Polymeric Nanoparticles for Biomedical Applications

Polymeric Nanoparticles for Biomedical Applications

Editors

Stephanie Andrade
Joana A. Loureiro
Maria João Ramalho

MDPI Basel • Beijing • Wuhan • Barcelona • Belgrade • Novi Sad • Cluj • Manchester

Editors

Stephanie Andrade
Department of Chemical Engineering
Univerity of Porto
Porto
Portugal

Joana A. Loureiro
Department of Chemical Engineering
Univerity of Porto
Porto
Portugal

Maria João Ramalho
Department of Chemical Engineering
Univerity of Porto
Porto
Portugal

Editorial Office
MDPI
St. Alban-Anlage 66
4052 Basel, Switzerland

This is a reprint of articles from the Special Issue published online in the open access journal *Polymers* (ISSN 2073-4360) (available at: www.mdpi.com/journal/polymers/special_issues/polym_nanoparticles_biomed_appli).

For citation purposes, cite each article independently as indicated on the article page online and as indicated below:

Lastname, A.A.; Lastname, B.B. Article Title. *Journal Name* **Year**, *Volume Number*, Page Range.

ISBN 978-3-7258-0152-7 (Hbk)
ISBN 978-3-7258-0151-0 (PDF)
doi.org/10.3390/books978-3-7258-0151-0

Contents

Stéphanie Andrade, Maria J. Ramalho and Joana A. Loureiro
Polymeric Nanoparticles for Biomedical Applications
Reprinted from: *Polymers* **2024**, *16*, 249, doi:10.3390/polym16020249 1

Mohammed Muqtader Ahmed, Md. Khalid Anwer, Farhat Fatima, Mohammed F. Aldawsari, Ahmed Alalaiwe, Amer S. Alali, et al.
Boosting the Anticancer Activity of Sunitinib Malate in Breast Cancer through Lipid Polymer Hybrid Nanoparticles Approach
Reprinted from: *Polymers* **2022**, *14*, 2459, doi:10.3390/polym14122459 5

Mohamed A. Alfaleh, Anwar M. Hashem, Turki S. Abujamel, Nabil A. Alhakamy, Mohd Abul Kalam, Yassine Riadi and Shadab Md
Apigenin Loaded Lipoid–PLGA–TPGS Nanoparticles for Colon Cancer Therapy: Characterization, Sustained Release, Cytotoxicity, and Apoptosis Pathways
Reprinted from: *Polymers* **2022**, *14*, 3577, doi:10.3390/polym14173577 20

Keelan Jagaran and Moganavelli Singh
Copolymer-Green-Synthesized Copper Oxide Nanoparticles Enhance Folate-Targeting in Cervical Cancer Cells In Vitro
Reprinted from: *Polymers* **2023**, *15*, 2393, doi:10.3390/polym15102393 39

Alessandra Teixeira Vidal-Diniz, Homero Nogueira Guimarães, Giani Martins Garcia, Érika Martins Braga, Sylvain Richard, Andrea Grabe-Guimarães and Vanessa Carla Furtado Mosqueira
Polyester Nanocapsules for Intravenous Delivery of Artemether: Formulation Development, Antimalarial Efficacy, and Cardioprotective Effects In Vivo
Reprinted from: *Polymers* **2022**, *14*, 5503, doi:10.3390/polym14245503 59

Shuo Li, Li Shi, Ting Ye, Biao Huang, Yuan Qin, Yongkang Xie, et al.
Development of Crosslinker-Free Polysaccharide-Lysozyme Microspheres for Treatment Enteric Infection
Reprinted from: *Polymers* **2023**, *15*, 1077, doi:10.3390/polym15051077 77

Daniela Predoi, Steluta Carmen Ciobanu, Simona Liliana Iconaru, Ştefan Ţălu, Liliana Ghegoiu, Robert Saraiva Matos, et al.
New Physico-Chemical Analysis of Magnesium-Doped Hydroxyapatite in Dextran Matrix Nanocomposites
Reprinted from: *Polymers* **2024**, *16*, 125, doi:10.3390/polym16010125 89

Iva Rezić
Nanoparticles for Biomedical Application and Their Synthesis
Reprinted from: *Polymers* **2022**, *14*, 4961, doi:10.3390/polym14224961 112

Md. Harun-Or-Rashid, Most. Nazmin Aktar, Md. Sabbir Hossain, Nadia Sarkar, Md. Rezaul Islam, Md. Easin Arafat, et al.
Recent Advances in Micro- and Nano-Drug Delivery Systems Based on Natural and Synthetic Biomaterials
Reprinted from: *Polymers* **2023**, *15*, 4563, doi:10.3390/polym15234563 131

polymers MDPI

Editorial

Polymeric Nanoparticles for Biomedical Applications

Stéphanie Andrade [1,2,*], Maria J. Ramalho [1,2,*] and Joana A. Loureiro [1,2,*]

1 LEPABE—Laboratory for Process Engineering, Environment, Biotechnology and Energy, Faculty of Engineering, University of Porto, Rua Dr. Roberto Frias, 4200-465 Porto, Portugal
2 ALiCE—Associate Laboratory in Chemical Engineering, Faculty of Engineering, University of Porto, Rua Dr. Roberto Frias, 4200-465 Porto, Portugal
* Correspondence: stephanie@fe.up.pt (S.A.); mjramalho@fe.up.pt (M.J.R.); jasl@fe.up.pt (J.A.L.)

Polymeric nanoparticles (NPs), utilized extensively in biomedical applications, have received increasing interest in the preceding years and today represent an established part of the nanotechnology field. Natural and synthetic polymers are versatile materials that offer several advantages, including biodegradability, biocompatibility, and non-toxicity. Encapsulating therapeutic molecules in polymeric NPs allows for sustained drug release, increasing their half-life, which is beneficial in terms of improving drug efficacy and safety, reducing unwanted side effects, and enhancing the acceptance and compliance of patients [1].

This Special Issue includes six research papers and eight reviews, with the most recent insights on advanced polymeric NPs and their current applications in healthcare, including the prevention, diagnosis, and treatment of diseases.

Because cancer is among the leading causes of premature death worldwide, numerous researchers have dedicated their investigations to curing this devasting disease by improving drug properties using polymer-based NPs. Ahmed et al. [2] designed sunitinib malate (SM)-loaded lipid polymer hybrid NPs (LPHNPs) for application in breast cancer therapy. LPHNPs were produced via the emulsion solvent evaporation technique, using lipoid90H and chitosan as the lipid and polymeric phases, respectively. Lecithin was also used as a stabilizer. The authors evaluated the effect of varying the amount of each component, and the selected formulation showed a monodisperse population of NPs with a mean diameter of 439 $\pm$ 6 nm, a positive zeta potential (34 $\pm$ 5 mV), and an encapsulation efficiency (EE) of 83 $\pm$ 5%. An in vitro release study revealed that SM was rapidly released in the first 6 h (approximately 70%), followed by sustained release over the following 42 h. The anti-breast cancer activity of the optimized SM-loaded NPs was then investigated in the MCF-7 breast cancer cell line using the MTT assay. The results indicated that the formulation showed a higher cell viability reduction than free SM.

Alfaleh et al. [3] also developed LPHNPs capable of encapsulating the natural compound apigenin for colon cancer therapy. NPs composed of poly(lactic-co-glycolic acid) (PLGA) and lipoid S PC-3 were prepared using nanoprecipitation. LPHNPs exhibited a mean size of 235 $\pm$ 12 nm, a polydispersity index (PDI) of 0.11 $\pm$ 0.04, a zeta potential of -5 ± 1 mV, and an EE of 55 $\pm$ 4%. An in vitro release study showed an initial burst release of apigenin in the first 8 h (around 30%), followed by sustained release until 72 h. The therapeutic efficacy of apigenin-loaded LPHNPs against colon cancer was assessed via flow cytometry. The results demonstrated that the nanosystem had more apoptotic activity than free apigenin. The apigenin-loaded NPs' anticancer efficacy was attributed to the reduction of signaling molecules (such as Bcl-2, BAX, NF-κB, and mTOR) involved in carcinogenic pathways.

Jagaran et al. [4] synthesized copper oxide NPs (CuONPs), containing a reporter gene (pCMV-Luc-DNA) for cervical cancer therapy. The NP surface was functionalized with two polymers, namely, polyethylene glycol (PEG) and chitosan. While PEG was used to extend the NPs' circulation time after systemic administration with reduced immunogenicity,

Citation: Andrade, S.; Ramalho, M.J.; Loureiro, J.A. Polymeric Nanoparticles for Biomedical Applications. *Polymers* **2024**, *16*, 249. https://doi.org/10.3390/polym16020249

Received: 8 January 2024
Accepted: 12 January 2024
Published: 15 January 2024

chitosan was employed due to the role its -NH_2 and -OH groups play in facilitating the conjugation of biomolecules. Furthermore, NPs' ability to targeting ability to cervical cancer cells was achieved by conjugating folic acid to their surface as folate receptors are overexpressed in these cells. Gene-loaded NPs presented a spherical shape, with mean diameters of 209 $\pm$ 10 nm and a positive zeta potential (about 35 mV). The nanocarrier (unloaded NPs) was revealed to be safe for human healthy embryonic kidney cells (HEK293) and cervical cancer cells (HeLa), while gene-loaded NPs showed significant transgene expression.

In turn, Vidal-Diniz et al. [5] prepared artemether-loaded polymeric nanocapsules (NCs) for malaria treatment. Three distinct formulations, composed of poly(D,L-lactide) (PLA), poly-ε-caprolactone (PCL), or PEG-PLA, were prepared via the polymer deposition method following solvent displacement with the aim of finding the formulation with the best physicochemical properties. Artemether-loaded PCL NCs showed the most promising characteristics, including mean sizes of 232 $\pm$ 3 nm, zeta potential of -49 ± 2 mV, EE of 92 $\pm$ 1%, and a monodisperse population. The in vitro release study revealed an initial burst release of artemether in the first 2 h (50%). After this period, drug release was negligible. The antimalarial activity of the artemether-loaded PCL NCs was then investigated in Swiss mice infected with *P. berghei*, the parasite responsible for causing malaria. After administering 1 or 4 daily doses of artemether-loaded NCs via the intravenous route, reduced parasitemia and increased mice survival were observed. Moreover, the encapsulation of artemether in PCL NCs reduced its cardiotoxicity.

Li et al. [6] prepared lysozyme-containing microspheres to treat enteric infections. The carboxymethyl starch and chitosan microspheres were produced via an electrostatic layer-by-layer self-assembly technique. The obtained loaded microspheres exhibited a spherical shape, with a mean diameter of 8 μm and an EE of 85%. In vitro lysozyme release under simulated gastrointestinal conditions was investigated, and the results showed a release of 29% in the simulated gastric fluid (2h), followed by a release reaching 97% in simulated intestinal fluid (during 6 h). The remaining amount of lysozyme was released in the next 4 h in simulated colonic fluid. After that, the antibacterial property of lysozyme-loaded microspheres against *Escherichia coli* (*E. coli*) than *Staphylococcus aureus* (*S. aureus*) was examined. Data revealed that the microspheres had more antibacterial activity against *E. coli* than *S. aureus*. Furthermore, the lysozyme's antibacterial activity was enhanced via its encapsulation within microspheres. In addition, the nanosystem was safe for human cells.

Predoi et al. [7] produced magnesium-doped hydroxyapatite in dextran matrix nanocomposites, using a coprecipitation technique to treat dental infections. The nanocomposites showed spherical morphology and an average diameter of 15 $\pm$ 2 nm. The in vitro antimicrobial activity of the nanocomposites was assessed, and the data revealed strong inhibitory activity against Gram-positive and Gram-negative bacteria. Additionally, the nanocomposites presented excellent biocompatibility towards human gingival fibroblast cells (HGF-1).

In addition to the research papers, I. Rezić [8] summarized the recent developments of NPs for different biomedical applications, including the treatment of inflammation, cancer, and infectious diseases, as well as implants, prosthetic, and theranostic devices. In addition to polymeric NPs, this article also covered metallic NPs. The author summarized the primary NP synthesis techniques and their toxicological effects.

Additionally, Harun-Or-Rashid et al. [9] conducted a literature review covering the latest advances in microparticles and NPs composed of natural and synthetic biomaterials for biomedical applications. The article covered a wide range of medical topics, ranging from the use of particles to treat health conditions to their application in disease diagnosis. The paper presented the benefits and limitations of using natural or synthetic biomaterial-based particles, as well as ongoing clinical trials for biomedical applications. The authors also discussed the advantages of combining biomaterials to address a majority of therapeutic needs.

The incorporation of NPs into hydrogels has been explored previously for the purpose of enhancing their individual beneficial properties. Nunes et al. [10] provided a systematic review of polymeric NP-loaded hydrogels for the treatment of several disorders, focusing on their evaluation in animal studies. The review validated the usefulness of polymeric NP-loaded hydrogels, particularly in reducing the frequency of drug administration.

Another approach covered in this Special Issue was the use of NPs in the innovative fields of tissue engineering and regenerative medicine by Phutane et al., who utilized polymeric nanofibers [11]. Nanofibrous scaffolds have been used as reinforcement to facilitate tissue regeneration. The authors highlighted the advantages and disadvantages of several natural and synthetic biodegradable polymers used to produce nanofibers. The functionalization of polymers to enhance cellular interaction and tissue regeneration as well as nanofibers' production techniques were discussed. Finally, the review article presented the recent developments of polymeric nanofibers in tissue engineering and regeneration, including neural, vascular, cartilage, bone, dermal, and cardiac tissues.

Even though nanotechnology has been extensively investigated for the treatment of several brain disorders, such as Alzheimer's and Parkinson's diseases, brain cancers, ischemic stroke, and epilepsy, the BBB significantly hinders effective drug-loaded NPs delivery to the brain. In this regard, NP functionalization with molecules capable of improving receptor-mediated transcytosis has been explored. The oligopeptide angiopep-2 has been used as a targeting ligand for this purpose. Habib et al. [12] reviewed the latest advances of angiopep-2-modified NPs (organic and inorganic) in diagnosing and treating brain disorders. Currently, the majority of angiopep-2-modified NPs are used to treat brain tumors, specifically glioblastoma.

Since cancer is a major medical concern, extensive research has been conducted to find a cure for this devasting disease. This led to numerous review articles in this field, focusing chiefly on the application of NPs. In the last year, polysaccharide-based carriers have been investigated to increase the therapeutic efficacy of anti-cancer drugs and reduce their toxicity. In addition to being non-toxic and biodegradable, these hydrophilic biopolymers are easily chemically changed to increase drug bioavailability. Yadav et al. [13] recently reviewed the recent progress in polysaccharide-based drug carriers for cancer therapy. The review highlighted the properties of a variety of polysaccharides nanocarriers, such as alginate, dextran, chitosan, and hyaluronic acid, as well as their biomedical applications.

The review of Caraway et al. [14] provided an overview of the main polymeric NPs' properties used in brain cancer therapy. The authors discussed the influence of NPs' composition, surface modification, and administration route in improving brain cancer treatment. Strategies for improving the targeting of NPs to the blood-brain barrier (BBB) and brain tumor cells were also presented.

Despite the abundance of articles demonstrating nanosystem effectiveness for cancer therapy in preclinical studies, only several have reached the market. This poor translation of nano-based medicines into clinical practice has been attributed to the poor targeting of nanosystems to tumors and heterogeneity of the enhanced permeability and retention (EPR) effect. The most promising strategies to improve the accumulation of NPs and macromolecules in tumors by enhancing the EPR effect have been reviewed by Ejigah et al. [15]. The authors also discussed the principles and challenges of the EPR effect.

Overall, the research and review articles published in this Special Issue represent a small portion of the global research on polymeric NPs for use in the medical field. It is our hope that the research produced will contribute to the progress of investigations in this area.

Funding: This work was financially supported by: LA/P/0045/2020 (ALiCE), UIDB/00511/2020 and UIDP/00511/2020 (LEPABE), funded by national funds through FCT/MCTES (PIDDAC); Project EXPL/NAN-MAT/0209/2021, funded by FEDER funds through COMPETE2020—Programa Operacional Competitividade e Internacionalização (POCI) and by national funds (PIDDAC) through FCT/MCTES. FCT supported MJR under the Scientific Employment Stimulus—Individual Call—(CEEC-IND/01741/2021) and SA (EXPL/NAN-MAT/0209/2021). This work has received funding from the European Union's Horizon 2020 research and innovation programme under grant agreement No 958174; and from national funds through FCT (M-ERA-NET3/0001/2021, https://doi.org/10.54499/M-ERA-NET3/0001/2021).

Acknowledgments: The Guest Editors would like to acknowledge all the authors and reviewers who contributed to this Special Issue.

Conflicts of Interest: The authors declare no conflicts of interest.

References

1. Ramalho, M.J.; Andrade, S.; Loureiro, J.A.; do Carmo Pereira, M. Nanotechnology to improve the Alzheimer's disease therapy with natural compounds. *Drug Deliv. Transl. Res.* **2020**, *10*, 380–402. [CrossRef] [PubMed]
2. Ahmed, M.M.; Anwer, M.K.; Fatima, F.; Aldawsari, M.F.; Alalaiwe, A.; Alali, A.S.; Alharthi, A.I.; Kalam, M.A. Boosting the anticancer activity of sunitinib malate in breast cancer through lipid polymer hybrid nanoparticles approach. *Polymers* **2022**, *14*, 2459. [CrossRef] [PubMed]
3. Alfaleh, M.A.; Hashem, A.M.; Abujamel, T.S.; Alhakamy, N.A.; Kalam, M.A.; Riadi, Y.; Md, S. Apigenin Loaded Lipoid–PLGA–TPGS Nanoparticles for Colon Cancer Therapy: Characterization, Sustained Release, Cytotoxicity, and Apoptosis Pathways. *Polymers* **2022**, *14*, 3577. [CrossRef] [PubMed]
4. Jagaran, K.; Singh, M. Copolymer-Green-Synthesized Copper Oxide Nanoparticles Enhance Folate-Targeting in Cervical Cancer Cells In Vitro. *Polymers* **2023**, *15*, 2393. [CrossRef] [PubMed]
5. Vidal-Diniz, A.T.; Guimarães, H.N.; Garcia, G.M.; Braga, É.M.; Richard, S.; Grabe-Guimarães, A.; Mosqueira VC, F. Polyester Nanocapsules for Intravenous Delivery of Artemether: Formulation Development, Antimalarial Efficacy, and Cardioprotective Effects In Vivo. *Polymers* **2022**, *14*, 5503. [CrossRef] [PubMed]
6. Li, S.; Shi, L.; Ye, T.; Huang, B.; Qin, Y.; Xie, Y.; Ren, X.; Zhao, X. Development of Crosslinker-Free Polysaccharide-Lysozyme Microspheres for Treatment Enteric Infection. *Polymers* **2023**, *15*, 1077. [CrossRef] [PubMed]
7. Predoi, D.; Ciobanu, S.C.; Iconaru, S.L.; Ţălu, Ş.; Ghegoiu, L.; Matos, R.S.; da Fonseca Filho, H.D.; Trusca, R. New Physico-Chemical Analysis of Magnesium-Doped Hydroxyapatite in Dextran Matrix Nanocomposites. *Polymers* **2024**, *16*, 125. [CrossRef] [PubMed]
8. Rezić, I. Nanoparticles for biomedical Application and their synthesis. *Polymers* **2022**, *14*, 4961. [CrossRef] [PubMed]
9. Harun-Or-Rashid, M.; Aktar, M.N.; Hossain, M.S.; Sarkar, N.; Islam, M.R.; Arafat, M.E.; Bhowmik, S.; Yusa, S.-I. Recent Advances in Micro-and Nano-Drug Delivery Systems Based on Natural and Synthetic Biomaterials. *Polymers* **2023**, *15*, 4563. [CrossRef]
10. Nunes, D.; Andrade, S.; Ramalho, M.J.; Loureiro, J.A.; Pereira, M.C. Polymeric nanoparticles-loaded hydrogels for biomedical applications: A systematic review on in vivo findings. *Polymers* **2022**, *14*, 1010. [CrossRef] [PubMed]
11. Phutane, P.; Telange, D.; Agrawal, S.; Gunde, M.; Kotkar, K.; Pethe, A. Biofunctionalization and Applications of Polymeric Nanofibers in Tissue Engineering and Regenerative Medicine. *Polymers* **2023**, *15*, 1202. [CrossRef] [PubMed]
12. Habib, S.; Singh, M. Angiopep-2-modified nanoparticles for brain-directed delivery of therapeutics: A review. *Polymers* **2022**, *14*, 712. [CrossRef] [PubMed]
13. Yadav, N.; Francis, A.P.; Priya, V.V.; Patil, S.; Mustaq, S.; Khan, S.S.; Alzahrani, K.J.; Banjer, H.J.; Mohan, S.K.; Mony, U. Polysaccharide-drug conjugates: A tool for enhanced cancer therapy. *Polymers* **2022**, *14*, 950. [CrossRef] [PubMed]
14. Caraway, C.A.; Gaitsch, H.; Wicks, E.E.; Kalluri, A.; Kunadi, N.; Tyler, B.M. Polymeric nanoparticles in brain cancer therapy: A review of current approaches. *Polymers* **2022**, *14*, 2963. [CrossRef] [PubMed]
15. Ejigah, V.; Owoseni, O.; Bataille-Backer, P.; Ogundipe, O.D.; Fisusi, F.A.; Adesina, S.K. Approaches to improve macromolecule and nanoparticle accumulation in the tumor microenvironment by the enhanced permeability and retention effect. *Polymers* **2022**, *14*, 2601. [CrossRef] [PubMed]

polymers

MDPI

Article

Boosting the Anticancer Activity of Sunitinib Malate in Breast Cancer through Lipid Polymer Hybrid Nanoparticles Approach

Mohammed Muqtader Ahmed [1,*], Md. Khalid Anwer [1], Farhat Fatima [1], Mohammed F. Aldawsari [1], Ahmed Alalaiwe [1], Amer S. Alali [1], Abdulrahman I. Alharthi [2] and Mohd Abul Kalam [3,4]

1 Department of Pharmaceutics, College of Pharmacy, Prince Sattam Bin Abdulaziz University, P.O. Box 173, Al-Kharj 11942, Saudi Arabia; m.anwer@psau.edu.sa (M.K.A.); f.soherwardi@psau.edu.sa (F.F.); moh.aldawsari@psau.edu.sa (M.F.A.); a.alalaiwe@psau.edu.sa (A.A.); a.alali@psau.edu.sa (A.S.A.)
2 Department of Chemistry, College of Science and Humanities, Prince Sattam Bin Abdulaziz University, P.O. Box 83, Al-Kharj 11942, Saudi Arabia; a.alharthi@psau.edu.sa
3 Nanobiotechnology Research Unit, College of Pharmacy, King Saud University, P.O. Box 2457, Riyadh 11451, Saudi Arabia; makalam@ksu.edu.sa
4 Department of Pharmaceutics, College of Pharmacy, King Saud University, Riyadh 11451, Saudi Arabia
* Correspondence: mo.ahmed@psau.edu.sa

Abstract: In the current study, lipid-polymer hybrid nanoparticles (LPHNPs) fabricated with lipoid-90H and chitosan, sunitinib malate (SM), an anticancer drug was loaded using lecithin as a stabilizer by employing emulsion solvent evaporation technique. Four formulations (SLPN1–SLPN4) were developed by varying the concentration of chitosan polymer. Based on particle characterization, SLPN4 was optimized with size (439 ± 5.8 nm), PDI (0.269), ZP (+34 ± 5.3 mV), and EE (83.03 ± 4.9%). Further, the optimized formulation was characterized by FTIR, DSC, XRD, SEM, and in vitro release studies. In-vitro release of the drug from SPN4 was found to be 84.11 ± 2.54% as compared with pure drug SM 24.13 ± 2.67%; in 48 h, release kinetics followed the Korsmeyer–Peppas model with Fickian release mechanism. The SLPN4 exhibited a potent cytotoxicity against MCF-7 breast cancer, as evident by caspase 3, 9, and p53 activities. According to the findings, SM-loaded LPHNPs might be a promising therapy option for breast cancer.

Keywords: sunitinib; lipoid 90H; chitosan; nanoparticles; breast cancer; caspase

Citation: Ahmed, M.M.; Anwer, M.K.; Fatima, F.; Aldawsari, M.F.; Alalaiwe, A.; Alali, A.S.; Alharthi, A.I.; Kalam, M.A. Boosting the Anticancer Activity of Sunitinib Malate in Breast Cancer through Lipid Polymer Hybrid Nanoparticles Approach. *Polymers* **2022**, *14*, 2459. https://doi.org/10.3390/polym14122459

Academic Editors: Stephanie Andrade, Joana A. Loureiro and Maria João Ramalho

Received: 23 April 2022
Accepted: 9 June 2022
Published: 16 June 2022

1. Introduction

Breast cancer is considered one of the leading types of cancer, surpassing lung cancer, as per worldwide cancer incidence in 2020 [1]. Breast cancer starts with the uncontrolled growth of breast cell in one or both sides. About one in eight women are diagnosed with breast cancer during their lifetime; the good news is that its curable if detected at an early stage [2]. Proliferation, apoptosis, angiogenesis, hypoxia, cancer stem cell activity, epithelial to mesenchymal transition (EMT), and metastasis are all related to the signaling system. Notch receptors and their ligands were shown to be overexpressed in breast cancer [3,4]. Signaling pathways upregulated leading to breast cancer include human epidermal growth factor receptor 2 (HER-2) tyrosine kinase pathway, a member of the ErbB family of transmembrane receptor tyrosine kinases [5]. The Hedgehog signaling pathway is also deregulated in breast cancer which is responsible for proper cell differentiation [6]. p53 mutation leads to aggressive diseases, such as breast cancer, playing a vital role in regulating the cell cycle, and apoptosis mutation of this gene causes cancer and shortens the overall survival. Another pathway actively involved in breast cancer is Phosphatase and tensin homolog (PTEN), reduced expression of which causes deceased formation of an enzyme phosphatase protein that acts as a tumor suppressor [7–9].

Sunitinib malate (SM) is a multiple tyrosine kinases inhibitor, used effectively in the cancer of the stomach, bowel, and esophagus, generally called gastrointestinal stromal tumors (GST), an abnormal proliferation of cells in gastrointestinal tract tissues [10,11]. It is a new vascular endothelial growth factor receptor practiced as a first-line therapy for advanced renal cell carcinoma [12] and a first-joint FDA-approved drug for these two indications. SM showed a promising activity against colorectal cancer, advanced non-small cell lung cancer (NSCLC) [13], hepatic cancer [14], and pancreatic neuroendocrine tumors (pNET) [15]. SM is one of the extensively studied antitumor agents in the breast cancer treatment with twenty-eight ongoing clinical trials, specifically sunitinib alone and in combinations. Sunhui et al.'s study showed sunitinib and curcumin have potential anticancer activity against breast cancer [16]. Sunitinib-loaded self-nanoemulsifying formulation has been developed with improved anticancer activity against the MCF-7 breast cancer cell line [17–19].

Moreover, breast cancer treatment failure could be surfaced due to poor drug solubility, low-bioavailability, permeability, cell uptake, drug resistance, and systemic toxicity. The cost of therapy and adverse drug effects could be lowered by adopting and aligning with new technologies. Nanotechnologies ensure it reduces adverse systemic toxic effects; thereby, the cost of therapy will also be reduced [20–22]. Nanocarriers enhance systemic drug circulation, and improve bioavailability, sustained release kinetics, and drug targeting at the receptor site. Enhanced permeability and retention effect facilitate the targeting of small molecules of nanosize and higher deposition of drug in cancer cells compared with the normal cells. Drug targeting involves the conjugate of chemotherapeutic-loaded nanocarrier with molecules that bind to the target (tumor) cell receptors [23–26]. Liposomes are thought to be a biocompatible vesicular structure with properties similar to biological membranes. Stability, low drug encapsulation, and burst drug release are the key concerns for vesicular systems. Polymer-based nanoparticles, on the other hand, are more stable than liposomes and also provide longer-lasting drug release. Synthetic (e.g., PLGA) and natural (e.g., chitosan) polymers are used to create polymeric nanoparticles. The preparation of lipid polymer hybrid nanoparticles (LPHNPs), which have both lipid and polymeric carriers, can address the constraints of both liposomes and polymeric nanoparticles. LPHNPs are next-generation core-shell nanostructures that are derived from both liposomes and polymeric nanoparticles (NPs), with a lipid coating encasing the polymer core loaded with drug that helps it to prolong systemic circulation and protect drug mitigation, and does not allow the water to obtain access into the drug-containing core. LPHNPs shows high entrapment, controlled release, cellular targeting, and serum stability. To the best of the author's knowledge, no study has been conducted or reported on LPHNPs of SM [27–29]. Therefore, to facilitate targeting of SM at the tumor cell with higher drug loading, nanocarriers with a conjugate of lipid and polymer were selected in order to achieve higher antineoplastic activity with reduced toxicity to the normal cell at the target region. The objective of the current study was to prepare SM-loaded lipid–polymer hybrid nanoparticles using lipoid-90, chitosan, and evaluated for particle size, drug release, and anticancer activity against MCF7 cell lines.

2. Materials and Methods

2.1. Materials

Sunitininb malate (SM) was purchased from "Mesochem Technology" Beijing, China. Chitosan and polyvinyl alcohol (PVA) were procured from Sigma Aldrich, St. Louis, MO, USA. Lipoid 90H was a generous gift from Lyon, France. Human breast cancer cell line (MCF-7 cells) with estrogen, progesterone, and glucocorticoid receptors was procured from the American Type Culture Collection (ATCC). All other chemicals used were of analytical grade, and Milli-Q water was used wherever needed.

2.2. Preparation of SM-loaded Lipid–Polymer Hybrid Nanoparticles

Lipid–polymer hybrid nanoparticles (LPH-NPs) were prepared by emulsification solvent evaporation technique [30]. Briefly, pure SM (20 mg) and lipoid 90H (40 mg) was dissolved in 10 mL of dichloromethane to obtain the organic phase. Separately, the aqueous phase was prepared by dissolving chitosan (25–100 mg) and soyalecithin (20 mg) in 0.5% *w*/*v* acetic acid (Table 1). Further, prepared organic phase was emulsified into aqueous phase (with rate of 0.3 mL/min) using probe sonication (ultrasonic processor, Fisher scientific, Waltham, MA, United States) at power 65%, on–off cycle 5 sec, for 3 min. Formed emulsion was kept on a magnetic stirrer (500 rpm) at room temperature overnight. After complete evaporation of dichloromethane, reduced volume was centrifuged (Hermle-Labortechnik, Z216MK, Wehingen, Germany) at 6000 rpm for 15 h to obtain the sediment. Sediment pellet was then washed with milli-Q water thrice and lyophilized (Millirock Technology, Kingston, NY, USA) and lyophilized LPHNPs were collected for further analysis (Figure 1).

Table 1. Composition of SM-loaded LPHNPs.

Composition (mg)	LPHNPs			
	SLPN1	SLPN2	SLPN3	SLPN4
Sunitinib	20	20	20	20
Lipoid 90H	40	40	40	40
Soyalecithin	20	20	20	20
Chitosan	25	50	75	100

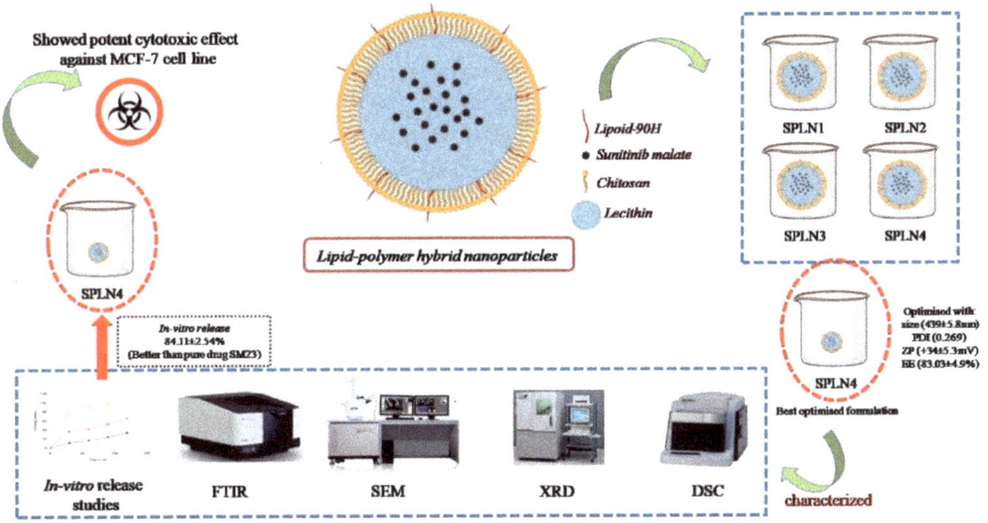

Figure 1. Schematic diagram of SLPNs synthesis.

2.3. Measurement of Particle Size, Polydispersity Index (PDI) and Zeta Potential (ZP)

Freshly prepared SM-loaded LPHNPs were diluted 200 times with milli-Q water, sonicated for 5 min, and transferred into plastic cuvette, then analyzed for particle size and PDI using Malvern Zetasizer (ZEN-3600, Malvern Instruments Ltd., Worcestershire, UK). Three measurements were carried out for 11 runs with 10 sec durations each run at 25 °C temperature. The same procedure was followed for ZP measurements as particle size except a glass electrode sample holder was used instead of a plastic cuvette [31]. Each sample measurement was performed in triplicate.

2.4. Percent Drug Entrapment Efficiency (%EE)

Percent entrapment efficiency (%EE) was measured indirectly [20]. Freshly prepared SM-loaded LPHNPs dispersion were subjected to the centrifugation at 10,000 rpm for 15 min. Supernatant was then collected, pre-filtered by syringe filter 45 µm, then suitably diluted with methanol and quantified for unentrapped drug. Aliquots were analyzed using UV/Visible spectroscopy at λ_{max} 250 nm [31] (Jasco V630 UV/Visible spectrophotometer, Tokyo, Japan). The %EE was calculated using the following equations:

$$\%EE = \frac{Initial\ SM\ added\ in\ LPHNPs - Free\ SM\ in\ supernatant}{Initial\ SM\ added\ in\ LPHNPs} \times 100 \quad (1)$$

2.5. Differential Scanning Calorimetry

Drug entrapment in the nanoparticles can be identified by differential scanning calorimetry (DSC). Samples 5 mg (pure SM, Lipoid 90H, chitosan, and optimized LPHNPs) (SLPN4) were packed in a hemispherical aluminium pan, separately. A pan filled with the sample was kept in a heating chamber against the blank. The temperature was raised from 50 °C to 350 °C at a rate of 20 °C/min; additionally, nitrogen gas was supplied at flow rate of 20 mL/min (Sinco 400, Seoul, Korea) [32]. Endothermic peaks were seen at the melting point of the sample; the temperature was then noted and studied.

2.6. FTIR Spectroscopy

FTIR spectra of SM, lipoid 90H, chitosan, and optimized LPHNPs (SLNP4) were taken using the KBr technique. The samples were mixed with crystalline KBr and the mixture was then compressed into transparent pellets using a hand-held compression machine. Thin transparent sample film enclosed in the die was fixed into the sample holder and scanned in the range of 400–4000 cm^{-1} (Jasco, V750, FTIR spectrophotometer, Tokyo, Japan). Peaks at the fingerprint region were interpreted, and additional or absent peaks were then studied for possible functional group interactions between drug and excipients. Spectrums were then collaged and presented for compatibility study [33].

2.7. XRD Diffraction Study

Sample (pure SM and SLPN4) were characterized by XRD to identify the nature of solid-state. X-ray diffractometer (Ultima IV, Rigaku Inc., Tokyo, Japan) was used with the following set parameters: Cu-kα radiation at 40 kV/40 mA, scan rate of 0.500°/min in the 0–60 (2θ) range, at a fixed monocromator (U4), attached with scintillation detector [33].

2.8. In-Vitro Release Studies

Comparative in vitro release studies of pure SM and optimized LPHNPs (SLPN4) were carried out using a dialysis bag (cut off mol wt. 12 kD) as per our previously reported method [30]. Briefly, pure SM and SLPN4 (equivalent to 20 mg of SM) were dissolved in pre-soaked dialysis containing 10 mL of phosphate buffer (pH 6.8), then dialysis bags were dipped into a beaker containing 40 mL of dissolution media and shaken on biological shaker (LBS-030S-Lab Tech, Kyonggi, Korea) at 100 rpm. At fixed time intervals (0.5, 1, 2, 3, 6, 12, 24, and 48 h), 1 mL of sample was withdrawn and compensated immediately with respective media to maintain sink condition, filtered, diluted, and analyzed at 250 nm [31]. Each sample was analyzed in triplicate. Further study of release mechanism was executed by fixing the release data into the following mathematical equations.

$$Qt = Q0 + k0t \text{ Zero order}$$

$$\log Qt = \log Q0 - kt/2.303 \text{ First order}$$

$$Qt = kHt\ 1/2 \text{ Higuchi model}$$

$$Mt/M\infty = ktn \text{ Korsmeyer Peppas model}$$

where, Qt and Q0 represents (SM dissolved in media overtime t), (initial amount of SM dissolved in media, i.e., equal to zero). K marked constants of models. Mt and M∞ are cumulative drug release at time t and infinite time, respectively, t is the release time and n denotes the diffusional exponent indicating release mechanism [30].

2.9. Morphology

The morphology and size of images of optimized LPHNPs (SLPN4) were viewed by Scanning Electron Microscopic (SEM) (Zeiss EVO LS10; Cambridge, UK). In a thin film coater under vacuum, the sample was homogeneously dispersed and coated with gold-metal (Quorum Q150R S, Lewes, East Sussex, UK). The pre-treated sample was then bombarded with an electron beam, resulting in the creation of secondary electrons known as auger electrons. Only the electrons scattered at ≥90 degrees were picked from this interaction between the electron beam and the specimen's atoms, and surface topography was obtained at 15 kV acceleration voltage and 7.58 K × magnification.

2.10. Cell Culture and Treatments

The Human breast cancer cell line (MCF7) was procured from the American Type Cutler Collection (Manassas, VA, USA). The cells were maintained in Dulbecco's Modified Eagle Medium (DMEM) with phenol red supplement with 10% Fetal Bovine Serum (FBS), with Penicillin (100 units/mL), Streptomycin (100 µg/mL), and Amphotericin B (250 ng/mL), Gibco® (New York, NY, USA). The cells were grown at 37 °C in 50 cm^2 tissue culture flasks in a 5% CO_2 humidified incubator. The cells were seeded into 96-well cell culture plates in DMEM.

2.11. MTT Assay on MCF7 Cells

To determine the dose dependent cell viability of MCF-7 cells, they were incubated with SM and SLPN4 ranging from 0.78 to 100 µg/mL (containing equivalent amount of SM drug) for 48h. The data presented demonstrate relative cell viability after the treatment, since MTT assay determines the viable cells through activity of mitochondria. This is primarily targeted through mitochondria-mediated apoptosis. Thus, this approach was adopted to determine the activity of SM and SLPN4. The IC_{50} values were calculated using Log (inhibitor) versus normalized response on variable slope by GraphPad Prism V-5.1 (San Diego, CA, USA).

2.12. Morphological Changes on MCF-7 Cells

The cytotoxic effect of pure STB and SLPN4 was also determined by visualizing the morphological changes in MCF7 cells [34]. The IC_{50} value of SM (10.79 µg/mL) equivalent to SLPN4 (8.24 µg/mL) was taken as the dose of treatment and morphological features were manifested by phase-contrast microscopy. Morphological features such as membrane blebbing, cell shrinkage, and necrosis were determined.

2.13. Caspase-3, Caspase-9 and p53 Assay by ELISA

Caspase-3, caspase-9, and p53 assay ELISA kits were used to measure caspase activity [35]. The MCF-7 cells (50,000 cell/well) were seeded in 96-well plates. The cells were cultured for 24 h at 37 °C in a humidified incubator with 5% CO_2. The SM, SLPN4-treated, and untreated control cells were then allowed to equilibrate at room temperature in 96-well plates. Each well of plate (SM, SLPN4-treated, and control) containing 100 µL of culture media received 100 µL of caspase-3 and 9 reagents. The plate was covered and the contents were stirred at 500 rpm for 30 s. After 30 min of incubation at room temperature, the optical density was measured at 405 nm using an an ELISA reader.

2.14. Stability Study

To analyze the change in formulation over storage or shelf life, a stability study was conducted for optimized LPHNPs (SLPN4). The formulation was sealed in a glass vial and stored for three months at 25 ± 0.5 °C/65 ± 5% RH and 40 ± 2 °C/75 ± 5% RH in a stability chamber, and physical appearance, particle size, PDI, ZP and entrapment were examined in samples obtained at 0, 1, 2, and 3 months [36,37].

2.15. Statistical Analysis

The experimental data were analyzed statistically using one way ANOVA followed by Tukey's multiple comparison test using SPSS 16 software (version 25.0; SPSS Inc., Chicago, IL, USA) ($p < 0.01$) was considered significant.

3. Results and Discussion

3.1. Measurement of Particle Size, Polydispersity Index (PDI) and Zeta Potential (ZP)

SM-loaded LPHNPs (SLPN1-SLPN4) were prepared by the single emulsification method. The developed LPHNPs were characterized for their particle size, PDI, and ZP and measured in the range of 218–439 nm, 0.269–0.504, and +18 to +34 mV, respectively (Figure 2). According to studies, nanoparticles with a size range of 40 to 400 nm are appropriate for ensuring long circulation duration and increased accumulation of drug in tumors with limited renal clearance [38,39]. The positive values of ZP was measured due to the amino group present on the surface of chitosan polymer [40,41]. From the results, it was observed that increase in concentration of chitosan in formulations increased the size of particles.

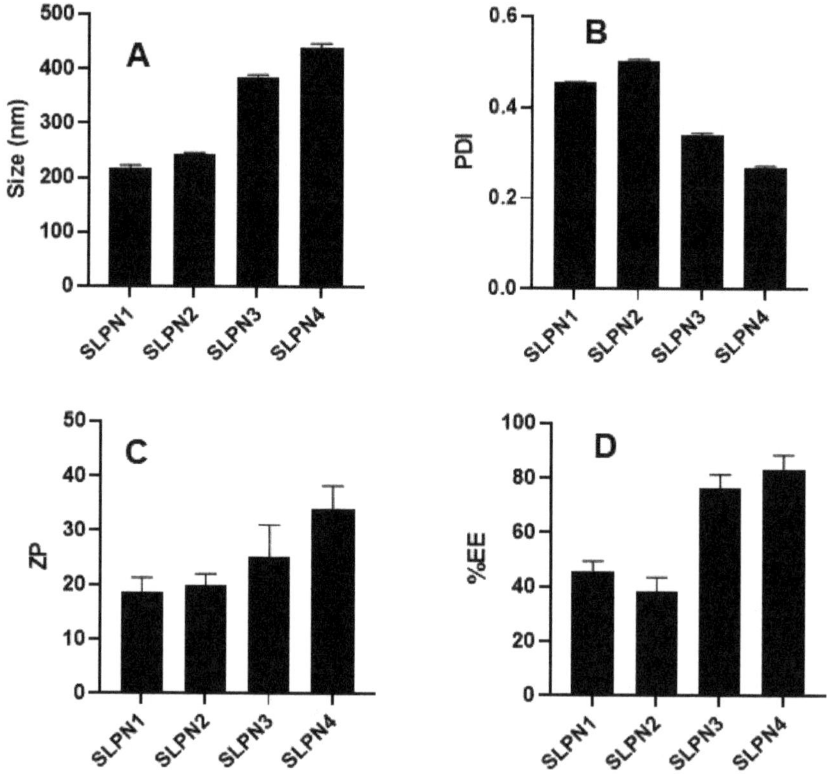

Figure 2. Particle size, PDI, ZP, and %EE of developed SM-loaded lipid polymer nanoparticles (SLPN1–SLPN4) tested with one way ANOVA followed by Tukey's multiple comparison between formulations. (**A**) Particle size—significant difference (** $p < 0.01$) among formulations. (**B**) PDI—significant difference (** $p < 0.001$) among formulations. (**C**) Zeta potential—results are not significant among (SLPN1 vs. SLPN2, SLPN1 vs. SLPN3, SLPN2 vs. SLPN3, and SLPN3 vs. SLPN4 formulations) and significant (** $p < 0.01$) between SLPN1 vs. SLPN4 and SLPN2 vs. SLPN4. (**D**) %EE—results are not significant among (SLPN1 vs. SLPN2 and SLPN3 vs. SLPN4 formulations) and the rest are significant (** $p < 0.01$).

3.2. Percent Drug Entrapment Efficiency (%EE)

The %EE of SM-loaded LPHNPs (SLPN1–SLPN4) were measured in the range of 45.71 ± 3.3–83.03 ± 4.9% (Figure 2). The highest drug entrapment (83.03 ± 4.9%) was measured in SLPN4, the large amount of chitosan (100 mg) used in this formulae expected to prevent leakage of drug from polymeric core, thereby improving the entrapment efficiency of drug [42]. Among the developed LPHNPs, SLPN4 was optimized with size (439 ± 5.8 nm), PDI (0.269), ZP (+34 ± 5.3 mV), and EE (83.03 ± 4.9%) and further evaluated.

3.3. Differential Scanning Calorimetry

DSC spectra of pure SM, Lipoid 90H, chitosan, and optimized LPHNPs (SLPN4) are presented in Figure 3. DSC studies confirmed the crystallinity and amorphocity nature of the sample, which indicates the encapsulation of drugs in nanoparticles [20,43]. Pure drug SM exhibited a sharp endothermic peak at 205 °C, which indicates its melting temperature [44,45]. The SM peak completely disappeared in DSC spectra of SLPN4, confirming SM entrapment in LPHNPs. Sharp endothermic and broad exothermic peaks of lipoid 90H and chitosan could be seen near 193 °C and 320 °C, respectively.

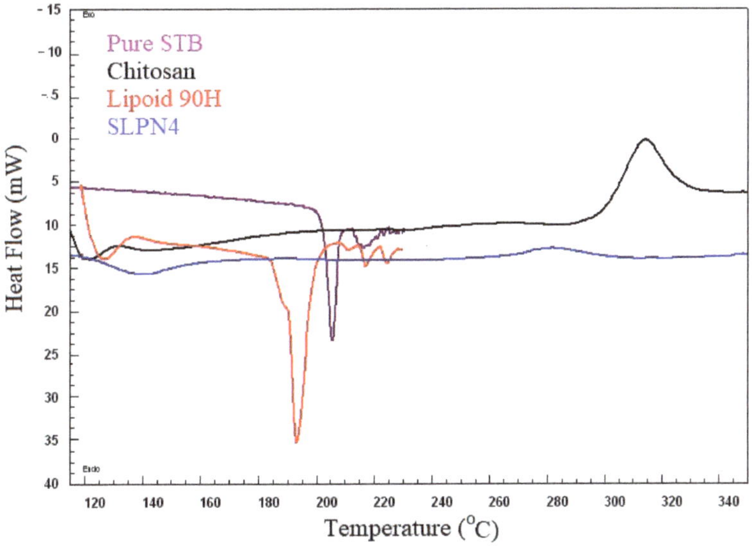

Figure 3. Comparative DSC spectra of pure SM, SLPN4, lipoid 90H, and chitosan.

3.4. FTIR Spectroscopy

Figure 4 indicates the FTIR spectra of pure SM, lipoid 90H, chitosan, and optimized LPHNPs (SLPN4) in the wavelength range of 400–4000 cm^{-1}. The FTIR spectra of pure SM showed many intense peaks at 3324 cm^{-1} for the acidic O-H, 2981 cm^{-1} for the acidic –CH=CH- (aryl) str, 2884 cm^{-1} for C-H (alkyl) str, 1635 cm^{-1} for the -NHCO str, and 1073 cm^{-1} bands correspond to the (C–F stretching) [41]. FTIR spectra of chitosan showed a broad peak at 3579 cm^{-1} (-OH str), 2873 cm^{-1} (CH_2 str) [46]. The FTIR spectra of phospholipon 90H exhibited characteristic peaks at 2934 cm^{-1} and 2861 cm^{-1} for –C-H- str present in long fatty acid chain, 1722 cm^{-1} for –C=O str, and 972 cm^{-1} for P=O str [47]. In the optimized LPHNPs (SLP4), peaks at 3324 cm^{-1}, 2981 cm^{-1}, and 2884 cm^{-1} were found to have disappeared. The FTIR peaks of SLPN4 represents no significant modifications in the functional peaks of the SM in the lipid–polymeric NPs matrix, thereby retaining drug's physicochemical properties and efficient chemical stability for the encapsulated SM in the fabricated nanocarriers. As there is no evidence of drug–polymers interaction, the selected lipid, polymer, and drug have the compatibility and suitability for the SM-loaded HNPs.

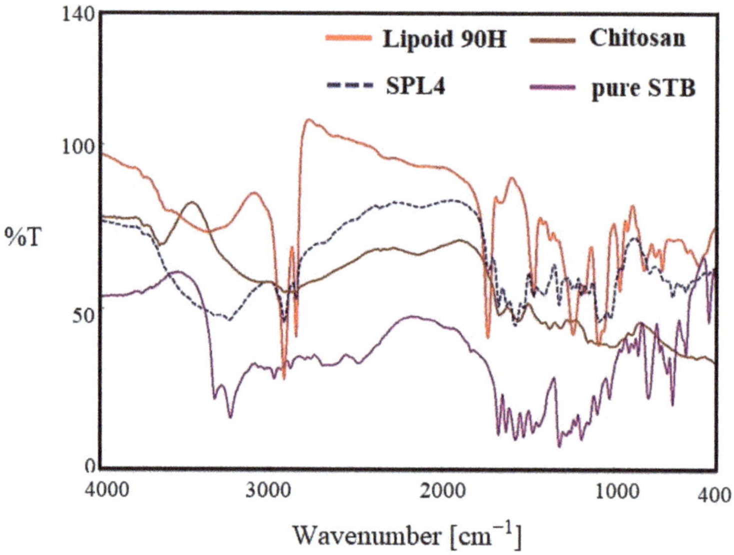

Figure 4. Comparative FTIR spectra of pure SM, SLPN4, lipoid 90H, and chitosan.

3.5. XRD Diffraction Study

Pure drug sunitinib (SM) showed intense X-ray diffractions at 13.3°, 19.5°, 22.4°, and 25.6° at 2θ, which represents the crystalline state of the drug [48]. However, the optimized LPHNPs (SLPN4) diffractogram also showed few peaks with reduced intensities, which are attributed to the amorphous state of drug in the nanoparticles due to the destruction of crystalline nature of SM and molecular dispersion of polymers, lipid, and drug (Figure 5).

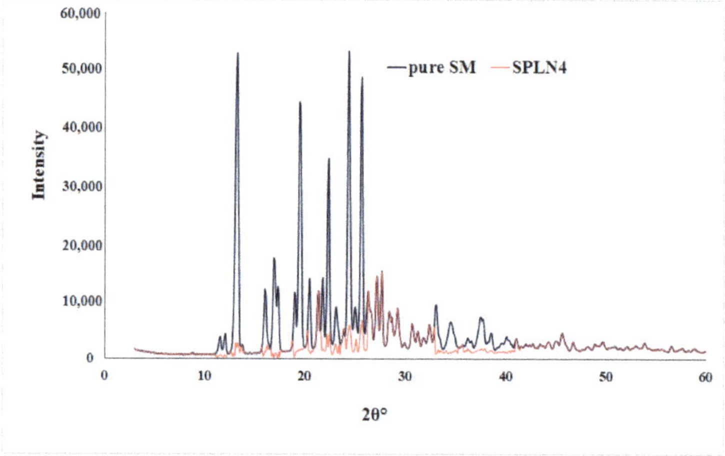

Figure 5. XRD spectra of pure SM and optimized LPHNPs (SLPN4).

3.6. In-Vitro Release Studies

In vitro drug release studies were conducted at pH 6.8, i.e., the pH of cancer cells [49]. A rapid drug release (98.45%) was exhibited by pure SM at the end of 6 h as compared with SLPN4. The optimized formulation (SLPN4) showed an initial rapid release of the drug for the first 6 h of study, followed by a sustained drug release. Initial rapid release could be due to the adsorbed drug over the polymer, which dissociated easily in the diffusion medium and released [20] (Figure 6).

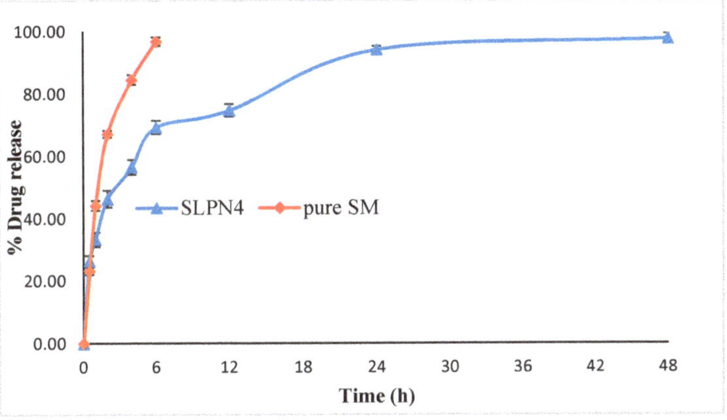

Figure 6. Comparative in vitro release profile of pure SM and optimized LPHNPs (SLPN4).

Release kinetics were assessed for optimized LPHNPs (SLPN4). The goodness of fit models was selected by evaluating R^2 value in the prediction of the release mechanism. The kinetics analysis of regression coefficient of all the four models used indicated R^2 values for zero order (0.612), first order (0.9359), Higuchi model (0.8494), and Korsmeyer–Peppas (0.9406) with diffusion coefficient n (0.271). The optimized LPHNPs (SLPN4) followed the Fickian diffusion ($n < 0.5$) and mechanism of release from the Korsmeyer–Peppas kinetic model [30].

3.7. Morphology

SEM images of optimized LPHNPs (SLPN4) are shown in the Figure 7. It was confirmed that optimized LPHNPs (SLNP4) were small and spherical in shape with aggregation, probably due to the presence of lipids. The size observed by SEM was approximately same as that measured by the DLS method.

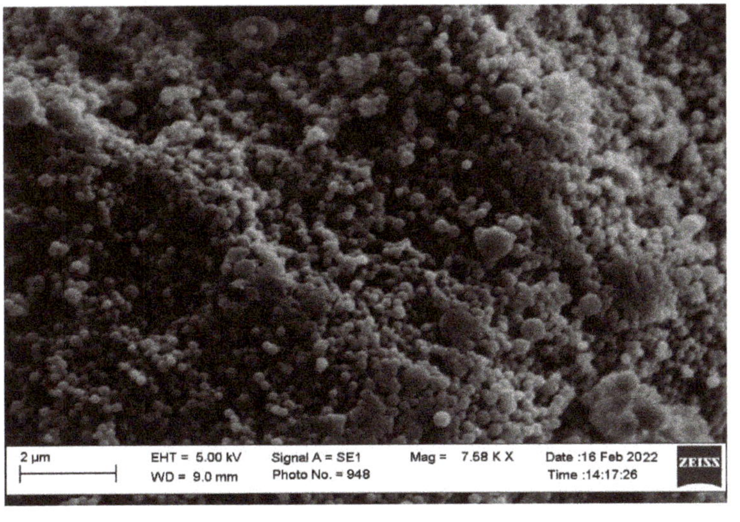

Figure 7. SEM images of optimized LPHNPs (SLPN4).

3.8. MTT Assay on MCF7 Cells

The MTT assay showed concentration-dependent reduction in cell viability for SM and optimized SM-loaded LPHNPs (SLPN4) against MCF-7 cell lines (Table 2 and Figure 8). The IC_{50} values for pure drug SM and SLP4 were found as 10.79 and 8.24 µg/mL for MCF-7 cells, respectively. The formulation SLP4 showed a significant reduction in cell viability (80.52%, 73.58%, 65.89%, 61.60%, 47.53%, 24.97%, 13.63%, and 6.79% at 0.8, 1.6, 3.1, 6.3, 12.5, 25, 50, and 100 µg/mL) in comparison with pure drug SM (92.05%, 85.26%, 75.35%, 67.49%, 54.04%, 32.94%, 15.24%, and 8.17% at 0.8, 1.6, 3.1, 6.3, 12.5, 25, 50, and 100 µg/mL), respectively, against MCF-7 cells. Based on the results of MTT assay, it was observed that SLP4 exhibited potential anticancer activity against breast cancer cell lines, probably due to the enhancement of the release of SM from the SLP4 formulation. SM-loaded LPHNPs (SLPN4) could be used as a potent carrier for the treatment of breast cancer.

Table 2. Percent cell viability against concentration.

Conc (µg/mL)	% Cell Viability	
	Pure SM	SLPN4
100.0	8.176 ± 0.457	6.793 ± 0.392
50.0	15.242 ± 1.188	13.636 ± 0.478
25.0	32.944 ± 0.305	24.976 ± 2.106
12.5	54.046 ± 2.463	47.537 ± 3.091
6.3	67.492 ± 1.627	61.607 ± 4.334
3.1	75.356 ± 0.885	65.893 ± 6.200
1.6	85.265 ± 2.041	73.585 ± 5.619
0.8	92.055 ± 0.763	80.511 ± 5.431
Control	100.000 ± 0.000	100.000 ± 0.000

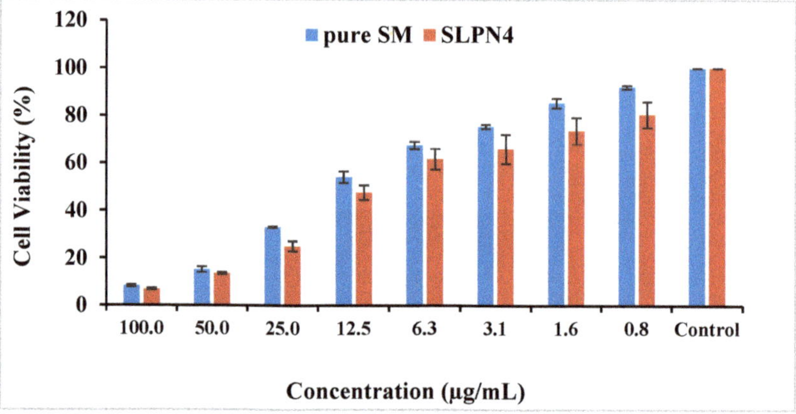

Figure 8. Cytotoxicity of pure SM and SLPN4 after 48 h incubation with MCF-7 cells at concentrations 0.8–100 µg/mL.

3.9. Morphological Changes on MCF-7 Cells

The morphological changes on MCF-7 cell lines after treatment of control, pure SM, and SLPN4 are presented in Figure 9, and after 24 h of incubation, SLPN4-treated cells evidenced maximum cell death in comparison with pure SM and control. The morphological changes observed in MCF-7 by SLPN4 were due to damage in cell organelles. The morphological changes were observed in MCF7 cells by SM as it is a tyrosine kinase inhibitor and has an anti-proliferative effect on breast cancer cell lines (MCF-7). The SM and SLPN4 treatments cause concentration-dependent cell growth suppression due to apoptosis, as made evident by Caspase-3, p53, and Caspase-9 levels in MCF7. These results are consistent and similar in previous reports [50,51].

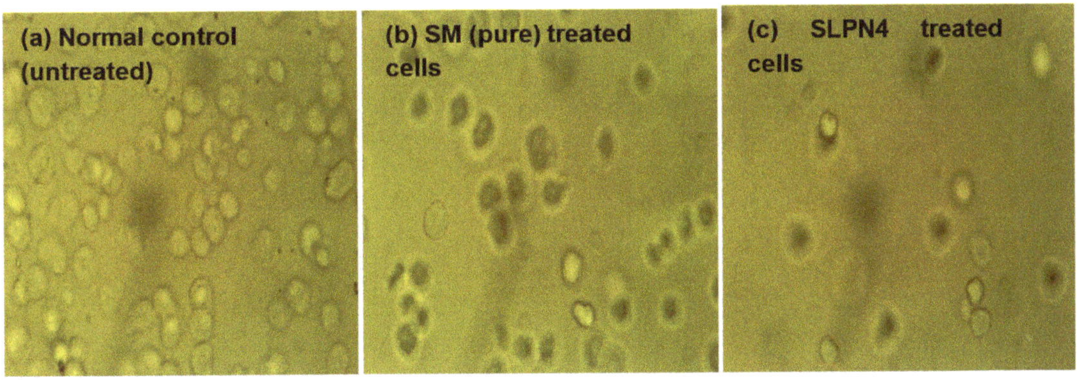

Figure 9. Morphological changes in MCF7 cells after exposure to Sunitinib-containing formulations. Untreated cells/control, cells incubated with pure SM, and SLPN4.

3.10. Caspase-3, Caspase-9 and p53 Assay by ELISA

Apoptosis is a type of programmed cell death that involves the disassembly of intracellular components while avoiding injury and inflammation to nearby cells [52,53]. The activation of caspase-3, -9, and p53 are mainly responsible for cancer cell apoptosis [54,55]. In this study, an increase in caspase 3, 9, and p53 production in pure SM- and SLPN4-treated MCF-7 cells was compared to the untreated control group to confirm apoptosis. When MCF-7 cells were exposed to pure SM and SLPN4, their caspase 3, 9, and p53 activity was found four, fourteen, and seven times higher compared with control (untreated) cells, respectively (Figure 10). Enhanced effectiveness of SM in LPHNPs suggests a possible reason for induction of apoptosis in cancer cells.

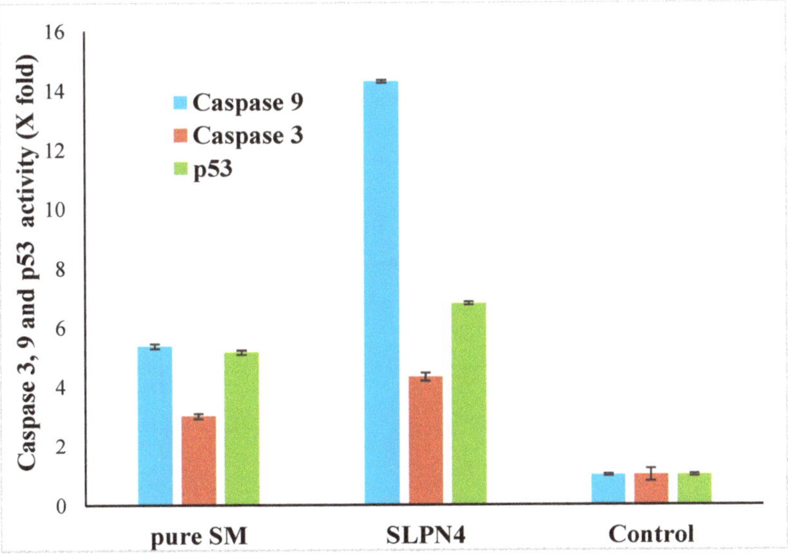

Figure 10. Activation of caspase 3, 9, and p53 in pure SM- and SLPN4-treated MCF7 cells was compared to the untreated control. SLPN4 ($p < 0.05$) vs. free SM and control group.

3.11. Stability Study

Following ICH guidelines, the stability of optimized LPHNPs (SLPN4) was assessed for three months in terms of particle size, PDI, ZP, and entrapment efficiency after storage. Table 3 shows the stability study parameters for LPHNPs (SLPN4), which were all within acceptable ranges, indicating that the developed formulation was stable for three months. A significant change (** $p < 0.01$) in particle size and PDI were noted when stored at 25 ± 0.5 °C/65 ± 5% RH and 40 ± 2 °C/75 ± 5% RH, which indicated that high temperature can lead to aggregation of the nanoparticles.

Table 3. Stability data of optimized LPHNPs (SLPN4).

Months	Storage Conditions	Particle Size	PDI	ZP (mV)	%EE
0	-	439 ± 5.8	0.269 ± 0.00052	+34 ± 5.3	83.03 ± 4.9
1	25 ± 0.5 °C/65 ± 5% RH	468 ± 4.5 **	0.267 ± 0.00052 ns	+33 ± 4.1 ns	82.42 ± 4.9 ns
2		472 ± 7.4 **	0.263 ± 0.00179 **	+31 ± 3.2 ns	78.11 ± 7.4 ns
3		476 ± 8.5 **	0.278 ± 0.00089 **	+29 ± 2.8 ns	77.28 ± 4.8 ns
1	40 ± 2 °C/75 ± 5% RH	467 ± 2.5 **	0.289 ± 0.00288 *	+31 ± 6.1 ns	81.45 ± 3.3 ns
2		473 ± 6.8 **	0.312 ± 0.00137 **	+24 ± 5.4 ns	74.32 ± 3.1 ns
3		476 ± 7.5 **	0.334 ± 0.00358 **	+21 ± 4.3 ns	71.31 ± 5.9 ns

Significant difference (** $p < 0.01$) compared with month 0; non-significant (ns) compared with month 0.

4. Conclusions

No research has been reported on SM as a breast cancer treatment using lipid–polymer nanoparticles. In vitro release and cell line investigations showed that the developed optimized SM-loaded LPHNPs (SLPN4) significantly enhanced the release and accessibility of SM at the breast cancer cells. Importantly, the potential cytotoxicity of SLPN4 on the MCF-7 breast cancer cell line was determined using the MTT test for cytotoxicity, ELISA activity of caspase-3, -9, and p53 in comparison to free SM or control. Thus, the developed formulation could provide an attractive nanoplatform for the treatment of breast cancer and may be the focus for the future chemotherapeutic investigations.

Author Contributions: Conceptualization, M.M.A. and M.K.A.; methodology, F.F.; software, A.S.A.; validation, M.A.K. and M.F.A.; investigation, F.F.; resources, A.A.; data curation, M.M.A.; writing—original draft preparation, M.K.A.; writing—review and editing, M.K.A. and A.I.A.; supervision, M.K.A.; project administration, M.M.A.; funding acquisition, M.M.A. All authors have read and agreed to the published version of the manuscript.

Funding: This research was funded by Deanship for Research & Innovation, Ministry of Education in Saudi Arabia, grant number IF/PSAU-2021/03/18862.

Institutional Review Board Statement: Not applicable.

Informed Consent Statement: Not applicable.

Data Availability Statement: The data presented in this study are available on request from the corresponding author.

Acknowledgments: The authors extend their appreciation to the Deanship for Research & Innovation, Ministry of Education in Saudi Arabia, for funding the research work.

Conflicts of Interest: The authors declare no conflict of interest.

References

1. Bray, F.; Ferlay, J.; Soerjomataram, I.; Siegel, R.L.; Torre, L.A.; Jemal, A. Global cancer statistics 2018: GLOBOCAN estimates of incidence and mortality worldwide for 36 cancers in 185 countries. *CA A Cancer J. Clin.* **2018**, *68*, 394–424. [CrossRef]
2. Yekedüz, E.; Dizdar, Ö.; Kertmen, N.; Aksoy, S. Comparison of Clinical and Pathological Factors Affecting Early and Late Recurrences in Patients with Operable Breast Cancer. *J. Clin. Med.* **2022**, *11*, 2332. [CrossRef]
3. Acar, A.; Simoes, B.M.; Clarke, R.B.; Brennan, K. A role for Notch signaling in breast cancer and endocrine resistance. *Stem Cells Int.* **2016**, *2016*, 2498764. [CrossRef]

4. Ling, H.; Sylvestre, J.R.; Jolicoeur, P. Notch1-induced mammary tumor development is cyclin D1-dependent and correlates with expansion of pre-malignant multipotent duct-limited progenitors. *Oncogene* **2010**, *29*, 4543–4554. [CrossRef]
5. Alladin, A.; Chaible, L.; Garcia Del Valle, L.; Sabine, R.; Loeschinger, M.; Wachsmuth, M.; Hériché, J.K.; Tischer, C.; Jechlinger, M. Tracking cells in epithelial acini by light sheet microscopy reveals proximity effects in breast cancer initiation. *eLife* **2020**, *9*, e54066. [CrossRef]
6. Sari, I.N.; Phi, L.; Jun, N.; Wijaya, Y.T.; Lee, S.; Kwon, H.Y. Hedgehog Signaling in Cancer: A Prospective Therapeutic Target for Eradicating Cancer Stem Cells. *Cells* **2018**, *7*, 208. [CrossRef]
7. Muñoz-Fontela, C.; Mandinova, A.; Aaronson, S.A.; Lee, S.W. Emerging roles of p53 and other tumour-suppressor genes in immune regulation. Nature reviews. *Immunology* **2016**, *16*, 741–750. [CrossRef]
8. Alimonti, A.; Carracedo, A.; Clohessy, J.G.; Trotman, L.C.; Nardella, C.; Egia, A.; Salmena, L.; Sampieri, K.; Haveman, W.J.; Brogi, E.; et al. Subtle variations in Pten dose determine cancer susceptibility. *Nat. Genet.* **2010**, *42*, 454–458. [CrossRef]
9. Antico Arciuch, V.G.; Russo, M.A.; Kang, K.S.; Di Cristofano, A. Inhibition of AMPK and Krebs cycle gene expression drives metabolic remodeling of Pten-deficient preneoplastic thyroid cells. *Cancer Res.* **2013**, *73*, 5459–5472. [CrossRef]
10. Wen, J.; Li, H.Z.; Ji, Z.G.; Jin, J. Human urothelial carcinoma cell response to Sunitinib malate therapy in vitro. *Cancer Cell Int.* **2015**, *15*, 26. [CrossRef]
11. Grimaldi, A.M.; Guida, T.; D'Attino, R.; Perrotta, E.; Otero, M.; Masala, A.; Cartenì, G. Sunitinib: Bridging present and future cancer treatment. *Ann. Oncol.* **2007**, *18*, vi31–vi34. [CrossRef] [PubMed]
12. Sangwan, S.; Panda, T.; Thiamattam, R.; Dewan, S.K.; Thaper, R.K. Novel Salts of Sunitinib an Anticancer Drug with Improved Solubility. *Int. Res. J. Pure Appl. Chem.* **2014**, *5*, 352–365. [CrossRef]
13. Papaetis, G.S.; Syrigos, K.N. Sunitinib: A multitargeted receptor tyrosine kinase inhibitor in the era of molecular cancer therapies. *BioDrugs.* **2009**, *23*, 377–389. [CrossRef] [PubMed]
14. Ramazani, F.; Hiemstra, C.; Steendam, R.; Kazazi-Hyseni, F.; Van Nostrum, C.F.; Storm, G.; Kiessling, F.; Lammers, T.; Hennink, W.E.; Kok, R.J. Sunitinib microspheres based on [PDLLA-PEG-PDLLA]-b-PLLA multi-block copolymers for ocular drug delivery. *Eur. J. Pharm. Biopharm.* **2015**, *95 (Pt B)*, 368–377. [CrossRef]
15. Raymond, E.; Dahan, L.; Raoul, J.L.; Bang, Y.J.; Borbath, I.; Lombard-Bohas, C.; Valle, J.; Metrakos, P.; Smith, D.; Vinik, A.; et al. Sunitinib malate for the treatment of pancreatic neuroendocrine tumors. *N. Engl. J. Med.* **2011**, *364*, 501–513. [CrossRef]
16. Chen, S.; Liang, Q.; Liu, E.; Yu, Z.; Sun, L.; Ye, J.; Shin, M.; Wang, J.; He, H. Curcumin/sunitinib co-loaded BSA-stabilized SPIOs for synergistic combination therapy for breast cancer. *J. Mater. Chem. B* **2017**, *5*, 4060–4072. [CrossRef]
17. Nazari-Vanani, R.; Azarpira, N.; Heli, H.; Karimian, K.; Sattarahmady, N. A novel self-nanoemulsifying formulation for sunitinib: Evaluation of anticancer efficacy. *Colloids Surfaces B Biointerfaces* **2017**, *160*, 65–72. [CrossRef]
18. Adams, V.R.; Leggas, M. Sunitinib malate for the treatment of metastatic renal cell carcinoma and gastrointestinal stromal tumors. *Clin. Ther.* **2007**, *29*, 1338–1353. [CrossRef]
19. Alshahrani, S.M.; Alshetaili, A.S.; Alalaiwe, A.; Alsulays, B.B.; Anwer, M.K.; Al-Shdefat, R.; Imam, F.; Shakeel, F. Anticancer Efficacy of Self-Nanoemulsifying Drug Delivery System of Sunitinib Malate. *AAPS PharmSciTech* **2018**, *19*, 123–133. [CrossRef]
20. Anwer, M.K.; Ahmed, M.M.; Ezzeldin, E.; Fatima, F.; Alalaiwe, A.; Iqbal, M. Preparation of sustained release apremilast-loaded PLGA nanoparticles: In vitro characterization and in vivo pharmacokinetic study in rats. *Int. J. Nanomed.* **2019**, *14*, 1587–1595. [CrossRef]
21. Jamil, A.; Aamir Mirza, M.; Anwer, M.K.; Thakur, P.S.; Alshahrani, S.M.; Alshetaili, A.S.; Telegaonkar, S.; Panda, A.K.; Iqbal, Z. Co-delivery of gemcitabine and simvastatin through PLGA polymeric nanoparticles for the treatment of pancreatic cancer: In-vitro characterization, cellular uptake, and pharmacokinetic studies. *Drug Dev. Ind. Pharm.* **2019**, *45*, 745–753. [CrossRef] [PubMed]
22. Anwer, M.K.; Al-Shdefat, R.; Ezzeldin, E.; Alshahrani, S.M.; Alshetaili, A.S.; Iqbal, M. Preparation, Evaluation and Bioavailability Studies of Eudragit Coated PLGA Nanoparticles for Sustained Release of Eluxadoline for the Treatment of Irritable Bowel Syndrome. *Front. Pharm.* **2017**, *8*, 844. [CrossRef] [PubMed]
23. Kaushik, N.; Borkar, S.B.; Nandanwar, S.K.; Panda, P.K.; Choi, E.H.; Kaushik, N.K. Nanocarrier cancer therapeutics with functional stimuli-responsive mechanisms. *J. Nanobiotechnol.* **2022**, *20*, 152. [CrossRef]
24. Anwer, M.K.; Mohammad, M.; Iqbal, M.; Ansari, M.N.; Ezzeldin, E.; Fatima, F.; Alshahrani, S.M.; Aldawsari, M.F.; Alalaiwe, A.; Alzahrani, A.A.; et al. Sustained release and enhanced oral bioavailability of rivaroxaban by PLGA nanoparticles with no food effect. *J. Thromb. Thrombolysis* **2020**, *49*, 404–412. [CrossRef]
25. Raza, F.; Zhu, Y.; Chen, L.; You, X.; Zhang, J.; Khan, A.; Khan, M.W.; Hasnat, M.; Zafar, H.; Wu, J.; et al. Paclitaxel-loaded pH responsive hydrogel based on self-assembled peptides for tumor targeting. *Biomater. Sci.* **2019**, *7*, 2023–2036. [CrossRef]
26. Raza, F.; Zafar, H.; You, X.; Khan, A.; Wu, J.; Ge, L. Cancer nanomedicine: Focus on recent developments and self-assembled peptide nanocarriers. *J. Mater. Chem. B* **2019**, *7*, 7639–7655. [CrossRef]
27. Anwer, M.K.; Iqbal, M.; Muharram, M.M.; Mohammad, M.; Ezzeldin, E.; Aldawsari, M.F.; Alalaiwe, A.; Imam, F. Development of Lipomer Nanoparticles for the Enhancement of *Drug* Release, Anti-microbial Activity and Bioavailability of Delafloxacin. *Pharmaceutics* **2020**, *12*, 252. [CrossRef]
28. Khan, M.M.; Madni, A.; Torchilin, V.; Filipczak, N.; Pan, J.; Tahir, N.; Shah, H. Lipid-chitosan hybrid nanoparticles for controlled delivery of cisplatin. *Drug Deliv.* **2019**, *26*, 765–772. [CrossRef]
29. Dong, W.; Wang, X.; Liu, C.; Zhang, X.; Zhang, X.; Chen, X.; Kou, Y.; Mao, S. Chitosan based polymer-lipid hybrid nanoparticles for oral delivery of enoxaparin. *Int. J. Pharm.* **2018**, *547*, 499–505. [CrossRef]

30. Anwer, M.K.; Ali, E.A.; Iqbal, M.; Ahmed, M.M.; Aldawsari, M.F.; Saqr, A.A.; Ansari, M.N.; Aboudzadeh, M.A. Development of Sustained Release Baricitinib Loaded Lipid-Polymer Hybrid Nanoparticles with Improved Oral Bioavailability. *Molecules* **2022**, *27*, 168. [CrossRef]
31. Joseph, J.J.; Sangeetha, D.; Gomathi, T. Sunitinib loaded chitosan nanoparticles formulation and its evaluation. *Int. J. Biol. Macromol.* **2016**, *82*, 952–958. [CrossRef] [PubMed]
32. Khuroo, T.; Verma, D.; Khuroo, A.; Ali, A.; Iqbal, Z. Simultaneous delivery of paclitaxel and erlotinib from dual drug loaded PLGA nanoparticles: Formulation development, thorough optimization and in vitro release. *J. Mol. Liq.* **2018**, *257*, 52–68. [CrossRef]
33. Alshetaili, A.S.; Anwer, M.K.; Alshahrani, S.M.; Alalaiwe, A.; Alsulays, B.B.; Ansari, M.J.; Imam, F.; Alshehri, S. Characteristics and anticancer properties of Sunitinib malate-loaded poly-lactic-co-glycolic acid nanoparticles against human colon cancer HT-29 cells lines. *Trop. J. Pharm. Sci.* **2018**, *17*, 1263. [CrossRef]
34. Alhakamy, N.A.; Fahmy, U.A.; Badr-Eldin, S.M.; Ahmed, O.A.A.; Asfour, H.Z.; Aldawsari, H.M.; Algandaby, M.M.; Eid, B.G.; Abdel-Naim, A.B.; Awan, Z.A.; et al. Optimized Icariin Phytosomes Exhibit Enhanced Cytotoxicity and Apoptosis-Inducing Activities in Ovarian Cancer Cells. *Pharmaceutics.* **2020**, *12*, 346. [CrossRef]
35. Md, S.; Alhakamy, N.A.; Alharbi, W.S.; Ahmad, J.; Shaik, R.A.; Ibrahim, I.M.; Ali, J. Development and Evaluation of Repurposed Etoricoxib Loaded Nanoemulsion for Improving Anticancer Activities against Lung Cancer Cells. *Int. J. Mol. Sci.* **2021**, *22*, 13284. [CrossRef] [PubMed]
36. Anwer, M.K.; Ahmed, M.M.; Aldawsari, M.F.; Alshahrani, S.; Fatima, F.; Ansari, M.N.; Rehman, N.U.; Al-Shdefat, R.I. Eluxadoline Loaded Solid Lipid Nanoparticles for Improved Colon Targeting in Rat Model of Ulcerative Colitis. *Pharmaceuticals* **2020**, *13*, 255. [CrossRef] [PubMed]
37. Sahu, A.R.; Bothara, S.B. Formulation and evaluation of phytosome drug delivery system of boswellia Serrata extract. *Int. J. Res. Med.* **2015**, *4*, 94–99.
38. Liechty, W.B.; Peppas, N.A. Expert opinion: Responsive polymer nanoparticles in cancer therapy. *Eur. J. Pharm. Biopharm.* **2012**, *80*, 241–246. [CrossRef]
39. Subhan, M.A.; Yalamarty, S.S.K.; Filipczak, N.; Parveen, F.; Torchilin, V.P. Recent Advances in Tumor Targeting via EPR Effect for Cancer Treatment. *J. Pers. Med.* **2021**, *11*, 571. [CrossRef]
40. Silva, M.M.; Calado, R.; Marto, J.; Bettencourt, A.; Almeida, A.J.; Gonçalves, L. Chitosan Nanoparticles as a Mucoadhesive Drug Delivery System for Ocular Administration. *Mar. Drugs* **2017**, *15*, 370. [CrossRef]
41. Alshetaili, A.S. Gefitinib loaded PLGA and chitosan coated PLGA nanoparticles with magnified cytotoxicity against A549 lung cancer cell lines. *Saudi J. Biol. Sci.* **2021**, *28*, 5065–5073. [CrossRef] [PubMed]
42. Zhang, L.I.; Zhang, L. Lipid–polymer hybrid nanoparticles: Synthesis, Characterization and Applications. *Nano Life* **2010**, *1*, 163–173. [CrossRef]
43. Anzar, N.; Mirza, M.A.; Anwer, M.K.; Khuroo, T.; Alshetaili, A.S.; Alshahrani, S.M.; Meena, J.; Hasan, N.; Talegaonkar, S.; Panda, A.K.; et al. Preparation, evaluation and pharmacokinetic studies of spray dried PLGA polymeric submicron particles of simvastatin for the effective treatment of breast cancer. *J. Mol. Liq.* **2018**, *249*, 609–616. [CrossRef]
44. Bhatt, P.; Narvekar, P.; Lalani, R.; Chougule, M.B.; Pathak, Y.; Sutariya, V. An in vitro Assessment of Thermo-Reversible Gel Formulation Containing Sunitinib Nanoparticles for Neovascular Age-Related Macular Degeneration. *AAPS PharmSciTech* **2019**, *20*, 281. [CrossRef] [PubMed]
45. Razmimanesh, F.; Sodeifian, G.; Sajadian, S.A. An investigation into Sunitinib malate nanoparticle production by US- RESOLV method: Effect of type of polymer on dissolution rate and particle size distribution. *J. Supercrit Fluids* **2021**, *170*, 105163. [CrossRef]
46. Varma, R.; Vasudevan, S. Extraction, Characterization, and Antimicrobial Activity of Chitosan from Horse Mussel Modiolus modiolus. *ACS Omega* **2020**, *5*, 20224–20230. [CrossRef]
47. Saoji, S.D.; Raut, N.A.; Dhore, P.W.; Borkar, C.D.; Popielarczyk, M.; Dave, V.S. Preparation and Evaluation of Phospholipid-Based Complex of Standardized Centella Extract (SCE) for the Enhanced Delivery of Phytoconstituents. *AAPS J.* **2016**, *18*, 102–114. [CrossRef]
48. Selic, L. New Crystal Form of Sunitinib Malate. European Patent Number EP2362873B1, 3 June 2015.
49. Arora, S.; Saharan, R.; Kaur, H.; Kaur, I.; Bubber, P.; Bharadwaj, L.M. Attachment of Docetaxel to Multiwalled Carbon Nanotubes for Drug Delivery Applications. *Adv. Sci. Lett.* **2012**, *17*, 70–75. [CrossRef]
50. Korashy, H.M.; Maayah, Z.H.; Al Anazi, F.E.; Alsaad, A.M.; Alanazi, I.O.; Belali, O.M.; Al-Atawi, F.O.; Alshamsan, A. Sunitinib Inhibits Breast Cancer Cell Proliferation by Inducing Apoptosis, Cell-cycle Arrest and DNA Repair While Inhibiting NF-κB Signaling Pathways. *Anticancer Res.* **2017**, *37*, 4899–4909. [CrossRef]
51. Maayah, Z.H.; El Gendy, M.A.; El-Kadi, A.O.; Korashy, H.M. Sunitinib, a tyrosine kinase inhibitor, induces cytochrome P450 1A1 gene in human breast cancer MCF7 cells through ligand-independent aryl hydrocarbon receptor activation. *Arch. Toxicol.* **2013**, *87*, 847–856. [CrossRef] [PubMed]
52. Ullah, I.; Khalil, A.T.; Ali, M.; Iqbal, J.; Ali, W.; Alarifi, S.; Shinwari, Z.K. Green-Synthesized Silver Nanoparticles Induced Apoptotic Cell Death in MCF-7 Breast Cancer Cells by Generating Reactive Oxygen Species and Activating Caspase 3 and 9 Enzyme Activities. *Oxid. Med. Cell. Longev.* **2020**, *2020*, 1215395. [CrossRef] [PubMed]
53. McIlwain, D.R.; Berger, T.; Mak, T.W. Caspase functions in cell death and disease. *Cold Spring Harb. Perspect. Biol.* **2013**, *5*, a008656. [CrossRef]

54. Boatright, K.M.; Renatus, M.; Scott, F.L.; Sperandio, S.; Shin, H.; Pedersen, I.M.; Ricci, J.E.; Edris, W.A.; Sutherlin, D.P.; Green, D.R.; et al. A unified model for apical caspase activation. *Mol. Cell.* **2003**, *11*, 529–541. [CrossRef]
55. Riedl, S.J.; Shi, Y. Molecular mechanisms of caspase regulation during apoptosis. *Nat. Rev. Mol. Cell Biol.* **2004**, *5*, 897–907. [CrossRef] [PubMed]

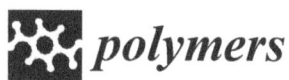

polymers

MDPI

Article

Apigenin Loaded Lipoid–PLGA–TPGS Nanoparticles for Colon Cancer Therapy: Characterization, Sustained Release, Cytotoxicity, and Apoptosis Pathways

Mohamed A. Alfaleh [1,2], Anwar M. Hashem [2,3], Turki S. Abujamel [2,4], Nabil A. Alhakamy [1], Mohd Abul Kalam [5], Yassine Riadi [6] and Shadab Md [1,*]

1 Department of Pharmaceutics, Faculty of Pharmacy, King Abdulaziz University, Jeddah 21589, Saudi Arabia
2 Vaccines and Immunotherapy Unit, King Fahd Medical Research Center, King Abdulaziz University, Jeddah 21589, Saudi Arabia
3 Department of Medical Microbiology and Parasitology, Faculty of Medicine, King Abdulaziz University, Jeddah 21589, Saudi Arabia
4 Department of Medical Laboratory Sciences, Faculty of Applied Medical Sciences, King Abdulaziz University, Jeddah 21589, Saudi Arabia
5 Nanobiotechnology Unit, Department of Pharmaceutics, College of Pharmacy, King Saud University, Riyadh 11451, Saudi Arabia
6 Department of Pharmaceutical Chemistry, College of Pharmacy, Prince Sattam Bin Abdulaziz University, Al-Kharj 11942, Saudi Arabia
* Correspondence: shaque@kau.edu.sa

Citation: Alfaleh, M.A.; Hashem, A.M.; Abujamel, T.S.; Alhakamy, N.A.; Kalam, M.A.; Riadi, Y.; Md, S. Apigenin Loaded Lipoid–PLGA–TPGS Nanoparticles for Colon Cancer Therapy: Characterization, Sustained Release, Cytotoxicity, and Apoptosis Pathways. *Polymers* **2022**, *14*, 3577. https://doi.org/10.3390/polym14173577

Academic Editor: Stephanie Andrade

Received: 9 June 2022
Accepted: 22 August 2022
Published: 30 August 2022

Abstract: Colon cancer (CC) is one of major causes of mortality and affects the socio-economic status world-wide. Therefore, developing a novel and efficient delivery system is needed for CC management. Thus, in the present study, lipid polymer hybrid nanoparticles of apigenin (LPHyNPs) was prepared and characterized on various parameters such as particle size (234.80 ± 12.28 nm), PDI (0.11 ± 0.04), zeta potential (−5.15 ± 0.70 mV), EE (55.18 ± 3.61%), etc. Additionally, the DSC, XRD, and FT-IR analysis determined drug entrapment and affinity with the selected excipient, demonstrating a promising drug affinity with the lipid polymer. Morphological analysis via SEM and TEM exhibited spherical NPs with a dark color core, which indicated drug entrapment inside the core. In vitro release study showed significant ($p < 0.05$) sustained release of AGN from LPHyNPs than AGN suspension. Further, the therapeutic efficacy in terms of apoptosis and cell cycle arrest of developed LPHyNPs against CC was estimated by performing flow cytometry and comparing its effectiveness with blank LPHyNPs and AGN suspension, which exhibited remarkable outcomes in favor of LPHyNPs. Moreover, the mechanism behind the anticancer attribute was further explored by estimating gene expression of various signaling molecules such as Bcl-2, BAX, NF-κB, and mTOR that were involved in carcinogenic pathways, which indicated significant ($p < 0.05$) results for LPHyNPs. Moreover, to strengthen the anticancer potential of LPHyNPs against chemoresistance, the expression of JNK and MDR-1 genes was estimated. Outcomes showed that their expression level reduced appreciably when compared to blank LPHyNPs and AGN suspension. Hence, it can be concluded that developed LPHyNPs could be an efficient therapeutic system for managing CC.

Keywords: hybrid nanoparticle; apigenin; sustained release; mTOR; apoptosis; colon cancer

1. Introduction

Colon cancer (CC) is among the leading causes of death worldwide, including lung cancer, breast cancer, and pancreatic cancer. In terms of diagnosis, it ranks 3rd (9.1%), whereas, in terms of the rate of mortality, it ranks 2nd (9.2%). As per the published report of 2020, approximately 1.9 million cases of CC were reported, and further global burden from CC may increase by 60%, i.e., ~2.2 million, by the end of 2035 [1]. Published reports have highlighted the fact that the level of incidence of CC is strongly correlated with the country's socio-economic development as well as the lifestyle of the population [2]. Coexisting metabolic syndrome, obesity, red meat consumption, alcohol, and sedentary lifestyle are some of the driving forces for the increased cases of CC. It was found that the majority of CC cases are associated with mutation-related polyposis MUTYH (MAP), Lynch syndrome (HNPCC), familial adenomatous polyposis (FAP), and hamartomatous polyposis syndromes. The mean age of diagnosis of disease ranges between 44 and 61 year and inflammatory bowel disease, Crohn's disease, as well as ulcerative colitis increase the risk of CC [3].

Cellular and molecular investigation showed the diverse role of multiple signaling pathways in its etiology. Among all, NF-κB/PI3K/Akt/mTOR/p53/BAX/JNK pathways have played a pivotal role [4]. NF-κB is one of the extensively explored proinflammatory transcription factors, but studies have shown its critical involvement in the etiology of colon and other types of cancer via increased proliferation, angiogenesis, metastasis, and reduced apoptosis [5]. In normal physiological conditions, NF-κB is present/restricted in the cytoplasm under the influence of the IKK complex via the IκB proteins. However, under the oncogenic stimulus, IKK gets activated, and IκB gets phosphorylated and degrades, leading to its nuclear translocation and subsequent initiation of the pro-oncogenic activity. Specifically, increased activity of Akt/PI3K pathways has been reported in CC. Akt is a critical regulator of tumor cell proliferation and survival and causes phosphorylation of various pro-oncogenic targets.

On the one hand, Akt causes nuclear translocation of NF-κB via phosphorylation of IKKα, while, on the other hand, it causes phosphorylation and activation of mTOR (mammalian target of rapamycin) [6]. Not only this, the direct relation between IKKα and mTOR were reported by Dan et al. 2007, where IKKα remains in association with TORC1 and controls the kinase activity of mTOR in tumorigenesis [7]. A significant role of NF-κB/Akt/mTOR pathways in tumorigenesis was confirmed by various preclinical studies where increased activity and level of PTEN (inhibitor of Akt) and rapamycin (inhibitor of mTOR) showed the anticancer effect [8]. Apart from the significant role of the NF-κB/mTOR pathway, reduced apoptosis via JNK/p53/BAX/Bcl-2 is also a decisive factor in the etiology of CC. In the CC and other types of cancer, reduced apoptosis is commonly reported, and any drug that stimulates or increases the apoptosis is considered a potential anticancer drug [9].

It is also important to note that various anticancer drugs were developed to manage and treat CC, but chemotherapeutic resistance is one of the leading factors for the poor clinical outcome of these drugs among the treated patients. One of the major reasons for the drug resistance in the CC is the increased expression of the multidrug resistance (MDR1) gene. MRD1 encodes the p-gp, also known as ATP-binding cassette subfamily B member 1 (ABCB1), and causes efflux of most anticancer drugs, leading to reduced bioavailability and drug resistance, and poor clinical outcomes [10]. Hence, in the present study, we designed and developed an apigenin (AGN) hybrid nanoparticle (HyNP) to manage and treat CC.

AGN is also known as 4′,5,7-trihydroxyflavone and belongs to the class of compounds known as flavones. AGN is commonly found as an active constituent in many Chinese traditional systems of medicines. It possesses significant antioxidant, anti-inflammatory, and antitumor potential in various preclinical studies. The first report on the anticancer potential of AGN was reported by Bart et al., in 1986, and since then, numerous research has been done to explore its anticancer potential in various types of cancers [11]. AGN exhibits an anticancer effect via apoptosis induction where it modulates the level of caspases, BAX, Bcl2, p53 [12], etc. AGN also modulates cell cycle progression via blocking the cycle arrest at G2/M or G0/G1 checkpoint. Moreover, AGN induces autophagy, blocks migration and invasion, and inhibits angiogenesis. Considering the molecular anticancer mechanism of AGN, it modulates the PI3K/AKT/mTOR signaling pathway, NF-κB/MAPK/ERK pathway, and Wnt/JNK pathway [13].

Despite being a potent anticancer molecule, AGN suffers from significant pharmacokinetic limitations. As per Biopharmaceutical Classification System, AGN has been categorized as a Class II drug. It has high permeability but low solubility [14]. In the pharmacokinetic study, it was found that upon administration of 60 mg/kg of AGN in rats, C_{max} was 1.33 ± 0.24 μg/mL, and $AUC_{0-t^{\circ}}$ was found to be 11.76 ± 1.52 μg h/mL [15]. Thus, to over the limitation, AGN-loaded lipid-polymer-HyNP (LPHyNPs) was fabricated and explored for anticancer potential in the in vitro model of CC.

It is further important to understand that lipid-based NPs, because of having an amphiphilic chain, easily get functionalized, become biocompatible, and possess increased duration of circulation, but at the same time suffers from the limitations of instability and loss of structural integrity [16]. So, polymeric NPs, which are one of the advanced drug delivery systems that enhanced stability and increased drug loading capacity was selected. Thus, attempts were made to combine both the drug delivery system and names as lipid–polymer HyNPs [17,18]. This drug delivery system possesses a polymeric core and lipidic shell where the lipid core lies on the outer side, retards the degradation of polymers via restraining inward diffusion of water. As per the report of Yu et al., 2018, LPHyNPs loaded with salinomycin showed improved pharmacokinetic attributes. Salzano et al., fabricated HyNPs and reported the controlled release of daunorubicin and lornoxicam [19,20]. Additionally, hybrid PLGA NPs was fabricated, in which a stabilizer was also added. One commonly used stabilizer is D-α-tocopherol polyethylene glycol 1000 succinate (vitamin E-TPGS). When PLGA as a polymer and TPGS as a stabilizer were used, p-gp efflux was reduced considerably, and improved pharmacological attributes could be achieved [21].

Thus, the present study was designed with novelty to provide promising therapeutic effects against CC. In this case, AGN-loaded lipid HyNPs of Vit-E-TPGS (LPHyNPs) was fabricated. Further, it was prepared and characterized on various parameters such as particle size, polydispersity index (PDI), zeta potential, and EE. Additionally, LPHyNPs was evaluated on DSC, XRD, and FT-IR. Morphological analysis and in vitro release study of the formulation was carried out. Next, the success of formulation toward anticancer potency against CC were performed via modulation NF-κB/mTOR/Bcl2/JNK/BAX/MDR1.

2. Experimental Methodology

2.1. Material

AGN, PLGA and D-α-tocopherol polyethylene glycol 1000 succinate (Vit E TPGS) were purchased from Sigma Aldrich, St. Louis, MO, USA. Lipoid SPC (hydrogenated phosphatidylcholine from soybean) was procured from LIPOID, (GmbH, Ludwigshafen Germany). Dichloromethane (DCM) and dimethyl sulfoxide (DMSO, 99.9%) were procured from Fischer Scientific (Loughborough, UK). A human colorectal cancer cell line was purchased from ATCC (Manassas, VA, USA). HCT 116 cells were grown in Dulbecco's modified Eagle medium (DMEM, Gibco, London, UK), which was supplemented with 10% fetal bovine serum (FBS), Pen Strep (5000 units/mL penicillin and 5000 μg/mL streptomycin). The Annexin V/PI apoptosis detection kit was procured from Invitrogen Corporation (Carlsbad, CA, USA).

2.2. Method of Preparation of LPHyNPs

LPHyNPs were prepared by nanoprecipitation method [22] with slight modification and in this case, solutions were designed in two-phase system. In phase one, 50 mg of PLGA (50:50), 100 mg of Lipoid S PC-3, and 5 mg of AGN (previously dissolved in 100 μL of DMSO) were dissolved in 5 mL of DCM. Whereas, in phase two, AGN/PLGA weight ratio was kept at 1:10 *w*/*w*, and the Lipoid: PLGA weight ratio was 2:1 *w*/*w*. Vit E-TPGS 1000 was dispersed in 10 mL of Milli-Q water at 0.5% *w*/*v* at heated to 70 °C. Then phase one solution was added drop-wise (at the rate of 1.5 mL/min) into the preheated phase two solution with magnetic stirring (500 rpm). The mixed solution was then homogenized (T25 digital Ultra-Turrax, IKA, UK) for 2 min (21,000 rpm), followed by magnetic stirring at 500 rpm for 4 h at 25 ± 1 °C for the complete evaporation of DCM. The final formulation was washed with Milli-Q water by ultracentrifugation at 30,000 rpm for 30 min (three cycles). Using the dialysis technique (Spectra/PorVR dialysis membrane), the prepared LPHyNPs were recovered and purified. The final LPHyNPs formulation (100 μL) was diluted 50-fold with Milli-Q water for particle characterization by dynamic light scattering (DLS) measurement, such as particle size, PDI, and zeta potential. Further, the formulations were lyophilized using mannitol (1%, *w*/*v*) as a cryoprotectant, and it was frozen at −80 °C and subjected to freeze-drying for further characterization.

2.3. *Characterization of Prepared LPHyNPs*

2.3.1. Determination of Particle Size, PDI and Zeta Potential

Particle size, PDI, and zeta potential measurements were carried out via the DLS technique using the Zetasizer Nano ZSP instrument (Malvern Instruments Ltd., Malvern, UK). The particle size and PDI of three individual collected samples were analyzed after 50-fold dilution in Milli-Q water at a temperature of 25 ± 0.5 °C [23].

2.3.2. Drug Entrapment and Loading Efficiency of Prepared LPHyNP

Drug entrapment (EE) and loading efficiency (DL) of prepared LPHyNP were analyzed in the supernatant (indirect method) of collected samples [24]. Approximately 5 mL of LPHyNPs was diluted in 5 mL of methanol to dissolve the drug and precipitate the PLGA and other excipients. The suspension was centrifuged (rpm for 20 min at 4 °C), and the supernatant was collected. To analyze the drug concentration in the collected supernatant, a 30 μL sample was injected into the HPLC-UV system. For this purpose, a chromatographic technique was developed, which contained a C_{18} analytical column (5 μm, 250 mm × 4.6 mm). The mobile phase was composed of acetonitrile and 0.1% formic acid at 55:45 (*v*/*v*), where pH was maintained at 7.4. The mobile phase was pumped isocratically at a 1 mL/min flow rate, and UV-detection was performed at 270 nm. The %EE and %DL were calculated as per the following equations:

%EE = [(Initial amount of drug − Amount of drug in supernatant)/Initial amount of drug] × 100

%DL = [(Initial amount of drug − Amount of drug in supernatant)/Initial amount of LPHyNPs] × 100

2.3.3. Differential Scanning Calorimeter (DSC) Analysis

A differential scanning calorimeter (DC-60 plus; Shimadzu, Japan) was used to investigate the thermal characteristics of AGN, LPHyNP, trehalose, PLGA, SPC-3, and TPGS samples. In this case, the HyNPs were dried overnight in a desiccator, and then powdered samples were sealed in aluminum pans with lids and heated from a temperature of 25–300 °C with a rate of 10 °C/min under nitrogen flow [17].

2.3.4. X-ray Diffraction Analysis

The samples of AGN, LPHyNPs, PLGA, SPC-3, trehalose, and Vit-E-TPGS were collected to determine their crystallinity. The degree of crystallization was analyzed using a high-resolution XRD (Maxima XRD-7000X, Shimadzu, Kyoto City, Japan) technology with an XRD scanning speed of 5–80°/min [25].

2.3.5. FT-IR Analysis

For FT-IR analysis, various collected samples such as AGN, LPHyNPs, PLGA, SPC-3, and Vit-E-TPGS were analyzed using an FT-IR spectrophotometer (Bruker 375 Tensor-27, Billerica, MA, USA). Additionally, dry samples were compressed as KBr pellets using an instrument pin. The selected transmission range was between a wave number of 4000–400 cm^{-1} [26].

2.3.6. Morphological Analysis

Scanning electron microscopy (SEM, ZEISS, Germany) and transmission electron microscopy (TEM, JEOL JEM 1010, Tokyo, Japan) were used to examine the nanostructure of the prepared LPHyNP. SEM is a high-resolution field emission scanning electron microscope that allows samples to be examined with an accelerating voltage of 15 kV [27]. Whereas, for TEM investigation, one drop of LPHyNP dispersion, which was prepared in Milli-Q water was dried on a copper grid. Uranyl acetate (2% w/w) was then used to stain the sample. The stained sample was air-dried to remove excess liquid media before being analyzed by TEM [28]. The length measurement of the particles was carried out with the help of ImageJ software (Version 1.53e, National institute of health, Bethesda, MD, USA). The particles are not exactly circular, so the length measurements were done at four different angles in such a manner that the shape is equally divided into two parts for each measurement. All measurements were transferred to the OriginPro software (Version 8.5.0, OriginLab Corporation, Northampton, MA, USA). The data were plotted as a histogram to get the bin worksheet. The bin worksheet generated a column plot using Bin centers vs. Bin counts. Finally, the plot was fitted using the nonlinear curve fit command of origin, and Gaussian fit was chosen to get the bell-shaped curve of Gaussian fit of counts.

2.4. *Study of In Vitro Release Pattern*

A dialysis bag with a molecular weight cut-off of 12,000 Da was used to perform in vitro drug release. In a nutshell, the LPHyNP and AGN suspensions were placed individually in the dialysis bag, knotted, and immersed in the release medium. The release media (50 mL) in the study was phosphate-buffered saline (PBS) of pH 7.8 with sodium lauryl sulfate (1%) as a solubilizer. Throughout the investigation, the system was kept at 37 °C in a shaker water bath. The samples were taken at 1, 2, 3, 4, 6, 8, 24, 48, and 72 h, and the AGN content was analyzed at given time intervals using the HPLC technique [29].

2.5. *Cell Viability Assay*

For the assessment of the cell viability assay, a colon cancer cell line (HCT-116) was used, and MTT proliferation kit (Sigma Aldrich, St Louis, MO, USA) was used. Initially, the cells were incubated at the temperature of 37 °C using a 96-well plate for 24 h where cells were grown and media was supplemented with 10% FBS, 1% Pen/strep, and 1 mM glutamine. The density of cells used in the process was 5×10^3 HCT-116 cells/well in a humidified CO_2 chamber. Cells were treated with the concentration of 6.25, 25, and 100 µg/mL of blank NPs, AGN suspension, and LPNHyNP and left for 24 h and then treated with MTT solution. Stock solution was prepared by dissolving 5 mg of MTT in 1 mL or 1000 µL of PBS. About 10 µL of the stock solution was used per well and kept for 4 h at 37 °C to develop formazan crystals. In the next step, excess culture media from each well was removed by washing and formazan crystals were solubilized by adding 100 µL DMSO for 20 min and the absorbance was recorded at 563 nm [30]. The same procedure mentioned above was used for a normal human cell line, i.e., HEK293 (the most widely used and readily available normal human cell line), to determine the cytotoxic effect of blank LPHyNPs, AGN suspension, and LPHyNPs formulations on a normal cell line. The formulations were treated in the same dose range as that used for the HCT-116 cell line.

2.6. *Cell Cycle and Apoptosis Analysis Using Flow Cytometry*

For the analysis of cell cycle, HCT-116 cells were properly fixed in 70% ethanol on ice for approximately 15 min. In the next step, HCT-cells were incubated in the binding buffer and propidium iodide at room temperature in a dark room for 20 min. Apoptosis assay was performed by using annexin V (FITC)/PI assay kit (K101-100, Biovision Inc., Milpitas, CA, USA) and performed as per the manufacturer's instruction. In brief, 1×10^5 cells/mL were allowed to get treated with HCT-116 and incubated for 24 h. Next, the cells were centrifuged, washed with the phosphate buffer, and resuspended using 500 µL of the buffer. Followed by this, 100 µL of resuspended cells were again incubated with the 5 μL of PI and Annexin-V at room temperature in a dark room for 15 min and analyzed using BD FACSCalibur reader, and data were analyzed using flow cytometer and flow system software, USA [31,32].

2.7. Gene Expression of the Various Carcinogenic Marker Using RT-PCR

For the estimation of BAX, Bcl-2, NF-κB, mTOR, JNK, and MD1, HCT-116 cells were initially treated with the various treatment groups at the concentration of IC50 of various samples and incubated in 96-well plated; blank NPs (124.77 μg/mL), AGN suspension (50.93 μg/mL), and LPNHyNPs (10.89 μg/mL). Concentration of IC50 was set in the present study because at this concentration maximum cytotoxicity was observed. In the RT-PCR study, to extract RNA, TRIzol reagent was used, and cDNA was synthesized. The expression level of the aforementioned genes was estimated using Rotor-Gene Q software and reported as fold change [30]. The primer sequence used in the study is shown in Table 1.

Table 1. The primer used in the RT-PCR.

Target	Primer Used
BAX	F: 5′-CTGCAGAGGATGATTGCCG-3′ R: 5′-TGCCACTCGGAAAAAGACCT-3′
Bcl-2	F: 5′-GACTTCGCCGAGATGTCCAG-3′ R: 5′-GAACTCAAAGAAGGCCACAATC-3′
mTOR	F: 5′-GCTTGATTTGGTTCCCAGGACAGT3 R: 5′-GTGCTGAGTTTGCTGTACCCATGT3′
JNK	F: 5′ -GTGT-GGAATCAAGCACCTTC-3′ R: 5′ -AGGCGTCATCATAAAACTCGTTC-3
NF-κB	F: 5′- CGCATCCAGACCAACAACA-3′ R: 5′- TGCCAGAGTTTCGGTTCAC-3′
MDR1	F: 5′-CCC ATC ATT GCA ATA GCA GG-3′ R: 5′-TGT TCA AAC TTC TGC TCC TGA-3′
β-actin	F: 5′-AGAGCTACGAGCTGCCTGAC-3′ R: 5′-AGCACTGTGTTGGCGTACAG-3′

2.8. Statistical Analysis

The experiments were carried out in triplicates, and the findings were given as mean ± standard deviation (SD). A one-way ANOVA followed by Tukey, multiple comparison tests was used to determine statistical significance, with a p-value of <0.05 considered significant.

3. Results and Discussion

3.1. Determination of Particle Size, PDI, and Surface Potential of LPHyNP

The DLS technique was used to determine the particle size, PDI, and zeta potential. The prepared blank LPHyNPs and drug-loaded LPHyNPs exhibited an average particle size of 200.26 ± 9.19 nm (Table 2) and 234.80 ± 12.28 nm (Figure 1a,b), respectively, which has been pondered as HyNPs. Simultaneously, the observed PDI of blank LPHyNPs and drug-loaded LPHyNPs were 0.34 ± 0.10 and 0.11 ± 0.04, respectively. These results were associated only with the addition of drug. When the drug was added, the particle size of NP was increased due to the entrapment of the drug, which expanded its particle size. On the contrary, the PDI of drug-loaded LPHyNPs decreased compared to blank LPHyNPs, which demonstrated a homogenous population of NP [33]. The average zeta potential of blank NPs and LPHyNPs were found to be −5.15 ± 0.70 mV and −4.14 ± 0.81, respectively (Figure 1c,d). Thus, the developed LPHyNPs was considered as a stable formulation.

Table 2. Result of characterization of blank LPHyNPs and drug-contained LPHyNPs.

Formulation	Particle Size (nm)	PDI	Zeta Potential (mV)	EE (%)	DL (%)
Blank LPHyNPs	200.26 ± 9.19	0.34 ± 0.10	−4.14 ± 0.81	-	-
LPHyNPs	234.80 ± 12.28	0.11 ± 0.04	−5.15 ± 0.70	55.18 ± 3.61	11.04 ± 0.72

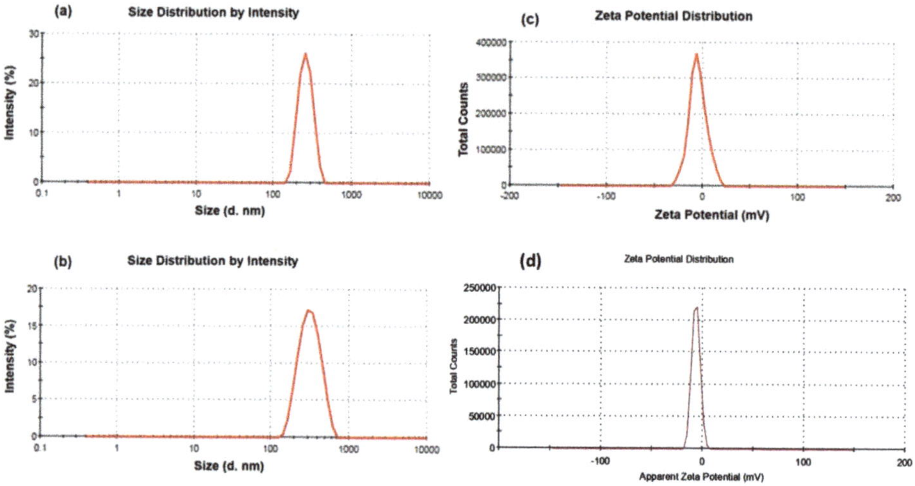

Figure 1. (**a**,**b**) Demonstrate particle size of blank LPHyNPs and drug-loaded LPHyNPs, and (**c**,**d**) show the zeta potential of blank LPHyNPs and drug-loaded LPHyNPs, respectively.

3.2. *Drug Entrapment and Loading Efficiency of Prepared LPHyNP*

The EE and DL of LPHyNPs were calculated to determine the drug concentration in the NP. The EE of LPHyNPs was recorded as 55.18 ± 3.61% (Table 2). In this case, the drug was generally entrapped inside the NP, which sustained the release of the drug from the LPHyNPs. Simultaneously, DL of LPHyNPs was recorded as 11.04 ± 0.72%, and this DL was due to AGN affinity toward the used polymer, i.e., PLGA.

3.3. *Differential Scanning Calorimeter (DSC) Analysis*

In this study, Figure 2 shows the DSC thermograms of AGN, LPHyNP, trehalose, PLGA, SPC-3, and TPGS. The thermogram of AGN showed a sharp endothermic peak at 320 °C, which demonstrated the crystallinity of drug molecules. Because of its glass transition temperature (Tg), the polymer shows an endothermic peak at 51.72 °C, whereas SPC-3 shows a gel–liquid crystalline phase transition at 43.28 °C, followed by a series of irregular peaks above 220 °C, indicating deterioration at higher temperatures [34]. The thermogram of drug-loaded LPHyNP demonstrated the distinctive peak of PLGA and TPGS at 40 °C, the short peak of trehalose at 270 °C, and the absence of AGN peak confirmed the encapsulation of the drug inside the NP.

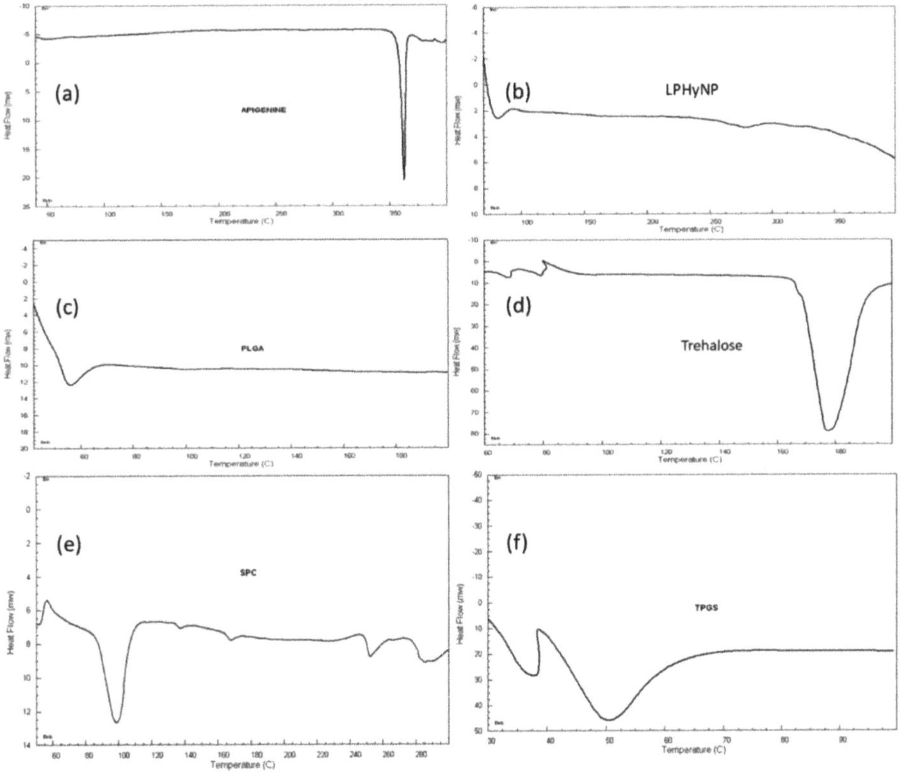

Figure 2. DSC thermogram of (**a**) AGN, (**b**) LPHyNP, (**c**) PLGA, (**d**) trehalose, (**e**) SPC-3, and (**f**) TPGS.

3.4. X-ray Diffraction Analysis

An XRD examination was carried out to determine the crystallinity of different collected samples, such as AGN, LPHyNP, PLGA, SPC-3, trehalose, and Vit-E-TPGS. Figure 3 shows the XRD diffractogram of these samples, in which Figure 3a revealed well-defined peaks of AGN in the examined range, which could reflect the AGN powder's crystalline composition. The lack of unique diffraction peaks in PLGA (Figure 3c) and SPC-3 (Figure 3d) can be attributed to their amorphous natures. On the other hand, the trehalose showed some of the well-defined peaks of the crystalline nature of the material (Figure 3e). Figure 3f shows some of the TPGS and Vit-E characteristic diffraction peaks in their physical mixing. Additionally, Figure 3b claims some of the AGN characteristic diffraction peaks with noisy PLGA and SPC-3 peaks. The amorphous form of the particles was confirmed by the XRD of the LPHyNP.

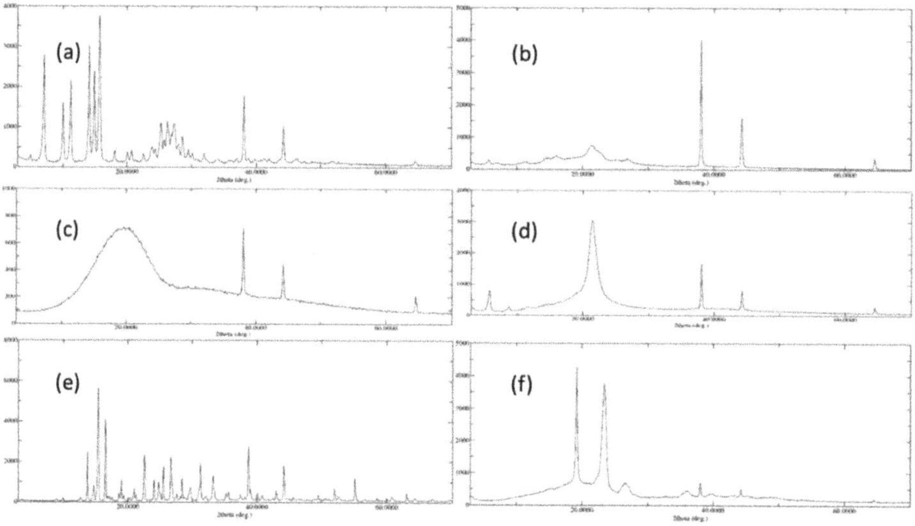

Figure 3. XRD diffractograms of (**a**) AGN, (**b**) LPHyNP, (**c**) PLGA, (**d**) SPC-3, (**e**) trehalose, and (**f**) TPGS.

3.5. FT-IR Analysis

To determine the comparative difference between free AGN and developed LPHyNPs, an FT-IR spectroscopy study was carried out, and spectra of various tested samples (AGN, LPHyNPs, PLGA, SPC-3, and Vit-E-TPGS) are shown in Figure 4. FT-IR spectra show the existence of specific functional groups in the AGN (Figure 4a). The characteristic bands of AGN were found at 3386.44, 1652.39, 1491.80, 1237.89, 1026.76, and 972.59 cm^{-1}. The absorption at 3386.44 cm^{-1} and 1652.39 cm^{-1} presented stretching vibration of the hydroxyl group (OH) and C=O stretching toward the lower vibrational frequencies. The absorption at 1491.80 cm^{-1} was attributed to CH_2 stretching vibration. The absorption at 1237.89 cm^{-1} (antisymmetric stretching) and 1026.76 cm^{-1} (C-O-P-O-C stretching) indicated polar head group vibration. The absorption peak at 971 cm^{-1} exhibited the antisymmetric N^+-CH_3 stretching vibrations [35]. The FT-IR spectra of developed LPHyNPs (Figure 4b) demonstrated a broad absorption peak at 3385.32 cm^{-1} due to the stretching vibrations of the OH group of PLGA and SPC-3. The stretching peak at 2920.81 cm^{-1} presented the stretching vibration of C-H. A peak at 1044.37 cm^{-1} was also observed, which showed the C-O stretching of TPGS. Hence, the FT-IR spectra of LPHyNP indicated promising drug entrapment in the polymer.

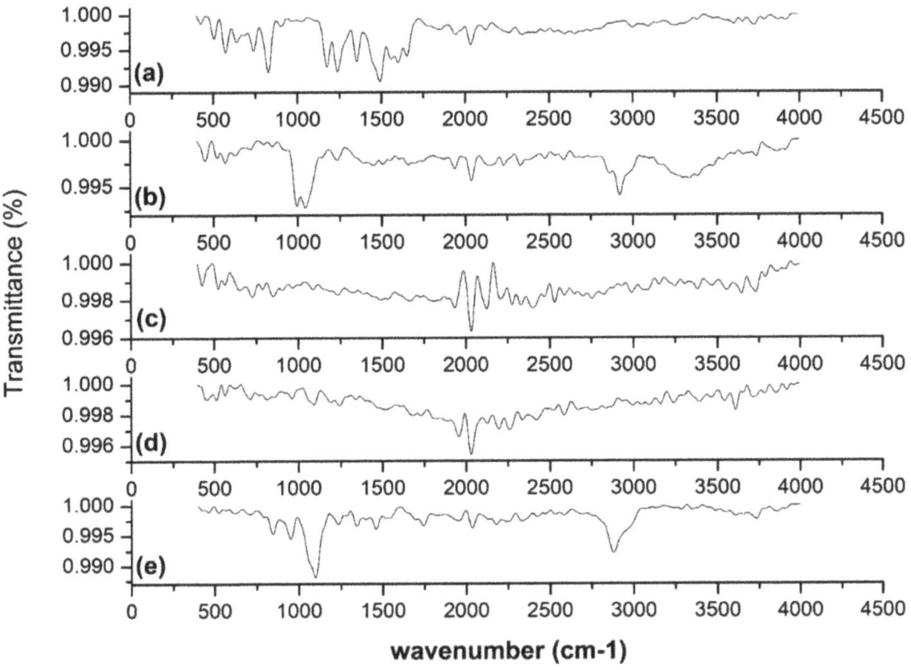

Figure 4. FT-IR spectra of (**a**) AGN, (**b**) LPHyNP, (**c**) PLGA, (**d**) SPC-3, and (**e**) TPGS.

3.6. Morphological Analysis

The morphological pattern of developed LPHyNP obtained via SEM study is depicted in Figure 5a. In the SEM image, LPHyNP exhibited a solid and asymmetrically shaped particle with a smooth surface. Concurrently, the TEM image (Figure 5b) also demonstrated asymmetrically shaped particles with more or less uniform size. The light gray spots in the TEM image indicated unreacted components of the formulation. Additionally, the size histogram (Figure 5c) of TEM showed the average particle size of optimized LPHyNP between 235 and 240 nm.

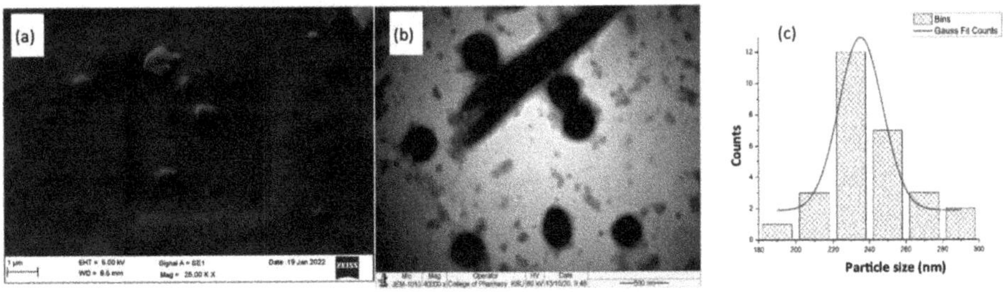

Figure 5. Morphological studies of LPHyNPs (**a**) SEM image, (**b**) TEM image, (**c**) size histogram of TEM.

3.7. Study of In Vitro Release Pattern

In vitro drug release study was performed using the dialysis bag method to analyze the release pattern of LPHyNPs and suspension of AGN, and the result is shown in Figure 6. The release of AGN from LPHyNPs was sufficient for 72 h. In contrast, AGN was released entirely within 24 h with a fast release pattern from the suspension of AGN. In the case of

AGN release from LPHyNPs, an initial burst release was observed for up to 8 h followed by sustained release till 72 h. Therefore, LPHyNPs released AGN significantly ($p < 0.05$) better than the suspension of AGN. Thus, the PLGA encapsulation of LPHyNPs controls the release and provides a sustained release pattern for more than 64 h. The encapsulation of lipophilic AGN in the core of HyNP-contained PLGA could be ascribed to such a sustained release behavior [36].

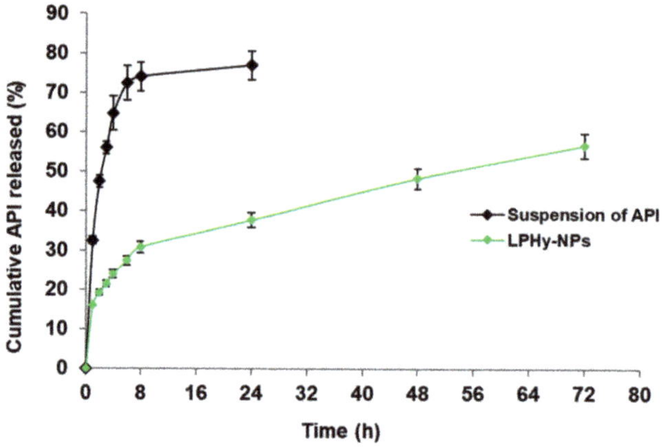

Figure 6. In vitro release of AGN from LPHyNP and suspension of AGN in PBS containing SLS (1%) as a solubilizer by dialysis bag method.

3.8. Cell Viability Analysis (MTT Assay)

MTT, or cell viability, is one of the extensively studied parameters during the initial phase of drug development in preclinical studies. MTT assay provides reliable and concrete information for the cell viability and proliferation property of the drug. In this process, when the drug is exposed to the MTT solution, the viable cells undergo fragmentation of DNA, get colorized, and absorbance is recorded. Upon quantifying the recorded absorbance, viability, as well as IC_{50} concentration in µg/mL, was calculated [37]. The lower the IC_{50} concentration, the more potential is the drug. Here, biosafety and cytotoxicity of blank LPHyNPs, AGN suspension, and LPHyNPs were evaluated on HEK293 and HCT-116. The prepared samples were tested in the same dose range as that tested for the colon cancer cell line, and the cytotoxicity was recorded as minimal and non-significant ($p < 0.05$), as shown in Supplementary Figure S1. Simultaneously, cell viability study was also performed on selected colon cancer cell line (HCT-116). In the current study, when HCT-116 cells were treated with the various drug formulation, and it was found that IC_{50} of blank LPHyNPs was 124.77 ± 7.01 µg/mL, AGN suspension showed IC_{50} of 50.93 ± 2.86 µg/mL, and LPHyNPs exhibited IC_{50} 10.89 ± 0.61 µg/mL, respectively (Figure 7). Thus, as discussed above, the observed results showed that LPNHyNPs was 5 times more potent than AGN suspension and 12 times more potent than blank LPHyNPs. So, it was concluded that the significantly higher anticancer potential of LPNHyNPs could be due to fabricated nanoformulation that exhibits superior drug penetration into the tumor cells. Moreover, the blank nanoformulation also showed anticancer activity due to the presence of Vit E-TPGS 1000 [38].

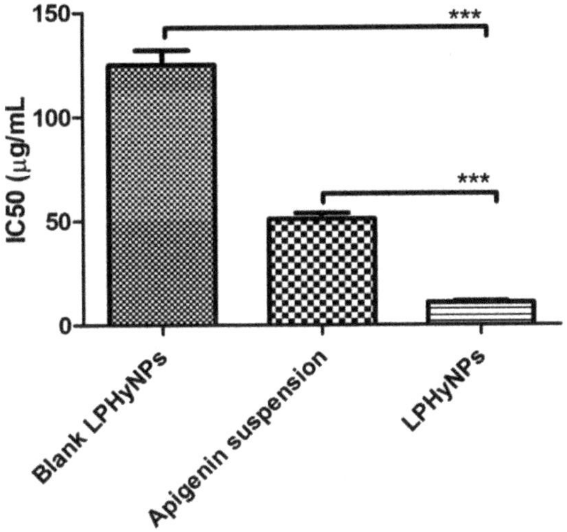

Figure 7. Showing the outcome of MTT assay for blank LPHyNPs, AGN suspension, and LPHyNPs. Values are presented as mean ± SD where $p < 0.001$) was represented by ***.

3.9. *Effect of LPHyNPs on Cell Cycle*

The uncontrolled division is one of the hallmarks of the initiation and progression of tumorigenesis. Although, in a normal homeostatic condition, the frequency of cell division is almost balanced by the apoptotic and necrotic events as an optimum cellular number and volume are critical for the smooth working of various vital organs. Multiple signaling pathways and molecules such as retinoblastoma protein (pRb), cyclin-dependent kinases (Cdk), CDK-activating kinase complex (CAK), etc., are needed for the continuous and balanced cell cycle [39]. However, in response to various oncogenic stimulus, disbalance occurs among these molecules, and an uncontrolled cell cycle occurs. Thus, more detailed and exploratory studies were performed to understand the multiple checkpoints, and likewise, a number of clinically used drugs were developed. At present, various FDA-approved drugs have been used that target the various cell cycle phases [40]. Thus, in the present study, we have also performed the cell cycle analysis to establish the anticancer effect of the developed formulation against colon cancer. Based on the outcome of cell cycle analysis, it was found that a maximum percentage of the cell of the control was present in the G0–G1 phase and exhibited proliferator activity. Moreover, no significant difference was found for the percent cell in the G0–G1 phase when treated with blank LPHyNPs, AGN suspension, and LPHyNPs.

In the present study, the cell population of LPHyNPs in the G2 phase is extremely higher than in the control and the AGN. As it is well-known that the accumulation of cells in the G2/M phases is a distinguishing feature of apoptosis and cell cycle arrest at the G2 checkpoint. Hence, the significant pro-apoptotic effect of LPHyNPs was confirmed and the result of the apoptosis assay was also validated. Hence, the finding of the current study indicates that the effect of LPHyNPs on activating apoptosis is related to an increased proportion of cells at the G2/M phases. Similar findings were observed by others. In other words, an increase in the cell population at the G2-M phase signifies that the relegation event is inhibited that follows DNA cleavage by topoisomerase II. This effect triggers the accumulation of double-strand breaks (loss of heterozygosity and chromosome rearrangements that result in the cell) in the genome after it is replicated, thus producing cell cycle arrest at the G2-M phase. Hence, increase in cell population by LPHyNPs in the G2-M phase signifies the potential of LPHyNPs to cause cell cycle arrest.

Furthermore, it is well established that the S phase is most crucial for the smooth continuation of the cell cycle, and in this phase only, the amount of DNA doubles. Thus, blockage of this phase is ideal for a potent anticancer drug. A maximum percent of cells were found in the S and G2-M phases. A significant difference was found between AGN suspension ($p < 0.05$) and LPHyNPs, as shown in Figure 8. Thus, the superiority of LPHyNPs was established in terms of increased apoptotic and necrotic activity. In a nutshell, LPHyNPs and AGN suspension arrest cell growth at G2/M phase, whereas blank NPs arrest cell growth at S phase and hence, the rationale for formulating LPHyNPs was justified.

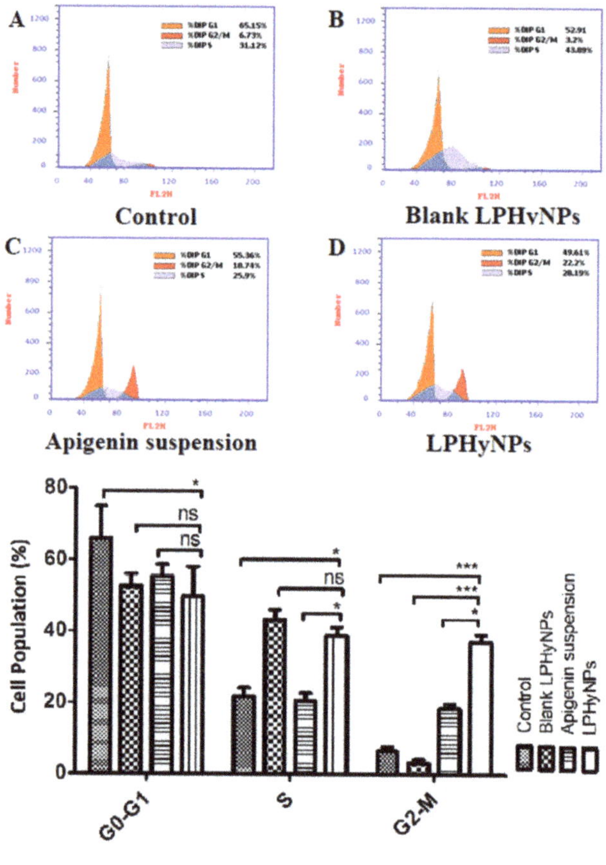

Figure 8. Figure 8 showing the outcome of cell cycle analysis using flow cytometry where effect of different groups such as (**A**) control, (**B**) Blank LPHyNPs, (**C**) apigenin and (**D**) LPHyNPs were evaluated. Values are presented as mean ± SD where $p < 0.05$ and $p < 0.001$ was represented by * and ***. 'ns' demonstrated non-significance differences between results.

3.10. Apoptosis Assay Using Flow Cytometry

Apoptosis is also known as programmed cell death, and it is one of the decisive parameters used to estimate the anticancer potential of a drug entity. Undoubtedly, apoptosis plays a major role in maintaining normal physiological function via maintaining the number and volume of cells [23]. However, when the normal pace of apoptosis is disturbed in response of carcinogenic stimulus, programmed cell death is abrogated, and uncontrolled cell division and survival of tumor cells take place, leading to the initiation and progression of the carcinogenic event. In line with this, numerous published pieces of evidence have shown a positive correlation between reduced apoptosis and progression of colon cancer. Thus, estimation of an apoptotic event or apoptotic cells provided a substantial idea about the progression of disease etiology and also helped in assessing the clinical out of anticancer drugs [41]. Moreover, the apoptotic assay is also regularly performed to validate the MTT assay outcome. The clinical relevance of apoptotic assay can be understood from the fact that many apoptosis-stimulating drugs are in different phases of the clinical trial, and many of the existing anticancer drugs act via stimulating apoptosis in the cancerous cells. Thus, looking into the clinical relevance of apoptosis, in the present study, we have also studied the apoptotic activity of various drug formulations using flow cytometry. Upon comparing the necrotic effect, it was found that control, blank LPHyNPs, and AGN suspension showed almost comparable effects. In contrast, LPHyNPs exhibited significantly superior necrotic effect (1.29%, 1.69%, 1.29% vs. 4.73%) as shown in Figure 9. When the late apoptotic activity was studied, it was found that LPHyNPs showed almost ~2 times higher apoptotic activity as compared to blank LPHyNPs (Q2: 4.22%, blank LPHyNPs) and AGN suspension (4.86%, AGN suspension) and 8.26% (LPHyNPs). Similarly, when the early apoptotic activity was estimated, we found that LPHyNPs showed almost ~2 times higher apoptotic activity as compared to blank LPHyNPs and AGN suspension that, on the other hand, showed similar apoptotic activity (Q4: 11.08%, blank LPHyNPs; 13.51%, AGN suspension) and 24.15% (LPHyNPs). Moreover, upon comparing the total apoptotic activity, it was found that AGN suspension exhibited 1.16 times higher apoptotic activity as compared to blank LPHyNPs, whereas LPHyNPs showed 2.19 and 1.89 times higher total apoptotic activity as compared to blank LPHyNPs and AGN suspension (Q4: 16.94%, blank LPHyNPs); 19.66% (AGN suspension) and 37.14% (LPHyNPs) as shown in Figure 9.

Based on the apoptotic activity it was found that blank NPs showed higher apoptotic activity as compared to control and blank NPs's apoptosis rate is due to its component "Vit E-TPGS 1000" that has proven anticancer property. Vit E-TPGS 1000 was added in blank NPs because of its solubilizer, emulsifier, stabilizer properties, absorption, permeation enhancer and its property to inhibit P-gp overexpression. Thus, in conclusion necrotic activity, early, late, and total apoptotic activity was significantly higher for LPHyNPs as compared AGN suspension. Hence, the LPHyNPs was found to be potential and effective in ameliorating the colonic cancerous cells by stimulating apoptotic activity.

3.11. Effect of LPHyNPs on the Gene Expression of Bcl-2, BAX, NF-κB, mTOR, JNK, and MDR-1

At present, the importance of apoptosis and modulation of various pro and anti-apoptotic genes are well established in the etiology of colon cancer. In a normal physiological condition, a balance is maintained between pro and anti-apoptotic genes. In contrast, during the cancerous phase, apoptotic activity is significantly reduced along with the reduction in expression of pro-apoptotic genes and elevation in the expression of anti-apoptotic genes. In the present study, we have already shown that overall apoptotic activity was reduced, and treatment in the LPHyNPs significantly elevates the apoptotic activity. Hence, we have tried to further investigate the mechanism and expression of various genes modulated upon exposure to different treatment groups. In general, apoptosis is regulated by proteins, namely caspase-3, caspase-6, cytochrome c, BAX, and Bcl-2 [42]. Numerous published evidences have shown increased expression level of Bcl-2 (anti-apoptotic gene) and reduced expression of BAX (pro-apoptotic gene), whereas, upon treatment with a potential anticancer drug, their level gets reversed, and anticancer effect is achieved. Bcl-2, an anti-apoptotic protein, mitigates apoptotic mechanism and supports the cancerous cell in undergoing uncontrolled cell division and survival. BAX, on the other hand, being a pro-apoptotic protein, gets attached to the anti-apoptotic proteins, sequesters their anti-apoptotic activity, stimulates the release of cytochrome C, and eventually promotes/stimulates apoptosis leading to the exhibition of anticancer effect [43]. In the present study, when blank LPHyNPs, AGN suspension, and LPHyNPs were treated with HCT-116 cells, LPHyNPs showed a significant reduction in the expression level of Bcl-2 and elevation in the expression of BAX when compared with AGN suspension ($p < 0.05$ for Bcl-2 and

0.01 for BAX) and blank LPHyNPs ($p < 0.01$ for Bcl-2 and $p < 0.001$ for BAX). Henceforth, the outcome of this study validated the rationale for formulating LPHyNPs as shown in Figure 10.

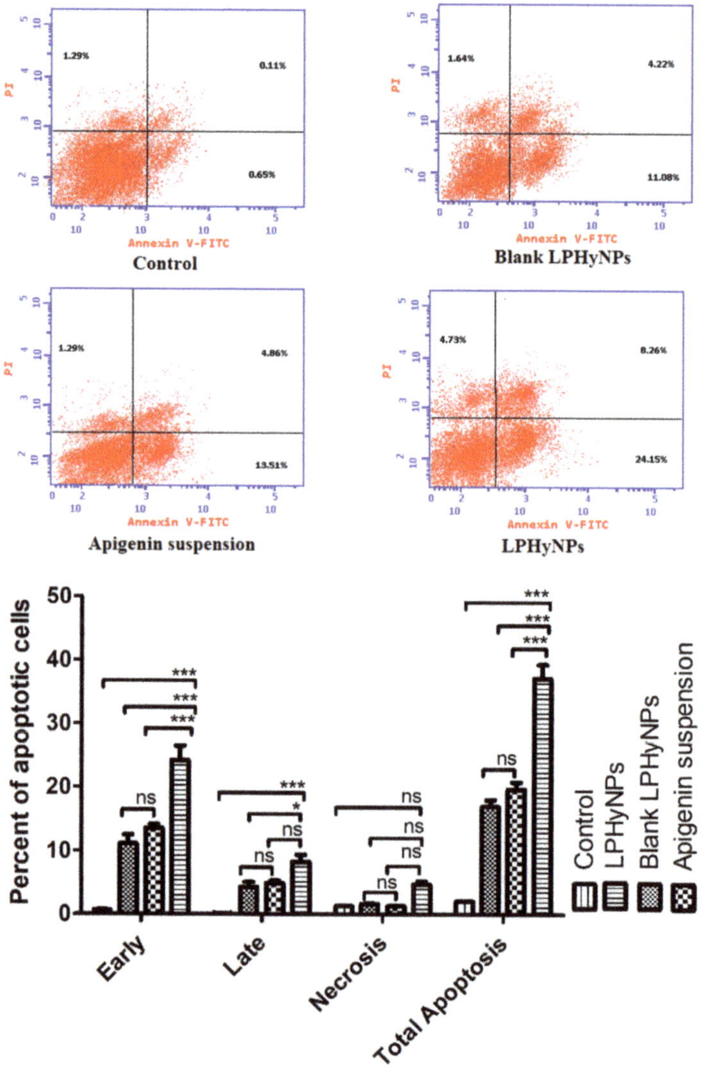

Figure 9. Showing the pro-apoptotic effect of various drug formulations such as control, Blank LPHyNPs, apigenin and LPHyNPs against colon cancer cell lines using flow cytometry. Values are presented as mean ± SD where $p < 0.05$ and $p < 0.001$ was represented by * and ***. 'ns' demonstrated non-significance differences between results.

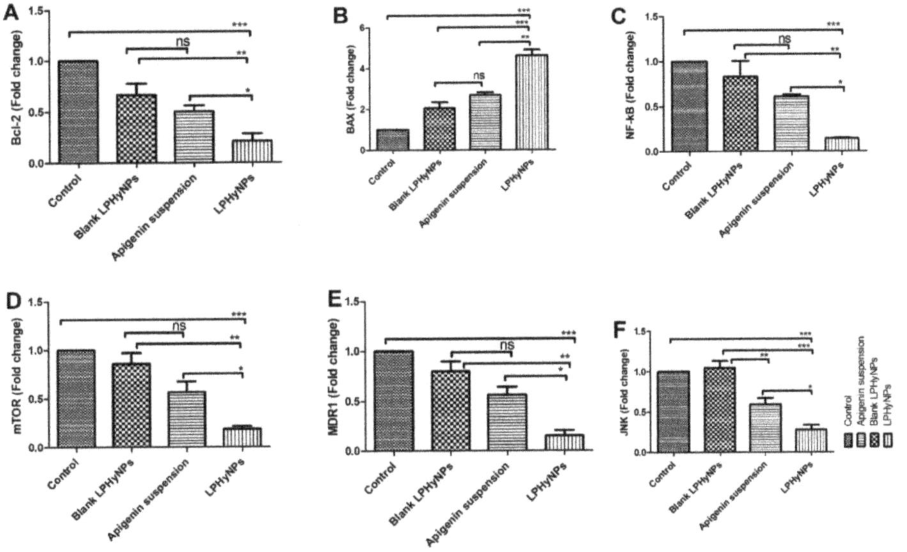

Figure 10. Showing the the outcome of RT-PCR on various formulations such as blank LPHyNPs, AGN suspension, and LPHyNPs. (**A**) represent mRNA expression of Bcl-2, (**B**) represent mRNA expression of BAX, (**C**) represent mRNA expression of NF-kB, (**D**) represent mRNA expression of mTOR, (**E**) represent mTNs expression of MDR1 and (**F**) represent mRNA expression of JNK. Values are presented as mean ± SD where * ($p < 0.05$: $p < 0.01$ and $p < 0.01$) was represented by *, **, and ***. 'ns' demonstrated non-significance differences between results.

Apart from the direct involvement of the apoptotic mechanism in the pathogenesis of colon cancer, the profound role of the inflammatory pathway has also been well established. Among various contributors of inflammation during carcinogenesis, NF-κB is one of the extensively studied transcription factors. During the normal physiological function, NF-κB remains sequestered in the cytoplasm associated with the ikB. Upon exposure to the carcinogenic stimulus, the inhibitory effect of ikB is withdrawn due to phosphorylation. NF-κB undergoes nuclear translocation and regulates the expression of inflammatory signaling molecules such as TNF-α, IL-6, IL-1β, COX-2, etc. Additionally, increased expression of NF-κB has been positively correlated with the enhanced angiogenesis, tumor invasion, and metastasis, where it modulates MMPs and VEGF activity [44]. Thus, a potent drug candidate ideally reduces NF-κB expression and exhibits a potent anticancer effect. We also analyzed the expression level of NF-κB upon treatment with the various formulation. The outcome of the study showed a significant reduction in the expression level of NF-κB when treated with LPHyNPs as compared to blank LPHyNPs ($p < 0.05$) and AGN suspension ($p < 0.01$), as shown in Figure 10.

Apart from NF-κB, mTOR is yet another extensively studied signaling molecule reported to have a potential pathogenic role in colon cancer. Undoubtedly, mTOR in a normal physiological process plays numerous biological functions via its cross-talk with PI3K/Akt/MAPK/NF-κB and p53, showing its involvement in the colon and other types of cancer via regulation of tumor cell cycle, survival, proliferation, invasion, and metastasis. Studies have shown that the dysregulated mTOR pathway is related to a gene mutation, leading to its increased expression and continuous hyperactivation. Moreover, the pathogenic role of the mTOR pathway was further confirmed in the studies where the use of its inhibitor exhibited a potent anticancer effect. The combination of mTOR inhibitor and conventional drugs is currently being investigated to achieve a superior clinical outcome in different types of cancer [45].

No doubt, the role of potent pharmacotherapy is needed at this time for the management and treatment of colon cancer. But at the same time, drug resistance is yet another hindrance in achieving the optimum clinical outcome. Many potent anticancer drugs fail to achieve desired clinical response just because of drug resistance. Recently, MDR-related genes such as MDR1 and JNK have gained the attention of researchers because of their role in promoting carcinogenic events [46]. c-Jun N-terminal kinase (JNK) is an important signaling molecule reported to be involved in the pathogenesis of colon and other types of cancer. Published evidence has reported the pro-carcinogenic role of these molecules. It acts in close proximity with NF-κB/mTOR/JAK-STAT signaling pathway and inhibits apoptosis, regulates autophagy, and eventually promotes tumor cell survival [47]. A more detailed study showed that JNK also plays a critical role in tumor evasion, modulating the expression of p21, p53 c-Myc, and drug resistance. MDR-1 is one of the important genes involved in the multidrug resistance of various anticancer drugs, and the exploratory studies have shown significant elevation in its expression in various types of cancer. Increased expression of MDR-1 has also been positively correlated with elevated Bcl-2 and JNK expression [48]. Hence, in the present study, we explored the anticancer effect of prepared LPHyNPs via the modulation of expression of mTOR, JNK, and MDR-1 genes. The outcome of the study showed a significant reduction in the expression level of mTOR, JNK, and MDR-1 when treated with LPHyNPs as compared to blank LPHyNPs ($p < 0.05$ for mTOR, MDR-1, and JNK) and AGN suspension (for mTOR, MDR-1, and JNK) as shown in Figure 10.

4. Conclusions

The outcome of the present study showed the successful preparation of LPHyNPs. In the current study, LPHyNPs was prepared and characterized on various parameters particle size, PDI, zeta potential, and EE. Further, prepared LPHyNPs was evaluated on DSC, XRD, and FT-IR, and the outcomes of the studies confirmed the entrapment and structure of NP. The in vitro release study provided data of the sustained release of AGN from LPHyNPs. The success and rationale of developing LPHyNPs against colon cancer were studied and validated via performing flow cytometry to estimate apoptotic activity, where a significant and superior anticancer effect was achieved compared to blank LPHyNPs and AGN suspension. The mechanism behind the anticancer attribute was further explored via estimation of gene expression of various signaling molecules such as Bcl-2, BAX, NF-κB, and mTOR that are involved in carcinogenic pathways. Moreover, to strengthen the anticancer potential of LPHyNPs against chemoresistance, the expression of JNK and MDR-1 genes were estimated. It was found that their expression level reduced considerably when compared to blank LPHyNPs and AGN suspension. However, more detailed investigational cellular and molecular studies are needed to provide substantial evidence to bring this developed formulation from bench to bedside.

Supplementary Materials: The following supporting information can be downloaded at: https://www.mdpi.com/article/10.3390/polym14173577/s1, Figure S1: Cytocompatibility study of blank LPHyNPs, AGN suspension, and LPHyNPs formulations on HEK293 normal cell line via MTT assay. Values are presented as mean ± SD (n = 3).

Author Contributions: Conceptualization, S.M.; data curation, N.A.A. and M.A.K.; formal analysis, A.M.H., N.A.A., M.A.K., Y.R. and S.M.; funding acquisition, A.M.H. and T.S.A.; investigation, M.A.A., A.M.H. and T.S.A.; methodology, M.A.A., A.M.H., T.S.A., N.A.A., M.A.K., Y.R. and S.M.; project administration, M.A.A. and T.S.A.; supervision, M.A.A.; writing—original draft, N.A.A., M.A.K. and Y.R.; writing—review and editing, S.M. All authors have read and agreed to the published version of the manuscript.

Funding: The authors extend their appreciation to the Deputyship for Research & Innovation, Ministry of Education in Saudi Arabia for funding this research work through the project number IFPRC-189-166-2020 and King Abdulaziz University, DSR, Jeddah, Saudi Arabia.

Institutional Review Board Statement: Not applicable.

Informed Consent Statement: Not applicable.

Data Availability Statement: The data presented in this study are available in article.

Conflicts of Interest: The authors declare no conflict of interest.

References

1. Xi, Y.; Xu, P. Global Colorectal Cancer Burden in 2020 and Projections to 2040. *Transl. Oncol.* **2021**, *14*, 101174. [CrossRef]
2. Condello, M.; Meschini, S. Role of Natural Antioxidant Products in Colorectal Cancer Disease: A Focus on a Natural Compound Derived from Prunus Spinosa, Trigno Ecotype. *Cells* **2021**, *10*, 3326. [CrossRef] [PubMed]
3. Sawicki, T.; Ruszkowska, M.; Danielewicz, A.; Niedźwiedzka, E.; Arłukowicz, T.; Przybyłowicz, K.E. A Review of Colorectal Cancer in Terms of Epidemiology, Risk Factors, Development, Symptoms and Diagnosis. *Cancers* **2021**, *13*, 2025. [CrossRef] [PubMed]
4. He, K.; Zheng, X.; Li, M.; Zhang, L.; Yu, J. MTOR Inhibitors Induce Apoptosis in Colon Cancer Cells via CHOP-Dependent DR5 Induction on 4E-BP1 Dephosphorylation. *Oncogene* **2016**, *35*, 148–157. [CrossRef]
5. Slattery, M.L.; Mullany, L.E.; Sakoda, L.; Samowitz, W.S.; Wolff, R.K.; Stevens, J.R.; Herrick, J.S. The NF-KB Signalling Pathway in Colorectal Cancer: Associations between Dysregulated Gene and MiRNA Expression. *J. Cancer Res. Clin. Oncol.* **2018**, *144*, 269–283. [CrossRef]
6. Tuli, H.S.; Sak, K.; Iqubal, A.; Garg, V.K.; Varol, M.; Sharma, U.; Chauhan, A.; Yerer, M.B.; Dhama, K.; Jain, M.; et al. STAT Signaling as a Target for Intervention: From Cancer Inflammation and Angiogenesis to Non-Coding RNAs Modulation. *Mol. Biol. Rep.* **2022**, 1–13. [CrossRef]
7. Dan, H.C.; Adli, M.; Baldwin, A.S. Regulation of Mammalian Target of Rapamycin Activity in PTEN-Inactive Prostate Cancer Cells by I Kappa B Kinase Alpha. *Cancer Res.* **2007**, *67*, 6263–6269. [CrossRef] [PubMed]
8. Pottier, C.; Fresnais, M.; Gilon, M.; Jérusalem, G.; Longuespée, R.; Sounni, N.E. Tyrosine Kinase Inhibitors in Cancer: Breakthrough and Challenges of Targeted Therapy. *Cancers* **2020**, *12*, 0731. [CrossRef]
9. Fatima, M.; Iqubal, M.K.; Iqubal, A.; Kaur, H.; Gilani, S.J.; Rahman, M.H.; Ahmadi, A.; Rizwanullah, M. Current Insight into the Therapeutic Potential of Phytocompounds and Their Nanoparticle-Based Systems for Effective Management of Lung Cancer. *Anticancer. Agents Med. Chem.* **2021**, *21*, 668–686. [CrossRef]
10. Linn, S.; Giaccone, G. MDR1/P-Glycoprotein Expression in Colorectal Cancer. *Eur. J. Cancer* **1995**, *31*, 1291–1294. [CrossRef]
11. Birt, D.F.; Walker, B.; Tibbels, M.G.; Bresnick, E. Anti-Mutagenesis and Anti-Promotion by Apigenin, Robinetin and Indole-3-Carbinol. *Carcinogenesis* **1986**, *7*, 959–963. [CrossRef]
12. Mabrouk Zayed, M.M.; Sahyon, H.A.; Hanafy, N.A.N.; El-Kemary, M.A. The Effect of Encapsulated Apigenin Nanoparticles on HePG-2 Cells through Regulation of P53. *Pharmaceutics* **2022**, *14*, 1160. [CrossRef]
13. Ahmed, S.A.; Parama, D.; Daimari, E.; Girisa, S.; Banik, K.; Harsha, C.; Dutta, U.; Kunnumakkara, A.B. Rationalizing the Therapeutic Potential of Apigenin against Cancer. *Life Sci.* **2021**, *267*, 118814. [CrossRef]
14. Zhang, J.; Liu, D.; Huang, Y.; Gao, Y.; Qian, S. Biopharmaceutics Classification and Intestinal Absorption Study of Apigenin. *Int. J. Pharm.* **2012**, *436*, 311–317. [CrossRef]
15. Ding, S.; Zhang, Z.; Song, J.; Cheng, X.; Jiang, J.; Jia, X. Enhanced Bioavailability of Apigenin via Preparation of a Carbon Nanopowder Solid Dispersion. *Int. J. Nanomed.* **2014**, *9*, 2327–2333. [CrossRef]
16. García-Pinel, B.; Porras-Alcalá, C.; Ortega-Rodríguez, A.; Sarabia, F.; Prados, J.; Melguizo, C.; López-Romero, J.M. Lipid-Based Nanoparticles: Application and Recent Advances in Cancer Treatment. *Nanomaterials* **2019**, *9*, 638. [CrossRef]
17. Silva, L.B.; Castro, K.A.D.F.; Botteon, C.E.A.; Oliveira, C.L.P.; da Silva, R.S.; Marcato, P.D. Hybrid Nanoparticles as an Efficient Porphyrin Delivery System for Cancer Cells to Enhance Photodynamic Therapy. *Front. Bioeng. Biotechnol.* **2021**, *9*, 679128. [CrossRef]
18. Hanafy, N.A.N.; Quarta, A.; Di Corato, R.; Dini, L.; Nobile, C.; Tasco, V.; Carallo, S.; Cascione, M.; Malfettone, A.; Soukupova, J.; et al. Hybrid Polymeric-Protein Nano-Carriers (HPPNC) for Targeted Delivery of TGFβ Inhibitors to Hepatocellular Carcinoma Cells. *J. Mater. Sci. Mater. Med.* **2017**, *28*, 120. [CrossRef]
19. Salzano, G.; Marra, M.; Porru, M.; Zappavigna, S.; Abbruzzese, A.; La Rotonda, M.I.; Leonetti, C.; Caraglia, M.; De Rosa, G. Self-Assembly Nanoparticles for the Delivery of Bisphosphonates into Tumors. *Int. J. Pharm.* **2011**, *403*, 292–297. [CrossRef]
20. Hanafy, N.A.; Dini, L.; Citti, C.; Cannazza, G.; Leporatti, S. Inihibition of Glycolysis by Using a Micro/Nano-Lipid Bromopyruvic Chitosan Carrier as a Promising Tool to Improve Treatment of Hepatocellular Carcinoma. *Nanomaterials* **2018**, *8*, 34. [CrossRef]
21. Wan, F.; Bohr, S.S.-R.; Kłodzińska, S.N.; Jumaa, H.; Huang, Z.; Nylander, T.; Thygesen, M.B.; Sørensen, K.K.; Jensen, K.J.; Sternberg, C.; et al. Ultrasmall TPGS-PLGA Hybrid Nanoparticles for Site-Specific Delivery of Antibiotics into Pseudomonas Aeruginosa Biofilms in Lungs. *ACS Appl. Mater. Interfaces* **2020**, *12*, 380–389. [CrossRef] [PubMed]
22. Ishak, R.A.H.; Mostafa, N.M.; Kamel, A.O. Stealth Lipid Polymer Hybrid Nanoparticles Loaded with Rutin for Effective Brain Delivery—Comparative Study with the Gold Standard (Tween 80): Optimization, Characterization and Biodistribution. *Drug Deliv.* **2017**, *24*, 1874–1890. [CrossRef] [PubMed]
23. Iqubal, M.K.; Iqubal, A.; Imtiyaz, K.; Rizvi, M.M.A.; Gupta, M.M.; Ali, J.; Baboota, S. Combinatorial Lipid-Nanosystem for Dermal Delivery of 5-Fluorouracil and Resveratrol against Skin Cancer: Delineation of Improved Dermatokinetics and Epidermal Drug Deposition Enhancement Analysis. *Eur. J. Pharm. Biopharm.* **2021**, *163*, 223–239. [CrossRef] [PubMed]

24. Mahmoudi, S.; Ghorbani, M.; Sabzichi, M.; Ramezani, F.; Hamishehkar, H.; Samadi, N. Targeted Hyaluronic Acid-Based Lipid Nanoparticle for Apigenin Delivery to Induce Nrf2-Dependent Apoptosis in Lung Cancer Cells. *J. Drug Deliv. Sci. Technol.* **2019**, *49*, 268–276. [CrossRef]
25. Md, S.; Abdullah, S.; Alhakamy, N.A.; Alharbi, W.S.; Ahmad, J.; Shaik, R.A.; Ansari, M.J.; Ibrahim, I.M.; Ali, J. Development, Optimization, and In Vitro Evaluation of Novel Oral Long-Acting Resveratrol Nanocomposite In-Situ Gelling Film in the Treatment of Colorectal Cancer. *Gels* **2021**, *7*, 276. [CrossRef]
26. Rimkiene, L.; Baranauskaite, J.; Marksa, M.; Jarukas, L.; Ivanauskas, L. Development and Evaluation of Ginkgo Biloba L. Extract Loaded into Carboxymethyl Cellulose Sublingual Films. *Appl. Sci.* **2021**, *11*, 270. [CrossRef]
27. Senthamarai Kannan, M.; Hari Haran, P.S.; Sundar, K.; Kunjiappan, S.; Balakrishnan, V. Fabrication of Anti-Bacterial Cotton Bandage Using Biologically Synthesized Nanoparticles for Medical Applications. *Prog. Biomater.* **2022**, *11*, 229–241. [CrossRef]
28. Almehmady, A.M.; Elsisi, A.M. Development, Optimization, and Evaluation of Tamsulosin Nanotransfersomes to Enhance Its Permeation and Bioavailability. *J. Drug Deliv. Sci. Technol.* **2020**, *57*, 101667. [CrossRef]
29. Alhakamy, N.A.; Ahmed, O.A.A.; Fahmy, U.A.; Md, S. Development and in Vitro Evaluation of 2-Methoxyestradiol Loaded Polymeric Micelles for Enhancing Anticancer Activities in Prostate Cancer. *Polymers* **2021**, *13*, 884. [CrossRef]
30. Md, S.; Alhakamy, N.A.; Aldawsari, H.M.; Husain, M.; Kotta, S.; Abdullah, S.T.; Fahmy, U.A.; Alfaleh, M.A.; Asfour, H.Z. Formulation Design, Statistical Optimization, and in Vitro Evaluation of a Naringenin Nanoemulsion to Enhance Apoptotic Activity in A549 Lung Cancer Cells. *Pharmaceuticals* **2020**, *13*, 152. [CrossRef]
31. Alhakamy, N.A.; Ahmed, O.A.A.; Fahmy, U.A.; Shadab, M. Apamin-Conjugated Alendronate Sodium Nanocomplex for Management of Pancreatic Cancer. *Pharmaceuticals* **2021**, *14*, 729. [CrossRef]
32. Mohamed, S.A.; Elshal, M.F.; Kumosani, T.A.; Aldahlawi, A.M.; Basbrain, T.A.; Alshehri, F.A.; Choudhry, H. L-Asparaginase Isolated from Phaseolus Vulgaris Seeds Exhibited Potent Anti-Acute Lymphoblastic Leukemia Effects In-Vitro and Low Immunogenic Properties In-Vivo. *Int. J. Environ. Res. Public Health* **2016**, *13*, 1008. [CrossRef]
33. Danaei, M.; Dehghankhold, M.; Ataei, S.; Hasanzadeh Davarani, F.; Javanmard, R.; Dokhani, A.; Khorasani, S.; Mozafari, M.R. Impact of Particle Size and Polydispersity Index on the Clinical Applications of Lipidic Nanocarrier Systems. *Pharmaceutics* **2018**, *10*, 57. [CrossRef]
34. Dina, F.; Soliman, G.M.; Fouad, E.A. Development and in Vitro/in Vivo Evaluation of Liposomal Gels for the Sustained Ocular Delivery of Latanoprost. *J. Clin. Exp. Ophthalmol.* **2015**, *6*, 16–19. [CrossRef]
35. Pawlikowska-Pawlęga, B.; Misiak, L.E.; Zarzyka, B.; Paduch, R.; Gawron, A.; Gruszecki, W.I. FTIR, 1H NMR and EPR Spectroscopy Studies on the Interaction of Flavone Apigenin with Dipalmitoylphosphatidylcholine Liposomes. *Biochim. Biophys. Acta—Biomembr.* **2013**, *1828*, 518–527. [CrossRef]
36. Liu, G.; Li, K.; Wang, H. Polymeric Micelles Based on PEGylated Chitosan-g-Lipoic Acid as Carrier for Efficient Intracellular Drug Delivery. *J. Biomater. Appl.* **2017**, *31*, 1039–1048. [CrossRef]
37. Iqubal, M.K.; Iqubal, A.; Anjum, H.; Gupta, M.M.; Ali, J.; Baboota, S. Determination of in Vivo Virtue of Dermal Targeted Combinatorial Lipid Nanocolloidal Based Formulation of 5-Fluorouracil and Resveratrol against Skin Cancer. *Int. J. Pharm.* **2021**, *610*, 121179. [CrossRef]
38. Kumbhar, P.S.; Nadaf, S.; Manjappa, A.S.; Jha, N.K.; Shinde, S.S.; Chopade, S.S.; Shete, A.S.; Disouza, J.I.; Sambamoorthy, U.; Kumar, S.A. D-α-Tocopheryl Polyethylene Glycol Succinate: A Review of Multifarious Applications in Nanomedicines. *OpenNano* **2022**, *6*, 100036. [CrossRef]
39. Ma, L.; Xu, G.B.; Tang, X.; Zhang, C.; Zhao, W.; Wang, J.; Chen, H. Anti-Cancer Potential of Polysaccharide Extracted from Hawthorn (Crataegus.) on Human Colon Cancer Cell Line HCT116 via Cell Cycle Arrest and Apoptosis. *J. Funct. Foods* **2020**, *64*, 103677. [CrossRef]
40. Sadeghi Ekbatan, S.; Li, X.-Q.; Ghorbani, M.; Azadi, B.; Kubow, S. Chlorogenic Acid and Its Microbial Metabolites Exert Anti-Proliferative Effects, S-Phase Cell-Cycle Arrest and Apoptosis in Human Colon Cancer Caco-2 Cells. *Int. J. Mol. Sci.* **2018**, *19*, 723. [CrossRef]
41. Pfeffer, C.M.; Singh, A.T.K. Apoptosis: A Target for Anticancer Therapy. *Int. J. Mol. Sci.* **2018**, *19*, 448. [CrossRef] [PubMed]
42. Lowe, S.W.; Lin, A.W. Apoptosis in Cancer. *Carcinogenesis* **2000**, *21*, 485–495. [CrossRef] [PubMed]
43. Watson, A.J.M. Apoptosis and Colorectal Cancer. *Gut* **2004**, *53*, 1701–1709. [CrossRef] [PubMed]
44. Xia, L.; Tan, S.; Zhou, Y.; Lin, J.; Wang, H.; Oyang, L.; Tian, Y.; Liu, L.; Su, M.; Wang, H.; et al. Role of the NFκB-Signaling Pathway in Cancer. *Onco. Targets. Ther.* **2018**, *11*, 2063–2073. [CrossRef]
45. Zou, Z.; Tao, T.; Li, H.; Zhu, X. MTOR Signaling Pathway and MTOR Inhibitors in Cancer: Progress and Challenges. *Cell Biosci.* **2020**, *10*, 31. [CrossRef]
46. Hasan, S.; Taha, R.; Omri, H. El Current Opinions on Chemoresistance: An Overview. *Bioinformation* **2018**, *14*, 80–85. [CrossRef]
47. Tournier, C. The 2 Faces of JNK Signaling in Cancer. *Genes Cancer* **2013**, *4*, 397–400. [CrossRef]
48. Elia, S.G.; Al-Karmalawy, A.A.; Nasr, M.Y.; Elshal, M.F. Loperamide Potentiates Doxorubicin Sensitivity in Triple-negative Breast Cancer Cells by Targeting MDR1 and JNK and Suppressing MTOR and Bcl-2: In Vitro and Molecular Docking Study. *J. Biochem. Mol. Toxicol.* **2022**, *36*, e22938. [CrossRef]

polymers

MDPI

Article

Copolymer-Green-Synthesized Copper Oxide Nanoparticles Enhance Folate-Targeting in Cervical Cancer Cells In Vitro

Keelan Jagaran and Moganavelli Singh *

Nano-Gene and Drug Delivery Laboratory, Discipline of Biochemistry, University of KwaZulu-Natal, Private Bag X54001, Durban 4000, South Africa; 215055447@stu.ukzn.ac.za
* Correspondence: singhm1@ukzn.ac.za; Tel.: +27-31-260-7170

Abstract: Cervical cancer is fast becoming a global health crisis, accounting for most female deaths in low- and middle-income countries. It is the fourth most frequent cancer affecting women, and due to its complexity, conventional treatment options are limited. Nanomedicine has found a niche in gene therapy, with inorganic nanoparticles becoming attractive tools for gene delivery strategies. Of the many metallic nanoparticles (NPs) available, copper oxide NPs (CuONPs) have been the least investigated in gene delivery. In this study, CuONPs were biologically synthesized using *Melia azedarach* leaf extract, functionalized with chitosan and polyethylene glycol (PEG), and conjugated to the targeting ligand folate. A peak at 568 nm from UV-visible spectroscopy and the characteristic bands for the functional groups using Fourier-transform infrared (FTIR) spectroscopy confirmed the successful synthesis and modification of the CuONPs. Spherical NPs within the nanometer range were evident from transmission electron microscopy (TEM) and nanoparticle tracking analysis (NTA). The NPs portrayed exceptional binding and protection of the reporter gene, pCMV-*Luc*-DNA. In vitro cytotoxicity studies revealed cell viability >70% in human embryonic kidney (HEK293), breast adenocarcinoma (MCF-7), and cervical cancer (HeLa) cells, with significant transgene expression, obtained using the luciferase reporter gene assay. Overall, these NPs showed favorable properties and efficient gene delivery, suggesting their potential role in gene therapy.

Keywords: cervical cancer; copper oxide; green synthesis; cytotoxicity; gene delivery; folate targeting

Citation: Jagaran, K.; Singh, M. Copolymer-Green-Synthesized Copper Oxide Nanoparticles Enhance Folate-Targeting in Cervical Cancer Cells In Vitro. *Polymers* **2023**, *15*, 2393. https://doi.org/10.3390/polym15102393

Academic Editors: Stephanie Andrade, Choon-Sang Park, Joana A. Loureiro and Maria João Ramalho

Received: 3 April 2023
Revised: 8 May 2023
Accepted: 18 May 2023
Published: 20 May 2023

1. Introduction

Cancer therapy is an ongoing challenge to mitigate the complexities associated with this dreaded disease, which is the second leading cause of death globally. Cancer accounts for an alarming 18.1 million new cases, with 9.6 million deaths annually. Cervical cancer poses a threat to women worldwide, being the fourth most frequent cancer affecting women, with 311,000 deaths worldwide and an estimated 570,000 cases in 2018 [1].

The human papillomavirus (HPV), a compromised immune system, and smoking are a few mentioned determinant conditions closely related to this disease [2,3]. Treatment options for cervical cancer depend on the cancer stage, the tumor's size and location, and the patient's health. Standard treatment options include radiation therapy, surgery, and chemotherapy, which increase the financial distress experienced by cancer patients, above and beyond the side effects caused. Hence, women in low- and middle-income countries portray a significantly higher mortality rate than those in high-income countries [3,4]. It is thus imperative to create innovative approaches to reduce this alarming statistic. Reducing the invasiveness of the current diagnostic and treatment procedures is urgently required.

The amalgamation of medicine and nanotechnology has become a highly efficient alternative to traditional chemotherapy. In addition, targeted gene/drug therapy and immunotherapy have attracted much attention. This has necessitated the formulation of safe and effective delivery systems. Nanoparticles (NPs) have gained popularity and have been investigated for the delivery and release of pharmacologically active agents such as genes or drugs [5]. Importantly, metal NPs have undergone the most extensive studies, with gold NPs being the most employed owing to their favorable properties [6]. Copper (Cu)-based NPs have not been thoroughly investigated as therapeutic carriers. Cu, a trace element and essential nutrient, is a good conductor of heat and electricity and possesses favorable antibacterial properties. Besides its use as a metabolic cofactor, Cu has been reported to be a metalloallosteric regulator and involved in signaling pathways [7]. Cu has been reported to have low toxicity, good biocompatibility, be readily available at lower costs, and be oxidation-resistant, making it an attractive prospect for delivering therapeutic drugs and genes [8,9]. Hence, exploiting Cu on the nanoscale may benefit therapeutic use. Furthermore, Cu-based NPs are relatively stable when bound to polymers, making them suitable for use in nanomedicine [10].

Compared to other noble metals, copper tends to become easily oxidized in the presence of air. Copper-based NPs portray unique properties in the nanoscale range. These NPs contain a single-phase tenorite with a narrow band gap of 1.7 eV [11]. They are thus used in various medical applications owing to their antiviral, antibacterial, and possibly anticancer properties [12–15]. CuONPs also have the potential to be used as MRI-ultrasound dual imaging contrast agents. However, CuONPs have not been extensively studied in cancer gene therapy compared to the other metallic NPs. Hence, this study was undertaken to assess their potential in gene delivery.

The study of green chemistry has significantly influenced the synthesis of metallic and inorganic NPs. NPs can be synthesized both chemically and biologically, with the biological method portrayed as being the least toxic approach. Various plant compounds or extracts have been used since they contain many secondary metabolites that enable them to reduce metallic ions and heavy metals. The leaves, stems, bark, fruit, and seeds have been more commonly used in NP synthesis [16]. Biological syntheses have been found to produce NPs rapidly with increased stability and varied shapes and sizes [17]. It has been reported that the reducing properties of plant material in NP synthesis depend on the antioxidant content of the specific part of the plant [16]. The green synthesis of metals or metal oxide NPs is advantageous due to the lack of toxic reducing agents such as sodium borohydride and hydrazine, which are also hazardous to the environment [18]. Due to the increased interest in waste minimization or elimination via implemented fundamental properties of green chemistry, the biological synthesis of NPs is a desirable approach.

In this study, the green synthesis approach uses the extract of the leaves of *Melia azedarach* (*M. azedarach*), or syringa, as it is commonly known. Syringa has been previously identified to contain polyphenolic compounds portraying remarkable antioxidant activities common to the Meliaceae family. This plant species has antimalarial, antifungal, antiviral, antibacterial, antifeedant, antifertility, and anticancer properties [19]. The extract of the syringa leaves has shown strong antiproliferative properties against cancer cells, both in vitro and in vivo, via the induction of tumor necrosis factor-α (TNF-α) and the light-chain 3(LC3)-11 autophagosomal proteins. This has been shown to suppress tumor growth and induce autophagy [20]. *M. azedarach* displays favorable biological activities, is readily available, and has excellent reduction capabilities. It has antioxidant, antimalarial, antihepatotoxic, antibacterial, antiparasitic, and antiulcer properties [21]. This is anticipated to enrich the CuONPs and produce a synergistic therapeutic effect. Owing to *M. azedarach*'s good biological properties, there has been substantive progress in its use in anticancer studies, warranting further studies, especially in gene therapy.

The success of gene therapy relies on the ability of NPs to compact, protect, and safely deliver the therapeutic gene to the desired target cell. The present study acquired this through surface modifications via the conjugation of cationic polymers. The advantages of using chitosan (Cs) as a functionalizing polymer for NPs include its good biodegradability, biocompatibility, minimal toxicity, and the presence of -NH2 and -OH groups in its backbone for the conjugation of biomolecules [22,23]. Folic acid, or folate (F), can be conjugated to NPs to enable receptor-mediated endocytosis via folate receptors (FRs) which are overexpressed in many human cancers, including those of the kidney, cervix, brain, breast, and lungs [24,25]. Hence, the folate-receptor-targeting of HeLa cells can achieve enhanced cellular uptake and transfection. Modification with polyethylene glycol (PEG) improves steric stabilization and prolongs systemic circulation with decreased immunogenicity [26,27].

Although some progress has been made using NPs in gene therapy, the Food and Drug Administration (FDA) has put forth several challenges. These include NP aggregation in physiological fluids, cellular internalization, biocompatibility, biodegradation, and endosomal escape [28]. Hence, further research investigating various NPs is imperative. The present study is a proof-of-concept study aimed at determining the efficiency of biologically synthesized folate–chitosan–copper oxide NPs (F-Cs-CuONPs) and PEG-F-Cs-CuONPs in the delivery of a reporter gene (pCMV-*Luc*-DNA) to cervical cancer (HeLa) cells. Importantly, no studies to date have used such green-synthesized CuONPs for targeted gene delivery to cervical cancer cells in vitro.

2. Materials and Methods

2.1. Materials

Copper nitrate ($Cu(NO_3)_2$, Mw: 187.56 g·mol^{-1}), sodium sulfide (Na_2S, Mw: 78.05 g·mol^{-1}), dimethyl sulfoxide (DMSO), 3-[4,5-dimethylthiazol-2-yl]-2.5-diphenyltetrazolium bromide (MTT), phosphate-buffered saline tablets (PBS, (140 mM NaCl, 10 mM phosphate buffer, 3 mM KCl)), PEG2000, ethidium bromide, bromophenol blue, xylene cyanol, sodium dodecyl sulfide (SDS), 1-ethyl-3-(3-dimethyl aminopropyl) carbodiimide (EDC), and *N*-hydroxysuccinimide (NHS) were sourced from Merck (Darmstadt, Germany). Chitosan (>75% DD, Mw: 25 kDa) and folic acid were purchased from Sigma-Aldrich (St. Louis, MO, USA), and the luciferase assay kit was supplied by Promega Corporation (Madison, WI, USA). Eagle's minimum essential medium (EMEM), antibiotic mixture (penicillin (10,000 U/mL)), streptomycin (10,000 µg/mL), amphotericin (25 µg/mL), and trypsin-versene-EDTA were obtained from Lonza BioWhittaker (Walkersville, MD, USA). Sterile fetal bovine serum (FBS) was purchased from GIBCO, Life Technologies LTD, Inchinnan, UK. Plasmid pCMV-*Luc* DNA (6.2 kbp) was obtained from Plasmid Factory (Bielefeld, Germany). Human breast adenocarcinoma (MCF-7), cervical carcinoma (HeLa), and embryonic kidney (HEK293) cells were originally purchased directly from the American Type Culture Collection (ATCC, Manassas, VA, USA). Ultrapure (18 Mohm) water (Milli-Q50, Millipore, France) was utilized in all assays, and all general reagents were of analytical grade.

2.2. Sample Collection and Preparation

2.2.1. Sample Collection

The *M. azedarach* (syringa) leaves were collected in January 2020 from the Durban area, KwaZulu-Natal, South Africa, coordinates: 29°50′48.195″ S; 31°0′34.233″ E. Young leaves were obtained from the tree (Figure 1) and placed in a sealed plastic bag. This allowed for the inhibition of transpiration and therefore the drying out of the leaves in transit to the laboratory. A *M. azedarach* L. (Meliaceae) specimen was deposited in the Ward Herbarium (voucher: K. Jagaran Collection Number 1), Westville Campus, UKZN. The species was identified and confirmed by the curator of the Ward Herbarium, School of Life Sciences, UKZN.

2.2.2. Preparation of *M. azedarach* Leaf Extract

The leaves of the *M. azedarach* plant were prepared as previously described [29]. The leaves were thoroughly washed with distilled water soon after collection and blotted dry using a paper towel. Approximately 10 g of the leaves were cut into small pieces, transferred to a beaker containing 40 mL of 18 Mohm water, and heated to between 80 and 90 °C, with constant stirring for 15 min, until a green-colored liquid solution was noted. The leaf extract was filtered (Whatmann Filter paper no. 5) and stored in a freezer at −4 °C until further use.

Figure 1. Photograph of *Melia azedarach*.

2.3. Synthesis and Chitosan (Cs) Functionalization of Copper Oxide Nanoparticles (CuONPs)

A copper nitrate (Cu $(NO_3)_2$) solution was prepared by dissolving 2 mg of $Cu(NO_3)_2$ in 5 mL of 18 Mohm water. The solution was heated to 60–70 °C with constant stirring, followed by the dropwise addition of 1 mL of the leaf extract until a light green color was visible. This indicated the formation of the CuONPs [30]. For chitosan (Cs) functionalization, Cs solution (1 g/mL in 1% acetic acid) was added to the nanoparticles with stirring at predetermined ratios of 1:1, 1:2.5, and 1:3.5 (*v*/*v*) for optimal binding [31]. A scheme for the above synthesis is illustrated in Figure 2.

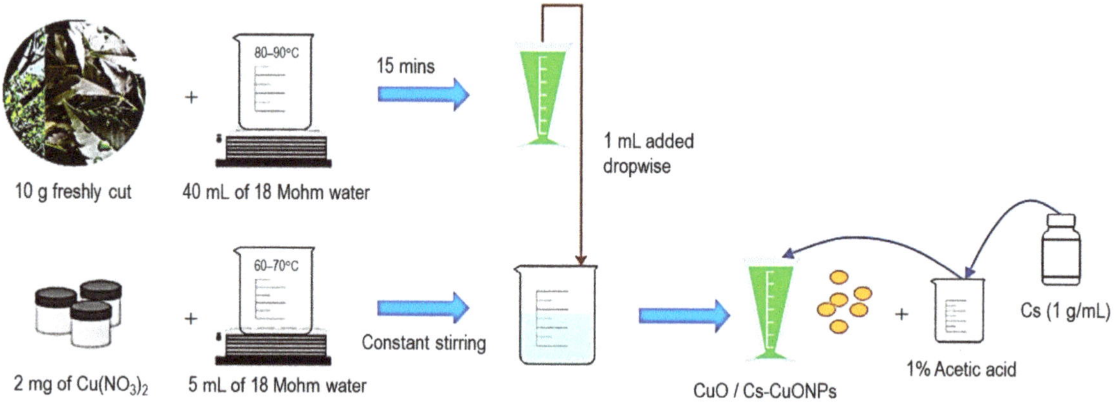

Figure 2. Schematic representation of the synthesis of CuO and Cs-CuONPs.

2.4. Formation of Folate–Chitosan–Copper Oxide (F-Cs-CuONPs) and PEG-F–Chitosan–Copper Oxide Nanoparticles (PEG-F-Cs-CuONPs)

To a solution of folic acid (10 mg/500 µL in DMSO), 15 mg of 1-ethyl-3-(3-dimethylamino propyl) carbodiimide (EDC) (0.19 g/mL), 12 mg of *N*-hydroxysuccinimide (NHS) (0.26 g/mL), and 5 mL of 18 Mohm water were added. The solution was covered with foil and stirred overnight at room temperature. To 2 mL of the previously prepared Cs solution, 0.5 mL of the folic acid solution was added. The CuONPs were added dropwise to the folate–chitosan (F-Cs) solution in a ratio of 1:3.5, forming the F-Cs-CuONPs [32].

Subsequently, PEGylation of the F-Cs-CuONPs was achieved by adding 2% Polyethylene glycol dropwise, followed by 2 h of stirring for successful conjugation. To remove unreacted products, dialysis (MWCO 12,000 -14,000 Da) was performed against 18 Mohm water [33].

2.5. Characterization of Nanoparticles and Nanocomplexes

2.5.1. UV-Vis Spectroscopy and X-ray Diffraction (XRD)

The absorption spectra of the CuO, Cs-CuO, F-Cs-CuO, PEG-Cs-CuO, and PEG-F-Cs-CuONPs were recorded using a UV-vis spectrophotometer (Jasco V-730, JASCO Corporation, Hachioji, Japan) over a wavelength range of 200–800 nm. The results were confirmed by those reported in the literature. X-ray diffraction (XRD) was conducted using a Malvern Panalytical Aeris diffractometer (Malvern, Worcestershire, UK).

2.5.2. Fourier-Transform Infrared Spectroscopy (FTIR)

FTIR spectroscopy of all NPs was performed using a PerkinElmer Spectrum 100 FTIR spectrophotometer (Shelton, CT, USA), fitted with a universal attenuated total reflection (ATR) sampling accessory. A single droplet of each colloidal nanoparticle (NP) was placed in the instrument stage, and the spectra were analyzed from 4000–400 cm^{-1} using 64 co-added scans at a resolution of 4 cm^{-1}.

2.5.3. Transmission Electron Microscopy (TEM)

TEM was used to determine the ultrastructural morphology of the CuO and its functionalized counterparts. Images were captured using a JEOL-JEM T1010 electron microscope and analyzed using iTEM Soft Imaging Systems (SIS) Mega view III (JEOL, Tokyo, Japan). One drop of the respective sample was placed onto a carbon-coated copper grid and air-dried at room temperature. After that, images were viewed, captured, and analyzed.

2.5.4. Nanoparticle Tracking Analysis (NTA)

Particle size and zeta (ζ) potential were assessed using NTA [34], operating at 24 V at 25 °C. A 1:100 dilution of the NPs in 18 Mohm water was used and evaluated using a Nanosight NS500 complete with NTA version 3.0 software (Malvern Instruments, Worcestershire, UK). The software utilized the Stokes–Einstein equation and Smoluchowski approximations to provide the particle size distribution and zeta potentials, respectively.

2.6. Nucleic Acid Binding Studies

2.6.1. Electrophoretic Mobility Shift Assay

Nanocomplex formation (Cs-CuO/pDNA, F-Cs-CuO/pDNA, PEG-Cs-CuO/pDNA, and PEG-F-Cs-CuO/pDNA) was assessed using the mobility shift electrophoresis assay [35]. The concentration of pCMV-*Luc* DNA (pDNA, 0.25 μg/μL) remained constant, while the amount of the NPs was gradually increased. The nanocomplexes were incubated for 1 h at 25 °C to ensure effective binding of the NP to the pDNA. This was followed by electrophoresis at 50 V for 90 min on a 1% agarose gel pre-stained with ethidium bromide (1 μg/mL) using BioRad mini-sub electrophoresis apparatus (BioRad, Richmond, VA, USA). Agarose gels were viewed, and images were captured using a Vacutec Syngene G: Box Bioimaging System (Syngene, Cambridge, UK). Three NP/pDNA binding ratios obtained were subsequently used, viz., sub-optimum, optimum, and supra-optimum binding ratios.

2.6.2. Nuclease Protection Assay

The serum nuclease protection assay used fetal bovine serum (FBS) to investigate the ability of the NPs to protect the bound pDNA from enzyme digestion [35]. The sub-optimum, optimum, and supra-optimum binding ratios (Section 2.6.1) were used. FBS to a final concentration of 10% was then added. A positive control (naked pDNA) and a negative control (FBS-treated pDNA) were included in the assay. The samples were incubated at 37 °C for 4 h, followed by adding EDTA (10 mM) to halt the enzyme reaction. After that, 0.5% SDS was introduced, and the reaction was allowed to proceed for 20 min at 55 °C to facilitate the release of the DNA. Following the incubation, agarose gel electrophoresis was conducted as previously described (Section 2.6.1).

2.6.3. Ethidium Bromide Intercalation Assay

This assay evaluated [35] the degree of complexation of the pDNA by the NPs. The relative fluorescence was measured using a Glomax® multi-detection system (Promega Biosystems, Sunnyvale, CA, USA). Approximately 2 µL of ethidium bromide was added to 100 µL of HBS in a well of a black 96-well plate. The fluorescence was measured at an excitation wavelength of 520 nm and an emission wavelength of 600 nm. This established a baseline fluorescence of 0%. Thereafter, 1.2 µg (4.8 µL) of pDNA was added, and the measured fluorescence was set as 100%. The NPs (1 µL at a time) were added, and the fluorescence measurements were recorded until a plateau in fluorescence was reached.

2.7. MTT Cell Viability Assay

The MTT cell viability assay was based on that originally described [36]. HeLa, HEK293, and MCF-7 cells at/near confluency were trypsinized and seeded into 48-well plates at a density of 2.5×10^5 cells/well. Cells were incubated at 37 °C for 24 h in 5% CO_2. Nanocomplexes at the sub-optimum, optimum, and supra-optimum (*w*/*w*) ratios obtained from the electrophoretic mobility shift assay were added to their respective wells in triplicate and incubated at 37 °C for 48 h. A positive control containing untreated cells was included. After incubation, the medium in each well was replaced with 200 µL fresh growth medium containing 20 µL MTT reagent (12 mM in PBS), and cells were incubated at 37 °C for 4 h. The growth medium infused with MTT was then removed, followed by the solubilization of the formazan crystals using 200 µL DMSO. Absorbance at 570 nm was then measured (Mindray MR-96A, Vacutec, Hamburg, Germany), and cell viability was calculated as follows:

$$\%\ \text{Cell Viability} = \frac{[\text{A570nm treated cells}]}{[\text{A570nm untreated cells}]} \times 100$$

2.8. Gene Expression Studies

Luciferase expression and receptor binding studies were conducted as previously reported [37,38]. Cells were seeded into 48-well plates and incubated as in Section 2.7. Following incubation, the medium was then replaced with 200 µL fresh growth medium. The nanocomplexes were added at the sub-optimum, optimum, and supra-optimum ratios, and cells were incubated at 37 °C for 48 h. The controls included naked pDNA and untreated cells. After the 48 h incubation, the cells were washed twice with 0.2 mL PBS and lysed with 80 µL reporter lysis buffer. The cells were then gently scraped off the wells, and the cell suspension was centrifuged at 12,000× *g* for 30 s using an Eppendorf 5418 centrifuge (Merck, Darmstadt, Germany). After that, 20 µL of each of the cell lysates was added to the respective wells in a white 96-well plate, followed by automatic injection of 100 µL of the luciferase assay reagent. Luminescence was measured using a Glomax® multi-detection system (Promega Biosystems, Sunnyvale, CA, USA). The relative light units obtained were normalized against the protein content measured using the standard Bicinchoninic acid assay (BCA) (Sigma, St. Louis, MO, USA). The readings were expressed as relative light units per milligram of total protein (RLU/mg protein). Each data point was averaged over three replicates ($n = 3$), and the results were presented as means ± SD. A competition assay was performed as above to confirm folate receptor uptake, except for adding a 50 times excess of free folic acid 30 min prior to nanocomplex addition.

2.9. Statistical Analysis

Statistical analysis was conducted via GraphPad Prism 9 and OriginLab 2021. Data are presented as a mean ± standard deviation (SD; $n = 3$). Statistical analysis among mean values was performed using a two-way multiple ANOVA. All statistics were performed using a 95% confidence interval (CI) and were considered to be significant at $p < 0.05$.

3. Results and Discussion

3.1. Characterization of Nanoparticles and Nanocomplexes

3.1.1. UV-Visible Spectroscopy and X-ray Diffraction (XRD)

The initial confirmation of the synthesis of the CuONPs was a visual color change from light blue to light yellow/green. This also indicated the excitation of the surface plasmon vibration of the CuONPs, allowing for their spectroscopic measurement. UV-vis spectroscopy (Figure 3) revealed a maximum absorbance peak at 258 nm for the CuONPs.

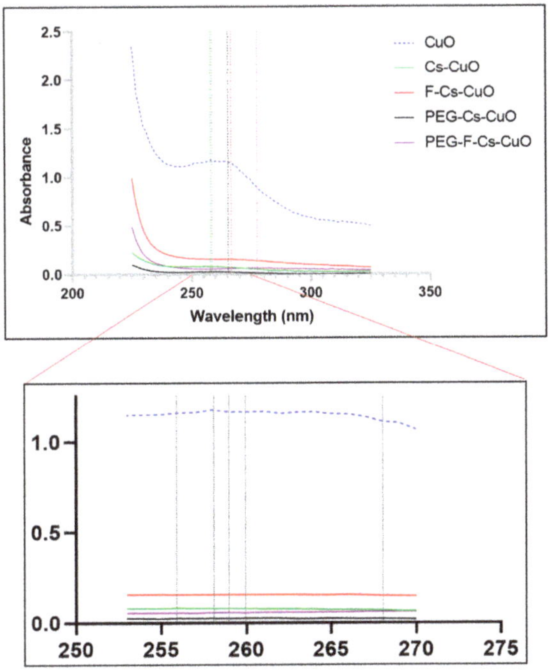

Figure 3. UV-vis spectra of CuONPs, their functionalized counterparts, and an amplified section of the UV-vis spectra showing the slight peak variations among the nanoparticles.

As observed in the literature, absorbance between 250 and 260 nm was noted and confirmed the synthesis of the CuONPs [39,40]. Upon the conjugation of chitosan, folate, and PEG, a slight shift in the λ_{max} was evidenced. The Cs-CuONPs peaked at 256 nm, while the F-Cs-CuONPs peaked at 266 nm. This shift is due to the successful modifications of the surface of the NPs [40]. Upon PEGylation, PEG-Cs-CuONPs and PEG-F-Cs-CuONPs had an SPR at 260 nm and 276 nm, respectively. Furthermore, the single visible peaks are evidence of a lack of by-products of the leaf extract used in the reduction process. An overall decline in the absorbance was noted due to the dilution of the original solution during the conjugation of chitosan, folate, and PEG.

The peaks seen in the XRD of the CuONPs (Supplementary Figure S1) were similar to those reported previously [41]. There is an agreement with the observed peaks upon the comparison of the data with JCPDS no. 00-048-1548. The strong peaks indicated the high crystalline state of the CuONPs [42].

3.1.2. Fourier-Transform Infrared Spectroscopy (FTIR)

Fourier-transform infrared spectroscopy plays an important role in the identification of the functional groups present in a molecule. It also serves as a validation for the synthesis of the CuONPs. Each chemical bond has a specific and unique energy absorption band.

Figure 4 portrays the spectrum generated following FTIR analysis. A clear, broad band at 3290.69 cm^{-1} indicated the OH- and C-O groups on the NP's surface. Stretching vibrations of the Cu-O bonds are characterized by the appearance of a narrower band at 1636.66 cm^{-1}. The peak at 585.85 cm^{-1} is ascribed to the formation of the CuONP.

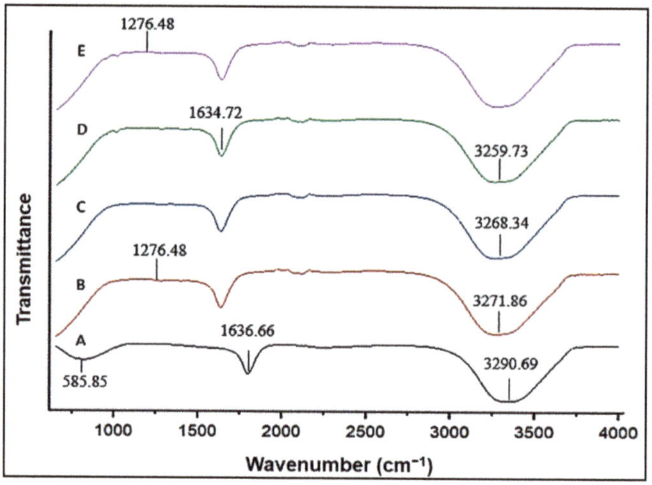

Figure 4. FTIR analysis of (A) CuONPs, (B) Cs-CuONPs, (C) PEG-Cs-CuONPs, (D) F-Cs-CuONPs, and (E) PEG-F-Cs-CuONPs.

The FTIR spectrum of the Cs-CuONPs (Figure 4) depicts a sharp peak at 1276.48 cm^{-1}. This peak indicates complex vibrations of the NHCO group (amide III bands) [43]. The spectrum shows structural changes owing to successful Cs conjugation. The distortion in the peaks at 1913.52 to 2304.16 cm^{-1} is due to the cross-coordination of the Cu ions and the Cs chains [44]. Following conjugation, the broad band at 3700–3100 cm^{-1} shifted to a lower frequency with a more asymmetrical appearance. This is due to the Cu ions complexing with the OH and NH_2 groups. A further reduction in frequency was noted for PEG-Cs-CuO (Figure 4), owing to the conjugation of PEG.

The subsequent addition of folic acid (Figure 4) produced a wider band at 1634.721 cm^{-1}, owing to the carboxylic acid group binding to the Cs-CuONPs. Furthermore, the band at 3259.729 cm^{-1} indicated the -NH and -OH groups of folic acid, correlating with the literature [45]. The peak at 1618 cm^{-1} was due to the conjugation of folic acid to the CuONPs [46]. A further peak at 1278.329 cm^{-1} for the PEG-F-Cs-CuONPs (Figure 4) was noted for the PEG-Cs-CuONPs (Figure 4), confirming the successful coupling of PEG to the NPs.

3.1.3. Transmission Electron Microscopy (TEM)

TEM provides a means of analyzing the ultrastructural morphology of the NPs together with their size and dispersity. Figure 5A shows that the CuONPs appear as smooth, spherical particles ranging from 48 nm to 69.5 nm, with an average size of 62.8 $\pm$ 12.8 nm (Table 1). Figure 5B depicts the Cs-CuONPs as clusters, maintaining their initial shape, with a larger size of 85.7 $\pm$ 26.9 nm (size range from 68.6 nm to 116.6 nm). The F-Cs-CuONPs (Figure 5C) demonstrated sizes ranging from 105.2 nm to 127.8 nm with an average size of 114.4 $\pm$ 11.9. nmTEM showed that the PEGylated NPs were larger, following the trend seen for the NTA (Section 3.1.4). In addition, the molecular weight of the PEG used may be a good indicator of the extent to which the NPs' size will increase, with a higher molecular weight possibly leading to larger NPs. The present study utilized PEG2000, which produced a significant size difference between the PEGylated and unPEGylated nanocomplexes. The PEG-Cs-CuONPs ranged from 87.3 nm to 98.2 nm, while the F-Cs-CuONPs were 125.5 nm to 159.5 nm. However, no significant difference was noted in the shape of the F-Cs-CuONPs, upon PEG addition, aside from the size increase.

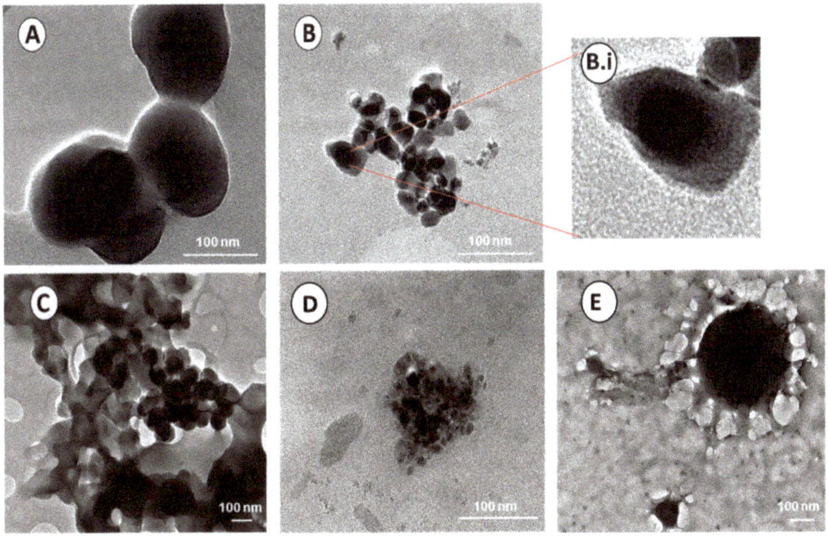

Figure 5. TEM micrographs of the nanoparticles. Bar scale = 100 nm. (**A**) CuONPs, (**B**) Cs-CuONPs, (**B.i**) Cs-CuONPs (zoomed), (**C**) F-Cs-CuONPs, (**D**) PEG-Cs-CuONPs, and (**E**) PEG-F-Cs-CuONPs.

Table 1. Particle size (nm) obtained from TEM imaging (n = 3).

Nanoparticles	Particle Size (nm) ± SE
CuO	62.8 ± 12.8
Cs-CuO	85.7 ± 26.9
F-Cs-CuO	114.4 ± 11.9
PEG-Cs-CuO	91.8 ± 9.4
PEG-F-Cs-CuO	119.3 ± 17.9

A thin coating around the NPs due to Cs coating was also observed (Figure 5B.i). The small size difference noted with that of CuONPs and Cs-CuONPs is in line with the slight red shift noted for UV-vis spectroscopy. The NPs showed minimal aggregation, which is not uncommon, and required sonication before treating the cells. This aggregation did not affect the stability of the NPs and their nanocomplexes, as evidenced by the zeta potentials obtained from the NTA analysis (Section 3.1.4).

3.1.4. Nanoparticle Tracking Analysis (NTA)

The biocompatibility of NPs depends on their size, shape, and charge. NTA combines laser light scattering microscopy and a charged-coupled device (CCD) camera to enable an accurate visualization of the NPs [47]. The analysis allows for the individual NPs to be tracked based on their Brownian motion in real-time, making NTA a very sensitive and robust analytical system. NTA accurately measured the size and zeta(ζ) potential of the NPs, as reported in Table 2. Although differences can be observed between the particle sizes obtained from the NTA and TEM, the general trend in the size increase was noted for the functionalized NPs. This difference in size may be attributed to the different sample preparation methods, with TEM using samples that were dried on a copper grid, while NTA utilized samples in aqueous solutions and measured the samples in real-time [31].

Table 2. Particle size (nm) and zeta potential (mV) of the nanoparticles.

Sample	Particle Size (nm) ± SE		ζ Potential (mV) ± SE	
	Nanoparticle	Nanocomplex (NP + pDNA)	Nanoparticle	Nanocomplex (NP + pDNA)
CuO	78.2 ± 20.7	-	−9 mV ± 0.1	-
Cs-CuO	103.9 ± 14.8	159.3 ± 6.7	45.3 mV ± 0.1	24.2 mV ± 0.1
F-Cs-CuO	128.0 ± 9.4	161.5 ± 9.5	55.1 mV ± 0.1	30.3 mV ± 0.2
PEG-Cs-CuO	148.8 ± 2.3	178.3 ± 3.7	42.3 mV ± 0.2	19.7 mV ± 0.1
PEG-F-Cs-CuO	197.1 ± 7.9	209.0 ± 9.8	55.5 mV ± 0.1	35.1 mV ± 0.1

Unmodified CuONPs exhibited the smallest size and a low negative ζ-potential (78.2 ± 20.7 nm; −9 ±0.1 mV). This negative charge was consistent with studies conducted by Sarkar et al. (2020) who observed a ζ potential of −2.67 mV. It has been proposed that the NPs' slight negative charge may be beneficial to prevent aggregation [48,49]. Upon modification with Cs, a notable increase in size and ζ-potential was seen (103.9 ± 14.8 nm; 45.3 ± 0.1 mV), which confirmed the successful chitosan conjugation. Furthermore, the positive charge indicates the presence of the cationic amino groups of Cs on the CuONPs. These results correlate to those obtained via UV-vis spectroscopy.

The conjugation of folate resulted in a further increase in the size and ζ-potential (128.0 ± 9.4 nm; 55.1 mV ± 0.1), confirming the successful binding of the folate to the NP. The positive charge is advantageous as it can support electrostatic interactions with the pDNA and anionic cellular membrane. The NPs portrayed a general trend in the increase in size and ζ-potential (mV) upon adding the polymer and ligand. This further confirmed that the Cs effectively stabilized the CuONPs and prevented aggregation.

The subsequent PEGylation of the Cs-CuONPs produced a slight decrease in ζ-potential. This decline may be due to the PEG chains masking the amino groups on the Cs. Furthermore, it has been reported that the molecular weight and density of PEG can increase the thickness of the PEG layer, resulting in a lowered ζ-potential, which could reach a neutral state [50]. Nanocomplexes containing PEG2000 have also been reported to be larger than those containing a lower-molecular-weight PEG [33]. However, the PEG-Cs-CuONPs and the PEG-F-Cs-CuONPs showed a further increase in the NP size, with improved ζ-potential. This increase in size further served as confirmation of successful PEG conjugation. Early research studies reported that nanomaterials functionalized with folate could utilize clathrin- and caveolae-independent endocytosis pathways for cellular uptake [51].

NPs that have ζ-potential values higher than +30 mV or lower than −30 mV are proposed to have excellent colloidal stability [31], as seen for the functionalized NPs and some of the nanocomplexes. The improved ζ-potential upon folate binding could have resulted in an altered PEG conformation on the NP, allowing for the availability of more of the cationic charges of the Cs. The changes in ζ-potential of the nanocomplexes were expected and confirmed the successful binding of the anionic DNA to the cationic NP. NPs with ζ-potentials not within the favorable range may experience some aggregation, since the attractive forces may exceed or equal the repulsive forces between the NPs. Although the ζ-potential for the PEG-Cs-CuO nanocomplex was reduced (19.7 mV), values around 20 mV have been reported to provide adequate stability [52]. However, this ζ-potential value proved to be sufficient to offer stability to the nanocomplex and facilitate favorable cellular uptake and transfection.

3.2. DNA Binding Studies

3.2.1. Electrophoretic Mobility Shift Assay

The principle of this assay works based on the cationic polymeric NPs electrostatically interacting with the negatively charged pDNA, forming nanocomplexes of varying charge ratios. Upon electroneutrality, the mobility of the pDNA into the gel is hindered, allowing for the determination of the point of complete condensation of the pDNA by the functionalized CuONPs (Figure 6). The white arrows in Figure 6 indicate the points of electroneutrality and the optimum binding (*w*/*w*) ratio. It can be seen that the F-Cs-CuONPs bind pDNA at a lower ratio (*w*/*w*) (0.4:1) compared to the Cs-CuONPs (0.6:1). A binding ratio of 0.3:1 (*w*/*w*) was obtained for both the PEG-Cs-CuONPs and PEG-F-Cs-CuONPs. There was some correlation with the NTA results, which showed an increase in the positive ζ-potential upon the conjugation of Cs, F, and PEG to the NPs.

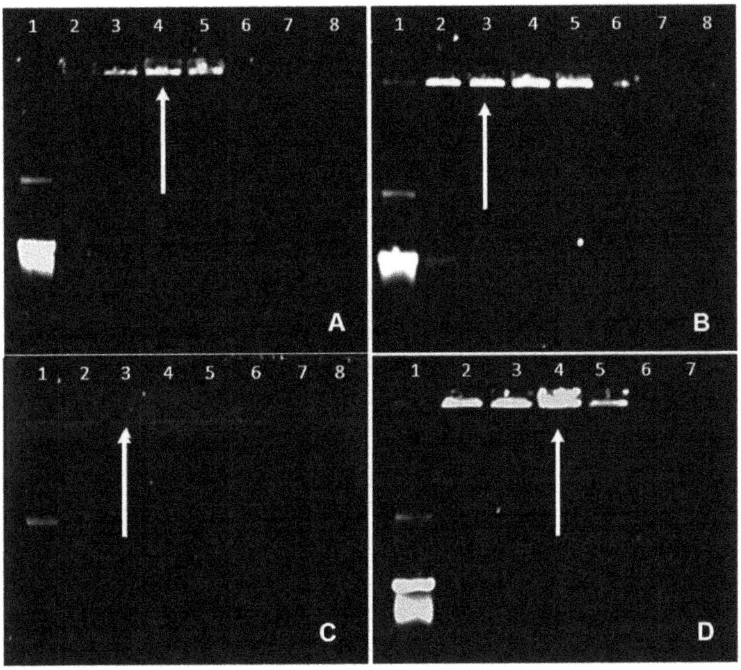

Figure 6. Electrophoretic mobility shift assay of the nanocomplexes. Lane 1 represents the positive control, consisting of naked pDNA (0.25 µg/mL). (**A**) Cs-CuONPs, lanes 1–5 (0, 0.2, 0.4, 0.6, 0.8 µg), (**B**) F-Cs-CuONPs, lanes 1–5 (0, 0.2, 0.4, 0.6, 0.8 µg), (**C**) PEG-Cs-CuONPs, lanes 1–5 (0, 0.1, 0.2, 0.3, 0.4, 0.5 µg), and (**D**) PEG-F-Cs-CuONPs, lanes 1–5 (0, 0.1, 0.2, 0.3, 0.4 µg). White arrows/yellow numbers indicate optimum binding ratios, preceded by the sub-optimum ratio and followed by the supra-optimum ratio.

The increased charge enhanced the complexation of the pDNA. The assay confirmed that these NPs have excellent pDNA binding capability. The inability of the DNA to migrate out of the wells at low ratios is in accordance with an early study [53] that showed the high compaction ability of chitosan. In some instances, such as in Figure 6C, the nanocomplexes formed are electroneutral and tend to float out of the wells. Hence, they are not seen as fluorescent bands in the wells. Similar results were seen before, such as in [37]. The low ratio of the F-Cs-CuONPs required in the binding of the pDNA further suggests that the covalently linked folate did not affect the DNA compaction ability of the chitosan.

3.2.2. Nuclease Protection Assay

A major intracellular challenge to a gene delivery system is the presence of various degradative enzymes, such as nucleases, resulting in the premature clearance of the genetic material in the bloodstream [54]. The carrier must protect the cargo during transportation for successful therapy to occur. The present assay assesses the ability of the NPs to protect the pDNA upon entering a simulated in vivo environment. Successful protection is visualized as distinct pDNA bands on the gel. This can be seen for all of the nanocomplexes (Figure 7). These bands confirm that the NPs can effectively protect the bound pDNA. The negative control (Figure 7A, lane 1) showed complete degradation of the naked pDNA.

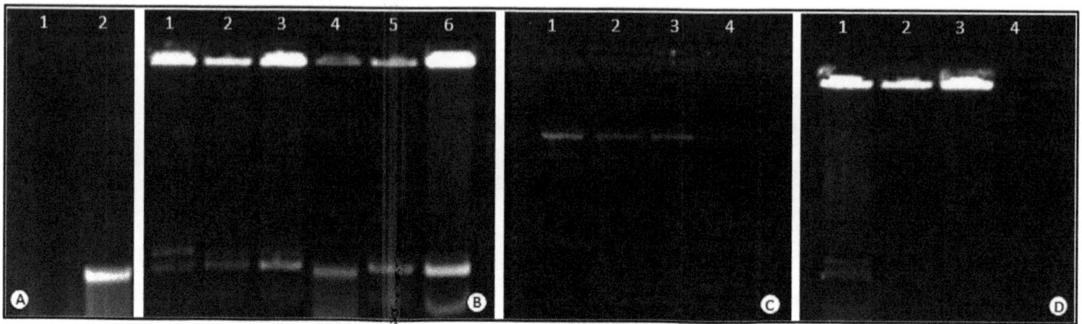

Figure 7. The serum nuclease protection assay. (**A**) Lane 1 contains the negative control (pDNA treated with FBS), while lane 2 contains the positive control of only naked pDNA (0.25 μg). (**B**) Lane 1–3: Cs-CuONPs (0.4, 0.6, 0.8 μg); lanes 4–6: F-Cs-CuONPs (0.2, 0.4, 0.6 μg). (**C**) Lanes 1–3: PEG-Cs-CuONPs (0.2, 0.3, 0.4 μg). (**D**) Lanes 1–3: PEG-F-Cs-CuONPs (0.2, 0.3, 0.4 μg). All nanocomplexes were complexed to pDNA (0.25 μg) and treated with 10% FBS.

In most cases, much of the DNA remained NP-bound in the wells, suggesting that the SDS used was insufficient to release the DNA. Alternate chemicals or increased SDS concentrations may be needed to obtain the complete release of the DNA. Nevertheless, this still confirmed the protection ability of the NPs. Similar results showing an incomplete release of the pDNA by SDS were noticed for metallic NPs [55].

3.2.3. Ethidium Bromide Intercalation Assay

Ethidium bromide is a cationic fluorescent dye that can intercalate between the DNA bases. The fluorescence is measured while increasing the concentration of cationic NPs, to determine the degree of compaction of the pDNA by the NPs. The complete complexation of the nanocomplex is indicated by a plateau in fluorescence readings. Figure 8A,B showed a steady decrease in fluorescence, correlating with an increase in NP concentration. The F-Cs-CuONPs produced a greater pDNA compaction of 82.4% compared to that of the Cs-CuONPs (71%).

Furthermore, the PEGylated nanocomplex showed increased compaction, as seen for PEG-Cs-CuONPs (84%) and PEG-F-Cs-CuONPs (95.4%). These results followed a similar trend to the electrophoretic mobility shift assay. The nanocomplexes must be compact enough to protect the nucleic acid cargo but not too tight to prevent disassociation of the nucleic acid from the nanoparticle. Overall, the binding of the NPs to the pDNA was sufficient to facilitate efficient gene expression, as observed in this study.

3.3. MTT Cell Viability Assay

The efficacy of gene therapy is dependent on its ability to target cancer cells while ensuring no adverse effects on healthy cells. The MTT assay was employed to determine the cell viability, measured in terms of the loss of metabolism in dead cells and the inability of the mitochondrial dehydrogenase enzyme to reduce MTT to formazan [37]. The cell viabilities (%) were expressed relative to the control of untreated cells (100%). The sub-optimum, optimum, and supra-optimum ratios obtained from the electrophoretic mobility shift assay (Section 3.2.1) were utilized. This was conducted to minimize the risk of any unconjugated CuONPs entering the cells and enabled the comparison between the three binding ratios, in order to ascertain which ratio produced the best transgene expression.

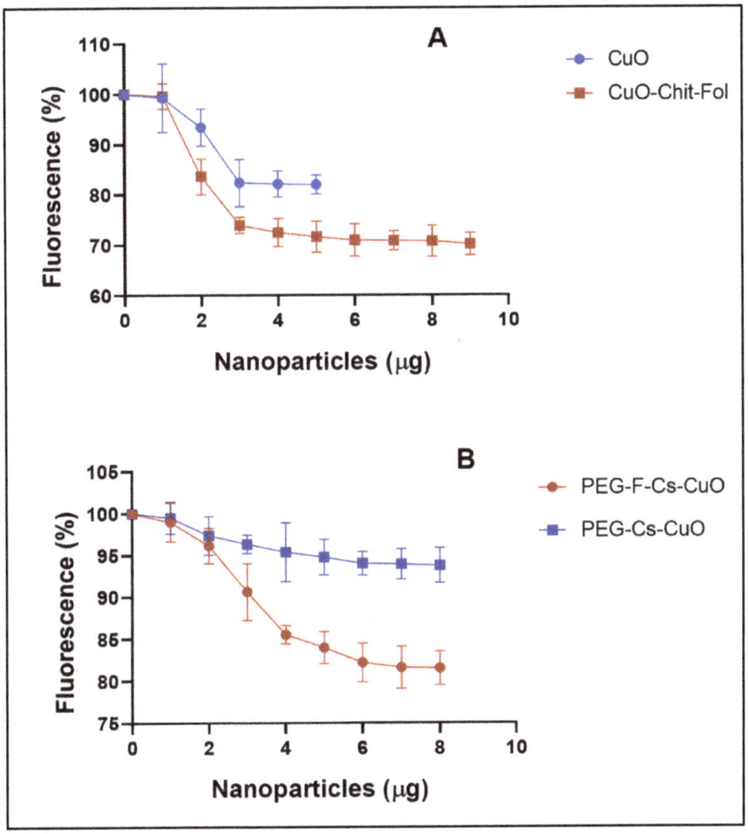

Figure 8. Ethidium bromide intercalation assay of the nanocomplexes. Incubation mixtures contained pCMV-*Luc* DNA (2.5 µg) in 100 µL HBS, with the addition of increasing amounts of the (**A**) Cs-CuONPs and F-Cs-CuONPs, and (**B**) PEG-Cs-CuONPs and PEG-F-Cs-CuONPs.

The promotion of cellular growth by the CuONPs was noted in the HeLa and MCF-7 cells, with a dose-dependent effect evident for the HEK293 cell (Figure 9A) with cell viability exceeding 70%. This is in keeping with Cus' natural properties of good biocompatibility and biodegradability [8]. It was proposed that an increased concentration of CuONPs would result in more significant cellular accumulation, resulting in enhanced stress and cell death [56]. A study by Fard et al. (2020) showed that CuONPs exhibited a dose-dependent cytotoxicity on peripheral blood cells, with a low concentration of CuONPs not significantly affecting the cell viability [57]. Overall, the nanocomplexes were well tolerated in all cells tested (HEK293, HeLa, and MCF-7), with cell viability >70% (Figure 9B–D). The increased cell viability may be due to the bound chitosan in the nanocomplexes, which can interact with the cell membranes and stimulate growth factors [58,59]. In addition, chitosan has been employed in tissue engineering [60], highlighting its suitability as a functionalizing polymer.

The targeted F-Cs-CuONPs portrayed a significantly high cell viability, especially at the optimum binding ratio in the HeLa cells, which was greater than that of the control. Similar high cell viabilities were noted in the HEK293 and MCF-7 cells. A study by Mansouri et al. (2006) revealed 79% cell viability in HEK293 cells using folate-targeted chitosan NPs. This affirms that folate does not affect the properties exhibited by chitosan but rather enhances them [53]. The additional conjugation of PEG to the functionalized NPs (Cs-CuO and F-Cs-

CuO) showed a further increase in viability in the HEK293 cells, and a moderate change in the viabilities of the HeLa and MCF-7 cells, compared to the unPEGylated nanocomplexes. A recent study reported that adding PEG to their NPs (Fe_3O_4) improved the NP stability and biocompatibility in 3T3 and HepG2 cells, preventing any toxic effects of the NPs due to their interactions with the cells or proteins [61]. Contrary to this, an earlier study using $MgFe_3O_4$ NPs noted that adding PEG increased cytotoxicity in breast cancer cells (SKBR-3). The authors attributed this to the possible interference of the function and structure of the cell membranes by the PEG molecules [62,63]. This study's overall low cytotoxicity is desirable, as it confirms the nanocomplexes' ability to be safe for gene delivery, especially regarding cervical cancer cells.

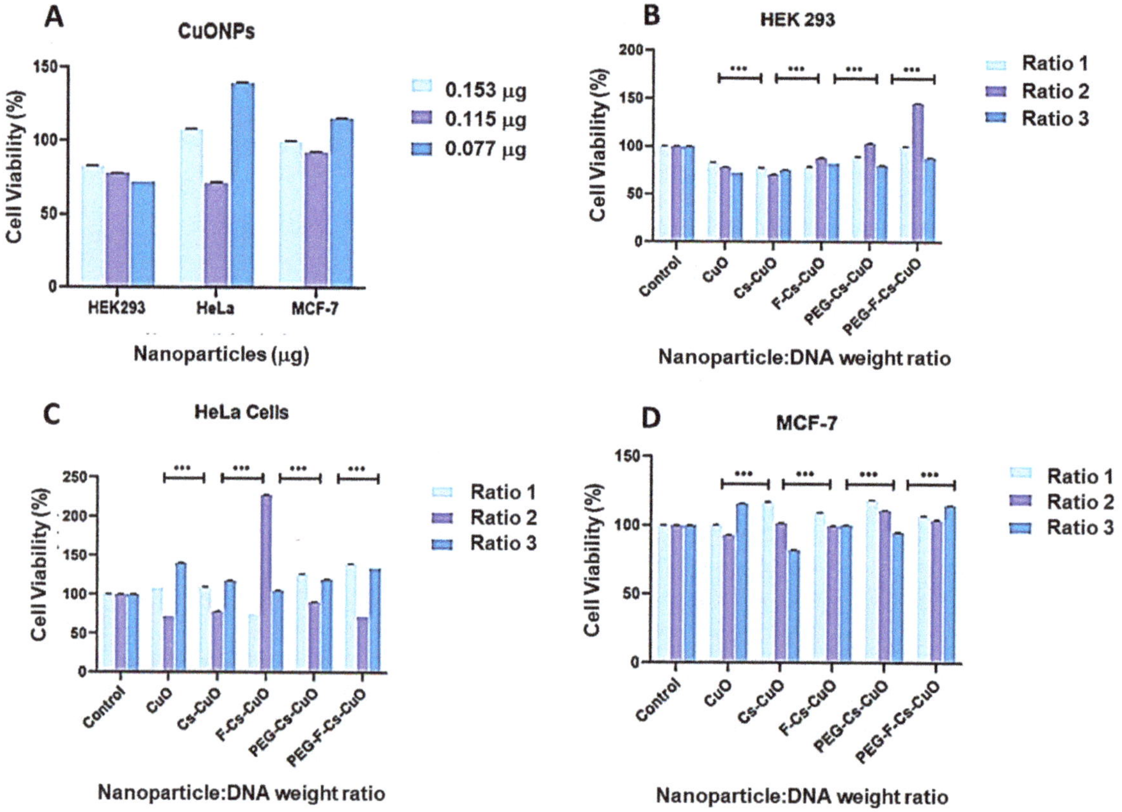

Figure 9. MTT assay portraying the cell viability of (**A**) CuONPs (0.153 µg; 0.115 µg; 0.077 µg), together with the nanoparticles in (**B**) HEK293, (**C**) HeLa, and (**D**) MCF-7 cells. The ratios are as follows: Cs-CuO (0.4:1; 0.6:1; 0.8:1), F-Cs-CuO (0.2:1; 0.4:1; 0.6:1), PEG-Cs-CuO (0.3:1; 0.4:1; 0.5:1), and PEG-F-Cs-CuO (0.2:1; 0.3:1; 0.4:1). Data are represented as means ± SD (n = 3). *** $p < 0.001$. was considered to be statistically significant.

3.4. Gene Expression Studies

This assay confirmed the potential of these Cu-based nanocomplexes in gene delivery and their ability to facilitate folate-receptor-mediated endocytosis (Figure 10). HEK293 cells served as control cells that were non-cancerous and lacking folate receptors. In the HEK293 cells, higher transgene expression was noted for the Cs-CuONPs and F-Cs-CuONPs compared to their PEGylated counterparts. This can be attributed to PEG shielding some of the cationic charges of Cs and the larger sizes of these PEG nanocomplexes. A key factor in determining transfection efficiency has been proposed to be the size of NPs [64]. NPs in the 120 to 150 nm range exhibit cellular internalization via clathrin- or caveolin-mediated endocytosis [65,66], with the optimum size limits being 50–100 nm [67]. NPs above (70–240 nm) and below (15–30 nm) have been reported to exhibit a decline in cellular uptake and transfection ability [68,69]. This is in keeping with the favorable transfection ability revealed by the F-Cs-CuONPs and Cs-CuONPs. This is further depicted in the difference between the transfection efficacy of PEG-Cs-CuONPs (91.8 $\pm$ 9.4 nm) and PEG-F-Cs-CuONPs (139.3 $\pm$ 17.9 nm), with the former showing greater transgene expression. A study using silver NPs noted that smaller particles might accumulate and contribute to lung damage compared to larger NPs [70].

The HeLa cells exhibited the highest transgene expression, followed by the MCF-7 and HEK293 cells. All nanocomplexes did perform almost equally well in the HeLa cells, suggesting that uptake occurred via various processes, including clathrin- or caveolae-mediated endocytosis and receptor-mediated endocytosis for the targeted nanocomplexes. Upon further examination using the competition assay, the PEG-F-Cs-CuO nanocomplexes showed cellular uptake via folate receptor mediation, as evidenced by the significantly lowered transgene expression obtained after the addition of excess folate which bound to and blocked the folate receptors overexpressed on the HeLa cells, preventing the entry of the PEG-F-Cs-CuO nanocomplexes. The size of the nanocomplexes was not the main determining factor in cellular internalization in this cell line, as the nanocomplex sizes varied. It was reported that different cell lines possessed different properties pertaining to their cell surfaces, resulting in a varied cellular uptake mechanism due to the range of NP sizes [71]. This study noted that uptake in the HeLa cells increased with an increase in NP size, while in the MCF-7 cells, a reduced uptake occurred with larger NPs. This was confirmed by the lack of significant reduction in the transgene expression in the PEGylated nanocomplexes compared to the unPEGylated nanocomplexes, as reported in previous studies [72,73].

The MCF-7 cells had fewer folate receptors on their cell surface than the HeLa cells, which was borne out via the significant differences in the transfection activities of the targeted nanocomplexes (PEG-F-Cs-CuO and F-Cs-CuO). Marshalek et al. (2016) compared the breast cancer cells MCF-7 and MDA-MB-231, and showed that the MCF-7 cells portrayed 1.76 times fewer folate receptors [74]. Another study examined the difference in folate receptors in the MCF-7 and HeLa cells utilizing nanoprobes, which revealed strong fluorescence in the HeLa cells, with very little fluorescence observed in the MCF-7 cells, validating the results obtained in the present study [75].

Overall, the transfection activity of all of the nanocomplexes declined in all of the cells in the competition assay. This could be due to the excess folate either blocking the receptors and preventing receptor-mediated endocytosis of the targeted nanocomplexes or their presence on the cell surface posing a challenge for the untargeted nanocomplexes to enter the cells via passive endocytosis, maropinocytosis, or micropinocytosis. Nevertheless, this study did confirm the uptake of the targeted nanocomplexes, especially the PEG-F-Cs-CuO nanocomplexes, via folate-receptor-mediated endocytosis. The addition of PEG did seem to improve the targeted delivery of these nanocomplexes. However, there is a need for further studies on the cellular uptake mechanisms involved and a need to improve the current system for enhanced targeted delivery.

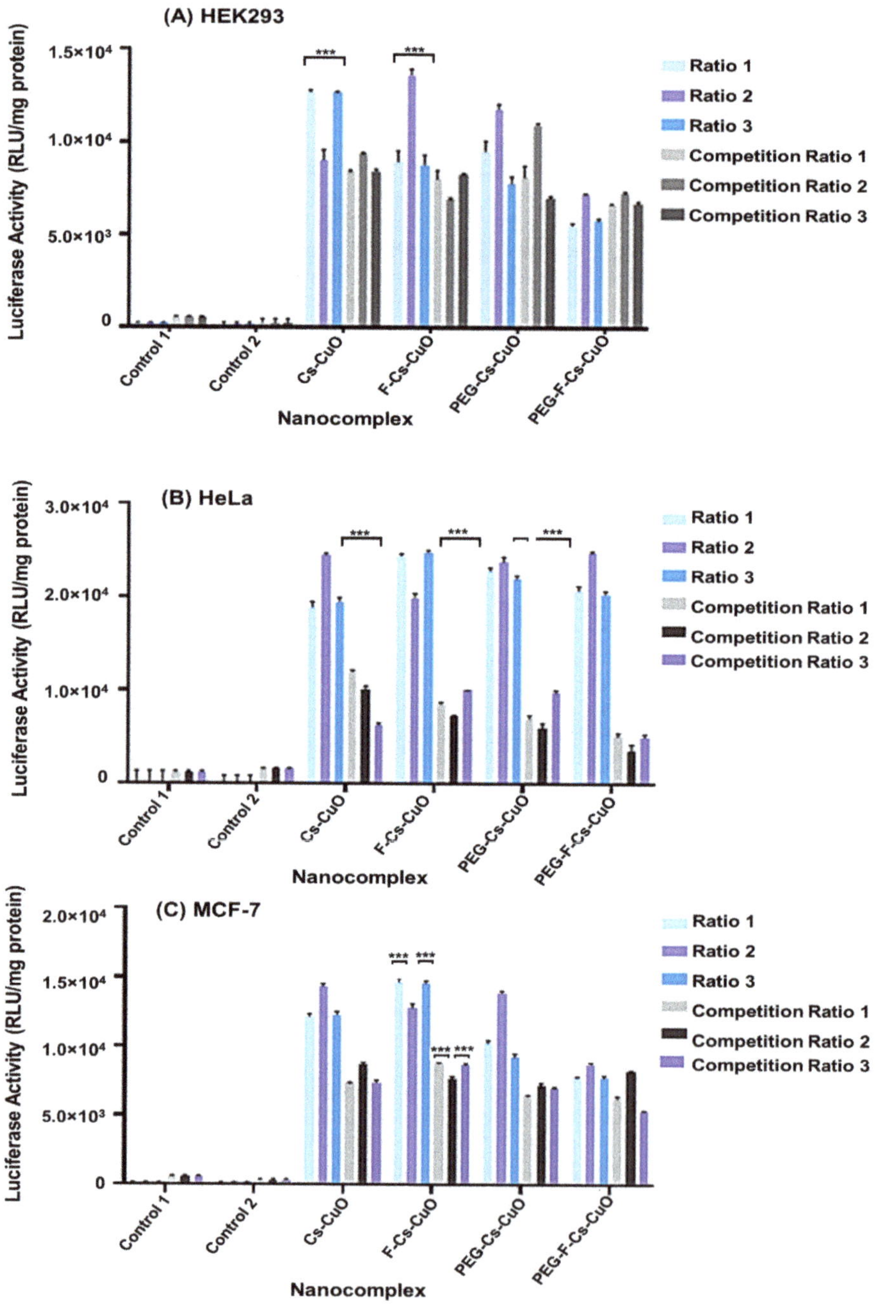

Figure 10. In vitro luciferase activity and competition studies in (**A**) HEK293, (**B**) HeLa, and (**C**) MCF-7 cells. Each column represents the mean ± SD (n = 3). Luciferase expression was measured in RLU/mg protein. *** $p < 0.001$ was considered to be statistically significant. The ratios were represented as follows: Cs-CuO (0.4:1; 0.6:1; 0.8:1), F-Cs-CuO (0.2:1; 0.4:1; 0.6:1), PEG-Cs-CuO (0.2:1; 0.3:1, 0.4:1), and PEG-F-Cs-CuO (0.2:1, 0.3:1, 0.4:1).

4. Conclusions

Holistically, this study demonstrated that the formulated CuONP-based nanocomplexes could safely and efficiently deliver the pDNA to cervical cancer cells in vitro. Our findings highlight the potential of folate-targeting for cancer therapy, and the benefits of the carrier systems' physiochemical properties for pharmaceutical research and future clinical applications. Hence, further investigations and optimizations are warranted to improve efficiency and to assess their in vivo applicability. Future studies could include optimizing the green synthesis approach by utilizing other biological materials, testing the nanocomplexes in a co-culture environment, i.e., with normal and cancer cells synergistically, and optimizing the receptor-mediated uptake in cervical cancer cells for enhanced transgene activity. Mechanistic studies to understand the cellular uptake and trafficking of the nanocomplexes in the cells would be beneficial. Overall, our study provides a robust tool for gene delivery studies and underscores the potential of gene therapy for cancer treatment. This study vaunts a non-invasive approach, laying the foundation for using other biological materials in NP synthesis. While CuONPs have been synthesized before via chemical, physical, and biological means, the novelty of this study lies in the gene delivery potential of the nanoconjugates produced, and the *M. azedarach* leaf extracts utilized, which acted synergistically with the NPs to achieve the desired therapeutic effect.

Supplementary Materials: The following supporting information can be downloaded at https://www.mdpi.com/article/10.3390/polym15102393/s1: Figure S1: XRD of copper oxide nanoparticles.

Author Contributions: Conceptualization, K.J. and M.S.; methodology, K.J.; software, K.J.; validation, M.S.; formal analysis, K.J.; investigation, K.J.; resources, M.S.; data curation, K.J.; writing—original draft preparation, K.J.; writing—review and editing, M.S.; visualization, K.J. and M.S.; supervision, M.S.; project administration, M.S.; funding acquisition, M.S. All authors have read and agreed to the published version of the manuscript.

Funding: Funding for this study was obtained from the National Research Foundation, South Africa (grant numbers 129263 and 120455).

Institutional Review Board Statement: Not applicable.

Data Availability Statement: All data presented in the study are included in the article. Further inquiries can be directed to the corresponding author.

Acknowledgments: The authors acknowledge the Nano-Gene and Drug Delivery group members for advice and technical support, and the curator (S Ramdhani) of the Ward Herbarium, UKZN, South Africa.

Conflicts of Interest: The authors declare no conflict of interest.

References

1. World Health Organization. Cervical Cancer. Available online: https://www.who.int/health-topics/cervical-cancer#tab=tab_1 (accessed on 20 February 2023).
2. Fang, J.H.; Yu, X.M.; Zhang, S.H.; Yang, Y. Effect of smoking on high-grade cervical cancer in women on the basis of human papillomavirus infection studies. *J. Cancer Res. Ther.* **2018**, *14*, 184–189. [CrossRef]
3. Venkatas, J.; Singh, M. Cervical Cancer: A meta-analysis, Therapy and future of Nanomedicine. *Ecancermedicalscience* **2020**, *14*, 1111. [CrossRef]
4. Small, W.; Bacon, M.; Bajaj, A.; Chuang, L.; Fisher, B.; Harkenrider, M.; Jhingran, A.; Kitchener, H.; Mileshkin, L.; Viswanathan, A.; et al. Cervical cancer: A global health crisis. *Cancer* **2017**, *123*, 2404–2412. [CrossRef]
5. Baetke, S.; Lammers, T.; Kiessling, F. Applications of nanoparticles for diagnosis and therapy of cancer. *Br. J. Radiol.* **2015**, *88*, 20150207. [CrossRef]
6. Bhattacharyya, S.; Kudgus, R.A.; Bhattacharya, R.; Mukherjee, P. Inorganic Nanoparticles in Cancer Therapy. *Pharm. Res.* **2011**, *28*, 237–259. [CrossRef]
7. Ge, E.J.; Bush, A.I.; Casini, A.; Cobine, P.A.; Cross, J.R.; DeNicola, G.M.; Dou, Q.P.; Franz, K.J.; Gohil, V.M.; Gupta, S.; et al. Connecting copper and cancer: From transition metal signalling to metalloplasia. *Nat. Rev. Cancer* **2022**, *22*, 102–113. [CrossRef]
8. Ingle, A.P.; Duran, N.; Rai, M. Bioactivity, mechanism of action, and cytotoxicity of copper-based nanoparticles: A review. *Appl. Microbiol. Biotechnol.* **2014**, *98*, 1001–1009. [CrossRef] [PubMed]

9. Wang, L. Synthetic methods of CuS nanoparticles and their applications for imaging and cancer therapy. *RSC Adv.* **2016**, *6*, 82596–82615. [CrossRef]
10. Chudobova, D.; Cihalova, K.; Kopel, P.; Melichar, L.; Ruttkay-Nedecky, B.; Vaculovicova, M.; Adam, V.; Kizek, R. Complexes of Metal-Based Nanoparticles with Chitosan Suppressing the Risk of Staphylococcus aureus and *Escherichia coli* Infections. In *Nanotechnology in Diagnosis, Treatment and Prophylaxis of Infectious Diseases*; Rai, M., Kon, K., Eds.; Academic Press: London, UK, 2015; pp. 217–232.
11. Tadjarodi, A.; Roshani, R. A green synthesis of copper oxide nanoparticles by mechanochemical method. *Curr. Chem. Lett.* **2014**, *3*, 215–220. [CrossRef]
12. Sankar, R.; Maheswari, R.; Karthik, S.; Shivashangari, K.; Ravikumar, V. Anticancer activity of Ficus religiosa engineered copper oxide nanoparticles. *Mater. Sci. Eng. C* **2014**, *44*, 234–239. [CrossRef] [PubMed]
13. Hang, X.; Peng, H.; Song, H.; Qi, Z.; Miao, X.; Xu, W. Antiviral activity of cuprous oxide nanoparticles against Hepatitis C Virus in vitro. *J. Virol. Meth.* **2015**, *222*, 150–157. [CrossRef] [PubMed]
14. Muhammad Imran Din, M.I.; Rehan, R. Synthesis, Characterization, and Applications of Copper Nanoparticles. *Anal. Lett.* **2017**, *50*, 50–62.
15. Jagaran, K.; Singh, M. Nanomedicines for COVID-19: Potential of Copper nanoparticles. *Biointerface Res. Appl.Chem.* **2021**, *11*, 10716–10728.
16. Olawale, F.; Oladimeji, O.; Ariatti, M.; Singh, M. Emerging Roles of green synthesized Chalcogen and Chalcogenide nanoparticles in Cancer theranostics. *J. Nanotechnol.* **2022**, *2022*, 6176610. [CrossRef]
17. Iravani, S. Green synthesis of metal nanoparticles using plants. *Green Chem.* **2011**, *13*, 2638. [CrossRef]
18. Černík, M.; Thekkae, P.V. Green synthesis of copper oxide nanoparticles using gum karaya as a biotemplate and their antibacterial application. *Int. J. Nanomed.* **2013**, *8*, 889–898. [CrossRef]
19. Vishnukanta, A.R. Melia Azedarach: A Phytopharmacological Review. *Phcog. Rev.* **2008**, *2*, 173–179.
20. Nerome, K.; Ito-Kureha, T.; Paganini, T.; Fukuda, T.; Igarashi, Y.; Ashitomi, H.; Ikematsu, S.; Yamamoto, T. Potent and broad anticancer activities of leaf extracts from *Melia azedarach* L. of the subtropical Okinawa islands. *Am. J. Cancer Res.* **2020**, *10*, 581–594.
21. Al-Marzoqi, A.H.; Imad Hadi Hameed, I.H.; Idan, S.A. Analysis of bioactive chemical components of two medicinal plants (*Coriandrum sativum* and *Melia azedarach*) leaves using gas chromatography-mass spectrometry (GC-MS). *Afr. J. Biotechnol.* **2015**, *14*, 2812–2830.
22. Cao, S.; Bi, Z.; Li, Q.; Zhang, S.; Singh, M.; Chen, J.D. Shape memory and antibacterial chitosan-based cryogel with hemostasis and skin wound repair. *Carbohyd. Polym.* **2023**, *305*, 120545. [CrossRef]
23. Wang, W.; Meng, Q.; Li, Q.; Liu, J.; Zhou, M.; Jin, Z.; Zhao, K. Chitosan derivatives and their application in biomedicine. *Int. J. Mol. Sci.* **2020**, *21*, 487. [CrossRef] [PubMed]
24. Cheung, A.; Bax, H.J.; Josephs, D.H.; Ilieva, K.M.; Pellizzari, G.; Opzoomer, J.; Bloomfield, J.; Fittall, M.; Grigoriadis, A.; Figini, M.; et al. Targeting folate receptor alpha for cancer treatment. *Oncotarget* **2016**, *7*, 52553–52574. [CrossRef]
25. Scaranti, M.; Cojocaru, E.; Banerjee, S.; Banerji, U. Exploiting the folate receptor α in oncology. *Nat. Rev. Clin. Oncol.* **2020**, *17*, 349–359. [CrossRef] [PubMed]
26. Suk, J.; Xu, Q.; Kim, N.; Hanes, J.; Ensign, L. PEGylation as a strategy for improving nanoparticle-based drug and gene delivery. *Adv. Drug Deliv. Rev.* **2016**, *99*, 28–51. [CrossRef]
27. Sercombe, L.; Veerati, T.; Moheimani, F.; Wu, S.Y.; Sood, A.K.; Hua, S. Advances and Challenges of Liposome Assisted Drug Delivery. *Front. Pharmacol.* **2015**, *6*, 286. [CrossRef] [PubMed]
28. Yin, H.; Kanasty, R.; Eltoukhy, A.; Vegas, A.; Dorkin, J.; Anderson, D. Non-viral vectors for gene-based therapy. *Nat. Rev. Genet.* **2014**, *15*, 541–555. [CrossRef]
29. Chinnasamy, G.; Chandrasekharan, S.; Bhatnagar, S. Biosynthesis of Silver Nanoparticles from *Melia azedarach*: Enhancement of Antibacterial, Wound Healing, Antidiabetic and Antioxidant Activities. *Int. J. Nanomed.* **2019**, *14*, 9823–9836. [CrossRef] [PubMed]
30. Arunkumar, B.; Jeyakumar, S.J.; Jothibas, M. A sol-gel approach to the synthesis of CuO nanoparticles using Lantana camara leaf extract and their photo catalytic activity. *Optik* **2019**, *183*, 698–705. [CrossRef]
31. Akinyelu, J.; Singh, M. Folate-tagged chitosan-functionalized gold nanoparticles for enhanced delivery of 5-fluorouracil to cancer cells. *Appl. Nanosci.* **2019**, *9*, 7–17. [CrossRef]
32. Rana, S.; Shetake, N.G.; Barick, K.C.; Pandey, B.N.; Salunke, H.G.; Hassan, P.A. Folic acid conjugated Fe3O4 magnetic nanoparticles for targeted delivery of doxorubicin. *Dalton Transact.* **2016**, *45*, 17401–17408. [CrossRef]
33. Daniels, A.N.; Singh, M. Sterically stabilized siRNA:gold nanocomplexes enhance c-MYC silencing in a breast cancer cell model. *Nanomedicine* **2019**, *14*, 1387–1401. [CrossRef]
34. Malloy, A. Count, Size and Visualize Nanoparticles. *Mater. Today* **2011**, *14*, 170–173. [CrossRef]
35. Singh, M. Assessing nucleic acid: Cationic nanoparticle interaction for gene delivery. In *Bio-Carrier Vectors*; Narayanan, K., Ed.; Springer: New York, NY, USA, 2021; Volume 2211, pp. 43–55.
36. Mosmann, T. Rapid colorimetric assay for cellular growth and survival: Application to proliferation and cytotoxicity assays. *J. Immunol. Methods* **1983**, *65*, 55–63. [CrossRef] [PubMed]
37. Maiyo, F.C.; Mbatha, L.S.; Singh, M. Selenium Nanoparticles in Folate-Targeted delivery of the pCMV-Luc DNA Reporter Gene. *Curr. Nanosci.* **2021**, *17*, 871–880. [CrossRef]

38. Naidoo, S.; Daniels, A.; Habib, S.; Singh, M. Poly-L-Lysine–Lactobionic Acid-Capped Selenium Nanoparticles for Liver-Targeted Gene Delivery. *Int. J. Mol. Sci.* **2022**, *23*, 1492. [CrossRef] [PubMed]
39. Kumar, B.; Smita, K.; Cumbal, L.; Debut, A.; Angulo, Y. Biofabrication of copper oxide nanoparticles using Andean blackberry (*Rubus glaucus* Benth.) fruit and leaf. *J. Saudi Chem. Soc.* **2017**, *21*, S475–S480. [CrossRef]
40. Haiss, W.; Thanh, N.T.; Aveyard, J.; Fernig, D.G. determination of size and concentration of gold nanoparticles from UV–Vis. spectra. *Anal. Chem.* **2007**, *79*, 4215–4221. [CrossRef]
41. Gamedze, N.; Mthiyane, D.M.N.; Babalola, O.; Singh, M.; Onwudiwe, D.C. Physico-chemical characteristics and cytotoxicity evaluation of CuO and TiO_2 nanoparticles biosynthesized using extracts of Mucuna pruriensutilis seeds. *Heliyon* **2022**, *8*, e10187. [CrossRef]
42. Velsankar, K.; Suganya, S.; Muthumari, P.; Mohandoss, S.; Sudhahar, S. Ecofriendly green synthesis, characterization and biomedical applications of CuO nanoparticles synthesized using leaf extract of *Capsicum frutescens*. *J. Environ. Chem. Eng.* **2021**, *9*, 106299.
43. Shigemassa, Y.; Matsuura, H.; Sashiwa, H.; Saimoto, H. Heavy metal contamination. *Int. J. Biol. Macromol.* **2006**, *18*, 237.
44. Qu, J.; Hu, Q.; Shen, K.; Zhang, K.; Li, Y.; Li, H.; Zhang, Q.; Wang, J.; Quan, W. The preparation and characterization of chitosan rods modified with Fe3þ by a chelation mechanism. *Carbohydr. Res.* **2011**, *346*, 822–827. [CrossRef] [PubMed]
45. Jalilian, A.R.; Hosseini-Salekdeh, S.L.; Mahmoudi, M.; Yousefnia, H.; Majdabadi, A.; Pouladian, M. Preparation and biological evaluation of radiolabeled-folate embedded superparamagnetic nanoparticles in wild-type rats. *J. Radioanal. Nucl. Chem.* **2011**, *287*, 119–127. [CrossRef]
46. Laha, D.; Pramanik, A.; Chattopadhyay, S.; Dash, D.K.; Roy, S.; Pramanik, P.; Karmakar, P. Folic acid modified copper oxide nanoparticles for targeted delivery in in vitro and in vivo systems. *RSC Adv.* **2015**, *5*, 68169. [CrossRef]
47. Filipe, V.; Hawe, A.; Jiskoot, W. Critical Evaluation of Nanoparticle Tracking Analysis (NTA) by NanoSight for the Measurement of Nanoparticles and Protein Aggregates. *Pharm. Res.* **2010**, *27*, 796–810. [CrossRef] [PubMed]
48. Sarkar, J.; Chakraborty, N.; Chatterjee, A.; Bhattacharjee, A.; Dasgupta, D.; Acharya, K. Green Synthesized Copper Oxide Nanoparticles Ameliorate Defence and Antioxidant Enzymes in Lens culinaris. *Nanomaterials* **2020**, *10*, 312. [CrossRef]
49. Sarkar, J.; Dey, P.; Saha, S.; Acharya, K. Mycosynthesis of selenium nanoparticles. *Micro Nano Lett.* **2011**, *6*, 599. [CrossRef]
50. Shi, L.; Zhang, J.; Zhao, M.; Tang, M.; Cheng, X.; Zhang, W.; Li, W.; Liu, X.; Peng, H.; Wang, Q. Effects of polyethylene glycol on the surface of nanoparticles for targeted drug delivery. *Nanoscale* **2021**, *13*, 10748–10764. [CrossRef]
51. Lu, T.; Low, P.S. Folate-mediated delivery of macromolecular anticancer therapeutic agents. *Adv. Drug Deliv. Rev.* **2002**, *54*, 675–693. [CrossRef]
52. Honary, S.; Zahir, F. Effect of Zeta Potential on the Properties of Nano-Drug Delivery Systems—A Review (Part 2). *Trop. J. Pharm. Res.* **2013**, *12*, 265–273.
53. Mansouri, S.; Cuie, Y.; Winnik, F.; Shi, Q.; Lavigne, P.; Benderdour, M.; Beaumont, E.; Fernandes, J. Characterization of folate-chitosan-DNA nanoparticles for gene therapy. *Biomaterials* **2006**, *27*, 2060–2065. [CrossRef]
54. Obata, Y.; Saito, S.; Takeda, N.; Takeoka, S. Plasmid DNA-encapsulating liposomes: Effect of a spacer between the cationic head group and hydrophobic moieties of the lipids on gene expression efficiency. *Biochim. Biophys. Acta Biomembr.* **2009**, *1788*, 1148–1158. [CrossRef] [PubMed]
55. Akinyelu, A.; Oladimeji, O.; Singh, M. Lactobionic Acid-Chitosan Functionalized Gold Coated Poly(lactide-co-glycolide) Nanoparticles for Hepatocyte Targeted Gene Delivery. *Adv. Nat. Sci. Nanosci. Nanotechnol.* **2020**, *11*, 045017. [CrossRef]
56. Alishah, H.; Pourseyedi, S.; Ebrahimipour, Y.S.; Mahani, S.E.; Rafiei, N. Green synthesis of starch-mediated CuO nanoparticles: Preparation, characterization, antimicrobial activities and in vitro MTT assay against MCF-7 cell line. *Rend. Fis. Acc. Lincei* **2017**, *28*, 65–71. [CrossRef]
57. Fard, M.Z.; Fatholahi, M.; Abyadeh, M.; Bakhtiarian, A.; Mousavi, S.E.; Falahati, M. The Investigation of the Cytotoxicity of Copper Oxide Nanoparticles on Peripheral Blood Mononuclear Cells. *Nanomed. Res. J.* **2020**, *5*, 364–368.
58. Boca, S.C.; Potara, M.; Gabudean, A.; Juhem, A.; Baldeck, P.L.; Astilean, S. Chitosan-coated triangular silver nanoparticles as a novel class of biocompatible, highly effective photothermal transducers for in vitro cancer cell therapy. *Cancer Lett.* **2011**, *311*, 131–140. [CrossRef] [PubMed]
59. Rajam, M.; Sivasami, P.; Rose, C.; Mandal, A.B. Chitosan nanoparticles as a dual growth factor delivery system for tissue engineering applications. *Int. J. Pharm.* **2011**, *410*, 145–152. [CrossRef]
60. Sivashankari, P.; Prabaharan, M. Prospects of chitosan-based scaffolds for growth factor release in tissueengineering. *Int. J. Biol. Macromol.* **2016**, *93*, 1382–1389. [CrossRef]
61. Ebadi, M.; Rifqi Md Zain, A.; Tengku Abdul Aziz, T.H.; Mohammadi, H.; Tee, C.A.T.; Rahimi Yusop, M. Formulation and Characterization of Fe_3O_4@PEG Nanoparticles Loaded Sorafenib; Molecular Studies and Evaluation of Cytotoxicity in Liver Cancer Cell Lines. *Polymers* **2023**, *15*, 971. [CrossRef]
62. Ramnandan, D.; Mokhosi, S.; Daniels, A.; Singh, M. Chitosan, Polyethylene glycol and Polyvinyl alcohol modified $MgFe_2O_4$ ferrite magnetic nanoparticles in Doxorubicin delivery: A comparative study in vitro. *Molecules* **2021**, *26*, 3893. [CrossRef]
63. Patil, U.; Adireddy, S.; Jaiswal, A.; Mandava, S.; Lee, B.; Chrisey, D. In vitro/in vivo toxicity evaluation and quantification of iron oxide nanoparticles. *Int. J. Mol. Sci.* **2015**, *16*, 24417. [CrossRef]
64. Zhu, M.; Nie, G.; Meng, H.; Xia, T.; Nel, A.; Zhao, Y. Physicochemical Properties Determine Nanomaterial Cellular Uptake, Transport, and Fate. *Acc. Chem. Res.* **2013**, *46*, 622–631. [CrossRef] [PubMed]

65. Rejman, J.; Oberle, V.; Zuhorn, I.; Hoekstra, D. Size-dependent internalization of particles via the pathways of clathrin-and caveolae-mediated endocytosis. *Biochem. J.* **2004**, *377*, 159–169. [CrossRef] [PubMed]
66. Panariti, A.; Miserocchi, G.; Rivolta, I. The effect of nanoparticle uptake on cellular behavior: Disrupting or enabling functions? *Nanotechnol. Sci. Appl.* **2012**, *5*, 87–100.
67. Foroozandeh, P.; Aziz, A.A. Insight into Cellular Uptake and Intracellular Trafficking of Nanoparticles. *Nanoscale Res. Lett.* **2018**, *13*, 339. [CrossRef] [PubMed]
68. Chithrani, B.; Chan, W. Elucidating the Mechanism of Cellular Uptake and Removal of Protein-Coated Gold Nanoparticles of Different Sizes and Shapes. *Nano Lett.* **2007**, *7*, 1542–1550. [CrossRef] [PubMed]
69. Lu, F.; Wu, S.; Hung, Y.; Mou, C. Size Effect on Cell Uptake in Well-Suspended, Uniform Mesoporous Silica Nanoparticles. *Small* **2009**, *5*, 1408–1413. [CrossRef]
70. Xu, L.; Wang, Y.-Y.; Huang, J.; Chen, C.-Y.; Wang, Z.-X.; Xie, H. Silver nanoparticles: Synthesis, medical applications, and biosafety. *Theranostics* **2020**, *10*, 8996–9031. [CrossRef]
71. Shang, L.; Nienhaus, K.; Nienhaus, G.U. Engineered nanoparticles interacting with cells: Size matters. *J. Nanobiotechnol.* **2014**, *12*, 5. [CrossRef]
72. Shan, Y.; Ma, S.; Nie, L.; Shang, X.; Hao, X.; Tang, Z.; Wang, H. Size-dependent endocytosis of single gold nanoparticles. *Chem. Comm.* **2011**, *47*, 8091–8093. [CrossRef]
73. Huang, K.; Ma, H.; Liu, J.; Huo, S.; Kumar, A.; Wei, T.; Zhang, X.; Jin, S.; Gan, Y.; Wang, P.C.; et al. Size-dependent localization and penetration of ultrasmall gold nanoparticles in cancer cells, multicellular spheroids, and tumors in vivo. *ACS Nano* **2012**, *6*, 4483–4493. [CrossRef]
74. Marshalek, J.P.; Sheeran, P.S.; Ingram, P.; Dayton, P.A.; Witte, R.S.; Matsunaga, T.O. Intracellular delivery and ultrasonic activation of folate receptor-targeted phase-change contrast agents in breast cancer cells in vitro. *J. Control. Release* **2016**, *243*, 69–77. [CrossRef] [PubMed]
75. Feng, D.; Song, Y.; Shi, W.; Li, X.; Ma, H. Distinguishing Folate-Receptor-Positive Cells from Folate-ReceptorNegative Cells Using a Fluorescence off–on Nanoprobe. *Anal. Chem.* **2013**, *85*, 6530–6535. [CrossRef] [PubMed]

polymers

MDPI

Article

Polyester Nanocapsules for Intravenous Delivery of Artemether: Formulation Development, Antimalarial Efficacy, and Cardioprotective Effects In Vivo

Alessandra Teixeira Vidal-Diniz [1,†], Homero Nogueira Guimarães [2], Giani Martins Garcia [1], Érika Martins Braga [3], Sylvain Richard [4,5,*], Andrea Grabe-Guimarães [1] and Vanessa Carla Furtado Mosqueira [1,*]

1 School of Pharmacy, Universidade Federal de Ouro Preto (UFOP), Campus Universitário Morro do Cruzeiro, Ouro Preto 35400-000, MG, Brazil
2 Department of Electrical Engineering, Federal University of Minas Gerais, Belo Horizonte 31270-901, MG, Brazil
3 Department of Parasitology, Institute of Biological Sciences, Universidade Federal de Minas Gerais (UFMG), Belo Horizonte 31270-901, MG, Brazil
4 CNRS, INSERM, Université de Montpellier, 34295 Montpellier, France
5 PhyMedExp, CHU Arnaud de Villeneuve 371, Avenue du Doyen Gaston Giraud, CEDEX 05, 34295 Montpellier, France
* Correspondence: sylvain.richard@inserm.fr (S.R.); mosqueira@ufop.edu.br (V.C.F.M.)
† Current address: Department of Natural Science, Instituto Federal de Minas Gerais (IFMG), Campus Congonhas, Ouro Preto 35400-000, MG, Brazil.

Citation: Vidal-Diniz, A.T.; Guimarães, H.N.; Garcia, G.M.; Braga, É.M.; Richard, S.; Grabe-Guimarães, A.; Mosqueira, V.C.F. Polyester Nanocapsules for Intravenous Delivery of Artemether: Formulation Development, Antimalarial Efficacy, and Cardioprotective Effects In Vivo. *Polymers* **2022**, *14*, 5503. https://doi.org/10.3390/polym14245503

Academic Editors: Stephanie Andrade, Joana A. Loureiro and Maria João Ramalho

Received: 1 November 2022
Accepted: 9 December 2022
Published: 15 December 2022

Abstract: Artemether (ATM) is an effective antimalarial drug that also has a short half-life in the blood. Furthermore, ATM is also cardiotoxic and is associated with pro-arrhythmogenic risks. We aimed to develop a delivery system enabling the prolonged release of ATM into the blood coupled with reduced cardiotoxicity. To achieve this, we prepared polymeric nanocapsules (NCs) from different biodegradable polyesters, namely poly(*D,L*-lactide) (PLA), poly-ε-caprolactone (PCL), and surface-modified NCs, using a monomethoxi-polyethylene glycol-*block*-poly(*D,L*-lactide) (PEG_{5kDa}-PLA_{45kDa}) polymer. Using this approach, we were able to encapsulate high yields of ATM (>85%, 0–4 mg/mL) within the oily core of the NCs. The PCL-NCs exhibited the highest percentage of ATM loading as well as a slow release rate. Atomic force microscopy showed nanometric and spherical particles with a narrow size dispersion. We used the PCL NCs loaded with ATM for biological evaluation following IV administration. As with free-ATM, the ATM-PCL-NCs formulation exhibited potent antimalarial efficacy using either the "Four-day test" protocol (ATM total at the end of the 4 daily doses: 40 and 80 mg/kg) in Swiss mice infected with *P. berghei* or a single low dose (20 mg/kg) of ATM in mice with higher parasitemia (15%). In healthy rats, IV administration of single doses of free-ATM (40 or 80 mg/kg) prolonged cardiac QT and QTc intervals and induced both bradycardia and hypotension. Repeated IV administration of free-ATM (four IV doses at 20 mg/kg every 12 h for 48 h) also prolonged the QT and QTc intervals but, paradoxically, induced tachycardia and hypertension. Remarkably, the incorporation of ATM in ATM-PCL-NCs reduced all adverse effects. In conclusion, the encapsulation of ATM in biodegradable polyester NCs reduces its cardiovascular toxicity without affecting its antimalarial efficacy.

Keywords: nanocapsules; QT interval; cardiotoxicity; artemether; malaria; self-assembled polymers; polylactide; drug delivery

1. Introduction

In 2020, nearly half of the world population was at risk of malaria with 241 million reported cases. Low immunity patients, such as infants, children under five years of age, pregnant women, and HIV/AIDS patients, are at the highest risk of developing severe malaria [1]. In the last 10 years, malaria has taken a higher toll on children from

Africa, accounting for about 80% of all malaria deaths in this region [2]. Severe malaria is a complicated condition that requires rapid intravenous treatment with fast-acting antimalarial drugs in hospitals. In this serious condition, the majority of the patients are infected by *Plasmodium falciparum*, and the gold standard therapy recommended is based on artemisinin derivatives [1–3].

Modern strategies used to treat severe malaria include the development of new therapeutic agents and also optimization of old drugs using new formulations [4,5]. Artemisinin and its derivatives, such as artemether (ATM), are among the most potent antimalarial drugs and have fast action against intraerythrocytic forms of the parasite in the blood. Furthermore, ATM is active against species and strains of chloroquine-resistant *Plasmodium*. ATM is also effective against *Schistosoma* spp. infections and is currently under evaluation for cancer treatment repurposing [6,7]. However, the therapeutic use and safety of ATM need improvement. Limiting issues have been identified. They include short plasma half-life [8,9], low oral bioavailability (~30%), chemical degradation, and high risk of toxicity, especially in the cardiovascular system [10]. Additionally, the ATM dosage for intramuscular (IM) injection is associated with poor patient compliance and higher risks of neurotoxicity [10]. In this sense, a safe formulation of ATM administered intravenously (IV) is urgently required to treat severe malaria. ATM is a methyl ether derivative of artemisinin and a lipophilic drug with a log P of 3.48 [10,11]. Thus, it is a suitable candidate to be associated with biodegradable nanometric carriers containing lipid reservoirs, allowing prolonged delivery of ATM into the blood.

The strategy of encapsulating existing drugs within nanocarriers for the treatment of malaria is one of the most promising approaches currently available [4,12,13]. Polymeric nanometric devices have demonstrated an outstanding ability to reduce the toxicity of drugs [14–17] and to control their release into the blood [11,18–20]. They act by modifying drug biodistribution and release rates in the heart [18,19,21,22]. Different polyesters can generate nanocarriers able to load high levels of lipophilic molecules [14,20]. Biodegradable and biocompatible polyesters, such as polycaprolactone [14,16], polylactides [14], polylactide-*co*-glycolide [23], and amphiphilic diblock polymers (PEG-PLA) [19,24,25], efficiently produce oil-core nanocapsules (NCs), where drugs with high lipophilicity can be encapsulated. In these biocompatible nanosystems, the drugs remain protected from degradation and they are released over a longer period into the blood with reduced toxicity even after IV administration of high doses [14,15,18,19,26]. ATM has been associated with different types of nanocarriers [21,27–29]. Among them, NCs have demonstrated their potential to reduce the cardiotoxicity of the ATM in vitro and in vivo when administered orally [21,22]. Nanoencapsulation of another antimalarial drug, halofantrine, has also reduced its cardiotoxicity in vivo [14].

The artemisinin derivatives, including ATM, can cause a prolonged cardiac QT interval and lead to potentially fatal cardiac arrhythmias [30–32]. Although the mechanism of the QT prolongation may be secondary to central nervous system toxicity [31], it has been demonstrated that there are also direct acute myogenic effects. Indeed, ATM lengthens the action potential of ventricular cardiomyocytes, explaining the QTc prolongation, and disrupts Ca^{2+} handling. Both effects promote pro-arrhythmogenic risks [21]. Interestingly, encapsulation of ATM in NCs can prevent these adverse effects when administered orally [21,22].

The encapsulation of ATM in NCs could be useful to improve its efficacy and reduce its cardiovascular toxicity when administered IV during severe malaria treatment. To investigate this strategy, we developed polymeric NCs from three different biodegradable polyesters. Next, we validated their efficacy to treat experimental malaria in vivo in *Plasmodium berghei*-infected mice. Finally, we assessed the cardiovascular toxicity following IV administration of ATM, free or loaded, in optimized PCL NCs (ATM-PCL-NCs).

2. Material and Methods

2.1. Drugs and REAGENTS

The National Malaria Control Program (Ministry of Health in Brazil) provided the ATM dissolved in oil solution for IM injection (PALUTHER® 80 mg/mL). We purchased ATM (dihydroartemisinin methyl ether), poly-ε-caprolactone (PCL) polymer Mn 42,500 g/mol, Poly(*D,L*-lactic) acid (PLA) Mn 30,000 g/mol and Mw 60,000 g/mol, Poloxamer 188 (Pluronic F68), and HPLC grade acetone from Sigma-Aldrich (Sigma-Aldrich Co., St. Louis, MO, USA). We synthesized and characterized diblock PEG-PLA polymer (Mn 45,000 g/mol, Đ 1.18, copolymerized with PEG 5000 g/mol, Đ:1.03), with approximately 10% *w*/*w* PEG as described [33]. Cargill (Berlin, Germany) generously donated soy lecithin with approximately 75% of phosphatidylcholine. Sasol (Hamburg, Germany GmBH) provided medium-chain triglyceride (Miglyol®810N). Synth (Rio de Janeiro, Brazil) supplied polyethylene glycol (PEG 300). The solvents were all of the analytical grade or HPLC grade. We purified the water used throughout the experiments with the MilliQ® system (Symplicity System 185, Millipore, Burlington, MA, USA).

2.2. Preparation of Nanocapsules and Intravenous Solution of Artemether

We prepared three NCs formulations (PCL, PLA, and PEG-PLA) by the polymer deposition method followed by solvent displacement [34]. We dissolved sixty milligrams of a polymer (PCL, PLA, or PEG-PLA) in 10 mL of an acetone solution containing 75 mg of Epikuron®170, and 250 μL of ATM oily solution (80 mg/mL in sesame oil for IM injection). We poured this organic solution into 20 mL of external aqueous phase under magnetic stirring using a syringe. The aqueous phase contained 75 mg of Poloxamer®188 for PCL and PLA NCs preparations and only water for PEG-PLA NCs preparation. After 10 min under agitation, we removed the solvents under reduced pressure in a rotary evaporator (Heildolph Rotary Evaporator Instruments, Schwabach, Germany) until a final 10 mL volume of NCs colloidal dispersion with 2 mg/mL of ATM. We prepared the unloaded NCs (blank-NCs) with the same method, replacing the ATM oily solution with 250 μL of Miglyol 810N oil. We obtained the ATM aqueous solution for IV injection by dissolving ATM powder in a dimethylacetamide/PEG 300/glucose 5 % (*w*/*v*) (2:4:96) solution. We protected free-ATM solution and ATM loaded in NCs from light exposure throughout the experiments. All the concentrations used for the different diluents were adequate for IV administration [35].

2.3. Nanocapsules Characterization

2.3.1. Hydrodynamic Diameter and Zeta Potential Determination

We determined the mean hydrodynamic diameter and the polydispersity index (PI) for the size distribution of NCs at 25 °C, using dynamic light scattering (DLS) technique for size and electrophoretic laser doppler anemometry for zeta potential determinations on the same equipment, a Zetasizer NanoZS equipment (Malvern Instruments, Malvern, UK). We analyzed the samples for size measurements after appropriate dilution in ultra-pure Milli-Q water and for zeta potential in 1mM NaCl solution at a final conductivity close to 1.2 ± 0.2 mS/cm^2. We expressed the reported values as the means ± standard deviations (SD) of at least three different batches of each NCs formulation.

2.3.2. Atomic Force Microscopy (AFM)

We performed a morphological analysis of the NCs by scanning probe microscopy in atomic force mode (AFM), using the equipment Multimode and Dimension 300, both monitored by NanoScope IIIa controller (Digital Instruments, Santa Barbara, Malvern, UK). We obtained the images in "tapping mode", using silicon probes of 228 μm length, with a resonance frequency of 75–98 kHz, force constant of 29–61 N/m, and a nominal tip radius of curvature of 5 to 10 nm. We obtained the images by depositing approximately 5 μL of NCs samples on a freshly cleaved mica surface. After deposition on mica, we dried the samples using argon flow. We executed the scan at a rate of 1 Hz with a resolution of

512 × 512 pixels. We performed sample analysis using the software "Analysis Section" of the equipment system. The values represent the mean ± SD derived from approximately 40 particle measurements.

2.3.3. Determination of Artemether Encapsulation

We determined the ATM content using an HPLC-validated method as described [21,22,36]. The system consisted of HPLC (Waters Alliance 2695, Waters Inc., Milford, MA USA) coupled to a UV detector at a wavelength of 216 nm. We used an RP-C18 Gemini Phenomenex column (150 mm × 4.6 mm) column with 0.5 μm of particle size, protected by a security guard column AJ0-4287 C18 (0.5 mm × 4.6 mm, 0.5 μm) (Phenomenex Inc., Torrance, CA, USA) at 30 °C. The mobile phase was acetonitrile/water (70:30 *v/v*) pumped at a flow rate of 1 mL/min (isocratic) and 50 μL of samples were injected, with a run time of 8 min each. The retention time was 6.35 min. We determined the total (100%) of ATM present in NCs suspension by diluting 400 μL of its in acetonitrile/ethanol (1:1). We assessed free ATM soluble in the aqueous external phase by ultrafiltration/centrifugation method in AMICON units (Microcon®, MWCO 100,000, Millipore®) of 400 μL suspension at 500 × *g* for 30 min. We calculated the ATM encapsulation yield (drug loading %) in nanostructures as the difference between the total amount of ATM in the final colloidal suspension after filtration through a 0.8-μm (Millex®-Millipore) membrane; and the amount of free soluble drug in the external aqueous phase of the colloidal suspension, divided by the total amount of ATM in the NCs dispersion. The encapsulation efficiency (EE%) takes into account the losses of the ATM in the total process, and it is calculated as a percentage of the amount of ATM truly encapsulated (400 μL) in suspension divided by the total of drug added to produce the formulation (400 μL) [37] using the equations below:

$$\text{Drug loading \%} = \frac{\text{total ATM in final NC dispersion} - \text{ATM in ultrafiltrate} \times 100}{\text{total ATM in the final NC dispersion}}$$

$$\text{Encapsulation efficiency \%} = \frac{\text{total ATM in the final NC dispersion} - \text{ATM in the ultrafiltrate} \times 100}{\text{total drug feed in the formulation}}$$

2.3.4. In Vitro Artemether Release from Nanocapsules

We determined the ATM solubility in water and PBS pH 7.4 using the HPLC method described above. The ATM thermodynamic solubility was 22.42 μg/mL in water and 14.54 μg/mL in PBS. In vitro release studies and assessment of ATM dissolution rate from NCs were conducted using the inverted dialysis method [38] in phosphate-buffered saline (PBS pH 7.4) under sink conditions (20% of saturation solubility) previously determined as 14 μg/mL at 37 °C [36]. Free-form ATM crystals (1.4 mg) or 350 μL of ATM-PCL-NCs (concentration of 4 mg/mL) were placed at time 0 in 500 mL of PBS release media at 37 °C in a thermostatic shaker bath containing seven dialysis sacks (Spectra Por 12,000–14,000 Da MWCO) with 1 mL de PBS. At each time interval (0, 5, 30, 60, 120, 360, 720, 1440 min), a dialysis sack was withdrawn simultaneously with a sample of release media (500 μL). The samples were diluted 1:1 with acetonitrile, then vortex mixed, centrifuged at 500× *g* for 5 min, and supernatant assayed by HPLC-UV to determine ATM concentrations using a validated method [21,36]. Three independent experiments were conducted; with each one tested in triplicate.

2.4. Experimental Animals

All procedures related to animal use conformed with the Ethical Principles of Animal Experimentation (Brazilian College of Animal Experimentation) and were approved by the UFOP Ethics Committee under number 03/2011.

Antimalarial efficacy in *Plasmodium berghei*-Infected Mice

Mice represent an ideal rodent model for *Plasmodium berghei* infection, which is fatal for these animals [39]. This experimental model simulates the human infection produced by *Plasmodium falciparum*, which is also lethal in non-treated humans. We used two treatment protocols to determine ATM efficacy in the mouse model. The first one was applied early after mice infection with high doses (total 40 mg and 80 mg/kg divided into 4 daily doses) following the "Four-day test" protocol [13,39]. This protocol aims to evidence efficacy and any general toxicity of the formulations in infected animals in repeated dose experiments. Efficacy was evaluated against chloroquine-sensitive *P. berghei* NK65-infected mice.

The second efficacy protocol consisted of the treatment of infected mice with established infection (15% blood parasitemia) with a single low dose of ATM (20 mg/kg IV) to better distinguish the formulation profile of the efficacy following time. An infective inoculum was prepared from an infected donor mouse with rising parasitemia (20%). Swiss female mice (18–22 g) were infected (IV) on day zero with 1×10^6 *P. berghei* parasitized red blood cells (RBC) in 0.2 mL of phosphate-buffered saline and randomly divided into groups. Then, the animals were treated on day 2, with only one IV dose of 20 mg/kg free-ATM solution or ATM-PCL-NCs. Control groups received only ATM vehicles or blank PCL NCs (blank-NCs). Thin blood smears were prepared with blood collected from the tail vein of all animals on days 3, 5, 7, 9, 14, 25, and 60 after infection. Parasitemia was measured in Giemsa-stained smears, counting at least 3000 RBC to determine the percentage of infected ones.

2.5. Determination of Cardiovascular Parameters and Protocols

The rat is the rodent model of choice to study cardiotoxicity using electrocardiogram (ECG) analysis and for measurements of arterial blood pressure. We used male Wistar rats (200–220 g). We anesthetized the animals with ketamine (100 mg/kg) and xylazine (14 mg/kg) mixture by intraperitoneal route. The femoral artery and vein were catheterized under anesthesia to allow recording of the AP and IV drug administration, respectively. AP was recorded using a disposable pressure transducer (TruWave, Edwards Life Sciences, Irvine, CA, USA) connected to a signal conditioning system. Limb lead II of ECG was continuously recorded using subcutaneous stainless steel needle electrodes connected by a shielded cable to a biopotential amplifier, with all the care related to the frequency response and the characteristics of the recorded signals [21]. The output signals of these systems were sampled at 1200 Hz by a 16-bit A/D conversion board (DaqBoard/2000, IOtech, Cleveland, OH, USA) and stored on a PC hard disk. We used two protocols to evaluate the cardiovascular safety of ATM in solution (free-ATM) or loaded in NCs (ATM-PCL-NCs) (Figure 1).

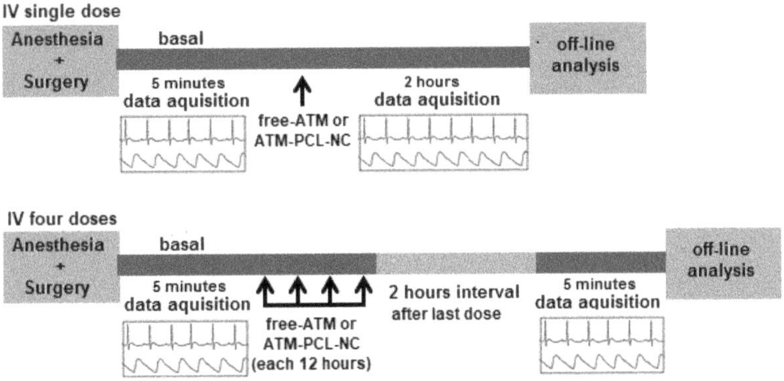

Figure 1. Experimental protocol of vehicle, blank-NC, free-ATM, and ATM-PCL-NC treatment for ECG and AP signal register.

Single IV dose: We recorded the ECG and AP signals continuously for 5 min before and after a single IV injection of the different formulations at 40 or 80 mg/kg. Thereafter, segmented data records of 30 s were performed every 5 min up to 30 min and every 15 min up to 2 h after the administration of the different formulations.

Four IV doses: The ECG and AP signals were recorded continuously for 5 min as basal controls of each animal. Then, the animals received four IV doses (ATM 20 mg/kg) of the different formulations every 12 h for 48 h. Thereafter, 2 h after the last injection, signal records of 5 min were obtained.

From the stored records analyzed offline, two seconds of segments (raw data) containing 6–12 heartbeats depending on the heart rate (HR) were extracted, and all the cardiovascular parameters were calculated as a mean value of these segments (filtered data). The cardiac parameters extracted from ECG records were the QT interval (interval between the beginning of the Q-wave and the end of the T-wave), RR interval (interval between two successive R-waves and used to obtain the HR = 60/RR), PR interval (interval between the beginning of the P-wave and the end of the R-wave), and QRS (interval from the beginning of the Q- wave to the end of the S-wave) interval. The QT interval was corrected to obtain QTc, using Fridericia's formula ($QTc = QT/(RR)^{1/3}$), to correct its HR dependence [22]. The cardiovascular parameters extracted from AP signals were systolic (SAP) and diastolic blood pressure (DAP). We calculated the percentual variation using the parameter value of each point related to the basal period, taken as a control of the same animal.

2.6. Statistical Analysis

We used the Kolmogorov–Smirnov method to determine whether continuous variables were normally distributed. We expressed the in vivo results as mean ± S.E.M. We performed statistical comparisons using ANOVA and Tukey's post-test. We accepted significance when $p < 0.05$.

3. Results

3.1. Nanocapsules Characterization and ATM Release Rate

Table 1 shows the physicochemical characterization of ATM NCs. The ATM encapsulation yield (drug loading %) was close to 100% for all types of NCs using the different polyesters. Blank-NCs for all formulations had an average size below 250 nm. However, after ATM association, the system increased in size and polydispersity. These larger sizes are related to the association of ATM with the oily core of the NCs. Loading of ATM in PLA-NCs increased the average particle size and polydispersity with increasing concentrations of ATM compared to the other polyester NC. On storage, PEG-PLA and PLA NCs showed a more unstable profile, and aggregations were observed seven days after preparation. Sizes larger than 300 nm are not suitable for IV administration.

Table 1. Characterization of the artemether-loaded nanocapsules.

Polymer/NCs	ATM mg/mL	Hydrodynamic Diameter (nm) [a]	Polydispersion Index [b]	Zeta Potential (mV)	Encapsulation Yield or Drug Loading (%)	Encapsulation Efficiency (%) [#]
Blank PCL	0	197.3 ± 0.8	0.138 ± 0.01	−56.2 ± 1.7	-	-
ATM-PCL 2	2	232.1 ± 2.7 *	0.266 ± 0.03	−49.3 ± 1.6 *	98.52 ± 0.3	91.81 ± 0.6
ATM-PCL 4	4	243.2 ± 4.7 *	0.278 ± 0.04	−41.9 ± 1.3 *	93.91 ± 0.5	80.03 ± 0.7
Blank PLA	0	256.7 ± 0.9	0.245 ± 0.05	−46.6 ± 1.3	-	-
ATM-PLA	0.5	251.9 ± 0.9	0.208 ± 0.07	−40.8 ± 2.1	ND	ND
ATM-PLA	1	301.5 ± 6.5 *	0.32 ± 0.15 *	−56.2 ± 0.6 *	90.22 ± 0.4	85.81 ± 0.8
ATM-PLA	2	328.7 ± 5.7 *	0.378 ± 0.41 *	−51.9 ± 2.3 *	87.11 ± 0.5	73.3 ± 0.7
Blank PEG-PLA	0	222.7 ± 3.4	0.17 ± 0,05	−54.4 ± 3.5	-	-
ATM PEG-PLA	1	296.6 ± 1.5 *	0.22 ± 0.05	−61.2 ± 1.2 *	94.5 ± 0.5	71.0 ± 2.8
ATM PEG-PLA	2	343.5 ± 4.7 *	0.58 ± 0.08 *	−63.8 ± 4.1 *	87.9 ± 1.2	74.8 ± 0.7

Values represent the mean ± standard deviation derived from 3 preparations; Refers to the freshly prepared formulation; # Refers to the total drug added to prepare the formulation (losses in the process). [a] Measured by dynamic light scattering. [b] Mean polydispersity index, (<0.3) monodispersed samples; The sizes and zeta potential values were compared using unpaired student's *t*-test. * Significantly different from the respective blank-nanocarrier ($p < 0.05$). ATM is artemether. PCL is polycaprolactone, PLA is homopolymer polylactide, and PEG-PLA is a *diblock* amphiphile polymer. ND: not determined.

The best formulation in terms of average size was ATM-PCL-NCs (Table 1 and Figure 2). The polydispersity index was less than 0.3 consistent with the size of monodispersed populations for PCL loaded with ATM at concentrations of 2 and 4 mg/mL. The mean size increased (by 35–45 nm) after loading with ATM. The association with ATM influenced the zeta potential in different ways depending on the polyester used in the formulation. However, PLA and PEG-PLA produced less stable particles with the highest polydispersity. Thus, the zeta potential was also affected by the heterogeneity of the system. In general, ATM influences the zeta potential, and it seems that part of the drug is localized on the surface of NCs. The PCL-NCs loaded with ATMs were the most stable in their physicochemical profile. They were selected for biological evaluation. Examination of the ATM-PCL-NCs morphology by AFM height and phase images showed monodispersed populations of particles. In contrast, AFM analysis showed that ATM-loaded PEG-PLA NCs were more polydispersed (Figure 2). The mean diameter extracted from the AFM analysis was greater than that measured by DLS, and the diameter/height ratio was greater than one in the topographical profile (Figure 2). This aspect is in agreement with the ability of NCs to deform under the pressure of the AFM tip. We also highlighted the higher heterogeneity of PEG-PLA NC in size dispersion by 3D AFM images of NCs in Figure 2D.

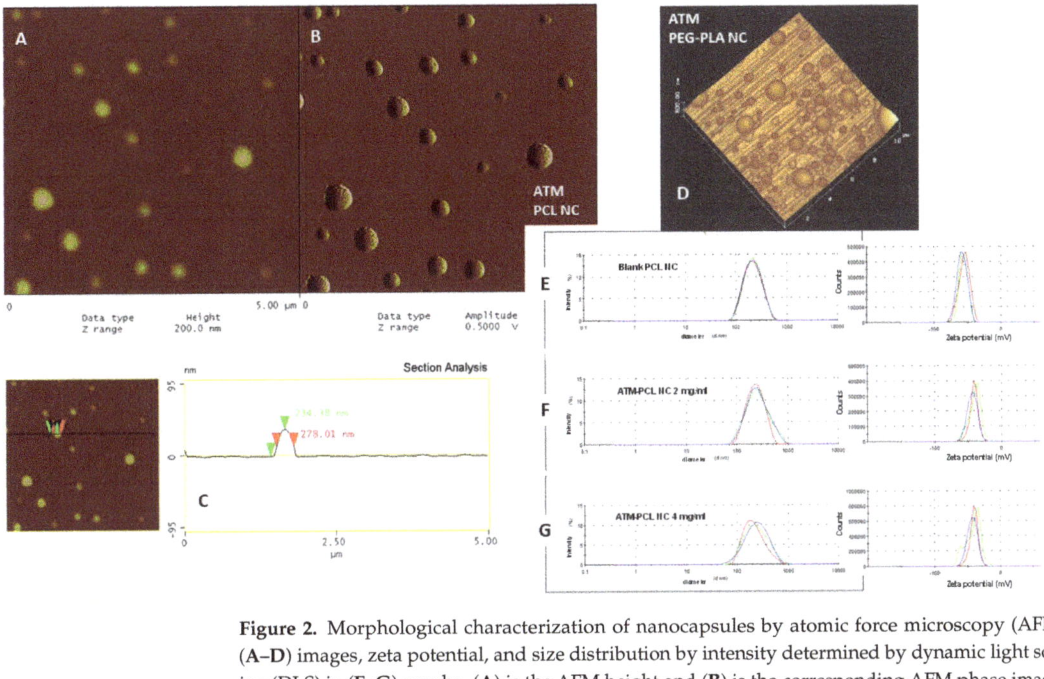

Figure 2. Morphological characterization of nanocapsules by atomic force microscopy (AFM) in (**A–D**) images, zeta potential, and size distribution by intensity determined by dynamic light scattering (DLS) in (**E–G**) graphs. (**A**) is the AFM height and (**B**) is the corresponding AFM phase images of ATM-PCL-NC 2 mg/mL and (**D**) is the 3D AFM image of ATM-PEG-PLA NC 2 mg/mL showing the size dispersion of spherical particles deposited under mica plates. Scheme (**C**) is the topographical profile of a selected NC in image (**A**) showing the measurement of diameter at half-height and the values measured by the equipment. In (**E**): blank-PCL-NC; (**F**): ATM-PCL-NC 2 mg/mL and (**G**): ATM-PCL-NC 4 mg/mL formulations.

Figure 3 shows the ATM dissolution profile in PBS 7.4 and the release profile of the ATM from PCL-NCs and PEG-PLA NCs determined by the reverse dialyze membrane method. The PEG-PLA showed a fast release of ATM (25%) in the first 5 min immediately after dilution in release media followed by a complete release within the next 6 h. This profile was similar to the free-ATM dissolution in this medium under *sink* conditions (20% of

ATM saturation solubility). In contrast, PCL-NCs showed a burst release of approximately 12% in the first minutes followed by a release of 50% of total ATM encapsulated for up to 2 h. The 50% amount was not released and was maintained associated with the nanocarrier even under *sink* conditions. We considered the profile of the ATM loaded in PCL-NCs to be the most suitable for the biological assessment of infection and cardiovascular toxicity models in vivo.

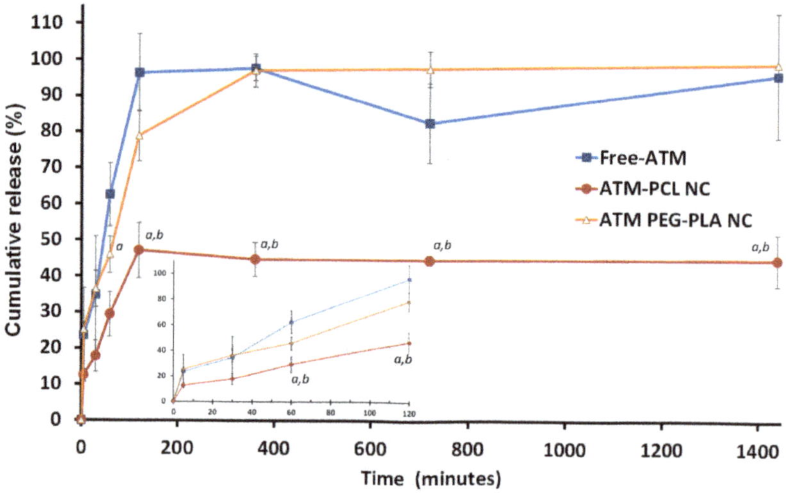

Figure 3. Profiles of the cumulative release of free-artemether (Free-ATM) and artemether from nanocapsules of polycaprolactone and PEG-PLA polymers following time in PBS pH 7.4 under *sink* conditions in 37 °C using inverted dialysis sac method. The insert represents the first 2 h. Data shown are means ± standard deviations of n = 3 independent experiments. The assay was made in triplicate (nine measurements in total). [a] is a mean statistically different from free-ATM and [b] statistically different from ATM-PEG-PLA NC. The values at each time point were compared using an unpaired student's *t*-test.

3.2. Antimalarial Efficacy in P. berghei-Infected Mice

The antimalarial efficacies of ATM-PCL-NCs and free-ATM were similar in Swiss mice infected with *P. berghei*. We treated mice with low parasitemia using 40 or 80 mg/kg divided into four daily doses delivered by the IV route. These are high doses and no systemic toxicity was observed in the infected mice. An untreated group was evaluated concurrently for each protocol. Control groups (IV solution vehicle, and blank-NCs) remained highly parasitized. All the animals died between day 5 to day 10 (Figure 4). The IV treatment with free ATM or ATM-PCL-NCs at 40 or 80 mg/kg/day for four consecutive days reduced parasitemia to very low levels, avoiding the progression of the infection, and increased mice survival (Figure 4). The success of the cure was verified in all groups treated with ATM for up to 10 days. No recrudescence occurred 60 days after treatment. Of note, there was no difference in survival profile and parasitemia levels in animals treated with blank NCs and glucose solution, indicating no antimalarial effect of blank-PCL-NCs. There was no difference between the two doses and formulations concerning parasitemia and survival after repeated doses (Figure 4).

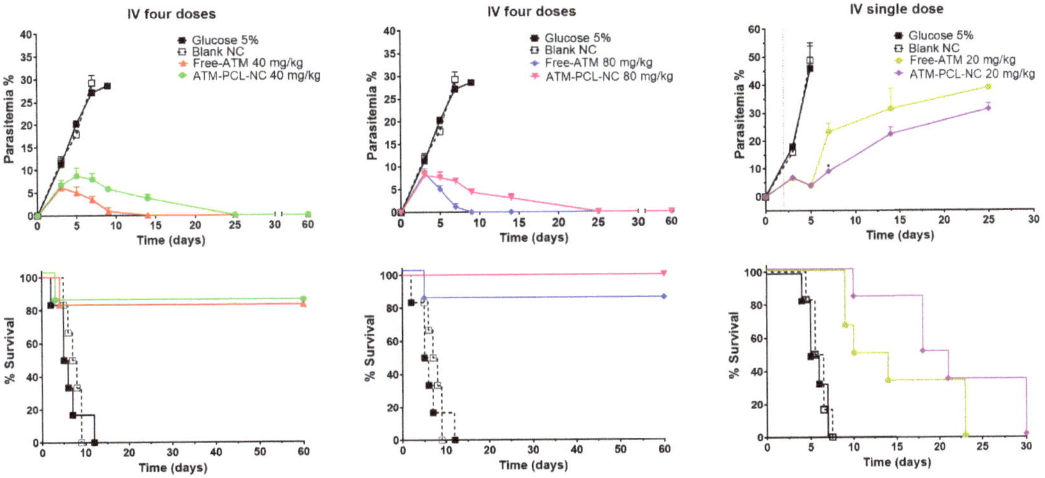

Figure 4. Efficacy of the artemether formulations represented in graphs of parasitemia (%) and survival (%) for both protocols (four-day-test) and single low dose (20 mg/kg) with established parasitemia (15%).

In the second treatment efficacy protocol (Figure 4-IV single dose), the aim was to distinguish the effect of the formulation with a low single dose of ATM in mice with established infection (parasitemia of 15%). ATM-PCL-NCs improved mice survival and reduced parasitemia more rapidly than single doses of the free-ATM solution (Figure 4).

3.3. *Determination of Cardiovascular Effects of ATM*

Figure 5 shows representative ECG signals before and after IV administration of blank NCs, free-ATM, and ATM-PCL-NCs in rats. IV administration of free-ATM increased the QT and QTc intervals of the ECG, occurring rapidly (between one to twenty minutes after delivery), which persisted until the end of the experiment (Figure 6). The effects were dose-dependent. Administration of ATM-PCL-NCs induced less prolongation of the QT and QTc intervals. In contrast, free-ATM or ATM-PCL-NCs did not affect the PR and QRS intervals of the ECG relative to the basal period. Additionally, we investigated free-ATM at 120 mg/kg in pilot experiments but, at this dose, the majority of animals died during the experiment with ECG alterations and arrhythmia (data not shown).

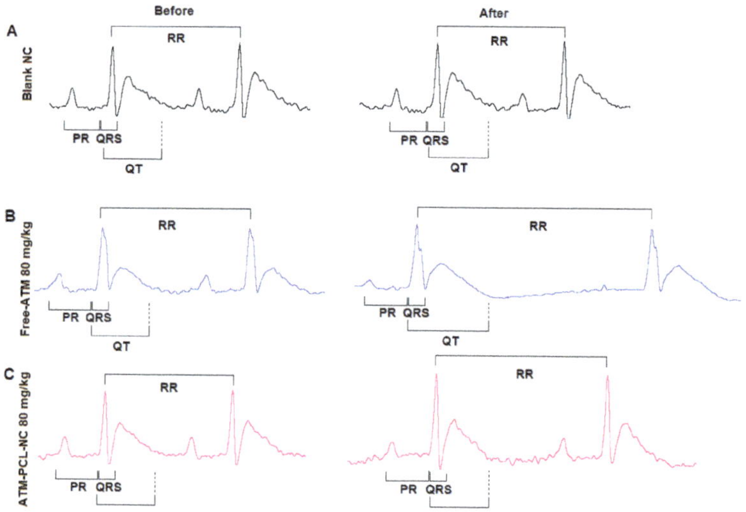

Figure 5. Representative ECG signal (lead II) at basal (before treatment) and after treatment with Blank-NC in (**A**), Free-ATM in (**B**) and ATM-PCL-NCs in (**C**), showing the ECG intervals analyzed. The RR interval was used to obtain the heart rate (HR).

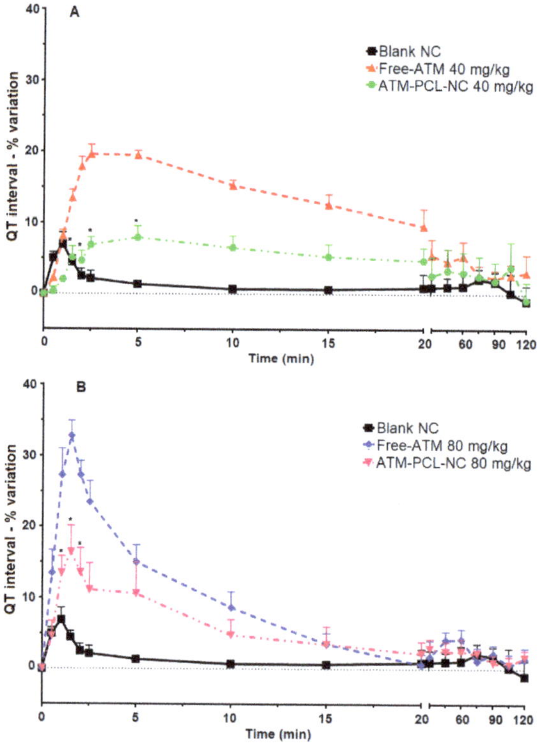

Figure 6. QT interval percentual variation from basal (before treatment) until two hours after IV single dose administration of blank-NC, free-ATM, and ATM-PCL-NC, both at 40 (**A**) and 80 mg/kg (**B**). * $p > 0.05$ compared to free-ATM administration. ANOVA followed by Tukey post-test.

Figure 7 shows the maximum changes in cardiovascular parameters measured before and after ATM administration of a single dose of 40 mg/kg (A) or 80 mg/kg (B), and of the four doses (20 mg/kg) in 48 h (C). After the single IV dose, the QT interval increased by 19.6% and 32.8%, respectively for 40 mg/kg and 80 mg/kg of free ATM. Encapsulation of ATM in NCs significantly prevented ATM-induced QT interval prolongation ($p < 0.05$), with approximately 60% and 50% of reduction after IV administration of 40 mg/kg or 80 mg/kg, respectively. Similarly, for the group receiving four repeated doses of 20 mg/kg free-ATM, the QT interval increased from 66.5 ms (basal) to 81.6 ms (two hours after the last dose of ATM-PCL-NCs). After the single dose, free-ATM reduced the blood pressure (SAP and DAP) (Figures 8 and 9A,B), and HR (Figure 7A,B), mainly from one to 20 min post-administration (Figures 8 and 9). The hypotension was severe with free-ATM (Figure 8). The decreases were, respectively for free-ATM doses of 40 mg/kg and 80 mg/kg, 17.8% and 41.4% for SAP, and 22.2 % and 49.9% for DAP. We observed no effect after the administration of the control solution or blank-NCs. The effects were weaker with the administration of ATM-PCL-NCs inducing a reduction of 9.8% and 13.8% for SAP and 8.5% and 16.4% for DAP for a single dose of 40 mg/kg or 80 mg/kg, respectively (Figure 9A and B). On the other hand, after the four doses protocol, free-ATM increased the HR (+76 bpm; Figure 7C), SAP (+26.1%), and DAP (+37.8%) (Figure 9C). The final absolute values of AP and HR in this group indicate substantial hypertension and tachycardia. These effects were reduced when ATM-PCL-NCs were administered, being only +4.4% for HR, and +3.6% and +5.1% for SAP and DAP, respectively. In summary, all cardiovascular alterations indicating cardiovascular toxicity of free-ATM were markedly reduced when the drug was administered as ATM-PCL-NCs (IV route). Encapsulation in our NCs formulation prevented alterations of the ECG parameters and blood pressure of ATM.

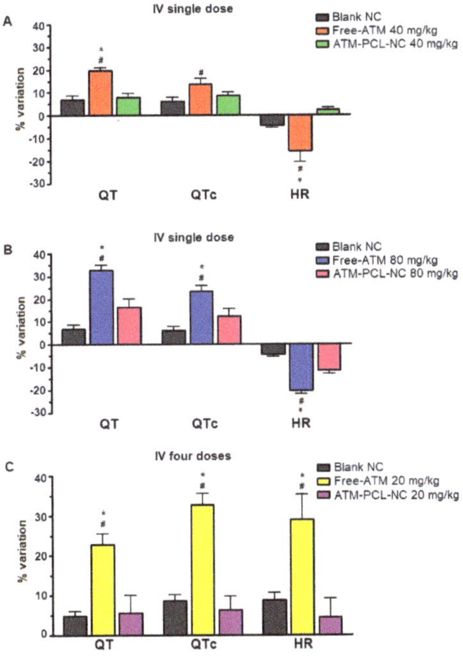

Figure 7. Maximal percentual variation of QT and QTc intervals and heart rate (HR) after IV administration of blank-NC, free-ATM, and ATM-PCL-NC, a single dose of 40 (**A**) and 80 mg/kg (**B**), and four doses (one every 12 h) of 20 mg/kg (**C**). * $p > 0.05$ compared to blank-NC and # $p > 0.05$ compared to ATM-PCL-NC administration. ANOVA followed by Tukey post-test.

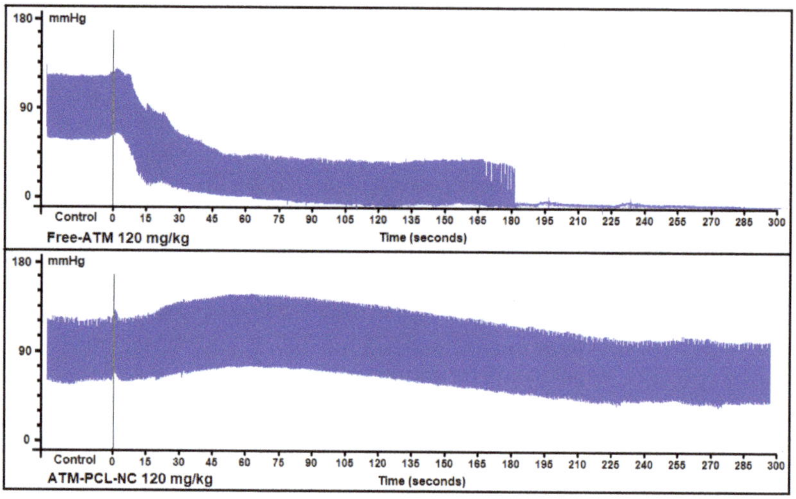

Figure 8. Representative signals of arterial blood pressure of anesthetized rats treated with free-ATM or ATM-PCL-NC, both at 120 mg/kg, showing the severe hypotension produced by the free form and the absence of hypotension with the NC formulation.

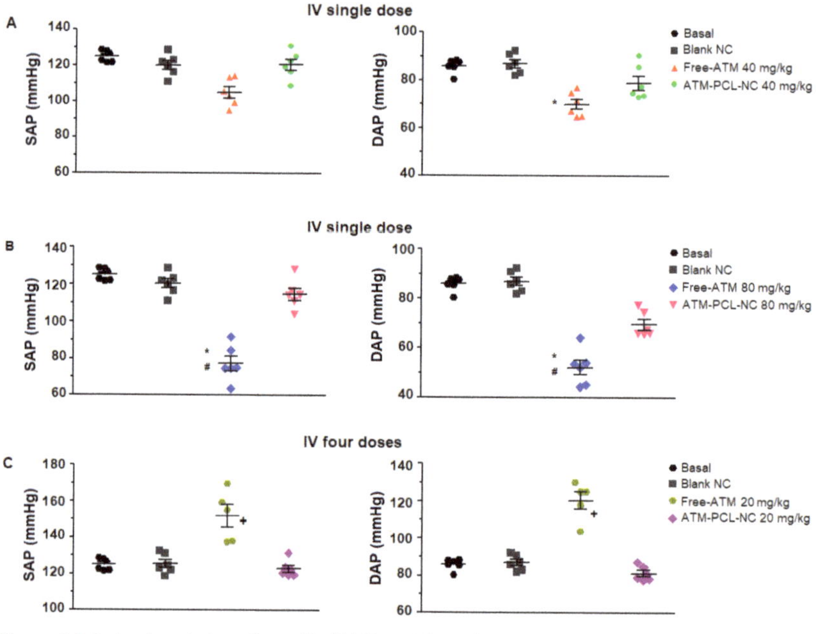

Figure 9. Maximal variation of systolic (SAP) and diastolic (DAP) pressure (mmHg) after IV administration of blank-NC, free-ATM, and ATM-PCL-NC, a single dose of 40 (**A**) and 80 mg/kg (**B**), and four doses (one every 12 h) of 20 mg/kg (**C**). * $p < 0.05$ compared to blank-NC and # $p < 0.05$ compared to ATM-PCL-NC administration, $^{+}$ $p < 0.05$ compared to basal, blank-NC, and ATM-PCL-NC. ANOVA followed by Tukey post-test.

4. Discussion

In this work, we sought to overcome the main limitations of ATM treatment by developing new formulations designed for intravenous administration for different therapeutical purposes. For example, to treat severe malaria successfully, rapid suppression of parasitemia using IV ATM administration as well as a sustained ATM release into the blood is highly desirable. In addition to the treatment of malaria, these formulations could also be useful for other uses of ATM in therapeutic repositioning strategies against schistosomiasis and cancer [6,7]. Difficulties concerning current treatment regimens include low water solubility of ATM, short half-life, fast degradation, and adverse effects. ATM is a lipophilic substance and, as such, NCs, with their oily core reservoir, make an ideal nanocarrier that can protect the ATM from degradation [26] and allow a controlled release after IV injection [19].

In this study, we demonstrate the development of NCs, prepared from different biodegradable polyesters, as intravenous formulations of ATM to treat severe malaria. These polymers are commercially available in high-purity grades with low dispersity and are approved for IV administration [24,25,40]. They can control the release of drugs into the blood as previously reported [19,22]. We prepared the NCs using a solvent displacement technique [34]. Advantages are the narrow distribution of NCs sizes and the use of well-defined and safe biodegradable polymers, such as polyesters, non-toxic excipients, and easily removable non-chlorinated solvents [24,25]. ATM has been encapsulated in PCL-NCs at high payloads, as much as 2 or 4 mg/mL of ATM, with an encapsulation efficiency of 92% and 80% respectively, in only 6 mg/mL polymer with a low drug interference at NC surface. This drug loading is promising compared with other lipophilic substances loaded into lipid nanocarriers [27–30,41–43]. For example, the optimized formulation of ATM in nanoemulsion was 0.4 mg/mL [42] and ATM showed 65% of encapsulation efficiency in nanoliposomes [44]. ATM in nanostructured lipid carriers with 87% of ATM encapsulation efficiency and 11% payload *w*/*w* were reported [41]. We found similar yields with ATM payload of 13% *w*/*w* in NC. Our data indicates a high affinity of the ATM for the oily core of PCL-NCs. The mean hydrodynamic diameters and the polydispersity indexes of both concentrations of ATM-PCL-NCs (2 and 4 mg/mL) are suitable for IV administration (Table 1). Particle size distribution must be strictly controlled in nanoparticulate formulations intended for intravascular delivery to prevent blood capillary occlusion [44]. ATM has already been associated with different types of lipid nanocarriers to be used in different routes of administration [27–30,41,42,44]. However, in these studies, no assessment of efficacy or cardiotoxicity upon intravenous administration was provided.

The morphology of polyester NCs prepared from PCL, PLA, and PEG-PLA polymers was characterized using different techniques as previously reported [19,20,26,45–47]. TEM and SEM techniques are useful, yet the shrinking and melting of these polymeric systems are frequently observed under electron beam incidence [18] as well as artifacts related to staining and coating [48]. The AFM technique in *tapping* mode has been employed recently to obtain information regarding the morphology [48,49], particle dispersion, stiffness, deformation [45,46], encapsulation [50], tip interaction with the surface [51], and surface porosity of nanospheres [52] and nanocapsules [18,45,46]. For these types of polymer nanocapsules, AFM provides much more additional physical information than electron microscopy. The AFM image analysis we performed indicates that NC, in contrast to hard nanospheres, can deform under tip pressure. We provide evidence in the AFM section analysis that the flattening of the ATM-PCL-NCs occurs under mica plates, based on the topographical profile of these NCs showing a diameter/height ratio higher than one (Figure 2C). This deformable property of the NCs structure, with an oily core surrounded by a polymeric wall-forming reservoir, is suitable for IV administration [44].

The negative value of PCL-NC zeta potential was influenced by ATM, indicating that part of the ATM was associated with the surface of the carrier. The presence of ATM at the NCs surface was confirmed by the ATM *burst* type release in PBS media in the first 5 min. However, as the NCs' surface charges are all very negative (>30 mV), we observed no impact on colloidal stability. The ATM-PCL-NC formulations were physically stable during the experimental period with no signs of aggregation, flocculation, or phase separation.

The hydrophobicity of the polyester also plays a role because PEG-PLA and PLA polymers with lower hydrophobicity were less efficient at encapsulating ATM in the nanocarrier. This behavior affects the release of ATM. PCL controls the release rate of ATM better than PEG-PLA under sink conditions, in line with the ability of this polymer to modify the release rate into blood (Figure 3). The size, polydispersity, and release profile of PCL NCs showed that this formulation was suitable for IV administration with a modified biodistribution allowing for sustained release of ATM and reduced toxicity.

The efficacy testing was conducted in a rodent model of malaria consisting of mice infected with *Plasmodium berghei,* which is one of the best models to simulate the fatal *Plasmodium falciparum* infection in humans. *Plasmodium berghei* produces an asynchronous infection in mice with high parasitemia that results in mortality if left untreated. We employed two protocols, one with low parasitemia and the other with high and established parasitemia. Importantly, the four-day test confirmed the anti-malarial efficacy of ATM-containing NC formulations in *P. berghei*-infected Swiss mice with no detectable adverse effects after four repeated doses. The IV administration of free and encapsulated ATM at total doses of 40 mg/kg and 80 mg/kg induced complete remission of parasitemia, with no recurrence of the infection for up to 60 days. Such efficacy has been reported for ATM combined with solid lipid nanoparticles [29], or when administered in NCs orally [22]. In a more stringent protocol at a single low dose (20 mg/kg) and high parasitemia (15%), ATM-PCL-NCs showed a faster reduction of parasitemia and a more sustained effect than the free-ATM solution delivered by IV injection.

Given the effectiveness of ATM in an encapsulated form on parasitemia of *P. berghei*-infected mice, we raised the legitimate question of whether, and to what extent, this property could induce cardiovascular toxicity. The determination of cardiotoxicity using QT interval prolongation is currently used, and mandatory, for decision-making during drug development [53,54]. Different antimalarial drugs are associated with decreases in both HR and AP as a cause of sinus bradycardia and fatal hypotension [55,56]. ECG recordings have shown that QT interval prolongation, a strong marker of potential cardiotoxicity [53], is associated with potentially lethal tachycardia known as *'torsade de pointes'* degenerating into ventricular fibrillation [55,56]. In our experiments, the two protocols used (either a single dose of 40 mg/kg or 80 mg/kg; or repeated administration of 20 mg/kg) prolonged the QT and the QTc interval.

The effects of ATM on HR and AP were more complex and rather counterintuitive. The two single doses (40 and 80 mg/kg) rapidly induced bradycardia and severe hypotension (decrease of SAP and DAP). In contrast, repeated administration of four doses of ATM at 20 mg/kg induced tachycardia and hypertension. Why this occurs is unclear. However, one possibility is that the treatment evokes a time-dependent impact on the sympatho-vagal balance with an immediate enhancing parasympathetic effect accounting for the bradycardia and hypotension. In this scheme, repeated episodes of ATM injection may activate a regulatory feedback mechanism to counteract vagal predominance and boost the sympathetic contribution, as indicated by tachycardia and hypertension. Although ATM has direct effects on the cardiomyocyte action potential, in line with the lengthening of the QTc interval, and abnormal Ca^{2+} handling [21], toxicity on the central nervous system has also been suggested [31]. Moreover, the targeting of membrane-delimited proteins involved in the electrical activity of neurons or vascular cells may also account for these observations.

A key outcome of our study was evidence of a safer cardiovascular profile of ATM at effective antiparasitic doses on *P. berghei* when ATM was released from the engineered NCs. Encapsulation of ATM in NCs reduced all adverse effects induced by either the two single doses (40 mg/kg and 80 mg/kg) or the four repeated doses (20 mg/kg) of ATM administered intravenously. This point is a novel finding concerning ATM encapsulated in nanocarriers. The modifications of cardiac parameters, such as the HR, QT, and QTc intervals, as well as blood pressure, were largely attenuated. Of note, while the adverse effects of ATM on HR and AP were the opposite, depending on the administration protocol,

NCs were efficient in preventing or minimizing them. Formulation design and administration route play a key role in drug plasma concentration. Although it appears that there is a discrepancy between the antiparasitic efficacy of ATM and the attenuation of its cardiovascular side effects provided by PCL-NCs, the differences may be inherent to the cell types (proliferative organisms versus differentiated mammalian cells). In particular, the activity against *P. berghei* possibly relates to the continuous blood release of an efficient antiparasitic drug sustaining the antiparasitic effect. The lower blood ATM dose, therefore, may have less systemic toxicity, particularly in the cardiovascular system. A fraction of free ATM, smaller than that observed following IV delivery of a high concentration of the free-ATM, may thus reach the cardiac tissue. Indeed, the toxicity of artemisinin derivatives has been correlated with drug concentration in the blood [10]. A modification of ATM biodistribution due to NC accumulation in the mononuclear phagocytic system after IV administration may also occur [57].

5. Conclusions

Our data show that the ATM-PCL-NCs maintain ATM antimalarial efficacy but have less associated cardiotoxicity than the free-ATM solution upon IV administration. The ATM-PCL-NCs exhibited the most favorable profile of ATM release compared to other engineered biodegradable polymers in our study. Furthermore, NC has an oily core where oil-soluble substances can be dissolved, resulting in high payloads. In this sense, the association of the ATM with NCs is a promising strategy for ATM IV administration. The modification of drug distribution using polymeric NCs appears as a strategy to be applied to other lipophilic, pro-arrhythmogenic drugs. This study also contributes to an important in vivo methodology to measure the cardiotoxicity of nanoparticulate drug carriers. In summary, the use of polymeric NCs represents a promising strategy for the production of safer IV treatments with a large plethora of cardiotoxic drugs.

Author Contributions: Conceptualization, A.T.V.-D., V.C.F.M. and A.G.-G.; methodology and formal analysis, A.T.V.-D., H.N.G., V.C.F.M., A.G.-G., É.M.B. and G.M.G.; investigation, A.T.V.-D., H.N.G., V.C.F.M., A.G.-G. and G.M.G.; resources, H.N.G., V.C.F.M. and A.G.-G.; data curation, A.T.V.-D., H.N.G., V.C.F.M. and A.G.-G.; writing—original draft preparation, A.T.V.-D., H.N.G., V.C.F.M., A.G.-G., G.M.G. and S.R.; writing—review and editing, A.T.V.-D., V.C.F.M., A.G.-G. and S.R.; visualization, A.T.V.-D., V.C.F.M., A.G.-G. and S.R.; supervision, V.C.F.M. and A.G.-G.; project administration and funding acquisition, V.C.F.M. and A.G.-G. All authors have read and agreed to the published version of the manuscript.

Funding: NANOBIOMG-Network funded by FAPEMIG, and the Brazilian Program for Malaria Control (MS, Brazil) for the PALUTHER® donation. This work was supported by CNPq grants, WHO-A0790 re-entry grant, NANOBIOMG-Network (#00007-14 and #40/11), and FAPEMIG grants. A bilateral CAPES-COFECUB Research Collaboration between Brazil and France (#768/13 and 978/20) also partially funded this work.

Institutional Review Board Statement: The study was conducted in accordance with the Declaration of Helsinki and approved by the Institutional Ethics Committee of Federal University of Ouro Preto and Ethical Principles of Animal Experimentation (Brazilian College of Animal Experimentation). The protocols were approved under number 03/2011).

Data Availability Statement: The data presented in this study are available on request from the corresponding author.

Acknowledgments: ATV-D thanks CAPES for the personal scholarship. A.G.-G. and V.C.F.M. also thank PPM FAPEMIG researcher grants and VCFM productivity fellowship from CNPq, Brazil. The authors thank J.M.C. Vilela and M.S. Andrade for the AFM analysis. We thank Chris Jopling (IGF Montpellier) for improving the English of the manuscript.

Conflicts of Interest: The authors report no conflict of interest. The authors alone are responsible for the content and writing of the paper.

References

1. World Health Organization. World Malaria Report. Geneva, Switzerland, 2021. License: CC BY-NC-SA 3.0 IGO 2021. Available online: https://www.who.int/news-room/fact-sheets/detail/malaria (accessed on 22 November 2022).
2. World Health Organization. *World Malaria Report 2020: 20 Years of Global Progress and Challenges*; WHO: Geneva, Switzerland, 2020.
3. Wells, T.N.C.; Van Huijsduijnen, R.H.; Van Voorhis, W.C. Malaria medicines: A glass half full? *Nat. Rev. Drug Discov.* **2015**, *14*, 424–442. [CrossRef] [PubMed]
4. Santos-Magalhães, N.S.; Mosqueira, V.C.F. Nanotechnology Applied to the Treatment of Malaria. *Adv. Drug Deliv. Rev.* **2010**, *62*, 560–575. [CrossRef] [PubMed]
5. Volpe-Zanutto, F.; Ferreira, L.T.; Permana, A.D.; Kirkby, M.; Paredes, A.J.; Vora, L.K.; Bonfanti, A.P.; Charlie-Silva, I.; Raposo, C.; Figueiredo, M.C.; et al. Artemether and lumefantrine dissolving microneedle patches with improved pharmacokinetic performance and antimalarial efficacy in mice infected with *Plasmodium yoelii*. *J. Control Release* **2021**, *333*, 298–315. [CrossRef] [PubMed]
6. Ho, W.E.; Peh, H.Y.; Chan, T.K.; Wong, W.S.F. Artemisinins: Pharmacological Actions beyond Anti-Malarial. *Pharmacol. Ther.* **2014**, *142*, 126–139. [CrossRef]
7. Andrews, K.T.; Fisher, G.; Skinner-Adams, T.S. Drug Repurposing and Human Parasitic Protozoan Diseases. *Int. J. Parasitol. Drugs Drug Resist.* **2014**, *4*, 95–111. [CrossRef]
8. van Agtmael, M.A.; Cheng-Qi, S.; Qing, J.X.; Mull, R.; van Boxtel, C.J. Multiple Dose Pharmacokinetics of Artemether in Chinese Patients with Uncomplicated Falciparum Malaria. *Int. J. Antimicrob. Agents* **1999**, *12*, 151–158. [CrossRef]
9. Hien, T.T.; Davis, T.M.E.; Chuong, L.V.; Ilett, K.F.; Sinh, D.X.T.; Phu, N.H.; Agus, C.; Chiswell, G.M.; White, N.J.; Farrar, J. Comparative Pharmacokinetics of Intramuscular Artesunate and Artemether in Patients with Severe Falciparum Malaria. *Antimicrob. Agents Chemother.* **2004**, *48*, 4234–4239. [CrossRef]
10. Medhi, B.; Patyar, S.; Rao, R.S.; Ds, P.B.; Prakash, A. Pharmacokinetic and Toxicological Profile of Artemisinin Compounds: An Update. *Pharmacology* **2009**, *84*, 323–332. [CrossRef]
11. Ameya, R.K.; Verma, M.; Karandikar, P.; Furin, J.; Langer, R.; Traverso, G. Nanotechnology approaches for global infectious diseases. *Nat. Nanotechnol.* **2021**, *16*, 369–384.
12. Melariri, P.; Kalombo, L.; Nkuna, P.; Dube, A.; Hayeshi, R.; Ogutu, B. Oral lipid-based nanoformulation of tafenoquine enhanced bioavailability and blood stage antimalarial efficacy and led to a reduction in human red blood cell loss in mice. *Int. J. Nanomed.* **2015**, *10*, 1493–1503. [CrossRef]
13. Mosqueira, V.C.F.; Loiseau, P.M.; Bories, C.; Legrand, P.; Devissaguet, J.P.; Barratt, G. Efficacy and Pharmacokinetics of Intravenous Nanocapsule Formulations of Halofantrine in *Plasmodium berghei*-Infected Mice. *Antimicrob. Agents Chemother.* **2004**, *48*, 1222–1228. [CrossRef] [PubMed]
14. Leite, E.A.; Grabe-Guimarães, A.; Guimarães, H.N.; Machado-Coelho, G.L.L.; Barratt, G.; Mosqueira, V.C.F. Cardiotoxicity Reduction Induced by Halofantrine Entrapped in Nanocapsule Devices. *Life Sci.* **2007**, *80*, 1327–1334. [CrossRef] [PubMed]
15. Garcia, G.M.; Roy, J.; Pitta, I.R.; Abdalla, D.S.P.; Grabe-Guimarães, A.; Mosqueira, V.C.F.; Richard, S. Polylactide Nanocapsules Attenuate Adverse Cardiac Cellular Effects of Lyso-7, a Pan-PPAR Agonist/Anti-Inflammatory New Thiazolidinedione. *Pharmaceutics* **2021**, *13*, 1521. [CrossRef]
16. Bulcão, R.P.; Freitas, F.A.; Venturini, C.G.; Dallegrave, E.; Durgante, J.; Göethel, G.; Cerski, C.T.S.; Zielinsky, P.; Pohlmann, A.R.; Guterres, S.S.; et al. Acute and Subchronic Toxicity Evaluation of Poly(ε-Caprolactone) Lipid-Core Nanocapsules in Rats. *Toxicol. Sci.* **2013**, *132*, 162–176. [CrossRef] [PubMed]
17. Branquinho, R.T.; Roy, J.; Farah, C.; Garcia, G.M.; Aimond, F.; Le Guennec, J.-Y.; Saude-Guimarães, D.A.; Grabe-Guimaraes, A.; Mosqueira, V.C.F.; de Lana, M.; et al. Biodegradable Polymeric Nanocapsules Prevent Cardiotoxicity of Anti-Trypanosomal Lychnopholide. *Sci. Rep.* **2017**, *7*, 44998. [CrossRef]
18. Garcia, G.M.; Oliveira, L.T.; da Rochia Pitta, I.; de Lima, M.D.C.A.; Vilela, J.M.C.; Andrade, M.S.; Abdalla, D.S.P.; Mosqueira, V.C.F. Improved Nonclinical Pharmacokinetics and Biodistribution of a New PPAR Pan-Agonist and COX Inhibitor in Nanocapsule Formulation. *J. Control Release* **2015**, *209*, 207–218. [CrossRef]
19. Branquinho, R.T.; Pound-Lana, G.; Marques Milagre, M.; Saúde-Guimarães, D.A.; Vilela, J.M.C.; Spangler Andrade, M.; de Lana, M.; Mosqueira, V.C.F. Increased Body Exposure to New Anti-Trypanosomal Through Nanoencapsulation. *Sci. Rep.* **2017**, *7*, 8429. [CrossRef]
20. Mosqueira, V.C.F.; Legrand, P.; Barratt, G. Surface-modified and conventional nanocapsules as novel formulations for parenteral delivery of halofantrine. *J. Nanosci. Nanotechnol.* **2006**, *6*, 3193–3202. [CrossRef]
21. Souza, A.C.M.; Grabe-Guimarães, A.; Cruz, J.D.S.; Santos-Miranda, A.; Farah, C.; Teixeira Oliveira, L.; Lucas, A.; Aimond, F.; Sicard, P.; Mosqueira, V.C.F.; et al. Mechanisms of Artemether Toxicity on Single Cardiomyocytes and Protective Effect of Nanoencapsulation. *Br. J. Pharmacol.* **2020**, *177*, 4448–4463.
22. Souza, A.C.M.; Mosqueira, V.C.F.; Silveira, A.P.A.; Antunes, L.R.; Richard, S.; Guimarães, H.N.; Grabe-Guimarães, A. Reduced cardiotoxicity and increased oral efficacy of artemether polymeric nanocapsules in *Plasmodium berghei*-infected mice. *Parasitology* **2018**, *145*, 1075–1083. [CrossRef]

23. Attili-Qadri, S.; Karra, N.; Nemirovski, A.; Schwob, O.; Talmon, Y.; Nassar, T.; Benita, S. Oral delivery system prolongs blood circulation of docetaxel nanocapsules via lymphatic absorption. *Proc. Natl. Acad. Sci. USA* **2013**, *110*, 17498–17503. [CrossRef] [PubMed]
24. Mitchell, M.J.; Billingsley, M.M.; Haley, R.M.; Wechsler, M.E.; Peppas, N.A.; Langer, R. Engineering precision nanoparticles for drug delivery. *Nat. Rev. Drug Discov.* **2021**, *20*, 101–124. [CrossRef] [PubMed]
25. Osorno, L.L.; Brandley, A.N.; Maldonado, D.E.; Yiantsos, A.; Mosley, R.J.; Byrne, M.E. Review of Contemporary Self-Assembled Systems for the Controlled Delivery of Therapeutics in Medicine. *Nanomaterials* **2021**, *11*, 278. [CrossRef]
26. Roy, J.; Oliveira, L.T.; Oger, C.; Galano, J.-M.; Bultel-Poncé, V.; Richard, S.; Guimaraes, A.G.; Vilela, J.M.C.; Andrade, M.S.; Durand, T.; et al. Polymeric Nanocapsules Prevent Oxidation of Core-Loaded Molecules: Evidence Based on the Effects of Docosahexaenoic Acid and Neuroprostane on Breast Cancer Cells Proliferation. *J. Exp. Clin. Cancer Res.* **2015**, *34*, 155–157. [CrossRef]
27. Prabhu, P.; Suryavanshi, S.; Pathak, S.; Patra, A.; Sharma, S.; Patravale, V. Nanostructured lipid carriers of artemether-lumefantrine combination for intravenous therapy of cerebral malaria. *Int. J. Pharm.* **2016**, *513*, 504–517. [CrossRef]
28. Shakeel, K.; Ahmad, F.J.; Harwansh, R.K.; Rahman, M.A. β-Artemether and Lumefantrine Dual Drug Loaded Lipid Nanoparticles: Physicochemical Characterization, Pharmacokinetic Evaluation and Biodistribution Study. *Pharm. Nanotechnol.* **2022**, *10*, 210–219. [CrossRef]
29. Aditya, N.P.; Patankar, S.; Madhusudhan, B.; Murthy, R.S.R.; Souto, E.B. Arthemeter-Loaded Lipid Nanoparticles Produced by Modified Thin-Film Hydration: Pharmacokinetics, Toxicological and in Vivo Anti-Malarial Activity. *Eur. J. Pharm. Sci.* **2010**, *40*, 448–455. [CrossRef]
30. Classen, W.; Altmann, B.; Gretener, P.; Souppart, C.; Skelton-Stroud, P.; Krinke, G. Differential Effects of Orally versus Parenterally Administered Qinghaosu Derivative Artemether in Dogs. *Exp. Toxicol. Pathol.* **1999**, *51*, 507–516. [CrossRef]
31. White, N.J. Cardiotoxicity of Antimalarial Drugs. *Lancet Infect. Dis.* **2007**, *7*, 549–558. [CrossRef]
32. Moskovitz, J.B.; Hayes, B.D.; Martinez, J.P.; Mattu, A.; Brady, W.J. Electrocardiographic Implications of the Prolonged QT Interval. *Am. J. Emerg. Med.* **2013**, *31*, 866–871. [CrossRef] [PubMed]
33. Pound-Lana, G.; Rabanel, J.-M.; Hildgen, P.; Mosqueira, V.C.F. Functional Polylactide via Ring-Opening Copolymerisation with Allyl, Benzyl and Propargyl Glycidyl Ethers. *Eur. Polym. J.* **2017**, *90*, 344–353. [CrossRef]
34. Fessi, H.; Puisieux, F.; Devissaguet, J.P.; Ammoury, N.; Benita, S. Nanocapsule Formation by Interfacial Polymer Deposition Following Solvent Displacement. *Int. J. Pharm.* **1989**, *55*, R1–R4. [CrossRef]
35. Li, P.; Zhao, L. Developing Early Formulations: Practice and Perspective. *Int. J. Pharm.* **2007**, *341*, 1–19. [CrossRef]
36. César, I.; Pianetti, G. Quantitation of Artemether in Pharmaceutical Raw Material and Injections by High-Performance Liquid Chromatography. *Braz. J. Pharm. Sci.* **2009**, *45*, 737–742. [CrossRef]
37. de Paula, C.S.; Tedesco, A.C.; Primo, F.L.; Vilela, J.M.C.; Andrade, M.S.; Mosqueira, V.C.F. Chloroaluminium Phthalocyanine Polymeric Nanoparticles as Photosensitisers: Photophysical and Physicochemical Characterisation, Release and Phototoxicity in Vitro. *Eur. J. Pharm. Sci.* **2013**, *49*, 371–381. [CrossRef]
38. Yu, M.; Yuan, W.; Li, D.; Schwendeman, A.; Schwendeman, S.P. Predicting drug release kinetics from nanocarriers inside dialysis bags. *J. Control Release* **2019**, *315*, 23–30. [CrossRef]
39. Peters, W.; Li, Z.L.; Robinson, B.L.; Warhurst, D.C. The Chemotherapy of Rodent Malaria, XL. The Action of Artemisinin and Related Sesquiterpenes. *Ann. Trop. Med. Parasitol.* **1986**, *80*, 483–489. [CrossRef]
40. Hanefeld, P.; Westedt, U.; Wombacher, R.; Kissel, T.; Schaper, A.; Wendorff, J.H.; Greiner, A. Coating of poly(p-xylylene) by PLA-PEO-PLA triblock copolymers with excellent polymer−polymer adhesion for stent applications. *Biomacromolecules* **2006**, *7*, 2086–2090. [CrossRef] [PubMed]
41. Wang, R.; Shi, G.; Chai, L.; Wang, R.; Zhang, G.; Ren, G.; Zhang, S. Choline and PEG dually modified artemether nano delivery system targeting intra-erythrocytic Plasmodium and its pharmacodynamics in vivo. *Drug Dev. Ind. Pharm.* **2021**, *47*, 454–464. [CrossRef] [PubMed]
42. Yang, Y.; Gao, H.; Zhou, S.; Kuang, X.; Wang, Z.; Liu, H.; Sun, J. Optimization and evaluation of lipid emulsions for intravenous co-delivery of artemether and lumefantrine in severe malaria treatment. *Drug Deliv. Res.* **2018**, *8*, 1171–1179. [CrossRef] [PubMed]
43. Shakeel, K.; Raisuddin, S.; Ali, S.; Imam, S.S.; Rahman, M.A.; Jain, G.K.; Ahmad, F.J. Development and in vitro/in vivo evaluation of artemether and lumefantrine co-loaded nanoliposomes for parenteral delivery. *J. Liposome Res.* **2017**, *29*, 35–43. [CrossRef]
44. Decuzzi, P.; Godin, B.; Tanaka, T.; Lee, S.-Y.; Chiappini, C.; Liu, X.; Ferrari, M. Size and Shape Effects in the Biodistribution of Intravascularly Injected Particles. *J. Control Release* **2010**, *141*, 320–327. [CrossRef] [PubMed]
45. Leite, E.A.; Vilela, J.M.C.; Mosqueira, V.C.F.; Andrade, M.S. Poly-Caprolactone Nanocapsules Morphological Features by Atomic Force Microscopy. *Microsc. Microanal.* **2005**, *11*, 48–51. [CrossRef]
46. Mosqueira, V.C.F.; Leite, E.A.; Barros, C.M.D.; Vilela, J.M.C.; Andrade, M.S. Polymeric Nanostructures for Drug Delivery: Characterization by Atomic Force Microscopy. *Microsc. Microanal.* **2005**, *11* (Suppl. 3), 36–39. [CrossRef]
47. Mosqueira, V.C.F.; Legrand, P.; Gulik, A.; Bourdon, O.; Gref, R.; Labarre, D.; Barratt, G. Relationship between complement activation, cellular uptake, and surface physicochemical aspects of novel PEG-modified nanocapsules. *Biomaterials* **2001**, *22*, 2967–2979. [CrossRef]
48. Falsafi, S.R.; Rostamabadi, H.; Assadpour, E.; Jafari, S.M. Morphology and microstructural analysis of bioactive-loaded micro/nanocarriers via microscopy techniques; CLSM/SEM/TEM/AFM. *Adv. Colloid Interface Sci.* **2020**, *280*, 102166. [CrossRef]

49. Takechi-Haraya, Y.; Ohgita, T.; Demizu, Y.; Saito, H.; Izutsu, K.-I.; Sakai-Kato, K. Current Status and Challenges of Analytical Methods for Evaluation of Size and Surface Modification of Nanoparticle-Based Drug Formulations. *AAPS PharmSciTech* **2022**, *23*, 150. [CrossRef]
50. Ural, M.S.; Dartois, E.; Mathurin, J.; Desmaële, D.; Collery, P.; Dazzi, A.; Deniset-Besseau, A.; Gref, R. Quantification of drug loading in polymeric nanoparticles using AFM-IR technique: A novel method to map and evaluate drug distribution in drug nanocarriers. *Analyst* **2022**, *147*, 5564. [CrossRef]
51. Nguyen, H.K.; Shundo, A.; Xiaobin Liang, X.; Yamamoto, S.; Tanaka, K.; Nakajima, K. Unraveling Nanoscale Elastic and Adhesive Properties at the Nanoparticle/Epoxy Interface Using Bimodal Atomic Force Microscopy. *ACS Appl. Mater. Interfaces* **2022**, *14*, 42713–42722. [CrossRef]
52. Sherief Essa, S.; Jean Michel Rabanel, J.-M.; Hildgen, P. Effect of aqueous solubility of grafted moiety on the physicochemical properties of poly(d,l-lactide) (PLA) based nanoparticles. *Int. J. Pharm.* **2010**, *388*, 263–273. [CrossRef]
53. Fermini, B.; Fossa, A.A. The Impact of Drug-Induced QT Interval Prolongation on Drug Discovery and Development. *Nat. Rev. Drug Discov.* **2003**, *2*, 439–447. [CrossRef] [PubMed]
54. EMA ICH E14 (R3) Clinical Evaluation of QT/QTc Interval Prolongation and Proarrhythmic Potential for Non-Antiarrhythmic Drugs—Questions Answers. Available online: https://www.ema.europa.eu/en/ich-e14-r3-clinical-evaluation-qt-qtc-interval-prolongation-proarrhythmic-potential-non (accessed on 29 October 2022).
55. Touze, J.E.; Heno, P.; Fourcade, L.; Deharo, J.C.; Thomas, G.; Bohan, S.; Paule, P.; Riviere, P.; Kouassi, E.; Buguet, A. The Effects of Antimalarial Drugs on Ventricular Repolarization. *Am. J. Trop. Med. Hyg.* **2002**, *67*, 54–60. [CrossRef] [PubMed]
56. Traebert, M.; Dumotier, B. Antimalarial Drugs: QT Prolongation and Cardiac Arrhythmias. *Expert Opin. Drug Saf.* **2005**, *4*, 421–431. [CrossRef] [PubMed]
57. Mosqueira, V.C.; Legrand, P.; Morgat, J.L.; Vert, M.; Mysiakine, E.; Gref, R.; Devissaguet, J.P.; Barratt, G. Biodistribution of Long-Circulating PEG-Grafted Nanocapsules in Mice: Effects of PEG Chain Length and Density. *Pharm. Res.* **2001**, *18*, 1411–1419. [CrossRef] [PubMed]

polymers

MDPI

Article

Development of Crosslinker-Free Polysaccharide-Lysozyme Microspheres for Treatment Enteric Infection

Shuo Li, Li Shi, Ting Ye, Biao Huang, Yuan Qin, Yongkang Xie, Xiaoyuan Ren * and Xueqin Zhao *

Zhejiang Provincial Key Laboratory of Silkworm Bioreactor and Biomedicine, College of Life Sciences and Medicine, Zhejiang Sci-Tech University, Hangzhou 310018, China
* Correspondence: xyren@zstu.edu.cn (X.R.); zhaoxueqin@zstu.edu.cn (X.Z.)

Abstract: Antibiotic abuse in the conventional treatment of microbial infections, such as inflammatory bowel disease, induces cumulative toxicity and antimicrobial resistance which requires the development of new antibiotics or novel strategies for infection control. Crosslinker-free polysaccharide-lysozyme microspheres were constructed via an electrostatic layer-by-layer self-assembly technique by adjusting the assembly behaviors of carboxymethyl starch (CMS) on lysozyme and subsequently outer cationic chitosan (CS) deposition. The relative enzymatic activity and in vitro release profile of lysozyme under simulated gastric and intestinal fluids were investigated. The highest loading efficiency of the optimized CS/CMS-lysozyme micro-gels reached 84.9% by tailoring CMS/CS content. The mild particle preparation procedure retained relative activity of 107.4% compared with free lysozyme, and successfully enhanced the antibacterial activity against *E. coli* due to the superposition effect of CS and lysozyme. Additionally, the particle system showed no toxicity to human cells. In vitro digestibility testified that almost 70% was recorded in the simulated intestinal fluid within 6 h. Results demonstrated that the cross-linker-free CS/CMS-lysozyme microspheres could be a promising antibacterial additive for enteric infection treatment due to its highest effective dose (573.08 μg/mL) and fast release at the intestinal tract.

Keywords: lysozyme; carboxymethyl starch; chitosan; layer-by-layer self assembly; enteric infection

Citation: Li, S.; Shi, L.; Ye, T.; Huang, B.; Qin, Y.; Xie, Y.; Ren, X.; Zhao, X. Development of Crosslinker-Free Polysaccharide-Lysozyme Microspheres for Treatment Enteric Infection. *Polymers* **2023**, *15*, 1077. https://doi.org/10.3390/polym15051077

Academic Editors: Stephanie Andrade, Joana A. Loureiro and Maria João Ramalho

Received: 3 February 2023
Revised: 18 February 2023
Accepted: 19 February 2023
Published: 21 February 2023

1. Introduction

Microbial infections, such as inflammatory bowel disease (IBD), threaten the health of human beings globally, and are associated with chronic inflammation, cancer, high mortality and so on [1]. The spread of antibiotic-resistant bacteria requires the development of new antibiotics or novel strategies for infection control [2,3]. Bioactive proteins extracted from natural products have caught more and more attention, due to their safety and few side effects compared with synthetic drugs [4–7]. Lysozyme, easily obtained from egg whites, is a glycoside hydrolase with high enzymatic specificity for the hydrolysis of β 1-4 glycosidic bonds in chitin or the peptidoglycan wall of fungal or Gram-positive bacteria, which can eventually induce cell lysis without involving antibiotic resistance [7]. Lysozyme is important for the resolution of inflammation at mucosal sites in the human gastrointestinal (GI) tract [8]. Thus, lysozyme has become a new antibacterial alternative for the treatment of IBD because of its innate activity [9], and its usage is limited by the harsh environment of the GI tract and its inherent limitations such as poor stability, low bioavailability and a narrow antimicrobial spectrum [10]. The development of immobilized lysozyme is an urgent need in medicine, the food industry and biotechnology.

The microencapsulation technique allows the isolation of active protein into solid matrices for the stabilization of active ingredients, enhancement of the bioavailability, responsive release, etc. [11]. Biopolymers, especially polysaccharides, are well-developed as the matrix to fabricate microspheres due to their excellent biocompatibility, biodegradability and low cost [12–17].Carboxymethyl starch (CMS) is a negatively charged ether derivative of starch with good water solubility and has become a suitable substrate for enzymatic immobilization. The presence of polar carboxylic functional groups in CMS grants the microspheres with mucoadhesive properties, pH sensitivity and digestion resistibility depending on the cross-linking and the degree of substitution [18–21]. However, the CMS-entrapped protein may be denatured by the proteolytic enzymes due to the rapid dissolution of CMS in an intestinal tract environment. Combining CMS with a second polymer further improves stability. In fact, chitosan (CS) as another polysaccharide can form complexes with CMS via electrostatic interaction for protein immobilization, particularly for oral administration [21–24]. For instance, CS/CMS-coated microspheres that were fabricated largely retained lysozyme under simulated stomach conditions and achieved the release of lysozyme under simulated intestinal conditions [23]. The CS/CMS polyelectrolyte complex enhanced the entrapment of bovine serum albumin (BSA) and release time (72 h) compared with CS-tripolyphosphate [24]. However, in most reports available, the protein-loaded CMS/CS preparation requires permanent chemical cross-linking or surfactants to obtain stable nanoparticles. On the other hand, the rapid release of antibacterial agents at the site is important for infection treatments. Considering the mucoadhesiveness and lysozyme triggered degradability of CS, we envisage that controlling the assembly behaviors of CMS on lysozyme, and in turn adjusting the outer CS deposition, can tailor the stability and release properties of lysozyme to enhance antimicrobial ability in enteric infections.

Herein, CMS and CS were employed as edible materials for lysozyme entrapment to treat enteric infection via layer-by-layer self-assembly without any surfactant and cross-linker. Different surface potential-contained CMS-lysozyme particles were first obtained and then deposited with different CS content to prepare CS/CMS-lysozyme microspheres. The relative enzymatic activity and in vitro release profile of lysozyme under simulated gastric and intestinal fluid were investigated. Finally, the biocompatibility and in vitro antibacterial activity of microspheres were also examined.

2. Materials and Methods

2.1. Materials

Lysozyme (from eggs, 70,000 U/mg), pancreatin, chitosan (CS, viscosity < 200 mPa.s) and carboxymethyl starch sodium (CMS, MW 50KD) with a substitution degree of 0.1046 were purchased from Aladdin Co., Ltd. (Shanghai, China). *Escherichia coli* O157:H7 (*E. coli*, ATCC 25922) and *Staphylococcus aureus* (*S. aureus*, ATCC25923) were bought from Sigma–Aldrich Corp., St. Louis, MO, USA. Human umbilical vein endothelial cells (hUVECs) were purchased from the China Center for Type Culture Collection. Lysozyme assay kit and bicinchoninic acid (BCA) protein assay kit were purchased from Jiancheng Bioengineering Institute (Nanjing, China). The Luria–Bertani broth (LB Broth), RPMI1640 culture medium, fetal bovine serum (FBS), Penicillin-Streptomycin solution (10 mg/mL) and Cell Counting Kit-8 (CCK-8) were purchased from Beijing LABLEAD Co., Ltd. (Beijing, China). Ultrapure water from a Milli-Q filtration system (Millipore Corp., Bedford, MA, USA) was used to prepare all solutions.

2.2. Preparation of Encapsulated Lysozyme Microspheres

t5 CS/CMS-lysozyme was prepared by layer-by-layer (LbL) assembly via ionic interaction of polysaccharides. Zeta-potential measurements and encapsulation efficiency were used to assess the optimum assembly condition. Typically, 0.1 g of lysozyme was firstly dissolved in 100 mL PBS (0.01 M, pH 3.0) to obtain a positively charged polyelectrolyte solution, then 0.3 g of CMS was mixed with the lysozyme solution gently for 2 h for the deposition of CMS on the lysozyme surface via electrostatic interaction. After centrifugation at 10,000 rpm for 10 min, the precipitate was washed with distilled water three times. Subsequently, 0.2 g of the above-obtained CMS-lysozyme microspheres was suspended in 1.6 mL of positively charged CS solution (1 mg/mL) and incubated at 25 °C for 1.5 h under stirring of 300 rpm. The CS solution was prepared in advance by dissolving

0.1 g chitosan in 100 mL 2% acetic acid solution (in PBS with 0.01 M, pH 4.0) and then pH was adjusted with hydrochloric acid (6 M). The resulting CS/CMS-lysozyme microspheres were centrifuged at 4000 rpm for 3 min and then washed with distilled water. Excessive lysozymes from all the washing solutions were collected for further determination of the lysozyme loading capacity.

The non-encapsulated lysozyme content was quantified using the micro-BCA protein assay kit by reading the absorbance at 562 nm. All experiments were performed in triplicate, and the drug loading capacity (LC) and drug encapsulation efficiency (EE) of the obtained microspheres were calculated according to the following formula:

$$\mathrm{LC}(\%, w/w) = \frac{W_{total\ lysozyme} - W_{free\ lysozyme}}{W_{total Ps}} \times 100\% \quad (1)$$

$$\mathrm{EE}(\%, w/w) = \frac{W_{totallysozyme} - W_{freelysozyme}}{W_{total\ lysozyme}} \times 100\% \quad (2)$$

where $W_{total\ lysozyme}$ is the initial amount of lysozyme; $W_{free\ lysozyme}$ is the amount of unloaded lysozyme measured in the supernatant and washing solution; and $W_{total Ps}$ is the mass of the initial microsphere powders added in the system.

2.3. *Microsphere Characterization*

The hydrodynamic sizes and zeta potentials were measured on a Nano-ZS Zetasizer dynamic light scattering (DLS) instrument (Malvern Instruments Ltd., Malvern, UK). The microstructure of the sample was measured using a field emission scanning electron microscope (SEM, ZEISS-ULTRA55; Hitachi Ltd., Tokyo, Japan) and the particle diameters were calculated using Nano Measurer 1.2 software. IR analysis and UV analysis were carried out on Nicolet 5700 FTIR spectrometer (Thermo Fisher Scientific, Waltham, MA, USA) using the KBr-disk method and Nanodrop 2000 UV–vis spectrometer (Thermo Fisher Scientific, Waltham, MA, USA), respectively.

2.4. *Enzymatic Activity of Lysozyme*

Lysozyme assay was performed using *M. lysodeikticus* cells. Briefly, 0.2 mL aliquot of diluted native lysozyme in phosphate buffer (pH 6.5) or the lysozyme released from CS/CMS-lysozyme was added to 2 mL suspension of *Micrococcus lysodeikticus* cells (0.1 mg/mL) and then incubated at 37 °C for 5 min. The turbidity of cell suspension was measured every 20 s over a 2 min period on a UV–visible spectrophotometer at 530 nm. The experiments were performed in triplicate, and the enzymatic activity was estimated as relative activity (% activity, U/U) based on a decrease in turbidity relative to the native lysozyme system.

2.5. *In Vitro Digestibility of Lysozyme Microspheres*

The in vitro releasing profile of lysozyme from CS/CMS-lysozyme microspheres was examined in pepsin-free simulated gastric fluid (SGF) and simulated intestinal fluid (SIF) with pancreatin. Simulated digestive fluids were prepared according to a previous method with little modification [20]. For the preparation of simulated gastric fluid (SGF), 36.5% (*w*/*w*) hydrochloric acid (7 mL) was diluted to 1 L with deionized water at pH 1.2. Similarly, simulated intestinal fluid (SIF) was prepared by dissolving 10 g pancreatin and 6.8 g potassium di-hydrogen phosphate into 1 L of distilled water. Then, the pH of the SIF solution was adjusted to 6.8 with 0.1 M sodium hydroxide solution. Simulated colonic fluid (SCF) at pH 7.2 was prepared by dissolving potassium di-hydrogen phosphate (3103 g) and di-potassium hydrogen phosphate (4355 g) in distilled water (1 L).

For the simulation study, microsphere-lysozyme complexes (6.25 mg) were added to 15 mL of the dissolution medium with a shaking speed of 100 rpm at 37 °C according to standard pharmacopeia methods [23]. The microspheres were firstly incubated for 2 h in SGF and then separated by centrifugation, followed by a 6 h treatment with SIF. Afterward, the medium was replaced with SCF for an additional 4 h. At periodic intervals, 1 mL of release medium was withdrawn and replaced with the same volume of fresh medium to maintain a constant volume. These experiments were performed in triplicate. The released lysozyme content was analyzed by UV–vis spectrophotometer at 278 nm. The beginning concentration was set as 100%. All experiments were performed in triplicate, and the cumulative release (%) was calculated using the equation below:

$$\text{Cumulative release } (\%,\ w/w) = \left(15C_i + \sum_{1}^{i-1} C_{i-1} \times 1\right)/M_0 \tag{3}$$

where M_0 is the initial mass of lysozyme in the samples and C_i is the concentration of lysozyme released at each sampling time point.

2.6. Antibacterial Activity

Gram-negative bacteria *Escherichia coli* (*E. coli*, ATCC 29425) and Gram-positive bacteria *Staphylococcus aureus* (*S. aureus*, ATCC25923) were cultured for 12 h in a 37 °C shaker. When the OD_{600} value of the cultured bacteria reached 0.7, CMS-lysozyme or CS/CMS-lysozymemicrospheres were added to 3 mL of the bacterial suspensions with a final concentration of 15 mg/mL. As a control, free lysozyme was also added to the suspensions to the same concentration as that of CMS-lysozyme or CS/CMS-lysozyme microspheres. After incubation at 37 °C for 20 h, 100 μL of 10-fold diluted suspensions from each test strain solution were spread onto agar plates and incubated for 20 h in shaker at 37 °C. The bacterial colonies were counted and the colony number of bacteria without treatment was used as the control. All experiments were performed in quintuplicate, and the antibacterial efficiency was calculated as follows:

$$\text{Antibacterial rate } (\%,\ \text{cfu/cfu}) = \frac{\text{cell numbers of control} - \text{cell numbers of samples}}{\text{cell numbers of control}} \times 100\% \tag{4}$$

2.7. Cytotoxicity

In vitro cytotoxicity was measured using a standard Cell Counting Kit-8 (CCK8) assay. The hCMEC cells were seeded into a 96-well plate at 5×10^3 cells/well and incubated for 24 h until the cell adhered to the wall. Then cells were treated with serum-free medium containing CS/CMS-lysozyme microspheres. After co-culturing for 24 h, cells were washed twice with PBS, and incubated with 10 μL CCK-8 solution for 1 h. The absorbance value (OD value) at 450 nm was measured by a microplate reader. All experiments were performed in octuplicate, and the relative cell viability (%) was expressed as a percentage relative to the untreated control cells.

2.8. Statistical Analysis

Statistical analysis was carried out using Origin 8.5 and Microsoft Excel and SPSS version 25.0. Statistical analysis was performed using two-tailed unpaired *t*-tests for comparison. The data were obtained from at least three independent experiments and expressed as means ± standard deviations (SD). Differences were statistically significant when the *p*-values were less than 0.05.

3. Results and Discussion

3.1. Preparation and Optimization of Encapsulation Conditions

The surface charge is one of the most important parameters tailoring the assembly behavior of polyelectrolytes during the fabrication of LBL self-assembly systems [20]. The dissociation degree of weak polyelectrolytes depends strongly on their content and the solution pH. Zeta-potential measurements were carried out to characterize the pH response behaviors of lysozyme, CS and CMS at different pH conditions. As shown in Figure 1a, CMS was nearly neutrally charged, then increased its surface negative charge with pH increase due to CMS deprotonation at the pH ranging from 1 to 6.8. CMS solution with the concentration of 3 mg/mL was found to have a maximum negative charge (−15.15 ± 2.788 mV) at pH 3, where the lysozyme molecules exhibited a positive charge (11.975 ± 1.402 mV) and smaller size (272.5 ± 6.6 nm). CMS and lysozyme were required with a relatively strong surface potential to fabricate compact structures of capsules via electrostatic interaction. When the pH was 3.0, the interaction between positively charged lysozyme molecules and negatively charged CMS molecules led to the formation of CMS-lysozyme LBL assembly capsules. As the pH value increased, the engineered CMS-lysozyme negatively charged because of CMS protonation, and exhibited a maximum opposite charge against CS at pH 4 (Figure 1b), which is the optimal pH for the assembly of the second layer. Oppositely charged polymers (CMS and CS) were electrostatically assembled directly onto the lysozyme surface to obtain CS/CMS-lysozyme microspheres. Furthermore, the substitution degrees of the CMS and the deacetylation degree of chitosan can influence their surface properties and the subsequent assembly behaviors of polyelectrolytes. To avoid the complexity of the self-assembly system, commercialized CMS and CS were selected for controlling the assembly behaviors of polyelectrolytes only by adjusting their deposition contents.

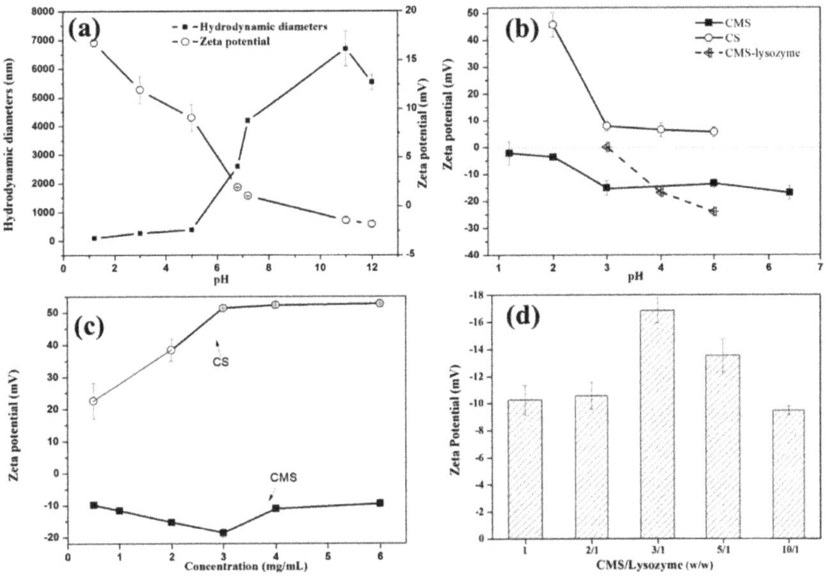

Figure 1. Optimization of preparation procedure according to DLS analysis: (**a**) zeta potential and size as a function of pH of lysozyme, (**b**) zeta potential as a function of pH of CMS, CS and CMS-lysozyme, (**c**) zeta potential as a function of the concentration of CMS and CS, and (**d**) zeta potential as a function of CMS-to-lysozyme mass ratio. Bars represent the corresponding standard deviations ($n = 3$).

In order to investigate the optimal encapsulation efficiency (EE), the assembly procedure was optimized by testing the key parameters such as the mass ratios of lysozyme/CMS and CMS-lysozyme/CS, CS concentration and incubation temperature. As shown in Figure 2, for the first layer, the EE was enhanced to reach a plateau with a maximum of 100% by increasing the mass ratio of CMS to lysozyme from 1:1 to 2:1, while the LC decreased significantly from 50% to 10% with the increase in the mass ratio between CMS and lysozyme from 1:1 to 10:1 (Figure 2a). Because the concentration of lysozyme in our experiment was fixed at a constant dosage of 1 mg/mL, increasing the ratio of CMS/lysozyme meant increasing the concentration of CMS. Thus, an increased CMS concentration gradient induced complete lysozyme entrapment, similar to the previous report [20]. At the same time, increased CMS/lysozyme microspheres induced LC to decrease. Considering that the highest negative charged (−16 mV, Figure 1d) CMS/lysozyme microspheres are beneficial for CS deposition, the ratio of 3/1 was selected as the optimal mass ratio of CMS/lysozyme microcapsule. For the second layer, with the increase in the CS concentration or the ratio between CS and CMS-lysozyme, both LC and EE decreased (Figure 2c,d). When the CS concentration was 1 mg/mL, EE and LC were 84.9% w/w, and 22.1% w/w, respectively.

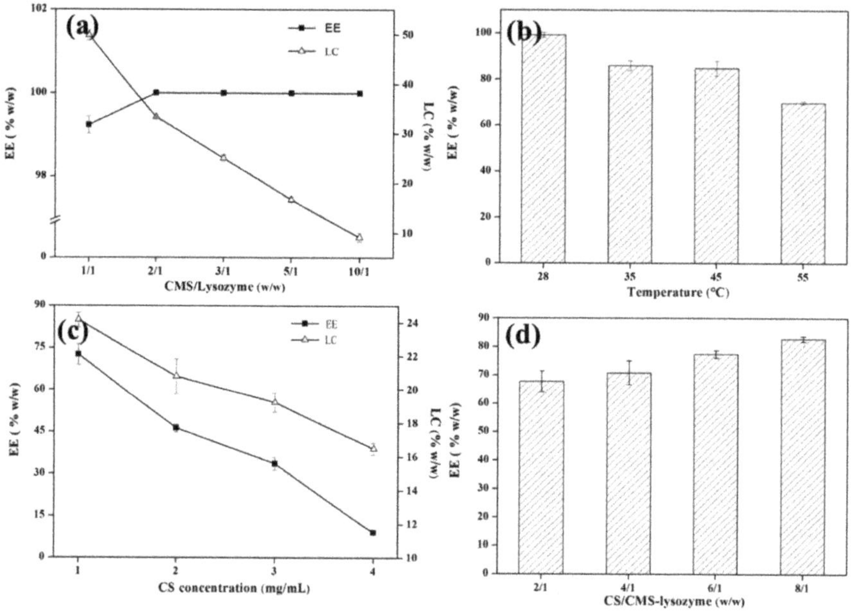

Figure 2. Optimization of encapsulation efficiency (EE) and loading capability (LC). EE and LC as a function of CMS-to-lysozyme mass ratio (**a**) and CS concentration (**c**). EE as a function of temperature (**b**,**d**) mass ratio of CS and CMS-lysozyme. Bars represent the corresponding standard deviations (n = 3).

3.2. Characterization of CS/CMS-Lysozyme Microspheres

The microstructure of the microspheres was detected by SEM. As shown in Figure 3, the microspheres exhibited a relatively spherical-shaped morphology and a relatively smooth surface with an average size of 8.09 μm.

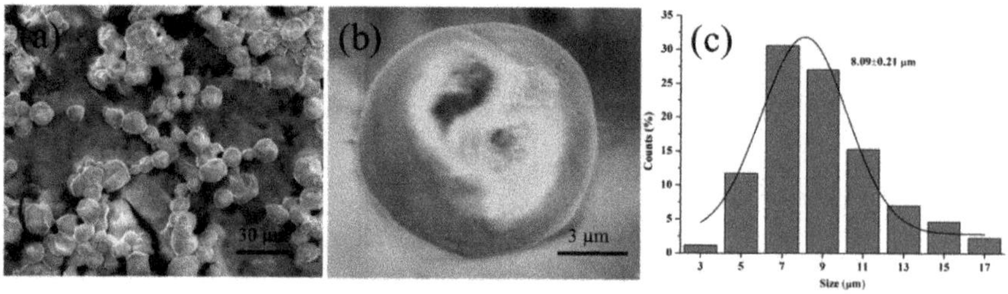

Figure 3. SEM morphological analysis of CS/CMS-lysozyme microspheres: (**a**) overview, (**b**) enlarged image of the individual capsule and (**c**) histograms of size distribution. The solid line in the size histograms is the simulation curve of Gaussian distribution. The data outlined refer to the most probable size.

The FT-IR analysis was performed to confirm the presence and possible interactions between CS/CMS and lysozyme. As shown in Figure 4a, the broad peaks centered at 3423 cm^{-1} for CS/CMS and 3487 cm^{-1} for CS/CMS-lysozyme microspheres came from CMS (3450 cm^{-1}) and CS (3443 cm^{-1}), likely due to the interaction of CMS with CS via hydrogen bonding. Lysozyme incorporation induced the O–H stretching peak of CS/CMS-lysozyme to decrease and shift to the lower wavenumber, which is similar to a previous report [16]. The peaks at 1650 cm^{-1} and 1531 cm^{-1} correspond to amide I (the stretching vibration of C=O conjugated peptide bond) and amide II (the secondary NH deformation), respectively. The characteristic peaks also appeared in the CS/CMS-lysozyme (1655 cm^{-1}, 1531 cm^{-1}) and the CS/CMS (1654 cm^{-1}, 1596 cm^{-1}) FTIR spectrum, indicating the amide appearance. The CS/CMS-lysozyme spectrum showed slight discrepancies compared with the CMS/CS. The band at 1596 cm^{-1} in the CS/CMS spectrum transformed into a shoulder one due to the entrapment of lysozyme. Due to the superposition effect of the polyelectrolyte complex between the amine group of CS (1650 cm^{-1}) and the carboxylate group of CMS (1601 cm^{-1}) [16], the intensive peak with a higher shift to 1655 cm^{-1} was observed in the CS/CMS-lysozyme spectrum, which demonstrates intermolecular interactions and good molecular compatibility between CMS and CS. Moreover, some characteristic peaks corresponding to compound CMS, CS and lysozyme appeared in gels, further confirming the presence of lysozyme within the CS/CMS.

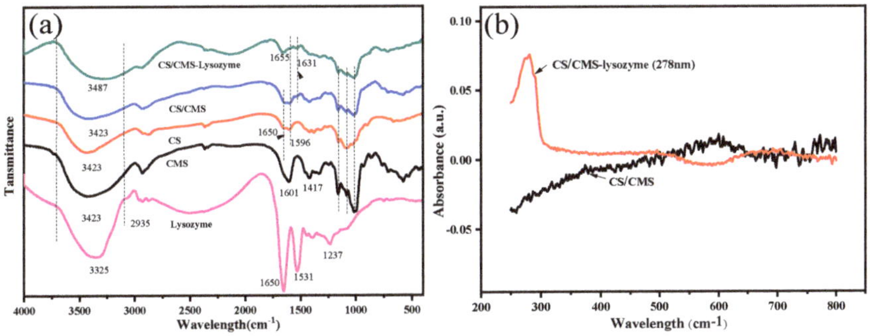

Figure 4. (**a**) FTIR spectra of lysozyme, CS, CMS and CS/CMS-lysozyme microspheres, (**b**) UV-visible absorption spectra of CS/CMS-lysozyme and hollow CS/CMS microspheres.

Monitoring wavelength changes of the UV–vis spectra can detect the formation of complexes between a protein and other macromolecules. Hence, UV–vis spectra were used to further analyze the microenvironment around microspheres. As shown in Figure 4b, CS/CMS had no absorbance from 800 cm^{-1} to 250 cm^{-1} wavelengths, while CS/CMS-lysozyme exhibited an obvious absorbance peak at 278 cm^{-1}. The slight 2 nm shift from 280 nm of native lysozyme, suggests the existence and combination of lysozyme with polysaccharides in the composite microspheres [25].

3.3. Enzymatic Activity of the Lysozyme

In vitro lytic activities on *M. lysodeikticus* cells of lysozymes released from microspheres are illustrated in Figure 5. The reactivity for released lysozyme was 107.4 ± 0.38% U/U, a bit higher than that of free lysozyme due to the enhancement of Na^{+} and K^{+} in the reaction solution [26,27]. Generally, lysozymes will aggregate together after release from the polymer matrix and their lytic activity may decrease due to the destabilization and unfolding of the protein [28]. It was also reported that polysaccharides and polyhydric alcohols can effectively prevent lysozymes from aggregation and inactivation [16]. Our study suggested that a mild preparation process minimized the above-mentioned defects of the polymer matrix and the CS/CMS matrix effectively protected the lysozyme.

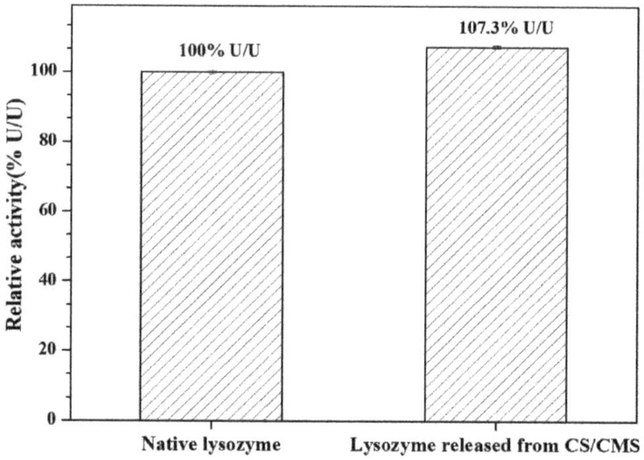

Figure 5. Relative activity of lysozyme released from CS/CMS. Bars represent the corresponding standard deviations (n = 3).

3.4. In Vitro Release Studies

As the intestine is generally considered as the main absorption site, in vitro digestion and release of lysozyme from chitosan-based formulations was performed under simulated digestive fluids to investigate their intestinal targeting during the formulation. The pH and ionic strength associated with local conditions in the stomach and small intestine fluids are quite different from each other. The protective effect of polysaccharide coating is of much importance to avoid degradation of lysozyme under the condition with high acid and enzymes in SGF. As shown in Figure 6, in simulated gastric digestion the release of lysozyme from CMS-coated microspheres exhibited a burst and more than 80% of lysozyme outflowed in 2 h owing to the protonation of carboxyl groups on the CMS backbone and the sequence detachment. In addition to the pH, pancreatin and the ionic strength affected release for the CMS-based carriers depending on the degree of substitution and crosslinking. However, coating the pH-responsive CS inhibited protein release at pH 2.0 and up to 70.8% was retained in CS/CMS through the molecule–molecule aggregation at extremely acidic pH. It suggested enhanced stability in gastric fluid and the possibility of using the second polysaccharide coating of CS as a gate material. After incubation in the SGF, the microspheres were separated and transferred to SIF (pH 7.2) with pancreatin. The release of lysozyme increased significantly due to the polysaccharide degradation by pancreatin. Almost all the lysozyme was released from CS/CMS-lysozyme and the cumulative release reached 96.7% w/w in 8 h. Then, in the SCF with pH 7.2, lysozyme was completely released in 2 h and the cumulative release rate reached 100%. The absolute amount of lysozyme that eventually reached the small intestine was 573.98 μg/mL within 6 h. The results indicate that CS/CMS-lysozyme exhibited a good digestion resistance in SGF and fast release in SIF.

The pH of simulated digestive fluid modulates ionic interactions of chitosan with oppositely charged CMS and consequently the properties of microspheres. It was reported that chitosan-based formulations remained intact in the low pH gastric environment but dissociated in the small intestine with higher pH [29]. The susceptibilities of microspheres in SIF present significant effects of pancreatin. CS-coated capsules are readily degradable by pancreatin in PBS [30]. Moreover, released lysozyme can also accelerate CS degradation [31]. These results indicated that CS/CMS-lysozyme exhibited a good resistance in SGF and fast release of lysozyme in SIF.

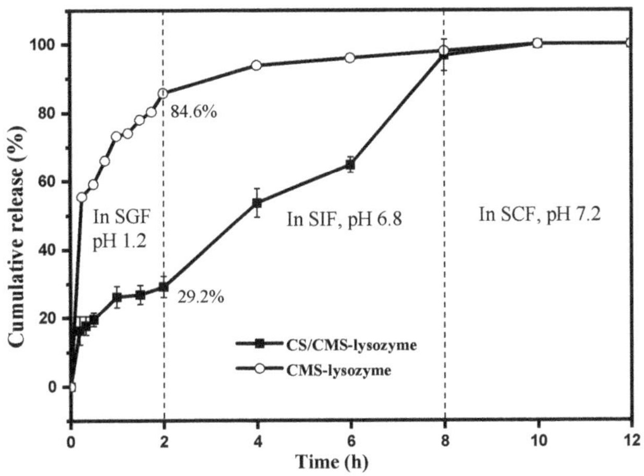

Figure 6. In vitro releasing kinetics of lysozyme from stabilized microspheres under stimulated digestive fluids. Bars represent the corresponding standard deviations ($n = 3$).

3.5. Antibacterial Analysis

Efficient antibacterial activity is essential for the antibacterial enzyme. The antimicrobial activity was quantified by counting the colony forming units (cfu) on the spread plate. As shown in Figure 7, free lysozyme has antibacterial ratios of 66.4 ± 10.4% cfu/cfu and 63.9 ± 10.2% cfu/cfu against *E. coli* and *S. aureus*, respectively. CS/CMS microspheres demonstrated a significantly enhanced antibacterial ability against *E. coli* (34.9% cfu/cfu) than *S. aureus* (14.2% cfu/cfu). Moreover, compared with the free lysozyme and hollow CS/CMS, the CS/CMS-lysozyme microspheres exhibited a better antibacterial ability against *E. coli* (97.4% cfu/cfu) than *S. aureus* (58.7% cfu/cfu) within 20 h of contact. This enhancement was attributed not only to the released lysozyme but also to a possible synergistic effect between chitooligomers and lysozyme obtained after chitosan hydrolysis [12,32,33].

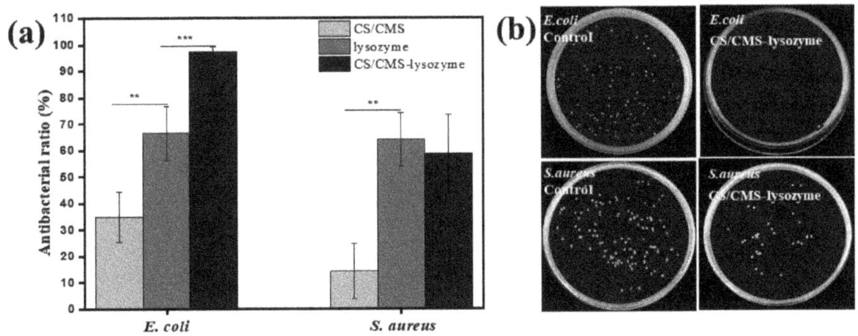

Figure 7. In vitro antibacterial activity: (**a**) antibacterial ratio against *E. coli* and *S. aureus*, (**b**) re-cultivated bacterial colonies on agar after bacteria dissociation from antibacterial tests. The data of antibacterial ratio were analyzed using a two-tailed unpaired Student's t-test. The error bars indicate means ± SD. **: $p < 0.01$ and ***: $p < 0.001$.

3.6. Biocompatibility

Cytotoxicity has always been a concern for biomaterials. The cytotoxicity was determined by CCK8 assay in a concentration-dependent manner. As shown in Figure 8, the cell viability is above 100% when the tested concentrations reached 0.875 g/L, suggesting significantly improved cell proliferation.

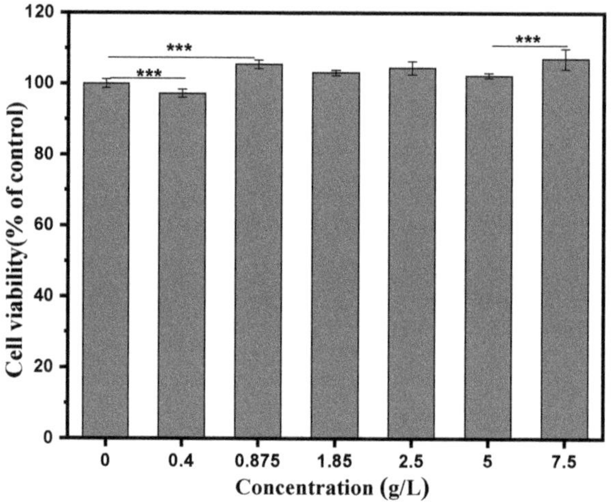

Figure 8. Cell viability of CS/CMS-lysozyme microspheres by CCK8 assay. The data of cell viability were analyzed using a two-tailed unpaired Student's *t*-test. The error bars indicate means ± SD and $n = 8$. ***: $p < 0.001$.

4. Conclusions

Here, crosslinker-free polysaccharide-lysozyme microspheres were constructed by adjusting the assembly behaviors of carboxymethyl starch (CMS) on lysozyme and subsequently outer cationic chitosan (CS) deposition. The UV–vis and FIRT-verified entrapment of lysozyme within the CS/CMS matrix was carried out successfully. The optimized CS/CMS-lysozyme microspheres had the highest loading efficiency of 84.9% w/w, and the relative activity of the released lysozyme from CS/CMS-lysozyme microspheres was 107.4 ± 0.38% U/U compared with free lysozyme, indicating a slight promotion potency on lysozyme activity. Furthermore, 67.5% w/w of cumulative release in SIF conferred that the CS/CMS double polysaccharide layers largely retained lysozyme within the matrix without a burst in the acid environment and achieved intestine-targeted release. Moreover, the antimicrobial activity assay indicated CS/CMS-lysozyme had a higher antibacterial activity especially towards *E. coli* owing to the synergistic effect between the membrane-attacking ability of CS and the cell wall-attacking ability of lysozymes. The cross-linker-free and pH-responsive CS/CMS-lysozyme microspheres can be a promising oral antibacterial additive for the treatment of intestinal infection.

Author Contributions: Conceptualization, X.R. and X.Z.; methodology, S.L., L.S. and Y.X.; validation, S.L. and Y.Q.; writing—original draft preparation, X.Z. and T.Y.; writing—review and editing, X.R. and X.Z.; supervision, B.H.; funding acquisition, T.Y., X.R. and X.Z. All authors have read and agreed to the published version of the manuscript.

Funding: This work was supported by funds from the Science Technology Department of Zhejiang Province (2017C32047), the Open Fund Project of State Key Laboratory Breeding Base of Marine Genetic Resource, MNR (HY201803) and the Zhejiang Provincial Natural Science Foundation (LY18C100002).

Institutional Review Board Statement: Not applicable.

Data Availability Statement: The data presented in this study are available on request from the corresponding author. The data are not publicly available due to confidentiality agreements.

Conflicts of Interest: The authors declare no conflict of interest for this work.

References

1. Mirsepasi-Lauridsen, H.C.; Vallance, B.A.; Krogfelt, K.A.; Petersen, A.M. Escherichia coli pathobionts associated with inflammatory bowel disease. *Clin. Microbiol. Rev.* **2019**, *32*, e00060-18. [CrossRef]
2. Su, C.; Hu, Y.; Song, Q.; Ye, Y.; Gao, L.; Li, P.; Ye, T. Initiated chemical vapor deposition of graded polymer coatings enabling antibacterial, antifouling, and biocompatible surfaces. *ACS Appl. Mater. Interfaces* **2020**, *12*, 18978–18986. [CrossRef]
3. Zhou, L.; Zhang, Y.; Ge, Y.; Zhu, X.; Pan, J. Regulatory mechanisms and promising applications of quorum sensing-inhibiting agents in control of bacterial biofilm formation. *Front. Microbiol.* **2020**, *11*, 589640. [CrossRef]
4. Chen, J.; Xu, L.; Zhou, Y.; Han, B. Natural products from actinomycetes associated with marine organisms. *Mar. Drugs* **2021**, *19*, 629. [CrossRef]
5. Chen, W.; Ye, K.; Zhu, X.; Zhang, H.; Si, R.; Chen, J.; Chen, Z.; Song, K.; Yu, Z.; Han, B. Actinomycin X2, an antimicrobial depsipeptide from marine-derived Streptomyces cyaneofuscatus applied as a good natural dye for silk fabric. *Mar. Drugs* **2021**, *20*, 16. [CrossRef]
6. Zhang, W.; Wei, L.; Xu, R.; Lin, G.; Xin, H.; Lv, Z.; Qian, H.; Shi, H. Evaluation of the antibacterial material production in the fermentation of Bacillus amyloliquefaciens-9 from whitespotted bamboo shark (Chiloscylliumplagiosum). *Mar. Drugs* **2020**, *18*, 119. [CrossRef]
7. Ferraboschi, P.; Ciceri, S.; Grisenti, P. Applications of lysozyme, an innate immune defense factor, as an alternative antibiotic. *Antibiotics* **2021**, *10*, 1534. [CrossRef]
8. Ragland, S.A.; Criss, A.K. From bacterial killing to immune modulation: Recent insights into the functions of lysozyme. *PLoSPathog.* **2017**, *13*, e1006512. [CrossRef]
9. Lee, M.; Kovacs-Nolan, J.; Yang, C.; Archbold, T.; Fan, M.Z.; Mine, Y. Hen egg lysozyme attenuates inflammation and modulates local gene expression in a porcine model of dextran sodium sulfate (DSS)-induced colitis. *J. Agric. Food Chem.* **2009**, *57*, 2233–2240. [CrossRef]
10. Wei, Y.; Wang, C.; Jiang, B.; Sun, C.; Middaugh, C. Developing biologics tablets: The effects of compression on the structure and stability of bovine serum albumin and lysozyme. *Mol. Pharmaceut.* **2019**, *16*, 1119–1131. [CrossRef]
11. Yan, C.; Kim, S.R.; Ruiz, D.R.; Farmer, J.R. Microencapsulation for Food Applications: A Review. *ACS Appl. Bio. Mater.* **2022**, *5*, 5497–5512. [CrossRef]
12. Aguanell, A.; Del Pozo, M.L.; Pérez-Martín, C.; Pontes, G.; Bastida, A.; Fernández-Mayoralas, A.; García-Junceda, E.; Revuelta, J. Chitosan sulfate-lysozyme hybrid hydrogels as platforms with fine-tuned degradability and sustained inherent antibiotic and antioxidant activities. *Carbohyd. Polym.* **2022**, *291*, 119611. [CrossRef]
13. Gennari, C.G.; Sperandeo, P.; Polissi, A.; Minghetti, P.; Cilurzo, F. Lysozyme mucoadhesive tablets obtained by freeze-drying. *J. Pharm. Sci.* **2019**, *108*, 3667–3674. [CrossRef] [PubMed]
14. Kim, S.; Fan, J.; Lee, C.S.; Lee, M. Dual functional lysozyme–chitosan conjugate for tunable degradation and antibacterial activity. *ACS Appl. Bio. Mater.* **2020**, *3*, 2334–2343. [CrossRef] [PubMed]
15. Hu, H.; Luo, F.; Zhang, Q.; Xu, M.; Chen, X.; Liu, Z.; Xu, H.; Wang, L.; Ye, F.; Zhang, K.; et al. Berberine coated biocomposite hemostatic film based alginate as absorbable biomaterial for wound healing. *Int. J. Biol. Macromol.* **2022**, *209*, 1731–1744. [CrossRef] [PubMed]
16. Wu, T.; Huang, J.; Jiang, Y.; Hu, Y.; Ye, X.; Liu, D.; Chen, J. Formation of hydrogels based on chitosan/alginate for the delivery of lysozyme and their antibacterial activity. *Food Chem.* **2018**, *240*, 361–369. [CrossRef] [PubMed]
17. Li, J.; Jin, H.; Razzak, M.A.; Kim, E.J.; Choi, S.S. Crosslinker-free bovine serum albumin-loaded chitosan/alginate nanocomplex for pH-responsive bursting release of oral-administered protein. *Biotechnol. Bioprocess Eng.* **2022**, *27*, 40–50. [CrossRef]
18. Zhang, B.; Tao, H.; Wei, B.; Jin, Z.; Xu, X.; Tian, Y. Characterization of different substitutedcarboxymethyl starch microgels and their interactions with lysozyme. *PLoS ONE* **2014**, *9*, e114634.
19. Zhang, Y.; Chi, C.; Huang, X.; Zou, Q.; Li, X.; Chen, L. Starch-based nanocapsules fabricated through layer-by-layer assembly for oral delivery of protein to lower gastrointestinal tract. *Carbohyd. Polym.* **2017**, *171*, 242–251. [CrossRef]
20. Zhang, Y.; Zhong, S.; Chi, C.; He, Y.; Li, X.; Chen, L.; Miao, S. Tailoring assembly behavior of starches to control insulin release from layer-by-layer assembled colloidal particles. *Int. J Biol. Macromol.* **2020**, *160*, 531–537. [CrossRef]
21. Pooresmaeil, M.; Namazi, H. Developments on carboxymethyl starch-based smart systems as promising drug carriers: A review. *Carbohyd. Polym.* **2021**, *258*, 117654. [CrossRef] [PubMed]
22. Boontawee, R.; Issarachot, O.; Kaewkroek, K.; Wiwattanapatapee, R. Foldable/expandable gastro-retentive films based on starch and chitosan as a carrier for prolonged release of resveratrol. *Curr. Pharm. Biotechnol.* **2022**, *23*, 1009–1018. [CrossRef] [PubMed]
23. Zhang, B.; Pan, Y.; Chen, H.; Liu, T.; Tao, H.; Tian, Y. Stabilization of starch-based microgel-lysozyme complexes using a layer-by-layer assembly technique. *Food Chem.* **2017**, *214*, 213–217. [CrossRef] [PubMed]
24. Quadrado, R.; Fajardo, A. Microparticles based on carboxymethyl starch/chitosan polyelectrolyte complex as vehicles for drug delivery systems. *Arab. J. Chem.* **2020**, *13*, 2183–2194. [CrossRef]
25. Xu, W.; Jin, W.; Wang, Y.; Li, J.; Huang, K.; Shah, B.; Li, B. Effect of physical interactions on structure of lysozyme in presence of three kinds of polysaccharides. *J. Food Sci. Technol.* **2018**, *55*, 3056–3064. [CrossRef] [PubMed]
26. Ghosh, G.; Panicker, L.; Barick, K. Protein nanoparticle electrostatic interaction: Size dependent counterions induced conformational change of hen egg white lysozyme. *Colloids Surf. B* **2014**, *118*, 1–6. [CrossRef] [PubMed]

27. Wawer, J.; Szociński, M.; Olszewski, M.; Piątek, R.; Naczk, M.; Krakowiak, J. Influence of the ionic strength on the amyloid fibrillogenesis of hen egg white lysozyme. *Int. J. Biol. Macromol.* **2019**, *121*, 63–70. [CrossRef]
28. Dyawanapelly, S.; Mehrotra, P.; Ghosh, G.; Jagtap, D.; Dandekar, P.; Jain, R. How the surface functionalized nanoparticles affect conformation and activity of proteins: Exploring through protein-nanoparticle interactions. *Bioorg. Chem.* **2019**, *82*, 17–25. [CrossRef]
29. Pathomthongtaweechai, N.; Muanprasat, C. Potential applications of chitosan-based nanomaterials to surpass the gastrointestinal physiological obstacles and enhance the intestinal drug absorption. *Pharmaceutics* **2021**, *13*, 887. [CrossRef]
30. Hoffmann, S.; Gorzelanny, C.; Moerschbacher, B.; Goycoolea, F. Physicochemical characterization of FRET-labelled chitosan nanocapsules and model degradation studies. *Nanomaterials* **2018**, *8*, 846. [CrossRef]
31. Li, X.; Hetjens, L.; Wolter, N.; Li, H.; Shi, X.; Pich, A. Charge-reversible and biodegradable chitosan-based microgels for lysozyme-triggered release of vancomycin. *J. Adv. Res.* **2023**, *43*, 87–89. [CrossRef] [PubMed]
32. Ceron, A.A.; Nascife, L.; Norte, S.; Costa, S.A.; do Nascimento, J.H.O.; Morisso, F.D.P.; Baruque-Ramos, J.; Oliveira, R.C.; Costa, S.M. Synthesis of chitosan-lysozyme microspheres, physicochemical characterization, enzymatic and antimicrobial activity. *Int. J. Biol. Macromol.* **2021**, *185*, 572–581. [CrossRef] [PubMed]
33. Wang, Y.; Li, S.; Jin, M.; Han, Q.; Liu, S.; Chen, X.; Han, Y. Enhancing the thermo-stability and anti-bacterium activity of lysozyme by immobilization on chitosan nanoparticles. *Int. J. Mol. Sci.* **2020**, *21*, 1635. [CrossRef] [PubMed]

polymers

MDPI

Article

New Physico-Chemical Analysis of Magnesium-Doped Hydroxyapatite in Dextran Matrix Nanocomposites

Daniela Predoi [1,*], Steluta Carmen Ciobanu [1,*], Simona Liliana Iconaru [1], Ştefan Ţălu [2], Liliana Ghegoiu [1], Robert Saraiva Matos [3], Henrique Duarte da Fonseca Filho [4] and Roxana Trusca [5]

[1] National Institute of Materials Physics, Atomistilor Street, No. 405A, P.O. Box MG 07, 077125 Magurele, Romania; simonaiconaru@gmail.com (S.L.I.); ghegoiuliliana@gmail.com (L.G.)
[2] The Directorate of Research, Development and Innovation Management (DMCDI), Technical University of Cluj-Napoca, 15 Constantin Daicoviciu St., 400020 Cluj-Napoca, Romania; stefan.talu@auto.utcluj.ro
[3] Amazonian Materials Group, Physics Department, Federal University of Amapá (UNIFAP), Macapá 68903-419, Amapá, Brazil; robert_fisic@unifap.br
[4] Laboratory of Synthesis of Nanomaterials and Nanoscopy (LSNN), Physics Department, Federal University of Amazonas-UFAM, Manaus 69067-005, Amazonas, Brazil; hdffilho@ufam.edu.br
[5] National Centre for Micro and Nanomaterials, University Politehnica of Bucharest, 060042 Bucharest, Romania; truscaroxana@yahoo.com
* Correspondence: dpredoi@gmail.com (D.P.); ciobanucs@gmail.com (S.C.C.)

Abstract: The new magnesium-doped hydroxyapatite in dextran matrix (10MgHApD) nanocomposites were synthesized using coprecipitation technique. A spherical morphology was observed by scanning electron microscopy (SEM). The X-ray diffraction (XRD) characterization results show hydroxyapatite hexagonal phase formation. The element map scanning during the EDS analysis revealed homogenous distribution of constituent elements of calcium, phosphor, oxygen and magnesium. The presence of dextran in the sample was revealed by Fourier transform infrared (FTIR) spectroscopy. The antimicrobial activity of the 10MgHAPD nanocomposites was assessed by in vitro assays using *Staphylococcus aureus* ATCC 25923, *Pseudomonas aeruginosa* ATCC 27853, *Streptococcus mutans* ATCC 25175, *Porphyromonas gingivalis* ATCC 33277 and *Candida albicans* ATCC 10231 microbial strains. The results of the antimicrobial assays highlighted that the 10MgHApD nanocomposites presented excellent antimicrobial activity against all the tested microorganisms and for all the tested time intervals. Furthermore, the biocompatibility assays determined that the 10MgHApD nanocomposites did not exhibit any toxicity towards Human gingival fibroblast (HGF-1) cells.

Keywords: biomedical applications; dextran; fractal features; hydroxyapatite; magnesium

Citation: Predoi, D.; Ciobanu, S.C.; Iconaru, S.L.; Ţălu, Ş.; Ghegoiu, L.; Matos, R.S.; da Fonseca Filho, H.D.; Trusca, R. New Physico-Chemical Analysis of Magnesium-Doped Hydroxyapatite in Dextran Matrix Nanocomposites. *Polymers* **2024**, *16*, 125. https://doi.org/10.3390/polym16010125

Academic Editors: Maria João Ramalho, Stephanie Andrade and Joana A. Loureiro

Received: 27 November 2023
Revised: 23 December 2023
Accepted: 28 December 2023
Published: 29 December 2023

1. Introduction

Dental caries is an expensive public health problem affecting up to 91% of adults (worldwide). The dental plaque formation usually occurs due to the microbial colonization of oral cavity surfaces [1]. Diet, age, oral hygiene routine, systemic and immune status are important factors that influence the apparition and development of dental plaque [2–4]. The excessive presence of sugar in the daily diet favors the apparition of dental caries [5].

Moreover, a diet rich in sugar favors the development of pathogens in the oral cavity (such as *Streptococcus mutans*), which leads to the formation of acidic and adherent biofilms that are difficult to combat. At the same time, these biofilms lead to a demineralization of dental enamel, thus favoring the apparition of caries [6]. Due to the nanotechnology progress in the medical field [7], materials that have in their composition zinc oxide, silver or magnesium ions were proposed as antibiofilm agents [8,9].

Dextran is a natural polysaccharide, and its use on humans is approved by FDA [10]. The efficiency of dextran-coated iron oxide nanozymes against oral biofilm development, together with their biocompatibility, was shown by Pratap et al. [11]. Moreover, the possibility of using calcium phosphate such as hydroxyapatite (Hap) as a dental filler was evaluated [12]. Studies show that the addition of Hap to dental composite leads to the reinforcement of cement mechanical properties [13]. HAp structure has the ability to allow a large number of substitutions with various ions, including magnesium, zinc, silver, etc. [14–18]. The addition of magnesium ions to the Hap structure is important due to its metabolic role in bone regeneration [19]. The Mg deficiency induces serious health problems, such as osteopenia/osteoporosis, because the lack of Mg in the body disturbs the activity of osteoblast cells [19,20]. For the development of biomaterials based on magnesium-doped hydroxyapatite, various synthesis routes were proposed: hydrothermal [21], precipitation [22], sol–gel [23], mechanochemical–hydrothermal [24], wet chemical [25], and microwave [26,27]. These approaches allow us to obtain nanomaterials with desired properties, such as morphology, dimension, biological properties, etc.

Polymeric nanoparticles have been used in many biomedical applications, such as drug delivery, tissue engineering, dentistry and imaging [28]. Usually, polymeric nanoparticles can be developed using various natural and/or synthetic polymers (e.g., polyethylene glycol (PEG), polylactic acid (PLA), poly(lactic-co-glycolic acid) (PLGA) and/or gelatin, alginate, albumin, chitosan, dextran etc.) [28]. Previous studies showed that polymeric nanoparticles exhibit a great potential for uses in drug delivery applications due to their biocompatibility and stability. Furthermore, polymeric nanoparticles were used for drug delivery and imaging applications due to their unique properties, such as high surface area to volume ratio and tunable size [29]. For example, in the work conducted by El-Meliegy et al. [30], the synthesis of a novel composite scaffold based on hydroxyapatite in dextran/chitosan polymeric matrix was reported. Their results highlighted that the presence of Hap nanoparticles in the polymeric matrix enhances the physicochemical properties of the obtained composite scaffolds [30]. In this context, Hap and dextran have been used for the development of polymeric nanoparticles with potential uses in various biomedical applications, such as drug delivery, tissue engineering, dentistry and imaging [31].

Shoba et al., in their work entitled "3D nano-bilayered spatially and functionally graded scaffold impregnated bromelain conjugated magnesium-doped hydroxyapatite nanoparticle for periodontal regeneration" revealed that scaffolds containing magnesium-doped hydroxyapatite possess an improved antibacterial activity and biocompatibility proving that such materials can be promising candidates for uses in periodontal tissue regeneration [32]. Moreover, in our previous antimicrobial study conducted on magnesium-doped hydroxyapatite suspension obtained by an adapted coprecipitation method, we underlined the efficacy of this biomaterial against *P. aeruginosa*, *S. aureus*, and *C. albicans* strains [33]. Also, we noticed that the antimicrobial activity of magnesium-doped hydroxyapatite is strongly correlated with the Mg concentration found in the samples. Therefore, a more efficient antimicrobial activity against Gram-positive strains (*B. subtilis* and *S. aureus*) was noticed in the case of an increased Mg concentration in the hydroxyapatite/chitosan composite samples when compared to those against Gram-negative strains [34]. Also, a study conducted on chitosan-coated magnesium-doped hydroxyapatite coatings highlighted the in vitro biocompatibility of the studied samples with the human fibroblast cell [34].

In the present study, we proposed the development of a novel biomaterial based on magnesium-doped hydroxyapatite in dextran matrix (10MgHApD). The research focused on the physicochemical characterization and antimicrobial evaluation of 10MgHApD nanocomposites with high potential to be applied in the dental field.

2. Materials and Methods

2.1. Materials

The synthesis of magnesium-doped hydroxyapatite ($Ca_{10-x}Mg_x(PO_4)_6(OH)_2$, $x_{Mg} = 0.1$ and [Ca + Mg]/P ratio equal with 1.67) in dextran matrix was conducted using ammonium hydrogen phosphate ($(NH_4)_2HPO_4$), calcium nitrate tetrahydrate ($Ca(NO_3)_2 \cdot 4H_2O$), magnesium nitrate hexahydrate ($Mg(NO_3)_2{\cdot}6H_2O$) purchased from Sigma-Aldrich (St. Louis, MO, USA) with a purity of 99.97%. Dextran ($H(C_6H_{10}O_5)_n$, Mr ~ 40,000) was also purchased from Sigma-Aldrich (St. Louis, MO, USA). In the synthesis double-distilled water was used.

2.2. Synthesis of Magnesium-Doped Hydroxyapatite in Dextran Matrix Nanocomposites

Magnesium-doped hydroxyapatite in dextran (10MgHApD) matrix nanocomposites were obtained by the coprecipitation technique [35]. In order to achieve this purpose, the atomic ratio of (Ca + Mg)/P was 1.67. A solution (300 mL) containing $(NH_4)_2{\cdot}HPO_4$ and 10% $H(C_6H_{10}O_5)_xOH$ (10 g) was stirred for 30 min at 40 °C. in air. A similar procedure was followed for the solutions (300 mL) containing $Ca(NO_3)_2{\cdot}4H_2O$ and $Mg(NO_3)_2{\cdot}6H_2O$. The solutions containing calcium and magnesium were dropped into the solution with dextran. The pH of synthesis was kept constant at 11 by adding NH_3. After 5 h of stirring after the end of dripping, the resulting suspension was centrifuged and redispersed in 10% solution of dextran (10 g at 100 mL of double-distilled water). The procedure was rehearsed five times. The resulting precipitate after the last centrifugation was redispersed in a 10% dextran solution and stirred for 12 h at 60 °C in air. The final suspension was centrifugated, and the last precipitate was dried at 40 °C (in air) and called 10MgHApD and afterwards analyzed in the present study.

2.3. Characterization Methods

2.3.1. Scanning Electron Microscopy

A scanning electron microscope (FEI Quanta Inspect F, FEI Company, Hillsboro, Oregon, United States) equipped with an energy-dispersive X-ray (EDS) attachment was used to study the morphology of 10MgHApD nanocomposites.

2.3.2. X-ray Diffraction

X-ray diffraction (XRD) was used to examine the magnesium-doped hydroxyapatite (10MgHAp), 10MgHApD nanocomposites and dextran. The equipment for XRD analysis was a Bruker D8 Advance diffractometer with CuKα (λ = 1.5418 Å) radiation (Bruker, Karlsruhe, Germany), equipped with a high-efficiency LynxEye™ 1D linear detector. The patterns were achieved in the 2θ range 20–60°. The step size was 0.02° and the dwell time was 5 s.

2.3.3. Fourier Transform Infrared Spectroscopy

The presence of functional groups was established by Fourier transform infrared (FTIR) spectroscopy in attenuated total reflectance (ATR) mode. A Perkin Elmer Spectrum BX II spectrometer (Perkin Elmer, Waltham, MA, USA) equipped with a Pike-MIRacle ATR head with diamond-ZnSe crystal plate, having a diameter of 1.8 mm (Pike Technologies, Madison, WI, USA) was used. The spectra were acquired in the 450–3800 cm^{-1} spectral range. The resolution was 4 cm^{-1} and represented the average of 32 individual scans.

2.3.4. Atomic Force Microscopy (AFM)

Detailed information regarding the morphology of the composites was achieved by atomic force microscopy (AFM) technique. For this purpose, the composite nanocomposites were pressed into pellets, and the surface topography of the pellet was studied using an instrument NT-MDT NTEGRA Probe Nano Laboratory (NT-MDT, Moscow, Russia). The measurements were performed at room temperature and in atmospheric conditions using semi-contact mode. Information about the morphology of the samples was obtained by recording AFM surface topographies on surface areas of 10 × 10, 5 × 5 and 3 × 3 μm^2 using a silicon NT-MDT NSG01 cantilever (NT-MDT, Moscow, Russia) with a 35 nm gold layer. Information about the roughness of the samples was also obtained by calculating the roughness parameter R_{RMS}. The recorded AFM data were processed using the 2.59 version of Gwyddion software (Department of Nanometrology, Czech Metrology Institute, Brno, Czech Republic) [36].

2.3.5. Monofractal and Multifractal Analysis

Herein, the monofractal parameters were performed evaluate the surface microtexture spatial complexity. The fractal dimension (*FD*) was computed using the Mandelbrot box-counting method [37], and the Hurts coefficient was obtained by applying the formula $H = (3 - FD)$ [37]. The Fractal succolarity (F_S) was determined using Equation (1) [38], where $T(k)$ represents the count of boxes of uniform sizes $T(n)$, $P_O(T(k))$ denotes the occupation percentage within each box, *PR* represents the occupation pressure, and p_c signifies the centroid's position (x,y) representing the applied pressure on the corresponding box. This equation provides a quantifiable measure of F_S, offering insights into the structural complexity and hole distribution across the analyzed surface [38].

$$F_s(T(k), dir) = \frac{\sum_{k=1}^{n} P_O(T(k)) \cdot PR(T(k), p_c)}{\sum_{k=1}^{n} PR(T(k), p_c)} \quad (1)$$

The topographic entropy was computed using Shannon entropy [39], as defined in Equation (2) [40]. In this equation, the term p_{ij} signifies the probability of pixels exhibiting discrepancies or not within the specified height range of the analyzed universe. The calculation of topographic entropy through Shannon entropy allows for a quantitative measure of the information content associated with the variability in pixel values across the analyzed topography [40].

$$TE = -\sum_{i=1}^{N} \sum_{j=1}^{N} p_{ij} \cdot log p_{ij} \quad (2)$$

On the other hand, the multifractal theory is a mathematical framework employed to characterize intricate objects or systems showcasing substantial variations in their properties across various scales [38]. Its application is prevalent in the analysis of 3D spatial patterns, time series, and diverse complex phenomena. Serving as an extension of fractal theory, which centers on objects with self-similarity—patterns recurring at different scales—multifractal theory broadens this perspective to encapsulate the multifaceted variability observed in complex systems [41]. We use the partition function (Equation (3)) for expressing the mass exponent (τ_q), where $p_i(\varepsilon) = r_i(\varepsilon) / \sum_{k=1}^{N(\varepsilon)} r_k(\varepsilon)$ represents the probability of occupancy of the i-th cell within a resolution grid ε. The $r_i(\varepsilon)$ term denotes the cumulative fluctuation of the height around the mean value within the i-th square [37].

$$Z(q, \varepsilon) = \sum_{i=1}^{N(\varepsilon)} p_i^q(\varepsilon) \sim \varepsilon^{\tau(q)} \quad (3)$$

The multifractal spectrum elucidates the variation in complexity associated with different exponents, such as Hölder exponent ($\alpha(q)$), and this relationship is expressed by Equation (4) [37].

$$f(\alpha(q)) = q \cdot \alpha(q) - \tau(q) \quad (4)$$

The mass exponent curve defines a relationship between moments of order (q) and generalized dimensions (D_q) and can be expressed according to Equation (5) [37].

$$D_q = \frac{\tau(q)}{(q-1)} \quad (5)$$

where $\alpha(q) = \frac{d\tau(q)}{dq}$ defines the connection between the ($\alpha(q)$) and the mass exponent ($\tau(q)$) for a given value of q.

2.3.6. In Vitro Antimicrobial Assays

The in vitro antimicrobial properties of 10MgHApD nanocomposites were studied with the aid of *Staphylococcus aureus* ATCC 25923, *Pseudomonas aeruginosa* ATCC 27853, *Streptococcus mutans* ATCC 25175, *Porphyromonas gingivalis* ATCC 33277 and *Candida albicans* ATCC 10231 microbial strains. The experiments were performed as previously described in [42]. For this purpose, 10MgHApD nanocomposites, as well as 10MgHAp and HAp nanoparticles, were exposed to 1.5 mL of microbial suspension of a standardized density equal to 5×10^6 CFU/mL (colony forming units/mL). *S. aureus*, *P. aeruginosa* and *C. albicans* microbial suspensions of a density of approximately 5×10^6 CFU/mL were prepared from 15 to 18 h. Solid cultures were grown in tryptone soy agar (TSA). Afterwards, the microbial suspensions were inoculated onto Muller Hinton agar (MHA) plates by swabbing. Afterwards, the suspensions were collected at different time intervals (24, 48 and 72 h) and incubated on Luria–Bertani (LB) agar medium for 24 h at 37 °C. *P. gingivalis* was grown on Brucella agar plates containing a blood agar base, yeast extract, glucose (4.5%) under anaerobic conditions (80% N_2, 10% H_2, 10% CO_2). The colonies were harvested and resuspended in a Brain Heart Infusion (BHI) broth (Difco). *S. mutans* were cultured from single colonies in BHI (Difco) in an aerobic atmosphere with 5% CO_2. The colonies were harvested and resuspended in BHI broth. The density of the microbial suspensions was adjusted by adding either *P. gingivalis* and *S. mutans* suspended in BHI broth or just pure BHI broth. The microbial suspensions were prepared as described above and then incubated for 24, 48 and 72 h, respectively with the 10MgHApD nanocomposites and 10MgHAp and HAp nanoparticles. As a positive control, a free microbial culture was assessed at the same time intervals.

The microbial suspensions were prepared in phosphate-buffered saline (PBS) and then incubated for 24, 48 and 72 h, respectively, with the 10MgHApD nanocomposites. The values of the CFU/mL were determined. The experiments were performed in triplicate and the data were presented as mean ± standard deviation (SD). The statistical analysis was performed using the ANOVA single-factor test.

2.3.7. In Vitro Biocompatibility Assay

The biocompatibility of the 10MgHApD nanocomposites was studied using a Human gingival fibroblasts (HGF-1) cell line. For this purpose, the cells were cultured using Dulbecco's Modified Eagle's Medium enriched with heat-inactivated fetal bovine at 37 °C in an atmosphere containing 95% air and 5% CO_2. The HGF-1 cells were seeded in culture plates and were allowed to adhere for 24 h. Afterwards, the cultured cells were incubated with 10MgHApD nanocomposites for 24, 48 and 72 h. An untreated cell culture was used as control. The cell viability of the HGF-1 cells was determined with the aid of the reduction assay MTT [3-(4,5dimethylthiazolyl)-2,5-diphenyltetrazolium bromide]. To achieve this, the cells were seeded in 96-well plates (5×10^4 cells/mL), incubated for 24 h, and then treated with the 10MgHApD nanocomposites. After 24, 48 and 72 h of incubation, the cells were washed using phosphate buffer saline (PBS) and incubated with 0.5 mg/mL MTT solution for 4 h. The HGF-1 cell viability was quantified by determining the optical density of the medium at 595 nm with the aid of a TECAN spectrophotometer. The percentage of the HGF-1 viable cells was quantified by rapport to the control sample, which was considered to have a viability of 100%.

3. Results and Discussion

3.1. X-ray Diffraction

To assess the magnesium incorporated in hydroxyapatite coated with dextran, XRD studies were conducted. In Figure 1, XRD patterns of the dextran, magnesium-doped hydroxyapatite in dextran matrix (10MgHApD) and magnesium-doped hydroxyapatite (10MgHAp) nanocomposites are presented. The reference hexagonal patterns of hydroxyapatite (JCPDS no. 09-0432) and dextran (JCPDS no. 063-1501) are also shown.

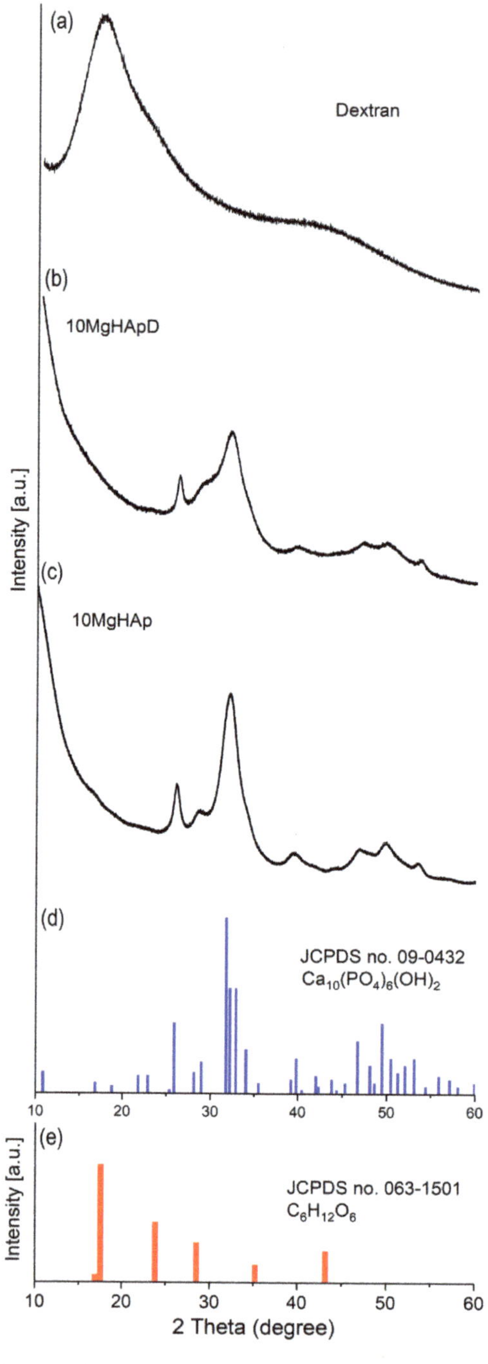

Figure 1. XRD spectrum of dextran (**a**), 10MgHApD nanocomposites (**b**) and 10MgHAp (**c**). The lines of the ICDD-PDF#9-432 reference file of hexagonal hydroxyapatite (**d**) and the ICDD-PDF#063-1501 reference file of dextran (**e**).

The diffraction pattern of 10MgHAp and 10MgHApD was similar to that of the reference hexagonal HAp pattern (ICDD-PDF#09-432). The diffraction pattern of both samples highlights the fact that the particles have nanometric dimensions (20.1 ± 2 nm for 10MgHAp and 13.6 ± 4 nm in the case of the 10MgHApD sample). It is observed that the diffraction pattern of 10MgHApD shows wider peaks than in the case of the 10MgHAp sample [43]. This behavior could also be caused by the presence of dextran. Due to its presence, dextran can lead to a decrease in crystallinity [44].

3.2. Scanning Electron Microscopy

The morphology of the as-synthesized 10MgHApD nanocomposite material is shown in Figure 2. Figure 2a represents the SEM micrograph at low resolution, while Figure 2b shows the SEM micrograph of 10MgHApD nanocomposites at high resolution. The micrograph shown in Figure 2b exhibits nanometric particles with a spherical shape. The average particle size calculated after measuring approximately 500 particles was 14.5 ± 2 nm (Figure 2d). The inset of Figure 2d presents the micrograph on which approximately 500 particles are numbered. Typical EDS patterns establish six prominent peaks which confirm the presence of magnesium, calcium, phosphor, oxygen and carbon, respectively (Figure 2c).

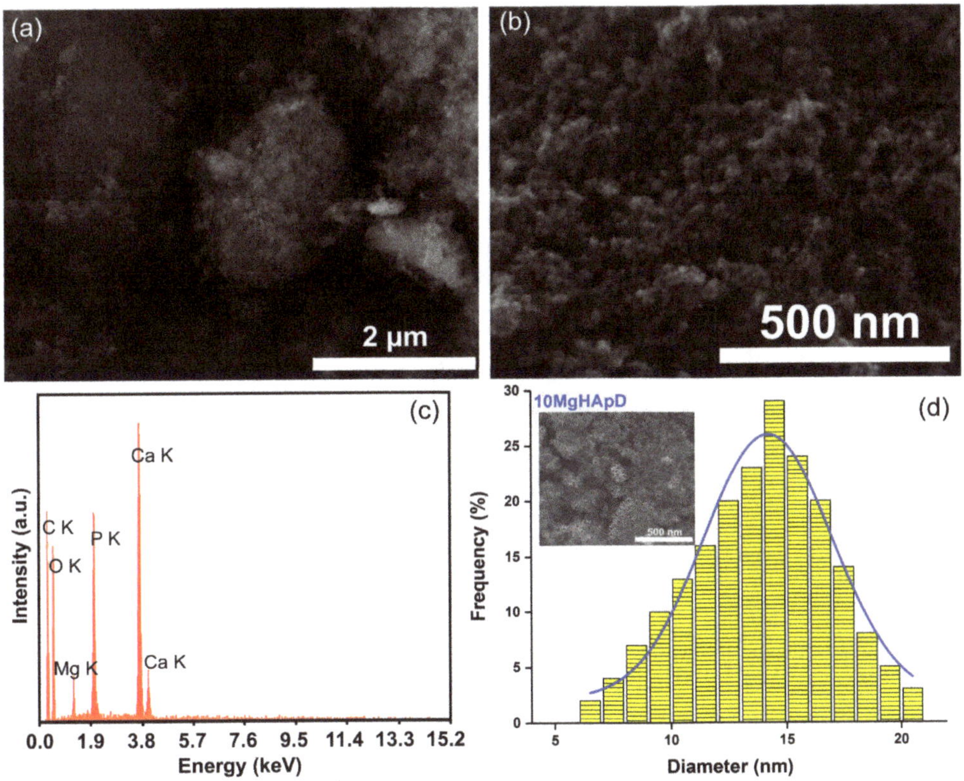

Figure 2. SEM micrograph at low (**a**) and high (**b**) resolution of 10MgHApD nanocomposites, EDS spectrum of 10MgHApD nanocomposites (**c**) and average particle size (**d**).

The element map scanning during the EDS analysis was conducted from the region revealed in Figure 2a. The results regarding the element map scanning during the EDS analysis of 10MgHApD nanocomposites are exhibited in Figure 3. The homogenous distribution of constituent elements Ca, P, O and Mg is observed. The C element is not presented because it is not conclusive (the C contribution has two sources, the carbon band and the dextran from the synthesized sample).

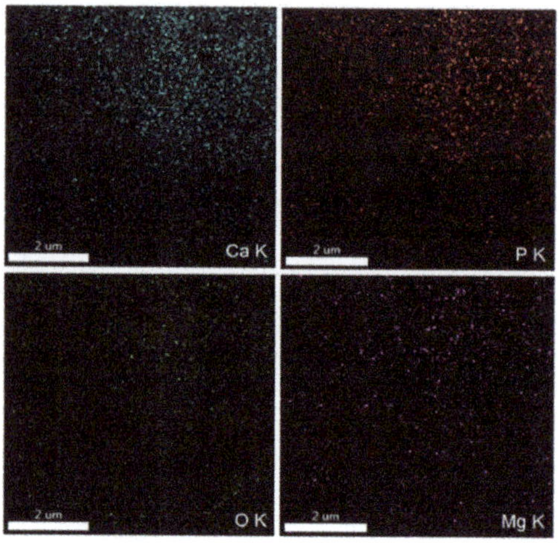

Figure 3. The EDS mapping of Ca, P, O and Mg in 10MgHApD nanocomposites.

3.3. Atomic Force Microscopy

The morphology of the 10MgHApD nanocomposites was further investigated using AFM technique. Information about the sample's morphology was obtained by recording AFM topographies on surface areas of 10 × 10, 5 × 5 and 3 × 3 μm^2 of the pellet surface topography obtained from the 10MgHApD nanocomposites nanocomposites. The results of the AFM studies are depicted in Figure 4a–f.

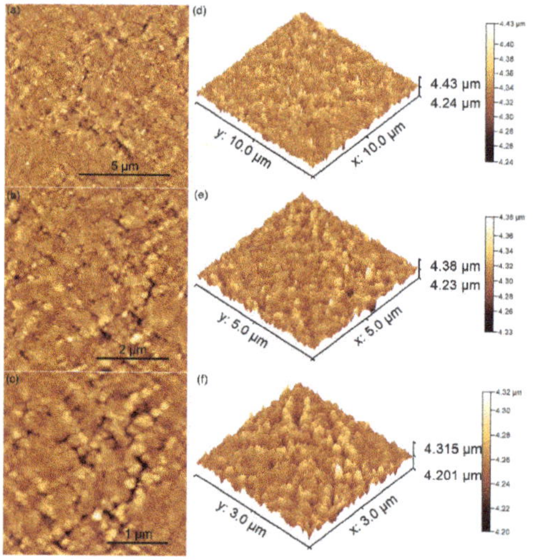

Figure 4. 2D AFM images of 10MgHApD pellet's topography recorded on an area of 10 × 10 μm^2 (**a**), 5 × 5 μm^2 (**b**), 3 × 3 μm^2 (**c**) and their corresponding 3D representations (**d–f**).

The AFM topography reveals a surface with slight irregularities resulting from the process of obtaining the pallets. The homogeneous distribution of the agglomerates formed by nanoparticles was also observed. The AFM topography obtained on the surface area of 3 × 3 μm^2 highlighted that the nanoparticles form agglomerates. In addition, the results suggested that the agglomeration of particles exhibited nanometric sizes. Both the 2D surface topography of the three investigated areas, as well as their 3D representation, emphasized that the nanocomposites present a uniform and homogenous morphology. Slight irregularities of the pellet surface could be observed. The roughness parameters, R_{RMS}, calculated for the areas of 10 × 10, 5 × 5 and 3 × 3 μm^2 of the pellet surface topography obtained from the 10MgHApD nanocomposites were 15.36, 13.56 and 10.91 nm, respectively. It can be seen that the obtained values are very close. The values of the roughness parameter suggest a homogeneous distribution of the agglomerates formed by nanoparticles on the surface of the pallets.

The investigation of coating surfaces using AFM has become a cornerstone for assessing nanoscale topographic changes. With its capability to map both topography and mechanical properties, AFM plays a pivotal role in advancing nanotechnology, characterizing biomaterials, and optimizing devices [37]. In this regard, we assessed the morphology and microtexture of 10MgHApD nanocomposites across various scales. The comprehensive view of the overall morphology of the 10MgHApD pellet's topography recorded on an area of 10 × 10 μm^2 and its 3D spatial configurations is depicted in Figure 5. The 3D topographic map covering dimensions of 10 × 10 μm^2 illustrates a relatively smooth surface with certain irregularities formed randomly along the surface following the process of obtaining the pallets. This behavior that appears on the surface of the pellets after the pressing process could be beneficial, contributing to the improvement of the adhesive properties of the surface [45] in order to develop applications in the biomedical field.

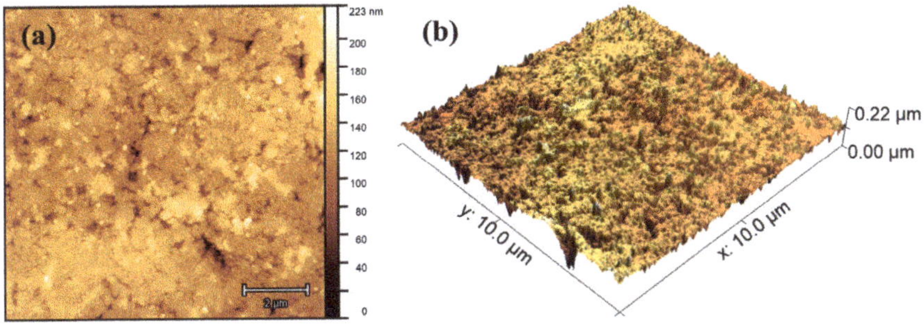

Figure 5. (**a**) 2D and (**b**) 3D representation of the atomic force microscopy (AFM) surface topography image of the 10MgHApD sample, with 10 × 10 μm scanned area.

To gain deeper insights into the 3D spatial configuration of the vertical growth profile of the investigated surface, we conducted a detailed analysis of its microtexture. This involved utilizing 3D AFM topographic maps with dimensions of 3 × 3 μm^2, as shown in Figure 6a,b. As it can be seen, the pellet surface appears nearly homogeneous and uniform, displaying a finely tuned vertical profile indicative of low topographic roughness. Additionally, the 10MgHApD particles are evenly distributed across the surface, showcasing sizes ranging from 50 to 200 nm. The average roughness (*Sa*) was computed to be 13 ± 0.2 nm, which is a markedly lower value than other values reported for different coatings, e.g., 980 nm [46] and 47 nm [47]. Such behavior was also observed for the other height ISO-based parameters: maximum peak height (*Sp*), maximum pit depth (*Sv*), and maximum height (*Sz*) (Table 1). Notably, the low roughness of the pellet surface suggests that the incorporation of MgHAp into the dextran matrix has a softening effect on the vertical profile of the investigated surface, with the particles being shaped and embedded by the polymer. Figure 6c shows the shape of the height distribution associated with the vertical profile of the pellet surface and its Abbot Firestone curve [39,48]. As observed, the height distribution of the investigated surface is centralized, a characteristic supported by the kurtosis value *Rku* ~3, signifying an almost perfectly platykurtic pattern [49]. Furthermore, the distribution is almost symmetric, which is supported by the skewness value *Rsk* ~0 [50] (Table 1). Additionally, the high quality of the investigated surface is also illustrated by the Abbot Firestone curve, characterized by its typical S shape. In this regard, the curve attains its peak value more rapidly at a specific relative height *z* value (in μm), providing evidence that the height distribution follows an almost normal behavior.

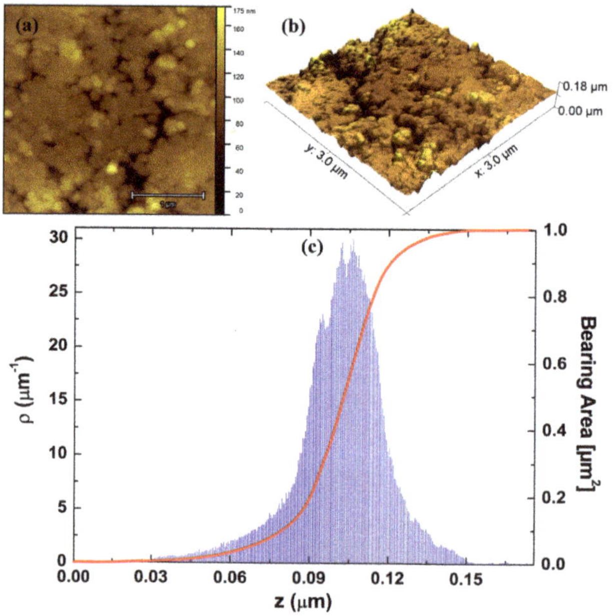

Figure 6. (**a**) 2D and (**b**) 3D AFM micrographs of 10MgHApD nanocomposites. (**c**) Height histogram and Abbot–Firestone curves from the respective image.

Table 1. Height parameters of 10MgHApD sample.

Sample	Parameters					
	Sa (nm)	*Ssk*	*Sku*	*Sp* (nm)	*Sv* (nm)	*Sz* (nm)
10MgHApD	13.0 ± 0.2	−0.7 ± 0.1	3.1 ± 0.3	68.2 ± 4.7	102.8 ± 0.8	180.9 ± 15.0

Minkowski Functionals (MFs) serve as geometric measurements employed to characterize and quantify the topological and morphological properties of geometric sets [51,52]. Their primary applications lie in morphological analysis, particularly in contexts such as image analysis and the study of porous materials [37]. The Minkowski volume (*V*) limit of the investigated surface, as depicted in Figure 7a, exhibits a characteristic S-like shape and approaches its minimum rapidly. This behavior indicates that the volume of material below a threshold is low, confirming the surface's exceptionally smooth vertical profile. The Boundary Minkowski (*S*) (Figure 7b) depicts a higher maximum value in 0.06, with a distribution of points similar to a normal curve. This suggests that the surface boundary of the sample is intricate, containing distinctive features in its contour. Finally, the Minkowski connectivity (χ) exhibits a typical minimum negative and a sharp positive maximum value, as shown in Figure 7c. This suggests that the investigated surface has homogeneous surface percolation. This behavior may be associated with a regular distribution of gaps along the pellet surface. Notably, these aspects of the nanocomposite morphology align with the observations made about its morphological and microtextural properties (Figure 6a,b).

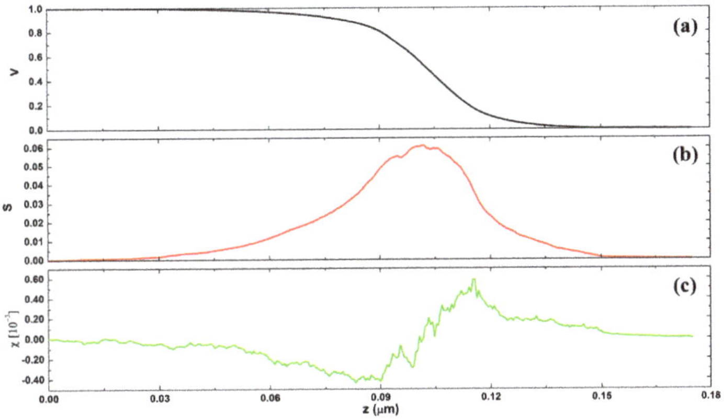

Figure 7. The MFs functionals of the surface of 10MgHApD obtained from AFM image for (**a**) Minkowski volume, (**b**) Minkowski boundary, and (**c**) Minkowski connectivity.

3.4. Monofractal Analysis

Analyzing the spatial complexity of surfaces using monofractal mathematics is crucial for deciphering intricate 3D patterns of polymer surfaces in nanoscale [53]. This approach offers a profound understanding of shapes, facilitating the optimization of industrial processes, material design, and environmental modeling. By unraveling the underlying geometry, monofractal mathematics emerges as a valuable tool for enhancing efficiency and fostering innovation across various domains, e.g., biological [54], thin films [55], and biomedical [56]. In our monofractal approach, we employed the Mandelbrot box-counting method [37] to obtain the fractal dimension, whose fit is shown in Figure 8. Remarkably, the smoothness of the nanocomposite surface is associated with relatively low spatial complexity (*FD* < 2.5) (Table 2). Despite this, the surface exhibits an *FD* value > 2, indicating the presence of topographic irregularities that give rise to long-range correlations. The relatively low spatial complexity of the nanocomposites is also linked to the presence of low dominant spatial frequencies, as indicated by the high value of the Hurst coefficient (*H* > 0.5) (Table 2). Thus, the arrangement of the topographic heights in the pellet surface microtexture promotes the development of a surface with low roughness and 3D spatial complexity characterized by low spatial frequencies in the topographic profile.

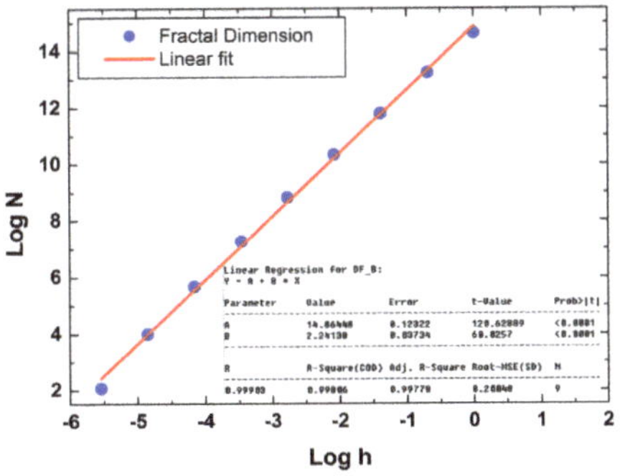

Figure 8. Representative fractal dimension determined using a cube counting method of 10MgHApD nanocomposites, obtained from AFM images shown in Figure 2.

Table 2. Measures of monofractal parameters of the surface of 10MgHApD nanocomposites.

Sample	Parameters			
	FD	*H*	*FS*	*E*
10MgHApD	2.243 ± 0.007	0.757 ± 0.007	0.361 ± 0.160	0.897 ± 0.006

Furthermore, we observed that the pellet surface is not highly porous ($FS < 0.5$), a characteristic that is not directly associated with the porous nature of the 10MgHApD structure. According to Ţălu et al. [38], a surface with ideal surface percolation has to display $FS = 0.5$, indicating a highly uniform hole distribution across the surface. In contrast, a more percolable surface is expected to exhibit $FS > 0.5$. In this regard, it is evident that the low roughness of the nanocomposites facilitated the development of a less porous surface, primarily attributed to the presence of dextran polymer in its structure. In addition, the average topographic entropy (E), a parameter linked to the uniformity of 3D spatial patterns in the distribution of topographic heights, was calculated to be 0.897 ± 0.006. This behavior implies the presence of more uniform spatial patterns ($E < 0.5$) than nonuniform ones ($E > 0.5$), indicating high surface quality and resistance of the coating. Moreover, surfaces with an E value ~1 tend to demonstrate homogeneous surface adhesion [40,57], which is advantageous for biological applications, including cell anchoring. In conclusion, our findings suggest that the 10MgHApD nanocomposites display monofractal behavior characterized by low spatial complexity, low surface percolation, and high topographic uniformity, largely attributed to the low surface roughness.

3.5. Multifractal Analysis

We conducted an in-depth examination of the pellet surface dynamics utilizing a multifractal methodology. In fact, it is recognized that monofractals have limitations as they solely rely on a single fractal dimension [37]. However, the analysis of multifractal behavior of a multifractal sample is crucial because it can address inhomogeneous surface complexity [37,57]. While the monofractal approach provides a global view of spatial complexity, multifractal analysis allows for a more refined description of local and regional variations using multiple scaling exponents [58,59]. This approach identifies specific scales of complexity, revealing nonlinear details and heterogeneities critical for full sample characterization, especially in systems where properties vary significantly at different spatial scales, e.g., roughness and surface isotropy. The analysis of 3D spatial patterns in the surface of 10MgHApD nanocomposites reveals unique multifractal properties, indicating significant structural complexity, as displayed in Figure 9. The multifractal parameters computed from the multifractal spectra of 10MgHApD nanocomposites are presented in Table 3. The multifractal spectra curve (Figure 9a), characterized by a downward concavity, serves as a hallmark of multifractal systems, with the maximum point corresponding to the Hausdorff dimension. This dimension offers insights into the three-dimensional complexity of the coating, highlighting structural variations at different scales. The spectra width ($\Delta\alpha = \alpha_{max} - \alpha_{min}$), calculated as 1.078, indicates a broad range of structural sizes within the coating. Higher values suggest substantial diversity in structural scales, contributing to the overall complexity of the material. Figure 9b illustrates a nonconstant relationship between the generalized dimensions Dq and the moments q, indicating high multifractality. This suggests that different regions of the nanocomposites exhibit varying degrees of fractal complexity. The mass exponent $\tau(q)$ versus q curve analysis (Figure 9a) confirms the multifractal nature of the nanocomposites. The nonlinear behavior of this curve underscores nontrivial variations in mass distribution across scales, adding an extra layer to understanding structural complexity. The fractal dimension difference ($\Delta f = f(\alpha_{max}) - f(\alpha_{min})$) was found to be 1.985 and emphasizes the disparity in fractal dimensions in different parts of the coating, highlighting marked variations in structural complexity. Notably, the low roughness of the nanocomposites promoted the formation of a microtexture with unique vertical growth dynamics marked by a strongly multifractal behavior. Comprehending the multifractality mechanism of the nanocomposites offers valuable insights for the design of advanced materials, bioengineering, and other fields where 3D surface architecture plays a pivotal role. Hence, these findings bear substantial implications, especially in the nanomaterial utilization in biomedical applications which demand meticulous control over structural properties, particularly in relation to their surface characteristics. Furthermore, the conjunction of multifractal behavior with microtextural properties, such as low percolation and high topographic uniformity in the nanocomposites, indicates that surface characteristics, including adhesion, friction, and resistance, are governed by distinctive and advantageous surface dynamics. This feature is assigned as beneficial for its application in the biological field.

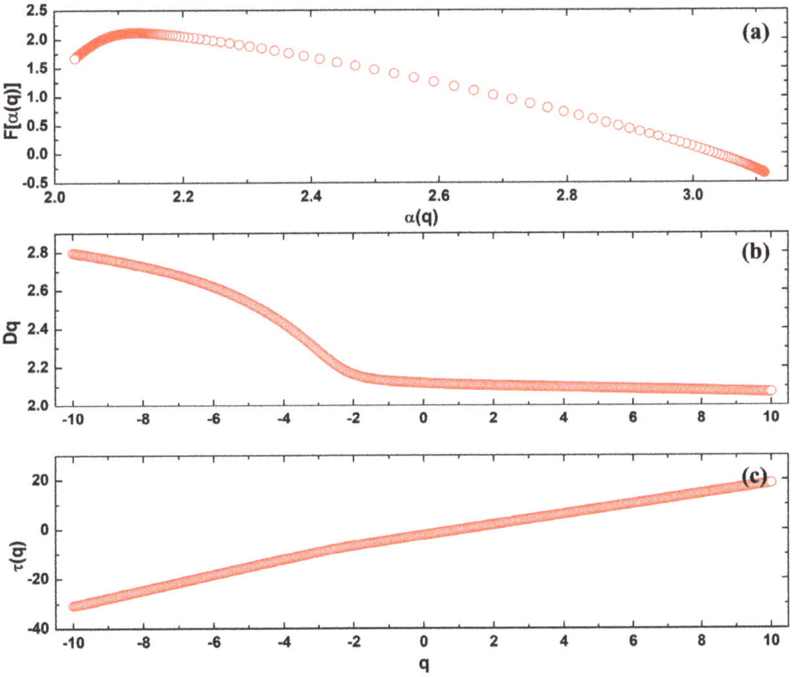

Figure 9. (**a**) Multifractal spectra (*f(α)* versus α), (**b**) generalized dimensions D_q, and (**c**) mass exponent *τ(q)*, as a function of the order of moments computed for 10MgHApD nanocomposites obtained using their 3D AFM topographical map.

Table 3. Multifractal parameters computed from the multifractal spectra of 10MgHApD nanocomposites.

Sample	Parameters					
	α_{max}	α_{min}	Δa	$f(\alpha_{max})$	$f(\alpha_{min})$	Δf
10MgHApD	3.109	2.031	1.078	−0.310	1.675	1.985

3.6. Fourier Transform Infrared Spectroscopy

The FTIR-ATR spectra of 10MgHApD nanocomposites, magnesium-doped hydroxyapatite (10MgHAp) and dextran powder as reference spectrum are presented comparatively in Figure 10a–i. According to previous studies [33,60], the vibration bands distinctive to HAp were observed together with the characteristic bands of dextran (Figure 10).

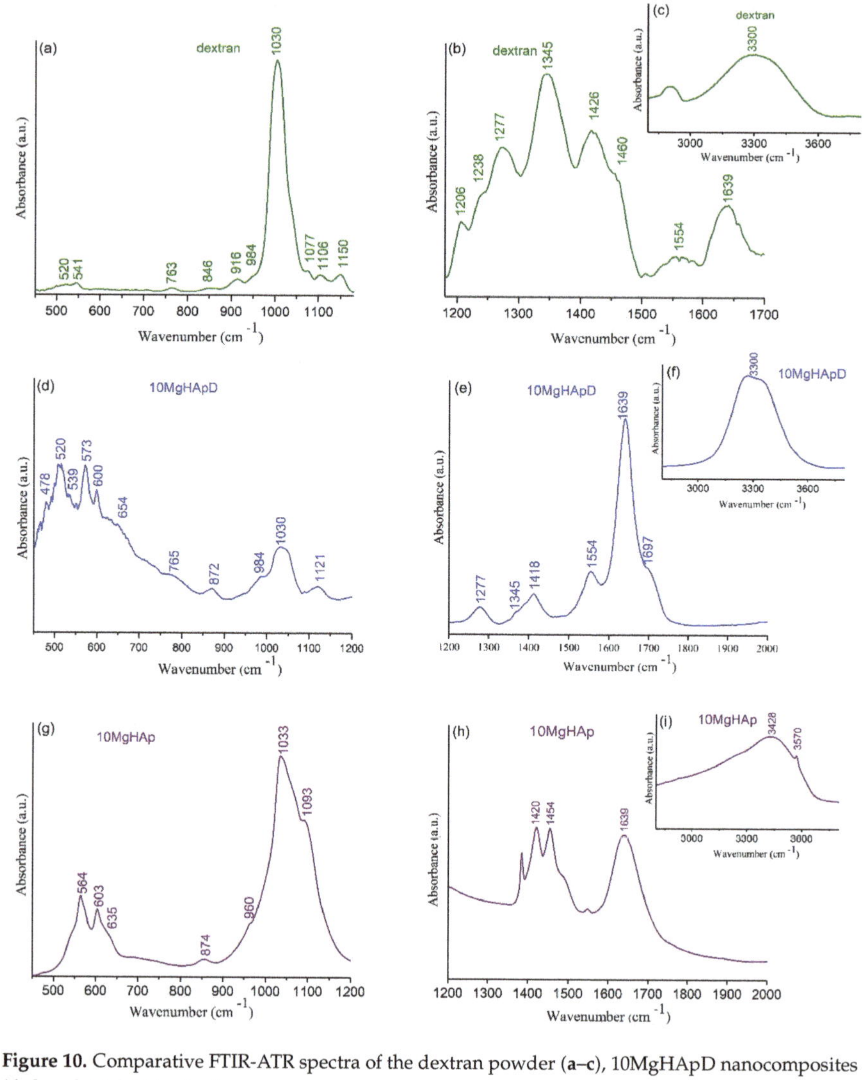

Figure 10. Comparative FTIR-ATR spectra of the dextran powder (**a–c**), 10MgHApD nanocomposites (**d–f**) and 10MgHAp (**g–i**).

For the magnesium-doped hydroxyapatite (10MgHAp) sample, the results of the FTIR studies are presented in Figure 10g–i. The dominant maxima observed in the FTIR spectra could be attributed mainly to the presence of the characteristic vibration of phosphate, hydroxyl groups (from HAp structure) and adsorbed water molecules. The maxima found between 500 and 620 cm^{-1} are characteristic of the ν_4 triply degenerated asymmetric stretching of phosphate groups [33]. Also, the maxima observed at 960 cm^{-1} are specific to ν_1 symmetric stretching of the phosphate group. The triply degenerated asymmetric stretching ν_3 maxima were observed between 980 and 1100 cm^{-1} spectral domain [33]. Moreover, the liberation and stretching vibration of hydroxyl groups are easily observed at 635 cm^{-1} and at around 3570 cm^{-1} [33]. In the FTIR spectra, we could also notice a maximum at around 1420 cm^{-1} and at 1454 cm^{-1}, which is usually attributed to the vibration of $CO_3{}^{2-}$ groups [33]. The presence of adsorbed water molecules in the studied sample is confirmed by the maxima located at around 1630 cm^{-1} (specific to the bending vibrations) and by the one observed at around 3428 cm^{-1} (specific to the stretching vibrations) of water molecules [33]. The presence of these intense and wide vibration bands confirms that the 10MgHAp samples are strongly hydrated.

The peak identified at about 465–470 cm^{-1} in FTIR spectra of 10MgHApD (Figure 10d) was assigned to characteristic stretching modes of O-H bands [61]. The bands at about 520–573 and 600 cm^{-1} (Figure 10d) were associated with ν_4 symmetric P-O stretching vibration of the PO_4^{3-} group [61,62]. Moreover, the peaks at 765, 846–872 and 984 cm^{-1} identified in Figure 10a,d represent the characteristic band sorption of dextran [63]. The formation of HAp (Figure 10d) is given by the observation of the broad band centered at 960–1121 cm^{-1} assigned to P-O asymmetric stretching of PO_4^{3-} [61,64]. On the other hand, the peaks that can be observed in 10MgHApD and dextran spectra in the range 1400–1426 cm^{-1} may be assigned to the dextran molecule ν(C–H) and δ(C–H) vibrational modes [63]. In agreement with recent studies [61,63], the presence of peaks at around 872 cm^{-1} (ν_2 bending vibrations) and 1418–1554 cm^{-1} (ν_3 asymmetric stretching vibrations) in the FTIR spectra of 10MgHApD was due to CO_3^{2-} groups (Figure 10e). The stronger peaks observed in the range of 846–1077 cm^{-1} in the FTIR spectra of dextran (Figure 10a) were also identified in the FTIR spectra of 10MgHApD and could be assigned to the stretching vibration of C–O–C [65]. Moreover, the bands in the range 1418–1460 cm^{-1} that can be assigned to C–O–H deformation vibration with contributions of O–C–O symmetric stretching vibration of the carboxylate group [65,66] were present in the FTIR spectra of the two analyzed samples (Figure 10b,e). It can be seen that both analyzed samples (10MgHApD and dextran) were strongly hydrated, as revealed by the intense bands at around 1630–1640 cm^{-1} appertaining to the bending vibrations of adsorbed water molecules (Figure 10b,e). The sample hydration is also confirmed and the intense vibration band is observed at around 3300 cm^{-1} that belongs to the stretching vibrations of water molecules (Figure 10c,f) [33]. Moreover, the results of the FTIR studies indicate that the presence of dextran in the sample induces a broadening and a slight displacement of the maxima associated with the functional groups. Therefore, from this point of view, the features revealed by FTIR measurements are in agreement with those provided by XRD studies.

3.7. Antimicrobial Assay

Nowadays, approximately two-thirds of the global population suffer from various dental affections, most encountered being tooth decay, which often leads to the apparition of lesions with various degrees of severity. Even though the surface decay could be easily treated, the tooth could become rapidly unhealthy due to inflammation or infection [67–69]. Recently, due to the emergence of microorganisms resistant to conventional therapies, studies regarding the antimicrobial effects of various types of possible antimicrobial agents, such as metallic ions (copper, iron, silver, magnesium, zinc), inorganic nanoparticles and natural polymers, were the focus of researchers due to the need of finding novel solutions having a wide-range action against common pathogens [70–74]. In this context, our study is focused on the development of novel biomaterials based on magnesium-doped hydroxyapatite in dextran matrix for biomedical and dental applications. Therefore, we have studied the antimicrobial activity of the 10MgHApD nanocomposites using *Staphylococcus aureus* ATCC 25923, *Pseudomonas aeruginosa* ATCC 27853, *Streptococcus mutans* ATCC 25175, *Porphyromonas gingivalis* ATCC 33277 and *Candida albicans* ATCC 10231 microbial strains. The in vitro antimicrobial assays were performed in triplicate and the results were presented graphically as mean ± SD. The results of the in vitro antimicrobial assays are depicted in Figure 11. The data suggested that the 10MgHApD nanocomposites exhibited strong inhibitory activity against all the tested microbial strains. In addition, the results of the in vitro antimicrobial experiments showed that the HAp nanoparticles promoted microbial cell development and proliferation. The results highlighted that the microbial cells CFU's values were higher even than for the control (C+) for all the tested microorganisms for all three incubation times (24, 48 and 72 h). Furthermore, the results also demonstrated that the 10MgHAp nanoparticles exhibited good antimicrobial activity against all the tested microorganisms for all the incubation time intervals. The data also emphasized that the incubation time played an important role in the antimicrobial activity of both 10MgHApD nanocomposites and 10MgHAp nanoparticles. The results showed that the 10MgHApD nanocomposites exhibited bactericidal activity against *P. aeruginosa, S. mutans* and *P. gingivalis* bacterial cells, as well as fungicidal activity against *C. albicans* fungal cells. The data also suggested that the bactericidal and fungicidal effects appear after 48 and 72 h of incubation, respectively. In addition, the results also highlighted that the samples were highly effective in reducing the CFU in the case of dental infection-related bacterial strains, *S. mutans and P. gingivalis,* thus revealing that these types of biocomposites could be successfully used in the development of novel application in dentistry. These results are in good agreement with previously reported data regarding the antimicrobial properties of hydroxyapatite doped with magnesium ions and also of composites based on doped hydroxyapatite in dextran matrix [47,75,76]. Moreover, the data reported by Salem et al. [67] suggested that there are tremendous benefits to employing materials based on magnesium ions in dental restorative applications.

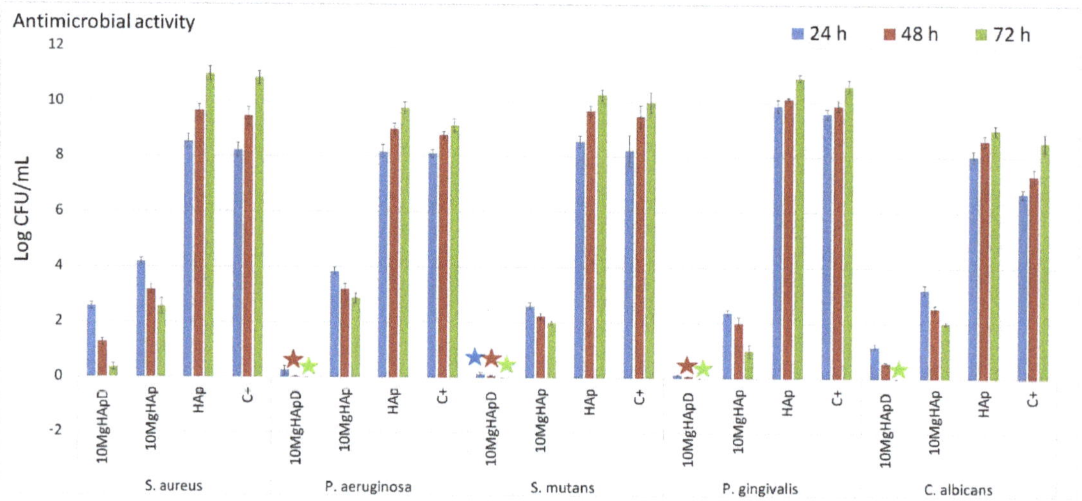

Figure 11. Antimicrobial assay of 10MgHApD nanocomposites, HAp and 10MgHAp nanoparticles against *S. aureus, P. aeruginosa, S. mutans, P. gingivalis* and *C. albicans* microbial strains. The results were considered statistically significant at * $p < 0.05$.

In their study, Salem et al. [68] showed that the presence of Mg^{2+} promoted an increase in the attachment rate, proliferation, differentiation, alkaline phosphatase activity, and mineralization, leading to a potential improvement of a pulp-capping material. These results, as well as the results obtained regarding the antimicrobial activity of 10MgHApD nanocomposites, suggest that magnesium-based composites might be employed in the development of novel future dental pulp-capping materials that could be used in regenerative endodontics applications. Magnesium is well known as a vital element that plays an important role in the physiological processes within the human body, from supporting muscle and nerve function to regulating blood pressure and contributing to bone health. Even though the exact mechanism is not yet fully understood, in recent years, scientific interest has expanded beyond magnesium's traditional roles, uncovering unexpected antimicrobial properties. The exact mechanism as to how exactly the magnesium ions exhibit antimicrobial activity still remains unclear, yet several suggestions have been explored in the scientific community. The antimicrobial activity was observed for the first time in early 1900 by Professor Pierre Delbet [77], who found out after numerous tests that a $MgCl_2$

solution was the most effective due to the fact that it was not toxic to the surrounding tissue, and it highly increased the leucocyte activity and phagocytosis. Later, Delbet [77], reported that the solution based on $MgCl_2$ proved to be efficient in the treatment of various diseases, including the ones related to various microorganisms. In recent years, studies have indicated that antibiotic efficacy could be notably enhanced in the presence of Mg^{2+} ions [78,79]. A prevailing hypothesis suggests that these divalent ions exert an influence on the membranes of bacterial cells. One of the primary proposed mechanisms by which magnesium ions demonstrate antimicrobial effects is through their capacity to interfere with microbial cellular functions. Due to the fact that it is a divalent cation, magnesium has the ability to compete with other metal ions, such as calcium and iron, for binding sites within microbial cells. This competition can disrupt crucial cellular processes, ultimately leading to compromising the growth and survival of bacteria and other microorganisms. In their study, Som et al., regarding "divalent metal ion-triggered activity of a synthetic antimicrobial in cardiolipin membranes" [80], reported that the divalent nature of the cation affected the curvature of the bacterial membrane, which left the bacteria more vulnerable, leading to an increase of the antibiotic's effects. In addition, Xie et al. [B5] also reported that the antimicrobial effects could be attributed to the fact that magnesium cations help permeabilize the membranes. Magnesium was also reported to be effective in the case of microorganism biofilms by destabilizing and disrupting them. Even though the exact mechanism as to how exactly the magnesium ions have the ability to delay the biofilm formation still remains unclear, several suggestions have been reported [80–83]. The studies suggested that magnesium ions possess the capability to interact directly with the cell membrane, potentially hindering the formation of biofilm. Another possibility could be attributed to their direct or indirect impact on the regulation of biofilm formation, leading to a delay in the process. Furthermore, a recent investigation illustrated the impact of Mg^{2+} ions against the development of Bacillus biofilm, highlighting a down-regulation of the expression of extracellular matrix genes by more than tenfold [84]. One other mechanism regarding the antimicrobial activity of magnesium composites is attributed to the presence of the existence of high concentrations of OH^- on their surface, which leads to an increase of the O^{2-} concentration, causing the destruction of the bacterial cell wall [85]. On the other hand, recently, dextran has gained attention in the scientific community for its diverse applications, particularly in the field of antimicrobial applications. One of the primary proposed antimicrobial mechanisms of dextran was reported to be the ability to inhibit biofilm formation. Dextran has been reported to be able to interfere with the initial stages of biofilm formation by preventing microbial adhesion to the surfaces. This behavior has been attributed to its hydrophilic nature and ability to modify surface properties. Furthermore, it has been reported that dextran's antimicrobial effects could be related to its ability to disrupt microbial cell membranes. In their studies, Amiri et al. [86] suggested that the polymer interacts with the bacterial membranes, causing destabilization and also an increased permeability. The mechanism responsible for this interaction is not yet fully elucidated, but it is believed that dextran's physical properties, such as molecular size and charge, could have an important role in compromising the integrity of microbial membranes. This disruption can lead to the leakage of cellular components, ultimately causing the bacterial cell's death. In addition to the direct effects of magnesium and dextran on microorganisms, both magnesium ions as well as dextran have been found to modulate the host immune response, enhancing the body's natural defense mechanisms. Dextran could stimulate the production of certain immune mediators, such as cytokines, and promote phagocytosis by immune cells [87,88].

Even though more complex biological studies should be performed to ensure the biological effects of magnesium and dextran on the dentin formation as well as the composite's antimicrobial activity and cytotoxic dosage, these preliminary results bring significant insight into the exquisite properties of these materials and their potential in future biomedical as well as dentistry applications.

3.8. In Vitro Biocompatibility Assay

Complementary information about the biological properties of the 10MgHApD nanocomposites was acquired by studying the biocompatibility of the nanocomposites using a Human Gingival Fibroblasts (HGF-1) cell line, which exhibits a typical fibroblast morphology. This particular cell line was isolated for the first time in 1989 from the gingiva of a 28-year-old white male patient. The toxicity of the 10MgHApD nanocomposites was evaluated by determining the cell viability of the HGF-1 cells after being exposed to the nanocomposites for three different time intervals. The results of the cell viability assays are depicted as a graphical representation in Figure 12.

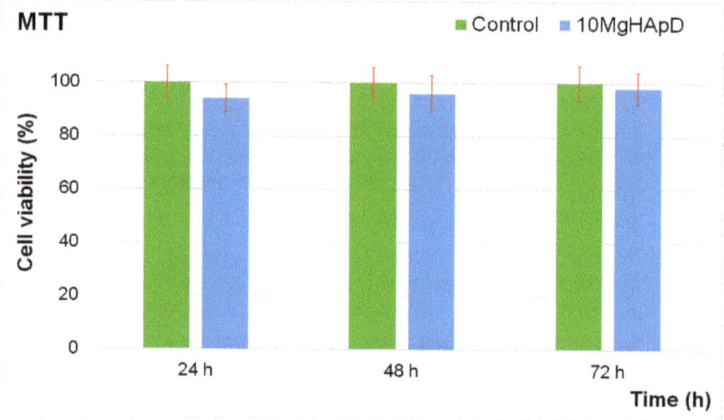

Figure 12. The graphical representation of the cell viability of HGF-1 cells exposed to 10MgHApD nanocomposites for 24, 48 and 72 h. The data are presented as mean ± standard deviation (SD) and are quantified as percentages of control (100% viability). The statistical analysis was performed using the ANOVA single-factor test and $p \leq 0.05$ was accepted as statistically significant.

The results of the MTT assay emphasized that the nanocomposites did not exhibit any significant toxicity against the HGF-1 cells for any of the incubation times when the data were acquired. The results showed that the viability was not significantly altered compared to the control after 24, 48 and 72 h of exposure with the 10MgHApD nanocomposites, which indicates a good biocompatibility of the samples. The results of the MTT reduction assay represented in Figure 12 showed that the cell viability of the HGF-1 cells exhibited values above 92% after the first 24 h of exposure to the 10MgHApD nanocomposites. Moreover, the findings also determined that the cell viability increased, reaching 96% and 98%, respectively, after 48 h and 72 h of exposure. These findings are in good agreement with other reported data regarding the biological properties of nanocomposites based on HAp, magnesium ions, and different biopolymers [47,89–92].

4. Conclusions

The scanning electronic microscopy (SEM) with energy-dispersive X-ray (EDS) methods in combination with X-ray diffraction (XRD) and Fourier transform infrared spectroscopy technique were applied in this study for the complex investigation of the structure and homogeneity of synthesized 10MgHApD sample. SEM analysis of 10MgHApD nanocomposites presented agglomerated particles with spherical morphology. The qualitative powder-XRD study revealed the nature of the hexagonal HAp. FTIR investigations demonstrated the presence of dextran in the 10MgHApD nanocomposites. Minkowski Functionals indicated nonconventional yet high-quality surface patterns. These findings deepen our understanding of biological surface interactions and have potential implications in materials and biomedicine, which can be further explored for practical applications. The in vitro antimicrobial assay of 10MgHApD nanocomposites against *S. aureus*, *P. aeruginosa*, *S. mutans*, *P. gingivalis* and *C. albicans* microbial strains emphasized that the samples exhibited a strong inhibitory effect on all the tested microbial cells. Moreover, the results also suggested that the 10MgHApD nanocomposite antimicrobial properties were influenced by both the incubation time and also the bacterial cells. In addition, the results demonstrated that 10MgHApD exhibited bactericidal activity against *P. aeruginosa*, *S. mutans*, *P. gingivalis* and *C. albicans* microbial strains. The in vitro cell viability assay also demonstrated that 10MgHApD exhibited good biocompatibility properties towards HGF-1 cells. The results obtained showed that this type of nanocomposite based on magnesium-doped hydroxyapatite in dextran matrix could be an effective antimicrobial agent that can be employed for the treatment of various oral diseases as well as dental caries. As a result, it is obviously seen that 10MgHApD nanocomposites have optimal properties for various dental field applications and they can probably be used in other medical or food applications.

Author Contributions: Conceptualization, D.P., S.C.C. and S.L.I.; methodology, S.L.I., S.C.C., Ş.Ţ. and D.P.; software, S.L.I., Ş.Ţ., R.S.M., H.D.d.F.F. and D.P.; validation, S.C.C., S.L.I., Ş.Ţ. and D.P.; formal analysis S.C.C., R.T., R.S.M., H.D.d.F.F. and L.G.; investigation, L.G., S.L.I., S.C.C., R.S.M., H.D.d.F.F., R.T. and D.P.; resources, S.L.I., Ş.Ţ. and D.P.; data curation, S.C.C., S.L.I., Ş.Ţ. and D.P.; writing—original draft preparation, D.P., S.C.C., Ş.Ţ., R.S.M., H.D.d.F.F. and S.L.I.; writing—review and editing, D.P., S.L.I., Ş.Ţ., R.S.M., H.D.d.F.F. and S.C.C.; visualization, L.G., D.P., S.L.I., R.S.M., H.D.d.F.F. and S.C.C.; supervision, R.S.M., H.D.d.F.F. and D.P. and S.C.C.; project administration, D.P.; funding acquisition, D.P. All authors have read and agreed to the published version of the manuscript.

Funding: This work was partially supported by the Romanian Ministry of Research and Innovation through the PN-III-P2-2.1-PED-2019-0868, contract number 467PED/2020 and by the Core Program of the National Institute of Materials Physics, granted by the Romanian Ministry of Research, Innovation and Digitalization through the Project PC1-PN23080101. Also, this research was supported by Contract No. T-IS 251801/04.05.2018 and Scientific Research Contract Nr.1/4.06.2020. The authors also thank CAPES (Coordenação de Aperfeiçoamento de Pessoal de Nível Superior–Código financeiro 001) and FAPEAM (Fundação de Amparo à Pesquisa do Estado do Amazonas, EDITAL N. 010/2021- CT&I ÁREAS PRIORITÁRIAS and EDITAL N. 013/2022-PRODUTIVIDADE EM CT&I) for the financial support, as well as the use of the infrastructure of the Analytical Center of Universidade Federal do Amazonas (UFAM). HDFF acknowledges funding support from CNPq Processo 306210/2022-3.

Institutional Review Board Statement: Not applicable.

Informed Consent Statement: Not applicable.

Data Availability Statement: Data are available on reasonable demand from the corresponding authors.

Acknowledgments: The authors would like to thank Monica Luminița Badea for assistance with the biological assays.

Conflicts of Interest: The authors declare no conflict of interest; The funders had no role in the design of the study; in the collection, analyses, or interpretation of data; in the writing of the manuscript; or in the decision to publish the results.

References

1. Kassebaum, N.J.; Bernabe, E.; Dahiya, M.; Bhandari, B.; Murray, C.J.; Marcenes, W. Global Burden of Untreated Caries: A Systematic Review and Metaregression. *J. Dent. Res.* **2015**, *94*, 650–658. [CrossRef] [PubMed]
2. Bowen, W.H.; Burne, R.A.; Wu, H.; Koo, H. Oral biofilms: Pathogens, matrix, and polymicrobial interactions in microenvironments. *Trends Microbiol.* **2018**, *26*, 229–242. [CrossRef] [PubMed]
3. Marsh, P.; Zaura, E. Dental biofilm: Ecological interactions in health and disease. *J. Clin. Periodontol.* **2017**, *44*, S12–S22. [CrossRef]
4. Bengtsson, U.G.; Hylander, L.D. Increased mercury emissions from modern dental amalgams. *Biometals* **2017**, *30*, 277–283. [CrossRef] [PubMed]
5. Lamont, R.J.; Koo, H.; Hajishengallis, G. The oral microbiota: Dynamic communities and host interactions. *Nat. Rev. Microbiol.* **2018**, *16*, 745–759. [CrossRef] [PubMed]
6. Pitts, N.B.; Zero, D.T.; Marsh, P.D.; Ekstrand, K.; Weintraub, J.A.; Ramos-Gomez, F.; Tagami, J.; Twetman, S.; Tsakos, G.; Ismail, A. Dental Caries. *Nat. Rev. Dis. Primers* **2017**, *3*, 17030. [CrossRef] [PubMed]
7. Hannig, M.; Hannig, C. Nanotechnology and its Role in Caries Therapy. *Adv. Dent. Res.* **2012**, *24*, 53–57. [CrossRef]

8. Gao, L.; Liu, Y.; Kim, D.; Li, Y.; Hwang, G.; Naha, P.C.; Cormode, D.P.; Koo, H. Nanocatalysts Promote Streptococcus mutans Biofilm Matrix Degradation and Enhance Bacterial killing to Suppress Dental Caries In Vivo. *Biomaterials* **2016**, *101*, 272–284. [CrossRef]
9. Hannig, M.; Hannig, C. Nanomaterials in Preventive Dentistry. *Nat. Nanotechnol.* **2010**, *5*, 565–569. [CrossRef]
10. Gibbons, R.J.; Banghart, S.B. Synthesis of Extracellular Dextran by Cariogenic Bacteria and its Presence in Human Dental Plaque. *Arch. Oral Biol.* **1967**, *12*, 11–23. [CrossRef]
11. Naha, P.C.; Liu, Y.; Hwang, G.; Huang, Y.; Gubara, S.; Jonnakuti, V.; Simon-Soro, A.; Kim, D.; Gao, L.; Koo, H.; et al. Dextran-Coated Iron Oxide Nanoparticles as Biomimetic Catalysts for Localized and pH-Activated Biofilm Disruption. *ACS Nano* **2019**, *13*, 4960–4971. [CrossRef] [PubMed]
12. Arcís, R.W.; López-Macipe, A.; Toledano, M.; Osorio, E.; Rodríguez-Clemente, R.; Murtra, J.; Fanovich, M.A.; Pascual, C.D. Mechanical properties of visible light-cured resins reinforced with hydroxyapatite for dental restoration. *Dent. Mater.* **2002**, *18*, 49–57. [CrossRef] [PubMed]
13. Chen, L.; Yu, Q.; Wang, Y.; Li, H.A.O. BisGMA/TEGDMA dental composite containing high aspect-ratio hydroxyapatite nanofibers. *Dent. Mater.* **2011**, *27*, 1187–1195. [CrossRef] [PubMed]
14. Laurencin, D.; Almora-Barrios, N.; de Leeuw, N.H.; Gervais, C.; Bonhomme, C.; Mauri, F.; Chrzanowski, W.; Knowles, J.C.; Newport, R.J.; Wong, A.; et al. Magnesium incorporation into hydroxyapatite. *Biomaterials* **2011**, *32*, 1826–1837. [CrossRef] [PubMed]
15. Sergi, R.; Bellucci, D.; Candidato Jr, R.T.; Lusvarghi, L.; Bolelli, G.; Pawlowski, L.; Candiani, G.; Altomare, L.; De Nardo, L.; Cannillo, V. Bioactive Zn-doped hydroxyapatite coatings and their antibacterial efficacy against *Escherichia coli* and *Staphylococcus aureus*. *Surf. Coat. Technol.* **2018**, *352*, 84–91. [CrossRef]
16. Stanić, V.; Janaćković, D.; Dimitrijević, S.; Tanasković, S.B.; Mitrić, M.; Pavlović, M.S.; Krstić, A.; Jovanović, D.; Raičević, S. Synthesis of antimicrobial monophase silver-doped hydroxyapatite nanopowders for bone tissue engineering. *Appl. Surf. Sci.* **2011**, *257*, 4510–4518. [CrossRef]
17. Laisney, J.; Chevallet, M.; Fauquant, C.; Sageot, C.; Moreau, Y.; Predoi, D.; Herlin-Boime, N.; Lebrun, C.; Michaud-Soret, I. Ligand-Promoted Surface Solubilization of TiO_2 Nanoparticles by the Enterobactin Siderophore in Biological Medium. *Biomolecules* **2022**, *12*, 1516. [CrossRef]
18. Laisney, J.; Rosset, A.; Bartolomei, V.; Predoi, D.; Truffier-Boutry, D.; Artous, S.; Bergé, V.; Brochard, G.; Michaud-Soret, I. TiO_2 nanoparticles coated with bio-inspired ligands for the safer-by-design development of photocatalytic paints. *Environ. Sci. Nano* **2021**, *8*, 297–310. [CrossRef]
19. Rondanelli, M.; Faliva, M.A.; Tartara, A.; Gasparri, C.; Perna, S.; Infantino, V.; Riva, A.; Petrangolini, G.; Peroni, G. An update on magnesium and bone health. *Biometals* **2021**, *34*, 715–736. [CrossRef]
20. Landi, E.; Logroscino, G.; Proietti, L.; Tampieri, A.; Sandri, M.; Sprio, S. Biomimetic Mg-substituted hydroxyapatite: From synthesis to in vivo behaviour. *J. Mater. Sci. Mater. Med.* **2008**, *19*, 239–247. [CrossRef]
21. Lim, G.K.; Wang, J.; Ng, S.C.; Gan, L.M. Formation of nanocrystalline hydroxyapatite in nonionic surfactant emulsions. *Langmuir* **1999**, *15*, 7472–7477. [CrossRef]
22. Tampieri, A.; Celotti, G.C.; Landi, E.; Sandri, M. Magnesium doped hydroxyapatite: Synthesis and characterization. *Key Eng. Mater.* **2004**, *264*, 2051–2054. [CrossRef]
23. Kalita, S.J.; Bhatt, H.A. Nanocrystalline hydroxyapatite doped with magnesium and zinc: Synthesis and characterization. *Mater. Sci. Eng. C* **2007**, *27*, 837–848. [CrossRef]
24. Suchanek, W.L.; Byrappa, K.; Shuk, P.; Riman, R.E.; Janas, V.F.; TenHuisen, K.S. Preparation of magnesium-substituted hydroxyapatite powders by the mechanochemical–hydrothermal method. *Biomaterials* **2004**, *25*, 4647–4657. [CrossRef] [PubMed]
25. Arul, K.T.; Kolanthai, E.; Manikandan, E.; Bhalerao, G.M.; Chandra, V.S.; Ramya, J.R.; Mudali, U.K.; Nair, K.G.M.; Kalkura, S.N. Green synthesis of magnesium ion incorporated nanocrystalline hydroxyapatite and their mechanical, dielectric and photoluminescence properties. *Mater. Res. Bull.* **2015**, *67*, 55–62. [CrossRef]
26. Padmanabhan, V.P.; Kulandaivelu, R.; Panneer, D.S.; Vivekananthan, S.; Sagadevan, S.; Lett, J.A. Microwave synthesis of hydroxyapatite encumbered with ascorbic acid intended for drug leaching studies. *Mater. Res. Innov.* **2020**, *24*, 171–178. [CrossRef]
27. Dhanalakshmi, C.P.; Vijayalakshmi, L.; Narayanan, V. Synthesis and characterization of poly(4-vinyl pyridine-co-styrene)/FHAP nanocomposite, and its biomedical application. *Appl. Nanosci.* **2013**, *3*, 373–382. [CrossRef]
28. Bolhassani, A.; Javanzad, S.; Saleh, T.; Hashemi, M.; Aghasadeghi, M.R.; Sadat, S.M. Polymeric nanoparticles: Potent vectors for vaccine delivery targeting cancer and infectious diseases. *Hum. Vaccines Immunother.* **2014**, *10*, 321–332. [CrossRef]
29. Rizvi, S.A.A.; Saleh, A.M. Applications of nanoparticle systems in drug delivery technology. *Saudi Pharm. J.* **2018**, *26*, 64–70. [CrossRef]
30. El-Meliegy, E.; Abu-Elsaad, N.; El-Kady, A.M.; Ibrahim, M.A. Improvement of physico-chemical properties of dextran-chitosan composite scaffolds by addition of nano-hydroxyapatite. *Sci. Rep.* **2018**, *8*, 12180. [CrossRef]
31. Kaushik, S. Polymeric and Ceramic Nanoparticles: Possible Role in Biomedical Applications. In *Handbook of Polymer and Ceramic Nanotechnology*; Hussain, C.M., Thomas, S., Eds.; Springer: Cham, Switzerland, 2021. [CrossRef]

32. Shoba, E.; Lakra, R.; Kiran, M.S.; Korrapati, P.S. 3D nano bilayered spatially and functionally graded scaffold impregnated bromelain conjugated magnesium doped hydroxyapatite nanoparticle for periodontal regeneration. *J. Mech. Behav. Biomed. Mater.* **2020**, *109*, 103822. [CrossRef] [PubMed]
33. Predoi, D.; Iconaru, S.L.; Predoi, M.V.; Stan, G.E.; Buton, N. Synthesis, Characterization, and Antimicrobial Activity of Magnesium-Doped Hydroxyapatite Suspensions. *Nanomaterials* **2019**, *9*, 1295. [CrossRef] [PubMed]
34. Sutha, S.; Dhineshbabu, N.R.; Prabhu, M.; Rajendran, V. Mg-doped hydroxyapatite/chitosan composite coated 316l stainless steel implants for biomedical applications. *J. Nanosci. Nanotechnol.* **2015**, *15*, 4178–4187. [CrossRef]
35. Ciobanu, C.S.; Iconaru, S.L.; Popa, C.L.; Motelica-Heino, M.; Predoi, D. Evaluation of samarium doped hydroxyapatite, ceramics for medical application: Antimicrobial activity. *J. Nanomater.* **2015**, *2015*, 849216. [CrossRef]
36. Gwyddion. Available online: http://gwyddion.net/ (accessed on 20 November 2023).
37. Ţălu, Ş. *Micro and Nanoscale Characterization of Three Dimensional Surfaces. Basics and Applications*; Napoca Star Publishing House: Cluj-Napoca, Romania, 2015.
38. Ţălu, Ş.; Abdolghaderi, S.; Pinto, E.P.; Matos, R.S.; Salerno, M. Advanced fractal analysis of nanoscale topography of Ag/DLC composite synthesized by RF-PECVD. *Surf. Eng.* **2020**, *36*, 713–719. [CrossRef]
39. Bulinski, A.; Dimitrov, D. Statistical Estimation of the Shannon Entropy. *Acta Math. Sin. Engl. Ser.* **2019**, *35*, 17–46. [CrossRef]
40. Matos, R.S.; Lopes, G.A.C.; Ferreira, N.S.; Pinto, E.P.; Carvalho, J.C.T.; Figueiredo, S.S.; Oliveira, A.F.; Zamora, R.R.M. Superficial Characterization of Kefir Biofilms Associated with Açaí and Cupuaçu Extracts. *Arab. J. Sci. Eng.* **2018**, *43*, 3371–3379. [CrossRef]
41. Brown, C.; Liebovitch, L. *Fractal Analysis*; SAGE Publications, Inc.: Thousand Oaks, CA, USA, 2010. [CrossRef]
42. Iconaru, S.L.; Prodan, A.M.; Turculet, C.S.; Beuran, M.; Ghita, R.V.; Costescu, A.; Groza, A.; Chifiriuc, M.C.; Chapon, P.; Gaiaschi, S.; et al. Enamel Based Composite Layers Deposited on Titanium Substrate with Antifungal Activity. *J. Spectrosc.* **2016**, *2016*, 4361051. [CrossRef]
43. Landi, E.; Tampieri, A.; Celotti, G.; Sprio, S. Densification behaviour and mechanisms of synthetic hydroxyapatites. *J. Eur. Ceram. Soc.* **2000**, *20*, 2377–2387. [CrossRef]
44. Unterweger, H.; Tietze, R.; Janko, C.; Zaloga, J.; Lyer, S.; Dürr, S.; Taccardi, N.; Goudouri, O.M.; Hoppe, A.; Eberbeck, D.; et al. Development and characterization of magnetic iron oxide nanoparticles with a cisplatin-bearing polymer coating for targeted drug delivery. *Int. J. Nanomed.* **2014**, *9*, 3659–3676. [CrossRef]
45. Nathanael, A.J.; Mangalaraj, D.; Ponpandian, N. Controlled growth and investigations on the morphology and mechanical properties of hydroxyapatite/titania nanocomposite thin films. *Compos. Sci. Technol.* **2010**, *70*, 1645–1651. [CrossRef]
46. Ahmadi, R.; Asadpourchallou, N.; Kaleji, B.K. In vitro study: Evaluation of mechanical behavior, corrosion resistance, antibacterial properties and biocompatibility of HAp/TiO_2/Ag coating on Ti6Al4V/TiO_2 substrate. *Surf. Interfaces* **2021**, *24*, 101072. [CrossRef]
47. Predoi, D.; Iconaru, S.L.; Predoi, M.V.; Motelica-Heino, M.; Buton, N.; Megier, C. Obtaining and Characterizing Thin Layers of Magnesium Doped Hydroxyapatite by Dip Coating Procedure. *Coatings* **2020**, *10*, 510. [CrossRef]
48. Schmähling, J.; Hamprecht, F.A. Generalizing the Abbott–Firestone curve by two new surface descriptors. *Wear* **2007**, *262*, 1360–1371. [CrossRef]
49. Matos, R.S.; Pinto, E.P.; Pires, M.A.; Ramos, G.Q.; Ţălu, Ş.; Lima, L.S.; da Fonseca Filho, H.D. Evaluating the roughness dynamics of kefir biofilms grown on Amazon cupuaçu juice: A monofractal and multifractal approach. *Microscopy* **2023**, dfad040. [CrossRef]
50. Ramos, G.Q.; Matos, R.S.; da Fonseca Filho, H.D. Advanced Microtexture Study of *Anacardium occidentale* L. Leaf Surface from the Amazon by Fractal Theory. *Microsc. Microanal.* **2020**, *26*, 989–996. [CrossRef]
51. Korpi, A.G.; Ţălu, Ş.; Bramowicz, M.; Arman, A.; Kulesza, S.; Pszczolkowski, B.; Jurečka, S.; Mardani, M.; Luna, C.; Balashabadi, P.; et al. Minkowski functional characterization and fractal analysis of surfaces of titanium nitride films. *Mater. Res. Express* **2019**, *6*, 086463. [CrossRef]
52. Zelati, A.; Mardani, M.; Rezaee, S.; Matos, R.S.; Pires, M.A.; Da Fonseca Filho, H.D.; Das, A.; Hafezi, F.; Rad, G.A.; Kumar, S.; et al. Morphological and multifractal properties of Cr thin films deposited onto different substrates. *Microsc. Res. Tech.* **2023**, *86*, 157–168. [CrossRef]
53. Ţălu, Ş.; Patra, N.; Salerno, M. Micromorphological characterization of polymer-oxide nanocomposite thin films by atomic force microscopy and fractal geometry analysis. *Prog. Org. Coat.* **2015**, *89*, 50–56. [CrossRef]
54. Salcedo, M.O.C.; Zamora, R.R.M.; Carvalho, J.C.T. Study fractal leaf surface of the plant species Copaifera sp. using the Microscope Atomic-Force-AFM. *Rev. ECIPerú* **2016**, *13*, 10–16. [CrossRef]
55. Ţălu, Ş.; Solaymani, S.; Bramowicz, M.; Naseri, N.; Kulesza, S.; Ghaderi, A. Surface micromorphology and fractal geometry of Co/CP/X (X = Cu, Ti, SM and Ni) nanoflake electrocatalysts. *RSC Adv.* **2016**, *6*, 27228–27234. [CrossRef]
56. Bullmore, E.; Barnes, A.; Bassett, D.S.; Fornito, A.; Kitzbichler, M.; Meunier, D.; Suckling, J. Generic aspects of complexity in brain imaging data and other biological systems. *Neuroimage* **2009**, *47*, 1125–1134. [CrossRef] [PubMed]
57. Silva, M.R.P.; Matos, R.S.; Pinto, E.P.; Santos, S.B.; Monteiro, M.D.S.; da Fonseca Filho, H.D.; Almeida, L.E. Advanced Microtexture Evaluation of Dextran Biofilms Obtained from Low Cost Substrate Loaded with Maytenus rigida Extract. *Mater. Res.* **2021**, *24*, 1–11. [CrossRef]
58. Shakoury, R.; Arman, A.; Ţălu, Ş.; Ghosh, K.; Rezaee, S.; Luna, C.; Mwema, F.; Sherafat, K.; Salehi, M.; Mardani, M. Optical properties, microstructure, and multifractal analyses of ZnS thin films obtained by RF magnetron sputtering. *J. Mater. Sci. Mater. Electron.* **2020**, *31*, 5262–5273. [CrossRef]

59. Pinto, E.P.; Matos, R.S.; Pires, M.A.; Lima, L.d.S.; Ţălu, Ş.; da Fonseca Filho, H.D.; Ramazanov, S.; Solaymani, S.; Larosa, C. Nanoscale 3D Spatial Analysis of Zirconia Disc Surfaces Subjected to Different Laser Treatments. *Fractal Fract.* **2023**, *7*, 160. [CrossRef]
60. Markovic, M.; Fowler, B.O.; Tung, M.S. Preparation and comprehensive characterization of a calcium hydroxyapatite reference material. *J. Res. Natl. Inst. Stand. Technol.* **2004**, *109*, 553–568. [CrossRef]
61. Mondal, S.; Mondal, B.; Dey, A.; Mukhopadhyay, S.S. Studies on processing and characterization of hydroxyapatite biomaterials from different bio wastes. *J. Miner. Mater. Charact. Eng.* **2012**, *11*, 55–67. [CrossRef]
62. Varma, H.K.; Babu, S. Synthesis of Calcium Phosphate Bioceramics by Citrate Gel Pyrolysis Method. *Ceram. Int.* **2005**, *31*, 109–114. [CrossRef]
63. Abifarin, J.K.; Obada, D.O.; Dauda, E.T.; Dodoo-Arhin, D. Experimental data on the characterization of hydroxyapatite synthesized from biowastes. *Data Brief* **2019**, *26*, 104485. [CrossRef]
64. Rocha, J.H.G.; Lemos, A.F.; Kannan, S.; Agathopoulos, S.; Ferreira, J.M.F. Hydroxyapatite scaffolds hydrothermally grown from aragonitic cuttlefish bones. *J. Mater. Chem.* **2005**, *15*, 5007–5011. [CrossRef]
65. Can, H.K.; Kavlak, S.; ParviziKhosroshahi, S.; Güner, A. Preparation, characterization and dynamical mechanical properties of dextran-coated iron oxide nanoparticles (DIONPs). *Artif. Cells Nanomed. Biotechnol.* **2018**, *46*, 421–431. [CrossRef] [PubMed]
66. Hradil, J.; Pisarev, A.; Babič, M.; Horák, D. Dextran-modified iron oxide nanoparticles. *China Particuology* **2007**, *5*, 162–168. [CrossRef]
67. Salem, R.M.; Zhang, C.; Chou, L. Effect of Magnesium on Dentinogenesis of Human Dental Pulp Cells. *Int. J. Biomater.* **2021**, *2021*, 6567455. [CrossRef] [PubMed]
68. NICDR. Dental Caries (Tooth Decay) in Adults (Age 20 to 64). 2014. Available online: http://www.nidcr.nih.gov.ezproxy.bu.edu/DataStatistics/FindDataByTopic/DentalCaries/DentalCariesAdults20to64.htm (accessed on 10 November 2023).
69. Flaxman, D.A.; Naghavi, M.; Lozano, R.; Michaud, C.; Ezzati, M.; Memish, Z.A. Years lived with disability (YLDs) for 1160 sequelae of 289 diseases and injuries 1990–2010: A systematic analysis for the global burden of disease study. *Lancet* **2012**, *380*, 2163–2196.
70. Graziani, G.; Boi, M.; Bianchi, M. A review on ionic substitutions in hydroxyapatite thin films: Towards complete biomimetism. *Coatings* **2018**, *8*, 269. [CrossRef]
71. LeGeros, R.Z. *Calcium Phosphates in Oral Biology and Medicine*; Karger: Basel, Switzerland, 1991.
72. Joo, L.; Ong, D.; Chanm, C.N. Hydroxyapatite and their use as coatings in dental implants: A review. *Crit. Rev. Biomed. Eng.* **1999**, *28*, 667–707.
73. Marques, C.F.; Olhero, S.; Abrantes, J.C.C.; Marote, A.; Ferreira, S.; Vieira, S.I.; Ferreira, J.M.F. Biocompatibility and antimicrobial activity of biphasic calcium phosphate powders doped with metal ions for regenerative medicine. *Ceram. Int.* **2017**, *43*, 15719–15728. [CrossRef]
74. Chung, R.-J.; Hsieh, M.-F.; Huang, C.-W.; Perng, L.-H.; Wen, H.-W.; Chin, T.-S. Antimicrobial effects and human gingival biocompatibility of hydroxyapatite sol-gel coatings. *J. Biomed. Mater. Res. B* **2006**, *76*, 169–178. [CrossRef]
75. Iconaru, S.L.; Predoi, M.V.; Motelica-Heino, M.; Predoi, D.; Buton, N.; Megier, C.; Stan, G.E. Dextran-Thyme Magnesium-Doped Hydroxyapatite Composite Antimicrobial Coatings. *Coatings* **2020**, *10*, 57. [CrossRef]
76. Monzavi, A.; Eshraghi, S.; Hashemian, R.; Momen-Heravi, F. In vitro and ex vivo antimicrobial efficacy of nano-MgO in the elimination of endodontic pathogens. *Clin. Oral Investig.* **2015**, *19*, 349–356. [CrossRef]
77. Delbet, P. *Politique Préventive du Cancer: Cytophylaxie*; Denoël: Paris, France, 1944.
78. Houlihan, A.J.; Russell, J.B. The effect of calcium and magnesium on the activity of bovicin HC5 and nisin. *Curr. Microbiol.* **2006**, *53*, 365–369. [CrossRef] [PubMed]
79. Khan, F.; Patoare, Y.; Karim, P.; Rayhan, I.; Quadir, M.A.; Hasnat, A. Effect of magnesium and zinc on antimicrobial activities of some antibiotics. *Pak. J. Pharm. Sci.* **2005**, *18*, 57–61. [PubMed]
80. Som, A.; Yang, L.; Wong, G.C.; Tew, G.N. Divalent metal ion triggered activity of a synthetic antimicrobial in cardiolipin membranes. *J. Am. Chem. Soc.* **2009**, *131*, 1510215103. [CrossRef] [PubMed]
81. Xie, Y.; Yang, L. Calcium and Magnesium Ions Are Membrane-Active against Stationary-Phase *Staphylococcus aureus* with High Specificity. *Sci. Rep.* **2016**, *6*, 20628. [CrossRef] [PubMed]
82. Gao, X.; Mukherjee, S.; Matthews, P.M.; Hammad, L.A.; Kearns, D.B.; Dann, C.E. Functional characterization of core components of the *Bacillus subtilis* cyclic-di-GMP signaling pathway. *J. Bacteriol.* **2013**, *195*, 4782–4792. [CrossRef] [PubMed]
83. Chen, Y.; Chai, Y.; Guo, J.H.; Losick, R. Evidence for cyclic Di-GMP-Mediated signaling in *Bacillus subtilis*. *J. Bacteriol.* **2012**, *194*, 5080–5090. [CrossRef]
84. Oknin, H.; Steinberg, D.; Shemesh, M. Magnesium ions mitigate biofilm formation of *Bacillus* species via downregulation of matrix genes expression. *Front. Microbiol.* **2015**, *6*, 907. [CrossRef]
85. Huang, L.; Li, D.-Q.; Lin, Y.-J.; Wei, M.; Evans, D.G.; Duan, X. Controllable preparation of nano-MgO and investigation of its bactericidal properties. *J. Inorg. Biochem.* **2005**, *99*, 986–993. [CrossRef]
86. Amiri, S.; Ramezani, R.; Aminlari, M. Antibacterial Activity of Dextran-Conjugated Lysozyme against *Escherichia coli* and *Staphylococcus aureus* in Cheese Curd. *J. Food Prot.* **2008**, *71*, 411–415. [CrossRef]
87. McCarthy, R.E.; Arnold, L.W.; Babcock, G.F. Dextran sulphate: An adjuvant for cell-mediated immune responses. *Immunology* **1977**, *32*, 963–974.

88. Ashique, S.; Kumar, S.; Hussain, A.; Mishra, N.; Garg, A.; Gowda, B.H.J.; Farid, A.; Gupta, G.; Dua, K.; Taghizadeh-Hesary, F. A narrative review on the role of magnesium in immune regulation, inflammation, infectious diseases, and cancer. *J. Health Popul. Nutr.* **2023**, *42*, 74. [CrossRef] [PubMed]
89. Predoi, D.; Ciobanu, S.C.; Iconaru, S.L.; Predoi, M.V. Influence of the Biological Medium on the Properties of Magnesium Doped Hydroxyapatite Composite Coatings. *Coatings* **2023**, *13*, 409. [CrossRef]
90. Bigi, A.; Foresti, E.; Gregorini, R.; Ripamonti, A.; Roveri, N.; Shah, J.S. The role of magnesium on the structure of biological apatites. *Calcif. Tissue Int.* **1992**, *50*, 439–444. [CrossRef] [PubMed]
91. Predoi, D.; Iconaru, S.L.; Predoi, M.V. Fabrication of Silver- and Zinc-Doped Hydroxyapatite Coatings for Enhancing Antimicrobial Effect. *Coatings* **2020**, *10*, 905. [CrossRef]
92. Luque-Agudo, V.; Fernández-Calderón, M.C.; Pacha-Olivenza, M.A.; Perez-Giraldo, C.; Gallardo-Moreno, A.M.; González-Martín, M.L. The role of magnesium in biomaterials related infections. *Colloids Surf. B* **2020**, *191*, 110996. [CrossRef]

polymers

MDPI

Review

Nanoparticles for Biomedical Application and Their Synthesis

Iva Rezić

Department of Applied Chemistry, Faculty of Textile Technology, University of Zagreb, 10000 Zagreb, Croatia; iva_rezic@net.hr or iva.rezic@ttf.hr

Abstract: Tremendous developments in nanotechnology have revolutionized the impact of nanoparticles (NPs) in the scientific community and, more recently, in society. Nanomaterials are by their definition materials that have at least one dimension in range of 1 to 100 nm. Nanoparticles are found in many types of different technological and scientific applications and innovations, from delicate electronics to *state-of-the-art* medical treatments. Medicine has recognized the importance of polymer materials coated with NPs and utilizes them widely thanks to their excellent physical, chemical, antibacterial, antimicrobial, and protective properties. Emphasis is given to their biomedical application, as the nanoscale structures are in the range of many biological molecules. Through this, they can achieve many important features such as targeted drug delivery, imaging, photo thermal therapy, and sensors. Moreover, by manipulating in a "nano-scale" range, their characteristic can be modified in order to obtain the desired properties needed in particular biomedical fields, such as electronic, optical, surface plasmon resonance, and physic-chemical features.

Keywords: biomedical application; nanoparticles; drug delivery; imaging; photo-thermal therapy; sensors

Citation: Rezić, I. Nanoparticles for Biomedical Application and Their Synthesis. *Polymers* **2022**, *14*, 4961. https://doi.org/10.3390/polym14224961

Academic Editors: Stephanie Andrade, Maria João Ramalho and Joana A. Loureiro

Received: 9 October 2022
Accepted: 9 November 2022
Published: 16 November 2022

1. Introduction

The current advancements in the field of biomedical application of nanoparticles is the result of the development of the synthesis and application of the engineered nanoparticles.

There is a wide variety of polymeric and metallic nanoparticles that are widely explored for possible biomedical applications. By this, such investigation is the focus of extensive research focused on the characterization and modification of intrinsic characteristics, including electronic, optical, and physicochemical properties, as well as surface plasmon resonance. All of the mentioned properties are changed during the modification of particular nanoparticle characteristics, for example, of their size, shape, or aspect ratio. [1].

Nanoparticles can be easily synthesized and modified so that they have novel electronic, optical, magnetic, medical, catalytic, and mechanical properties. Such a powerful modification results in a high surface-to-volume ratio and quantum size effect, which depend greatly on their size, structure, and shape. By deposition of targeted nanoparticles in polymers, new materials foreseen as biomedical devices are made.

Such materials are found in woven and nonwoven medical materials, polymers, and in other applications. The list of NPs applied on textile and polymer materials is presented in Table 1.

It is estimated that, among all different engineered nanoparticles, silver nanoparticles have the largest degree of commercialization [2]. A great diversity of synthetic bottom-up and top-down methods is used for developing and producing nanoparticles of various chemical composition, size, and shape [3]. However, most of those methods are wet chemical processes based on solution-phase colloidal chemical reactions and often harsh conditions. Except for those widely applied routine procedures, NPs can be produced by enzyme-mediated reactions.

The production of NPs by enzymes offers many advantages: reactions are performed at ambient temperature, moderate pH values are sufficient for an effective enzymatic reactions, the control of reactions is easy, produced NPs can be easily combined with other organic or heat sensitive materials (e.g., proteins), the crystalline phase of the produced NPs can be different from the one(s) obtained by conventional methods and lead to new products, and the morphology of NPs can be controlled by enzymatic reaction engineering. One of the most important features of the nanoparticles is their small size. It is reported that, by comparing the size of novel engineered nanomaterials with well-known biological nanostructures, such as the DNA double-helix with a diameter of 2 nm or Mycoplasma bacteria with a length of 200 nm, nanoparticles cover this exact region. Through this, it is possible to engineer new colloidal nanoparticles in order to become *"living nanodrugs"* that mimic bacteria. Moreover, *"robotic nanodrugs"* could be designed to manipulate molecular events in human or animal bodies [4].

Table 1. Nanoparticles in biomedical applications [2,3].

NPs	Diameter	NPs	Diameter	NPs	Diameter
Ag	1.5–350 um	Eu_2O_3	30–58 nm	Pr_6O_{11}	15–30 nm
Al	18 or 80 nm	Fe	25–250 nm	Si	30–70 nm
Au	50–150 nm	Gd_2O_3	15–80 nm	SiO_2	15–80 nm
B_2O_3	40–80 nm	In_2O_3	30–50 nm	Sm_2O_3	15–55 nm
$BaSO_4$	1–5 um	$In(OH)_3$	20–70 nm	SnO_2	45–60 nm
C	3–400 nm	La_2O_3	15–30 nm	$SrTiO_3$	100 nm
CeO_2	15–105 nm	$Li_4Ti_5O_{12}$	20–60 nm	Ti	30–50 nm
Co	28 nm	MgO	20–100 nm	W	50 nm
Cr	50 nm	$Mg(OH)_2$	15 nm	Y_2O_3	20–40 nm
Cu	25–500 nm	Mn_2O_3	30–60 nm	YbF_3	40–80 nm
Dy_2O_3	30 or 55 nm	Mo	70 nm	Zn	80–130 nm
Er_2O_3	20–53 nm	Ni	20–50 nm	ZrC	30–60 nm

Interaction of NPs with living organisms can be monitored at different levels, from molecular/cellular to higher, tissue systematic levels. Moreover, their interaction will depend on different routes of administration that include inhalation or intranasal, -venous, -dermal, -muscular, and others.

Interactions are under investigation of many interdisciplinary fields and scientists from both a toxicological and therapeutic point of view in the development of a new generation of nano-immuno-therapeutic drugs using in vitro and in vivo preclinical trials. It has to be emphasized, however, that although primary interaction of NPs with immune cells is related to the physical and chemical properties of the NPs, direct conclusions on the mechanisms cannot be foreseen [4,5].

Man-made polymers are synthesized in a controlled manner and final products contain well-defined characteristics such as defined molar mass, architecture, hydrophobic or hydrophilic properties, crystallinity, and functional groups in the structure of the molecules. Special interest is also given to gold nanoparticles (Au NPs) with or without a polymeric coating. This focus originates from the fact that Au NPs exhibit two very valuable advantages that make them the "stars" among a wide variety of NPs for bio-applications: biocompatibility and ease of surface modifications by a wide range of molecules (Figure 1). Figure 1 presents the illustration of the mechanism in which the photo-induced reactions enables the release of biologically active nanoparticles (gold nanospheres).

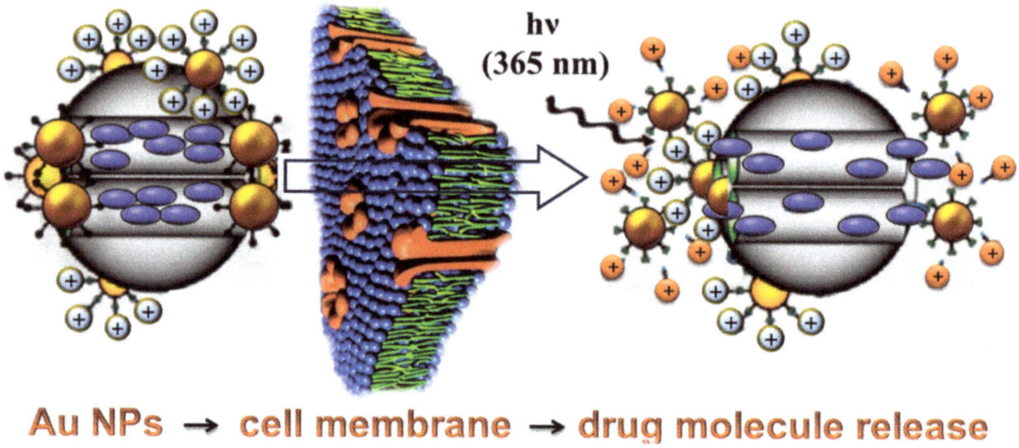

Figure 1. Schematic illustration of the photo-induced intracellular controlled release of gold nanospheres. Upon UV irradiation, the photo-labile linker on the gold NPs is cleaved, changing the surface charge of gold NPs from positive to negative. The charge repulsion between the gold NPs then uncaps the mesopores and allows the release of the drugs for cancer therapy [5].

The most relevant synthetic methods for producing Au NPs are as follows: seed-mediated method for producing nano-rods and nano-prisms in a two-step process (Zsigmondy nuclear method), thermal decomposition (polyol synthesis) using polyvinylpyrrolidone (PVP) as a surface-capping agent, template mediated synthesis for non-spherical Au NPs in which Au is generated in situ and shaped into a morphology complementary to a template (like channels within porous materials, organic surfactants or block polymer, and biological molecules such as DNA or viruses), and galvanic replacement reaction introduced by Brenner and Riddell comprising the spontaneous reduction of metal ions to metallic particles and films in the absence of an external electric field. In contrast to those methods, enzymatically catalyzed reactions are used to achieve environmentally friendly and cost-effective conditions for the production of NPs. Among a great variety of Au NPs that can be prepared in a controlled way with a high yield and reproducibility, especially important are nanorods, prisms, shells, cages, and hollow nanostructures like stars. Gold shells are spherical NPs consisting of a dielectric core (silica, polystyrene, or sodium sulfide) covered by a thin layer of gold.

Applications of Au NPs can be categorized into different technological areas (energy, environment, information technologies, and bio-applications) and are based on their outstanding properties [6], such as, for example, optical property to absorb and scatter light with extraordinary efficiency [7] (Genet and Ebbesen, 2007). Recent applications cover the application of Au NPs in photovoltaic devices and environmental conversion of CO into CO_2, as well as in the design of antennas, lenses, and resonators [5]. Important medical applications of Au NPs today are present in drug delivery [8]. Those are proposed for the use of laser irradiation, pH change, ionic concentration change, and similar reactions that trigger a drug release (Figure 1). Because of their biocompatibility, these NPs can be used with a near-infrared laser source to thermally destroy cancer tissue without significant damage to surrounding healthy tissue. Therefore, gold nanoshells are already in advanced stages of Food and Drug Administration (FDA) clinical trials. Except for drug delivery, medical usage of Au NPs includes medicine diagnosis and therapy, imaging of tumor cells, and other applications listed in Table 2.

Table 2. Imaging and biomedical capabilities of Au NPs [5].

	Modality	Nanoparticle/Agent
Imaging	Optical scattering PET, SPECT CT	Au nanoshells, nanorodes, nanocages Radioisotope ^{198}Au Au NPs
Therapeutic actuation	Phototermal Photoacoustic Chemotherapy Gene therapy	Au nanoshells, nanocages NIR-absorbing Au NPs Au NPs loaded with anticancer drugs Au NPs loaded with RNA, DNA

Although there is a wide variety of methods for the characterization of NPs in different textile and polymer samples [9], not all of them can reveal the particle size, morphology, and size distribution [2,3]. Among the mostly used methods are scanning electron microscopy (SEM), transmission electron microscopy (TEM), photon correlation spectroscopy (PSC), surface area analysis (BET), and X-ray diffraction (XRF) peak broadening analysis [10]. Characterization methodology is crucial for targeting of special nanoparticles for desired application (Table 2). However, specific targeting using desired ligands can be hard to achieve, and the comparison between targeted and non-targeted NPs obtained similar results [4]. Moreover, the interaction of NPs with biological fluids at the contact point site of drug administration (oral, intranasal, or intravenous) makes the problem even more complex. Some strategies are implemented to manipulate this effect, such as PEG nanoparticles that have the ability to avoid interactions with other molecules. Secondly, there are NPs coated with cell membranes that are able to reduce their blood clearance and, thirdly, there are NPs that are able to drain lymph nodes. All of those properties influence the effectiveness of NPs in reaching their target. More importantly, understanding of the functioning of our immune system provides innovative ideas for the application of immunotherapies through other routes of administration (intraperitoneal injection of NPs or subcutaneous administration of NPs for vaccination purposes) [4]. Furthermore, it has also been shown that certain NP compositions exhibit per se, without the addition of any drug, immunostimulant or immunosuppressive properties [5]. Many examples are provided, from the hyaluronic acid, PLGA, chitosan NPs, or porous silicon micro particles—all of these have shown the ability to act as self-adjuvants or stimulate M1 or Th1 responses.

There are many advantages of nanoparticles. Firstly, the advantages of NPs are their bioactive properties, good attachment to other polymeric materials, not complex mechanisms of formation of composite and bio-composite materials, as well as simple steps in the modification and surface functionalization of different carriers with them. However, a strong hydration shell, precipitation and turbid solution, surface aggregation, dispersive composition, size, and other properties are only a few disadvantages that make them very hard to operate and use in technological and biomedical applications. Moreover, the disadvantages of nanomaterials are the consequence of their small dimensions. Because of this, they easily infiltrate through the cell walls of skin and lung cells. Even worse, they can cross the blood–brain barrier.

2. Polymeric Nanoparticles in Biomedical Application

Polymeric nanoparticles (NPs) contain a core with an inner filling that can contain dyes, other inorganic nano-particles, or drug molecules. The outer shell contains usually hydrophilic polymers. Such an outer shell can be created in such a manner that additional functional units are combined with it [10].

A biomedical application of polymeric nanoparticles is presented in the following parts of the manuscript.

2.1. Polymeric Nanoparticles in the Treatment of Inflammation

The World Health Organization (WHO) has recognized the most abundant chronic diseases of global human population as the following: Alzheimer's disease, cardiovascular diseases, cancer, chronic obstructive pulmonary disease, and diabetes (type 2).

Chronic inflammation is recognized as the underlying cause of most chronic diseases, in which the immune system is activated for long periods, resulting in various pro-inflammatory cytokines that induce damage to organisms [4]. Owing to the lack of specificity, conventional therapies that involve non-steroidal and anti-inflammatory drugs have numerous side effects, while at the same time, the application of anti-inflammatory drugs results in low bioavailability. In contrast, specially designed nanocarriers offer possibilities of overcoming such obstacles through the optimization of site-specific delivery and improvements in the solubility of the drugs. This so-called nano-carrier designs is adapted from the microenvironment of the inflamed tissue. The microenvironment has many characteristics, including increased permeability, acidic pH value, and a high presence of reactive oxygen species (ROS) [4,6].

2.2. Polymeric Nanoparticles in the Treatment of Cancer

Cancer is the cause of more than 10 million deaths per year, which makes it the second leading cause of death around the world. One of the major drawbacks of conventional medical approaches and therapies is the difficulty in discriminating cancer cells from healthy cells. Nevertheless, there are other distinctive properties of cancer cells, useful in providing new targeting strategies. Firstly, there is an extracellular environment of cancer that is acidic. Secondly, the temperature of the cancer cell seems to be higher than that of the environment. Thirdly, the cancer-associated enzymes and the surface molecules expressed on cancer cells, together with the hypoxic conditions and reductive oxygen species (ROS), are specific properties of such diseases.

The answer to such a specific regime is polymeric NPs that are able to be easily modified and manipulated to a desired size and surface architecture. Moreover, as the addition to the effect of the passive targeting approach, particles are prepared in range of desired size between 10 and 200 nm. Such a strategy enables improvements in localization and drug delivery uptake in cancer-recognizing molecules for active targeting.

There is a wide variety of different polymeric NP delivery systems of anticancer drugs investigated for targeting different cancers. Furthermore, the curcumin can be combined with efficient cancer-targeting delivery systems using appropriate biocompatible polymers (PLGA, lecithin, chitosan silk fibroin, and Eudragit®). In comparison with free curcumin investigations, the results obtained with curcuma in nano-formulation achieved a huge increase in bioavailability as a result of better adhesion to cancer cells in the colon.

2.3. Polymeric Nanoparticles in the Treatment of Infectious Diseases

Microorganisms are more and more resistant to antibiotics. The World Health Organization presented data and outlined this problem as one of the worst threats to public health [11].

Various materials that patients in hospitals come into contact with (such as bedding, towels, sheets, bandages, catheters, and other hospital equipment) can present sources of danger if they are not sterile, so there is an imperative that technology aims to develop such materials that will have protective and preventive action against antibiotic-resistant microorganisms [12].

Staphylococcus aureus can cause various infections, especially infections on skin, soft tissues, bones, and blood vessels, which is the most common consequence of postoperative surgery. Some strains of *S. aureus* can cause various specific symptoms including toxic shock syndrome. The first strains of resistant microorganisms appeared in the 1960s. Initially, they appeared mainly in hospitals, but over the past decade, the appearance of MRSA has significantly expanded to several countries around the world (WHO 2014, Figure 2). *S. aureus* mortality decreased significantly between 1981 and 2000, but, as incidence rates doubled, the total number of deaths increased, emphasizing the need for further preventive measures and the development of new materials (Benfield et al., 2007). In hospitals, open-wound patients, inhaled devices, and those with impaired immune systems have a higher risk of infections from a wider population. MRSA infections after surgery are rare, but may occur in wounds, chests, or bloodstream (bacteriemia). MRSA infections occur in 1 to 33% of cases in persons undergoing surgery and infections can endanger life and cause a prolongation of hospital stay.

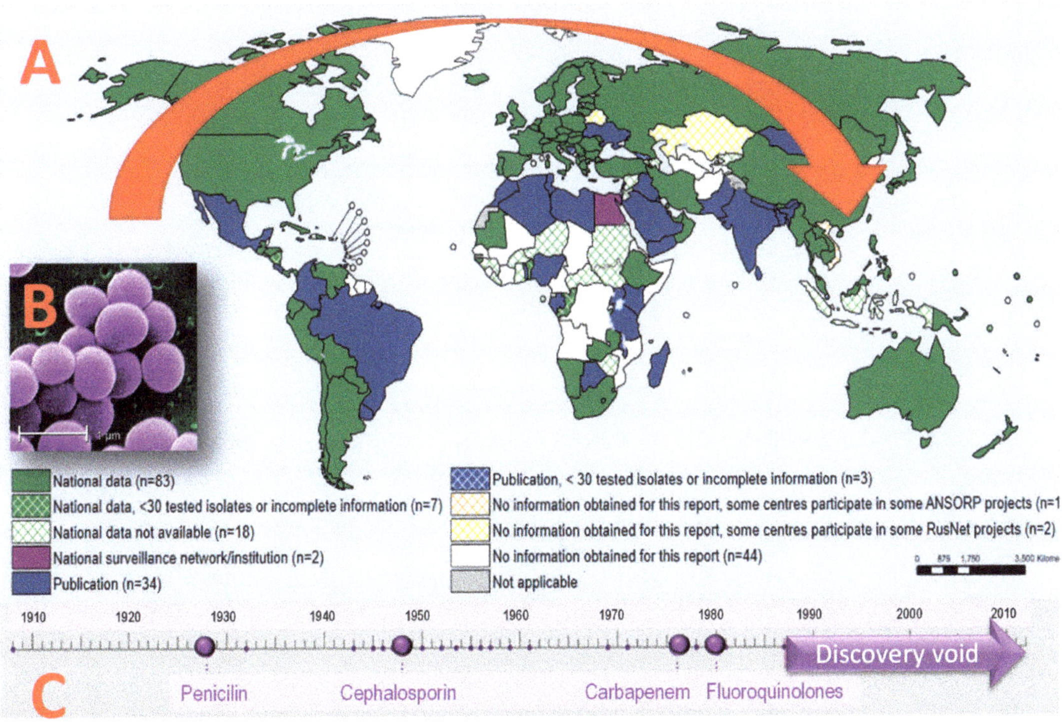

Figure 2. (**A**) Collected data on the number of infections by *S. Aureus* (MRSA), WHO 2014; (**B**) SEM microphotographs of *Staphylococcus aureus* [13]; (**C**) discovery void in new antibacterial drugs during the last 30 years.

The first wave of strain *S. aureus* resistance to penicillin appeared in the mid-1940s; therefore, since then, *S. aureus* can be divided into strains *resistant (MRSA)* or *sensitive (MSSA)* to methicillin. MRSA infection results in 50% higher probability of death in hospital compared with MSSA infection [14] The most frequent incidence of MRSA infection in hospitals has led to greater use of other antibiotics such as vancomycin, resulting in vancomycin-resistant *S. aureus (VRSA)* microorganisms. The first such case was described in Japan in 1996 [15] and afterwards in England, France, and the United States. Furthermore, *S. aureus* causes many types of infections:

- Skin (McCaig et al., 2006, [16]);
- Food poisoning (Wieneke et al., 1993 [17]);
- Soft tissues, bones (Sheehy et al., 2010 [18]);
- Bacteriemia–blood vessels (Khatib et al., 2009 [19]);
- Infective endocarditis (Fowler et al., 2005 [20]);
- Infections–pneumonie [21].

Untreated bacteriemia of *S. aureus* has a mortality rate of 80% [19,22]; most patients do not survive the first year [23] and, in Western European countries, very often patients in hospitals do not receive adequate therapy at the time [22]. A major problem in hospitals is non-sterile materials, because MRSA is easily transmitted through contaminated hands, clothing, or nesterious medical supplies after contact with a patient infected with MRSA either directly (after contact with a patient, blood, tissue fluid, secretions, and excretions) or indirectly (through contaminated instruments, objects, equipment, surfaces, and similar materials). The spread of MRSA can be prevented by the usage of disposable gloves, capes, and masks, but this is not always feasible. Therefore, new materials need to be developed that are active against microorganisms resistant to antibiotics. In the development of new antimicrobial materials, as well as in the functionalization and modification of existing ones, the application of nanoparticles of metal and metal oxides plays a very important role. The nanoparticles have excellent new medical, mechanical, optical, magnetic, catalytic, and electronic properties owing to a specific surface that is highly dependent on their size, structure, and shape. The global demand for nanoparticles of metal and metal oxide is projected to grow to 1700 tons in 2020 [5].

Nanoparticles have strong antibacterial activity on a wide range of Gram-positive and Gram-negative bacteria, including strains resistant to antibiotics [24]. They have great application because of their chemical stability, catalytic activity, and high conductivity. Rezić et al. have shown that silver nanoparticles have higher antibacterial activity owing to their high surface-to-volume ratio, which ensures better contact with microorganisms [4]. Furthermore, Al-Dhabi et al. have demonstrated visible antimicrobial activity against pathogenic wound infections such as *Bacillus subtilis*, *Enterococcus faecalis*, *Staphylococcus epidermidis*, multidrug-resistant *Staphylococcus aureus*, and *Escherichia coli* [25]. Silver nanoparticles with their own antimicrobial activity in combination with antibiotics (such as penicillin G, amoxicillin, erythromycin, clindamycin, or vancomycin) enhance the action of antibiotics in the treatment of resistant *Staphylococcus aureus* and *Escherichia coli* infections [26].

The interaction between nanoparticles and microorganisms is complex. Two interrelated mechanisms are described in the literature: (1) membrane potential disturbance and (2) production of reactive oxygen species (ROS), where nano-particles act as nanocatalysts [27,28]. Microbial membrane damage occurs when nanoparticles electrostatically bind to the surface of the bacterium, resulting in changes in the membrane wall and membrane potential and loss of integrity. Reactive oxygen species (ROS) are the cytotoxicity of nanoparticles in vivo and in vitro conditions where oxidative stress on the surface damages the integrity of microbial membranes by lipid peroxidation, further damaging the protein and enzyme function and damaging DNA and RNA. In some cases, ROS is induced by either visible or UV light and photocatalytic nanoparticle toxicity, such as, for example, TiO_2 nanoparticles, which, under UV light, cause lipid peroxidation, respiratory dysfunction, and death of methicillin-resistant *Staphylococcus aureus* cells [29]. Other mechanisms of toxicity of bacterial cell nanoparticles include direct inhibition of specific essential enzymes, induction of nitrogen reactive species [30], and induction of programmed cell death or apoptosis [31]. Table 3 shows the activity of silver nanoparticles of various dimensions on microorganisms.

Polymers containing metal nanoparticles and metal oxides have antimicrobial effects [32], but their effects on many pathogenic microorganisms have not yet been sufficiently explored. The main antibiotic groups currently used affect three bacterial targets: cell membrane synthesis, translation, and DNA replication [33]. Unfortunately, bacterial resistance can occur on each of these antimicrobial effects. The main mode of nanoparticle effect is direct contact with the bacterial membrane, without the need for penetration of the membrane and the wall, which opens the possibility for the microorganisms to grow more resistant to them. Today, silver nanoparticles are used in coatings on many materials such as surgical instruments, catheters or masks, orthopedics and dentistry (as additives for bone and dental materials), diagnostics (for increasing sensitivity of bio-detection), and ultra-susceptible clinical tests for the diagnosis of infarction myocardium and fluorescence detection of RNA [34] (Table 4).

Table 3. Antimicrobial activity of silver nanoparticles on microorganisms [3,9,32].

Biological System	Effect	Size, nm	Biological System	Effect	Size, nm
Acinetobacter baumannii	Antibacterial	5–30	*Micrococcus luteus*	Growth inhibition	~20–~30
Acinetobacter	Growth inhibition	~20–100	*multidrug resistant Pseudomonas aeruginosa*	Inhibition of biofilm formation	20–30
Aeromonas	Growth inhibition	~20–100	*multidrug-resistant Escherichia coli*	Antibacterial	~6
Aspergillus foetidus	Antifungal	104.9	*multidrug-resistant Klebsiella pneumoniae*	Antibacterial	~6
Aspergillus fumigatus	Antifungal	5–30	*Staphylococcus aureus*	Bactericide	~10–~100
Aspergillus niger	Antifungal	4.75–8.31	*Paenibacillus koreensis*	Antibacterial	~10
Aspergillus oryzae	Antifungal	104.9	*Penicillium*	Antifungal	>42
Aspergillus	Antifungal	7–20	*Peptostreptococcus*	Antibacterial	~100
Aspergillus terreus	Antifungal	Nije specificirano	*Pichia pastoris*	Antifungal	9–10
Bacillus cereus	Antibacterial	120	*Propionibacterium acnes*	Antibacterial	20–70
Bacillus megaterium	Antibacterial	24	*Proteus mirabilis*	Cytotoxic	Not specified
Bacillus mycoides	Bactericide	16	*Proteus vulgaricus*	Antibacterial	50–70
Bacillus pumulis	Bactericide	Not specified	*Pseudomonas aeruginosa*	Antibacterial	1–12
Bifidobacterium	Antibacterial	~100	*Rhizoctonia bataticola*	Antifungal	~2–40
Bordetella pertussis Candida albicans	Growth inhibition	~20–~30	*Saccharomyces*	Antifungal	~15
Candida albicans ATCC 10239	Antifungal	~20–45	*Salmonella paratyphi*	Antibacterial	63–90
Candida glabrata ATCC90030	Antifungal	Not specified	*Scedosporium JAS1*	Antifungal	Not specified
Candida tropicalis	Antifungal	Not specified	*Serratia marcescens*	Antibacterial	Not specified
Citrobacter	Growth inhibition	~20–100	*Shigella flexneri MTCC 1475*	Antibacterial	17–29
Cryptococcus neoformans	Antifungal	50–70	*Staphylococcus aureus*	Antibacterial	1–12
Eschericia coli MTCC 443	Bactericide	8–12	*Staphylococcus aureus ATCC 25923*	Antibacterial	5–25
Enterobacter aerogenes	Antibacterial	8–12	*Staphylococcus aureus ATCC BAA-1721*	Antibacterial	~89
Enterococcus faecalis	Antibacterial	~89	*Staphylococcus epidermidis*	Antibacterial	20–70
Fusarium oxysporum	Antifungal	104.9	*Streptococcus*	Antibacterial	15
Klebsiella pneumoniae	Antibacterial	~6	*Trichopyton rubrum*	Antifungal	~15
Lactobacillus acidophilus	Bactericide	16	*Vibrio cholerae*	Antibacterial	5–25
Listeria monocytogenes	Antibacterial	17–29	*Xanthomonas campestris*	Antibacterial	~54
MRSA	Growth inhibition	198–595	*Yersinia enterocolitica*	Antibacterial	-

Table 4. Application of silver nanoparticles in medicine items [35].

Area	Application of Silver Nanoparticles
Anesteziology	Coating on breathing masks, endotracheal tubes for mechanical ventilation assistance
Cardiology	Coating on the tracking catheter
Stomatology	Adhesives in dental materials, silver-filled SiO_2 nanocomposite resins
Diagnostics	Ultra-sensitive and ultra-fast platform for clinical tests for myocardial infarction diagnosis, fluorescence detection of RNA
Drug release	Remote laser induced opening microcapsules
Eye care	Contrasts on contact lenses
Visualisation	Nomenclatures for marking cells
Neurosurgery	Coating of the catheter for the drainage of cerebrospinal fluid
Orthopedics	Bone cement additives, joint replacement implants, orthopedic socks
Drugs	Treatments for dermatitis, ulcerative colitis and acne, HIV-1 inhibition
Surgery	Coats in medical textiles—surgical suits and masks
Urology	Plastered on surgical nets for pelvic reconstruction

In addition to the pharmaceutical industry, nanoparticles of silver are used in the cosmetic industry thanks to their strong antibacterial and anti-inflammatory properties, in cosmetic products such as deodorants, anti-aging creams, and so on. Nanotechnology can solve the problem of bio-film formation by using an antibacterial active surface with a combination of ZnO and MgO nanoparticles that is activated in the dark and is effective against *MRSA* species [36].

Considering the new solutions, it is not surprising that resistance to nanoparticles by microorganisms has already occurred; Panaček et al. demonstrated that gram-negative bacteria *of Escherichia coli 013, Pseudomonas aeruginosa CCM 3955* and *E. coli CCM 3954* can develop resistance to silver nanoparticles after repeated exposure and resistance results from the production of adhesive flagellin protein, which activates the aggregation of nanoparticles [37].

Research on antimicrobial nanoparticle coatings has also been applied in the space program: recently, the *International Space Station* tested a silver and ruthenium coating that showed excellent properties against Gram-negative and Gram-positive bacterial strains, including pathogenic bacteria with multiple MRSA resistance, *Enterococcus faecalis, Staphylococcus epidermidis, E. coli* pathogens (ESTEC), *Pseudomonas aeruginosa, Acinetobacter baumannii,* and *Legionella.* The International Space Station is an extremely closed area where bacteria develop special antibiotic defense mechanisms by developing a thicker cell wall or highly express virulent genes. Moreover, the microgravity and cosmic radiation can increase the virulence of microorganisms and transform them into potential pathogens. In addition, these conditions also reduce the immune defense of astronauts, which, when linked to psychological stress associated with space flights, makes them much more prone to infections [38].

2.4. Polymeric Nanoparticles in Implants and Prosthetic Devices

One of the most common uses of NPs is in implantable devices. Implantable devices must possess specific requirements, including good biocompatibility, tissue affinity, corrosion resistance and particularly antibacterial property. There is a wide variety of different classical medical devices, implants, and prosthetics: catheters, dental implants, pacemakers, prostheses, and others. Such materials come in direct and prolonged contact with biological tissue. Therefore, their biocompatibility is a critical property that limits their application.

Frequently reported problems are related to toxicity, allergic reactions, inflammation, and sometimes bacterial infections [39].

2.5. Polymeric Nano-Particles as Theranostic Devices

Therapeutics systems simultaneously act as therapeutic and a diagnostic agent. They are designed to enable target-specific drug delivery while at the same time monitoring the release of the drugs. Moreover, they can monitor the progression of the disease progression while at the same time delivering the treatment response. Although traditional diagnostic imaging (magnetic resonance imaging, computed tomography, and angiography) operates with inorganic contrast agents, organic fluorophores represent a promising alternative, as they show a good safety profile.

2.6. Other Applications

There are many other applications, from formulations created for skin, hair, nails, or for other formulations. For other applications, there are different sizes of nanoparticles (from the range of 50 to 500 nm), as those enable the penetration through the skin into the organism. Through this property, the development of so-called "transdermal drugs" was enabled [39]. Polymers used as carriers are chitosan, albumin, PLGA, or PLA.

In addition, inorganic nanoparticles are useful in antibacterial coating; thus, for example, TiO_2 coated with silver-doped hydroxyapatite or silver-coated collagen proved to possess antibacterial activity useful for catheters and dental implants effective against *S. aureus* and *P. aeruginosa* [40].

3. Synthesis of Nanoparticles for Biomedical Application

The first attempt of synthesis of metallic particles by enzymes was performed by McConnel and Frajola (1961), when the synthesis of carbonate apatite was carried out in the presence of carbon anhydrase. Today, the advents and developments in nanotechnology resulted in a wide range of microorganisms and their enzymes, which can be utilized in the production of NPs.

There are three chemical synthetic pathways for synthesis of nanoparticles: the Turkevich method, Brust–Schiffrin method, and seed growth method.

John Turkevich reported his method in 1951 using Na-citrate as a reducing agent and stabilizer for creating spheric NPs of size of 10 to 20 nm. Similarly, the Brust–Schiffrin method from 1994 enabled formation of 1.5–5.2 nm NPs, using thiole at an ambient condition. Lastly, the seed growth method enabled excellent control of size and shape.

In contrast to conventional chemical synthetic pathways, there are many green methods that use plant extracts or microorganisms in the synthesis of NPs. New methods are developed in order to avoid application and huge usage of unwanted toxic chemicals that are usually used in synthesis of NPs. Those approaches are known as *green methods* in which plant extracts and microorganisms are used for the synthesis of AuNPs as substitutions for toxic reagents.

Although a reduction step is also involved in this process, this methodology is known as green synthesis owing to the utilization of herbal extracts. Another advantage of this procedure is the fact that NPs prepared using this methodology do not require a further functionalization agent [39].

Some examples of green synthesis of biomedically applicable nanoparticles are presented in Table 5.

Table 5. Nanoparticles synthesized by microorganisms [41–44].

Microorganisms	Nanoparticles	Temperature, °C	Size (nm)	Shape
Rhodococcus species	Au	37	5–15	sphere
Shewanella oneidensis	Au	30	12 ± 5	sphere
Plectonemaboryanum	Au	25–100	<10–25	cube
Plectonema boryanum	Au	25	10–600	octahedral
Escherichia coli	Au	37	20–30	triangle
Yarrowia lipolytica	Au	30	15	triangle
Rhodopseudomonas capsulate	Au	30	10–20	sphere
Brevibacterium casei	Au, Ag	37	10–50	sphere
Trichoderma viride	Ag	27	5–40	sphere
Phaenerochaete chrysosporium	Ag	37	50–200	pyramidal
Bacillus cereus	Ag	37	4–5	sphere
Lactobacillus species	Ba, TiO_3	25	20–80	tetragonal
Fusarium oxysporum	CdSe	10	9–15	sphere
Escherichia coli	Cd, Te	37	2.0–3.2	sphere
Fusariumoxysporum	$CdCO_3$, $PbCO_3$	27	120–200	sphere
Lactobacillus	CdS	25–60	4.9 ± 0.2	sphere
Pyrobaculum islandicum	Cr, Mn, U, Tc	100	different	sphere
Shewanella oneidensis	Fe_3O_4	28	40–50	hexagonal
Enterobacter species	Hg	30	2–5	sphere
Desulfovibrio desulfuricans	Pd	25	50	sphere
Shewanella algae	Pt	25	5	-
Saccharomyces cerevisiae	Sb_2O_3	25–60	2–10	sphere
Shewanella species	Se	30	181 ± 40	sphere
Fusarium oxysporum	$SrCO_3$	27	10–50	needle like
Fusarium oxysporum	TiO_2	300	6–13	sphere
Fusarium oxysporum	ZrO_2	25	3–11	sphere

There are several biopolymers used in the synthesis of nanoparticles: PVP, PEG, polyvinyl alcohol/, -/acryl amide, -/ally amine, -/phenols, and -/methyl amino ethyl metha crylate. In addition, different surfactants can be used in this reaction [39].

The NPs can also be prepared using biopolymers. Such a procedure can be performed using a single biopolymer or their combination with other reducing agents.

Some of the useful biopolymers are as follows: pectin, glucose, dextrin, amino-dextrin, gum Arabic, starch, chitosan, proteins, gelatin, collagen, casein, tyrosine, alpha-amylase, aspartic acid, tyrosine, tryptophan, glutamic acid, sodium glutamate, cystine, and L-leucine, among others [39] The organic reagents used are as follows: amino alcohols, glycerol, luminol, nitriloacetic acid, and sodium rhodizonate, enabling the reduction of gold NPs from $HAuCl_4$.

Enzymatic synthesis of nanoparticles at moderate temperatures will be investigated in this project, because it is of prime interest for the industry; that is, mild manufacturing conditions will ensure low energy consumption, economical production, less manpower, and compatibility with other fabrication processes. Combinations of NPs with proteins allow medical usage of NPs as therapies for effective cancer treatment. Moreover, if the synthesis of NPs is carried out in enzyme-mediated reactions, chemical and physical properties of products can differ from the properties of NPs produced without the presence of enzymes, providing wide variety of their possible novel applications. By understanding this mechanism, the reactions can be guided in the direction of production of targeted nanoparticles with desired applicable properties.

Enzymes are protein catalysts that are characterized by having a high degree of specificity for their substrates. They catalyze specific chemical reactions: hydrolysis, reduction-oxidation, phase transition, elimination of specific functional groups, and isomerisation. Although the enzymatic reactions naturally occur in living organisms, thanks to technological development, they can now be utilized in bioreactors in different scientific and industrial laboratories with the goal of producing different chemical compounds (drugs and detergent products), as well as nanoparticles [45].

Urease is a very interesting enzyme for producing NPs. It can be isolated from different microorganisms (*Aspergillus niger*, *Proteus mirabilis*, *Bacillus subtilis*, and *Aerobacter aerogenes*) and applied for the production of NPs of different morphologies like thin films, hollow micro- and nano-spheres, nanotubes, 2D patterns, 3D replicas of biominerals, and others [43]. Therefore, the *urease*-mediated synthesis of NPs is an attractive research field that enables the possibilities of producing NPs for a wide variety of applications: in cancer therapy, for producing materials of special purposes, in food packing, geotextiles, and many others [46].

Methods for producing NPs by *urease*-catalyzed reactions can be divided into two main categories: (i) reactions in which the precipitants of metal ions are produced by enzymatic reactions (metal ions in solutions precipitate into oxides, hydroxides, carbonates, or phosphates), and (ii) reactions in which enzymes directly interact with metal-containing substrates to produce metallic materials [42].

Reactions using *urease* are usually very simple and obtained NPs can be formed as hydroxides, carbonates, hydratized oxides, or other chemical species. For example, if a small amount of *urease* is diluted in a solution containing calcium chloride and urea, the *urease* will hydrolyze urea into ammonia (carbonic acid will also be produced).

In the same reaction, calcium ions react with the carbonate at room temperature, and calcium carbonate is produced [47,48]:

$$H_2NCONH_2 + 2\,H_2O \rightarrow 2\,NH_3 + H_2CO_3 \quad (1)$$

$$Ca^{2+} + CO_3{}^{2-} \rightarrow CaCO_3 \quad (2)$$

The crystallinity, size, and mechanism of growth of produced NPs are governed by reaction parameters (e.g., concentrations, stirring, pH, temperature, pressure, and ultrasound effects). In this project, the kinetic of such enzymatically catalyzed reactions will be thoroughly investigated by different classical instrumental methods. In addition, *beyond-state-of-the-art* analytical and bioanalytical methods will be implemented for revealing the mechanisms of preselected reactions. Unuma et al. (2004) applied *urease* immobilized on alginate gel particles to produce porous aluminum NPs [43].

In this reaction, after reaction (1), the following reaction occurred:

$$Al^{3+} + 3NH_3 + 3H_2O \rightarrow Al(OH)_3 + 3\,NH_4^+ \quad (3)$$

Under high temperatures, aluminum hydroxide produced in reaction (3) converts into Al_2O_3 with a particle size of 1–2 mm (Unuma et al., 2004). Porous NPs of iron oxide are produced by a similar reaction:

$$Fe^{3+} + xOH^- + nH_2O \rightarrow FeO_{y/2}(OH)_{3-y} \times mH_2O \quad (4)$$

In contrast to the process of producing aluminum NPs (reaction 3), process (4) also utilizes the polymer matrices molecules, which degrade after drying and calcification in order to obtain porous iron NP structures [49]. In the reaction of *urease* with ammonia and carbonates, many other metallic NPs can be produced: magnetite [50], iron hydroxide [51], aluminum sulfate [52], calcium carbonate [47,53], strontium and barium carbonate [48], calcium phosphate [54], yttrium hydroxide [55], magnesium oxide, and zinc oxide, in the form of nano-shells (Figure 3).

Figure 3. Synthesis of ZnO nanoshells using the *urease* enzyme as a catalytic template. In the first step, NH_3 is generated by the hydrolysis of urea. This adjusts the pH value around the enzyme. In the second step, Zn^{2+} is added. The growth of nanoshells is governed by the local pH value at the enzyme/solution interface (De la Rica and Matsui, 2008).

Targeted products' metallic NPs have biological and medical activity, antimicrobial, water-repellent, and other protective and polymer-applicable properties. Owing to the special characteristics of the mechanism of its reaction, catalytic *urease* processes can be used for the synthesis of Al, Au, Ca, Cr, Fe, Ge, Pd, Pt, Ti, Y, Zn, and Zr nanoparticles. Some of those NPs (Ca, Ge, Pd, and Pt) are currently not applied in medical materials, while others (Zn and Zr) are applied only as oxides (Table 1). The produced NPs are subsequently deposited onto a polymer substrate in the form of a thin film. The application of such a coated protective material is foreseen for medical applications.

Urease is not the only enzyme that can be used. In addition, four oxidoreductases can be utilized: *cellobiose dehydrogenase* (CDH), *glucose dehydrogenase*, *glucose oxidase*, and *laccase*. These enzymes have many applications in biotechnology and are used for producing metallic NPs. For example, *glucose oxidase* extracted from *Aspergillus niger* can be used for production of different nanoparticles (e.g., Au, Sn, and others) [53]. The usage of CDH for producing gold nanoparticles has already been investigated and reported in the literature [54]. Their results showed that CDH acts as an electron donor and reduces gold from $[AuCl_4]^-$ complex into metallic gold nanoparticle by direct electron transfer (DET) at pH 5 or by mediated electron transfer (MET) at pH 7. The limiting factor for NPs' growth is the ability of CDH to interact with the growing Au NPs. This is an important fact, useful for controlling the direction of the reaction. For example, the ability of enzymes to control the rate of nucleation of NPs is the key step in using them as plasmatic nano-sensors.

Having all of this in mind, Au NPs should be the first targeted metal ion for biomedical applications [56,57].

4. Characterization of Biomedical Active Nano-Particles

There are many different methods for revealing the mechanisms of metallic NPs' biomedical application: *"beyond-state-of-the-art"* instrumental methods such as gas-phase electrophoretic mobility molecular analyzer (GEMMA) [58–61], parallel differential mobility analyzer (PDMA) [62], and liquid chromatography/capillary electrophoresis coupled to MALDI and ESI mass spectrometry (LC/CE-MS/MS), as well as classical spectroscopic methods (graphite furnace atomic absorption spectrometry (GF-AAS), inductively coupled plasma optical emission spectrometry (ICP-OES) [63–66], scanning electron microscopy equipped with the EDX detector (SEM-EDX) [67,68], thermal gravimetric analysis (TGA), Fourier transform infrared spectroscopy (FTIR), UV/VIS spectroscopy, and liquid chromatography (HPLC)), which is convenient to provide important information on mechanisms of reactions and obtaining products of desired crystallinity, size, and shape [69–72]. The identification of the newly isolated enzymes can be performed by MALDI-TOF/TOF-MS and LC-MS/MS methods.

Special focus and attention should be given to the determination of the kinetics of the systems studied, as well as to the revealing of the mechanisms of reaction by *"beyond-state-of-the-art-techniques"* GEMMA and PDMA coupled to capillary electrophoresis (CE), which were invented and developed by prof. Allmaier et al., for the analysis of nanosize materials. To the best of our knowledge, the application of GEMMA and PDMA to monitor enzymatic in situ reactions and to follow both of the enzymes and produce nanoparticles, i.e., the size-separated nanoparticles are collected and can be investigated by other techniques as SEM or AFM, is a new, *beyond-state-of-the-art* concept in the field.

Although different *state-of-the-art* techniques can be today utilized for the in situ measurement of nanoparticles [73–77], such attempts cannot provide insight into the particle growth. In future work, on the other hand, the focus should be shifted to understanding the mechanism of NPs' growth during the reaction by monitoring reactants and products (enzymes and occurred metallic nanoparticles). Having the possibility to obtain a deep insight into the mechanism of enzymatically guided reactions is crucial, as contradictory observations and conclusions are currently reported in highly ranked scientific journals (*Nature* and *Angewandte Chemie*). For example, B. Sharma et al. (2013) concluded that, during *urease*-guided reaction of the formation of hollow ZnO nanoparticles, the enzyme covers the outer shell (Sherma et al., 2013). This conclusion is in contradiction to the interpretation of De la Rica and Matsui (2008), who reported that the enzyme (Figure 3) fulfills the nanoparticles' inner core [78].

Widely used NPs include oxides of Al, Sb, Bi, Ce, Cu, In/Sn, Fe, Mg, Mn, Ni, Si, Ti, Y, Zn, and Zr, thanks to their novel electronic, optical, magnetic, medical, catalytic, and mechanical properties resulting from the high surface-to-volume ratio and quantum size effect. The primary target of biomedical application research should be to modify and control protective materials using newly prepared NPs (e.g., Ag, Al, Au, Cr, Ge, Fe, Pd, Pt, Ti, Y, Zn, and Zr NPs). Polymer materials coated with thin films of NPs have special protective properties, such as preventing oxygen, moisture, and microorganism permeability. The process of thin film deposition is very important, as it influences graininess and the structure of the deposit. In other words, the deposit can comprise grains of different sizes and its structure may suffer from defects in connectivity [79]. There are several thin film deposition technologies that can be applied: sputtering, sol–gel deposition, spray coating, chemical vapor deposition, spray pirolysis, cathode electro deposition, anodic conversion, and conventional evaporation. The properties of thin metal films with nanoholes smaller than the light wavelength have recently attracted large amounts of interest, owing to the remarkable observation of extraordinary light transmission through metallic sub-wavelength hole arrays, which does not obey classical optical theory [80]. Therefore, several methodologies should be tested including the sol–gel process of dip coating [81–84], for combining different NPs and different polymer materials [85,86], in order to create metal NP films of different thicknesses that are most efficient for biomedical applications (Figure 4) [87].

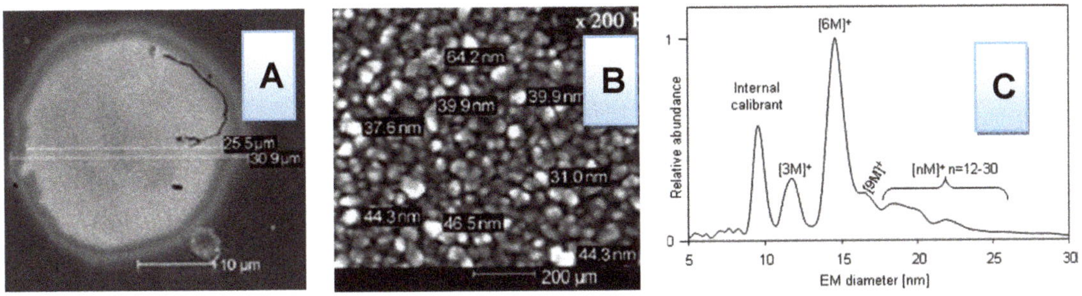

Figure 4. Results of investigation into biomedically active NPs: (**A**,**B**) SEM-magnified images of gold NPs obtained by *cellobiose dehydrogenase* treatment of $[AuCl_4]^-$ (Malel et al., 2010); (**C**) GEMMA spectrum of non-covalent homo-complex of *Jackbean urease* (Allmaier et al., 2008).

The important nanoparticles in biomedical applications are inorganic and organic nanoparticles. Among inorganic species, gold nanoparticles and silver NPs are today the most investigated and prominent species.

The advantages of nanoparticles that are important in biomedical application, as well as their synthesis, have been described in this work. However, it has to be emphasized that such nanomaterials can be very toxic. The level of their toxicity will depend on their physical, chemical, and mechanical (morphological) properties. At this point, toxic effects of nanoparticles should be considered before the determination of the most appropriate synthesizing pathway for producing nanoparticles. By optimizing this step, the desired size, shape, composition, and other most important parameters should be varied in order to prove that the result will not have any adverse toxic effects.

Therefore, during the synthesis of NPs by engineering the desired outcome properties, the toxicity can be avoided and minimized. For example, the parameters such as the size and surface area, as well as their shape, aspect ratio, surface coating, crystalline, and dissolution, as well as agglomeration, are dominant parameters. Such parameters will influence the capability of NPs to begin the formation of reactive species (ROS) and all other mechanisms important in cytotoxicity, genotoxicity, and neurotoxicity processes.

There are some identified toxic mechanisms. Those are detected through the induction of the formation of reactive species and cytotoxicity to cells, as well as genotoxic and neurotoxic effects. For example, smaller nanoparticles had a higher rate of acute toxicity in in vivo experiments. In contrast to this field, the area of investigation of polymeric nano-sized materials is much broader and enables a huge variety of different targeting materials.

However, the advantages of nanomaterials also provide interesting applications in technological and industrial areas. What are the future aspects of the nanomaterials? Firstly, there are carbon nanotubes, which are light materials with special physical properties, because of which they can revolutionize the design of cars by enabling the possibility to conduct electricity and heat. Secondly, there are nano-robots. It is likely that such devices are the next stage in miniaturization and, as such, they can lead to the production of nanorobots that can help cells of living organisms in their daily tasks [88]. Last, but not least, many authors describe nano and micro vehicles that can be operated, moved, and guided to move on particular surfaces; this application is important in many different aspects of life.

5. Conclusions

The field of nanomaterial science is growing so rapidly that it has already resulted in the implementation of materials with nanoparticles in many different innovations. It can be estimated that the application of biomedically active NPs in combination with different polymers in the form of thin films, nanocapsules, or some other form will result in producing new materials with changed physical, chemical, and mechanical properties. For example, properties such as anti-microbial activity, water repellence, and low oxygen permeability can be exploited for biomedically active surfaces. On the other hand, hollow NPs that can release encapsulated molecules can be used in targeted drug delivery. In addition to this, polymers coated with NPs could be beneficial for catheters, prostatic materials, and others. Although there are many review articles with similar topics, many similar articles are focused on only one nanoparticle [89–100]. In contrast to the work that can be found in the recent literature, this manuscript covered many different nanoparticles, from inorganic silver, gold, to organic (polymeric) species. More importantly, the most important aspect of this work is the fact that not only the possible biomedical application of nanoparticles is presented, but also their toxicological effect. As this can be minimized by influencing the parameters of the synthesis of such materials, only an integral approach can bring benefits to the desired nanomaterials and their real application.

However, the application of newly created biomedically active materials will depend on their toxicity, environmental impact, and costs of the final product. Significant transfer of knowledge among interdisciplinary researchers is expected in the following years, which should lead to solving particular problems in analytical chemistry, material sciences, and nanotechnology related to biomedically active NPs.

Funding: This research was funded by the Croatian Science Foundation grant number IP-2019-04-1381 (project under title 'Antibacterial coating for biodegradable medicine materials ABBAMEDICA'). Any opinions, findings, and conclusions or recommendations expressed in this material are those of the authors and do not necessarily reflect the views of the Croatian Science Foundation.

Institutional Review Board Statement: Not applicable.

Informed Consent Statement: Not applicable.

Data Availability Statement: Not applicable.

Conflicts of Interest: The author declares no conflict of interest. The funders had no role in the design of the study; in the collection, analyses, or interpretation of data; in the writing of the manuscript; or in the decision to publish the results.

References

1. Khursheed, R.; Dua, K.; Vishwas, S.; Gulati, M.; Jha, N.K.; Aldhafeeri, G.M.; Alanazi, F.G.; Goh, B.H.; Gupta, G.; Paudel, K.R.; et al. Biomedical applications of metallic nanoparticles in cancer: Current status and future perspectives. *Biomed. Pharmacother.* **2022**, *150*, 112951. [CrossRef]
2. Rezić, I. Determination of engineered nanoparticles on textiles and in textile wastewaters. *TrAC Trends Anal. Chem.* **2011**, *30*, 1159–1167. [CrossRef]
3. Rezić, I. Engineered Nanoparticles in Textiles Textile Wastewaters. In *Analysis Risk of Nanomaterials in Environmental Food Samples*; Farre, M., Barcelo, D., Eds.; Elsevier: Chennai, India, 2012; Volume 59, pp. 235–264.
4. Anfray, C.; Mainini, F.; Andón, F.T. Nanoparticles for immunotherapy. *Front. Nanosci.* **2020**, *16*, 265–306.
5. Jesus, M.D.L.F.; Grazu, V. *Nanobiotechnology: Inorganic Nanoparticles vs Organic Nanoparticles, Frontiers of Nanoscience*; Palmer, R.E., Ed.; Elsevier: Oxford, UK, 2012; Volume 4.
6. Lopez-Sanchez, J.A.; Dimitratos, N.; Hammond, C.; Brett, G.L.; Kesavan, L.; White, S.; Miedziak, P.; Tiruvalam, R.; Jenkins, R.L.; Carley, A.F.; et al. Facile removal of stabilizer-ligands from supported gold nanoparticles. *Nat. Chem.* **2011**, *3*, 551–556. [CrossRef] [PubMed]
7. Genet, C.; Ebbesen, T.W. Light in tiny holes. *Nature* **2007**, *445*, 39–46. [CrossRef] [PubMed]
8. Ding, Y.; Jiang, Z.; Saha, K.; Kim, C.S.; Kim, S.T.; Landis, R.F.; Rotello, V.M. Gold Nanoparticles for Nucleic Acid Delivery. *Mol. Ther.* **2014**, *22*, 1075–1083. [CrossRef] [PubMed]
9. Rezić, I.; Zeiner, M.; Steffan, I. Determination of 28 selected elements in textiles by axially viewed inductively coupled plasma optical emission spectrometry. *Talanta* **2011**, *83*, 865–871. [CrossRef] [PubMed]

10. Hinterwirth, H.; Wiedmer, S.K.; Moilanen, M.; Lehner, A.; Allmaier, G.; Waitz, T.; Lindner, W.; Lämmerhofer, M. Comparative method evaluation for size and size-distribution analysis of gold nanoparticles. *J. Sep. Sci.* **2013**, *36*, 2952–2961. [CrossRef]
11. World Health Organization. Antimicrobial Resistance Global Report on Surveillance. Available online: https://www.who.int/drugresistance/documents/surveillancereport/en/ (accessed on 2 October 2022).
12. Van Hal, S.J.; Jensen, S.O.; Vaska, V.L.; Espedido, B.A.; Paterson, D.L.; Gosbell, I.B. Predictors of mortality in Staphylococcus aureus bacteremia. *Clin. Microbiol. Rev.* **2012**, *25*, 362–386. [CrossRef]
13. Arduino, J.M. DRPH Photo Credit: Janice Haney Carr Center for Disease Control Prevention's Public Health Image Library PHIL, ID #11157. Available online: https://phil.cdc.gov/Details.aspx?pid=11156 (accessed on 2 October 2022).
14. Hanberger, H.; Walther, S.; Leone, M.; Barie, P.S.; Rello, J.; Lipman, J.; Marshall, J.C.; Anzueto, A.; Sakr, Y.; Pickkers, P.; et al. Increased mortality associated with meticillin-resistant Staphylococcus aureus (MRSA) infection in the Intensive Care Unit: Results from the EPIC II study. *Int. J. Antimicrob. Agents* **2011**, *38*, 331–335. [CrossRef]
15. Hiramatsu, K.; Aritaka, N.; Hanaki, H.; Kawasaki, S.; Hosoda, Y.; Hori, S.; Fukuchi, Y.; Kobayashi, I. Dissemination in Japanese hospitals of strains of Staphylococcus aureus heterogeneously resistant to vancomycin. *Lancet* **1997**, *350*, 1670–1673. [CrossRef]
16. McCaig, L.F.; McDonald, L.C.; Mandal, S.; Jernigan, D.B. Staphylococcus aureus–associated skin and soft tissue infections in ambulatory care. *Emerg. Infect. Dis.* **2006**, *12*, 1715. [CrossRef] [PubMed]
17. Wieneke, A.A.; Roberts, D.; Gilbert, R.J. Staphylococcal food poisoning in the United Kingdom, 1969–90. *Epidemiol. Infect.* **1993**, *110*, 519–531. [CrossRef]
18. Sheehy, S.; Atkins, B.; Bejon, P.; Byren, I.; Wyllie, D.; Athanasou, N.; Berendt, A.; McNally, M. The microbiology of chronic osteomyelitis: Prevalence of resistance to common empirical anti-microbial regimens. *J. Infect.* **2010**, *60*, 338–343. [CrossRef] [PubMed]
19. Khatib, R.; Johnson, L.B.; Sharma, M.; Fakih, M.G.; Ganga, R.; Riederer, K. Persistent Staphylococcus aureus bacteremia: Incidence and outcome trends over time. *Scand. J. Infect. Dis.* **2009**, *41*, 4–9. [CrossRef] [PubMed]
20. Fowler, V.G.; Miro, J.M.; Hoen, B.; Cabell, C.H.; Abrutyn, E.; Rubinstein, E.; Corey, G.R.; Spelman, D.; Bradley, S.F.; Barsic, B.; et al. Staphylococcus aureus endocarditis: A consequence of medical progress. *JAMA* **2005**, *293*, 3012–3021. [CrossRef]
21. Tong, S.Y.C.; Davis, J.S.; Eichenberger, E.; Holland, T.L.; Fowler, V.G., Jr. Staphylococcus aureus Infections: Epidemiology, Pathophysiology, Clinical Manifestations, and Management. *Clin. Microbiol. Rev.* **2015**, *28*, 603–661. [CrossRef]
22. Ammerlaan, H.; Seifert, H.; Harbarth, S.; Brun-Buisson, C.; Torres, A.; Antonelli, M.; Kluytmans, J.; Bonten, M. Study on European Practices of Infections with Staphylococcus aureus (SEPIA) Study Group Adequacy of Antimicrobial Treatment and Outcome of *Staphylococcus aureus* Bacteremia in 9 Western European Countries. *Clin. Infect. Dis.* **2009**, *49*, 997–1005. [CrossRef]
23. Gotland, N.; Uhre, M.L.; Mejer, N.; Skov, R.; Petersen, A.; Larsen, A.R.; Benfield, T.; Danish Staphylococcal Bacteremia Study Group. Long-term mortality and causes of death associated with Staphylococcus aureus bacteremia. A matched cohort study. *J. Infect.* **2016**, *73*, 346–357. [CrossRef]
24. Díez-Pascual, A.M. Antibacterial Activity of Nanomaterials. *Nanomaterials* **2018**, *8*, 359. [CrossRef]
25. Al-Dhabi, N.A.; Mohammed Ghilan, A.K.; Arasu, M.V. Characterization of silver nanomaterials derived from marine Streptomyces sp. al-dhabi-87 and its in vitro application against multidrug resistant and extended-spectrum beta-lactamase clinical pathogens. *Nanomaterials* **2018**, *8*, 279. [CrossRef] [PubMed]
26. Wijnhoven, S.W.; Peijnenburg, W.J.; Herberts, C.A.; Hagens, W.I.; Oomen, A.G.; Heugens, E.H.; Roszek, B.; Bisschops, J.; Gosens, I.; Van De Meent, D.; et al. Nano-silver—A review of available data and knowledge gaps in human and environmental risk assessment. *Nanotoxicology* **2009**, *3*, 109–138. [CrossRef]
27. Blecher, K.; Nasir, A.; Friedman, A. The growing role of nanotechnology in combating infectious disease. *Virulence* **2011**, *2*, 395–401. [CrossRef] [PubMed]
28. Huh, A.J.; Kwon, Y.J. "Nanoantibiotics": A new paradigm for treating infectious diseases using nanomaterials in the antibiotics resistant era. *J. Control. Release* **2011**, *156*, 128–145. [CrossRef]
29. Shah, M.A.; Ahmad, T. *Principles of Nanoscience and Nanotechnology*; Alpha Science International: London, UK; Oxford, UK, 2010; pp. 10–20.
30. Pelgrift, R.Y.; Friedman, A.J. Nanotechnology as a therapeutic tool to combat microbial resistance. *Adv. Drug Deliv. Rev.* **2013**, *65*, 1803–1815. [CrossRef]
31. Abramovitz, I.; Wisblech, D.; Zaltsman, N.; Weiss, E.I.; Beyth, N. Intratubular Antibacterial Effect of Polyethyleneimine Nanoparticles: An *Ex Vivo* Study in Human Teeth. *J. Nanomater.* **2015**, *2015*, 1–5. [CrossRef]
32. Stoimenov, P.K.; Klinger, R.L.; Marchin, G.L.; Klabunde, K.J. Metal Oxide Nanoparticles as Bactericidal Agents. *Langmuir* **2002**, *18*, 6679–6686. [CrossRef]
33. Magiorakos, A.P.; Srinivasan, A.; Carey, R.B.; Carmeli, Y.; Falagas, M.E.; Giske, C.G.; Harbarth, S.; Hindler, J.F.; Kahlmeter, G.; Olsson-Liljequist, B.; et al. Multidrug-resistant, extensively drug-resistant prug-resistant bacteria: An international expert proposal for interim stard definitions for acquired resistance. *Clin. Microbiol. Infect.* **2012**, *183*, 268–281. [CrossRef]
34. Varner, K.E.; El-Badawy, A.; Feldhake, D.; Venkatapathy, R. *State-of-the-Science Review: Everything Nanosilver and More*; US Environmental Protection Agency: Washington, DC, USA, 2010.
35. Kesić, A.; Horozić, E. Prednosti i neželjeni efekti primjene metalnih nanočestica i metalnih oksida u farmaceutskim i kozmetičkim preparatima. *Šesti Međunarodni Kongr. Biomed. I Geonauke-Utic. Zivotn. Sred. Na Zdr. Ljud.* **2016**, *1*, 218–225.

36. Sehmi, S.K.; Noimark, S.D.; Pike, S.; Bear, J.C.; Peveler, W.J.; Williams, C.K.; Shaffer, M.; Allan, E.; Parkin, I.P.; MacRobert, A.J. Enhancing the Antibacterial Activity of Light-Activated Surfaces Containing Crystal Violet and ZnO Nanoparticles: Investigation of Nanoparticle Size, Capping Ligand, and Dopants. *ACS Omega* **2016**, *1*, 334–343. [CrossRef]
37. Panáček, A.; Kvítek, L.; Smékalová, M.; Večeřová, R.; Kolář, M.; Röderová, M.; Dyčka, F.; Šebela, M.; Prucek, R.; Tomanec, O.; et al. Bacterial resistance to silver nanoparticles and how to overcome it. *Nat. Nanotechnol.* **2017**, *13*, 65–71. [CrossRef] [PubMed]
38. Grohmann, E. Available online: https://prof.beuth-hochschule.de/grohmann/ (accessed on 30 September 2022).
39. Shkodra-Pula, B.; Vollrath, A.; Schubert, U.S.; Schubert, S. Polymer-based nanoparticles for biomedical applications. In *Frontiers of Nanoscience*; Elsevier: Amsterdam, The Netherlands, 2020; Volume 16, pp. 233–252. ISBN 978-0-08-102828-5.
40. Franco, D.; Calabrese, G.; Guglielmino, S.P.P.; Conoci, S. Metal-Based Nanoparticles: Antibacterial Mechanisms and Biomedical Application. *Microorganisms* **2022**, *10*, 1778. [CrossRef] [PubMed]
41. Unuma, H.; Hirose, Y.; Ito, M.; Watanabe, K. Preparation of the Precursor of Porous Alumina Particles using Immobilized Urease in Alginate Gel Templates. *J. Ceram. Soc. Jpn.* **2004**, *112*, 409–411. [CrossRef]
42. Unuma, H.; Matsushima, Y.; Kawai, T. Enzyme-mediated synthesis of ceramic materials. *J. Ceram. Soc. Jpn.* **2011**, *119*, 623–630. [CrossRef]
43. Li, X.; Xu, H.; Chen, Z.-S.; Chen, G. Biosynthesis of Nanoparticles by Microorganisms and Their Applications. *J. Nanomater.* **2011**, *2011*, 1–16. [CrossRef]
44. Mandal, D.; Bolander, M.E.; Mukhopadhyay, D.; Sarkar, G.; Mukherjee, P. The use of microorganisms for the formation of metal nanoparticles and their application. *Appl. Microbiol. Biotechnol.* **2006**, *69*, 485–492. [CrossRef]
45. Stoddart, A. Enzyme catalysis. *Nat. Mater.* **2012**, *11*, 910. [CrossRef]
46. Chen, L.; Shen, Y.; Xie, A.; Huang, B.; Jia, R.; Guo, R.; Tang, W. Bacteria-Mediated Synthesis of Metal Carbonate Minerals with Unusual Morphologies and Structures. *Cryst. Growth Des.* **2008**, *9*, 743–754. [CrossRef]
47. Sondi, I.; Matijević, E. Homogeneous Precipitation of Calcium Carbonates by Enzyme Catalyzed Reaction. *J. Colloid Interface Sci.* **2001**, *238*, 208–214. [CrossRef]
48. Sondi, I.; Matijević, E. Homogeneous Precipitation by Enzyme Catalyzed Reactions. II. Strontium Barium Carbonates. *Chem. Mater.* **2003**, *15*, 1322–1326. [CrossRef]
49. Zhao, J.; Sekikawa, H.; Kawai, T.; Unuma, H. Ferrimagnetic magnetite hollow microspheres prepared via enzymatically precipitated iron hydroxide on a urease-bearing polymer template. *J. Ceram. Soc. Jpn.* **2009**, *117*, 344–346. [CrossRef]
50. Kawashita, M.; Takayama, Y.; Kokubo, T.; Takaoka, G.H.; Araki, N.; Hiraoka, M. Enzymatic Preparation of Hollow Yttrium Oxide Microspheres for In Situ Radiotherapy of Deep-Seated Cancer. *J. Am. Ceram. Soc.* **2006**, *89*, 1347–1351. [CrossRef]
51. Hamaya, T.; Takizawa, T.; Hidaka, H.; Horikoshi, K. A new method for the preparation of magnetic alginate beads. *J. Chem. Eng. Jpn.* **1993**, *26*, 223–224. [CrossRef]
52. Kato, S.; Makino, T.; Unuma, H.; Takahashi, M. Enzyme Catalyzed Preparation of Hollow Alumina Precursor Using Emulsion Template. *J. Ceram. Soc. Jpn.* **2001**, *109*, 369–371. [CrossRef]
53. Kato, S.; Unuma, H.; Ota, T.; Takahashi, M. Homogeneous Precipitation of Hydrous Tin Oxide Powders at Room Temperature Using Enzymatically Induced Gluconic Acid as a Precipitant. *J. Am. Ceram. Soc.* **2004**, *83*, 986–988. [CrossRef]
54. Yeom, B.; Char, K. Nanostructured $CaCO_3$ Thin Films Formed on the Urease Multilayers Prepared by the Layer-by-Layer Deposition. *Chem. Mater.* **2009**, *22*, 101–107. [CrossRef]
55. Kawai, T.; Sekikawa, H.; Unuma, H. Preparation of hollow hydroxyapatite microspheres utilizing poly(divinylbenzene) as a template. *J. Ceram. Soc. Jpn.* **2009**, *117*, 340–343. [CrossRef]
56. Kawashita, M.; Sadaoka, K.; Kokubo, T.; Saito, T.; Takano, M.; Araki, N.; Hiraoka, M. Enzymatic preparation of hollow magnetite microspheres for hyperthermic treatment of cancer. *J. Mater. Sci. Mater. Med.* **2006**, *17*, 605–610. [CrossRef] [PubMed]
57. Malel, E.; Ludwig, R.; Gorton, L.; Madnler, D. Local Deposition of Au Nanpoarticles by Direct Electron Transfer through Cellobiose Dehydrogenase. *Chem. A Eur. J.* **2010**, *16*, 11697–11706. [CrossRef]
58. Rodrigues-Lorenzo, L.; Rica, R.; Alvarez-Puebla, R.A.; Liz-Marzan, L.M. Plasmonic nanosensors with inverse sensitivity by means of enzyme guided crystal growth. *Nat. Mater.* **2012**, *11*, 604–607.
59. Allmaier, G.; Bacher, G.; Kaufman, S.L.; Szymanski, W.W. Molecular mass determination of noncovalent complexes biopolymers—Proteins carbohydrates–by applying a gas-phase electrophoretic mobility mass analyzer. *J. Aerosol Sci.* **1999**, *30*, S303–S304. [CrossRef]
60. Allmaier, G.; Laschober, C.; Reischl, G.P.; Szymanski, W.W. Austrian Patent Surgical Cutting and Fastening Instrument with Closure Trigger Locking Mechanism. Patent 502 207 A1 2007 02, 15 February 2007.
61. Allmaier, G.; Laschober, C.; Szymanski, W.W. 2008 Nano ES GEMMA nad PDMA, New Tools for the Analysis of Nanobioparticles–Protein Complexes, Lipoparticles, Viruses. *J. Am. Soc. Mass Spectrom.* **2008**, *19*, 1062–1068. [CrossRef]
62. Kallinger, P.; Weiss, V.; Lehner, A.; Allmaier, G.; Szymanski, W. Analysis hling of bio-nanoparticles environmental nanoparticles using electrostatic aerosol mobility. *Particuology* **2013**, *11*, 14–19. [CrossRef]
63. Rezić, I.; Rezić, T.; Bokić, L. Optimization of the TLC separation of seven amino acids. *J. Planar Chromatogr. Mod. TLC* **2007**, *20*, 173–177. [CrossRef]
64. Rezić, I.; Majdak, M.; Kirin, P.; Vinceković, M.; Jurić, S.; Sopko, S.; Vlahoviček Kahlina, K.; Somogyi Škoc, M. Functionalized Microcapsules with Silver for Medical Textiles. In Proceedings of the 14th Scientific-Professional Symposium Textile Science and Economy: Book of Abstracts, Tekstilno-tehnološki fakultet, Zagreb, Croatia, 26 January 2022.

65. Rezić, I. Prediction of the surface tension of surfactant mixtures for detergent formulation using Design Expert software. *Mon. Chem.-Chem. Mon.* **2011**, *142*, 1219–1225. [CrossRef]
66. Rezić, I.; Somogyi Škoc, M.; Pokrovac, I.; Ljoljić Bilić, V.; Kosalec, I. Characterization of materials with nanoparticles and their biological and environmental impact. In Proceedings of the 141ST Iastem International Conference/International Academy of Science, Technology, Engineering and Management, Vienna, Austria, 25–26 September 2018; pp. 54–57.
67. Rezić, I.; Ćurković, L.; Ujević, M. Simple methods for characterization of metals in historical textile threads. *Talanta* **2010**, *82*, 237–244. [CrossRef] [PubMed]
68. Rezić, I.; Ćurković, L.; Ujević, M. Study of microstructure and corrosion kinetic of steel guitar strings in artificial sweat solution. *Mater. Corros.* **2010**, *61*, 524–529. [CrossRef]
69. Rezić, I.; Steffan, I. ICP-OES determination of metals present in textile materials. *Microchem. J.* **2007**, *85*, 46–51. [CrossRef]
70. Rezić, I. Optimization of ultrasonic extraction of 23 elements from cotton. *Ultrason. Sonochem.* **2009**, *16*, 63–69. [CrossRef]
71. Rezić, I.; Rolich, T. Artificial neural networks in chromatography spectroscopy. In *Artificial Neural Networks*; Nova Science Publishers: New York, NY, USA, 2011; pp. 123–138.
72. Rezić, I. Thin layer chromatographic monitoring of sonolytic degradation of surfactants in wastewaters. *J. Planar Chromatogr. Mod. TLC* **2019**, *26*, 96–101. [CrossRef]
73. Reich, E.S. Ultimate upgrade for US synchrotron. *Nature* **2013**, *501*, 148–149. [CrossRef] [PubMed]
74. Schmitt, B.; Brönnimann, C.; Eikenberry, E.F.; Gozzo, F.; Hörmann, C.; Horisberger, R.; Patterson, B. Mythen detector system. *Nucl. Instrum. Methods Phys. Res. A* **2003**, *501*, 267–272. [CrossRef]
75. Brönnimann, C.; Baur, R.; Eikenberry, E.F.; Kohout, S.; Lindner, M.; Schmitt, B.; Horisberger, R. A pixel read-out chip for the PILATUS project. *Nucl. Instrum. Methods Phys. Res. A* **2001**, *465*, 235–272. [CrossRef]
76. Petkov, V.; Peng, Y.; Williams, G.; Huang, B.; Tomalia, D.; Ren, Y. Structure of gold nanoparticles suspended in water studied by x-ray diffraction and computer simulations. *Phys. Rev. B* **2005**, *72*, 195402. [CrossRef]
77. Jacques, S.D.M.; Di Michiel, M.; Kimber, S.A.J.; Yang, X.; Cernik, R.J.; Beale, A.M.; Billinge, S.J.L. Pair distribution function computed tomography. *Nat. Commun.* **2013**, *4*, 2536. [CrossRef] [PubMed]
78. Sherma, B.; Mani, S.; Sarma, T.K. Biogenic Growth of Alloys Core Shell Nanostructures Using Urease as a Nanoreactor at Ambient Conditions. *Nat. Sci. Rep.* **2013**, *3*, 1–8. [CrossRef]
79. De la Rica, R.; Matsui, H. Urease as a Nanoreactor for Growing Crystalline ZnO Nanoshells at Room Temperature. *Angewte Chem. Int. Ed.* **2008**, *47*, 5415–5417. [CrossRef]
80. Granquist, C.G. Transparent conductors as solar energy materials: A panoramic review. *Sol. Energy Mater. Sol. Cells* **2007**, *91*, 1529–1598. [CrossRef]
81. Ebbesen, T.W.; Lezec, H.J.; Ghaemi, H.F.; Thio, T.; Wolff, P.A. Extraordinary optical transmission through sub-wavelength hole arrays. *Nature* **1998**, *391*, 667–669. [CrossRef]
82. Martinaga Pintarić, L.; Somogi Škoc, M.; Ljoljić Bilić, V.; Pokrovac, I.; Kosalec, I.; Rezić, I. Synthesis, modification and characterization of antimicrobial textile surface containing ZnO nanoparticles. *Polymers* **2020**, *12*, 1210. [CrossRef]
83. Rezić, I.; Majdak, M.; Bilić, V.L.; Pokrovac, I.; Martinaga, L.; Škoc, M.S.; Kosalec, I. Development of Antibacterial Protective Coatings Active against MSSA and MRSA on Biodegradable Polymers. *Polymers* **2021**, *13*, 659. [CrossRef] [PubMed]
84. Rezić, I.; Somogyi Škoc, M.; Majdak, M.; Jurić, S.; Stracenski, K.S.; Vinceković, M. Functionalization of Polymer Surface with Antimicrobial Microcapsules. *Polymers* **2022**, *14*, 1961. [CrossRef] [PubMed]
85. Rezić, I.; Škoc, M.S.; Majdak, M.; Jurić, S.; Stracenski, K.S.; Vlahoviček-Kahlina, K.; Vinceković, M. ICP-MS Determination of Antimicrobial Metals in Microcapsules. *Molecules* **2022**, *27*, 3219. [CrossRef] [PubMed]
86. Vukoja, D.; Vlainić, J.; Ljolić Bilić, V.; Martinaga, L.; Rezić, I.; Brlek Gorski, D.; Kosalec, I. Innovative Insights into In Vitro Activity of Colloidal Platinum Nanoparticles against ESBL-Producing Strains of Escherichia coli and Klebsiella pneumoniae. *Pharmaceutics* **2022**, *14*, 1714. [CrossRef]
87. Rezić, I.; Kracher, D.; Oros, D.; Mujadžić, S.; Anđelini, M.; Kurtanjek, Ž.; Ludwig, R.; Rezić, T. Application of Causality Modelling for Prediction of Molecular Properties for Textile Dyes Degradation by LPMO. *Molecules* **2022**, *27*, 6390. [CrossRef]
88. Egbuna, C.; Parmar, V.K.; Jeevanandam, J.; Ezzat, S.M.; Patrick-Iwuanyanwu, K.C.; Adetunji, C.O.; Khan, J.; Onyeike, E.N.; Uche, C.Z.; Akram, M.; et al. Toxicity of Nanoparticles in Biomedical Application: Nanotoxicology. *J. Toxicol.* **2021**, *2021*, 1–21. [CrossRef]
89. Blanco-Andujar, C.; Thanh, N.T. Synthesis of nanoparticles for biomedical applications. *Annu. Rep. Sect. A Inorg. Chem.* **2010**, *106*, 553–568. [CrossRef]
90. Uthaman, A.; Lal, H.; Thomas, S. Fundamentals of Silver Nanoparticles and Their Toxicological Aspects. In *Polymer Nanocomposites Based on Silver Nanoparticles. Engineering Materials*; Lal, H.M., Thomas, S., Li, T., Maria, H.J., Eds.; Springer: Cham, Switzerland, 2021; pp. 1–24.
91. Dikshit, P.K.; Kumar, J.; Das, A.K.; Sadhu, S.; Sharma, S.; Singh, S.; Gupta, P.K.; Kim, B.S. Green synthesis of metallic nanoparticles: Applications and limitations. *Catalysts* **2021**, *11*, 902. [CrossRef]
92. De Silva, C.; Nawawi, N.M.; Abd Karim, M.M.; Abd Gani, S.; Masarudin, M.J.; Gunasekaran, B.; Ahmad, S.A. The mechanistic action of biosynthesised silver nanoparticles and its application in aquaculture and livestock industries. *Animals* **2021**, *11*, 2097. [CrossRef]

93. Xu, L.; Yi-Yi, W.; Huang, J.; Chun-Yuan, C.; Zhen-Xing, W.; Xie, H. Silver nanoparticles: Synthesis, medical applications and biosafety. *Theranostics* **2020**, *10*, 8996–9031. [CrossRef]
94. Almatroudi, A. Silver nanoparticles: Synthesis, characterisation and biomedical applications. *Open Life Sci.* **2020**, *15*, 819–839. [CrossRef] [PubMed]
95. Barani, H.; Mahltig, B. Microwave-assisted synthesis of silver nanoparticles: Effect of reaction temperature and precursor concentration on fluorescent property. *J. Clust. Sci.* **2020**, *1*, 1–11. [CrossRef]
96. Patil, A.H.; Jadhav, S.A.; More, V.B.; Sonawane, K.D.; Patil, P.S. Novel one step sonosynthesis and deposition technique to prepare silver nanoparticles coated cotton textile with antibacterial properties. *Colloid J.* **2019**, *81*, 720–727. [CrossRef]
97. Kuntyi, O.; Shepida, M.; Sozanskyi, M.; Sukhatskiy, Y.; Mazur, A.; Kytsya, A.; Bazylyak, L. Sonoelectrochemical synthesis of silver nanoparticles in sodium polyacrylate solution. *Biointerface Res. Appl. Chem.* **2021**, *11*, 12202–12214.
98. Seku, K.; Gangapuram, B.R.; Pejjai, B.; Kadimpati, K.K.; Golla, N. Microwave-assisted synthesis of silver nanoparticles and their application in catalytic, antibacterial and antioxidant activities. *J. Nanostruct. Chem.* **2018**, *8*, 179–188. [CrossRef]
99. Tola, O.H.; Oluwole, O.I.; Omotayo, A.B. Synthesis and characterization of silver nanoparticles from ecofriendly materials: A review. *Int. J. Eng. Res. Technol.* **2020**, *9*, 782–795.
100. Naganthran, A.; Verasoundarapandian, G.; Khalid, F.E.; Masarudin, M.J.; Zulkharnain, A.; Nawawi, N.M.; Karim, M.; Abdullah, C.A.C.; Ahmad, S.A. Synthesis, Characterization and Biomedical Application of Silver Nanoparticles. *Materials* **2022**, *15*, 427. [CrossRef]

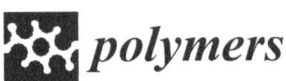

polymers

MDPI

Review

Recent Advances in Micro- and Nano-Drug Delivery Systems Based on Natural and Synthetic Biomaterials

Md. Harun-Or-Rashid [1], Most. Nazmin Aktar [1], Md. Sabbir Hossain [2], Nadia Sarkar [2], Md. Rezaul Islam [2], Md. Easin Arafat [2], Shukanta Bhowmik [1] and Shin-ichi Yusa [1,*]

[1] Department of Applied Chemistry, Graduate School of Engineering, University of Hyogo, 2167 Shosha, Himeji 671-2280, Hyogo, Japan; rashid685@diu.edu.bd (M.H.-O.-R.); nazmin29-846@diu.edu.bd (M.N.A.); shukantabhowmik@nstu.edu.bd (S.B.)

[2] Department of Pharmacy, Faculty of Allied Health Sciences, Daffodil International University, Dhaka 1207, Bangladesh; sabbir29-1219@diu.edu.bd (M.S.H.); nadiasarkar181@gmail.com (N.S.); rezaul29-1301@diu.edu.bd (M.R.I.); easin29-1217@diu.edu.bd (M.E.A.)

* Correspondence: yusa@eng.u-hyogo.ac.jp

Abstract: Polymeric drug delivery technology, which allows for medicinal ingredients to enter a cell more easily, has advanced considerably in recent decades. Innovative medication delivery strategies use biodegradable and bio-reducible polymers, and progress in the field has been accelerated by future possible research applications. Natural polymers utilized in polymeric drug delivery systems include arginine, chitosan, dextrin, polysaccharides, poly(glycolic acid), poly(lactic acid), and hyaluronic acid. Additionally, poly(2-hydroxyethyl methacrylate), poly(*N*-isopropyl acrylamide), poly(ethylenimine), dendritic polymers, biodegradable polymers, and bioabsorbable polymers as well as biomimetic and bio-related polymeric systems and drug-free macromolecular therapies have been employed in polymeric drug delivery. Different synthetic and natural biomaterials are in the clinical phase to mitigate different diseases. Drug delivery methods using natural and synthetic polymers are becoming increasingly common in the pharmaceutical industry, with biocompatible and bio-related copolymers and dendrimers having helped cure cancer as drug delivery systems. This review discusses all the above components and how, by combining synthetic and biological approaches, micro- and nano-drug delivery systems can result in revolutionary polymeric drug and gene delivery devices.

Keywords: natural polymers; synthetic polymers; drug delivery; copolymers; biomimetic

Citation: Harun-Or-Rashid, M.; Aktar, M.N.; Hossain, M.S.; Sarkar, N.; Islam, M.R.; Arafat, M.E.; Bhowmik, S.; Yusa, S.-i. Recent Advances in Micro- and Nano-Drug Delivery Systems Based on Natural and Synthetic Biomaterials. *Polymers* **2023**, *15*, 4563. https://doi.org/10.3390/polym15234563

Academic Editors: Maria João Ramalho, Stephanie Andrade and Joana A. Loureiro

Received: 31 October 2023
Revised: 22 November 2023
Accepted: 26 November 2023
Published: 28 November 2023

1. Introduction

Nanomedicine uses nanotechnology to improve healthcare by manufacturing medication nanocarriers with the enhanced permeability and retention (EPR) effect. These nanocarriers may passively or actively target tumour tissues, resulting in more effective treatment. In the past two decades, pharmaceutical research and development has focused on developing drug delivery vehicles. A better drug transporter is biocompatible, biodegradable, non-toxic, and delivers active substances to the action site. Until recently, liposomes, organic and inorganic nanoparticles, and hydrogels were considered promising drug carriers [1,2]. Due to their abundance in nature and unique properties, including sustainability, biocompatibility, and biodegradability, natural biopolymers are gaining favour in drug delivery systems (DDSs) [3,4]. Living cells build natural biopolymers by covalently bonding monomeric units to form large molecular-weight molecules. Polysaccharides and proteins are the most common natural biopolymers, which include chitosan, cellulose, dextran, starch, pectin, collagen, gelatin, fibronectin, elastin, keratin, actin, myosin, etc. Polymers made up of long chains of amino acid residues (proteins) exhibit a broad range of physical and chemical characteristics. Oligopeptides are linear chains of 20–30 amino acids. Nondegradable materials used in biomedical applications can be replaced by synthetic biodegradable polymers. The majority of synthetic biodegradable polymers used today as common materials and even in the biomedical industry belong to the polyester family. In most cases, poly(glycolide), poly(lactide), poly(caprolactone), and their copolymers are part of applications and research. Various synthetic polymers based on poly(amino ester) (PAE) are suggested as candidates for gene and drug delivery owing to their pH responsiveness, which contributes to efficient delivery performance [5].

This review addresses natural and manmade/synthetic polymers, their desirable qualities, ongoing clinical trials, and their limitations in regard to drug delivery systems. Moreover, we discuss how combining two or more biomaterials with enhanced capabilities, such as copolymers, polymer–polymer blends, or composites, may meet most therapeutic needs. The existing literature is reviewed, and a scaffold design, commercial viability, and manufacturing methods are also examined.

2. Data Source and Search Strategy

We comprehensively searched online publications in the WOSCC (Web of Science Core Collection) database on 25 September 2023. The publication timeframe ranged from an unspecified date to 25 September 2023. To prevent bias stemming from daily database updates, we specifically looked into articles on a single day. Our focus was on English-language publications, with only "articles" and "reviews" being considered. Prior to assessing the relevance of the literature to the theme of natural and synthetic biomaterials in drug delivery as well as ongoing clinical research, we eliminated publications that did not meet the specified language and article-type criteria by evaluating their titles and abstracts.

3. Nano-Based Drug Delivery Systems

3.1. Fundamentals of Nanotechnology-Based Drug Design Methodologies

Nanomedicine is a subspecialty of medicine that makes use of nanoscale materials, such as biocompatible nanoparticles [6] and nanorobots [7], to carry out a variety of biological tasks, such as diagnostics [8], transportation [9], sensing [10], and actuation [11]. Before the conventional method of formulating vaccinations was established, medicines with very low aqueous solubility presented a number of difficulties in terms of biopharmaceutical delivery. These difficulties included restricted bio-access on oral intake, a lower diffusion capacity into the outer membrane, larger intravenous (IV) dosage requirements, immunogenicity, and unwanted side effects. It is possible that if nanotechnology was included in the pharmaceutical distribution process, all of these problems would be eliminated. Drug design at the nanoscale has been the subject of extensive research and is currently the most cutting-edge technology in the field of nanoparticle applications. This is due to the fact that nanoscale drug design offers an abundance of benefits, including the capability to modify properties such as solubility, drug release profiles, diffusivity, bioavailability, and immunogenicity. Because of this, it may be feasible to devise more comfortable administration techniques that have less toxicity, fewer side effects, better biodistribution, and a longer pharmaceutical life cycle [12].

DDSs were developed to either direct therapeutic chemicals to a particular location in a more concentrated fashion or to disperse therapeutic chemicals to a certain area in a more manageable way. Self-assembly is defined as the process through which the assembly of components results in the spontaneous emergence of well-defined forms or patterns [13]. Endocytosis and absorption via a system of mononuclear phagocytes are also very important [14]. As a result of the hydrophobic properties of a structure, medicines may be injected into an internal cavity. When nanostructure components are guided to specific locations, the amount of medication that is expected to be released is achieved despite the low concentration of the medicine that is kept in a hydrophobic environment. Both passive and active methods of medication administration are viable options for nanostructured DDSs. As a result of the hydrophobic nature of drugs, they are often extensively absorbed into the interior cavity of a structure. When nanostructure materials are directed to certain locations, pharmaceuticals are kept in an environment that is hydrophobic, and the appropriate quantity of medication is released [14]. To facilitate distribution more easily, supplied medications are conjugated to the transporter nanostructure material as quickly as is practically possible. If a medication is released from its nanocarrier system at the incorrect moment, it will not reach the target it was designed for, and it will rapidly dissociate from the carrier, resulting in a reduction in its bioactivity and effectiveness [15].

Another essential feature of a DDS is drug targeting, wherein nanoparticles or nanoformulations make up the DDS and are split up into active and passive categories. In order to successfully target antibodies and peptides to receptor complexes expressed in a particular area, several DDSs are coupled with antibodies and peptides. Active targeting involves the use of antibodies and peptides in combination with pharmaceutical delivery methods to connect to receptor structures that are expressed in the target site. The produced drug carrier complex is directed to the target site by affinity or binding, which is governed by pH, temperature, molecular size, and shape. This process takes place while the complex travels through circulation. The receptors that are located on the cell membranes, the lipid components that are located on the cell membranes, and the antigens or proteins that are located on the surface of the cells are the primary targets in the body [16]. The majority of nanotechnology-assisted DDSs are now under investigation for use in the treatment and prevention of cancer.

3.2. Diagnostic, Detection, and Imaging Applications of Biopolymeric Nanoparticles

The combination of treatment and diagnosis is known as theragnostic therapy, and it is used widely in cancer treatment [17]. When used in theragnostic therapy, nanoparticles have the potential to improve the disease diagnosis, treatment localization, stage, and treatment response. In addition, nanoparticles have the capability of transporting tumour-targeted therapeutic medication, which may then be released into the body in response to either an internal or external stimulus [18]. Chitosan, also known as chitin, is a biomaterial that has distinctive qualities, such as biological uses and functional groups [19]. Chitosan is being employed to encapsulate or coat a broad variety of nanoparticles, which will result in a diversity of particles with various activities that may be used in the detection and diagnosis of illness [20]. Using oleic acid-coated iron oxide nanoparticles encapsulated in oleic acid-conjugated chitosan (oleyl-chitosan), Lee et al. [18] investigated the accretion of nanoparticles in tumour cells via the EPR effect in vivo for analytical reasons. This was done using oleic-acid-coated iron oxide nanoparticles for magnetic resonance imaging (MRI). In vivo, experiments of cyanine-5-attached oleyl-chitosan nanoparticles demonstrated high signal intensity and recovery in tumour tissues using both techniques.

Yang et al. [19] reported nanoparticles with improved 5-aminolevulinic (5-ALA) release in the cell lysosome; the nanoparticles were produced by physically connecting alginate to folic acid-modified chitosan. The nanoparticles were extremely effective for light-mediated colon cancer (CC) cell detection. According to the results, the modified nanoparticles were readily endocytosed by CC cells via a folate receptor-based endocytosis mechanism. Due to the usage of deprotonated alginate, there was a reduction in the binding strength between 5-ALA and chitosan. As a consequence, the charged 5-ALA was delivered to the lysosome, which led to an accumulation of protoporphyrin IX for photodynamic detection inside the cells. Researchers found that chitosan-based nanoparticles coupled with alginate and folic acid are good vectors for delivering 5-ALA to CC cells while also allowing for endoscopic fluorescence monitoring.

Cathepsin B (CB) is an enzyme that plays an essential role in the detection of metastasis because of its close connection to the metastatic process and its prevalence in the pericellular regions where it takes place. Ryu et al. [20] produced fluorogenic peptide and tumour-targeting glycol chitosan nanoparticles, which were incorporated on the surface of a CB-sensitive nanoprobe. The resulting nanoprobe was spherical, with a diameter of 280 nm, and did not glow when placed in biological environments. In three different rat metastatic models, a CB-sensitive nanoprobe was used in conjunction with non-invasive imaging to differentiate between metastatic and healthy cells.

Another example of a biopolymeric molecule is hyaluronic acid (HA). This glycosaminoglycan, which is biocompatible and has a negatively charged ion, is found in the extracellular matrix [21]. As a result of the interaction between the receptor and the linker, HA has the potential to bind to the CD44 receptor, which is overexpressed in certain cancer cells. As a consequence of this, HA-modified nanoparticles have the potential to assist in the diagnosis and treatment of cancer [22–24]. Dopamine-modified HA was used to coat the surface of iron oxide nanoparticles in a study carried out by Wang and co-workers [25]. Siyue et al. [26] created HA nanoparticles with varying diameters using varying degrees of hydrophobic HA replacement. Nanoparticles were given to animals afflicted with malignancies, and the treatment responses were analyzed. For the purpose of the early detection and targeted treatment of CC, the same research group developed a potent and flexible thermostatic system based on PEG-conjugated HA nanoparticles (P-HA-NPs). To determine whether or not the nanoparticles were effective against cancer, they were first chemically coupled to a near-fluorescent dye, cyanine 5.5 (Cy5.5), and then encapsulated with the anticancer medication irinotecan (IRT). The therapeutic potential of P-HA-containing nanoparticles was then investigated using a variety of CC animal models. The near-infrared fluorescence imaging technology was successfully used to scan tiny and early-stage malignancies as well as CCs that were implanted in the liver. This was accomplished after an IV injection of fluorescent dye-associated nanoparticles (Cy5.5-P-HA-NPs). The remarkable ability of drug-containing nanoparticles (IRT-P-HA-NP) to target tumours resulted in a considerable reduction in the growth of malignancies while causing no damage to the body as a whole. Cy5.5-P-HA-NPs are a potential tool for concurrently analysing many aspects of the healing process [27].

The natural polymer alginate, which is obtained from brown seaweed, has been subjected to intensive research for the purpose of determining whether or not it could be used in the medical field. This is due to the fact that alginate possesses a number of desirable qualities, such as low toxicity, compatibility, and an easy gelling ability, when divalent cations are present. Baghbani et al. [28] employed alginate-stabilized perfluorohexane (PFH) nanodroplets to administer doxorubicin (DOX). In addition to determining its potential therapeutic utility, they subsequently examined the sensitivity of the nanodroplets to ultrasonography and imaging. Treatment with ultrasound-assisted therapy using DOX-loaded PFH nanodroplets also showed significant promise in breast cancer rat models. The degree to which the tumour was able to be broken down was an indicator of how well the treatment worked. Podgórna et al. [29] produced gadolinium nanogels (GdNGs) for MRI scanning and the loading of hydrophilic pharmaceuticals. The diameter of the gadolinium alginate nanogels was an average of 110 nm, and their stability lasted for sixty days. In MRI imaging, gadolinium combinations are often used as positive contrast agents due to the inherent paramagnetic characteristics of gadolinium. The spin-lattice relaxation time (T_1) was dramatically shortened due to the presence of GdNGs in comparison to controls. As a consequence of this, alginate nanogels have the potential to be used in the medical industry as contrast enhancers.

It is believed that dextran, a non-toxic neutral polymer, was the first exopolysaccharide used in medicine by bacteria. Dextran is a peculiar substance since it is safe for human consumption, does not cause any harm, and can be broken down by biological processes. The cancer treatment known as photodynamic therapy eliminates cancer cells while sparing the surrounding healthy tissue from damage. Ding et al. [30] built a nanoparticulate multifunctional composite system for near-infrared (NIR) imaging and MRI. They did this by encapsulating Fe_3O_4 nanoparticles in dextran nanoparticles and then coupling the dextran nanoparticles to redox-responsive chlorine 6 (C6). The redox biological reaction that occurs as a result of the nanoparticles generates a fluorescent signal that has an "on/off" characteristic, which enables accurate tumour imaging. In addition, improvements in magnetic targeting capabilities led to an increase in the effectiveness of photodynamic therapy, both in vitro and in vivo. The production of theragnostic nanoparticles and glioma cells was carried out by Ali et al. [31] using C6 mice. In order to produce these particles, gadolinium oxide nanoparticles were first coated with either paclitaxel (PTX) or folic acid (FA)-conjugated dextran. The MTT test was used to evaluate both the chemotherapeutic effects of PTX on C6 glioma cells and the bioprotective advantages of a dextran coating. Due to the paramagnetic properties of the gadolinium nanoparticles, the nanoparticles were able to penetrate C6 tumour cells through a process known as receptor-mediated endocytosis. Additionally, the nanoparticles displayed enhanced contrast MRI concentration-dependent activity. When it came to inhibiting cell growth, uncoated gadolinium nanoparticles were shown to be less efficient than their multifunctional counterparts. As a consequence of this, drugs that are paramagnetic and chemotherapeutic may be produced by utilizing theragnostic nanoparticles containing both FA and PTX.

4. Drug Delivery Using Synthetic Polymers

4.1. *Poly(2-hydroxyethyl methacrylate)*

Poly(2-hydroxyethyl methacrylate) (PHEMA) hydrogel for intraocular lens components was synthesized via solution polymerization using 2-hydroxyethyl methacrylate (HEMA) with cross-linking agents, such as ethyleneglycoldimethacrylate and triethyleneglycoldimethacrylate [32]. In cancer research, PHEMA (Figure 1) is often used to wrap cell-culture flasks to reduce the amount of cell adhesion and increase the creation of spheroids. There are two older alternatives to PHEMA: agar and agarose gels [33]. For the purpose of developing drug delivery systems, equilibrium swelling, structural characterization, and solute transports in swollen PHEMA gels cross-linked with tripropyleneglycol diacrylate (TPGDA) were investigated across a large TPGDA concentration range [34]. In order to get a better understanding of the mechanism of drug–polymer interaction and its influence on the drug-release behaviour of controlled-release polymeric devices, the physical and chemical features of pilocarpine extracted from PHEMA hydrogels were investigated. PHEMA hydrogels are often employed in the field of biomedical implants. PHEMA, which is biocompatible, is also a potential coating for ventricular catheters [35,36] because of its high hydrophilicity, which provides resistance to protein fouling.

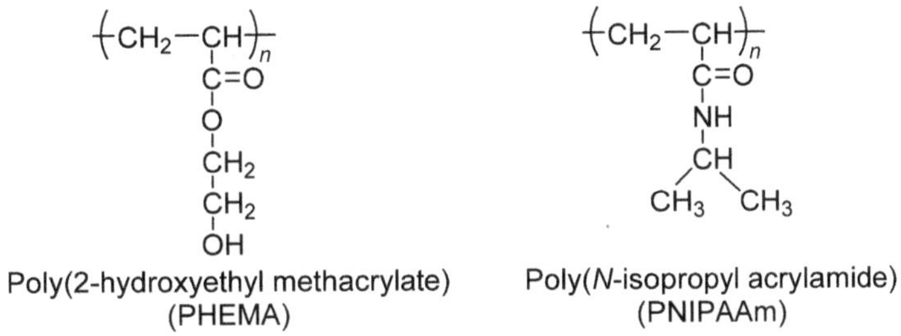

Figure 1. Chemical structure of poly(2-hydroxyethyl methacrylate) (PHEMA) and poly(*N*-isopropyl acrylamide) (PNIPAAm).

4.2. *Poly(N-isopropyl acrylamide)*

Poly(*N*-isopropyl acrylamide) (PNIPAAm) dissolves in water at a low temperature; however, it cannot dissolve in high-temperature water (Figure 1). In the 1960s, temperature-sensitive polymers were a prevalent research subject [37]. The lower critical solution temperature (LCST) of thermo-sensitive PNIPAAm was 32 °C. To determine the thermodynamic properties of the system, the phase diagram and amount of heat absorbed during phase separation were used in the calculation [38]. While carrying out radical polymerization of *N*-isopropyl acrylamide (NIPAAm), it is common practice to make use of the initiator known as azobis(isobutyronitrile). Thermo-responsive polymers have a variety of applications in biological and medical fields, including medicine and gene transfer [39]. An investigation was conducted into the temperature dependence of the swelling of cross-linked poly(*N*,*N*-alkyl substituted acrylamides) in water. Thermo-sensitivity of water swelling has been linked to the delicate balance of hydrophilic and hydrophobic groups on polymer chains, which is determined by the size, conformation, and mobility of alkyl side-chain groups [40]. This balance is regulated by the size of the alkyl side-chain groups. This method has been used to create hydrophilic and hydrophobic coatings on a cell culture surface of PNIPAAm-grafted polymers in a reversible manner [41].

NIPAAm was copolymerized with acrylic acid (AAc) to develop temperature- and pH-sensitive hydrogels. The influence of polyelectrolytes on the LCST of temperature/pH-sensitive hydrogels was investigated within the pH range of the swelling ratio. In the same conditions and in the presence of poly(allyl amine) (PAA) as a polyelectrolyte, an investigation into the swelling ratio of hydrogels was carried out [42]. Some of the discussed subjects in the study were pH, redox responsiveness, hypoxia sensitivity, and other tumour microenvironmental-sensitive nanoparticle in situ stimuli.

4.3. *Poly(ethylenimine)*

When the pH is low, the compound known as poly(ethylenimine) (PEI) is soluble in hot water, ethanol, and chloroform. The compound is impervious to the solubility of cold water, acetone, benzene, and ethyl ether. Branched PEI (BPEI) is a polymer with repeating units composed of ethylene diamine groups. Aziridine ring-opening polymerization led to the formation of BPEI. BPEI is a cationic polymer that contains primary, secondary, and tertiary amino groups (Figure 2). Such types of water-soluble polymers having a high density of amines are one of the most promising cationic vectors for gene delivery. Hence, constructing nanocarriers that contain PEI has attracted considerable research efforts in gene therapy because of the synergetic effects of PEI molecules (for their efficient transfection) and the multi-functionality of nanoparticles in delivery. BPEI can be prepared with a divergent synthesis method from an ethylenediamine core. Day by day, highly branched polymers have also attracted growing interest in various fields, especially in drug delivery, gene delivery, and diagnosis [43,44]. Major applications of BPEI include the stable combination with other positively charged particles, layer-by-layer construction of nanoparticle surfaces, binding to negatively charged substrates or larger particles, and colour engineering.

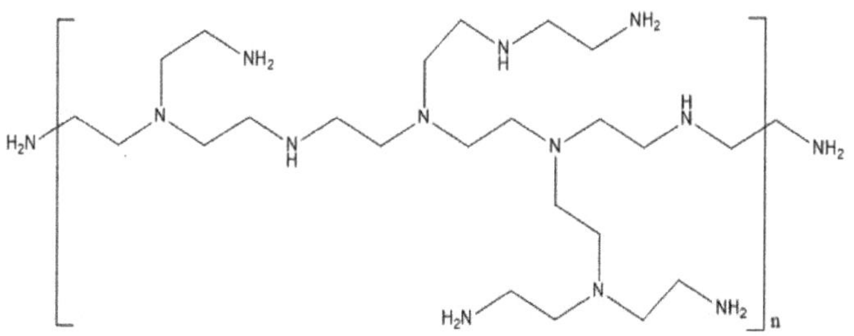

Polyethylenimine (PEI)

Figure 2. Chemical structure of branched poly(ethylenimine) (BPEI) for polymeric drug delivery.

4.4. *Dendritic Polymers*

Dendrimers are three-dimensional structures that have a high bifurcation level, are monodisperse, and have clear boundaries. Effective drug delivery systems may be characterized by their spherical form, which makes it possible for them to be quickly functionalized in a regulated manner [45]. The path to dendrimer formation can be convergent (the dendrimer develops inwards from the exterior) or divergent (the dendrimer expands outwards from its centre) [46]. Dendritic polymers are distinguished by their large population of terminal functional groups as well as their low solution or melt viscosity and great solubility (Figure 3). The scale of dendritic polymer synthesis processes may be regulated and varied, as can their branching and overall utility.

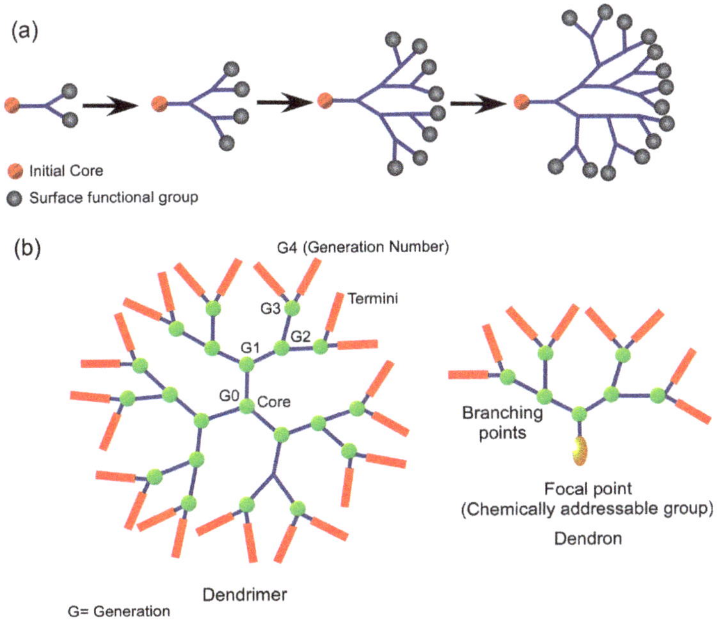

Figure 3. (**a**) Schematic design for divergent dendrimer manufacturing for drug delivery; (**b**) drug delivery dendrimer and dendron chemical structures.

Dendrimers are a kind of polymer that is classified under the dendritic polymer family. Other types of polymers included in this family are linear, cross-linked, and branched polymers. The design and production of biocompatible dendrimers, in addition to their use in a range of bioscience sectors that include drug delivery, immunology, and vaccine generation, have been the primary focus of dendrimer research [47,48]. Dendrimers are a form of reducible polymer that has also been investigated for their potential to effectively transport genes [49].

4.5. Biodegradable and Bioabsorbable Polymers

If an implant is only going to be needed for a short amount of time, the best choice for drug carriers would be bio-absorbable drug delivery devices [50]. Aliphatic polyesters are examples of synthetic biodegradable polymers. Some examples of aliphatic polyesters are PGA and PLA. PCL and polydioxanone are the polymers used as bio-absorbable drug delivery devices the majority of the time. Many other types of polymers have been developed, including polyesters, poly(ortho ester)s, polyanhydrides, and biodegradable polycarbonates (Figure 4) [51,52]. There are several types of biodegradable polymers, including hydroxy acid, polyanhydride, polyamide, poly (ester amide), polyphosphoester, poly(alkyl cyanoacrylate), PHA, and natural sugars such as chitosan. In drug delivery systems, synthetic biodegradable polymers are favoured over natural biodegradable polymers [53–55]. This is due to the fact that natural biodegradable polymers are immunogenic.

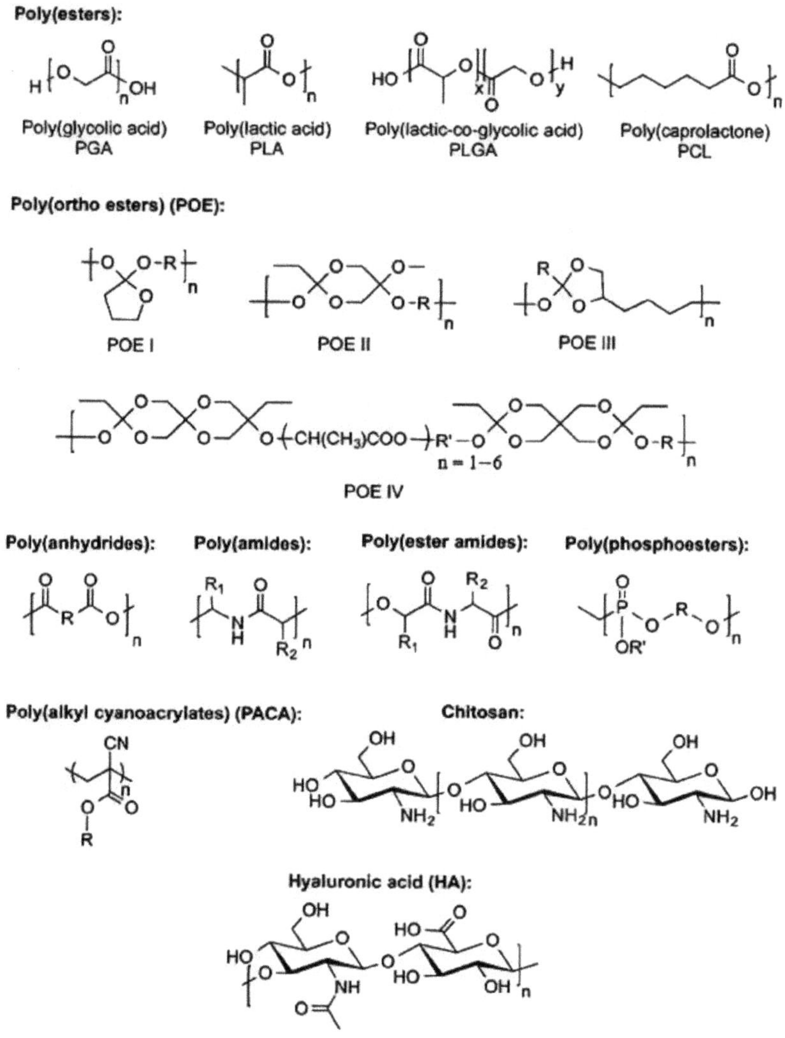

Figure 4. Biodegradable polymers for polymeric drug delivery with typical monomer units (Reproduced from [51]).

5. Drug Delivery Methods Using Nanoparticles

5.1. Polymeric Micelles

Polymer-based micelles are nanostructures composed of amphiphilic block copolymers that may self-assemble into a core-shell structure when placed in an aqueous solution. While the hydrophilic shell renders the whole system water soluble and stabilizes the core, the hydrophobic core may include hydrophobic medications, such as camptothecin, docetaxel, and PTX. These nanostructures have a promising future in the field of hydrophobic drug delivery because their interior core shape makes it possible for drugs to be absorbed, which results in higher stability and bioavailability [56].

The production of polymeric micelles may be accomplished by direct polymer dissolution in a solvent, solvent evaporation, and dialysis methods [57]. Micelle creation is affected by a number of factors, including the size of the hydrophobic chain on the amphiphilic molecule, the concentration of amphiphiles, the temperature, and the kind of solvent system [58]. The process of micelle creation starts when the concentration of amphiphilic molecules reaches a critical level, which is also referred to as the critical micelle concentration (CMC). Because of their diminutive size, amphiphilic molecules are only capable of existing on their own in trace quantities [59]. During direct dissolution, the copolymer and pharmaceuticals interact with one another on their own in an aqueous environment, resulting in a drug that is packed with micelles. In the method known as solvent evaporation, the needed medication and the copolymer are first dissolved in a volatile organic solvent. Next, the drug in the solution and the copolymer in the organic solvent are mixed in a dialysis bag and dialyzed, making use of micelle formation [60].

Stimuli that promote penetrability and the hold effect include monoclonal antibodies connected to the corona of the micelle or a carefully targeted ligand molecule complexed to the surface of the micelle [61]. The use of polymeric micelles may be beneficial when it comes to the administration of cancer treatments [58] and the delivery of eye medications (Figure 5) [62]. For the treatment of progressive vitreoretinopathy, polymeric micelles were encased in nanoparticles produced using the micellization of poly(ethylene glycol)-*block*-poly(bisphenol A carbonate). ARPE-19 cells were not affected by the cytotoxicity of the nanoparticles, which had a diameter of 55 nm. When compared to free medicines, the micellar formulation demonstrated a significant increase in the ability to suppress cell growth, adhesion, and translocation [63]. After appropriate treatments are administered, polymeric micelles are commonly injected into the posterior ocular tissues via the transscleral channel [61].

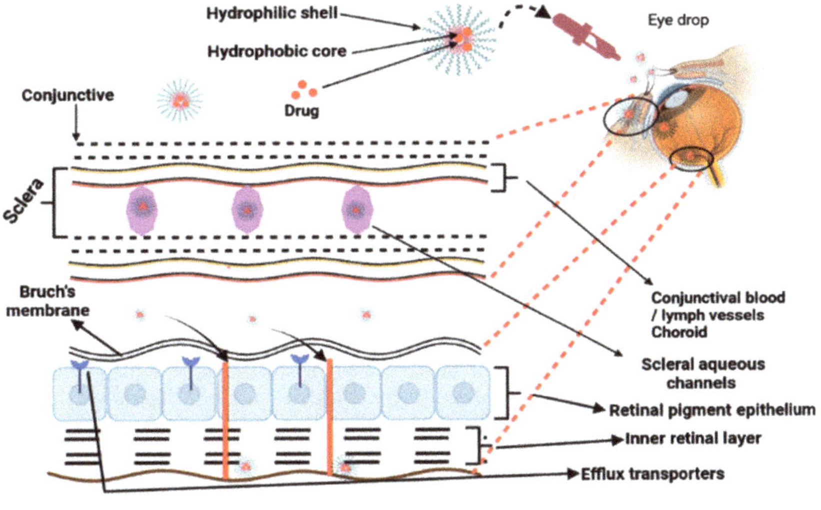

Figure 5. Polymeric micelles employed to reach posterior ocular tissues via the transscleral channel following topical treatment.

5.2. Dendrimers

Dendrimers for oral medication delivery have received the most attention because they are water soluble and can pass through epithelial tissue [64]. Because dendrimers include amine groups, their potential as medicinal agents are rather restricted. The cationic or negatively charged nature of dendrimers results in them often being changed in an effort to lessen or remove their toxicity. The following is a list of procedures used to load medicines into dendrimers: encapsulation, electrostatic contact, and covalent conjugation, which are all included in simple encapsulation [65]. Dendrimers are primarily responsible for the delivery of drugs through two different pathways: (a) in vivo degradation of the covalent bonding of the drug dendrimer as a result of suitable enzymes or a favorable environment that can cleave the bonds and (b) drug discharge as a result of changes in the physical environment, such as pH, temperature, and so on. Dendrimers have the potential for use in-drug administration through transdermal, oral, ophthalmic, pulmonary, and targeted methods.

Jain et al. [66] showed that folate-attached poly(*L*-lysine) dendrimer-encapsulated DOX is a promising cancer prevention drug carrier model due to its pH-dependent drug discharge, target selectivity, antiangiogenic ability, and anticancer potential. The use of DOX-folate conjugated poly(*L*-lysine) dendrimers resulted in a 121.5% increase in DOX concentration in the tumour. Using folate-conjugated polypropylene imine dendrimers, Kaur et al. [67] developed a pH-sensitive methotrexate (MTX) nanocarrier for cancer cell targeting and anticancer therapy via folate-conjugated poly(propylene imine) (PPI) dendrimers (FAPPI). FAPPI is considered a pH-sensitive DDS. In vitro studies with MCF-7 cell lines demonstrated substantial release, enhanced cell uptake, and mild cytotoxicity [68]. In addition to these results, the generated formulations, which were methotrexate MTX-loaded and folic acid-conjugated generation 5 PPI, were preferentially taken up by tumour cells compared to the free drug.

5.3. Inorganic Nanoparticles

Inorganic nanoparticles include gold, silver, iron oxide, and silica nanoparticles. A very small number of inorganic nanoparticles have been approved for use in therapeutic applications, but the vast majority are currently undergoing testing in clinical studies. Surface plasmon resonance (SPR) is only found in silver and gold nanoparticles; liposomes, dendrimers, and micelles do not possess this property. Among inorganic particles, gold and silver nanoparticles provide a wide range of advantages, the most notable of which is their excellent biocompatibility and flexible surface functionalization. There is very little evidence to suggest that paracellular transport and transcytosis really take place in living organisms [68] despite the fact that these processes have been postulated. Although paracellular and transcytosis transport and absorption have been proposed as possibilities, little is known for certain about how these processes work in the body. The surfaces of gold nanoparticles may have drugs conjugated to them through ionic or covalent bonding as well as through physical absorption. These drugs can then be delivered and controlled through the use of biological stimuli or light activation [69]. In recent years [70], the use of metallic nanoparticles has become more common in a variety of medical applications, including bioimaging, biosensors, target/sustained drug delivery, hyperthermia, and photoablation. By adding functional groups, these nanoparticles have the potential to interact with antibodies, medicines, and other ligands, hence increasing their value for use in biological applications [71]. Zinc oxide, titanium oxide, platinum, selenium, gadolinium, palladium, and cerium dioxide nanoparticles are also all gaining increased interest.

In order to facilitate the release of ornidazole, Prusty and Swain [72] developed an interconnected and spongy polyacrylamide/dextran nano-hydrogel hybrid system. This system had covalently bonded silver nanoparticles and obtained a 98.5% success rate in vitro. Another study used laser pyrolysis to create iron oxide nanoparticles, which were then coated with violamycine B1 and antracyclinic antibiotics [73]. These nanoparticles were tested in MCF-7 cells for cytotoxicity and anti-proliferation properties and compared to commercially available iron oxide nanoparticles.

5.4. Nanocrystals

Nanocrystals are defined as pure solid pharmaceutical particles with dimensions of fewer than one thousand nanometres. The performance of nanocrystal suspensions in thin liquid media may be significantly enhanced by the use of a surfactant component known as nano-suspension. Water, or other aqueous or non-aqueous media, such as liquid PEG and oils, are used as the dispersion mechanism [74]. Nanocrystals offer unique properties, such as enhanced saturation solubility, speedier dissolution, and improved adherence to surface and cell membranes. Both bottom-up and top-down methods used to produce nanocrystals are examined here. Sono-crystallization, precipitation, high gravity-controlled precipitation technology, multi-inlet vortex mixing methods, and the limited impinging liquid jet precipitation technique are all components of the top-down approach [75]. This method is somewhat pricey since it requires the use of an organic solvent that is subsequently thrown away. Grinding and homogenization steps take place under intense pressure during the bottom-up process. Milling, high-pressure homogenization, and precipitation are the three processes most often used when attempting to create nanocrystals. Nanocrystals increase medication absorption by increasing solubility, suspension rate, and the ability to keep the intestinal wall in place. Cinaciguat nanocrystals coated in chitosan microparticles were the vehicles that Ni et al. [76] used to deliver a hydrophobic medication to the lungs. Swelling and muco-adhesive properties of the polymers were used in the production of nanoparticles for continuous medication release. They found that sickness may limit the effectiveness of inhalation, which suggests that more research is required to prove if this method is promising [77].

5.5. Quantum Dots

Quantum dots (QDs) are semiconductor nanocrystals that range in diameter from 2 to 10 nm and feature size-dependent optical characteristics that include absorbance and photoluminescence [78]. QDs have attracted a lot of interest in the field of nanomedicine since they emit in the NIR region (650 nm), which is a particularly desirable property in biomedical imaging due to decreased tissue absorption and light scattering at this wavelength [79]. Biocompatible multifunctional graphene oxide QDs with a brilliant magnetic nanoplatform were created by Shi et al. [79] for detecting and identifying individual liver cancer tumour cells (glypican-3-expressing Hep G2). The researchers claim that by combining an anti-glypican-3 antibody with a nanoplatform, they are able to selectively eliminate Hep G2 hepatocellular cancer cells from contaminated blood samples. Coating QD antibodies with norbornene-displaying polyimidazole ligands was the method that Ahmad et al. [80] used to develop a new fluorophore for intravital cytometric imaging. This fluorophore was utilized to highlight hematopoietic stem and progenitor cells in bone marrow in vivo. Additionally, a single light source has the potential to excite QDs of various sizes and/or compositions, which may lead to a wide spectrum of colour emission [81]. In terms of multiplex imaging, we feel that QDs have a very bright future. In the field of medicine, QDs have been the subject of substantial research as potential targets for targeted medication administration, sensors, and bioimaging. Recent years have seen a proliferation of studies on the use of QDs as contrast agents in in vivo imaging [82].

As a parenteral multifunctional system, Olerile et al. [83] created a theragnostic system based on co-loaded QDs and anti-cancer drugs in nanostructured lipid carriers. This system is intended to be administered intravenously. A tumour growth suppression rate of 77.9% and an encapsulating efficacy of 80.7% were achieved by the spherical nanoparticles. According to the findings of the study, the procedure has the potential to be utilized to target and detect H22 tumour cells. Cai et al. [84] created pH-responsive ZnO QDs coated with PEG and HA acid. These QDs were stable in physiological environments and targeted cells that expressed the HA receptor CD44. This nanocarrier was used to investigate both short-term and long-term DOX release. The physiological pH was used to load DOX into the nanocarrier by either combining it with Zn^{2+} ions or conjugating it to PEG. The appearance of DOX in tumour cells only occurred after the ZnO QDs had been degraded in an acidic environment inside the cell. The researchers found that increasing the amount of DOX and ZnO QDs in treatment increased its anticancer effect.

5.6. Protein and Carbohydrate Nanoparticles

The term "natural biopolymers" refers to polysaccharides and proteins that originate chitosan from biological sources, such as plants, animals, and microbes [85]. Protein-based nanoparticles are an attractive choice due to their ability to metabolize drugs and other ligands as well as their biodegradability and ease of functionalization for drug attachment. In order to produce natural biopolymers, water-soluble proteins, such as bovine and human serum albumin, and insoluble proteins, such as zein and gliadin, are used [86]. Coacervation and desolvation, emulsion and solvent extraction, complicated coacervation, and electro-spraying are the industrial processes that are most often used to create such particles. The targeting mechanism of protein nanoparticles may be strengthened and improved by including targeting ligands. Targeting ligands are identifiers for certain cell and tissue types via a process known as chemical modification [85]. Polysaccharides, also known as polycarbohydrates, are the most abundant carbohydrates found in food; polysaccharides are constructed from sugar units linked to one another by *O*-glycosidic bonds. Polysaccharides have the potential to exhibit a broad variety of physical and chemical properties depending on their monomer composition and biological source [86]. One of the most significant difficulties of using polysaccharides in nanomedicine is the fact that these molecules are prone to oxidation (degradation) at high temperatures (far over their melting point), which are often necessary in industrial procedures. In addition, the vast majority of polysaccharides are water soluble, which restricts their applicability in nanomedicine applications, such as tissue engineering [87]. In aqueous environments, polymer chain crosslinking has been shown to be an effective method for ensuring the structural integrity of the polysaccharide chain. The many different sources of polysaccharides that have been exploited in nanomedicine are shown in Figure 6. These biopolymers are used in nanomedicine and drug delivery due to their adaptability and specific properties. These properties include their ability to originate from soft gels, flexible fibers, and hard shapes, which can make them porous or non-porous, and their high similarity to components of the extracellular matrix, which may help them avoid immunological reactions [88]. Although not much research has been done on these nanoparticles, the fact that they are made of biocompatible materials indicates that there is a significant amount of promise for their use in future drug delivery techniques. Bovine serum albumin was artificially produced by Yu et al. [89], who then studied the capacity of the protein to bind to and infiltrate the cochlea and middle ear of guinea pigs. They have also looked into the loading capacity and release behaviours of nanoparticles that have the potential to be used as drug transporters. Their goal was to determine whether or not these nanoparticles could provide greater bio-suitability, increased drug loading capacity, and a well-ordered discharge mechanism.

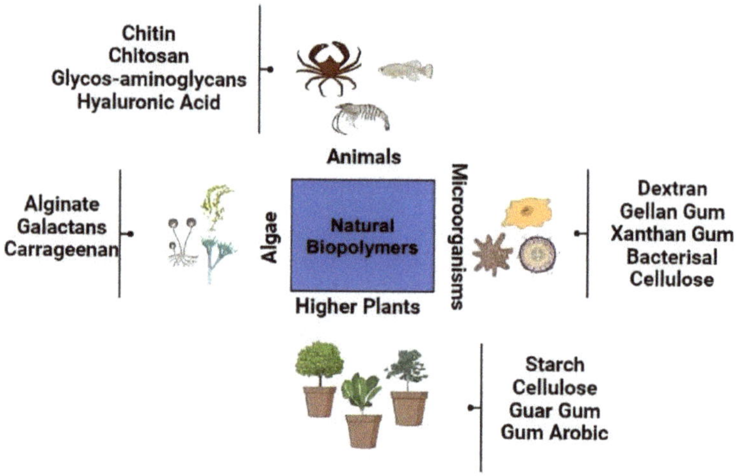

Figure 6. Natural biopolymers for nanomedicine applications from various sources. Natural biopolymers could come from higher plants, animals, microorganisms, and algae.

6. Natural Polymers for Drug Delivery

6.1. Chitosan Derivative

A DDS regulates the distribution of pharmaceuticals inside a living organism in terms of both temporal and geographical parameters [90]. The objective of the DDS is to provide the appropriate quantity of medicine at the appropriate time and location in order to maximize bioavailability while simultaneously minimizing expenditures and adverse effects [91]. Within a DDS, the fields of medicine, engineering (including materials, mechanics, and electronics), and pharmaceuticals all interact with one another. The medication, drug carrier, and related delivery mechanisms are all a part of the research, as are any physical or chemical adjustments that were made to either the drug or the carrier [92]. Chitosan is derived from the chitin via deacetylation (Figure 7).

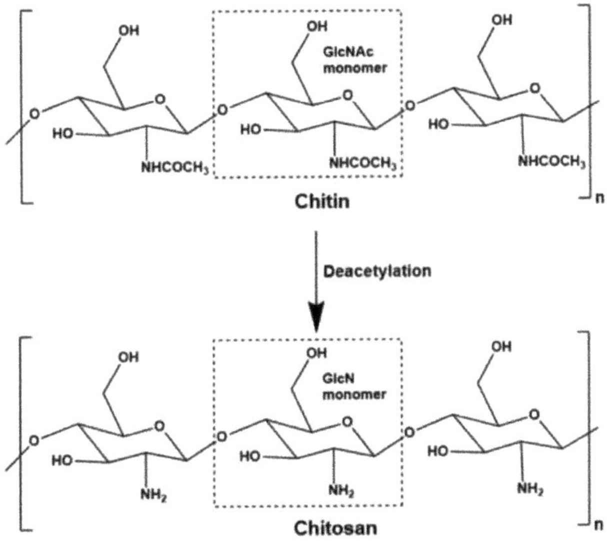

Figure 7. Mechanism of conversion of chitin to chitosan. (Reproduced from [93]).

6.1.1. Carrier for Deliveries

Chitosan and its derivatives are often dispersed by the use of micelles, gels, microspheres, and nanoparticles [94]. Microspheres have particle sizes that range from 1 to 500 μm, while nanoparticles have particle sizes that are less than 100 nm. Because of their diminutive size, nanoparticles are able to traverse a variety of biological barriers, paving the way for the targeted delivery of pharmaceuticals [95]. A micelle is a two-part structure that has exceptional stability, tissue permeability, and long-term drug release [96]. Micelles are characterized by their capacity to release drugs in a controlled manner. By producing a self-assembling micelle, amphiphilic chitosan has the potential to enhance the solubility of fat-soluble medications as well as their biological activity and the precision with which they may be administered [97]. A gel is a material that has undergone polymerization in three dimensions and has a crosslinked network structure. Gels have the ability to retain a significant amount of water. Gels that are flexible and malleable may boost chemical activity and dispersibility in biological fluids while also regulating stability, biodegradability, and biodegradability [98]. The mucoadhesion and permeability of gels are far higher than those of nanoparticles, and they also have the potential to transport very small molecules, all of which contribute to their use in a wide range of biological applications.

6.1.2. Controlled Drug Delivery

Drug control and long-term release are two prospective study subjects [99]. Because certain medications have a short half-life and are swiftly absorbed, their plasma concentration drops. As a consequence, greater therapeutic dosages are necessary to maintain the plasma balance, increasing patient discomfort. Scientists are working hard to develop new pharmaceutical delivery methods that provide therapeutically acceptable medication concentrations [100]. Quick, constant, and long-lasting pharmacological release allows for rapid effectiveness while also offering long-term advantages [101]. A continuous, progressive release of chitosan-derivative nanoparticles is straightforward to accomplish, improving bioavailability and therapeutic effectiveness while decreasing negative effects [102]. While proteins are a suggested therapy for a variety of illnesses due to their accuracy and biocompatibility, protein treatments have certain disadvantages [103]. Proteins are quickly digested by enzymes, have minimal epithelial permeability in the intestine, and are poorly absorbed via the mouth. Protein medicines have limited utility due to these properties [104].

6.1.3. Chitosan-Derivative Nanoparticles for Polypeptide Delivery

Strong hydrogen bonds and static electricity are required for the generation of peptide-loaded nanoparticles from chitosan-derivative nanoparticles interacting with peptides. Peptides that are not coated with nanoparticles are superior to their free counterparts in terms of temperature stability and the ability to be modified in vitro [105]. According to researchers, insulin-loaded, fatty acid-modified quaternary ammonium chitosan nanoparticles displayed a 98% encapsulation efficiency and loading capacity, which is superior to free insulin. This was shown by the ability of the nanoparticles to encapsulate insulin. When using insulin in a pill form, it is necessary to take into account the digestive and absorption capacities of the gastrointestinal tract [106]. Fucoidan (FD) is an amino acid that plays a role in the control of blood sugar. Trimethyl chitosan (TMC) and FD are responsible for the creation of nanoparticles that are then injected with insulin. TMC/FD nanoparticles are sensitive to pH, which allows them to block the breakdown of insulin while simultaneously boosting insulin cellular transport across the intestinal barrier [107]. In terms of insulin delivery, it has been shown that chitosan nanoparticles that have been treated with glycerol monocaprylate are comparable to TMC/FD nanoparticles [108].

6.1.4. Chitosan-Derivative Nanoparticles for Gene Delivery

In terms of solubility, biodegradability, biocompatibility, nontoxicity, and transfection rate, nonviral vectors constructed of chitosan-derivative nanoparticles have performed better than chitosan nanoparticles [109]. Chitosan nanoparticles were also less toxic. It is essential for cancer treatments that genes have the capacity to change signaling networks and improve chemotherapy-induced tumour suppression [110]. Nucleic acids that are not packed tightly are unable to pass across cell membranes and are quickly nuclease-degraded [111]. It is possible for TMC to be modified in such a way as to prevent serum genes from being degraded by nucleases [112,113]. DOX and CA were covalently bonded to methoxy PEG-modified TMC (mPEG-TMC) to form mPEG-tetracyclododecene (TCD) nanoparticles, which are more effective against tumours than DOX and plasmid DNA on their own [114]. Studies conducted in vitro have shown that the presence of *O*-carboxymethyl chitosan nanoparticles inhibits the migration of tumour cells [115]. The cell transfection rates of poly(amino ester) and thiolated *O*-carboxymethyl chitosan composite nanoparticles with loaded genes were found to be greater than those of the former alone [116]. By altering chitosan derivatives, the target ligand may also be used to enhance targeted gene delivery. It has also been proven that the use of target ligands may increase tumour-specific delivery and cellular absorption and decrease adverse effects [117].

6.2. Alginate Derivatives

Alginate is a block copolymer composed of (1,4)-linked-*D*-mannuronate and *L*-guluronate monomers. It has been shown to be biocompatible and has the ability to produce hydrogels under conditions that are considered to be moderately physiologic. This has led to a great deal of interest in the application of alginate in biomedical research. Alginate is especially valuable in the distribution of medicines because its breakdown can be regulated, it is simple to chemically modify, and it has self-healing capabilities. Although previous chemical modifications of alginate for the administration of drugs have been documented, the objective of this study was to develop a method that maintains the inherent biocompatibility and non-toxic rapid crosslinking capabilities of alginate when combined with a divalent cation, such as calcium. This was accomplished by developing a technique that preserves these properties. There have been several applications for alginate-based scaffolds, including bone and tissue regeneration [118], wound healing [119], and drug delivery platforms [120–123]. Although alginate hydrogel microbeads may expand under physiological conditions, they are frequently solid and slow to break down [124], rendering them unsuitable for regulated therapeutic drug release in the gastrointestinal tract. Alginate hydrogel microbeads may expand under physiological conditions, and we have hypothesized that alginate microbeads containing basic functional groups will behave differently than those containing acidic functional groups when exposed to acidic or basic solutions [125–128]. As a consequence, modifying the release rate may be accomplished by elevating the alginate polymer capacity for swelling and/or its water-soluble properties at a certain pH. In research on hydrogel formation and features based on chemical and physical crosslinking polymers, the network structure and permeability of the material are often two of the primary factors [129]. In contrast, modifications to the vicinal hydroxyl group include the oxidation of alcohols to dialdehyde and the reductive amination of the oxidized alginate. Chemical adjustments to alginate include esterification [130], the Ugi reaction [131,132], and amidation [133]. The Ugi four-component condensation of an aldehyde, an amine, a carboxylic acid, and an isocyanide permits the production of α-aminoacyl amide derivatives in a short period of time. The Ugi reaction is exothermic and usually complete within minutes of adding the isocyanide. The products of the Ugi reaction may exhibit a broad range of substitution patterns and are peptidomimetics with potential medicinal uses. This reaction is thus very important for generating compound libraries for screening purposes. This was the first study to modify alginate by oxidizing the vicinal dialcohol backbone of the alginate to improve and better control the release rate of potential therapeutic substances in the gastrointestinal tract. Although a previous study modified alginate using three different bioactive peptide sequences (GRGDYP, GRGDSP, and KHIFSDDSSE) in combination with 8% periodate oxidized alginate, this was the first study to use the modified alginate [134].

6.3. Xanthan Gum (XG)

Xanthan gum (XG) is a polysaccharide with a high molecular weight that is produced via spontaneous fermentation. In 1961, the United States Department of Agriculture discovered the bacterium Xanthomonas campestris on cabbage plants. The bacteria were able to produce an extracellular polysaccharide that had favorable rheological characteristics. As a direct result of this finding, several significant advancements have been made in the production of polysaccharides. XG was the most widely used and widely accessible commercially produced microbial polysaccharide at the time [135]. In a wide range of different cosmetic and medicinal goods, XG is included as a component that serves the purpose of emulsifying and suspending [136].

6.4. Cellulose

One of the oldest plentiful biopolymers, cellulose is a potential material for nano-DDSs due to its cheap cost, high biodegradability, and remarkable biocompatibility [137]. In order to be effective, pharmaceutical delivery techniques need excipients that are not only bioavailable, but also biocompatible and biodegradable. Excipients based on cellulose and its derivatives have been the subject of a significant amount of study due to their green and natural qualities as well as their one-of-a-kind encapsulating and binding properties. Cellulose and its derivatives are widely used in controlled and long-term DDSs. Additionally, cellulose and its derivatives may change the solubility or gelling behaviour of drugs in a number of different applications, which results in a wide diversity of methods for regulating drug release patterns [138,139].

6.5. Cyclodextrin Derivative

PTX is an anticancer medication that has shown promise in treating a variety of different types of cancer. Because of its low water solubility and tendency to re-crystallize after being diluted, PTX is often manufactured with the assistance of co-solvents, such as Cremophor EL®, during the commercial production process. Amphiphilic cyclodextrins are chosen oligosaccharides for the administration of anticancer medications because they have the potential to spontaneously form nanoparticles and do not need a surfactant or co-solvent. Polycationic amphiphilic cyclodextrins have recently been produced as effective gene-delivery vehicles [140]. These cyclodextrins have been generated in the form of nanoplexes.

6.6. Hyaluronic Acid, Poly(Glycolic Acid), and Poly(Lactic Acid)

For the purpose of delivering drugs specifically to tumours, docetaxel (DCT) nanoparticles loaded with poly(*D*,*L*-lactide-*co*-glycolide) (PLGA) were employed to produce an HA acid–ceramide (HACE) nanostructure [141]. DCT-loaded PLGA nanoparticles were implanted into an HACE nanostructure to create nanoparticles with a limited size distribution and a negative zeta potential. This was accomplished using the DCT/PLGA/HACE formulation. PLA and PLGA have been used in the construction of vaccine, medicine, and gene delivery systems that are both safe and effective using manufacturing techniques that are well-described and easy to replicate [142,143]. With the help of these polymers, a wide range of nano- and micro-particulates may be produced. Table 1 summarized typical examples of natural and synthetic biomaterials.

Table 1. List of natural and synthetic biomaterials.

Natural Biomaterials	Synthetic Biomaterials
Amylose	Cellulose acetate phthalate
Cellulose	Poly(vinyl acetate)
Chitin	Hydroxy acetate phthalate
Chitosan	Phthalate 55
Pectin	Phthalate 50
Alginate	Eudragit L 100
Dextran	Eudragit RS 30 D
Cyclodextrin	Hydroxy ethyl cellulose
Arginine	Poly(diethyl siloxane) (PDES)
Guar Gum	Poly(methyl hydrogen siloxane) (PMHS)
poly(glycolic acid)	Poly(glycolic acid) (PGA)
poly(lactic acid)	Poly(acrylic acid) (PAAc)
Hyaluronic acid	Poly(lactic acid) (PLA)
Heparin	Poly(lactic acid-*co*-glycolic acid) (PLGA)
Chondroitin sulphate	Poly(2-hydroxyethyl methacrylate)
Agarose	Poly(*N*-isopropyl acrylamide)
Gellan	Polycaprolactone (PCL)
Keratin	Poly(ethylenimine)s
Silk fibroin	Dendritic polymers
Collagen	Poly(*N*,*N*-diethylacrylamide) (PDEAAm)
Gelatin	Poly(ethylene oxide) (PEO)
Fibronectin	Poly(ethylene glycol) (PEG)

Table 1. *Cont.*

Natural Biomaterials	Synthetic Biomaterials
Laminins	Poly(2-(methacryloyloxy)ethyl phosphorylcholine)
Elastin	Poly(methyl methacrylate)
Glycosaminoglycan	Poly(maleic anhydride)
Ovomucin	Poly(methacrylate)
Lactoferrin	Poly(vinylacetaldiethylaminoacetate) (PVD)
Sericin	Poly(2-acrylamido 2-methylpropane sulfonate)

7. A Revolutionary Nano-Biomaterial for Biomedical Purposes

Nanotechnology has shown promise in the medical field with regard to the optimization and production of one-of-a-kind nanoparticles. When it comes to biological applications, nanoparticles stand out as particularly useful in the detection, monitoring, and therapy of diseases or damage to tissue [144]. This capability is attributed to a wide range of physicochemical and biological features [145]. Polyester-based polymer nanoparticles have found use in the field of regenerative medicine, such as in imaging agents, systems for the administration of medicine, and multifunctional intelligent structures [146–150]. Nanoparticles that are connected to scaffolds, which are three-dimensional structures that support cells, have the potential to proliferate and heal damaged tissue. Poly(*L-co-D,L*-lactic acid-*co*-trimethylene carbonate) (PLDLA-*co*-TMC) is a high molecular mass polyester that is used in tissue engineering [151,152]. This material is biocompatible, biodegradable, and bioresorbable. When a polymer is combined with amphiphilic block copolymers, the efficiency of the initial material is preserved in terms of drug encapsulation and controlled release. Non-ionic block copolymers have several useful properties, including the ability to coat, stabilize, and self-assemble, as well as the ability to regulate the release of drugs. PEO-PPO-PEO is an abbreviation for poly(ethylene oxide) (PEO) and poly(propylene oxide) (PPO) [153–155]. When the physicochemical features of polymers are coupled, it is much easier to generate unique nanoparticles with attributes that are beneficial from a therapeutic standpoint. The characteristics of the final polymer nanoparticles are determined by the precursor polymer parameters as well as the preparation conditions, such as the amount of polymer and its molecular weight, the type of solvent, the addition sequence and phase concentration, the drip rate, and the amount of surfactant.

The solvent displacement technique, which is also known as nanoprecipitation, is one of the many methods that can be used to produce nanoparticles. This method has the potential to produce nanoparticles that are effective in encapsulating hydrophobic drugs at a low cost, with a simple production process, and with a low level of complexity. The degree of miscibility between organic (also known as the internal phase) and aqueous (also known as the external phase) solvents is what governs the nucleation, development, and aggregation of nanoparticles [156]. The Marangoni [157,158] and Ouzo effects are two hypothesized components of the molecular interfacial environment that are responsible for the generation of solutions. The Marangoni effect encourages the processing of nanoparticles by increasing flow, diffusion, and surface tension fluctuations at the interface between the solvent and non-solvent (water) [159]. Nanoparticle processing occurs when metastable liquid dispersions spontaneously emulsify through liquid–liquid nucleation. When processing polymer nanoparticles, it is essential to have an understanding of the effect of molecular interface variables and physicochemical parameters on the process [160]. Quality is achieved by design and process optimization, which includes the design of trials for product development backed by the industry. As a direct result of this, producing perfect nanoparticles and developing a strategy that can be replicated on a large scale will be simple tasks. Statistical methods for analysing nanoparticle formation processes, such as the factorial 23 and Box–Behnken designs [161], have been shown to be useful in evaluating the effects of independent variables on responses and forecasting changes in NP physical–chemical properties, such as hydrodynamic diameter, polydispersity index, and zeta potential [162].

8. Consideration of General Mechanisms

8.1. Tissue-Targeting Design, Surface Functionalization, and Controlled Release

Passive (increased accumulation owing to passive physiological variables) and active (application of ligands to a specific target) diffusion are the two methods that may be used for concentrating medications to a particular area of interest [163]. The surface modification of drug carriers with bioactive compounds that interact with cell receptors and that are adsorbed, coated, conjugated, or connected to them demonstrates a preference for a certain cell or tissue type, which may increase medicine absorption. The surface modification of drug carriers with bioactive compounds can be done in a number of different ways. It is possible to apply modified coatings (for instance, ones that include albumin and chitosan) in order to limit the enzymatic breakdown that occurs in the gastrointestinal tract and plasma [164].

Both non-antibody ligands, such as lectins (carbohydrates that are adapted for cell surfaces) and monoclonal antibodies, have been put through their paces in this type of research [165]. Recent research has shown that small molecules or peptides that function as agonists/subtracts or antagonist inhibitors for overexpressed receptors on the cell surface of particular organs have shown potential for concentrated drug delivery [166,167]. There are a number of considerations that need to be made, one of which is the use of targeting ligands, which have the potential to boost distribution to secondary target sites in tissues other than those that are initially targeted [168]. On the other hand, ligands that are not antibodies have the drawback of not being selective in their binding [169]. Immunoconjugates pose concerns about immunogenicity and reticuloendothelial system (RES) retention [170]. Coating the carrier surface is an option for modifying the lipophilicity and hydrophilicity profile, lowering the rate at which immune cells are absorbed, and enhancing cell identification, e.g., the synergy between the distribution and signaling of antibodies. Within minutes after receiving an IV injection, nanoparticles were eliminated from the plasma as a result of opsonization, which was followed by phagocytosis carried out by RES cells. It is possible that surface ligands might assist in reducing opsonization. PEG is a hydrophilic polymer that has been shown to increase plasma protein resistance while simultaneously reducing serum aggregation brought on by ions and proteins [171,172]. Reduced immune responses may be achieved by preventing phagocytes from opsonizing and detecting the invading pathogen. PEG may also inhibit enzymes from accessing dendrimer scaffolds, which results in a reduction in the rate at which dendrimer is broken down [173]. PEG-coated in vivo nanoparticles and liposomes extend the amount of time it takes for an individual's blood to circulate, from minutes to hours [174,175]. The surface density of PEG, the length of its chain, and its capacity to inhibit hepatic absorption all contribute to the efficiency of the compound. On the other hand, PEG carriers are designed to infiltrate cells, and the presence of PEG may prevent the carrier from interacting negatively with cells in some circumstances. PEG carriers are meant to enter cells. PEGylated nanocarrier systems have also been connected to the accelerated blood clearance (ABC) phenomenon, which causes increased accumulation in the liver and spleen after repeated injections [176,177]. This accumulation occurs because PEGylated nanocarrier systems are lipid-coated. It has been proven that the ABC phenomenon activates an immune response known as the ABC response. This reaction causes greater accumulation in the liver and spleen after repeated injections, which is the phenomenon known as ABC.

8.2. Simultaneously Encapsulated Drugs for Combined Therapy

Simultaneously encapsulated drugs are drug delivery devices that can simultaneously administer many medications to deliver efficient chemotherapeutic drugs while suppressing the P-glycoprotein (P-gP) efflux pathway. P-gP is considered an obstacle to the successful pharmacotherapy of cancers because this protein pumps the drugs out of the cells. Consequently, P-gP overexpression is one of the main mechanisms behind decreased intracellular drug accumulation and the development of multidrug resistance in human multidrug-resistant cancers. PLGA nanoparticles were loaded with vincristine sulfate and verapamil hydrochloride at the same time and, by overcoming tumour insensitivity, boosted the therapeutic index. As a result, the same approach was used to administer DOX and cyclosporine-A [178,179]. According to current research, PLGA-PEG interacts with P-gP [180] to potentially improve system efficiency. When constructing these systems, the features of the pharmaceuticals to be encapsulated must be considered. Hydrophobic drugs, for example, can be encased in hydrophobic polymers [181]. This limitation may be solved by utilizing novel polymers, such as PLA-PEG-PLA or PCL-PEG [182], which have been successfully employed to encapsulate retinoic acid and calf-thymus DNA [183]. Another way to improve therapy is to use two different drug release rates (such as in cancer treatment). A PTX and a C6-ceramide were encapsulated in a bespoke mixture of poly(lactic-*co*-glycolic acid) and poly(β-amino ester) (PLGA/PβAE) [184] to effectively overcome cancer treatment resistance mechanisms.

8.3. Carrier Distribution

RES, which are found largely in the liver and spleen, are the most important barriers to carrier systems [185]. This is because of their inclination to internalize and withdraw themselves from systemic circulation. The concentration of PLGA NPs was greatest in the liver (40%), followed by the kidney (26%), the heart (12%), and the brain (13%), with just a small proportion present in the plasma [186]. PLGA/PbAE yielded findings that were comparable [187]. The method of administration has no effect on the distribution pattern since lymphatic clearance occurs after intraperitoneal (IP) injection and applies to all kinds of charged particles [188]. Additionally, carrier lipophilicity affects cell absorption by causing more hydrophilic particles to be expelled more rapidly [189]. This has an effect on how well the cell can take up the carrier. The nanoparticle charge surface and route of administration were studied using 10 nm gold nanoparticles that had been functionalized with a variety of groups, aiming for distinct zeta potentials (neutral, negative, positive, and zwitterionic). IV and IP administrations were utilized. Following IV administration, the peak plasma concentration of positively charged particles was ten times lower than before, and both negatively and positively charged particles were eliminated from the body more quickly [190,191]. Following IP injection, only a minute amount of both positively and negatively charged particles could be identified in the sample [192]. According to these data, the circulation was enhanced by neutral and zwitterionic nanoparticles. The significant discrepancies in bioavailability may be explained by the opsonization of nanoparticles with antibodies for detection by local macrophages [193] as well as by a similar discovery in dendrimers [194]. Nanocarriers that can be manipulated have the potential to deliver medications to specific organs and tissues. Dendrimer branch size may be adjusted to affect dendrimer dispersion and removal. As a consequence of this, dendrimers could avoid renal clearance using a cut-off that is between 40 and 60 kDa, just like the G7 [195,196]. G1 to G5 dendrimers are quickly evacuated to the kidneys and bladder, whereas G3 to G7 dendrimers are regularly recognized in circulation [197]. G8 dendrimers may be found in the lymph nodes, and G9 dendrimers are identified in the liver. As mentioned earlier, PEG plays a role in determining how evenly the carrier is distributed throughout the body. As the molecular weight of the PEGylated dendrimers rises, uptake from the injection site into the lymph becomes a considerable contribution to the total absorption profile. This indicates innovative drug delivery options as well as increased lymphatic system imaging agents [198,199].

9. Drug Delivery Using Mucoadhesive Hydrogels

9.1. Mucoadhesive Biomaterials

In the early 1980s, Nagai [200] came up with a new way to treat aphthae at the local level with an adhesive tablet. This sparked a lot of interest in developing mucoadhesive

DDSs. Nagai also discovered that using a mucoadhesive polymer enhanced the absorption of peptides when administered nasally [201]. There has been a lot of pioneering work in this field, such as the development of mucoadhesive creams based on poly(acrylic acid) (PAAc) [202] and poly(methyl methacrylate) [203]. The majority of early research on mucoadhesion was done with traditional polymers in the form of tablets [204], powders [205,206], or films [207]. The promising results observed during the creation of these early mucoadhesive formulations suggested that mucoadhesion should be investigated further. The advantages of mucoadhesive devices over traditional drug delivery methods should certainly be exploited.

Biomaterials, such as synthetic and natural polymers, have been used in the creation of innovative mucoadhesive medicinal devices to date. PAAc and cellulose derivatives make up the majority of currently used synthetic mucoadhesive polymers. Polymers that are seminatural, such as chitosan, gellan carrageenan, and pectin, are also included in the list of seminatural mucoadhesive polymers. There are also other synthetic mucoadhesive polymers, such as poly(*N*-vinyl pyrrolidone) (PNVP) and PVA [208]. The most widely used polymers for the preparation of hydrogels are listed in Table 2.

Table 2. Monomers are extensively utilized to make hydrogels for medical and pharmacological purposes.

Abbreviation	Monomer
HEMA	2-Hydroxyethyl methacrylate
HEEMA	2-Hydroxyethoxyethyl methacrylate
HDEEMA	2-Hydroxydiethoxy methacrylate
MEMA	2-Methoxyethyl methacrylate
EEMA	2-Ethoxyethyl methacrylate
MPC	2-(Methacryloyloxy)ethyl phosphorylcholine
EGDMA	Ethylene glycol dimethacrylate
NVP	*N*-Vinyl-2-pyrrolidine
NIPAAm	*N*-Isopropyl acrylamide
VAc	Vinyl acetate
AAc	Acrylic acid
MAAc	Methacrylic acid
HPMA	*N*-(2-Hydroxypropyl) methacrylamide
EG	Ethylene glycol
PEGA	PEG acrylate
PEGMA	PEG methacrylate
PEGDA	PEG diacrylate
PEGDMA	PEG dimethacrylate

At the molecular level, the chemical structures of already-existing mucoadhesive polymers could be improved. Kali et al. [209] suggested the incorporation of thiol groups into polymers in order to enhance mucoadhesion. As a result of the formation of disulfide bonds with cysteine-rich regions in mucins, polycarbophil-cysteine conjugates exhibit a great deal of mucoadhesion. The discovery of lectins was yet another groundbreaking achievement. Proteins known as lectins possess a great degree of selectivity in the carbohydrates to which they bind. The second generation of mucoadhesive uses lectin-conjugated polymers to attach to receptors on epithelial cell surfaces. Because of their capacity to bind to cellular structures, these polymers are often referred to as cytoadhesives. In spite of the fact that this polymer property differentiates it from mucoadhesive polymers, it should be noted that in order for the polymer to reach the epithelial cell membrane, it must first diffuse through the whole mucus layer. As mentioned earlier, one of the functions of this mucus layer is to serve as a protective barrier for the cells that lie underneath it. In some parts of the body, the mucus coat could be up to 400 μm thick. In order for cytoadhesive polymers to be able to reach the epithelial surface, they must first be able to selectively adhere to the cell membrane and then spread through the mucus layer.

The theory of diffusion is now being used in an effort to provide an explanation for some of the additional molecular changes that are utilized in mucoadhesive polymers. In 1963 [210], Voyutskii proposed the diffusion hypothesis as an explanation for the adherence of rubbery polymers. According to this idea, a chemical potential gradient leads the polymer chains to extend beyond the initial barrier whenever two rubbery polymers come near enough to one another. A polymer–mucus system is formed when polymer chains form and interweave with mucin glycoproteins. This results in the formation of a polymer–mucus system. In the context of mucoadhesion, Peppas and colleagues [211] suggested the interdiffusion hypothesis. They reasoned that increasing chain interpenetration would lead to an improvement in mucoadhesion. Minghao et al. [212] used ATR/FTIR spectroscopy to investigate the mucin interpenetration that occurred at the PAAc/mucin interface. According to their results, the amount of mucin present in the polymer made from PAAc increased as more time passed. Nicholas and Mikos et al. [213,214] supported the use of sticky promoters in order to improve chain interpenetration and, as a result, polymer mucoadhesion. Their reasoning was based on the notion of chain interpenetration. The injection or grafting of adhesion promoters may be done into the surface of the matrix.

Morello et al. [215] used a theoretical approach in their research on the interpenetration of free chains in mucoadhesion. They found that the length of the chain as well as the percentage of gel volume impacted the mobility of the distributed chains. Figure 8 illustrates how free polymer chains may be grafted onto the backbone of a hydrogel in order to assist in the ability of the hydrogel to adhere to other surfaces. When free adhesion promoters are brought into direct contact with mucus in a mucoadherent device, this causes the chains to scatter, which results in a concentration gradient being created in the device. Effective interactions, such as hydrogen bonding and physical entanglements, manifest themselves at the point of contact [216]. According to Haimhoffer et al. [217] the mucoadhesive properties of the polymer were improved by the incorporation of free PEG chains into crosslinked PHEMA particles. These chains served as adhesion promoters. It was hypothesized that free PEG chains moving over the interface were the cause of the problem.

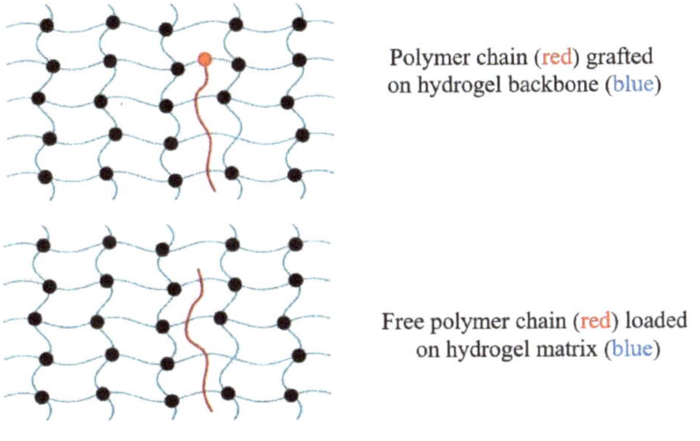

Figure 8. Schematic illustration of polymer chains used as adhesion promoters in hydrogel matrixes.

The black-filled circles in Figure 8 indicate cross linking of the polymers. Crosslinking changes a liquid polymer into a "solid" or "gel" by restricting movement. There are two types of crosslinking: physical and chemical. Physical crosslinking may not be permanent, but chemical or permanent hydrogels are formed by the covalent crosslinking of polymers. The hydrogel backbone may be grafted with polymer chains, or the network can be inserted freely. Sahlin and Peppas [218] used near-field FTIR microscopy to investigate the diffusion of free PEG chains through a PAAc hydrogel. The ability of linear PEG chains to improve mucoadhesion via interpenetration in hydrogels was first established in 1997 [219]. Tethered polymer chains may also be used to make adhesion promoters. For diffusion and penetration, the tether is chemically linked to one end of the hydrogel surface and left unattached on its opposite end. While in close contact with mucus, grafted chains are able to expand across the interface because of a concentration gradient. Covalent bonds between the backbone hydrogel structure and the adhesion promoter chains prevent them from being lost. The polymer chains of the adhesion promoter may be able to cross mucus and connect tissue to a mucoadhesive device [220,221]. Further uses for surface-anchored polymers have been discussed. For example, the single-chain mean field (SCMF) theory was used by Huang et al. [222] to investigate tethered polymer gel–gel adhesion. Theoretical developments assisted in the creation of novel tethered polymer chain biomaterials. Analysis of interactions between PEG chains and mucin glycoproteins was performed using a surface-force instrument. PEG-tethered structures might also be used to develop new mucoadhesive drug delivery systems in the future, according to their results.

9.2. Mucoadhesive Medication Delivery with Hydrogels

Water-soluble and water-insoluble systems coexist in mucoadhesive polymers. Mucoadhesive polymers that are water soluble are typically linear or branched polymeric molecules. The rate at which they dissolve determines how long they stay in the water. Swellable networks having a crosslinked chemical structure are known as water-insoluble polymer networks. The residence period of water-insoluble mucoadhesive biomaterials is determined by mucus turnover or cell desquamation. When exposed to water or physiological fluids, hydrogels form three-dimensional polymer networks [223,224]. Hydrophilic homopolymers or copolymers are used to make these networks. The crosslinks in the chemical structure keep them from dissolving and give them a distinct physical integrity. Hydrogels have been widely employed in medical and pharmaceutical fields because they better imitate genuine tissue than any other synthetic biomaterial. Because of their high-water content and flexible structure, hydrogels are biocompatible [225,226]. Contact lenses, biosensor membranes, prosthetic skin, artificial heart linings, and drug-delivery devices have all been made using hydrogels [227–232]. The network architecture of a hydrogel determines the characteristics of a drug delivery system. Three essential parameters may be used to understand the network structure of hydrogels:

- ✓ The swelling ratio, including the mass swelling ratio and the volume swelling ratio;
- ✓ The polymer volume fraction in the swollen state;
- ✓ The number-average molecular weight between cross-links (M_c);
- ✓ The network mesh size.

The swollen polymer volume fraction is used to determine the quantity of fluid that the hydrogel can absorb and hold. Molecular weight differences between crosslinks dictate the degree of crosslinking. Since polymerization is a stochastic process, only average M_c values may be derived. The distance between crosslinks or connections is determined by the mesh size, which regulates the amount of available drug diffusion space between macromolecular chains. Hydrogels may be either neutral or ionic depending on the kind of charge in their pendant groups. The swelling of hydrogels may also be caused by the surrounding environment. Recent years have seen a surge in interest in hydrogels that are responsive to body chemistry [233,234]. Many factors, including temperature, electromagnetic radiation, and pH, may affect the swell ability of hydrogels (Figure 9).

pH-sensitive hydrogels, which feature swelling behaviour and a three-dimensional architecture that are affected by the external environmental pH, employ acidic or basic pendant groups. Some chemical groups ionize as the pH and ionic strength of the environment change, causing structural changes in hydrogels. Because medicine may be delivered to particular parts of the body while simultaneously being shielded from potentially harmful biological circumstances, these qualities make pH-sensitive hydrogels ideal for use in the development of DDSs. As a medication delivery and protein control strategy, complexation hydrogels have also been previously studied.

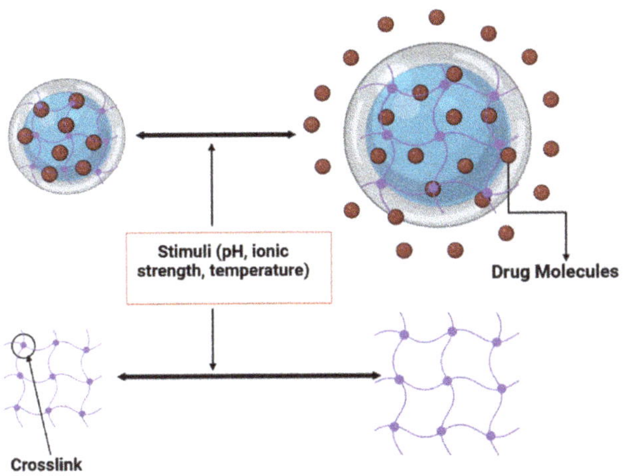

Figure 9. Physiologically sensitive hydrogel: certain external events cause the drug to swell and release.

At present, bacterial cellulose is gaining more interest to researchers due to it having a wide variety of current and potential future applications. Due to its many unique properties, it has been used in the food industry, the medical field, commercial and industrial products, and other technical areas. In the medical industry, bacterial cellulose-based hydrogels are attractive materials for wound dressing applications due to their hydrophilic properties, purity, ability to maintain appropriate moisture balance, and flexibility in conforming to any contour of the wound, forming a tight barrier between the wound and the environment, thus preventing bacterial infections. It also found its place in tissue engineering applications because of its biocompatibility, non-toxic effects, porous structure, and good mechanical strength [235,236].

The hydrogel is distinguished by the joining of chemical groups from several polymer chains. Hydrogen bonding [237–242] is one of the interactions that results in these chemical links between macromolecular chains. A wide range of monomers have been used to make therapeutic hydrogel biomaterials. Hydrogel topologies may be created by combining different monomers. Table 2 lists some of the monomers often utilized to make hydrogels for medical and pharmaceutical applications. Polymer chains are fused together to form the hydrogel [242]. Hydrogen bonding is a method through which macromolecular chains form chemical bonds. To create hydrogel biomaterials for therapeutic reasons, many monomers have been utilized. Hydrogels for medical and pharmaceutical applications may be made from any of the monomers mentioned in Table 2. It is incredible that so many different kinds of hydrogel structures can made, each with unique physical and chemical properties. To achieve precise drug delivery goals, the molecular design of customized biomaterials has exceeded the strategy of converting "off the shelf" polymeric materials for consumer usage to therapeutic use. As shown in Table 3, there are a number of hydrogel polymers that might be used in the administration of medicines under strict supervision. There are hydrogels that are able to cling and stick to mucosal surfaces, making the medication delivery device persist longer. The mucoadhesive qualities of a biomaterial are determined by the chemical composition and topology of the hydrogel, as previously stated.

Table 3. Common hydrogels used in the pharmaceutical field for the preparation of controlled drug delivery systems.

Hydrogel Polymer	Notes
Biodegradable Hydrogels	
Poly(glycolic acid) (PGA)	
Poly(lactic acid) (PLA)	
PLA-*b*-PGA	
PLA-*b*-PEG	
Chitosan	
Dextran	
Non-biodegradable hydrogels	
Poly(2-hydroxyethyl methacrylate) (PHEMA)	
Poly(vinyl alcohol) (PVA)	
Poly(*N*-vinyl pyrrolidone) (PNVP)	
Poly(ethylene-*co*-vinyl-acetate) (PEVAc)	
Poly(acrylamide) (PAAm)	
Poly(acrylic acid) (PAAc)	pH-responsive
Poly(methacrylic acid) (PMAAc)	pH-responsive
Poly(*N*,*N*-diethylaminoethyl methacylate) (PDEAMEA)	pH-responsive
Poly(*N*,*N*-dimethylaminoethyl methacrylate) (PDMAEMA)	pH-responsive
Poly(methacrylic acid)-*graft*-poly(ethylene glycol) (PMAAc-*g*-PEG))	Complexing hydrogels
Poly(acrylic acid)-*graft*-poly(ethylene glycol) (PAAc-*g*-PEG))	Complexing hydrogels
Poly(*N*-isopropyl acrylamide) (PNIPAAm)	Temperature-responsive
Poly(NIPAAm-*co*-AAc)	pH/temperature-responsive
Poly(NIPAAm-*co*-MAAc)	pH/temperature-responsive

9.3. Mechanisms of Drug Release from Mucoadhesive Hydrogels

To deliver therapeutic pharmaceuticals, a mucoadhesive hydrogel may provide suitable drug release rates and locate and hold the pharmaceutical device in a particular location of the body for an extended period of time. In comparison to typical medicinal devices, mucoadhesive hydrogels provide a number of advantages. Understanding how a medication releases over time is crucial in the science of drug distribution. To construct novel pharmacological formulations and evaluate experimental data, mathematical models are needed [243]. The majority of theoretical models used to analyse pharmaceutical transport rely on simple diffusion equations. Controlled drug release systems are available in a wide range of forms since medication may be administered in a number of ways [244].

- DDSs that are controlled through diffusion;
- Chemically controlled DDSs;
- Swelling-controlled DDSs.

Each of these systems will be addressed in this section. A quick review of diffusion fundamentals is also included, as diffusion plays a significant part in all of the systems.

9.3.1. Diffusion Fundamentals

Drug molecules must pass through the polymer bulk to be released from a mucoadhesive hydrogel. Using mass transportation concepts, we can explain the occurrence of diffusion phenomena. Drug molecules are transferred from a hydrogel matrix to an external environment using Fick's equations of diffusion. The first and second of Fick's diffusion laws are represented in Equations (1) and (2), respectively. The Fickian rules are shown in a flat manner:

$$J = -D\frac{dC}{dx} \quad (1)$$

$$\frac{\partial C}{\partial t} = D\frac{\partial^2 C}{\partial x^2} \quad (2)$$

where c and J represent concentration and mass flow, respectively, and x and t represent the independent variables of location and time, respectively. The diffusion coefficient (D) is the unit of measurement. Equation (1) of Fick's first law of diffusion is used when the steady state has been achieved or when the drug concentration within the diffusion volume does not fluctuate over time. This may be shown by using a zero-order drug release model. Equation (2) of Fick's second equation of diffusion is used to describe diffusion when a steady state has not been achieved or the concentration within the diffusion volume fluctuates with time. In this equation, ∂C is the change in C, ∂t is the change in t, and ∂x is the change in x.

9.3.2. Drug Delivery Systems with Diffusion Control

There are two types of diffusion-controlled systems: monolithic and reservoir. In monolithic devices, the medication is tightly attached to the polymer, which possesses rate-controlling properties. When the polymer has been dissolved or distributed, two monolithic systems may be formed. No matter how long a medicine stays in the system, zero-order kinetics cannot be applied to therapeutic compounds. Reservoir devices have a polymeric membrane that controls the pace at which the medication is delivered. Medications are transported from the polymer core to the polymer membrane by dissolution at one interface and diffusion caused by a change in thermodynamic activity. Using Fick's first law, Equation (1) depicts the movement of drugs. If the thermodynamic activity of the drug in the reservoir is constant and infinite sink conditions are maintained, the release rate of the drug will be constant and predictable. To achieve zero-order kinetics, the medication will be administered at a constant rate. Fick's first law may be recast to predict drug release rates from planar devices, as shown in following Equation:

$$\frac{\mathrm{d}M_t}{\mathrm{d}t} = \frac{ADK\Delta C}{t} \quad (3)$$

where $\mathrm{d}M_\mathrm{t}/dt$ represents the rate of drug release (mass/time), A is the membrane surface area, D is the diffusion coefficient, K is the partition coefficient of the drug in the membrane, and ΔC is the concentration gradient. Similarly, the restated Fick's first law may be used to calculate drug release rates from cylindrical delivery devices:

$$\frac{\mathrm{d}M_t}{\mathrm{d}t} = \frac{2\pi hDK\Delta C}{\ln(r_0 - r_1)} \quad (4)$$

where r_0 is the interior radius, r_1 is the exterior radius, and h is the height. It is possible to analyse drug release rates from spherical delivery systems using a rederived version of Fick's first law:

$$\frac{\mathrm{d}M_t}{\mathrm{d}t} = 4DKC\frac{r_0 r_1}{(r_0 - r_1)} \quad (5)$$

According to these equations, the shape of the delivery system may be used to control the release of pharmaceuticals. If you know the polymer structure and the partition coefficient of the system, you may determine how much medicine will be released through the membrane as well as the thickness of the membrane.

9.3.3. Drug Delivery Systems with Swelling Control

When water enters a hydrophilic polymer matrix, it causes the device to swell, which in turn controls the amount of medication released. A glassy polymer matrix first distributes the drug in an even manner. The medication is held in place by glassy polymers that are almost impenetrable. Drugs cannot disperse because of the glassy nature of the polymer. A rubbery phase is formed when the polymer matrix expands in the presence of water or biological fluids. Inner glass and exterior rubber phases are both visible. Drug compounds may be more difficult to detect if they go through the rubbery phase (Figure 10). In order to control the quantity of medication discharged, the pace and location of the rubbery front need to be adjusted. Swelling-controlled devices may disperse medication through diffusion and polymer chain relaxations at the glassy–rubbery interface [245]. Ritger et al. [246] provided a simple equation for evaluating diffusion and macromolecular relaxation:

$$\frac{M_t}{M_\infty} = kt^n \tag{6}$$

where M_t/M_∞ is the fraction of drugs released, M_t is the amount of drug released over time t, M_∞ is the amount of drug at the equilibrium state, k is the rate constant, and n is the diffusional exponent characteristic of the release mechanism. This exponential equation [243,246] for drug release from polymeric matrixes was developed after both Fickian and non-Fickian diffusional behaviour were researched and the implications on exponent n of these two types of diffusion were investigated. Dispersion coefficient values were shown to be connected with drug delivery systems that were either flat, cylindrical, or spherical. Table 4 summarizes these results.

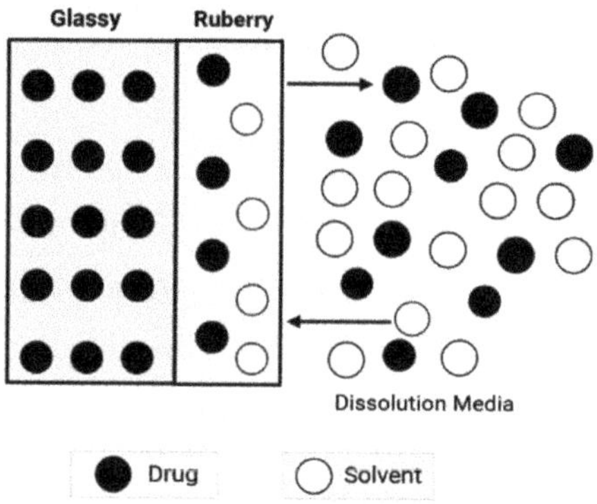

Figure 10. Depiction of a drug delivery system whose swelling is controlled. The polymer matrix swells and the rubbery phase develops when water comes into contact with the glassy hydrogel. Drug molecules can scatter out after passing through the rubbery phase.

Table 4. Different controlled release devices may be used to acquire the diffusional exponent, n, as well as the associated drug release processes [246].

Diffusion Exponent, n			Drug Release Mechanism
Thin Film	Cylindrical Sample	Spherical Sample	
0.5	0.45	0.43	Fickian diffusion
$0.5 < n < 1.0$	$0.45 < n < 1.0$	$0.43 < n < 1.0$	Anomalous transport (non-Fickian diffusion)
1.0	1.0	1.0	Case II transport (zero-order release)

10. Natural and Synthetic Biomaterials to Deliver Extracellular Vesicles (EVs)

Extracellular vesicles (EVs) are nanosized membranous structures derived from cells and can be classified into subclasses, such as exosomes, microvesicles, and apoptotic bodies. Exosomes, formed through the inward budding of endosomes within multivesicular bodies (MVBs), are released when MVBs fuse with the plasma membrane (Figure 11). Researchers often categorize EV subclasses based on physical characteristics, biochemical composition, conditions, or cell of origin due to challenges in capturing live images of EV release [247,248]. EVs play various roles in physiological and pathological processes, including suppressing inflammation, modulating cellular function, regenerating tissue injuries, and influencing the immune system. The specific functions of EVs depend on their origin, and when taken up by target cells, they release contents such as proteins, lipids, and genetic material, leading to changes in gene expression. The passage emphasizes the importance of studying reliable markers for EV subtypes to establish a consensus on nomenclature. Additionally, it highlights the potential benefits of using EVs in bioengineering, as their contents allow for the regulation of phenotype, function, and immune cell homing. Specific EV cargoes related to positive therapeutic outcomes are explored in subsequent sections [249].

There are two main types of biopolymers that are utilized in bioengineering: natural and synthetic. Natural biomacromolecules, such as silk fibroin, collagen, gelatin, chitosan, and hyaluronic acid, are explored, along with widely studied synthetic biopolymers, such as PEG, PCL, PLGA, and PLLA (Figure 11). Each type has its benefits and challenges; natural biomaterials may vary and have potential issues with mechanical stability, while synthetic biomaterials lack native tissue structure and may pose toxicity risks [250].

Here, we will highlight some natural and synthetic biopolymers for therapeutic EV delivery. Sodium alginate is a linear polysaccharide derived from brown seaweed, specifically a derivative of alginic acid composed of α-1-guluornic and 1,4-linked-β-D-mannuronic monomers. Alginate-based hydrogels loaded with EVs have been investigated for various therapeutic purposes, such as healing diabetic wounds [251], regenerating peripheral nerves [252], and addressing myocardial infarction [253]. Silk fibroin (SF) is a hydrophobic protein derived from the Bombyx mori silkworm known for its self-assembling properties that create strong and resilient materials. Cunnane and colleagues [254] investigated the impact of extracellular vesicles derived from human adipose-derived mesenchymal stem cells on vascular cells in an in vitro setting. Their findings revealed that the use of these EVs led to a dose-dependent enhancement in both the proliferation and migration of smooth muscle cells and endothelial cells. Chitosan is a cationic polysaccharide derived from chitin and composed of di-glucose amine and *N*-acetyl glucose amine groups. Scaffolds made from chitosan have been employed for the administration of EVs to enhance the reparation of bone defects, address corneal diseases, facilitate the healing of skin wounds, and treat injuries to articular cartilage. Wu and colleagues [255] specifically created chitosan-based thermosensitive hydrogels loaded with small EVs derived from bone mesenchymal stem cells to expedite the processes of osteogenesis and angiogenesis. Collagen, the predominant protein in mammals, is formed with three interwoven α-chains. Scaffolds made from collagen have been applied for therapeutic objectives, such as bone and endometrium regeneration. Xin and colleagues [256] developed a collagen scaffold incorporating exosomes derived from umbilical cord-derived mesenchymal stem cells to facilitate endometrial regeneration in a rat model of endometrial damage. HA is a linear polysaccharide composed of repeating units of *D*-glucuronic acid and *N*-acetyl-*D*-glucosamine. Scaffolds based on hyaluronic acid have been employed for delivering EVs as a treatment for tendon repair and injuries to cartilage in osteoarthritis. K. Song and colleagues [257] isolated exosomes from tendon-derived stem cells, loaded an HA scaffold with these exosomes, and investigated the therapeutic effects of this system for tendon repair. Gelatin, a commonly employed natural biopolymer in the fields of regenerative medicine and tissue engineering, has found application in scaffolds for bone regeneration. In their study, Man and colleagues [258] enhanced the osteoinductive potency of osteoblast-derived EVs through epigenetic modification using the histone deacetylase inhibitor Trichostatin A. PEG is an FDA-approved polymer known for its hydrophilic and flexible characteristics and deemed safe for biomedical applications. PEG hydrogels have been employed to deliver exosomes, aiding in the healing of cutaneous wounds [259]. PCL, an aliphatic polyester characterized by its linear nature, hydrophobic nature, high mechanical strength, and biocompatibility, is also biodegradable. Wei and colleagues explored the potential of heparin-functionalized vascular PCL grafts to improve their anti-thrombogenic properties. The researchers manufactured tubular PCL grafts through electrospinning, introduced heparin modifications to the grafts, and incorporated small extracellular vesicles derived from mesenchymal stem cells by immersing the scaffolds in an EV solution [260]. PLGA is a copolymer that shares similarities with PCL, being a biocompatible, biodegradable, and flexible biopolymer. Research has explored the use of PLGA scaffolds loaded with EVs to enhance the treatment of bone defects and chronic kidney disease. Ko and colleagues [261] specifically developed a PLGA-based scaffold for the delivery of EVs derived from stem cells, aiming to promote kidney regeneration. PLLA undergoes degradation through nonenzymatic hydrolysis, and its resulting by-products are eliminated through regular cell metabolism. Swanson and colleagues devised a biodegradable delivery platform based on PLLA to regulate the release of exosomes from microspheres, aiming to enhance craniofacial bone healing. In their approach, they employed PLGA and PEG triblock copolymer microspheres to encapsulate and control the timed release of exosomes derived from human dental pulp stem cells [262]. PLA is both biocompatible and biodegradable through hydrolysis and enzymatic activity, and it exhibits high hydrophobicity. In their work, Gandolfi and colleagues sought to create a mineral-doped, PLA-based scaffold that could be functionalized with EVs to enhance the osteogenic commitment of human adipose-derived mesenchymal stem cells. The researchers observed that mineral-doped PLA scaffolds effectively adsorbed red-labeled exosomes derived from human adipose mesenchymal stem cells [263].

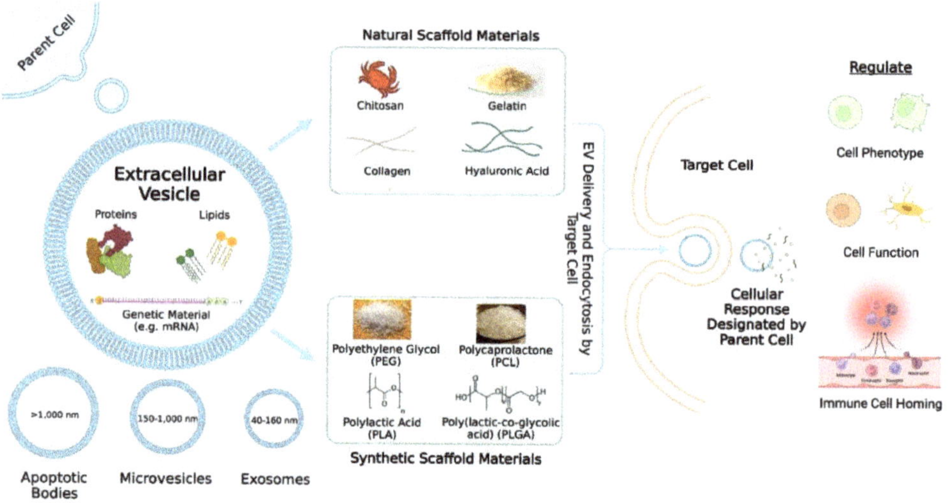

Figure 11. An overview of extracellular vesicle (EV) delivery via scaffolds. The contents and categories of the extracellular vesicles as well as what the extracellular vesicles may regulate are described. Common natural and synthetic biomaterials for scaffold fabrication are highlighted (Reproduced form [264]).

11. Pharmaceutical Applications

Pharmaceuticals based on nanotechnology have grown in popularity and have had a significant impact on the pharmaceutical industry. Compared to other industries, nanoparticle technology is a significant amount of the nanotech pharma business. The electro-spraying method addresses scalability, reproducibility, efficient encapsulation, and other potential nanoparticle production needs. Pharmaceuticals electro-sprayed with and without polymer-carriers have been employed in a wide range of applications. Drug release quality is improved by using biodegradable polymer-carriers to delay the release of encapsulated medications. In pharmaceutical applications, electro-spraying is a popular approach for generating nanoparticles [265]. Polymers have played an integral role in the advancement of drug delivery technology by providing a controlled release of therapeutic agents in constant doses over long periods, cyclic dosage, and tuneable release of both hydrophilic and hydrophobic drugs. The types of polymers used in the pharmaceutical industry must meet the same safety requirements, including sterilizability, biocompatibility, processability, fluid compatibility, and an optimum balance of mechanical properties tailored for the given application.

11.1. Brain Delivery

To sustain homeostasis and other vital tasks, the brain is a fragile neuronal organ system that requires a steady supply of fuels, gases, and nutrients. The blood–brain barrier (BBB), which is a type of vasculature in the central nervous system that acts as a physical barrier, is responsible for a number of problems. The BBB makes it more difficult for therapeutic drugs to reach the brain and spinal cord [266]. Antibiotics, anticancer drugs, and neuropeptides are among the types of pharmaceuticals that cannot get past endothelial capillaries and into the brain. For central nervous system (CNS)-related disease therapy, several pharmacological delivery systems and procedures have been devised. However, the majority of these procedures have been shown to be intrusive and lack target selectivity. Regardless, all prior medication delivery systems have been developed by trial and error. These are nearly always used to deliver a small number of medications with good structure–activity correlations, drug-receptor linkages, and structure-transport ties [267]. When it comes to allowing medications to pass through by diffusion or active transport, the BBB is nonselective, posing significant challenges for CNS drug development. The brain, on the other hand, quickly absorbs glucose and fat/lipid soluble medications. Due to their fat-insoluble nature, certain drugs, contrary to popular belief, are difficult to transfer into the brain. Pharmacological availability at potentially deadly doses is reduced because the brain capillary endothelium only distributes a small amount of medication [268]. CDs are employed in pharmaceutical applications for a variety of reasons, including improving medication bioavailability. Current CD-based therapies have been explored and potential future uses provided [269]. Additionally, various carrier materials are continually being developed to overcome the limits of therapeutic medications [270]. Cyclodextrins have been identified as attractive candidates due to their capacity to impact the physical, chemical, and biological features of guest molecules through the formation of inclusion complexes [271].

11.2. Mucosal Drug Delivery

The mucus layer works as a barrier when drugs are administered to mucosal surfaces despite the fact that it is typically ignored. The mucus gel layer that surrounds the mucosal epithelia forms a barrier [272] and has been shown to be an important component of the mucus layer owing to its water content of roughly 83% [273]. The apical location of a cell is connected to its mucus layer, which acts as a barrier for the cells below [274]. Several investigations have shown that chitosan has excellent mucoadhesive properties, which have been researched extensively. A variety of therapeutic chitosan derivatives have also been tested for their ability to adhere to the mucous membrane, with the purpose of identifying how structural modifications impact mucoadhesion [275,276]. Several processes will be discussed in this review, including ionic interactions with mucin chains and the hydration state of chitosan. Three-dimensional printing has been used to build single- and multi-layered medicine delivery systems in addition to its most common usage in oral delivery. Additionally, oral mucosa-based administration is possible through buccal and sublingual routes. Squamous cells lining the buccal cavity have a surface area of 50 cm^2 and an average thickness of 500–800 μm. Sublingual epithelium on the bottom of the mouth may potentially absorb drugs. The lower thickness of non-keratinized sublingual epithelium (100–200 μm) and its substantial vascularization frequently result in better drug penetration and early onset when compared to the buccal route.

11.3. Pulmonary Drug Delivery

Pulmonary medicine distribution is a prominent topic in research because the lungs may absorb medications for both local and systemic administration. Respiratory epithelial cells are in charge of both the production of airway-lining fluid and the regulation of airway tone. A non-invasive delivery of therapeutic medicines may be achieved via pulmonary administration, which has a high permeability, a large absorptive surface area (about 70–140 m^2 in human adults with an exceptionally thin absorptive mucosal membrane), and a good blood supply [277–279]. Chlorofluorocarbons propellants are also being phased out of industrial and residential applications across the world, which has sparked the creation of novel substitutes. Various amendments to the Montreal Protocol on Substances that deplete the ozone layer as well as to the original text outline the timescale for this project. As a consequence of these critical modifications, pharmaceutical companies and academics were able to develop new methods of administering drugs through the pulmonary route. Respiratory infection drugs and systemic sickness treatments are separate medication categories [280]. In this second state, the lungs are solely regarded as a route for systemic pharmaceutical administration.

11.4. Skin Drug Delivery

Due to physicochemical constraints, the protective function of human skin restricts the types of permeants that may pass through it. Passive skin administration necessitates the use of a lipophilic drugs with a molecular weight below 500 Da. Only a small number of commercially available medications fulfill these strict standards for percutaneous administration. Active and passive strategies have emerged in recent years to improve drug delivery. The passive strategy involves improving the formulation or the medicinal carrier in order to increase skin permeability. Drugs having molecular weights larger than 500 Da, on the other hand, do not benefit from passive techniques. For active techniques, physical or mechanical methods of distribution have been demonstrated to be more effective in the long run [281,282].

12. Ongoing Clinical Trials on Natural and Synthetic Biomaterials

This section contains human clinical trials that are currently ongoing with natural and synthetic biomaterials. Here, we highlighted the present stage of a clinical trial of the selected biomaterials. All clinical information we collected from ClinicalTrials.gov websites. Currently, there are several clinical studies ongoing with natural and synthetic biomaterials and their formulations, summarized below in Table 5. Chitosan, a modified carbohydrate polymer obtained from chitin present in various natural sources, has a wide array of applications in the fields of medicine and pharmaceuticals. Its main approved purpose is serving as a biomaterial in medical devices. Additionally, chitosan and its derivatives play essential roles in pharmaceuticals, functioning as excipients, drug carriers, or therapeutic agents. At the moment, chitosan and its composition are in different clinical phases to treat different diseases, such as periodontitis, periodontal pocket, periodontal diseases, periodontal inflammation, osteoarthritis, and vaginal bleeding. Cellulose is a complex polysaccharide that forms the principal component of the plant cell wall and is the most abundant biopolymer in the biological world. Cellulose and microcrystalline cellulose are currently in the early clinical phase to mitigate some diseases, such as long COVID-19, sarcoidosis, and pulmonary diseases. With a collagen sponge and hydrolysed collagen-based supplements, different researchers are trying to find a new way to treat alveolar osteitis, pain (acute and chronic), and knee osteoarthritis. Different formulations of dextran and gelatin are now applied to treat pain (acute and chronic) and knee osteoarthritis, but some of them are in the early clinical phase, and few of the formulations are in phase 4 clinical trials.

Table 5. Natural and synthetic biomaterials-based formulations or devices currently in clinical trials [283].

No.	Clinical Trial Phase	Participants	Composition or Device	Disease or Conditions	Estimated Study Completion Date
1	Phase 4	130	Chitosan	Vaginal bleeding, loop electrosurgical excision	31 January 2024
2	N/A	40	4% Chitosan gel (pH 3.48).	Periodontitis, periodontal pocket, periodontal diseases, periodontal inflammation	31 December 2024
3	Phase 2	48	α-Mangostin hydrogel film with chitosan alginate base	Recurrent aphthous stomatitis	20 December 2024
4	Phase 4	104	KiOmedine® CM-chitosan (KiOmed Pharma, Herstal, Belgium)	Osteoarthritis, knee	20 February 2024
5	Phase 3	150	Microcrystalline cellulose	long COVID-19	1 November 2025
6	Phase 2	40	Cellulose	Sarcoidosis, pulmonary	1 September 2024
7	Phase 2	40	Collagen sponge	Alveolar osteitis	15 November 2023 (Just completed)
8	N/A	80	Hydrolysed collagen-based supplement	Pain (acute and chronic), knee osteoarthritis	June 2024
9	Phase 4	90	Iron dextran injection	cCKD-chronic kidney disease	31 December 2024
10	N/A	100	Dextran 40	Type 2 diabetes mellitus, diabetic kidney disease	1 September 2026
11	N/A	30	Gelatin sponge-loaded apoptotic vesicle complex (Kuaikang®, Hengshui Kuaikang Medical Device Co. Ltd., Hengshui, China)	Third molar extraction	30 December 2024
12	N/A	50	Gelatin sponge stabilization with suture and cyanoacrylate	Wound heal	5 April 2024
13	Phase 1 and 2	26	18F-OP-801 (18F hydroxyl dendrimer)	Amyotrophic lateral sclerosis (ALS)	30 June 2023 (Just completed)
14	Phase 4	480	Poly(ethylene glycol) Losenatide	Type2 diabetes mellitus	31 December 2025
15	Phase 4	102	Poly(ethylene glycol)s	Hepatic encephalopathy	30 December 2023
16	N/A	40	Poly(ethylene glycol) (PEG) mediated fusion	Peripheral nerve injuries	1 October 2024
17	Phase 2	150	Poly(ethylene glycol) (PEG) recombinant human erythropoietin injection	Renal anaemia	
18	N/A	40	3D printed upper-limb prosthesis (made of poly(lactic acid))	Amniotic band syndrome, upper extremity deformities, congenital	31 August 2024
19	N/A	40	PDLLA (Poly(*D*,*L*-lactic acid)	Pigmentation, pigmentation disorder	31 December 2023
20	N/A	50	Gana X (Poly(*L*-lactic acid))	Buttocks volume loss	November 2024
21	N/A	150	Poly(*L*-lactic acid) [PLLA-SCA, Sculptra®, Dermik Laboratories, New Jersey, USA]	Skin laxity	November 2023

Table 5. *Cont.*

No.	Clinical Trial Phase	Participants	Composition or Device	Disease or Conditions	Estimated Study Completion Date
22	N/A	20	Surgical implantation of the Polycaprolactone (PCL) breast scaffold with autologous fat grafting	Breast implant revision, congenital breast defect correction	20 June 2026
23	N/A	60	PCL based breast implants Lifesil	Breast augmentation, mammoplasty	30 May 2024
24	N/A	50	Hydrogel injection (Hydroxy ethyl cellulose)	Osteoarthritis, knee pain	31 July 2024

Besides natural biomaterials, synthetic biomaterials are also gaining interest from researchers to treat different types of diseases. Dendrimer, polyethylene glycol, polylactic acid, polycaprolactone, and hydroxy ethyl cellulose are now in different clinical trial phases to treat different diseases, such as amyotrophic lateral sclerosis, type 2 diabetes mellitus, pigmentation disorder, congenital breast defect correction, and osteoarthritis, respectively.

In the below table, we have mentioned some natural and synthetic biomaterials that are currently in the clinical trial; in the table, we highlighted some points: (1) clinical trial phase; (2) participants number; (3) composition or device; (4) diseases or conditions; (5) estimated study completion date.

13. Future Direction and Perspectives

More than 1500 patents have been registered in this sector in the last two decades and dozens of clinical investigations have been completed [284]. Natural and synthetic biomaterials seem to be the ideal paradigm for therapy, diagnostics, and generation of drug delivery carriers, as detailed in the sections above. With nanoparticles, it will be easier than other techniques to deliver a precise dose of medicine to cancerous or tumorous cells without interfering with the normal physiology of healthy cells. However, material homogeneity, drug loading, and release capabilities need to be further researched. This review covers vast area to the delivery and diagnosis of medicines using metal-based nanoparticles and synthetic and natural biopolymers. One of the most promising areas of research is the use of precious metals, such as gold and silver, in diagnostic and therapeutic procedures. A major source of excitement in this review is nanoparticles, which seem to be able to penetrate soft tumour tissues, making the tumour accessible to radiation-based heat therapy for targeted eradication.

Natural and synthetic biomaterials are rapidly attracting the attention of researchers who want to use them to administer drugs in a manner that is both precise and regulated. In the near future, these materials are expected to earn greater scientific legitimacy. Despite widespread recognition of their long-term potential, nanomedicine and nano-DDSs have yet to have a significant influence on the healthcare system, notably in cancer treatment and diagnostics. Due to the fact that this field has only been studied for two decades, there are many key fundamental qualities that have yet to be discovered. One of the most important future study areas will be the development of basic indicators of sick tissues, such as vital biological markers, that permit precise targeting without interrupting the normal cellular process. With an increasing understanding of disease mechanisms and the discovery of nanomaterial-subcellular scale indicators, new diagnostic and therapeutic avenues will open up by expanding the use of the nanomedicine field. Because of this, nanomedicine will benefit from an increased understanding of disease molecular signatures.

New studies are needed to maximize the potential of nanomedicine, including animal studies and cross-disciplinary research. The growing need for accurate medications and diagnostics has led to the development of nanorobots and nanodevices for tissue diagnostics and healing systems. However, nanomedicine has drawbacks, such as acute and chronic toxicity effects on humans and the environment, and the regulation of nanomedicine will change with advancements in applications. Nanotechnology has become a significant component in increasing the effectiveness of natural bioactive chemicals, such as berberine, curcumin, resveratrol, and quercetin. Nanocarriers made with gold, silver, and other metals; titanium dioxide polymers; solid lipid nanoparticles; micelles; and superparamagnetic iron oxide nanoparticles have been identified to boost their effectiveness. Natural biomaterials, such as biodegradable, biocompatible, renewable, and low-toxicity biomaterials, are in high demand. Advanced biopolymer research issues include crosslinking biopolymers to make them more resistant to industrial processing and biological matrixes. Polymeric nanoparticles have been produced using solvent evaporation, emulsion polymerization, and surfactant-free emulsion polymerization. Nanomedicine research has seen a spike in interest in integrating therapy and diagnostics, as shown by the cancer disease model. Advances in nanomedicine have allowed for better diagnosis and treatment of illnesses by combining diagnostic and therapeutic processes.

Author Contributions: Conceptualization, project administration, funding acquisition, and supervision, S.-i.Y.; writing—original draft preparation, M.H.-O.-R., M.N.A., N.S., M.R.I., M.E.A., M.S.H., S.B. and S.-i.Y.; writing—review and editing, M.H.-O.-R. and S.-i.Y. All authors have read and agreed to the published version of the manuscript.

Funding: This study was partially supported by KAKENHI grants (21H02005, 21H05027, 21H05535, and 23H04088) from the Japan Society for the Promotion of Science (JSPS), JSPS Bilateral Joint Research Projects (JPJSBP120203509 and JPJSBP120223510), the Cooperative Research Program of Network Joint Research Center for Materials and Devices (20234041), and MEXT Promotion of Distinctive Joint Research Center Program (JPMXP0621467946).

Institutional Review Board Statement: Not applicable.

Data Availability Statement: Data will be made available on request.

Conflicts of Interest: The authors declare no conflict of interest.

Abbreviation

Abbreviation	Definition
DDS	Drug delivery system
PAE	Poly(amino ester)
IV	Intravenous
5-ALA	5-Aminolevulinic
CC	Colon cancer
CB	Cathepsin B
HA	Hyaluronic acid
Cy5.5	Cyanine 5.5
IRT	Irinotecan
PFH	Perfluorohexane
DOX	Doxorubicin
GdNGs	Gadolinium nanogels
MRI	Magnetic resonance imaging
NIR	Near infrared
C6	Chlorine 6
PTX	Paclitaxel
FA	Folic acid
PHEMA	Poly(2-hydroxyethyl Methacrylate)
HEMA	2-Hydroxyethyl methacrylate
TPGDA	Tripropyleneglycol diacrylate
PNIPAAm	Poly(*N*-isopropyl acrylamide)
LCST	Lower critical solution temperature
AAc	Acrylic acid

PAA	Poly(allyl amine)
PEI	Poly(ethylenimine)
BPEI	Branched poly(ethylenimine)
QDs	Quantum dots
FD	Fucoidan
TMC	Trimethyl chitosan
TCD	Tetracyclododecene
CMC	Critical micelles concentration
FAPPI	Folate acid conjugated poly(propylene imine)
MTX	Methotrexate
XG	Xanthan gum
DCT	Docetaxel
PLGA	poly(*D*,*L*-lactide-*co*-glycolide)
HACE	HA acid–ceramide
PLA	Poly(lactic acid)
PEO	Poly(ethylene oxide)
PPO	Poly(propylene oxide)
RES	Reticuloendothelial system
ABC	Accelerated blood clearance
P-gP	P-glycoprotein
IP	Intraperitoneal
PNVP	Poly(*N*-vinyl pyrrolidone)
BBB	Blood-brain barrier
CNS	Central nervous system
EVs	Extracellular vesicles
MSCs	Mesenchymal stem cells
MVBs	Multivesicular bodies

References

1. Patra, J.K.; Das, G.; Fraceto, L.F.; Campos, E.V.R.; Rodriguez-Torres, M.D.P.; Acosta-Torres, L.S.; Diaz-Torres, L.A.; Grillo, R.; Swamy, M.K.; Sharma, S.; et al. Nano based drug delivery systems: Recent developments and future prospects. *J. Nanobiotechnol.* **2018**, *16*, 71. [CrossRef] [PubMed]
2. Daraba, O.M.; Cadinoiu, A.N.; Rata, D.M.; Atanase, L.I.; Vochita, G. Antitumoral drug-loaded biocompatible polymeric nanoparticles obtained by non-aqueous emulsion polymerization. *Polymers* **2020**, *12*, 1018. [CrossRef]
3. Jacob, J.; Haponiuk, J.T.; Thomas, S.; Gopi, S. Biopolymer based nanomaterials in drug delivery systems: A review. *Mater. Today Chem.* **2018**, *9*, 43–55. [CrossRef]
4. Gopi, S.; Amalraj, A.; Sukumaran, N.P.; Haponiuk, J.T.; Thomas, S. Biopolymers and Their Composites for Drug Delivery: A Brief Review. *Macromol. Symp.* **2018**, *380*, 1800114. [CrossRef]
5. Rodriguez-Galan, A.; Franco, L.; Puiggali, J. Degradable Poly(ester amide)s for Biomedical Applications. *Polymers* **2011**, *3*, 65–99. [CrossRef]
6. Giri, G.; Maddahi, Y.; Zareinia, K. A Brief Review on Challenges in Design and Development of Nanorobots for Medical Applications. *Appl. Sci.* **2021**, *11*, 10385. [CrossRef]
7. Al-Arif, S.; Quader, N.; Shaon, A.M.; Islam, K.K. Sensor based autonomous medical nanorobots: A cure to demyelination. *J. Sel. Areas Nanotechnol.* **2011**, *2*, 1–7.
8. Mamani, J.B.; Borges, J.P.; Rossi, A.M.; Gamarra, L.F. Magnetic Nanoparticles for Therapy and Diagnosis in Nanomedicine. *Pharmaceutics* **2023**, *15*, 1663. [CrossRef]
9. Kim, J.; Cho, H.; Lim, D.K.; Joo, M.K.; Kim, K. Perspectives for Improving the Tumor Targeting of Nanomedicine via the EPR Effect in Clinical Tumors. *Int. J. Mol. Sci.* **2023**, *24*, 10082. [CrossRef] [PubMed]
10. Dhaliwal, A.; Zheng, G. Improving accessibility of EPR-insensitive tumor phenotypes using EPR-adaptive strategies: Designing a new perspective in nanomedicine delivery. *Theranostics* **2019**, *9*, 8091–8108. [CrossRef]
11. Herea, D.D.; Zară-Dănceanu, C.M.; Lăbușcă, L.; Minuti, A.E.; Stavilă, C.; Ababei, G.; Tibu, M.; Grigoraș, M.; Lostun, M.; Stoian, G.; et al. Enhanced Multimodal Effect of Chemotherapy, Hyperthermia and Magneto-Mechanic Actuation of Silver-Coated Magnetite on Cancer Cells. *Coatings* **2023**, *13*, 406. [CrossRef]
12. Lu, H.; Wang, J.; Wang, T.; Zhong, J.; Bao, Y.; Hao, H. Recent progress on nano- structures for drug delivery applications. *J. Nanomater.* **2016**, *2016*, 5762431. [CrossRef]
13. Blanco, E.; Shen, H.; Ferrari, M. Principles of nanoparticle design for overcoming biological barriers to drug delivery. *Nat. Biotechnol.* **2015**, *33*, 941–951. [CrossRef]
14. Kumari, A.; Kumar, V.; Yadav, S. Nanotechnology: A tool to enhance therapeutic values of natural plant products. *Trends Med. Res.* **2012**, *7*, 34–42.

15. Guanyou, L.; Miqin, Z. Ligand Chemistry in Antitumor Theranostic Nanoparticles. *Acc. Chem. Res.* **2023**, *56*, 1578–1590.
16. Haozhe, H.; Xindan, Z.; Lihua, D.; Minwen, Y.; Yonglai, L.; Jiajia, X.; Jun, W.; Xintao, S. Molecular imaging nanoprobes for theranostic applications. *Adv. Drug Deliv. Rev.* **2022**, *186*, 114320.
17. Baroni, S.; Argenziano, M.; La Cava, F.; Soster, M.; Garello, F.; Lembo, D.; Cavalli, R.; Terreno, E. Hard-Shelled Glycol Chitosan Nanoparticles for Dual MRI/US Detection of Drug Delivery/Release: A Proof-of-Concept Study. *Nanomaterials* **2023**, *13*, 2227. [CrossRef]
18. Lee, C.M.; Jang, D.; Kim, J.; Cheong, S.J.; Kim, E.M.; Jeong, M.H.; Kim, S.H.; Kim, D.W.; Lim, S.T.; Sohn, M.H.; et al. Oleyl-Chitosan nanoparticles based on a dual probe for Optical/MR imaging in vivo. *Bioconjug. Chem.* **2011**, *22*, 186–192. [CrossRef]
19. Yang, S.J.; Lin, F.H.; Tsai, H.M.; Lin, C.F.; Chin, H.C.; Wong, J.M.; Shieh, M.J. Alginate-folic acid-modified chitosan nanoparticles for photodynamic detection of intestinal neoplasms. *Biomaterials* **2011**, *32*, 2174–2182. [CrossRef]
20. Ryu, J.H.; Na, J.H.; Ko, H.K.; You, D.G.; Park, S.; Jun, E.; Yeom, H.J.; Seo, D.H.; Park, J.H.; Jeong, S.Y. Non-invasive optical imaging of cathepsin B with activatable fluorogenic nanoprobes in various metastatic models. *Biomaterials* **2014**, *35*, 2302–2311. [CrossRef]
21. Juhaščik, M.; Kováčik, A.; Huerta-Ángeles, G. Recent Advances of Hyaluronan for Skin Delivery: From Structure to Fabrication Strategies and Applications. *Polymers* **2022**, *14*, 4833. [CrossRef]
22. Puluhulawa, L.E.; Joni, I.M.; Elamin, K.M.; Mohammed, A.F.A.; Muchtaridi, M.; Wathoni, N. Chitosan–Hyaluronic Acid Nanoparticles for Active Targeting in Cancer Therapy. *Polymers* **2022**, *14*, 3410. [CrossRef] [PubMed]
23. Uthappa, U.T.; Suneetha, M.; Ajeya, K.V.; Ji, S.M. Hyaluronic Acid Modified Metal Nanoparticles and Their Derived Substituents for Cancer Therapy: A Review. *Pharmaceutics* **2023**, *15*, 1713. [CrossRef]
24. Tang, H.; Zhang, Z.; Zhu, M.; Xie, Y.; Lv, Z.; Liu, R.; Shen, Y.; Pei, J. Efficient Delivery of Gemcitabine by Estrogen Receptor-Targeted PEGylated Liposome and Its Anti-Lung Cancer Activity In Vivo and In Vitro. *Pharmaceutics* **2023**, *15*, 988. [CrossRef] [PubMed]
25. Wang, G.; Gao, S.; Tian, R.; Miller-Kleinhenz, J.; Qin, Z.; Liu, T.; Li, L.; Zhang, F.; Ma, Q.; Zhu, L. Theranostic hyaluronic acid-iron micellar nanoparticles for magnetic-field-enhanced in vivo cancer chemotherapy. *Chem. Med. Chem.* **2018**, *13*, 78–86. [CrossRef] [PubMed]
26. Ma, S.; Kim, J.H.; Chen, W.; Li, L.; Lee, J.; Xue, J.; Liu, Y.; Chen, G.; Tang, B.; Tao, W.; et al. Cancer Cell-Specific Fluorescent Prodrug Delivery Platforms. *Adv. Sci.* **2023**, *10*, 2207768. [CrossRef]
27. Weng, Y.; Yang, G.; Li, Y.; Xu, L.; Chen, X.; Song, H.; Zhao, C.-X. Alginate-based materials for enzyme encapsulation. *Adv. Colloid. Interface Sci.* **2023**, *318*, 102957. [CrossRef]
28. Baghbani, F.; Moztarzadeh, F.; Mohandesi, J.A.; Yazdian, F.; Mokhtari-Dizaji, M. Novel alginate-stabilized doxorubicin-loaded nanodroplets for ultrasounic theranosis of breast cancer. *Int. J. Biol. Macromol.* **2016**, *93*, 512–519. [CrossRef]
29. Podgórna, K.; Szczepanowicz, K.; Piotrowski, M.; Gajdošová, M.; Štěpánek, F.; Warszyński, P. Gadolinium alginate nanogels for theranostic applications. *Coll. Surf. B* **2017**, *153*, 183–189. [CrossRef]
30. Ding, Z.; Liu, P.; Hu, D.; Sheng, Z.; Yi, H.; Gao, G.; Wu, Y.; Zhang, P.; Ling, S.; Cai, L. Redox-responsive dextran based theranostic nanoparticles for near-infrared/magnetic resonance imaging and magnetically targeted photodynamic therapy. *Biomater. Sci.* **2017**, *5*, 762–771. [CrossRef]
31. Ali, M.M.; Brown, S.L.; Snyder, J.M. Dendrimer-Based Nanomedicine (Paramagnetic Nanoparticle, Nanocombretastatin, Nanocurcumin) for Glioblastoma Multiforme Imaging and Therapy. *Nov. Aproaches Cancer Study* **2021**, *6*, 609–614. [CrossRef] [PubMed]
32. Kobryń, J.; Raszewski, B.; Zięba, T.; Musiał, W. Modified Potato Starch as a Potential Retardant for Prolonged Release of Lidocaine Hydrochloride from Methylcellulose Hydrophilic Gel. *Pharmaceutics* **2023**, *15*, 387. [CrossRef] [PubMed]
33. Nayak, P.; Bentivoglio, V.; Varani, M.; Signore, A. Three-Dimensional In Vitro Tumor Spheroid Models for Evaluation of Anticancer Therapy: Recent Updates. *Cancers* **2023**, *15*, 4846. [CrossRef] [PubMed]
34. Simeonov, M.; Kostova, B.; Vassileva, E. Interpenetrating Polymer Networks of Poly(2-hydroxyethyl methacrylate) and Poly(*N*,*N*-dimethylacrylamide) as Potential Systems for Dermal Delivery of Dexamethasone Phosphate. *Pharmaceutics* **2023**, *15*, 2328. [CrossRef] [PubMed]
35. Hou, X.; Li, J.; Hong, Y.; Ruan, H.; Long, M.; Feng, N.; Zhang, Y. Advances and Prospects for Hydrogel-Forming Microneedles in Transdermal Drug Delivery. *Biomedicines* **2023**, *11*, 2119. [CrossRef] [PubMed]
36. Hanak, B.W.; Hsieh, C.Y.; Donaldson, W.; Browd, S.R.; Lau, K.S.; Shain, W. Reduced cell attachment to poly(2-hydroxyethyl methacrylate)-coated ventricular catheters in vitro. *J. Biomed. Mater. Res. B Appl. Biomater.* **2018**, *106*, 1268–1279. [CrossRef]
37. Liu, Z.; Zhang, S.; Gao, C.; Meng, X.; Wang, S.; Kong, F. Temperature/pH-Responsive Carboxymethyl Cellulose/Poly(*N*-isopropyl acrylamide) Interpenetrating Polymer Network Aerogels for Drug Delivery Systems. *Polymers* **2022**, *14*, 1578. [CrossRef]
38. Szewczyk-Łagodzińska, M.; Plichta, A.; Dębowski, M.; Kowalczyk, S.; Iuliano, A.; Florjańczyk, Z. Recent Advances in the Application of ATRP in the Synthesis of Drug Delivery Systems. *Polymers* **2023**, *15*, 1234. [CrossRef]
39. Luo, S.; Lv, Z.; Yang, Q.; Chang, R.; Wu, J. Research Progress on Stimulus-Responsive Polymer Nanocarriers for Cancer Treatment. *Pharmaceutics* **2023**, *15*, 1928. [CrossRef]
40. Jente, V.; Tomáš, S.; Valentin, V.J.; Yann, B.; Joachim, F.R.; Van, G.; Richard, H. Poly(*N*-allyl acrylamide) as a Reactive Platform toward Functional Hydrogels. *ACS Macro. Lett.* **2023**, *12*, 79–85.
41. Sarabia-Vallejos, M.A.; Cerda-Iglesias, F.E.; Pérez-Monje, D.A.; Acuña-Ruiz, N.F.; Terraza-Inostroza, C.A.; Rodríguez-Hernández, J.; González-Henríquez, C.M. Smart Polymer Surfaces with Complex Wrinkled Patterns: Reversible, Non-Planar, Gradient, and Hierarchical Structures. *Polymers* **2023**, *15*, 612. [CrossRef] [PubMed]

42. Yoo, M.K.; Sung, Y.K.; Lee, Y.M.; Cho, C.S. Effect of polyelectrolyte on the lower critical solution temperature of poly(*N*-isopropyl acrylamide) in the poly(NIPAAm-*co*-acrylic acid) hydrogel. *Polymer* **2000**, *41*, 5713–5719. [CrossRef]
43. Zhu, Y.; Liu, C.; Pang, Z. Dendrimer-Based Drug Delivery Systems for Brain Targeting. *Biomolecules* **2019**, *9*, 790. [CrossRef] [PubMed]
44. Corchero, J.L.; Favaro, M.T.P.; Márquez-Martínez, M.; Lascorz, J.; Martínez-Torró, C.; Sánchez, J.M.; López-Laguna, H.; de Souza Ferreira, L.C.; Vázquez, E.; Ferrer-Miralles, N.; et al. Recombinant Proteins for Assembling as Nano-and Micro-Scale Materials for Drug Delivery: A Host Comparative Overview. *Pharmaceutics* **2023**, *15*, 1197. [CrossRef] [PubMed]
45. Madaan, K.; Kumar, S.; Poonia, N.; Lather, V.; Pandita, D. Dendrimers in drug delivery and targeting: Drug-dendrimer interactions and toxicity issues. *J. Pharm. Bioallied Sci.* **2014**, *6*, 139–150.
46. Noriega-Luna, B.; Godínez, L.A.; Rodríguez, F.J.; Rodríguez, A.; Larrea, G.Z.L.D.; Sosa-Ferreyra, C.F.; Mercado-Curiel, R.F.; Manríquez, J.; Bustos, E.B. Applications of Dendrimers in Drug Delivery Agents, Diagnosis, Therapy, and Detection. *J. Nanomater.* **2014**, *2014*, 39. [CrossRef]
47. Sahiner, M.; Yilmaz, A.S.; Demirci, S.; Sahiner, N. Physically and Chemically Crosslinked, Tannic Acid Embedded Linear PEI-Based Hydrogels and Cryogels with Natural Antibacterial and Antioxidant Properties. *Biomedicines* **2023**, *11*, 706. [CrossRef]
48. Cai, X.; Dou, R.; Guo, C.; Tang, J.; Li, X.; Chen, J.; Zhang, J. Cationic Polymers as Transfection Reagents for Nucleic Acid Delivery. *Pharmaceutics* **2023**, *15*, 1502. [CrossRef]
49. Tarach, P.; Janaszewska, A. Recent Advances in Preclinical Research Using PAMAM Dendrimers for Cancer Gene Therapy. *Int. J. Mol. Sci.* **2021**, *22*, 2912. [CrossRef]
50. Ortega, M.Á.; Guzmán Merino, A.; Fraile-Martínez, O.; Recio-Ruiz, J.; Pekarek, L.; Guijarro, L.G.; García-Honduvilla, N.; Álvarez-Mon, M.; Buján, J.; García-Gallego, S. Dendrimers and Dendritic Materials: From Laboratory to Medical Practice in Infectious Diseases. *Pharmaceutics* **2020**, *12*, 874. [CrossRef]
51. Sung, Y.K.; Kim, S.W. Recent advances in polymeric drug delivery systems. *Biomater. Res.* **2020**, *24*, 12. [CrossRef]
52. Assad, H.; Assad, A.; Kumar, A. Recent Developments in 3D Bio-Printing and Its Biomedical Applications. *Pharmaceutics* **2023**, *15*, 255. [CrossRef]
53. Lee, T.S.; Bee, S.T. A practical guide for the processing, manufacturing and applications of PLA. In *Polylactic Acid*, 2nd ed.; Plastics design library; Elsevier: Hoboken, NJ, USA, 2019; pp. 53–95. ISBN 9780128144732.
54. Heller, J.; Barr, J.; Yng, S.; Abdellauoi, K.S.; Gurny, R. Poly(ortho esters): Synthesis, characterization, properties and uses. *Adv. Drug Deliv. Rev.* **2002**, *54*, 1015–1039. [CrossRef]
55. Kumar, N.; Langer, R.S.; Domb, A.J. Polyanhydrides: An overview. *Adv. Drug Deliv. Rev.* **2002**, *54*, 889–910. [CrossRef]
56. Kamaly, N.; Yameen, B.; Wu, J.; Farokhzad, O.C. Degradable controlled-release polymers and polymeric nanoparticles: Mechanisms of controlling drug release. *Chem. Rev.* **2016**, *116*, 2602–2663. [CrossRef]
57. Mustafai, A.; Zubair, M.; Hussain, A.; Ullah, A. Recent Progress in Proteins-Based Micelles as Drug Delivery Carriers. *Polymers* **2023**, *15*, 836. [CrossRef] [PubMed]
58. Xu, W.; Ling, P.; Zhang, T. Polymeric micelles, a promising drug delivery system to enhance the bioavailability of poorly water-soluble drugs. *J. Drug Deliv.* **2013**, *2013*, 340315. [CrossRef] [PubMed]
59. Kotta, S.; Aldawsari, H.M.; Badr-Eldin, S.M.; Nair, A.B.; YT, K. Progress in Polymeric Micelles for Drug Delivery Applications. *Pharmaceutics* **2022**, *14*, 1636. [CrossRef] [PubMed]
60. Devarajan, P.V.; Jain, S. *Targeted Drug Delivery: Concepts and Design*; Springer: Berlin, Germany, 2015; ISBN 978-3-319-11354-8.
61. Hu, Q.; Lu, Y.; Luo, Y. Recent advances in dextran-based drug delivery systems: From fabrication strategies to applications. *Carbohydr. Polym.* **2021**, *264*, 117999. [CrossRef] [PubMed]
62. Wakaskar, R.R. Polymeric micelles for drug delivery. *Int. J. Drug Dev. Res.* **2017**, *9*, 1–2.
63. Mandal, A.; Bisht, R.; Rupenthal, I.D.; Mitra, A.K. Polymeric micelles for ocular drug delivery: From structural frameworks to recent preclinical studies. *J. Control. Release* **2017**, *248*, 96–116. [CrossRef]
64. Kesharwani, P.; Xie, L.; Banerjee, S.; Mao, G.; Padhye, S.; Sarkar, F.H.; Iyer, A.K. Hyaluronic acid-conjugated polyamidoamine dendrimers for targeted delivery of 3,4-difluorobenzylidene curcumin to CD44 overexpressing pancreatic cancer cells. *Coll. Surf. B* **2015**, *136*, 413–423. [CrossRef] [PubMed]
65. Kesharwani, P.; Jain, K.; Jain, N.K. Dendrimer as nanocarrier for drug delivery. *Progr. Polym. Sci.* **2014**, *39*, 268–307. [CrossRef]
66. Jain, K.; Gupta, U.; Jain, N.K. Dendronized nanoconjugates of lysine and folate for treatment of cancer. *Eur. J. Pharm. Biopharm.* **2014**, *87*, 500–509. [CrossRef] [PubMed]
67. Kaur, A.; Jain, K.; Mehra, N.K.; Jain, N. Development and characterization of surface engineered PPI dendrimers for targeted drug delivery. *Artif. Cells Nanomed. Biotechnol.* **2017**, *45*, 414–425. [CrossRef]
68. Choi, S.J.; Lee, J.K.; Jeong, J.; Choy, J.H. Toxicity evaluation of inorganic nanoparticles: Considerations and challenges. *Mol. Cell Toxicol.* **2013**, *9*, 205–510. [CrossRef]
69. Kong, F.Y.; Zhang, J.W.; Li, R.F.; Wang, Z.X.; Wang, W.J.; Wang, W. Unique roles of gold nanoparticles in drug delivery, targeting and imaging applications. *Molecules* **2017**, *22*, 1445. [CrossRef] [PubMed]
70. Volkov, Y. Quantum dots in nanomedicine: Recent trends, advances and unresolved issues. *Biochem. Biophys. Res. Commun.* **2015**, *468*, 419–427. [CrossRef]

71. Linkova, N.; Diatlova, A.; Zinchenko, Y.; Kornilova, A.; Snetkov, P.; Morozkina, S.; Medvedev, D.; Krasichkov, A.; Polyakova, V.; Yablonskiy, P. Pulmonary Sarcoidosis: Experimental Models and Perspectives of Molecular Diagnostics Using Quantum Dots. *Int. J. Mol. Sci.* **2023**, *24*, 11267. [CrossRef]
72. Prusty, K.; Swain, S.K. Nano silver decorated polyacrylamide/dextran nanohydrogels hybrid composites for drug delivery applications. *Mater. Sci. Eng.* **2018**, *85*, 130–141. [CrossRef]
73. Marcu, A.; Pop, S.; Dumitrache, F.; Mocanu, M.; Niculite, C.; Gherghiceanu, M.; Lungu, C.; Fleaca, C.; Ianchis, R.; Barbut, A. Magnetic iron oxide nanoparticles as drug delivery system in breast cancer. *Appl. Surf. Sci.* **2013**, *281*, 60–65. [CrossRef]
74. Junyaprasert, V.B.; Morakul, B. Nanocrystals for enhancement of oral bioavailability of poorly water-soluble drugs. *Asian J. Pharm. Sci.* **2015**, *10*, 13–23. [CrossRef]
75. Du, J.; Li, X.; Zhao, H.; Zhou, Y.; Wang, L.; Tian, S.; Wang, Y. Nanosuspensions of poorly water-soluble drugs prepared by bottom-up technologies. *Int. J. Pharm.* **2015**, *495*, 738–749. [CrossRef]
76. Ni, R.; Zhao, J.; Liu, Q.; Liang, Z.; Muenster, U.; Mao, S. Nanocrystals embedded in chitosan-based respirable swellable microparticles as dry powder for sustained pulmonary drug delivery. *Eur. J. Pharm. Sci.* **2017**, *99*, 137–146. [CrossRef]
77. McNamara, K.; Tofail, S.A. Nanoparticles in biomedical applications. *Adv. Phys.* **2017**, *2*, 54–88. [CrossRef]
78. Xu, G.; Zeng, S.; Zhang, B.; Swihart, M.T.; Yong, K.T.; Prasad, P.N. New generation cadmium-free quantum dots for biophotonics and nanomedicine. *Chem. Rev.* **2016**, *116*, 12234–12327. [CrossRef]
79. Shi, Y.; Pramanik, A.; Tchounwou, C.; Pedraza, F.; Crouch, R.A.; Chavva, S.R.; Vangara, A.; Sinha, S.S.; Jones, S.; Sardar, D. Multifunctional biocompatible graphene oxide quantum dots decorated magnetic nanoplatform for efficient capture and two-photon imaging of rare tumor cells. *ACS Appl. Mater. Interfaces* **2015**, *7*, 10935–10943. [CrossRef]
80. Ahmad, J.; Garg, A.; Mustafa, G.; Ahmad, M.Z.; Aslam, M.; Mishra, A. Hybrid Quantum Dot as Promising Tools for Theranostic Application in Cancer. *Electronics* **2023**, *12*, 972. [CrossRef]
81. Zheng, F.F.; Zhang, P.H.; Xi, Y.; Chen, J.J.; Li, L.L.; Zhu, J.J. Aptamer/graphene quantum dots nanocomposite capped fluorescent mesoporous silica nanoparticles for intracellular drug delivery and real-time monitoring of drug release. *Anal. Chem.* **2015**, *87*, 11739–11745. [CrossRef] [PubMed]
82. Huang, C.L.; Huang, C.C.; Mai, F.D.; Yen, C.L.; Tzing, S.H.; Hsieh, H.T.; Ling, Y.C.; Chang, J.Y. Application of paramagnetic graphene quantum dots as a platform for simultaneous dual-modality bioimaging and tumor targeted drug delivery. *J. Mater. Chem. B* **2015**, *3*, 651–664. [CrossRef]
83. Olerile, L.D.; Liu, Y.; Zhang, B.; Wang, T.; Mu, S.; Zhang, J.; Selotlegeng, L.; Zhang, N. Near-infrared mediated quantum dots and paclitaxel co-loaded nanostructured lipid carriers for cancer theragnostic. *Coll. Surf. B* **2017**, *150*, 121–130. [CrossRef] [PubMed]
84. Cai, X.; Luo, Y.; Zhang, W.; Du, D.; Lin, Y. pH-Sensitive ZnO quantum dots- doxorubicin nanoparticles for lung cancer targeted drug delivery. *ACS Appl. Mater. Interfaces* **2016**, *8*, 22442–22450. [CrossRef] [PubMed]
85. Balaji, A.B.; Pakalapati, H.; Khalid, M.; Walvekar, R.; Siddiqui, H. Natural and synthetic biocompatible and biodegradable polymers. In *Biodegradable and Biocompatible Polymer Composites: Processing, Properties and Applications*; Shimpi, N.G., Ed.; Woodhead Publishing series in composites science and engineering; Woodhead Publishing: Duxford, UK, 2017; pp. 3–32. ISBN 9780081009703.
86. Bassas-Galia, M.; Follonier, S.; Pusnik, M.; Zinn, M. Natural polymers: A source of inspiration. In *Bioresorbable Polymers for Biomedical Applications*; Elsevier: New York, NY, USA, 2017; pp. 31–64. ISBN 9780081002629.
87. Lohcharoenkal, W.; Wang, L.; Chen, Y.C.; Rojanasakul, Y. Protein nanoparticles as drug delivery carriers for cancer therapy. *BioMed. Res. Int.* **2014**, *2014*, 180549. [CrossRef] [PubMed]
88. Cardoso, M.J.; Costa, R.R.; Mano, J.F. Marine origin polysaccharides in drug delivery systems. *Mar. Drugs* **2016**, *14*, 34. [CrossRef] [PubMed]
89. Yu, Z.; Yu, M.; Zhang, Z.; Hong, G.; Xiong, Q. Bovine serum albumin nanoparticles as controlled release carrier for local drug delivery to the inner ear. *Nanoscale Res. Lett.* **2014**, *9*, 343. [CrossRef] [PubMed]
90. Wang, B.; Wang, S.; Zhang, Q.; Deng, Y.; Li, X.; Peng, L.; Zuo, X.; Piao, M.; Kuang, X.; Sheng, S.; et al. Recent advances in polymer-based drug delivery systems for local anesthetics. *Acta Biomater.* **2019**, *96*, 55–67. [CrossRef] [PubMed]
91. Ewart, D.; Peterson, E.J.; Steer, C.J. A new era of genetic engineering for autoimmune and inflammatory diseases. Semin. *Arthritis Rheum.* **2019**, *49*, e1–e7.
92. Shamsi, M.; Mohammadi, A.; Manshadi, M.K.D.; Sanati-Nezhad, A. Mathematical and computational modeling of nano-engineered drug delivery systems. *J. Control. Release* **2019**, *307*, 150–165. [CrossRef]
93. Pal, K.; Sarkar, P.; Anis, A.; Wiszumirska, K.; Jarzębski, M. Polysaccharide-Based Nanocomposites for Food Packaging Applications. *Materials* **2021**, *14*, 5549. [CrossRef]
94. Su, C.; Liu, Y.; Li, R.; Wu, W.; Fawcett, J.P.; Gu, J. Absorption, distribution, metabolism and excretion of the biomaterials used in nanocarrier drug delivery systems. *Adv. Drug Deliv. Rev.* **2019**, *143*, 97–114. [CrossRef]
95. Jiang, W.Z.; Cai, Y.; Li, H.Y. Chitosan-based spray-dried mucoadhesive microspheres for sustained oromucosal drug delivery. *Powder Technol.* **2017**, *312*, 124–132. [CrossRef]
96. Rassu, G.; Gavini, E.; Jonassen, H.; Zambito, Y.; Fogli, S.; Breschi, M.C.; Giunchedi, P. New chitosan derivatives for the preparation of rokitamycin loaded microspheres designed for ocular or nasal administration. *J. Pharm. Sci.* **2009**, *98*, 4852–4865. [CrossRef] [PubMed]

97. Wang, F.; Zhang, Q.; Li, X.; Huang, K.; Shao, W.; Yao, D.; Huang, C. Redox-responsive blend hydrogel films based on carboxymethyl cellulose/chitosan microspheres as dual delivery carrier. *Int. J. Biol. Macromol.* **2019**, *134*, 413–421. [CrossRef]
98. Chu, L.; Zhang, Y.; Feng, Z.; Yang, J.; Tian, Q.; Yao, X.; Zhao, X.; Tan, H.; Chen, Y. Synthesis and application of a series of amphipathic chitosan derivatives and the corresponding magnetic nanoparticle-embedded polymeric micelles. *Carbohydr. Polym.* **2019**, *223*, 114966. [CrossRef] [PubMed]
99. Qu, G.; Hou, S.; Qu, D.; Tian, C.; Zhu, J.; Xue, L.; Ju, C.; Zhang, C. Self-assembled micelles based on *N*-octyl-*N*′-phthalyl-*O*-phosphoryl chitosan derivative as an effective oral carrier of paclitaxel. *Carbohydr. Polym.* **2019**, *207*, 428–439. [CrossRef] [PubMed]
100. Cuggino, J.C.; Blanco, E.R.O.; Gugliotta, L.M.; Alvarez Igarzabal, C.I.; Calderon, M. Crossing biological barriers with nanogels to improve drug delivery performance. *J. Control. Release* **2019**, *307*, 221–246. [CrossRef]
101. Li, S.; Hu, L.; Li, D.; Wang, X.; Zhang, P.; Wang, J.; Yan, G.; Tang, R. Carboxymethyl chitosan-based nanogels via acid-labile ortho ester linkages mediated enhanced drug delivery. *Int. J. Biol. Macrmol.* **2019**, *129*, 477–487. [CrossRef]
102. Wang, J.; Xu, M.; Cheng, X.; Kong, M.; Liu, Y.; Feng, C.; Chen, X. Positive/negative surface charge of chitosan based nanogels and its potential influence on oral insulin delivery. *Carbohydr. Polym.* **2016**, *136*, 867–874. [CrossRef]
103. Bulbul, Y.E.; Eskitoros-Togay, S.M.; Demirtas-Korkmaz, F.; Dilsiz, N. Multi-walled carbon nanotube-incorporating electrospun composite fibrous mats for controlled drug release profile. *Int. J. Pharm.* **2019**, *568*, 118513. [CrossRef]
104. Ozlu, B.; Kabay, G.; Bocek, I.; Yilmaz, M.; Piskin, A.K.; Shim, B.S.; Mutlu, M. Controlled release of doxorubicin from polyethylene glycol functionalized melanin nanoparticles for breast cancer therapy: Part I. Production and drug release performance of the melanin nanoparticles. *Int. J. Pharm.* **2019**, *570*, 118613. [CrossRef]
105. Gajendiran, M.; Jo, H.; Kim, K.; Balasubramanian, S. In vitro controlled release of tuberculosis drugs by amphiphilic branched copolymer nanoparticles. *J. Ind. Eng. Chem.* **2019**, *77*, 181–188. [CrossRef]
106. Safdar, R.; Omar, A.A.; Arunagiri, A.; Regupathi, I.; Thanabalan, M. Potential of Chitosan and its derivatives for controlled drug release applications—A review. *J. Drug Deliv. Sci. Technol.* **2019**, *49*, 642–659. [CrossRef]
107. Bajracharya, R.; Song, J.G.; Back, S.Y.; Han, H.-K. Recent Advancements in Non-Invasive Formulations for Protein Drug Delivery. *Comput. Struct. Biotechnol. J.* **2019**, *17*, 1290–1308. [CrossRef]
108. Lee, S.H.; Song, J.G.; Han, H.K. Development of pH-responsive organic-inorganic hybrid nanocomposites as an effective oral delivery system of protein drugs. *J. Control. Release* **2019**, *311*, 74–84. [CrossRef]
109. Du, Z.; Liu, J.; Zhang, T.; Yu, Y.; Zhang, Y.; Zhai, J.; Huang, H.; Wei, S.; Ding, L.; Liu, B. A study on the preparation of chitosan-tripolyphosphate nanoparticles and its entrapment mechanism for egg white derived peptides. *Food Chem.* **2019**, *286*, 530–536. [CrossRef]
110. Rekha, M.R.; Sharma, C.P. Synthesis and evaluation of lauryl succinyl chitosan particles towards oral insulin delivery and absorption. *J. Control. Release* **2009**, *135*, 144–151. [CrossRef] [PubMed]
111. Tsai, L.C.; Chen, C.H.; Lin, C.W.; Ho, Y.C.; Mi, F.L. Development of multifunctional nanoparticles self-assembled from trimethyl chitosan and fucoidan for enhanced oral delivery of insulin. *Int. J. Biol. Macromol.* **2019**, *126*, 141–150. [CrossRef]
112. Trivedi, A.; Hoffman, J.; Arora, R. Gene therapy for atrial fibrillation—How close to clinical implementation? *Int. J. Cardiol.* **2019**, *296*, 177–183. [CrossRef]
113. Gollomp, K.L.; Doshi, B.S.; Arruda, V.R. Gene therapy for hemophilia: Progress to date and challenges moving forward. *Transfus. Apher. Sci.* **2019**, *58*, 602–612. [CrossRef]
114. Gallego, I.; Villate-Beitia, I.; Martinez-Navarrete, G.; Menendez, M.; Lopez-Mendez, T.; Soto-Sanchez, C.; Zarate, J.; Puras, G.; Fernandez, E.; Pedraz, J.L. Non-viral vectors based on cationic niosomes and minicircle DNA technology enhance gene delivery efficiency for biomedical applications in retinal disorders. *Nanomedicine* **2019**, *17*, 308–318. [CrossRef]
115. Kochhar, S.; Excler, J.L.; Bok, K.; Gurwith, M.; McNeil, M.M.; Seligman, S.J.; Khuri-Bulos, N.; Klug, B.; Laderoute, M.; Robertson, J.S.; et al. Brighton Collaboration Viral Vector Vaccines Safety Working, G. Defining the interval for monitoring potential adverse events following immunization (AEFIs) after receipt of live viral vectored vaccines. *Vaccine* **2019**, *37*, 5796–5802. [CrossRef]
116. Mashal, M.; Attia, N.; Martinez-Navarrete, G.; Soto-Sanchez, C.; Fernandez, E.; Grijalvo, S.; Eritja, R.; Puras, G.; Pedraz, J.L. Gene delivery to the rat retina by non-viral vectors based on chloroquine-containing cationic niosomes. *J. Control. Release* **2019**, *304*, 181–190. [CrossRef] [PubMed]
117. Massaro, M.; Barone, G.; Biddeci, G.; Cavallaro, G.; Di Blasi, F.; Lazzara, G.; Nicotra, G.; Spinella, C.; Spinelli, G.; Riela, S. Halloysite nanotubes-carbon dots hybrids multifunctional nanocarrier with positive cell target ability as a potential non-viral vector for oral gene therapy. *J. Colloid. Interface Sci.* **2019**, *552*, 236–246. [CrossRef] [PubMed]
118. Javan, B.; Atyabi, F.; Shahbazi, M. Hypoxia-inducible bidirectional shRNA expression vector delivery using PEI/chitosan-TBA copolymers for colorectal Cancer gene therapy. *Life Sci.* **2018**, *202*, 140–151. [CrossRef] [PubMed]
119. Jaiswal, S.; Dutta, P.K.; Kumar, S.; Koh, J.; Pandey, S. Methyl methacrylate modified chitosan: Synthesis, characterization and application in drug and gene delivery. *Carbohydr. Polym.* **2019**, *211*, 109–117. [CrossRef] [PubMed]
120. Mallick, S.; Song, S.J.; Bae, Y.; Choi, J.S. Self-assembled nanoparticles composed of glycol chitosan-dequalinium for mitochondria-targeted drug delivery. *Int. J. Biol. Macromol.* **2019**, *132*, 451–460. [CrossRef]
121. Tang, Y.; Liu, Y.; Xie, Y.; Chen, J.; Dou, Y. Apoptosis of A549 cells by small interfering RNA targeting survivin delivery using poly-β-amino ester/guanidinylated O-carboxymethyl chitosan nanoparticles. *Asian J. Pharm. Sci.* **2020**, *13*, 121–128. [CrossRef] [PubMed]

122. Wen, L.; Hu, Y.; Meng, T.; Tan, Y.; Zhao, M.; Dai, S.; Yuan, H.; Hu, F. Redox-responsive polymer inhibits macrophages uptake for effective intracellular gene delivery and enhanced cancer therapy. *Colloids Surf. B* **2019**, *175*, 392–402. [CrossRef]
123. Lin, J.T.; Liu, Z.K.; Zhu, Q.L.; Rong, X.H.; Liang, C.L.; Wang, J.; Ma, D.; Sun, J.; Wang, G.H. Redox-responsive nanocarriers for drug and gene co-delivery based on chitosan derivatives modified mesoporous silica nanoparticles. *Colloids Surf. B* **2017**, *155*, 41–50. [CrossRef]
124. Augst, A.D.; Kong, H.J.; Mooney, D.J. Alginate hydrogels as biomaterials. *Macromol. Biosci.* **2006**, *6*, 623–633. [CrossRef]
125. Smidsrød, O.; Skja, G. Alginate as immobilization matrix for cells. *Trends Biotechnol.* **1990**, *8*, 71–78. [CrossRef]
126. Chen, C.Y.; Ke, C.J.; Yen, K.C.; Hsieh, H.C.; Sun, J.S.; Lin, F.H. 3D Porous Calcium-Alginate Scaffolds Cell Culture System Improved Human Osteoblast Cell Clusters for Cell Therapy. *Theranostics* **2015**, *5*, 643–655. [CrossRef] [PubMed]
127. Doniparthi, J.; Chappidi, S.R.; Bhargav, E. *Alginate Based Micro Particulate Systems for Drug Delivery, Alginate Biomaterial: Drug Delivery Strategies and Biomedical Engineering*; Springer Nature: Singapore, 2023; pp. 19–59. ISBN 978-981-19-6937-9.
128. Alvarez-Lorenzo, C.; Blanco-Fernandez, B.; Puga, A.M.; Concheiro, A. Crosslinked ionic polysaccharides for stimuli-sensitive drug delivery. *Adv. Drug Deliv. Rev.* **2013**, *65*, 1148–1171. [CrossRef] [PubMed]
129. Li, Y.; Xu, Z.; Wang, J.; Pei, X.; Chen, J.; Wan, Q. Alginate-based biomaterial-mediated regulation of macrophages in bone tissue engineering. *Int. J. Biol. Macromol.* **2023**, *230*, 123246. [CrossRef] [PubMed]
130. Darrabie, M.D.; Kendall, W.F.; Opara, E.C. Effect of alginate composition and gelling cation on micro-bead swelling. *J. Microencapsul.* **2006**, *23*, 29–37. [CrossRef] [PubMed]
131. Patil, J.S. Hydrogel system: An approach for drug delivery modulation. *Adv. Pharmacoepidemiol. Drug Saf.* **2015**, *5*, e135.
132. Dalheim, M.Ø.; Vanacker, J.; Najmi, M.A.; Aachmann, F.L.; Strand, B.L.; Christensen, B.E. Efficient functionalization of alginate biomaterials. *Biomaterials* **2016**, *80*, 146–156. [CrossRef]
133. Yang, J.S.; Xie, Y.J.; He, W. Research progress on chemical modification of alginate: A review. *Carbohydr. Polym.* **2011**, *84*, 33–39. [CrossRef]
134. Bu, H.; Kjøniksen, A.L.; Elgsaeter, A.; Nyström, B. Interaction of unmodified and hydrophobically modified alginate with sodium dodecyl sulfate in dilute aqueous solution: Calorimetric, rheological, and turbidity studies. *Colloids Surf. A Physicochem. Eng.* **2006**, *278*, 166–174. [CrossRef]
135. Gomez, C.G.; Chambat, G.; Heyraud, A.; Villar, M.; Auzély-Velty, R. Synthesis and characterization of a β-CD-alginate conjugate. *Polymer* **2006**, *47*, 8509–8516. [CrossRef]
136. Pandey, S.; Mishra, S.B. Graft copolymerization of ethyl acrylate onto xanthan gum, using potassium peroxydisulfate as an initiator. *Int. J. Biol. Macromol.* **2011**, *49*, 527–535. [CrossRef]
137. Rana, V.; Rai, P.; Tiwary, A.K.; Singh, R.S.; Kennedy, J.F.; Knill, C.J. Modified gums: Approaches and applications in drug delivery. *Carbohydr. Polym.* **2011**, *83*, 1031–1047. [CrossRef]
138. Adepu, S.; Ramakrishna, S. Controlled Drug Delivery Systems: Current Status and Future Directions. *Molecules* **2021**, *26*, 5905. [CrossRef] [PubMed]
139. Dai, L.; Si, C. Recent advances on cellulose-based nano-drug delivery systems: Design of prodrugs and nanoparticles. *Curr. Med. Chem.* **2019**, *26*, 2410–2429. [CrossRef] [PubMed]
140. Sun, B.; Zhang, M.; Shen, J.; He, Z.; Fatehi, P.; Ni, Y. Applications of cellulose-based materials in sustained drug delivery systems. *Curr. Med. Chem.* **2019**, *26*, 2485–2501. [CrossRef] [PubMed]
141. Varan, G.; Benito, J.M.; Mellet, C.O.; Bilensoy, E. Development of polycationic amphiphilic cyclodextrin nanoparticles for anticancer drug delivery. *Beilstein J. Nanotechnol.* **2017**, *8*, 1457–1468. [CrossRef] [PubMed]
142. Elmowafy, E.M.; Tiboni, M.; Soliman, M.E. Biocompatibility, biodegradation and biomedical applications of poly(lactic acid)/poly(lactic-*co*-glycolic acid) micro and nanoparticles. *J. Pharm. Investig.* **2019**, *49*, 347–380. [CrossRef]
143. Yiye, L.; Coates, G.W. Pairing-Enhanced Regioselectivity: Synthesis of Alternating Poly(lactic-*co*-glycolic acid) from Racemic Methyl-Glycolide. *J. Am. Chem. Soc.* **2023**, *145*, 22425–22432.
144. Andrade, A.L.; Fabris, J.D.; Pereira, M.C.; Domingues, R.Z.; Ardisson, J.D. Preparation of composite with silica-coated nanoparticles of iron oxide spinels for applications based on magnetically induced hyperthermia. *Hyperfine Interact.* **2012**, *218*, 71–82. [CrossRef]
145. Vieira, S.; Vial, S.; Reis, R.L.; Oliveira, J. Nanoparticles for bone tissue engineering. *Biotechnol. Prog.* **2017**, *3*, 590–611. [CrossRef]
146. Hickey, J.W.; Santos, J.L.; Williford, J.M.; Mao, H.Q. Control of polymeric nanoparticle size to improve therapeutic delivery. *J. Control. Release* **2015**, *219*, 536–547. [CrossRef] [PubMed]
147. Banik, B.L.; Fattahi, P.; Brown, J.L. Polymeric nanoparticles: The future of nanomedicine. *Wiley Interdiscip. Rev. Nanomed. Nanobiotechnol.* **2016**, *8*, 271–299. [CrossRef] [PubMed]
148. Capasso, P.U.; Maraldi, M.; Manfredini, N.; Moscatelli, D. Zwitterionic polyester-based nanoparticles with tunable size, polymer molecular weight, and degradation time. *Biomacromolecules* **2018**, *19*, 1314–1323. [CrossRef] [PubMed]
149. Sousa, F.; Fonte, P.; Cruz, A.; Kennedy, P.J.; Pinto, I.M.; Sarmento, B. Polyester-based nanoparticles for the encapsulation of monoclonal antibodies. *Methods Mol. Biol.* **2018**, *1674*, 239–253. [PubMed]
150. Fonte, P.; Sousa, F. Sarmento B/ Polyester-Based Nanoparticles for Delivery of Therapeutic Proteins. *Methods Mol. Biol.* **2018**, *1674*, 255–274.
151. Rijt, S.V.; Habibovic, P. Enhancing regenerative approaches with nanoparticles. *J. R. Soc. Interface* **2017**, *129*, 20170093. [CrossRef] [PubMed]

152. Motta, A.C.; Duek, E.A.D.R. Synthesis and characterization of a novel terpolymer based on *L*-lactide, *D,L*-lactide and trimethylene carbonate. *Mat. Res.* **2014**, *17*, 619–626. [CrossRef]
153. Messias, A.D.; Martins, K.F.; Motta, A.C.; Duek, E.A.D.R. Synthesis, characterization, and osteoblastic cell culture of poly(*L-co-D,L*-lactide-*co*-trimethylene carbonate) scaffolds. *Int. J. Biomater.* **2014**, *2014*, 501789. [CrossRef]
154. Cardoso, T.P.; Ursolino, A.P.S.; Casagrande, P.D.M.; Caetano, E.B.; Mistura, D.V.; Duek, E.A.D.R. In vivo evaluation of porous hydrogel pins to fill osteochondral defects in rabbits. *Rev. Bras. Ortop.* **2016**, *52*, 95–102. [CrossRef]
155. Wang, N.; Guan, Y.; Yang, L.; Jia, L.; Wei, X.; Liu, H.; Guo, C. Magnetic nanoparticles (MNPs) covalently coated by PEO-PPO-PEO block copolymer for drug delivery. *J. Colloid. Interface Sci.* **2013**, *395*, 50–57. [CrossRef]
156. Miladi, K.; Sfar, S.; Fessi, H.; Elaissari, A. Nanoprecipitation Process: From Particle Preparation to In Vivo Applications. In *Polymer Nanoparticles for Nanomedicines*; Springer Intern. Publishing: Berlin, Germany, 2016; pp. 17–53.
157. Wissink, J.; Herlina, H. Surface-temperature-induced Marangoni effects on developing buoyancy-driven flow. *J. Fluid. Mech.* **2023**, *962*, A23. [CrossRef]
158. Mahsa, M.; Mozhdeh, S.; Hossein, H.S. Thermally driven Marangoni effects on the spreading dynamics of droplets. *Int. J. Multiph. Flow.* **2023**, *159*, 104335.
159. Beck-Broichsitter, M.; Erik, R.; Tobias, L.; Xiaoying, W.; Thomas, K. Preparation of nanoparticles by solvent displacement for drug delivery: A shift in the "ouzo region" upon drug loading. *Eur. J. Pharm. Sci.* **2010**, *41*, 244–253. [CrossRef] [PubMed]
160. Schubert, S.; Delaney, J.J.T.; Schubert, U.S. Nanoprecipitation and nanoformulation of polymers: From history to powerful possibilities beyond poly(lactic acid). *Soft Matter.* **2011**, *7*, 1581–1588. [CrossRef]
161. Prakobvaitayakit, M.; Nimmannit, U. Optimization of polylactic-*co*-glycolic acid nanoparticles containing itraconazole using 23 factorial design. *Pharm. Sci. Tech.* **2003**, *4*, 565–573. [CrossRef] [PubMed]
162. Jan, A.T.; Azam, M.; Siddiqui, K.; Ali, A.; Choi, I.; Haq, Q.M.R. Heavy Metals and Human Health: Mechanistic Insight into Toxicity and Counter Defense System of Antioxidants. *Int. J. Mol. Sci.* **2015**, *16*, 29592–29630. [CrossRef] [PubMed]
163. Jabeen, N.; Muhammad, A. Polysaccharides based biopolymers for biomedical applications: A review. *Polym. Adv. Technol.* **2023**, e6203. [CrossRef]
164. Mondal, A.; Nayak, A.K.; Chakraborty, P.; Banerjee, S.; Nandy, B.C. Natural Polymeric Nanobiocomposites for Anti-Cancer Drug Delivery Therapeutics: A Recent Update. *Pharmaceutics* **2023**, *15*, 2064. [CrossRef] [PubMed]
165. Haisheng, H.; Yi, L.; Jianping, Q.; Quangang, Z.; Zhongjian, C.; Wei, W. Adapting liposomes for oral drug delivery. *Acta. Pharm. Sin. B* **2019**, *9*, 36–48.
166. Allen, T.M. Ligand-Targeted Therapeutics in Anticancer Therapy. *Nat. Rev. Cancer* **2002**, *2*, 750–763. [CrossRef]
167. Riaz, M.K.; Riaz, M.A.; Zhang, X.; Lin, C.; Wong, K.H.; Chen, X.; Zhang, G.; Lu, A.; Yang, Z. Surface Functionalization and Targeting Strategies of Liposomes in Solid Tumor Therapy: A Review. *Int. J. Mol. Sci.* **2018**, *19*, 195. [CrossRef] [PubMed]
168. Millard, M.; Yakavets, I.; Zorin, V.; Kulmukhamedova, A.; Marchal, S.; Bezdetnaya, L. Drug delivery to solid tumors: The predictive value of the multicellular tumor spheroid model for nanomedicine screening. *Int. J. Nanomed.* **2017**, *12*, 7993–8007. [CrossRef] [PubMed]
169. Yoo, J.; Park, C.; Yi, G.; Lee, D.; Koo, H. Active Targeting Strategies Using Biological Ligands for Nanoparticle Drug Delivery Systems. *Cancers* **2019**, *11*, 640. [CrossRef] [PubMed]
170. Yu, X.; Yu-Ping, Y.; Dikici, E.; Deo, S.K.; Daunert, S. Beyond Antibodies as Binding Partners: The Role of Antibody Mimetics in Bioanalysis. *Annu. Rev. Anal. Chem.* **2017**, *10*, 293–320. [CrossRef] [PubMed]
171. Kukowska-Latallo, J.F.; Candido, K.A.; Cao, Z.; Nigavekar, S.S.; Majoros, I.J. Nanoparticle Targeting of Anticancer Drug Improves Therapeutic Response in Animal Model of Human Epithelial Cancer. *Cancer Res.* **2005**, *65*, 5317–5324. [CrossRef] [PubMed]
172. Bennewitz, M.F.; Saltzman, W.M. Nanotechnology for Delivery of Drugs to the Brain for Epilepsy. *Neurotherapeutics* **2009**, *6*, 323–336. [CrossRef] [PubMed]
173. Taratula, O.; Garbuzenko, O.B.; Kirkpatrick, P.; Pandya, I.; Savla, R. Surface-Engineered Targeted PPI Dendrimer for Efficient Intracellular and Intratumoral siRNA Delivery. *J. Control. Release* **2009**, *140*, 284–293. [CrossRef]
174. Zhu, S.; Hong, M.; Zhang, L.; Tang, G.; Jiang, Y.; Pei, Y. PEGylated PAMAM Dendrimer-Doxorubicin Conjugates: In Vitro Evaluation and In Vivo Tumor Accumulation. *Pharm. Res.* **2010**, *27*, 161–174. [CrossRef]
175. Park, J.; Fong, P.M.; Lu, J.; Russell, K.S.; Booth, C.J.; Saltzman, W.M.; Fahmy, T.M. PEGylated PLGA nanoparticles for the improved delivery of doxorubicin. *Nanomedicine* **2009**, *5*, 410–418. [CrossRef]
176. Garcia-Garcia, E.; Andrieux, K.; Gil, S.; Couvreur, P. Colloidal carriers and blood–brain barrier (BBB) translocation: A way to deliver drugs to the brain? *Int. J. Pharm.* **2005**, *298*, 274–292. [CrossRef]
177. Chen, D.; Liu, W.; Shen, Y.; Mu, H.; Zhang, Z. Effects of a novel pH-sensitive liposome with cleavable esterase-catalyzed and pH-responsive double smart mPEG lipid derivative on ABC phenomenon. *Int. J. Nanomed.* **2011**, *6*, 2053–2061. [CrossRef] [PubMed]
178. Martinho, N.; Damgé, C.; Reis, C.P. Recent Advances in Drug Delivery Systems. *J. Biomater. Nanobiotech.* **2011**, *2*, 510–526. [CrossRef]
179. Song, X.; Zhao, Y.; Wu, W.; Bi, Y.; Cai, Z. PLGA nanoparticles simultaneously loaded with vincristine sulfate and verapamil hydrochloride: Systematic study of particle size and drug entrapment efficiency. *Int. J. Pharm.* **2008**, *350*, 320–329. [CrossRef]
180. Ke, W.; Zhao, Y.; Huang, R.; Jiang, C.; Pei, Y. Enhanced Oral Bioavailability of Doxorubicin in a Dendrimer Drug Delivery System. *J. Pharmacol. Sci.* **2008**, *97*, 2208–2216. [CrossRef] [PubMed]

181. Geldenhuys, W.; Mbimba, T.; Bui, T.; Harrison, K.; Sutariya, V. Brain-targeted delivery of paclitaxel using glutathione-coated nanoparticles for brain cancers. *J. Drug Target.* **2011**, *19*, 837–845. [CrossRef] [PubMed]
182. Su, C.W.; Chiang, C.S.; Li, W.M.; Hu, S.H.; Chen, S.Y. Multifunctional nanocarriers for simultaneous encapsulation of hydrophobic and hydrophilic drugs in cancer treatment. *Nanomedicine* **2014**, *9*, 1499–1515. [CrossRef] [PubMed]
183. Hammady, T.; El-Gindy, A.; Lejmi, E.; Dhanikula, R.S.; Moreau, P.; Hildgen, P. Characteristics and properties of nanospheres co-loaded with lipophilic and hydrophilic drug models. *Int. J. Pharm.* **2009**, *369*, 185–195. [CrossRef]
184. Amani, A.; Kabiri, T.; Shafiee, S.; Hamidi, A. Preparation and Characterization of PLA-PEG-PLA/PEI/DNA Nanoparticles for Improvement of Transfection Efficiency and Controlled Release of DNA in Gene Delivery Systems. *Iran. J. Pharm. Res.* **2019**, *18*, 125–141.
185. Van Vlerken, L.E.; Duan, Z.; Little, S.R.; Seiden, M.V.; Amiji, M.M. Biodistribution and Pharmacokinetic Analysis of Paclitaxel and Ceramide Administered in Multifunctional Polymer-Blend Nanoparticles in Drug Resistant Breast Cancer Model. *Mol. Pharm.* **2008**, *5*, 516–526. [CrossRef]
186. Reis, C.P.; Neufeld, R.J.; Ribeiro, A.J.; Veiga, F.; Nanoencapsulation, I. Methods for preparation of drug-loaded polymeric nanoparticles. *Nanomedicine* **2006**, *2*, 8–21. [CrossRef]
187. Semete, B.; Booysen, L.; Lemmer, Y.; Kalombo, L.; Katata, L. In vivo evaluation of the biodistribution and safety of PLGA nanoparticles as drug delivery systems. *Nanomedicine* **2010**, *6*, 662–671. [CrossRef] [PubMed]
188. Fields, C.J.; Cheng, E.; Quijano, C.; Weller, N.; Kristofik, N.; Duong, C.; Hoimes, M.E.; Egan, W.M. Saltzman, Surface modified poly(β amino ester)-containing nanoparticles for plasmid DNA delivery. *J. Control. Release* **2012**, *164*, 41–48. [CrossRef] [PubMed]
189. Arvizo, R.R.; Miranda, O.R.; Moyano, D.F.; Wal-den, C.A.; Giri, K. Modulating Pharmacokinetics, Tumor Uptake and Biodistribution by Engineered Nanoparticles. *PLoS ONE* **2011**, *6*, e24374. [CrossRef] [PubMed]
190. Testa, B.; Crivori, P.; Reist, M.; Pierre-Alain, C. The influence of lipophilicity on the pharmacokinetic behavior of drugs: Concepts and examples. Perspect. *Drug Discov. Des.* **2000**, *19*, 179–211.
191. Xin, X.C.; Nabisab, M.M.; Shaukat, A.M.; Abdul, S.J.; Awais, A.; Mohammad, K.; Rashmi, W.; Abdullah, E.C.; Rama, R.K.; Siddiqui, M.T.H.; et al. A review on the properties and applications of chitosan, cellulose and deep eutectic solvent in green chemistry. *J. Ind. Eng. Chem.* **2021**, *104*, 362–380.
192. Shoyaib, A.A.; Archie, S.R.; Karamyan, V.T. Intraperitoneal Route of Drug Administration: Should it Be Used in Experimental Animal Studies? *Pharm. Res.* **2020**, *37*, 12. [CrossRef] [PubMed]
193. Zheng, W.; Xue, F.; Zhang, M.; Wu, Q.; Yang, Z.; Ma, S.; Liang, H.; Wang, C.; Wang, Y.; Ai, X.; et al. Charged Particle (Negative Ion)-Based Cloud Seeding and Rain Enhancement Trial Design and Implementation. *Water* **2020**, *12*, 1644. [CrossRef]
194. Zolnik, B.S.; González-Fernández, Á.; Sadrieh, N.; Dobrovolskaia, M.A. Minireview: Nanoparticles and the Immune System. *Endocrinology* **2010**, *151*, 458–465. [CrossRef]
195. Gustafson, H.H.; Holt-Casper, D.; Grainger, D.W.; Ghandehari, H. Nanoparticle uptake: The phagocyte problem. *Nano. Today* **2015**, *10*, 487–510. [CrossRef]
196. Sadekar, S.; Linares, O.; Noh, G.J.; Hubbard, D.; Ray, A.; Janát-Amsbury, M.; Peterson, C.M.; Facelli, J.; Ghandehari, H. Comparative pharmacokinetics of PAMAM-OH dendrimers and HPMA copolymers in ovarian tumor-bearing mice. *Drug Deliv. Transl. Res.* **2013**, *3*, 260–271. [CrossRef]
197. Lee, C.C.; MacKay, J.A.; Fréchet, J.M.; Szoka, F.C. Designing dendrimers for biological applications. *Nat. Biotechnol.* **2005**, *23*, 1517–1526. [CrossRef]
198. Chenthamara, D.; Subramaniam, S.; Ramakrishnan, S.G.; Krishnaswamy, S.; Essa, M.M.; Lin, F.H.; Qoronfleh, M.W. Therapeutic efficacy of nanoparticles and routes of administration. *Biomater. Res.* **2019**, *23*, 20. [CrossRef]
199. Ryan, G.M.; Kaminskas, L.M.; Bulitta, J.B.; McIntosh, M.P.; Owen, D.J.; Porter, C.J. EGylated polylysine dendrimers increase lymphatic exposure to doxorubicin when compared to PEGylated liposomal and solution formulations of doxorubicin. *J. Control. Release* **2013**, *172*, 128–136. [CrossRef] [PubMed]
200. Nagai, T. Adhesive topical drug delivery system. *J. Control. Release* **1985**, *2*, 121–134. [CrossRef]
201. Yermak, I.M.; Davydova, V.N.; Volod'ko, A.V. Mucoadhesive Marine Polysaccharides. *Mar. Drugs* **2022**, *20*, 522. [CrossRef] [PubMed]
202. Keldibekova, R.; Suleimenova, S.; Nurgozhina, G.; Kopishev, E. Interpolymer Complexes Based on Cellulose Ethers: Application. *Polymers* **2023**, *15*, 3326. [CrossRef]
203. Alvani, M.; Bahri Najafi, R.; Mahmood, M.; Fazel, N. Preparation and pharmaceutical evaluation of Mucoadhesive Buccal film extracts of petroselinum for pharyngitis symptoms. *Med. Sci. J. Islam. Azad Univesity-Tehran Med. Branch* **2022**, *32*, 256–263. [CrossRef]
204. Račić, A.; Krajišnik, D. Biopolymers in Mucoadhesive Eye Drops for Treatment of Dry Eye and Allergic Conditions: Application and Perspectives. *Pharmaceutics* **2023**, *15*, 470. [CrossRef]
205. Lee, E.; Park, H.Y.; Kim, S.W.; Sun, Y.; Choi, J.H.; Seo, J.; Jung, Y.P.; Kim, A.J.; Kim, J.; Lim, K. Enhancing Supplemental Effects of Acute Natural Antioxidant Derived from Yeast Fermentation and Vitamin C on Sports Performance in Triathlon Athletes: A Randomized, Double-Blinded, Placebo-Controlled, Crossover Trial. *Nutrients* **2023**, *15*, 3324. [CrossRef]
206. Sharma, P.; Joshi, R.V.; Pritchard, R.; Xu, K.; Eicher, M.A. Therapeutic Antibodies in Medicine. *Molecules* **2023**, *28*, 6438. [CrossRef]

207. Brito-Casillas, Y.; Caballero, M.J.; Hernández-Baraza, L.; Sánchez-Hernández, R.M.; Betancort-Acosta, J.C.; Wägner, A.M. Ex vivo evaluation of adhesive strength and barrier effect of a novel treatment for esophagitis. *Gastroenterol. Y Hepatol.* **2023**, *46*, 455–461. [CrossRef] [PubMed]
208. Song, S.Y.; Ahn, M.S.; Mekapogu, M.; Jung, J.A.; Song, H.Y.; Lim, S.H.; Jin, J.S.; Kwon, O.K. Analysis of Floral Scent and Volatile Profiles of Different Aster Species by E-nose and HS-SPME-GC-MS. *Metabolites* **2023**, *13*, 503. [CrossRef] [PubMed]
209. Kali, G.; Fürst, A.; Efiana, N.A.; Dizdarević, A.; Bernkop-Schnürch, A. Intraoral Drug Delivery: Highly Thiolated κ-Carrageenan as Mucoadhesive Excipient. *Pharmaceutics* **2023**, *15*, 1993. [CrossRef] [PubMed]
210. Voyutskii, S.S.; Vakula, V.L. The role of diffusion phenomena in polymer-to-polymer adhesion. *J. Appl. Polym. Sci.* **1963**, *7*, 475–491. [CrossRef]
211. Gurny, R.; Meyer, J.M.; Peppas, N.A. Bioadhesive intraoral release systems: Design, testing and analysis. *Biomaterials* **1984**, *5*, 336–340. [CrossRef]
212. Minghao, Z.; Yuanyuan, Q.; Jinlong, L.; Lijun, Y.; Qingrong, H.; Xin, J. Charge characteristics of guar gums on the three-stage interaction mechanism with mucin to improve the mucoadhesion ability. *Food Hydrocoll.* **2023**, *145*, 109107.
213. Nicholas, A.P.; Daniel, A.C. Impact of Absorption and Transport on Intelligent Therapeutics and Nano-scale Delivery of Protein Therapeutic Agents. *Chem. Eng. Sci.* **2009**, *64*, 4553–4565.
214. Mikos, A.G.; Peppas, N.A. *Bioadhesion: Possibilities and Future Trends: First International Joint Workshop of the Association for Pharmaceutical Technology (APV) and the Controlled Release Society (CRS)*; CRC Press (Taylor & Francis): Boca Raton, FL, USA, 1990; pp. 31–37.
215. Morello, G.; De Iaco, G.; Gigli, G.; Polini, A.; Gervaso, F. Chitosan and Pectin Hydrogels for Tissue Engineering and In Vitro Modeling. *Gels* **2023**, *9*, 132. [CrossRef]
216. Georgiev, N.I.; Bakov, V.V.; Anichina, K.K.; Bojinov, V.B. Fluorescent Probes as a Tool in Diagnostic and Drug Delivery Systems. *Pharmaceuticals* **2023**, *16*, 381. [CrossRef]
217. Haimhoffer, Á.; Dossi, E.; Béresová, M.; Bácskay, I.; Váradi, J.; Afsar, A.; Rusznyák, Á.; Vasvári, G.; Fenyvesi, F. Preformulation Studies and Bioavailability Enhancement of Curcumin with a 'Two in One' PEG-β-Cyclodextrin Polymer. *Pharmaceutics* **2021**, *13*, 1710. [CrossRef]
218. Jabbari, E.; Peppas, N.A. Use of ATR-FTIR to study interdiffusion in polystyrene and poly(vinyl methyl ether). *Macromolecules* **1999**, *26*, 2175–2186. [CrossRef]
219. Osorno, L.L.; Brandley, A.N.; Maldonado, D.E.; Yiantsos, A.; Mosley, R.J.; Byrne, M.E. Review of Contemporary Self-Assembled Systems for the Controlled Delivery of Therapeutics in Medicine. *Nanomaterials* **2021**, *11*, 278. [CrossRef] [PubMed]
220. De Gennes, P.G. *Soft Interfaces*; Cambridge University Press: Cambridge, UK, 1997; ISBN 9780511628764.
221. Marques, C.; Leal-Júnior, A.; Kumar, S. Multifunctional Integration of Optical Fibers and Nanomaterials for Aircraft Systems. *Materials* **2023**, *16*, 1433. [CrossRef] [PubMed]
222. Huang, Y.; Szleifer, I.; Peppas, N.A. Gel–gel adhesion by tethered polymers. *J. Chem. Phys.* **2001**, *114*, 3809–3816. [CrossRef]
223. Riley, R.G.; Green, K.L.; Smart, J.D. The gastrointestinal transit profile of ^{14}C-labelled poly(acrylic acids): An in vivo study. *Biomaterials* **2001**, *22*, 1861–1867. [CrossRef]
224. Lowman, A.M.; Peppas, N.A. *Hydrogels, Encyclopedia of Controlled Drug Delivery*; Wiley: New York, NY, USA, 1999; pp. 197–418.
225. Peppas, N.A.; Huang, Y.; Torres-Lugo, M.; Ward, J.H.; Zhang, J. Physicochemical foundations and structural design of hydrogels in medicine and biology. *Annu. Rev. Biomed. Eng.* **2000**, *2*, 9–29. [CrossRef]
226. Peppas, N.A.; Bures, P.; Leobandung, W.; Ichikawa, H. Hydrogels in pharmaceutical formulations. *J. Pharm. Biopharm.* **2000**, *50*, 27–46. [CrossRef]
227. Dziubla, T.D.; Lowman, A.M.; Peppas, N.A. Evaluation of Poly(ethylene glycol)-Based Copolymers for Contact Lenses. *Trans. Soc. Biomater.* **2001**, *27*, 232.
228. Peppas, N.A.; Ratner, B.D.; Hoffman, A.S.; Schoen, F.J.; Lemons, J.E. *Biomaterials Science: An Introduction to Materials in Medicine*; Academic Press: New York, NY, USA, 2004.
229. Peppas, N.A.; Langer, R. New challenges in biomaterials. *Science* **1994**, *263*, 1715–1720. [CrossRef]
230. Park, K. *Controlled Drug Delivery: Challenges and Strategies*; American Chemical Society: Washington, DC, USA, 1997; ISBN 0-8412-3418-3.
231. Brannon-Peppas, L.; Peppas, N.A. *Polymer Science of Controlled Release Systems*; Controlled Release Systems KSSD: Istanbul, Turkey, 2002; pp. 1–40.
232. Peppas, N.A.; Wood, K.M.; Blanchette, J.O. Hydrogels for oral delivery of therapeutic proteins. *Biol. Ther.* **2004**, *4*, 881–887. [CrossRef]
233. Peppas, N.A.; Zhang, J. *Diffusional Behavior in pH- and Temperature Sensitive Interpenetrating Polymeric Networks Used in Drug Delivery, Biomaterials and Drug Delivery Systems towards the New Millennium*; Hanrimwon: Seoul, Republic of Korea, 2000; pp. 87–96.
234. Peppas, N.A. *Kinetics of Smart Hydrogels, Reflexive Polymers and Hydrogels: Understanding and Designing Fast-Responsive Polymeric Systems*; CRC Press (Taylor & Francis): Boca Raton, FL, USA, 2004; pp. 99–113.
235. Horue, M.; Silva, J.M.; Berti, I.R.; Brandão, L.R.; Barud, H.D.S.; Castro, G.R. Bacterial cellulose-based materials as dressings for wound healing. *Pharmaceutics* **2023**, *15*, 424. [CrossRef] [PubMed]

236. Ciolacu, D.E.; Nicu, R.; Suflet, D.M.; Rusu, D.; Darie-Nita, R.N.; Simionescu, N.; Ciolacu, F. Multifunctional Hydrogels Based on Cellulose and Modified Lignin for Advanced Wounds Management. *Pharmaceutics* **2023**, *15*, 2588. [CrossRef] [PubMed]
237. Lowman, A.M.; Morishita, M.; Kajita, M.; Nagai, T.; Peppas, N.A. Oral delivery of insulin using pH-responsive complexation gels. *J. Pharm. Sci.* **1999**, *88*, 933–937. [CrossRef] [PubMed]
238. Foss, A.C.; Goto, T.; Morishita, M.; Peppas, N.A. Development of acrylic-based copolymers for oral insulin delivery. *J. Pharm. Biopharm.* **2004**, *57*, 163–169. [CrossRef]
239. Torres-Lugo, M.; Garcia, M.; Record, R.; Peppas, N.A. pH-Sensitive Hydrogels as Gastrointestinal Tract Absorption Enhancers: Transport Mechanisms of Salmon Calcitonin and Other Model Molecules Using the Caco-2 Cell Model. *Biotechnol. Prog.* **2002**, *28*, 612–616. [CrossRef]
240. Lowman, A.M.; Peppas, N.A. Molecular analysis of interpolymer complexation in graft copolymer networks. *Polymer* **2000**, *41*, 73–80. [CrossRef]
241. Kim, B.; Peppas, N.A. Analysis of molecular interactions in poly(methacrylicacid-*g*-ethyleneglycol) hydrogels. *Polymer* **2003**, *44*, 3701–3707. [CrossRef]
242. Lowman, A.M. *Complexing Polymers in Drug Delivery, Handbook of Pharmaceutical Controlled Release Technology*; Marcel Dekker: New York, NY, USA, 2000; pp. 89–98.
243. Narasimhan, B.; Peppas, N.A. *The Role of Modeling Studies in the Development of Future Controlled Release Devices, Controlled Drug Delivery: Challenges and Strategies*; American Chemical Society: Washington, DC, USA, 1997; pp. 529–557.
244. Heller, J. *Use of Polymers in Controlled Release of Active Agents, Controlled Drug Delivery, Fundamentals and Applications*; Marcel Dekker: New York, NY, USA, 1987; pp. 179–212.
245. Burnette, R.R. *Theory of Mass Transfer, Controlled Drug Delivery, Fundamentals and Applications*; Marcel Dekker: New York, NY, USA, 1987; pp. 95–138.
246. Ritger, P.L.; Peppas, N.A. A simple equation for description of solute release I. Fickian and non-fickian release from non-swellable devices in the form of slabs, spheres, cylinders or discs. *J. Control. Release* **1987**, *5*, 23–36. [CrossRef]
247. Fonseka, P.; Marzan, A.L.; Mathivanan, S. Introduction to the Community of Extracellular Vesicles. In *New Frontiers: Extracellular Vesicles*; Mathivanan, S., Fonseka, P., Nedeva, C., Atukorala, I., Eds.; Subcellular Biochemistry; Springer International Publishing: Cham, Switzerland, 2021; pp. 3–18. ISBN 978-3-030-67170-9.
248. Rai, A.; Fang, H.; Fatmous, M.; Claridge, B.; Poh, Q.H.; Simpson, R.J.; Greening, D.W. A Protocol for Isolation, Purification, Characterization, and Functional Dissection of Exosomes. *Methods Mol. Biol.* **2021**, *2261*, 105–149.
249. Garcia-Martin, R.; Wang, G.; Brandão, B.B.; Zanotto, T.M.; Shah, S.; Kumar Patel, S.; Schilling, B.; Kahn, C.R. MicroRNA Sequence Codes for Small Extracellular Vesicle Release and Cellular Retention. *Nature* **2022**, *601*, 446–451. [CrossRef]
250. Mallick, S.P.; Singh, B.N.; Rastogi, A.; Srivastava, P. Design and Evaluation of Chitosan/Poly(l-Lactide)/Pectin Based Composite Scaffolds for Cartilage Tissue Regeneration. *Int. J. Biol. Macromol.* **2018**, *112*, 909–920. [CrossRef] [PubMed]
251. Zhang, Y.; Zhang, P.; Gao, X.; Chang, L.; Chen, Z.; Mei, X. Preparation of Exosomes Encapsulated Nanohydrogel for Accelerating Wound Healing of Diabetic Rats by Promoting Angiogenesis. *Mater. Sci. Eng. C* **2021**, *120*, 111671. [CrossRef] [PubMed]
252. Yang, Z.; Yang, Y.; Xu, Y.; Jiang, W.; Shao, Y.; Xing, J.; Chen, Y.; Han, Y. Biomimetic Nerve Guidance Conduit Containing Engineered Exosomes of Adipose-Derived Stem Cells Promotes Peripheral Nerve Regeneration. *Stem Cell Res. Ther.* **2021**, *12*, 442. [CrossRef] [PubMed]
253. Lv, K.; Li, Q.; Zhang, L.; Wang, Y.; Zhong, Z.; Zhao, J.; Lin, X.; Wang, J.; Zhu, K.; Xiao, C.; et al. Incorporation of Small Extracellular Vesicles in Sodium Alginate Hydrogel as a Novel Therapeutic Strategy for Myocardial Infarction. *Theranostics* **2019**, *9*, 7403–7416. [CrossRef]
254. Cunnane, E.M.; Lorentz, K.L.; Ramaswamy, A.K.; Gupta, P.; Mandal, B.B.; O'Brien, F.J.; Weinbaum, J.S.; Vorp, D.A. Extracellular Vesicles Enhance the Remodeling of Cell-Free Silk Vascular Scaffolds in Rat Aortae. *ACS Appl. Mater. Interfaces* **2020**, *12*, 26955–26965. [CrossRef]
255. Wu, D.; Qin, H.; Wang, Z.; Yu, M.; Liu, Z.; Peng, H.; Liang, L.; Zhang, C.; Wei, X. Bone Mesenchymal Stem Cell-Derived SEV-Encapsulated Thermosensitive Hydrogels Accelerate Osteogenesis and Angiogenesis by Release of Exosomal MiR-21. *Front. Bioeng. Biotechnol.* **2022**, *9*, 829136. [CrossRef]
256. Xin, L.; Lin, X.; Zhou, F.; Li, C.; Wang, X.; Yu, H.; Pan, Y.; Fei, H.; Ma, L.; Zhang, S. A Scaffold Laden with Mesenchymal Stem Cell-Derived Exosomes for Promoting Endometrium Regeneration and Fertility Restoration through Macrophage Immunomodulation. *Acta Biomater.* **2020**, *113*, 252–266. [CrossRef]
257. Man, K.; Barroso, I.A.; Brunet, M.Y.; Peacock, B.; Federici, A.S.; Hoey, D.A.; Cox, S.C. Controlled Release of Epigenetically-Enhanced Extracellular Vesicles from a GelMA/Nanoclay Composite Hydrogel to Promote Bone Repair. *Int. J. Mol. Sci.* **2022**, *23*, 832. [CrossRef]
258. Nikhil, A.; Kumar, A. Evaluating Potential of Tissue-Engineered Cryogels and Chondrocyte Derived Exosomes in Articular Cartilage Repair. *Biotechnol. Bioeng.* **2022**, *119*, 605–625. [CrossRef]
259. Kwak, G.; Cheng, J.; Kim, H.; Song, S.; Lee, S.J.; Yang, Y.; Jeong, J.H.; Lee, J.E.; Messersmith, P.B.; Kim, S.H. Sustained Exosome-Guided Macrophage Polarization Using Hydrolytically Degradable PEG Hydrogels for Cutaneous Wound Healing: Identification of Key Proteins and MiRNAs, and Sustained Release Formulation. *Small* **2022**, *18*, 2200060. [CrossRef]

260. Wei, Y.; Wu, Y.; Zhao, R.; Zhang, K.; Midgley, A.C.; Kong, D.; Li, Z.; Zhao, Q. MSC-Derived SEVs Enhance Patency and Inhibit Calcification of Synthetic Vascular Grafts by Immunomodulation in a Rat Model of Hyperlipidemia. *Biomaterials* **2019**, *204*, 13–24. [CrossRef] [PubMed]
261. Ko, K.-W.; Park, S.-Y.; Lee, E.H.; Yoo, Y.-I.; Kim, D.-S.; Kim, J.Y.; Kwon, T.G.; Han, D.K. Integrated Bioactive Scaffold with Polydeoxyribonucleotide and Stem-Cell-Derived Extracellular Vesicles for Kidney Regeneration. *ACS Nano* **2021**, *15*, 7575–7585. [CrossRef] [PubMed]
262. Swanson, W.B.; Zhang, Z.; Xiu, K.; Gong, T.; Eberle, M.; Wang, Z.; Ma, P.X. Scaffolds with Controlled Release of Pro-Mineralization Exosomes to Promote Craniofacial Bone Healing without Cell Transplantation. *Acta Biomater.* **2020**, *118*, 215–232. [CrossRef] [PubMed]
263. Gandolfi, M.G.; Gardin, C.; Zamparini, F.; Ferroni, L.; Esposti, M.D.; Parchi, G.; Ercan, B.; Manzoli, L.; Fava, F.; Fabbri, P.; et al. Mineral-Doped Poly(L-Lactide) Acid Scaffolds Enriched with Exosomes Improve Osteogenic Commitment of Human Adipose-Derived Mesenchymal Stem Cells. *Nanomaterials* **2020**, *10*, 432. [CrossRef]
264. Leung, K.S.; Shirazi, S.; Cooper, L.F.; Ravindran, S. Biomaterials and Extracellular Vesicle Delivery: Current Status, Applications and Challenges. *Cells* **2022**, *11*, 2851. [CrossRef]
265. Sridhar, R.; Ramakrishna, S. Electrosprayed nanoparticles for drug delivery and pharmaceutical applications. *Biomatter.* **2013**, *3*, 24281. [CrossRef]
266. Kinoshita, M. Targeted Drug Delivery to the Brain Using Focused Ultrasound. Top. Magn. Reson. *Imaging* **2006**, *17*, 209–215.
267. Jones, A.R.; Shusta, E.V. Blood-brain barrier transport of therapeutics via receptor-mediation. *Pharma. Res.* **2007**, *24*, 1759–1771. [CrossRef]
268. Dadparvar, M.; Wagner, S.; Wien, S.; Kufleitner, J.; Worek, F.; Von Briesen, H.; Kreuter, J. HI 6 human serum albumin nanoparticles-development and transport over an in vitro blood-brain barrier model. *Toxicol. Lett.* **2011**, *206*, 60–66. [CrossRef]
269. Tiwari, G.; Tiwari, R.; Rai, A.K. Cyclodextrins in delivery systems: Applications. *J. Pharm. Bioallied Sci.* **2010**, *2*, 72–79. [CrossRef]
270. Zuccari, G.; Alfei, S. Development of Phytochemical Delivery Systems by Nano-Suspension and Nano-Emulsion Techniques. *Int. J. Mol. Sci.* **2023**, *24*, 9824. [CrossRef]
271. Hussein, N.R.; Omer, H.K.; Abdelbary, M.A.E.; Ahmed, W. Chapter 15—Advances in nasal drug delivery systems. In *Advances in Medical and Surgical Engineering*; Academic Press: Cambridge, MA, USA, 2020; pp. 279–311.
272. Hall, D.J.; Khutoryanskaya, O.V.; Khutoryanskiy, V.V. Developing synthetic mucosa-mimetic hydrogels to replace animal experimentation in characterization of mucoadhesive drug delivery systems. *Soft Matter* **2011**, *7*, 9620–9623. [CrossRef]
273. Zhu, Q.; Talton, J.; Zhang, G.; Cunningham, T.; Wang, Z.; Waters, R.C.; Kirk, J.; Eppler, B.; Klinman, D.M.; Sui, Y.; et al. Large intestine-targeted, nanoparticle-releasing oral vaccine to control genitorectal viral infection. *Nat. Med.* **2012**, *18*, 1291–1296. [CrossRef]
274. Lawson, L.B.; Norton, E.B.; Clements, J.D. Defending the mucosa: Adjuvant and carrier formulations for mucosal immunity. *Curr. Opin. Immunol.* **2011**, *23*, 414–420. [CrossRef] [PubMed]
275. Sandri, S.R.G.; Bonferoni, M.C.; Ferrari, F.; Mori, M.; Caramella, C. The role of chitosan as a mucoadhesive agent in mucosal drug delivery. *J. Drug Deliv. Sci. Technol.* **2012**, *22*, 275–284. [CrossRef]
276. Lam, J.K.W.; Cheung, C.C.K.; Chow, M.Y.T.; Harrop, E.; Lapwood, S.; Barclay, S.I.G.; Wong, I.C.K. Transmucosal drug administration as an alternative route in palliative and end-of-life care during the COVID-19 pandemic. *Adv. Drug Deliv. Rev.* **2020**, *160*, 234–243. [CrossRef]
277. Wang, C.; Chu, C.; Ji, X.; Luo, G.; Xu, C.; He, H.; Yao, J.; Wu, J.; Hu, J.; Jin, Y. Biology of Peptide Transporter 2 in Mammals: New Insights into Its Function Structure and Regulation. *Cells* **2022**, *11*, 2874. [CrossRef] [PubMed]
278. Mikihisa, T.; Shiori, K.; Nanako, K.; Masashi, K.; Ryoko, Y. Effect of Corticosteroids on Peptide Transporter 2 Function and Induction of Innate Immune Response by Bacterial Peptides in Alveolar Epithelial Cells. *Biol. Pharm. Bull.* **2022**, *45*, 213–219.
279. Banat, H.; Ambrus, R.; Csóka, I. Drug combinations for inhalation: Current products and future development addressing disease control and patient compliance. *Int. J. Pharm.* **2023**, *643*, 123070. [CrossRef]
280. Smola, M.; Vandamme, T.; Sokolowski, A. Nanocarriers as pulmonary drug delivery systems to treat and to diagnose respiratory and non-respiratory diseases. *Int. J. Nanomed.* **2008**, *3*, 1–19.
281. Czajkowska-Kosnik, A.; Szekalska, M.; Winnicka, K. Nanostructured lipid carriers: A potential use for skin drug delivery systems. *Pharmacol. Rep.* **2019**, *71*, 156–166. [CrossRef] [PubMed]
282. Brown, M.B.; Martin, G.P.; Jones, S.A.; Akomeah, F.K. Dermal and transdermal drug delivery systems: Current and future prospects. *Drug Deliv.* **2006**, *13*, 175–187. [CrossRef] [PubMed]
283. Clinical Trials.Gov Website. Available online: https://classic.clinicaltrials.gov/ (accessed on 18 November 2023).
284. Pandit, A.; Zeugolis, D.I. Twenty-five years of nano-bio-materials: Have we revolutionized healthcare? *Nanomedicine* **2016**, *11*, 985–987. [CrossRef] [PubMed]

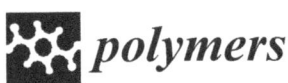

polymers

MDPI

Review

Polymeric Nanoparticles-Loaded Hydrogels for Biomedical Applications: A Systematic Review on In Vivo Findings

Débora Nunes [1,2], Stéphanie Andrade [1,2,†], Maria João Ramalho [1,2,†], Joana A. Loureiro [1,2,*] and Maria Carmo Pereira [1,2,*]

1 LEPABE—Laboratory for Process Engineering, Environment, Biotechnology and Energy, Faculty of Engineering, University of Porto, Rua Dr. Roberto Frias, 4200-465 Porto, Portugal; deborasspinola@hotmail.com (D.N.); stephanie@fe.up.pt (S.A.); mjramalho@fe.up.pt (M.J.R.)
2 ALiCE—Associate Laboratory in Chemical Engineering, Faculty of Engineering, University of Porto, Rua Dr. Roberto Frias, 4200-465 Porto, Portugal
* Correspondence: jasl@fe.up.pt (J.A.L.); mcsp@fe.up.pt (M.C.P.)
† These authors contributed equally to this work.

Abstract: Clinically available medications face several hurdles that limit their therapeutic activity, including restricted access to the target tissues due to biological barriers, low bioavailability, and poor pharmacokinetic properties. Drug delivery systems (DDS), such as nanoparticles (NPs) and hydrogels, have been widely employed to address these issues. Furthermore, the DDS improves drugs' therapeutic efficacy while reducing undesired side effects caused by the unspecific distribution over the different tissues. The integration of NPs into hydrogels has emerged to improve their performance when compared with each DDS individually. The combination of both DDS enhances the ability to deliver drugs in a localized and targeted manner, paired with a controlled and sustained drug release, resulting in increased drug therapeutic effectiveness. With the incorporation of the NPs into hydrogels, it is possible to apply the DDS locally and then provide a sustained release of the NPs in the site of action, allowing the drug uptake in the required location. Additionally, most of the materials used to produce the hydrogels and NPs present low toxicity. This article provides a systematic review of the polymeric NPs-loaded hydrogels developed for various biomedical applications, focusing on studies that present in vivo data.

Keywords: nanomaterials; polymers; drug release; local delivery; administration routes; thermosensitive hydrogel; cancer therapy; chronic wound treatment

Citation: Nunes, D.; Andrade, S.; Ramalho, M.J.; Loureiro, J.A.; Pereira, M.C. Polymeric Nanoparticles-Loaded Hydrogels for Biomedical Applications: A Systematic Review on In Vivo Findings. *Polymers* **2022**, *14*, 1010. https://doi.org/10.3390/polym14051010

Academic Editors: Mikyung Shin and Paolo Ferruti

Received: 2 February 2022
Accepted: 1 March 2022
Published: 2 March 2022

1. Introduction

The physiological barriers of the human body challenge the traditional delivery of drugs, limiting drug access to the desired organs and tissues. Furthermore, the efficacy and retention of drugs in the target tissue are affected by their bioavailability, pharmacokinetic, and pharmacodynamic parameters. These challenges lead to the need for higher dosages and more frequent administrations to reach effective treatment doses, which induces undesirable side effects and toxicity in the other tissues [1].

Over the last years, due to considerable advances in the nanotechnology field, several nanomaterials have been developed as drug delivery systems (DDS). These systems have emerged to overcome the drawbacks of drug administration by improving the drug's solubility and bioavailability, decreasing drug degradation, and extending the drug's half-life [2,3]. DDS can provide a specific and targeted therapy, reducing the required dose to achieve a therapeutic effect. DDS capabilities make therapies more accurate, effective, and less invasive by preventing systemic toxicity and unwanted side effects [2–5]. Among the potential DDS approaches under development, the design of nanoparticles (NPs), hydrogels, and, more recently, NPs-loaded hydrogel (NLH) systems for drug release applications are gaining attention.

The growing advances in the use of nanomaterials led to the need to establish guidelines and regulations for their usage in medicine. Updates in the future are expected to guarantee the quality, effectiveness, and security of these DDS for human use [6]. Besides, scale-up manufacturing is another important condition for clinical use and commercialization of DDS. The clinical application of the DDS faces some concerns regarding their production. When it comes to large-scale production, the procedures should ensure the preservation of the DDS physicochemical characteristics for the desired application since these properties directly affect the efficiency, safety, and drug delivery capabilities of the developed DDS. Simultaneously, the methods must be scalable, reproducible, and cost-effective [7]. Good manufacturing practices verification is required to ensure conformance in large-scale production [6]. In addition, to evaluate whether DDS large-scale production influences clinical performance, comprehensive quality controls of drug carriers are essential [8]. Finally, clinical trials are mandatory to determine the ratio between benefits and risks of the developed DDS [6].

As in vitro studies can only give a partial indication of potential toxicity, in vivo studies are fundamental to evaluate the efficacy and safety of these DDS [4]. In this sense, this review aims to discuss the most recently developed polymeric NLH systems for biomedical applications focusing on their in vivo performance for different administration routes, including parenteral and topical administration. Polymeric NLH systems integrate NPs into a hydrogel, synergistically combining their individual functionalities and benefits. The resulting system presents increased performance and can be successfully employed as DDS for biomedical applications, including tissue engineering, bacterial applications (e.g., wounds or eye infections), and contact lenses [9,10].

Several review papers focusing on the use of NPs combined with hydrogels have been published in recent years owing to the importance of this subject for the scientific community [11–18]. Despite the considerable content provided, while some of these works only focus on NLH production methods, others focus on applying NLH systems to treat specific disorders. Thus, the current work provides the first systematic review of this area, comprising all the research evidence published in the period between 2011 and 2021. The research methodology of this systematic review is described in Section 2, which presents the protocol used to select the articles. Section 3 introduces polymeric NLH systems, and Section 4 describes the routes of drugs administration, divided into sub-sections, with a summary of the outcomes of the selected works for each route. Finally, the discussion and conclusion are presented in Section 5, summarizing the developed work and the perspective based on the obtained findings.

2. Research Methodology

This systematic review is based on the PRISMA (Preferred Reporting Items for Systematic Reviews and Meta-Analyses) reporting guide [19], a process that identifies, evaluates, and interprets all published research relevant to the study topic. An initial literature search of polymeric NPs combined with hydrogels was performed. Databases PubMed, Science Direct, Google Scholar, Scopus, and Web of Science were consulted between November 2021 and December 2021, considering "hydrogel", "polymeric nanoparticles", "nanoparticles", "nanogels", and "gel" as key terms for all databases. Articles were then collected based on their main themes and relation to the current article's scope.

The inclusion criteria applied was polymeric NLH developed for biomedical applications, published in the period between 2011 and 2021. Only articles in English were included. As exclusion criteria, studies that were repeated in different databases, studies published in journals that were not Quartile (Q) 1 or Q2, review articles, and research that did not evaluate the in vivo performance of the developed polymeric NLH. As the journal's quartile was an exclusion criterion, all the considered studies were published in indexed journals of renown, indicating good paper quality.

Based on the applied research strategy, 119 articles were found, including 33 articles in PubMed, 20 articles in ScienceDirect, 37 articles in Google Scholar, 19 articles in Scopus, and 10 articles in Web of Science. After duplication analysis, 54 studies were eliminated, leaving 65 articles. Inclusion and exclusion criteria were applied, removing 28 articles that did not meet the research requirements. After full-text article evaluation, 7 were excluded by lack of information or corresponding to a hydrogel formed by the NPs themselves, rather than NLH, leaving 30 articles for the qualitative synthesis. This methodology is summarized in Figure 1.

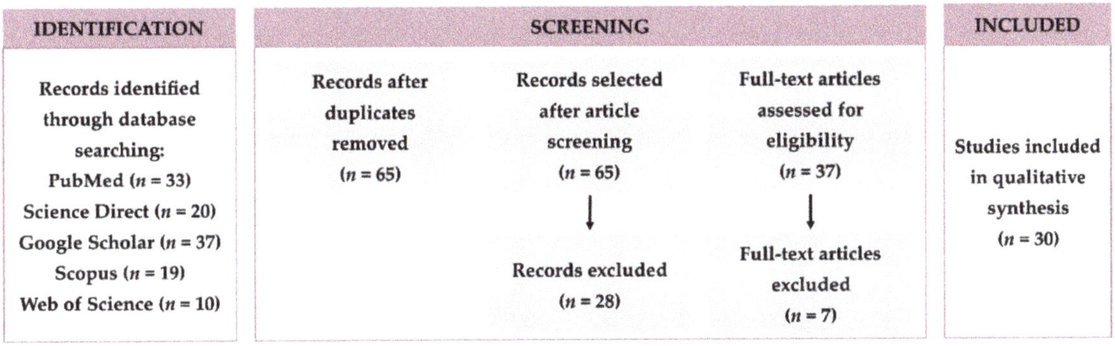

Figure 1. Schematic representation of the applied methodology.

The expected outcome was identifying an improved activity of NPs and hydrogels combined as DDS. In in vivo experimental studies, the description of the results was mainly based on the performance of polymeric NLH as DDS to explore new therapeutic strategies. The following data were extracted from the included studies: NPs material, hydrogel material, NPs and hydrogel production methods, description of the drug-loaded into polymeric NLH systems, biomedical application, route of administration, in vitro and in vivo studies and results; and are then discussed under further sub-sections.

3. NPs-Loaded Hydrogel System

NPs are colloidal structures designed and produced to transport drugs across biological barriers. Their optimal size range is approximately between 100 and 200 nm. They protect drugs from degradation, increasing their half-life, improving drugs' bioavailability, and providing a sustained and localized release [13]. Moreover, their surface could be modified for targeted delivery, reducing drugs' toxicity resultant from the systemic spread and, consequently, side effects [20]. Among them, polymeric NPs have been widely chosen as DDS in several biomedical applications due to their high stability, water solubility, biocompatibility, biodegradability, and non-immunogenicity [21,22]. Another advantage of these NPs is their capacity to encapsulate both hydrophilic and hydrophobic drugs [20]. Moreover, their loading capacity, drug release kinetics, and biological performance could be regulated by adjusting their composition or surface charge. However, these NPs are quite predisposed to premature burst release of drugs [13,22], their permanency on the target site until complete drug release is unpredictable [13,23]. When in contact with the biological environment, NPs can present instability or clearance by the immune system [5].

Hydrogels are three-dimensional porous structures produced with hydrophilic polymers through physical or chemical cross-linking methods [24] and can be prepared from a wide range of natural and synthetic polymers. Natural polymers include alginate, chitosan, gelatin, collagen, hydroxypropyl methylcellulose (HPMC), and hyaluronic acid (HA); in contrast, synthetic polymers could be polyacrylamide (PAM), poly(hydroxyethyl methacrylate) (PHEMA), polyvinylpyrrolidone (PVP), poly(vinyl alcohol) (PVA), poly(ethylene glycol) (PEG), and poly-ε-caprolactone (PCL) [25]. Hydrogels are materials with great solute permeability, and a high-water retention capability [9,25–27]. Depending on the polymers employed in their production, hydrogels can be biocompatible, biodegradable, and present minimal toxicity. They can encapsulate molecules in an effective amount, protecting and releasing them over time while increasing their local concentration and reducing their toxicity in the remaining tissues [28]. Moreover, hydrogels kinetics can respond to biological, chemical, or physical external stimuli. Biological triggers include antibodies or enzymes; chemical factors comprising pH, type of ions, or organic solvents; and physical stimuli include temperature, light, and magnetic or electrical fields [25]. Thermo-sensitive hydrogels are the most commonly used variety of hydrogels for medical use due to their sol-gel transition behavior at body temperature. While at room temperature hydrogels are in the form of an aqueous suspension that can be easily injected; at body temperature, the solution rapidly transits to a stable gel [25,29]. Their administration is minimally invasive since most of them can be administered without the need for a surgical procedure [11]. Besides, thermo-sensitive hydrogels manage to mold themselves perfectly to the shape of the place where they are administered, creating a drug depot for a localized and sustained release [1]. On the other hand, hydrogels have a few drawbacks limiting their application as DDS. Hydrogels have weak mechanical properties; for example, their mechanical strength decreases after swelling [19]. Drug release from hydrogels depends on their network structure, rearrangement, size, materials employed in their production, and drugs' physicochemical properties [26]. Most of the time, hydrogels present an initial burst release of drugs when in contact with the release medium due to their high-water content [15,30]. Another problem of hydrogels' hydrophilic nature is the integration of poorly soluble drugs, which are rapidly released through diffusion [13,31].

Individual limitations of NPs and hydrogels could be addressed with their combination into a single platform (Figure 2). As NPs can be physically or covalently integrated into hydrogels, the development of these DDS has emerged, taking advantage of their benefits synergistically combined in one system [26].

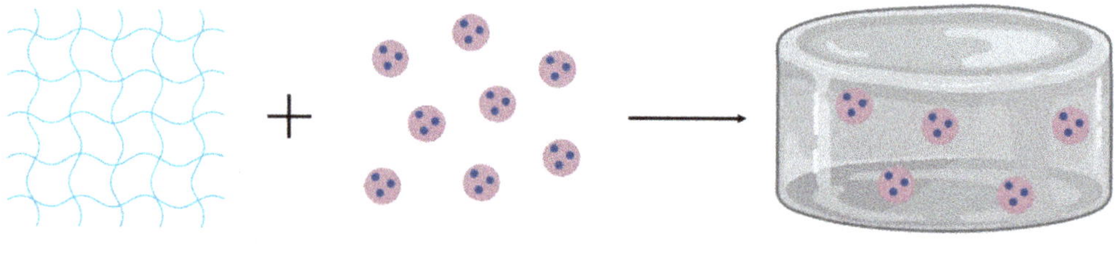

Figure 2. Schematic representation of the combination of drug-loaded NPs and a hydrogel as DDS.

Besides the biocompatibility, biodegradability, and non-toxicity of both NPs and hydrogels as individual DDS, their combination provides benefits that none of them could achieve independently (Table 1).

NPs and hydrogels can both provide multiple drug loading. When combined, the hydrogel protects the NPs from degradation, prevents their aggregation [33], and promotes the local delivery of drugs [15,30]. Incorporating NPs into hydrogels can improve hydrogel mechanical properties, such as strength, stiffness, and degradation, in a concentration-dependent relationship [13]. NPs may also act as a hydrogel crosslinking and fortify the stimuli-responsive behavior of the hydrogel [12]. Although NPs and hydrogels can improve drugs' bioavailability and release them over time [13], their combination forms a depot at the administration site for prolonged local drug retention [32]. This dual delivery system provides a double encapsulation of drugs, regulates drug release kinetics, and prevents the initial burst release. It can also encapsulate both hydrophobic and hydrophilic drugs [13,17].

These benefits work together to minimize side effects and systemic toxicity [13] and improve the therapeutic effect of treatment and patients' compliance [15,30].

Table 1. Benefits of polymeric NPs, hydrogels, and polymeric NLH as DDS. (+) and (–) indicate advantages and disadvantages, respectively.

	Polymeric NPs	Hydrogel	Polymeric NLH	Refs.
Multiple drug loading	+	+	Improved	[17,20,28]
Hydrophobic drugs loading	+	–	Maintained	[13,17,20,31]
Controlled and sustained release	+	+/–	Improved	[13,28]
Drug bioavailability improvement	+	+	Improved	[13]
Targeted drug delivery	+	–	Maintained	[20]
Local retention of drug	–	+	Maintained	[1,15,30,32]
Stimuli-responsive behavior	+	+	Improved	[12,25,29]

4. Routes of Drugs Administration

The administration route determines the effectiveness of the treatment and depends on the drug characteristics, biomedical application, and patients' compliance. The routes of drugs administration are classified as enteral, parenteral, topical, or inhalation (Figure 3) based on the drug's administration site.

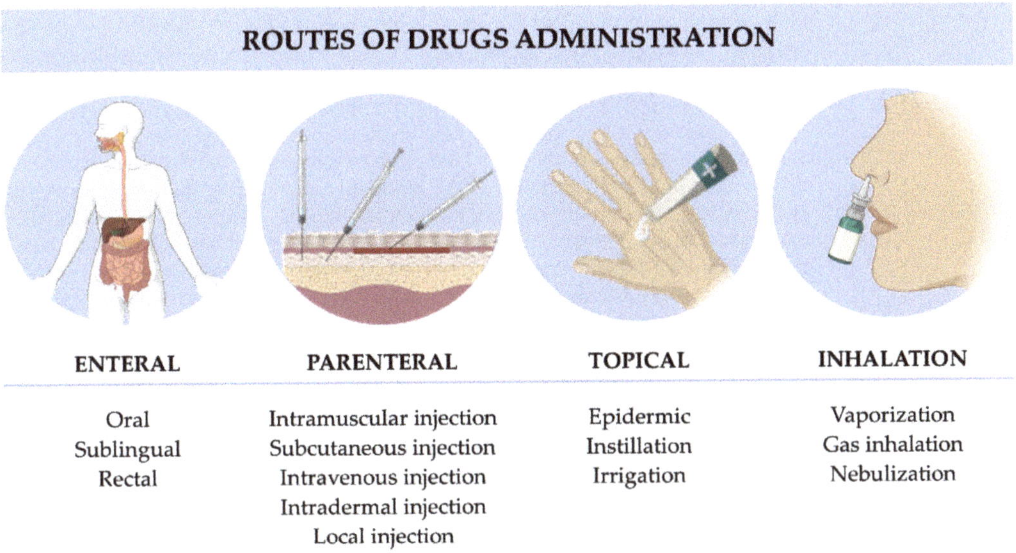

Figure 3. Classification of routes of drugs administration.

In the enteral route, drugs are absorbed by the gastrointestinal tract and comprise oral, sublingual/buccal, and rectal administration. Oral administration of drugs is the most frequently used administration route in ambulatory [34]. This route is convenient for frequent treatments that require repeated intakes. Besides the easy application, safety, convenience, and economic advantages of this administration, some limitations compromise the therapeutic effect. Due to first-pass metabolism and poor drug pharmacokinetics and pharmacodynamic properties, such as low oral absorption, low bioavailability, and enzymatic degradation [34], the drug concentration is highly reduced until they reach their action site. Another limitation of this route of administration is the poor targeting of drugs [35], which slows their onset of action and induces systemic toxicity.

For the parenteral route, drugs directly reach the systemic circulation and are usually administered with an injection. This route is preferable for drugs with poor solubility and stability in the gastrointestinal tract. The commonly injections forms are (a) intravenous, where the drugs are directly administered to the systemic circulation for a rapid drug effect; (b) intramuscular, where the drugs are directly administered to the muscle, that due to its high blood circulation, provides a higher absorption of drugs; and (c) subcutaneous, where the drugs are administered in the subcutaneous tissue that, due to its reduced vascularization compared to intramuscular injection, provides a slower absorption of drugs, suitable for treatments that require frequent administrations. However, drugs can also be administered through other less common injection forms, such as intradermal, intraperitoneal, intrathecal, or local injections [34]. The limitations of these routes include systemic toxicity and subsequent side effects depending on the drugs [34].

The topical route comprises the administration of drugs at the skin and mucous membranes, such as eyes (through instillation or irrigation) or vagina. The topical application through the skin is not only painless and non-invasive but is also simple to apply, resulting in high patient compliance. It also allows increased amounts of drugs at the target site. The topical route avoids the first-pass metabolism, which improves drugs bioavailability and treatments efficacy [36]. The drawbacks of this route depend on each mucosa. The skin has an acidic pH (4.2–5.6), requires a high degree of hydration, and presents a barrier at the most outermost layer that could limit drugs' permeability [36]. Due to the aqueous environment of eyes and their tear drainage, drug loss is frequent, which induces low retention of drugs at the ocular tissue. Furthermore, the drug's low bioavailability contributes to an ineffective treatment [34]. Vaginal administration efficacy is reduced by degradative enzymes and vaginal fluid, decreasing the drugs' bioavailability and residence time [36].

The inhalation route comprises the nasal administration of drugs through vaporization, gas inhalation, or nebulization and could be applied for both local and systemic therapies. The drugs are rapidly absorbed into the respiratory epithelium due to their large surface area and huge permeability [37]. Then, due to respiratory epithelium's high irrigation, drugs could be directly diffused to the bloodstream, avoiding the first-pass metabolism and enzymatic degradation, which increases their bioavailability [34]. Besides, this route provides a target delivery in the case of pulmonary problems. Compared with other administration routes, inhalation provides fewer systemic side effects. However, the drug's efficacy can be conditioned by the size of the inhaled molecules, physiology of the patient's respiratory system, mucosal turnover, and nasal epithelium that reduces the nasal residence time of drugs [34,36,37].

Given the limitations of current medicines, NLH systems can be advantageous for therapies that require a local, frequent, and long-term administration of drugs. The most appropriate route of administration is chosen based on the biomedical application. The following sections provide and discuss the most recently developed polymeric NLH for the various administration routes, including subcutaneous, local, ocular, and topical.

4.1. *Parenteral Administration of Polymeric NPs-Loaded Hydrogels*

4.1.1. Subcutaneous Administration

Drug solutions are quickly dispersed through neighboring tissues in subcutaneous administration and consequently eliminated. Most of the time, regular administrations are required to maintain the drug levels in the blood and achieve the desired therapeutic effect. Furthermore, subcutaneous administration is associated with high dosages of drugs to reach therapeutic levels, leading to undesired side effects [38]. Considering this, polymeric NLH systems have been investigated for therapies involving drugs' subcutaneous administration

(Table 2). These systems ensure a controlled and sustained drug release, significantly decreasing their frequency of administration. Besides, NLH systems can maintain the therapeutic levels of drugs, thus increasing their efficacy [39].

Table 2. Summary of polymeric NLH administered subcutaneously for biomedical applications.

NPs Material	Hydrogel Material	Loaded Cargo	NPs Production Method	Hydrogel Crosslinking Nature	Biomedical Application	Main Conclusions	Ref.
PHBHHx	Chitosan	Insulin	Single-emulsion solvent-evaporation	Physical (β-GP)	Diabetes	NLH increased insulin bioavailability and prolonged hypoglycemic effect	[40]
PLGA	PCL–PEG–PCL	ICG and l-Arginine	Double-emulsion solvent-evaporation	Chemical	Various types of cancer	NLH increased ICG and l-Arg concentration and retention at the tumor site, inhibiting tumor growth and regression of the established tumors	[28]
Micelles of PEG-phenylboronic acid-polycarbonate	P(Bor)5-PEG-P(Bor)5 and P(Gu)5-PEG-P(Gu)5	BTZ	Film hydration	Chemical	Cancer (myeloma)	BTZ-loaded NLH enhanced anti-cancer activity by decreasing the tumor size and inhibiting its progression	[41]

Peng et al. (2013) proposed poly(3-hydroxybutyrate-co-3-hydroxyhexanoate) (PHBHHx) NPs incorporated in a thermosensitive hydrogel for an ultralong sustained release of insulin for diabetes management [40]. The hydrogel was prepared with chitosan and β-glycerophosphate disodium salt (β-GP) as a crosslinking agent, and their concentrations were optimized to obtain gelation at 37 °C. In vitro studies proved that the incorporation of the NPs did not affect the hydrogel's gelation temperature, gelation time, or degradation. The performed in vitro release studies revealed the significant role of the NPs-hydrogel combination in prolonging insulin release, compared with free insulin or insulin-loaded hydrogel. Further, to evaluate the in vivo performance, animal experiments were performed with diabetic transgenic mice. The animals were divided into three groups and subcutaneously injected with control hydrogel, insulin-loaded hydrogel (4 IU/kg), and insulin-loaded NLH (6 IU/kg) separately. Compared to the other groups, the animal group administrated with the insulin-loaded NLH presented a prolonged hypoglycemic effect, supporting in vitro release studies that showed that the combination of NPs with hydrogel exhibited a more sustained insulin release. The authors concluded that the NLH prolonged the insulin release and the hypoglycemic effect for more than 5 days after a single subcutaneous administration, which allows for a reduction in the frequency of insulin administration.

Subcutaneous applications of these NLH systems have also been widely explored for other biomedical applications by increasing drugs' bioavailability and improving therapies, such as in the case of cancer treatments, as proposed by Sun et al. (2021) [28]. The authors developed an injectable thermosensitive hydrogel containing poly(lactic-glycolic acid) (PLGA) NPs as a synergic approach to destroy the tumor's extracellular matrix. The hydrogel was composed of PCL–PEG–PCL, and PLGA NPs were used to encapsulate booth photosensitizer indocyanine green (ICG) and nitric oxide donor l-arginine (l-Arg). The ICG is a photosensitizer able to produce reactive oxygen species under infrared light irradiation, which will cause apoptotic cell death and, simultaneously, the oxidation of l-Arg, which will generate nitric oxide and inhibit cancer cell proliferation. The in vitro studies showed that the incorporation of the NPs into the hydrogel did not change the sol-gel transition behavior of the hydrogel. The authors also verified that loading the NPs in the hydrogel increased the ICG and l-Arg concentration and retention at the tumor site, decreasing their toxicity in the other tissues. In vitro experiments were performed with human mammary carcinoma cells to evaluate the cellular uptake, cytotoxicity activity, and apoptosis effect of NLH, compared with NPs in solution and ICG in solution. Cellular uptake assay revealed efficient endocytosis of NPs by cancer cells. In vitro apoptosis showed higher apoptosis of cancer cells for the NLH formulation than NPs in solution and ICG in solution. The in vivo studies with mammary carcinoma-bearing transgenic mice showed tumor growth inhibition and regression of the established tumors when the animals were subcutaneously administered with the NLH.

For multiple myeloma therapy, Lee et al. (2018) proposed the administration of bortezomib (BTZ) through its encapsulation into NPs, and the incorporation of those NPs in a hydrogel [41]. The NPs were produced with a triblock copolymer of PEG-phenylboronic acid-polycarbonate. The hydrogel was prepared with P(Bor)5-PEG-P(Bor)5 and P(Gu)5-PEG-P(Gu)5 polymers and optimized to obtain suitable properties for injection and BTZ-loaded NPs delivery. The incorporation of NPs in the hydrogel increased the storage moduli but did not affect the injectability. In vitro release was investigated at pH 7.4 and 5.8, to mimic the extracellular and endolysosomal environments, respectively. Results showed a faster BTZ release in the acid environment than pH 7.4. A multiple myeloma xenograft mouse model was used to evaluate the in vivo anti-cancer activity of the hydrogel. Animals were injected subcutaneously at approximately 1 cm away from the tumor site (0.8 mg/kg BTZ, 150 µL). Results showed a hydrogel degradability of more than 14 days and an excellent tolerance to the formulation. The anti-cancer efficacy was investigated by comparing animals injected with BTZ-loaded NPs incorporated in hydrogel and BTZ-loaded NPs in solution. Mice treated with BTZ-loaded NLH formulation showed smaller tumor size and inhibition of tumor progression.

4.1.2. Local Administration

Local administration was envisaged to allow for the accumulation of drugs at the target tissue, enhancing treatment efficacy while minimizing systemic toxicity. Thus, several polymeric NLH have been proposed for local drug delivery (Table 3) since they can reduce drugs' clearance and increase their retention time [42]. By achieving higher drug concentrations at the target tissue, hydrogels can prolong the therapeutic activity of drugs and reduce their systemic toxicity. Thermosensitive hydrogels attract particular attention for local implantation due to their ability to increase viscosity above physiological temperatures, allowing for their injection in the liquid form and further gelation in situ [38].

Table 3. Summary of polymeric NLH locally administered for biomedical applications.

NPs Material	Hydrogel Material	Loaded Cargo	NPs Production Method	Hydrogel Crosslinking Nature	Biomedical Application	Main Conclusions	Ref.
Ethyl cellulose	Chitosan	Carboplatin	Double-emulsion solvent-evaporation	Physical (dibasic sodium phosphate)	Various types of cancer	NLH reduced systemic toxicity, increased drug concentration at the tumor site, and improved anti-tumor activity	[43]
PCL-PEG-PCL	Pluronic F-127	Norcantharidin	Thin-film dispersion	Chemical	Cancer (hepatocellular carcinoma)	NLH provided high anti-tumor activity, with inhibition of the implanted tumors growth and prolonged the survival time of the tumor-bearing mice	[44]

Table 3. *Conts.*

NPs Material	Hydrogel Material	Loaded Cargo	NPs Production Method	Hydrogel Crosslinking Nature	Biomedical Application	Main Conclusions	Ref.
PLA	PCL and Pluronic 10R5	Oxaliplatin and Tannic acid	Double-emulsion solvent-evaporation	Chemical (Sn(Oct)2)	Cancer (colorectal peritoneal carcinoma)	NPs incorporation in the hydrogel allowed for a sustained release in vivo, improving tumor growth inhibition while reducing systemic toxicity	[30]
pBAE	PAMAM crosslinked with dextran aldehyde	siRNA	Self-assembly	Chemical	Breast cancer	The formulation exhibited a sustained and controlled release in vivo, but the therapeutic effect was not improved	[45]
PCL-PEG and DOTAP	Pluronic F127	Deguelin	Film hydration	Chemical	Bladder cancer	NLH acts as a drug depot, allowing for sustained local drug delivery	[46]
PUR	Poloxamer 407	BODIPY (mimic)	Nanoprecipitation	Physical (saline solution)	Glioblastoma	NPs incorporation in the hydrogel increased drug retention time in the tumor tissue, without systemic toxicity	[32]
PLGA	PEGDA and HA	Paclitaxel	Single-emulsion solvent-evaporation	Physical (UV light)	Lung cancer	Incorporating the drug-loaded NPs into the hydrogel improved in vivo tumor growth inhibition	[26]
PLGA-PEG	Pluronic F-127, Pluronic F-68, HPMC, MC and SA	Paclitaxel	Single-emulsion solvent-evaporation	Chemical	Pancreatic cancer	NPs incorporation in the hydrogel increased drug retention time in the tumor tissue, improving tumor growth inhibition.	[47]
PLGA	Gelatin	Andrographolide	Single-emulsion solvent-evaporation	n.d.	Osteoarthritis	NPs incorporation in the hydrogel increases the retention time in the joint, maintaining a sustained release for over 8 weeks	[48]
PLGA	Alginate and PAM	TGF-β3	Nanoprecipitation	Chemical (PAAm crosslinker)	Tissue Regeneration (cartilage)	TGF-β3-loaded NLH induces the formation of new cartilage tissue	[49]
PLGA	Keratin	EGF and bFGF	Double-emulsion solvent-evaporation	Physical (hydrogen peroxide)	Intracerebral hemorrhage (iron overload)	The NLH improved stem cell differentiation and accelerated neurological recovery in vivo	[50]
Chitosan	Collagen	Insulin	Ionic gelation	Chemical (EDC)	Tissue regeneration (peripheral nerve)	Collagen hydrogel has tissue regeneration ability, but the incorporation of insulin-loaded NPs enhances the effect	[51]

Table 3. *Conts.*

NPs Material	Hydrogel Material	Loaded Cargo	NPs Production Method	Hydrogel Crosslinking Nature	Biomedical Application	Main Conclusions	Ref.
Chitosan	Alginate and Chitosan	Berberine	Ionic gelation	Physical (β-GP)	Spinal cord injury	The hydrogel containing berberine-loaded NPs and stem cells exhibited a higher tissue regeneration ability	[52]
Ag-Lignin	Pectin and PAA	None	Self-assembly	Chemical	Wound healing	NPs incorporation in the hydrogel improves wound healing ability-enhancing the formation of mature tissue	[53]

n.d.: not defined.

Thus, thermosensitive hydrogels have been widely explored for intratumoral drug delivery. Chemotherapy is associated with several serious side effects, and local delivery by hydrogels can overcome those. For instance, Thakur et al. (2020) developed an in situ injectable NLH for carboplatin delivery, a drug used to treat various cancers, such as breast, cervix, colon, lung, prostate, among others [43]. Carboplatin was entrapped in ethyl cellulose NPs to overcome the toxicity issues of the drug. Then, the loaded NPs were incorporated in a thermosensitive hydrogel composed of chitosan. The experimental parameters were optimized to obtain the ideal gelation time, temperature, and syringeability for the NLH. The authors also verified an improvement in carboplatin in vitro release profile, with a sustained and controlled release for drug-loaded NLH, compared with drug-loaded hydrogel, drug-loaded NPs in solution, or pure drug. In vitro cytotoxicity was performed with the drug-loaded NLH and carboplatin commercial formulations against six different human cancer cells: cervix, colon, prostate, lung, osteosarcoma, and normal breast epithelial cells. For all cell lines, no cytotoxicity was observed. Furthermore, in vitro cellular uptake study was performed in human osteosarcoma cells, whose results proved that the NPs could penetrate those cells. Further, studies were performed with carcinoma mice models to evaluate the in vivo efficacy of the developed NLH. The animals were injected intraperitoneally with the drug-loaded NLH at 123 mg/kg. The results revealed a significant improvement in the drug's pharmacokinetic profile, which leads to a reduction of systemic toxicity. The authors concluded that the drug-loaded NLH provides a great anti-tumor activity with a decrease in tumor size, justified by the higher NPs' uptake by cancer cells and the NLH's ability to release the drug in a sustained and controlled manner for 7 days.

An approach for hepatocellular carcinoma treatment was proposed by Gao et al. (2021) [44]. The authors developed an injectable thermosensitive hydrogel composed of Pluronic F-127 for intratumoral administration. The hydrogel was co-loaded with norcantharidin-loaded NPs and doxorubicin, two widely used anti-cancer drugs that potentiate the treatment efficacy when administrated together. The NPs were prepared with poly(ε-caprolactone)-poly(ethylene glycol)-poly(ε-caprolactone) (PCL-PEG-PLC) and used to encapsulate norcantharidin due to its poor water solubility and low bioavailability. The in vitro release studies proved that hydrogel containing loaded NPs released their content in a controlled and sustained way compared with free drugs or drug-loaded NPs in solution. In vitro cellular studies were performed with a human hepatoma cell line. To evaluate the NPs uptake by tumor cells, NPs were loaded with a fluorescent molecule, coumarin-6, and the NPs uptake by tumor cells was efficiently achieved. In the presence of drug-loaded NLH, a significant decrease in the proliferative activity of tumor cells was observed. The in vivo potential of the drug-loaded NLH was evaluated in a hepatoma tumor-bearing mice model after intratumoral administration. The results showed significant tumor growth inhibition and an extension of the survival time of the tumor-bearing mice compared with free drugs.

Ren et al. (2019) developed a thermosensitive hydrogel composed of PCL and Pluronic 10R5 to treat colorectal peritoneal carcinoma [30]. Poly(lactic acid) (PLA) NPs loaded with oxaliplatin, the clinically used chemotherapeutic agent for this tumor, were incorporated in the hydrogel. To potentiate the oxaliplatin's activity, tannic acid, a natural compound with anti-cancer properties and lower side effects, was also entrapped in the NPs. The authors verified that while tannic acid showed a similar in vitro release profile from NPs both in solution and in the hydrogel, NPs incorporation slowed the release of oxaliplatin. After intraperitoneal injection in healthy mice, the hydrogel underwent self-gelation at physiological temperature and exhibited a slow degradation rate for 20 days, allowing for a sustained and controlled drug release. Further, a tumor mice model was established by injection of colon cancer cells in the peritoneal cavity of the animals. The incorporation of drug-loaded NPs in the hydrogel improved tumor growth inhibition and increased mice survival compared with the free drugs combination. The NLH also proved to decrease drug toxic effects associated with systemic administration.

Segovia et al. (2015) proposed oligopeptide-terminated poly(β-aminoester) (pBAE) NPs for the small interfering RNA (siRNA) delivery for gene silencing in breast cancer therapy [45]. Since these NPs exhibit rapid degradation limiting their ability to maintain a sustained delivery, these were incorporated in a hydrogel composed of polyamidoamine (PAMAM) cross-linked with dextran aldehyde. In in vitro release studies, the authors verified that NPs' degradation rate decreased with their immobilization in the hydrogel, increasing stabilization. This experiment also confirmed that NPs incorporation in the hydrogel did not affect the hydrogel properties, such as its degradation rate. The siRNA delivery by NLH enhanced gene silencing ability in human breast cancer cells due to a higher transfection activity than the hydrogel containing free siRNA or siRNA-NPs in solution. In vivo studies were performed in mice bearing tumors in the mammary fat pad implanted with loaded-NLH with disk-shapes (6 mm diameter, 3 mm thick). These studies revealed that NPs exhibited a more sustained and controlled release when intratumorally injected embedded in the hydrogel than in suspension. Although the prepared formulation proved to be safe, the authors did not verify a significant therapeutic effect in the treated animals.

Men et al. (2012) developed NPs composed of pegylated PCL and 1,2-dioleoyl-3-trimethylammonium-propane (DOTAP) for the delivery of deguelin for bladder cancer therapy [46]. Deguelin is a natural compound with anticancer activity but exhibits neurotoxicity. The authors proposed its local delivery to decrease toxic effects by incorporating the loaded NPs in a thermosensitive hydrogel composed of Pluronic F127. In vitro release studies revealed that the hydrogel is gradually degraded, allowing for the sustained release of deguelin. Additionally, in vitro studies also showed that this hydrogel can adhere to the mucous membrane of the bladder wall, thus increasing the drug residence time. Further, to evaluate the ability of the developed NLH formulation in delivering drug cargo to the bladder, mice were intravesically administered with 2 mg/kg hydrogel. The hydrogel was administered in the liquid form, and since its gelation temperature is 25 °C, this formed a gel in situ inside the bladder. The NPs incorporated in the hydrogel were marked with a fluorescent dye (coumarin 6) to allow their visualization. Fluorescent NPs in solution were used as control. The authors verified that although NPs can enhance the drug permeability into the bladder, NPs incorporation in the hydrogel is advantageous since, due to its bioadhesive property, the hydrogel is not eliminated during urination, allowing for a prolonged local drug release without systemic toxicity.

Brachi et al. (2020) proposed polyurethane (PUR) NPs loaded in a Poloxamer 407 thermosensitive hydrogel for delivery to glioblastoma tumors [32]. Local administration can be particularly advantageous to treat brain disorders, such as brain tumors, once NPs can circumvent the blood–brain barrier that poses a significant obstacle for brain drug delivery. The hydrogel formulation was optimized, and it was verified that gelation time decreases with the increase of poloxamer concentration. A polymer concentration of 25 wt.% was chosen to obtain a gelation time of 4 min at physiological temperature. NP incorporation in the hydrogel proved to be advantageous by slowing the cargo release and preventing the burst effect. To assess the ability of the hydrogel to increase the NPs retention time in the tumor, intracranial xenografts were established in immunocompromised mice. BODIPY fluorophore was encapsulated in the NPs to allow for their real-time biodistribution evaluation, and 2.5 mg hydrogels containing the NPs were administered to the animals by intratumoral injection. A group of animals was treated with NPs in solution as a control. NPs incorporation in the hydrogel increased the retention time of NPs at the tumor site, thus avoiding their migration from the tumor and covering larger areas of the tumor compared with NPs in suspension. The authors concluded that these NPs are a promising tool for glioblastoma drug delivery to increase drug accumulation in the tumor without systemic toxicity.

PLGA NPs are biocompatible, Food and Drug Administration (FDA) approved [54], and safe for repeat-dose exposure in vivo [55], being widely explored for incorporation in hydrogels for drug delivery. Wang et al. (2021) developed a hydrogel composed of poly (ethylene glycol) diacrylate (PEGDA) and HA for the local delivery of paclitaxel for lung cancer therapy [26]. The authors encapsulated the drug into PLGA NPs and incorporated them in the hydrogel. Aiming for the in situ ultraviolet (UV) polymerization of the hydrogel at the tumor site, a radical photoinitiator for the UV curing was added to the formulation, and its effect on the hydrogel gelation time, swelling rate, and degradation rate were evaluated. The authors verified that increasing the concentration of the photoinitiation led to a decrease in the hydrogel gelation time, degradation rate, and pore size due to a reduction of the water content. Incorporating the natural polymer HA in the formulation increased the hydrogel's swelling rate, leading to higher NPs' loading capacity. In vitro release studies showed that paclitaxel encapsulation in the PLGA NPs allows for a slow and sustained drug release. PLGA NPs have similar release profiles in the hydrogel or in solution. Mice bearing subcutaneous tumor xenografts were treated with paclitaxel-loaded NPs in solution or incorporated in the hydrogel (5 mg/kg). After local injection, the regions of tumors were irradiated with UV light to allow the in situ hydrogel polymerization. The obtained results depicted that incorporating the drug-loaded NPs into the hydrogel improved tumor growth inhibition since the hydrogel can retain the NPs at the tumor site avoiding elimination by the lymphatic system into the systemic circulation.

Shen et al. (2015) also proposed PLGA NPs-loaded thermosensitive hydrogels for the local delivery of paclitaxel for cancer therapy, in this case, to treat pancreatic cancer [47]. In this work, the hydrogel was composed of HPMC, methyl cellulose (MC), sodium alginate (SA), and Pluronic F-127 and F-68. Its degradation speed and gelation time at physiological temperature were optimized to ensure that the hydrogel remained permeable in the tumor tissue with a slow erosion rate. In vitro release studies revealed that incorporating the NPs into the hydrogel significantly delayed paclitaxel release, preventing the burst release. In vitro studies with drug-resistant pancreatic tumor cells suggested that paclitaxel encapsulation in the NLH increases cell uptake due to the ability of the NPs to circumvent the p-glycoprotein pump and the ability of F-127 to increase the cell membrane' fluidity. Additionally, in vitro studies in a 3D model composed of tumor cells supported in agarose/collagen scaffold were performed. Results showed that drug elimination rates decreased when NPs were incorporated in the hydrogel, leading to higher drug concentrations in the tissue and, consequently, to more efficient inhibition of cell regrowth. Real-time imaging studies in subcutaneous tumor-bearing mice depicted that, after intratumoral injection, NLH remained near the injection site, while control NPs (without hydrogel) were distributed evenly throughout the tumor tissue. NPs incorporated in the hydrogel proved to be more efficient in inhibiting tumor growth than NPs in solution or free paclitaxel, with no toxicity to healthy organs.

Hydrogels containing drug-loaded PLGA NPs were also widely explored for other biomedical applications to enhance therapeutic efficacy by increasing drug-residence time. Kulsirirat et al. (2021) proposed a gelatin-based hydrogel to deliver andrographolide for osteoarthritis therapy [48]. Andrographolide is a natural compound with anti-inflammatory properties but has inadequate absorption, distribution, metabolism, and excretion properties. For the preparation of the NPs, PLGA polymers with different molecular weights and end groups were used. The choice of the adequate formulation was based on the drug encapsulation efficiency. The chosen NPs formulation was incorporated in the hydrogel. In vitro release studies revealed that andrographolide release from the hydrogel is slower when the compound is entrapped in the selected PLGA NPs than the free compound. Healthy mice were treated with different formulations by intra-articular injection, hydrogel containing a free fluorescent dye, dye-loaded NPs, and dye-loaded NLH. Real-time biodistribution of dye was evaluated using a non-invasive in vivo imaging system. The obtained results showed that incorporating NPs into the hydrogel allowed for a long-term sustained release, increasing the retention time in the target tissue maintaining a constant fluorescence intensity in the joint over 8 weeks. The authors concluded that this hydrogel is a suitable approach for the local management of osteoarthritis.

Saygili et al. (2021) developed a hydrogel composed of alginate and PAM for the in situ delivery of the transforming growth factor beta-3 (TGF-β3) for cartilage regeneration [49]. The growth factor was entrapped in PLGA NPs, and the experimental parameters were optimized to yield NPs with suitable dimensions. Three hydrogel formulations were prepared, one containing empty PLGA NPs, the other containing TGF-β3-loaded PLGA NPs, and control hydrogel without NPs. The mechanical strength of the hydrogels was increased when the incorporation of the NPs was performed. The thermal degradation of the hydrogels was not affected by the NPs incorporation. In vitro biodegradability and stability studies showed that the hydrogels retain their mechanical stability under different temperatures and humidity conditions over 3 months, exhibiting a biodegradation rate of 3.5% (*w*/*w*) per week. The three hydrogel formulations proved to be biocompatible in vitro using mice cells. The hydrogels' performance was further evaluated in a cartilage defect rat model. Cylindrical cartilage defects (1.5 mm in diameter and 1.5 mm in depth) were created in healthy rats, and then the prepared hydrogels were implanted into the defect site. Control animals were left untreated. Animals treated with TGF-β3-NLH showed enhanced tissue repair with newly formed tissue with chondrogenic differentiation and cell proliferation without excessive inflammation. The authors concluded that the developed NLH is suitable for cartilage regeneration due to its ability to mimic the extracellular matrix structure.

Hydrogels have gained popularity for tissue repair and regeneration applications since these can mimic the extracellular matrix, regulate cell processes, and form new tissue. The porous structure of hydrogels promotes cell attachment and proliferation and allows for the diffusion of crucial nutrients [56]. Therefore, other groups have explored PLGA NPs incorporated in hydrogels for tissue regeneration applications. For example, Gong et al. (2020) [50] developed PLGA NPs incorporated in hydrogel for tissue repair after hemorrhagic injury. The proposed hydrogel was composed of keratin and contained bone marrow mesenchymal stem cells. The PLGA NPs incorporated in the hydrogel were loaded with epidermal (EGF) and basic fibroblast (bFGF) growth factors to promote stem cell differentiation. An iron-chelator was also entrapped in the hydrogel to regulate iron levels since iron overload can decrease the success of stem cell therapy. The hydrogel was prepared using keratin with different molecular weights to assemble an outer shell composed of low-molecular-weight keratin and the iron chelator. The stem cells and PLGA NPs were incorporated in the inner core consisting of high-molecular-weight keratin. The outer shell of the hydrogel exhibited a faster in vitro degradation rate than the inner core, which is advantageous by allowing chelation of iron at a faster rate while slowing the release of the growth factors for stem cells differentiation. The authors also verified that the growth factors were released slower when the NPs were loaded in the hydrogel, which allowed avoiding of the burst release effect. In vitro cytotoxicity studies revealed that the hydrogels are biocompatible at a concentration below 100 mg/L, causing no harmful effects to the stem cells. After injecting animals intracranially with the hydrogel, a neurobehavioral evaluation was conducted. The authors verified that the developed formulation improved stem cell differentiation and accelerated neurological recovery.

Other materials have been proposed for tissue regeneration applications, such as chitosan NPs. Ai et al. (2019) proposed collagen hydrogel containing insulin-loaded chitosan NPs to promote the regeneration of the sciatic nerve caused by traumatic injury or some diseases [51]. Although clinically used to regulate blood glucose, recent evidence has shown that insulin possesses neurotrophic activities. Because collagen is abundantly distributed in the sciatic nerve and other peripheral nerves, it was chosen for hydrogel formation. The developed hydrogel exhibited a mean pore size (75–235 μm) suitable for cell attachment and migration. In vitro studies revealed that after incorporating insulin-loaded NPs, the hydrogel's degradation rate was decreased, allowing for a controlled and sustained release of insulin for 14 days. The prepared formulation did not induce hemolysis in vitro, thus proving the good human blood compatibility of the used materials. Additionally, the insulin-NLH did not show any in vitro cytotoxicity to rat glial cells. Therefore, the authors proceed to animal experiments to evaluate the in vivo regeneration potential of the developed hydrogel. A sciatic nerve injury was created in healthy rats, and 0.5 mL of hydrogel formulations (with or without insulin-loaded NPs) were injected at the lesion site. A group of control animals received no treatment. Histological analysis of untreated animals revealed poor fiber arrangement and damages, such as edema, disintegration of the myelin sheath, degenerated nerve fibers, and fibrosis. In contrast, animals treated with insulin-loaded NLH showed regenerated tissue resembling the normal sciatic nerve. Animals treated with control hydrogel without NPs exhibited regenerated tissue, although to a lower extent than those treated with the hydrogel-containing NPs. Both hydrogel formulations proved to decrease the muscle weight loss in the injured limb and recover motor and sensory functions, with NLH being more efficient. The authors concluded that the synergic effect between collagen hydrogel and insulin-loaded NPs could enhance peripheral nerve regeneration.

Chitosan NPs have also been explored for tissue repair after spinal cord injury. Mahya et al. (2012) developed a hybrid hydrogel composed of alginate and chitosan [52]. A natural compound, berberine, was encapsulated into the chitosan NPs to enhance its ability to promote the growth and reconstruction of damaged neuronal cells. In vitro studies revealed that incorporating the NPs into the hydrogel prolonged the berberine release three-fold, preventing the burst release. The incorporation of the NPs also altered the mechanical properties of the hydrogel by increasing the degree of crosslinking, leading to an increased elastic modulus. The NPs presence also increased the degradation rate of the hydrogel. The degradation of the scaffolds in vivo allows for the formation of new tissue. Stem cells were also incorporated in the scaffold formulation to potentiate the neurodegeneration ability of this hydrogel. In vitro studies revealed that the cells remained viable after their entrapment in the hydrogel. For the in vivo evaluation, healthy rats were submitted to a laminectomy surgery to induce a moderate spinal cord contusion injury. Then, hydrogels were implanted at the lesion sites. The biological performance of different hydrogel formulations was evaluated using a hydrogel with control NPs (without natural compound), a hydrogel with berberine-loaded NPs, and a hydrogel with berberine-loaded NPs and stem cells. A group of animals was left untreated as a control. The authors verified that the berberine-loaded NLH induced successful spinal cord injury recovery contrary to the control hydrogel. Additionally, the hydrogel containing both berberine-NPs and stem cells exhibited the better recovery of sensory and motor functions, suggesting that combination therapy with stem cells is a promising approach for repairing spinal cord injury.

Furthermore, for tissue repair, Gan et al. (2019) have proposed an approach using Ag-Lignin NPs [53]. Ag-Lignin NPs exhibit bactericidal activity due to their ability to generate free radicals, advantageous for wound repair. The NPs were loaded in a hydrogel composed of pectin and polyacrylic acid (PAA) to potentiate their therapeutic effect. The formulations were physicochemically characterized, and the authors verified that Ag-Lignin NPs improved the mechanical properties of the hydrogel and triggered its self-gelation at room temperature, not requiring UV or thermal treatments that could damage tissues. The adhesiveness properties of the hydrogel to different surfaces were evaluated in vitro, and it was verified that the hydrogel maintained good adhesion to porcine skin after 28 days. The NLH exhibited high antibacterial activity in vitro and in vivo, inhibiting the growth of both Gram-negative and Gram-positive bacteria. The biological effect of the hydrogel performance was evaluated in a wound rat model. A skin wound was created in the dorsal area of the animals, and the animals were treated with 30 μg NLH or control hydrogel without NPs. The hydrogels were implanted in the wound site, and control animals were left untreated. Although rats treated with control hydrogel exhibited healed wounds, the NLH showed better healing with higher portions of mature tissue with collagen fibers. Besides its potential for wound healing, this NLH system could also be applied for bone or cartilage repair.

4.2. Topical Administration of Polymeric NPs-Loaded Hydrogels

4.2.1. Ocular Administration by Instillation

Due to eye complexity (Figure 4), eye barriers, and their physiological functions, the conventional ocular administration of drugs has several shortcomings in terms of low bioavailability and reduced activity due to a low permanence and permeability at the target site [57].

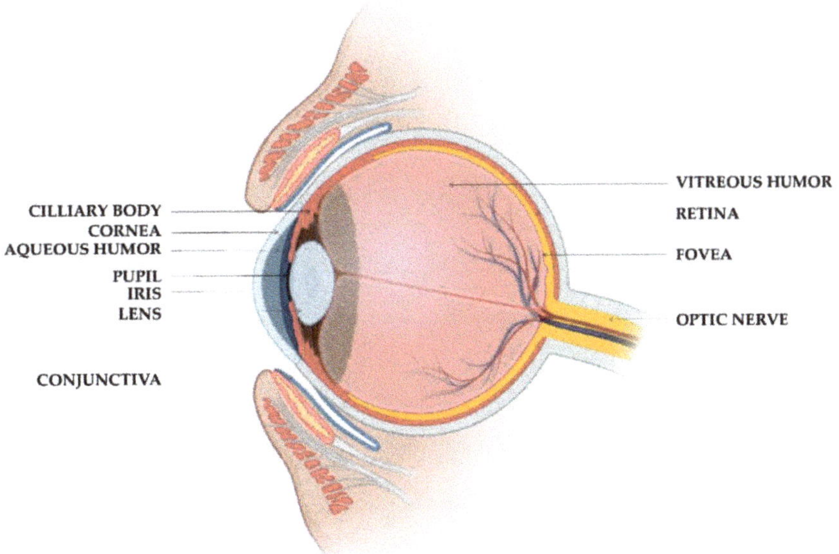

Figure 4. Schematic representation of the human eye's anatomy.

High dosages and frequent application of drugs are required to achieve the therapeutic drug concentrations, which can induce eye tissue injury, unwanted side effects, and low patient compliance. In the last years, polymeric NLH have been explored as DDS for ocular administration, as summarized in Table 4. These systems can increase the drugs' bioavailability and retain drugs at the ocular tissues for an extended period of time and release them in a controlled and sustained manner, enhancing the therapeutic effect [58]. Furthermore, with hydrogels, a reduced frequency of administration is required to achieve the therapeutic drug concentration, increasing patients' compliance.

Table 4. Summary of polymeric NLH ocular administered for biomedical applications.

NPs Material	Hydrogel Material	Loaded Cargo	NPs Production Method	Hydrogel Crosslinking Nature	Biomedical Application	Main Conclusions	Ref.
PLGA	PAMAM	Brimonidine and Timolol maleate	Single-emulsion solvent-evaporation	Chemical	Glaucoma	NLH provided a controlled release of drugs, reduction of IOP, and higher concentrations of drugs at the target site.	[59]
PLGA	Chitosan and Gelatin	Curcumin and Latanoprost	Single-emulsion solvent-evaporation	Physical (β-GP)	Glaucoma	The loaded-NLH reduced the oxidative stress effect that causes glaucoma, provided an anti-inflammatory effect.	[58]
PLGA	Chitosan and Gelatin	Levofloxacin and Prednisolone acetate	Single-emulsion solvent-evaporation	Physical (β-GP)	Anti-inflammatory treatment following surgery	Incorporating NPs into a hydrogel, a longstanding anti-inflammatory and anti-bacterial treatment were obtained, reducing side effects.	[60]
PLGA	Carbomer 934	Pranoprofen	Solvent displacement	Chemical	Anti-inflammatory treatment following surgery	NLH provided therapy with improved anti-inflammatory effects and edema reduction.	[61]
Chitosan	Carbopol 974P	Gentamycin	Ionotropic gelation	Chemical	Ophthalmic bacterial infections	NLH increased drug contact time in the cornea, extended release, and excellent antimicrobial properties.	[62]
Chitosan	Chitosan or its derivatives (TSOH)	5-fluorouracil	Self-assembly	Physical (β-GP)	Several ophthalmic diseases	NLH increased drug bioavailability and prolonged drug retention at the cornea.	[63]

Different approaches have been developed for glaucoma treatment. For example, Yang et al. (2012) proposed an NLH system to deliver two antiglaucoma drugs, timolol maleate and brimonidine [59]. The authors developed PLGA NPs as drugs carrier integrated into a PAMAM-based hydrogel. This hydrogel presents low viscosity, allowing for its administration in the liquid form as eye instillation. In vitro release studies were performed, showing that the NPs could release both drugs simultaneously for 28 days, while NPs integrated into the hydrogel released them for 35 days. Additionally, studies with human corneal epithelial cells were performed to evaluate the NLH cytotoxicity and the cellular uptake of the NPs. The results proved the non-toxicity of the NLH formulation, posteriorly corroborated with an in vivo analysis. The NPs uptake was about seven times higher when embedded in the hydrogel than in solution. Furthermore, the authors proceeded to animal experiments with ocular normotensive rabbits to evaluate the in vivo therapeutic activity of the developed hydrogel. A topical drop was administrated in the right eye, while the other eye was used as control, and the intraocular pressure (IOP), a parameter that in high values is associated with glaucoma development, was measured for seven days in both eyes. When NPs were embedded in the hydrogel, results showed an extended and controlled release of drugs, reduced IOP, and higher concentrations of drugs in aqueous humor, cornea, and conjunctiva for 4 days. The authors concluded that the NLH could retain drugs at the target tissue, significantly reducing the frequency of administration and improving the patients' compliance.

Cheng et al. (2019) also proposed PLGA NPs loaded in a thermosensitive chitosan-gelatin-based hydrogel as an eye drop formulation for glaucoma treatment [58]. Recently, oxidative stress in the ocular tissue has been associated with glaucoma. To overcome that, the authors proposed the administration of curcumin, an antioxidant and anti-inflammatory natural compound, encapsulated into NPs. Then, loaded-NPs were incorporated in the hydrogel together with latanoprost. Hydrogel in vitro release study revealed a prolonged release of curcumin-loaded NPs and latanoprost for 7 days. Under oxidative stress, an in vitro cell viability assay was performed with human ocular connective tissue. In the presence of the hydrogel containing loaded NPs, the cell damage caused by the oxidative stress decreased. Additionally, the anti-inflammatory effect of NLH formulation was confirmed by the considerable regulation of the inflammation-related genes compared with control hydrogel. Rabbit corneal epithelial cells were used to evaluate the in vitro biocompatibility of the developed NLH, and the results showed no damaging effect on cells. Thus, in vivo studies were performed with albino rabbits, instilling one eye with the loaded NLH (50 μL) and using the untreated eye as a control. After 7 days, the animals did not present indications of inflammation, confirming the biocompatibility results obtained in vitro.

The same authors developed an approach with the same materials to deliver two drugs with anti-inflammatory properties for inflammatory treatment after intraocular surgery [60]. In this work, the authors proposed levofloxacin encapsulated into PLGA NPs. Thus, NPs were simultaneously encapsulated in the hydrogel with prednisolone acetate. An in vitro release study was performed with drug-loaded NLH formulation, whose results showed a sustained release of drugs for 7 days. The levofloxacin-loaded NPs and prednisolone acetate were both incorporated in the hydrogel, and rabbit corneal epithelial cells were used to evaluate the in vitro cell viability and the anti-inflammatory and anti-bacterial properties of the NLH. Tumor necrosis factor-alpha was used to induce cells' inflammation. Then, cells were treated in vitro with blank solution, prednisolone acetate loaded-hydrogel, levofloxacin loaded-NPs, or drugs-loaded NLH. The results showed a considerable down-regulation in the expression of inflammation-related genes for all the tested formulations, compared to the blank solution, the levofloxacin-loaded NLH being the most promising. The loaded NLH performance was further evaluated in treating infected corneas of an ex vivo rabbit model of *Staphylococcus aureus* keratitis. The corneas were treated with 50 μL formulation for 24 h, and the results confirmed the anti-inflammation and anti-bacterial properties of the hydrogel. The NLH sustains drugs at the cornea, which increases the ocular bioavailability of drugs, and reduces the frequency of administration. Furthermore, the extended and controlled release of drugs provides a longstanding anti-inflammatory and anti-bacterial treatment, reducing side effects.

Abrego et al. (2015) also used PLGA NPs for the anti-inflammatory treatment after ocular surgery [61]. Pranoprofen is a non-steroidal anti-inflammatory drug commonly used after ocular surgery but with drawbacks in ocular bioavailability, stability, and solubility. PLGA NPs were used to encapsulate pranoprofen to increase its absorptivity in the ocular barrier. Then, two formulations of carbomer 934-based hydrogel with loaded-NPs were prepared for ocular administration: containing 0% and 1% azone as permeation enhancer agent. Both NLH formulations (with and without azone) decreased drug release by about 50% for 24 h compared with free drug or commercial eye drops. In vitro permeation and retention studies through the cornea were performed with ex vivo corneas of healthy rabbits using NLH formulations (with and without azone), free drug solution, and commercial eye drops. The results showed that both formulations increased the drug permeation and retention in the cornea. As expected, the NLH prepared with azone presented better results proving its permeation enhancer properties. The ocular tolerance of the hydrogel was evaluated in vivo through a single instillation of 50 μL of each NLH. Ocular lesions involving animals' corneas, irises, and conjunctivae were analyzed. Both formulations were considered safe since no indications of opacity, inflammation degree, congestion, swelling, or discharge were observed. In this experiment, the authors also demonstrated the anti-inflammatory efficacy of NLH with azone and its ability to reduce edema, being a suitable approach for ocular administration of pranoprofen.

Ocular administration with polymeric NLH systems can also be advantageous to treat bacterial conjunctivitis. The treatment's efficacy is hindered by low bioavailability and poor permeability of antibiotics in the ocular tissue. Due to this, higher concentrations of antibiotics are prescribed, which increases bacterial resistance [64]. To overcome this problem, Alruwaili et al. (2020) prepared gentamycin-loaded chitosan NPs integrated into a Carbopol 974P-based hydrogel for bacterial conjunctivitis treatment [62]. The incorporation of NPs in the hydrogel was optimized to obtain suitable properties for ocular delivery and evaluated in terms of clarity, pH, gelling ability, and rheological behavior. Both in vitro and ex vivo studies were performed with NPs in solution, gentamycin-loaded NLH, and commercial eye drops of gentamycin. The NPs-hydrogel system significantly prolonged the release of the drug. The authors verified that both NPs and NLH slowed the drug release compared with the commercial solution, with the NLH exhibiting the slower release. Cornea tissue collected from the goat eye was used for ex vivo studies to evaluate the mucoadhesive strength, ocular permeation, ocular tolerance, and antimicrobial properties of the hydrogel formulation. Outcomes revealed a high mucoadhesive strength indicating a greater retention capacity of the system in the cornea. As the ocular permeation was higher for the NPs in solution, the author suggested that the hydrogel network decreased the permeation. Furthermore, results indicated that the NLH significantly improved the drug's antimicrobial activity. All the findings indicate that the NLH was suitable for ocular administration of gentamycin by increasing the corneal contact time and extending the drug release. Furthermore, the antimicrobial properties of hydrogel enhance the effectiveness of bacterial conjunctivitis treatment.

Fabiano et al. (2019) proposed a chitosan-based hydrogel and NPs prepared with chitosan derivatives to promote 5-fluorouracil transcorneal administration, a drug widely used for several ophthalmic diseases [63]. NPs were prepared with four different chitosan and its derivatives: chitosan, quaternary ammonium-Ch conjugate (QA-Ch), S-protected derivative thereof (QA-Ch-S-pro), and a sulfobutyl chitosan derivative (SB-Ch). The NPs were optimized to increase the ocular bioavailability of 5-fluorouracil. The hydrogels were prepared with chitosan, loaded-NPs, and β-glycerophosphate disodium salt as a gelling agent. In vitro 5-fluorouracil release was performed with NPs in solution and NLH. All the prepared NPs revealed long drug retention and a sustained release, suitable for corneal delivery. In vitro results of the NLH demonstrated a slower release for hydrogel combined with SB-Ch NPs. SB-Ch is negatively charged and interacts electrostatically with the positive charge of chitosan hydrogel. The rheological analysis further validated this interaction, which revealed a robust mucoadhesive role influenced by NPs charge. In vivo studies were performed with albino rabbits, instilled in one eye with 50 μL of each formulation, using the other untreated eye as a control. NLH were able to increase the bioavailability of the drug and prolong its retention time at the cornea, compared to control. No signs of corneal edema were observed for all the tested formulations. The authors concluded that the formulation with SB-Ch NPs is the most efficient in prolonging the retention of the drug at the cornea, constituting a more effective treatment.

4.2.2. Epidermic Administration

Skin is the largest organ and the first barrier of the human body against external influences [65,66]. Due to this constant exposure to the external environment, skin is constantly subject to wounds, which in turn are quite predisposed to inflammation or infection that make their healing difficult [67]. Beyond that, for patients with other comorbidities, such as diabetes, inflammation could be uncontrollable, and wound healing could turn into a chronic medical problem [68]. Nowadays, the treatments applied to wounds protect the wounds from pathogens and incorporate active pharmaceutical agents to help with the healing process by treating the existent inflammation/infections [69]. Topical administration of drugs is usually used for external treatments and corresponds to the most non-invasive form of drug delivery. However, drug penetration in skin tissue is limited by drugs' properties, such as high molecular weight or reduced bioavailability [25], which require high dosages of drugs. Furthermore, a frequent application is needed, especially in the case of chronic wound healing, which demands high patient compliance that most of the time is not achieved. These limitations compromise the treatments and could be addressed using polymeric NLH as DDS, as proposed by several authors (Table 5). Besides the NPs improving drugs' bioavailability, protecting them from degradation, they could also be modified to enhance tissue penetration [70]. Hydrogels offer a favorable environment for wound healing with appropriate humidity, oxygen levels [71], and temperature to the tissue [69]. Hydrogels can also absorb wound exudates and provide impermeability to bacteria, preventing infection [68]. By incorporating NPs loaded with drugs, hydrogels offer a controlled release of drugs after application and increased retention of drugs in the wound area [25], improving the therapeutic effect.

Table 5. Summary of polymeric NLH epidermally administered for biomedical applications.

NPs Material	Hydrogel Material	Loaded Cargo	NPs Production Method	Hydrogel Crosslinking Nature	Biomedical Applications	Main Conclusions	Ref.
PLGA	Pluronic F-127	Platelet lysate	Double-emulsion solvent-evaporation	Chemical	Wound healing	NLH accelerates wound closure by promoting the cell migration and proliferation of fibroblasts	[23]
HA	Gelatin and methacryloyl (GelMA)	miR-223 5p mimic	n.d.	Chemical	Wound healing	Promotion of wound healing by initiating the resolution of the inflammatory phase and stimulating the formation of new vascularized skin tissue	[68]
PEG 4000	Carbopol	Simvastatin	Nanoprecipitation	Chemical	Wound healing	Acceleration of the wound healing by forming a normal epithelial layer and mature collagen fibers, with minimal inflammatory cell infiltration	[72]
Polydopamine	Xanthan gum and Konjac glucomannan	-	Nanoprecipitation	Chemical	Wound healing	NLH significantly accelerates the healing of wounds by reducing the inflammatory response and promoting vascular reconstruction	[73]
Chitosan	HA, pullulan and PVA	Cefepime	Ionic gelation	Physical (sodium tripolyphosphate)	Wound healing	Accelerates the wound healing process by inhibiting Gram-positive and Gram-negative bacteria growth, with no cytotoxicity against a human cell line	[66]
PAMAM	PAM	Platensimycin	Double-emulsion solvent-evaporation	Physical (PEG di-methacrylate)	Wound healing and subcutaneous bacterial infections	Accelerates wound closure and treats subcutaneous infections by exhibiting antibacterial activity	[67]
PLGA	Acrylamide, PEG dimethacrylate and PVA	Ciprofloxacin	Double-emulsion solvent-evaporation	Physical (PEG di-methacrylate)	Bacterial infections	The bioadhesive NLH showed superior adhesion and antibiotic retention under high shear stress, with no skin toxicity	[74]

Table 5. *Cont.*

NPs Material	Hydrogel Material	Loaded Cargo	NPs Production Method	Hydrogel Crosslinking Nature	Biomedical Applications	Main Conclusions	Ref.
RRR-α-tocopheryl succinate-grafted-ε-polylysine	Silk fibroin	Curcumin	Self-assembly	Chemical	Psoriasis	NPs incorporation in the hydrogel improved the therapeutic effect of curcumin by inhibiting skin inflammation	[75]
Eudragit L 100	Carbopol 934 and argan oil	Ibuprofen	Nanoprecipitation	Physical (glutaraldehyde)	Rheumatoid arthritis	Incorporating the drug-loaded NPs into the hydrogels improved the anti-inflammatory effect of ibuprofen compared to the commercially available ibuprofen cream	[65]

n.d.: not defined.

Bernal-Chávez et al. (2020) developed a Pluronic F-127 thermo-responsive hydrogel containing PLGA NPs loaded with a platelet lysate to treat chronic wounds [23]. The hydrogel formulation was optimized, and the gelation temperature was 32 °C, allowing the formulation to solidify in situ and be retained in the wound area during the platelet lysate release period. In vitro release studies revealed that lysate-loaded NPs incorporated into the hydrogel significantly prolonged its release from 12 to 24 h. The therapeutic efficacy of the NLH was evaluated in healthy mice with a full-thickness cutaneous wound in the dorsal region with a diameter of 3 mm. In vivo results demonstrate that the topical application of platelet lysate-loaded NPs incorporated into the hydrogel accelerates wound closure by promoting fibroblasts' cell migration and proliferation, proving to be more efficient than platelet lysate in solution.

Saleh et al. (2019) proposed a gelatin methacryloyl-based hydrogel containing HA NPs loaded with miR-223 5p mimic (miR-223) to promote wound healing [68]. miR-223 regulates the expression of anti-inflammatory and proinflammatory markers. The developed NLH showed a controlled release of miR-223 over 48 h. In vitro results confirmed that the hydrogel containing miR-223-loaded NPs was successfully internalized by murine macrophages and induced their polarization by decreasing the expression levels of pro-inflammatory markers. The therapeutic efficacy of the system was evaluated in male mice with wounds on the dorsum. According to the visual inspection of the wounds, the percentage of wound closure induced by the hydrogel containing miR-223-loaded NPs was substantially higher than the unloaded hydrogel and the free miR-223. Histological analysis of tissue samples validated these results, revealing that the miR-223-loaded NLH promotes wound healing by initiating the resolution of the inflammatory phase and stimulating the formation of new vascularized skin tissue.

A hydrogel loaded with polymeric NPs was also proposed by Aly et al. (2019) to be topically used on skin wounds [72]. PEG NPs were prepared to encapsulate simvastatin, a molecule that promotes the wound healing process. The hydrogel made of Carbopol showed a controlled release of simvastatin in vitro. Furthermore, an ex vivo permeation study revealed that 69% of the drug permeated through the skin. The therapeutic efficacy of the hydrogel containing loaded NPs was assessed in healthy rats with 8 mm excisional skin wounds on their backs. The wounds treated with the drug-NLH showed a more accentuated reduction in the wound area than those receiving the control hydrogel. In addition, the wounds treated with simvastatin-loaded NLH appeared contracted without hard or abnormal tissue on their surface, unlike control wounds, which showed a hard and abnormal tissue on the surface. The histopathological examination of the medicated wound confirmed the efficacy of the NLH in accelerating wound healing by forming a normal epithelial layer and mature collagen fibers with minimal inflammatory cell infiltration. In contrast, granulation tissue formation with massive inflammatory cell infiltration was observed in the control wound. Hair follicle growth was observed at the end of the treatment indicating good tissue regeneration.

Because open wounds provide an access point for microorganisms, these are particularly vulnerable to bacterial infections. This can result in significant wound inflammation, compromising wound healing and ultimately resulting in death [69]. Thus, keeping a wound from becoming infected is crucial for proper wound healing. Zeng et al. (2021) synthesized a hydrogel composed of xanthan gum and konjac glucomannan to heal bacteria-infected wounds [73]. The hydrogel was loaded with polydopamine NPs. In vitro near-infrared photothermal antibacterial experiments indicate that the NLH has a broad-spectrum antibacterial activity against Gram-negative (*Escherichia coli)* and Gram-positive (*Staphylococcus aureus*) bacteria. While bacteria incubated with the buffer and the unloaded hydrogel had smooth and intact membranes, bacteria cultured with the loaded hydrogel showed many wrinkles and disruptions. The authors further evaluated the in vitro blood compatibility of the hydrogel using mouse-derived fibroblasts. After 3 days of incubation, the hydrogel showed no cytotoxicity to the healthy cells. To further examine the suitability of the NLH for the topical application in skin tissue repair, male rats with a round skin wound of 8 mm diameter on their backs were used. Next, an *Escherichia coli* suspension was added to the wound surfaces to create bacteria-infected wounds. The developed NLH exhibited a rapid wound shape adaptability. The obtained results demonstrate that the NLH significantly accelerates wound healing compared with the control groups. The wound sites of the animals treated with NLH, were histopathologically analyzed. Results revealed greater regularity of both epithelium and connective tissue, improved collagen deposition, promotion of vascular angiogenesis, and new hair follicles in the wound sites of the animals treated with NLH. In addition, the expression of pro-inflammatory cytokines was significantly lower in the animals receiving NLH than the buffer and unloaded hydrogel groups.

Shafique et al. (2021) also devised a hydrogel with antibacterial properties for wound healing. The hydrogel was composed of HA, pullulan, and PVA, containing cefepime loaded into chitosan NPs [66]. Cefepime is a water-insoluble hydrophobic drug that acts as a parenteral antibiotic against Gram-positive and Gram-negative bacteria. The prepared NLH showed good stability and swelling capacity, allowing for the sustained release of cefepime. The antibacterial activity of cefepime-NLH was evaluated against *Staphylococcus aureus*, *Pseudomonas aeruginosa*, and *Escherichia coli*. It was found to inhibit both Gram-positive and Gram-negative bacteria growth. The in vitro cytocompatibility was also assessed against a human fibrosarcoma cell line, showing no cytotoxicity. Authors evaluated the therapeutic efficacy of the cefepime-loaded NLH in healthy rats with a wound on the back of 1 cm^2 area. The drug-NLH showed a greater potential to accelerate the wound healing process with a wound closure rate of 100% after 14 days, compared to cefepime in solution, with a wound closure rate of 80% at the same period. The data also demonstrated the ability of the blank hydrogel to accelerate wound healing, with a 90% wound closure rate following 14 days of the application. This is related to the HA's ability to facilitate the migration and proliferation of fibroblasts as well as absorb water, keeping the wound bed wet. Furthermore, the degradation products of HA promote tissue regeneration by meeting the nutritional needs of the surrounding cells. Moreover, pullulan provides energy to the cells, regenerating the skin through glucose consumption. The healing process of the developed NLH was proved by the presence of hair follicles, sweat, and sebaceous glands and the absence of inflammatory cells at the wound site.

A promising approach to treat bacterial infections present either subcutaneously or in open wounds was proposed by Wang et al. (2021) [67]. To this end, a PAM hydrogel containing PAMAM NPs was prepared for the controlled release of platensimycin, a natural antibiotic with great potential to treat infections caused by *Staphylococcus aureus*. The in vitro release data suggest that incorporating the platensimycin-loaded NPs into the hydrogel provided a more controlled release behavior than the free drug and platensimycin-loaded NPs. Furthermore, the in vitro antibacterial activity study revealed that the hydrogel containing platensimycin-loaded NPs could completely inhibit the *Staphylococcus aureus* growth, and to a greater extent than the free platensimycin and the drug-loaded NPs, without inducing toxicity to murine macrophage cells. In the first approach, the authors evaluated the in vivo activity of the drug-loaded NLH against the subcutaneous *Staphylococcus aureus* infection. The buffer, free drug, drug-loaded NPs, blank NLH, and the hydrogel containing the drug-loaded NPs were subcutaneously administered in the backs of male mice, and the number of colonies in each group was quantified. While the free platensimycin and the blank NLH exhibited no antibacterial activity, the drug-loaded NPs, loaded or not in the hydrogel, displayed high anti-staphylococcus activity. Further in vivo data revealed that the NLH can remain in situ for 24 h, unlike the free NPs, suggesting that the hydrogel can improve platensimycin's therapeutic activity against bacterial infections by extending its residence time. On the second approach, the authors evaluated the drug-loaded NLH ability in promoting the healing of bacteria-infected wounds following topical application. For this purpose, *Staphylococcus aureus* was used to infect wounds of 10 mm diameter formed on the back of healthy rats. The platensimycin-loaded NLH showed a significantly higher wound healing rate on the ninth day when compared to the other groups. The NLH's antibacterial activity was assessed after two days by quantifying the remaining colonies in the wound. The hydrogel containing drug-loaded NPs exerted the strongest antibacterial activity in infected wounds than the free drug and the drug-loaded NPs, implying that the NLH is more advantageous for treating bacteria-infected wounds due to the sustained drug release.

In addition to open wounds, bacteria can enter the body through other routes, including mouth, eyes, nose, or urogenital openings. Frequently, bacterial infections occur at sites with high shear forces, which facilitate bacterial adhesion and prevent effective drug accumulation [76]. Thus, developing formulations that can withstand strong shear forces and work at infection sites involving the shear flow of biological fluids is particularly desirable. To address this issue, Zhang et al. (2016) designed a bioadhesive NLH for the local delivery of ciprofloxacin, a broad-spectrum antibiotic [74]. Ciprofloxacin-loaded PLGA NPs were embedded in a hydrogel composed of acrylamide, PEG dimethacrylate, and PVA. Dopamine methacrylamide, a catechol moiety responsible for the adhesion of marine mussels to various surfaces, was included in the hydrogel network. The authors started to compare the in vitro ciprofloxacin-loaded NLH release profile to that of ciprofloxacin-loaded hydrogel (without NPs). While the drug showed a burst release profile from the blank hydrogel, the NLH showed a gradual drug release profile, highlighting the benefit of introducing NPs into the hydrogel to allow the controlled and sustained ciprofloxacin release. Under the flow environment, the in vitro antibacterial efficacy of the formulations (free drug, drug-NPs, blank hydrogel, and drug-loaded NLH) was investigated. The ciprofloxacin-NLH completely inhibited the formation of an *Escherichia coli* bacterial film, unlike the other formulations. The authors then tested the adhesive capabilities of the hydrogel and its ability to hold the embedded NPs under flow conditions, using distinct surfaces including a bacterial film, a mammalian cell monolayer, and a shaved mouse skin tissue. The retention of NPs in the bioadhesive hydrogel was quantified and revealed that the totality of the NPs remained on the three biological surfaces. In contrast, a small quantity of NPs in the non-adhesive hydrogel was retained. Lastly, the NLH's skin toxicity was investigated in healthy mice following its daily topical application for 7 days, and the results showed no skin reaction or toxicity.

Aiming to improve current psoriasis' therapy, Mao et al. (2017) [75] designed a silk fibroin hydrogel containing RRR-α-tocopheryl succinate-grafted-ε-polylysine NPs for the topical delivery of curcumin for psoriasis' treatment. The in vitro release study revealed that the encapsulation of curcumin into the polymeric NPs induced a slower release profile with no evident initial burst release, unlike the free curcumin. The incorporation of the NPs in the hydrogel resulted in a longer-lasting release of the natural compound. The in vivo skin penetration of the curcumin-loaded hydrogel, with and without NPs, was investigated by applying the formulations topically on the back of mice skin (1 cm^2 area). A high curcumin skin permeation ability was observed when the NPs were included in the hydrogel. The in vivo therapeutic effect of the two formulations was evaluated on a psoriatic mice model and compared to the clobetasol's activity, a topical corticosteroid used on the psoriasis therapy. Clobetasol-treated mice exhibited the most improvement in psoriatic symptoms, including erythema, thickness, and scaling of the back skins. Although to a lesser extent, curcumin-loaded NPs also improved the psoriatic symptoms, which were further enhanced by incorporating the NPs into the hydrogel. The histological examination confirmed the visual observation as a decrease in thicknesses of the dermis and epidermis—associated with psoriasis—was detected. Significantly few leukocytic infiltrations were noticed in the curcumin-loaded NPs treated group. This suppression was even more evident in the curcumin-loaded NLH group. Finally, the effect of the curcumin-loaded hydrogel, with and without NPs, on inflammatory cytokine levels was evaluated and compared to clobetasol, a corticosteroid used to treat psoriasis. Curcumin-loaded NLH inhibited the expression of inflammatory cytokines to the same degree as the commercially available clobetasol but to a greater extent than the curcumin-loaded hydrogel without NPs.

The topical application of hydrogels containing polymeric NPs has been used to manage other health conditions, such as autoimmune disorders. Rheumatoid arthritis therapy was recently addressed by Khan et al. (2021) [65] by developing a pH-responsive NLH to deliver ibuprofen, a nonsteroidal anti-inflammatory drug widely employed to treat this condition. Eudragit L 100 polymer was chosen to produce the NPs due to its pH-responsive dissolution profile, as it particularly releases drugs at pH 6.8, a pH found in inflamed tissues. The NPs were incorporated into a Carbopol 934 hydrogel containing argan oil as a permeation enhancer agent. The in vitro ibuprofen release data at pH 6.8 indicate that the NPs caused a more sustained release pattern than the free drug, which was further sustained by incorporating the NPs into the hydrogel. On the other hand, both formulations induced a residual drug release at the physiological pH of the skin (5.5). The ex vivo mice skin permeability study revealed that the hydrogel containing ibuprofen-loaded NPs showed a higher permeation of the NPs than a commercially available ibuprofen cream. The presence of argan oil in the formulation further increased the skin permeability of the NPs 14-fold. The in vivo safety of the ibuprofen-loaded NLH was assessed in healthy mice. The formulation caused no visual signs of skin toxicity, with the histological analysis showing no harm to skin tissues. Further, behavioral experiments and biochemical analysis demonstrated the in vivo therapeutic efficacy of the hydrogel in both acute and chronic inflammatory pain mice models, which was significantly improved compared to the group treated with the marketed cream. The prepared formulation was also able to inhibit the inflammatory processes and oxidative stress. In addition, both bone erosion and soft-tissue edema in the ankle joints of the treated mice were considerably reduced in the group treated with the hydrogel containing ibuprofen-loaded NPs.

4.2.3. Vaginal Administration

Local vaginal therapy is a non-invasive drug administration route used to create a local pharmacological impact while avoiding systemic exposure. However, the presence of degradative enzymes and vaginal fluid, which reduces medication bioavailability and residence time, decreases the efficacy of vaginal administration. The vaginal fluid also induces a leak of drugs, requiring repeated applications, that may result in low patient compliance [36]. Polymeric NLH systems for vaginal drug administration are promising to overcome the highlighted drawbacks. The mucoadhesive ability of hydrogels provides a high interaction with vaginal tissue, favoring formulation permanence on the vaginal mucosa. This capability lengthens drug residence time, maximizing pharmacological activity. The combination of NPs and hydrogels provides a high concentration of drugs at the site of action and controlled drug release rates, which decreases the need for frequent applications [77].

Recently, Zimmermann et al. (2021) [78] produced an NLH system to treat vulvovaginal candidiasis. The approach containing gellan gum hydrogel and PCL NPs was used for diphenyl diselenide delivery. Diphenyl diselenide exhibit a wide range of biological effects, including antioxidant, anti-inflammatory, and antifungal activity against *Candida* spp. In vitro assays performed against various *Candida* species confirmed the diphenyl diselenide's antifungal activity and revealed that the encapsulation of the drug into the NPs did not affect its therapeutic activity. The in vivo efficacy of the formulations (hydrogel containing free or encapsulated drug) was assessed in a mice model of vulvovaginal candidiasis. The formulations were topically administered once a day for 7 days, and the total fungal burden was quantified. Treatment with both formulations reduced the fungal load, validating the diphenyl diselenide's antifungal activity previously demonstrated in vitro. Additionally, the hydrogel containing drug-loaded NPs presented better pharmacological efficacy than the hydrogel containing free diphenyl diselenide, and the control group treated with Nystatin cream, a commercially available antifungal medicine.

5. Discussion and Conclusions

Until this date, the use of polymeric NLH has been studied for different routes of administration, including parenteral and topical administration. As demonstrated in the reported works, the described systems validate the usefulness of polymeric NLH as DDS for various biomedical applications. Based on the in vivo findings, polymeric NLH systems have shown the ability to regulate drug release kinetics and increase the drug's bioavailability, prolonging the therapeutic effect while minimizing systemic spread and toxicity. The innovative integration of NPs in hydrogels has resulted in a new second generation of DDS with improved performance and characteristics. The target and localized drug delivery offered by NPs could be enhanced through simultaneous use of hydrogels, demonstrating superior capabilities on local drug retention. The presented studies have highlighted the crucial role of polymeric NLH in the reduction of drugs administration frequency, resulting in enhanced patient compliance. Polymeric NLH appear to be a promising strategy to improve the treatment of various diseases by boosting drug delivery efficiency. Two different approaches of polymeric NLH could be employed, NPs loaded into unloaded hydrogels or NPs loaded into the hydrogel, which contains an additional drug. That way, a synergic effect of different drugs could also be obtained.

Due to the extensive availability of materials to produce NPs and hydrogels, many polymeric NLH systems can be generated for a broad range of potential biomedical applications. PLGA is undoubtedly the most explored polymeric material to produce NPs to be incorporated in hydrogels, accounting for about 33% of the works reported above. PLGA is a versatile polymer that can be customized in terms of biodegradation and release rates, which is ideal for drug delivery. PLGA is also biocompatible, biodegradable, and FDA-approved for biomedical use [8]. Chitosan NPs have also been widely studied and were reported in 15% of the works described in this review. Additionally, chitosan is also the most popular material for hydrogel production (18% of the works reported here). This popularity is given by the physicochemical and biological properties of chitosan, which is useful for biomedical applications. In addition to being a natural polymer, it is also non-toxic, biodegradable, and biocompatible. As described previously by Fabiano et al. (2019) [63], chitosan has a cationic characteristic that allows it to interact with the negatively charged cell membranes of microorganisms, conferring to this polymer powerful antimicrobial and antibacterial effects [79]. Other materials have also been explored for hydrogels production, with a particular focus on surfactants, such as Pluronic F-127 and F-68 (12%). Pluronic block copolymers are commonly used in the manufacturing of hydrogels due to their thermosensitive properties. Furthermore, these copolymers can improve the pharmacokinetics and pharmacodynamics of DDS [80].

Although many DDS have considerable in vitro and in vivo drug delivery benefits, scaling up production is still a challenge. Despite that, clinically available strategies of NPs and hydrogels as individual DDS could already be found on the market. The polymeric NPs that have resulted in commercial products are Abraxane, a paclitaxel delivery in a suspension of albumin particles, for breast cancer treatment [81]; Lupron Depot, Nutropin Depot, and Vivitrol, injectable medicines for extended drugs release; and Accurins, for targeted cancer therapy [8]. There is an exhaustive list of commercial products based on hydrogels for biomedical applications. Some examples of commercial products are Metrogel

Vagina, a vaginal drug delivery form of the synthetic antibacterial agent, metronidazole; Voltaren Gel, a topical application form of diclofenac used to reduce pain and inflammation; Astero, a topical application form indicated for wounds, first and second-degree burns, post-surgical incisions, cuts, and abrasions; Ocusert, an ocular administration system used for glaucoma treatment; among others [82].

However, their combination as a single DDS is recent, and no clinical trials are concluded nowadays. Just one experiment reporting the use of silk particles distributed in a hydrogel as a tissue filler is in an ongoing clinical trial (NCT04534660). That combination of particles with a hydrogel, called SMI-01, is an injectable formulation applied in the deep dermis, and it is used to correct moderate to severe wrinkles and folds. Furthermore, the researchers are testing the effect of that technology to correct age-related volume deficiency in the midface through either a subcutaneous, supraperiosteal, or both, injection of SMI-01. This feasibility study is multicentered (two clinical sites), unblinded, with no control group, performed to access the preliminary safety and effectiveness of SMI-01 as a tissue filler. The administration of NLH takes place on day 1 and, if desired, on day 30, with the primary safety and effectiveness evaluation taking place at the second month. Extended follow-up examinations will be conducted on subjects for 24 months after the first administration. This clinical trial is still in recruitment status, so there are no reported data on its effectiveness.

Thus, even with the promising advantages of this innovative type of DDS, more research is needed before it reaches the market.

Author Contributions: Conceptualization, methodology, formal analysis, investigation, validation and writing—original draft preparation, D.N., S.A., M.J.R. and J.A.L.; writing—review and editing and visualization, D.N., S.A., M.J.R., J.A.L. and M.C.P.; resources, supervision, project administration and funding acquisition, J.A.L. and M.C.P. All authors have read and agreed to the published version of the manuscript.

Funding: This work was financially supported by: Base Funding—LA/P/0045/2020 (ALiCE), and UIDP/00511/2020 (LEPABE), funded by national funds through the FCT/MCTES (PIDDAC); Project 2SMART—engineered Smart materials for Smart citizens, with reference NORTE-01-0145-FEDER-000054, supported by Norte Portugal Regional Operational Program (NORTE 2020), under the PORTUGAL 2020 Partnership Agreement, through the European Regional Development Fund (ERDF); FCT supported D.N. (UI/BD/150946/2021), S.A. (SFRH/BD/129312/2017 and COVID/BD/151869/2021) and J.A.L under the Scientific Employment Stimulus—Institutional Call—(CEECINST/00049/2018); and Prize Maratona da Saúde for Cancer Research.

Institutional Review Board Statement: Not applicable.

Informed Consent Statement: Not applicable.

Data Availability Statement: Not applicable.

Conflicts of Interest: The authors declare no conflict of interest.

References

1. Bellotti, E.; Schilling, A.L.; Little, S.R.; Decuzzi, P. Injectable thermoresponsive hydrogels as drug delivery system for the treatment of central nervous system disorders: A review. *J. Control. Release* **2020**, *329*, 16–35. [CrossRef]
2. El-Sayed, A.; Kamel, M. Advances in nanomedical applications: Diagnostic, therapeutic, immunization, and vaccine production. *Environ. Sci. Pollut. Res.* **2019**, *27*, 19200–19213. [CrossRef] [PubMed]
3. Kargozar, S.; Mozafari, M. Nanotechnology and Nanomedicine: Start small, think big. *Mater. Today Proc.* **2018**, *5*, 15492–15500. [CrossRef]
4. Nobile, L.; Nobile, S. Recent advances of nanotechnology in medicine and engineering. *AIP Conf. Proc.* **2016**, *1736*, 20058. [CrossRef]
5. Saxena, S.K.; Nyodu, R.; Kumar, S.; Maurya, V.K. Current Advances in Nanotechnology and Medicine. *NanoBioMedicine* **2020**, 3–16. [CrossRef]
6. Souto, E.B.; Silva, G.F.; Dias-Ferreira, J.; Zielinska, A.; Ventura, F.; Durazzo, A.; Lucarini, M.; Novellino, E.; Santini, A. Nanopharmaceutics: Part I—Clinical Trials Legislation and Good Manufacturing Practices (GMP) of Nanotherapeutics in the EU. *Pharmaceutics* **2020**, *12*, 146. [CrossRef]
7. Zhao, F.; Yao, D.; Guo, R.; Deng, L.; Dong, A.; Zhang, J. Composites of Polymer Hydrogels and Nanoparticulate Systems for Biomedical and Pharmaceutical Applications. *Nanomaterials* **2015**, *5*, 2054–2130. [CrossRef]

8. Operti, M.C.; Bernhardt, A.; Grimm, S.; Engel, A.; Figdor, C.G.; Tagit, O. PLGA-based nanomedicines manufacturing: Technologies overview and challenges in industrial scale-up. *Int. J. Pharm.* **2021**, *605*, 120807. [CrossRef]
9. Wang, K.; Hao, Y.; Wang, Y.; Chen, J.; Mao, L.; Deng, Y.; Chen, J.; Yuan, S.; Zhang, T.; Ren, J.; et al. Functional Hydrogels and Their Application in Drug Delivery, Biosensors, and Tissue Engineering. *Int. J. Polym. Sci.* **2019**, *2019*, 1–14. [CrossRef]
10. Wahid, F.; Zhao, X.J.; Jia, S.R.; Bai, H.; Zhong, C. Nanocomposite hydrogels as multifunctional systems for biomedical applications: Current state and perspectives. *Compos. Part B Eng.* **2020**, *200*, 108208. [CrossRef]
11. Mellati, A.; Hasanzadeh, E.; Gholipourmalekabadi, M.; Enderami, S.E. Injectable nanocomposite hydrogels as an emerging platform for biomedical applications: A review. *Mater. Sci. Eng. C* **2021**, *131*, 112489. [CrossRef] [PubMed]
12. Biondi, M.; Borzacchiello, A.; Mayol, L.; Ambrosio, L. Nanoparticle-Integrated Hydrogels as Multifunctional Composite Materials for Biomedical Applications. *Gels* **2015**, *1*, 162–178. [CrossRef] [PubMed]
13. Jiang, Y.; Krishnan, N.; Heo, J.; Fang, R.H.; Zhang, L. Nanoparticle–hydrogel superstructures for biomedical applications. *J. Control. Release* **2020**, *324*, 505–521. [CrossRef] [PubMed]
14. Meis, C.M.; Grosskopf, A.K.; Correa, S.; Appel, E.A. Injectable Supramolecular Polymer-Nanoparticle Hydrogels for Cell and Drug Delivery Applications. *JoVE* **2021**, *168*, e62234. [CrossRef]
15. Gao, W.; Zhang, Y.; Zhang, Q.; Zhang, L. Nanoparticle-Hydrogel: A Hybrid Biomaterial System for Localized Drug Delivery. *Ann. Biomed. Eng.* **2016**, *44*, 2049–2061. [CrossRef]
16. Gonçalves, C.; Pereira, P.; Gama, M. Self-Assembled Hydrogel Nanoparticles for Drug Delivery Applications. *Materials* **2010**, *3*, 1420–1460. [CrossRef]
17. Desfrançois, C.; Auzély, R.; Texier, I. Lipid Nanoparticles and Their Hydrogel Composites for Drug Delivery: A Review. *Pharmaceuticals* **2018**, *11*, 118. [CrossRef]
18. Mauri, E.; Negri, A.; Rebellato, E.; Masi, M.; Perale, G.; Rossi, F. Hydrogel-Nanoparticles Composite System for Controlled Drug Delivery. *Gels* **2018**, *4*, 74. [CrossRef]
19. Welch, V.; Petticrew, M.; Petkovic, J.; Moher, D.; Waters, E.; White, H.; Tugwell, P.; Atun, R.; Awasthi, S.; Barbour, V.; et al. Extending the PRISMA statement to equity-focused systematic reviews (PRISMA-E 2012): Explanation and elaboration. *J. Clin. Epidemiol.* **2015**, *70*, 68–89. [CrossRef]
20. Begines, B.; Ortiz, T.; Pérez-Aranda, M.; Martínez, G.; Merinero, M.; Argüelles-Arias, F.; Alcudia, A. Polymeric Nanoparticles for Drug Delivery: Recent Developments and Future Prospects. *Nanomaterials* **2020**, *10*, 1403. [CrossRef]
21. Adhikari, C. Polymer nanoparticles-preparations, applications and future insights: A concise review. *Polym. Technol. Mater.* **2021**, 1–29. [CrossRef]
22. Mitchell, M.J.; Billingsley, M.M.; Haley, R.M.; Wechsler, M.E.; Peppas, N.A.; Langer, R. Engineering precision nanoparticles for drug delivery. *Nat. Rev. Drug Discov.* **2020**, *20*, 101–124. [CrossRef]
23. Bernal-Chávez, S.A.; Alcalá-Alcalá, S.; Cerecedo, D.; Ganem-Rondero, A. Platelet lysate-loaded PLGA nanoparticles in a thermo-responsive hydrogel intended for the treatment of wounds. *Eur. J. Pharm. Sci.* **2020**, *146*, 105231. [CrossRef] [PubMed]
24. Hoffman, A.S. Hydrogels for biomedical applications. *Adv. Drug Deliv. Rev.* **2012**, *64*, 18–23. [CrossRef]
25. Chang, D.; Park, K.; Famili, A. Hydrogels for sustained delivery of biologics to the back of the eye. *Drug Discov. Today* **2019**, *24*, 1470–1482. [CrossRef]
26. Wang, Y.; Li, Q.; Zhou, J.-E.; Tan, J.; Li, M.; Xu, N.; Qu, F.; Chen, J.; Li, J.; Wang, J.; et al. A Photopolymerized Semi-Interpenetrating Polymer Networks-Based Hydrogel Incorporated with Nanoparticle for Local Chemotherapy of Tumors. *Pharm. Res.* **2021**, *38*, 669–680. [CrossRef] [PubMed]
27. Madduma-Bandarage, U.S.K.; Madihally, S.V. Synthetic hydrogels: Synthesis, novel trends, and applications. *J. App. Polym. Sci.* **2021**, *138*, 50376. [CrossRef]
28. Sun, Z.; Wang, X.; Liu, J.; Wang, Z.; Wang, W.; Kong, D.; Leng, X. ICG/l-Arginine Encapsulated PLGA Nanoparticle-Thermosensitive Hydrogel Hybrid Delivery System for Cascade Cancer Photodynamic-NO Therapy with Promoted Collagen Depletion in Tumor Tissues. *Mol. Pharm.* **2021**, *18*, 928–939. [CrossRef]
29. Boffito, M.; Gioffredi, E.; Chiono, V.; Calzone, S.; Ranzato, E.; Martinotti, S.; Ciardelli, G. Novel polyurethane-based thermosensitive hydrogels as drug release and tissue engineering platforms: Design and in vitro characterization. *Polym. Int.* **2016**, *65*, 756–769. [CrossRef]
30. Ren, Y.; Li, X.; Han, B.; Zhao, N.; Mu, M.; Wang, C.; Du, Y.; Wang, Y.; Tong, A.; Liu, Y.; et al. Improved anti-colorectal carcinomatosis effect of tannic acid co-loaded with oxaliplatin in nanoparticles encapsulated in thermosensitive hydrogel. *Eur. J. Pharm. Sci.* **2018**, *128*, 279–289. [CrossRef]
31. Cao, D.; Zhang, X.; Akabar, M.D.; Luo, Y.; Wu, H.; Ke, X.; Ci, T. Liposomal doxorubicin loaded PLGA-PEG-PLGA based thermogel for sustained local drug delivery for the treatment of breast cancer. *Artif. Cells Nanomed. Biotechnol.* **2019**, *47*, 181–191. [CrossRef] [PubMed]
32. Brachi, G.; Ruiz-Ramírez, J.; Dogra, P.; Wang, Z.; Cristini, V.; Ciardelli, G.; Rostomily, R.C.; Ferrari, M.; Mikheev, A.M.; Blanco, E.; et al. Intratumoral injection of hydrogel-embedded nanoparticles enhances retention in glioblastoma. *Nanoscale* **2020**, *12*, 23838–23850. [CrossRef] [PubMed]
33. Thoniyot, P.; Tan, M.J.; Karim, A.A.; Young, D.J.; Loh, X.J. Nanoparticle–Hydrogel Composites: Concept, Design, and Applications of These Promising, Multi-Functional Materials. *Adv. Sci.* **2015**, *2*, 1400010. [CrossRef] [PubMed]

34. Ghasemiyeh, P.; Mohammadi-Samani, S. Hydrogels as Drug Delivery Systems; Pros and Cons. *Trends Pharm. Sci.* **2019**, *5*, 7–24. [CrossRef]
35. Li, J.; Mooney, D.J. Designing hydrogels for controlled drug delivery. *Nat. Rev. Mater.* **2016**, *1*, 16071. [CrossRef]
36. Malik, D.S.; Mital, N.; Kaur, G. Topical drug delivery systems: A patent review. *Expert Opin. Ther. Patents* **2015**, *26*, 213–228. [CrossRef]
37. Patil, J.S.; Sarasija, S. Pulmonary drug delivery strategies: A concise, systematic review. *Lung India* **2012**, *29*, 44–49.
38. Huang, H.; Qi, X.; Chen, Y.; Wu, Z. Thermo-sensitive hydrogels for delivering biotherapeutic molecules: A review. *Saudi Pharm. J.* **2019**, *27*, 990–999. [CrossRef]
39. Rafael, D.; Melendres, M.M.R.; Andrade, F.; Montero, S.; Martinez-Trucharte, F.; Vilar-Hernandez, M.; Durán-Lara, E.F.; Schwartz, S., Jr.; Abasolo, I. Thermo-responsive hydrogels for cancer local therapy: Challenges and state-of-art. *Int. J. Pharm.* **2021**, *606*, 120954. [CrossRef]
40. Peng, Q.; Sun, X.; Gong, T.; Wu, C.-Y.; Zhang, T.; Tan, J.; Zhang, Z.-R. Injectable and biodegradable thermosensitive hydrogels loaded with PHBHHx nanoparticles for the sustained and controlled release of insulin. *Acta Biomater.* **2012**, *9*, 5063–5069. [CrossRef]
41. Lee, A.L.Z.; Voo, Z.X.; Chin, W.; Ono, R.J.; Yang, C.; Gao, S.; Hedrick, J.L.; Yang, Y.Y. Injectable Coacervate Hydrogel for Delivery of Anticancer Drug-Loaded Nanoparticles in vivo. *ACS Appl. Mater. Interfaces* **2018**, *10*, 13274–13282. [CrossRef] [PubMed]
42. Chao, Y.; Chen, Q.; Liu, Z. Smart Injectable Hydrogels for Cancer Immunotherapy. *Adv. Funct. Mater.* **2019**, *30*. [CrossRef]
43. Thakur, S.; Singh, H.; Singh, A.; Kaur, S.; Sharma, A.; Singh, S.K.; Kaur, S.; Kaur, G.; Jain, S.K. Thermosensitive injectable hydrogel containing carboplatin loaded nanoparticles: A dual approach for sustained and localized delivery with improved safety and therapeutic efficacy. *J. Drug Deliv. Sci. Technol.* **2020**, *58*, 101817. [CrossRef]
44. Gao, B.; Luo, J.; Liu, Y.; Su, S.; Fu, S.; Yang, X.; Li, B. Intratumoral Administration of Thermosensitive Hydrogel Co-Loaded with Norcantharidin Nanoparticles and Doxorubicin for the Treatment of Hepatocellular Carcinoma. *Int. J. Nanomed.* **2021**, *16*, 4073–4085. [CrossRef] [PubMed]
45. Segovia, N.; Pont, M.; Oliva, N.; Ramos, V.; Borrós, S.; Artzi, N. Hydrogel Doped with Nanoparticles for Local Sustained Release of siRNA in Breast Cancer. *Adv. Healthc. Mater.* **2015**, *4*, 271–280. [CrossRef] [PubMed]
46. Men, K.; Liu, W.; Li, L.; Duan, X.; Wang, P.; Gou, M.; Wei, X.; Gao, X.; Wang, B.; Du, Y.; et al. Delivering instilled hydrophobic drug to the bladder by a cationic nanoparticle and thermo-sensitive hydrogel composite system. *Nanoscale* **2012**, *4*, 6425–6433. [CrossRef] [PubMed]
47. Shen, M.; Xu, Y.-Y.; Sun, Y.; Han, B.-S.; Duan, Y.-R. Preparation of a Thermosensitive Gel Composed of a mPEG-PLGA-PLL-cRGD Nanodrug Delivery System for Pancreatic Tumor Therapy. *ACS Appl. Mater. Interfaces* **2015**, *7*, 20530–20537. [CrossRef]
48. Kulsirirat, T.; Sathirakul, K.; Kamei, N.; Takeda-Morishita, M. The in vitro and in vivo study of novel formulation of andrographolide PLGA nanoparticle embedded into gelatin-based hydrogel to prolong delivery and extend residence time in joint. *Int. J. Pharm.* **2021**, *602*, 120618. [CrossRef]
49. Saygili, E.; Kaya, E.; Ilhan-Ayisigi, E.; Saglam-Metiner, P.; Alarcin, E.; Kazan, A.; Girgic, E.; Kim, Y.-W.; Gunes, K.; Eren-Ozcan, G.G.; et al. An alginate-poly(acrylamide) hydrogel with TGF-β3 loaded nanoparticles for cartilage repair: Biodegradability, biocompatibility and protein adsorption. *Int. J. Biol. Macromol.* **2021**, *172*, 381–393. [CrossRef]
50. Gong, Y.; Wang, Y.; Qu, Q.; Hou, Z.; Guo, T.; Xu, Y.; Qing, R.; Deng, J.; Wang, B.; Hao, S. Nanoparticle encapsulated core-shell hydrogel for on-site BMSCs delivery protects from iron overload and enhances functional recovery. *J. Control. Release* **2020**, *320*, 381–391. [CrossRef]
51. Ai, A.; Behforouz, A.; Ehterami, A.; Sadeghvaziri, N.; Jalali, S.; Farzamfar, S.; Yousefbeigi, A.; Ai, A.; Goodarzi, A.; Salehi, M.; et al. Sciatic nerve regeneration with collagen type I hydrogel containing chitosan nanoparticle loaded by insulin. *Int. J. Polym. Mater. Polym. Biomater.* **2018**, *68*, 1133–1141. [CrossRef]
52. Mahya, S.; Ai, J.; Shojae, S.; Khonakdar, H.A.; Darbemamieh, G.; Shirian, S. Berberine loaded chitosan nanoparticles encapsulated in polysaccharide-based hydrogel for the repair of spinal cord. *Int. J. Biol. Macromol.* **2021**, *182*, 82–90. [CrossRef]
53. Gan, D.; Xing, W.; Jiang, L.; Fang, J.; Zhao, C.; Ren, F.; Fang, L.; Wang, K.; Lu, X. Plant-inspired adhesive and tough hydrogel based on Ag-Lignin nanoparticles-triggered dynamic redox catechol chemistry. *Nat. Commun.* **2019**, *10*, 1–10. [CrossRef] [PubMed]
54. Ramalho, M.J.; Loureiro, J.A.; Gomes, B.; Frasco, M.F.; Coelho, M.A.N.; Pereira, M.C. PLGA nanoparticles for calcitriol delivery. In Proceedings of the 2015 IEEE 4th Portuguese Meeting on Bioengineering, ENBENG, Porto, Portugal, 26–28 February 2015.
55. Fonseca-Gomes, J.; Loureiro, J.A.; Tanqueiro, S.R.; Mouro, F.M.; Ruivo, P.; Carvalho, T.; Sebastião, A.M.; Diógenes, M.J.; Pereira, M.C. In vivo bio-distribution and toxicity evaluation of polymeric and lipid-based nanoparticles: A potential approach for chronic diseases treatment. *Int. J. Nanomed.* **2020**, *15*, 8609. [CrossRef]
56. Spang, M.T.; Christman, K.L. Extracellular matrix hydrogel therapies: In vivo applications and development. *Acta Biomater.* **2018**, *68*, 1–14. [CrossRef] [PubMed]
57. Gaudana, R.; Ananthula, H.K.; Parenky, A.; Mitra, A.K. Ocular drug delivery. *AAPS J.* **2010**, *12*, 348–360. [CrossRef] [PubMed]
58. Cheng, Y.-H.; Ko, Y.-C.; Chang, Y.-F.; Huang, S.-H.; Liu, C.J.-L. Thermosensitive chitosan-gelatin-based hydrogel containing curcumin-loaded nanoparticles and latanoprost as a dual-drug delivery system for glaucoma treatment. *Exp. Eye Res.* **2018**, *179*, 179–187. [CrossRef] [PubMed]
59. Yang, H.; Tyagi, P.; Kadam, R.S.; Holden, C.A.; Kompella, U.B. Hybrid Dendrimer Hydrogel/PLGA Nanoparticle Platform Sustains Drug Delivery for One Week and Antiglaucoma Effects for Four Days Following One-Time Topical Administration. *ACS Nano* **2012**, *6*, 7595–7606. [CrossRef]

60. Cheng, Y.-H.; Chang, Y.-F.; Ko, Y.-C.; Liu, C.J.-L. Development of a dual delivery of levofloxacin and prednisolone acetate via PLGA nanoparticles/thermosensitive chitosan-based hydrogel for postoperative management: An in-vitro and ex-vivo study. *Int. J. Biol. Macromol.* **2021**, *180*, 365–374. [CrossRef]
61. Abrego, G.; Alvarado, H.; Souto, E.B.; Guevara, B.; Bellowa, L.H.; Parra, A.; Calpena, A.C.; Garcia, M.L. Biopharmaceutical profile of pranoprofen-loaded PLGA nanoparticles containing hydrogels for ocular administration. *Eur. J. Pharm. Biopharm.* **2015**, *95*, 261–270. [CrossRef]
62. Alruwaili, N.K.; Zafar, A.; Imam, S.S.; Alharbi, K.S.; Alotaibi, N.H.; Alshehri, S.; Alhakamy, N.A.; Alzarea, A.I.; Afzal, M.; Elmowafy, M. Stimulus Responsive Ocular Gentamycin-Ferrying Chitosan Nanoparticles Hydrogel: Formulation Optimization, Ocular Safety and Antibacterial Assessment. *Int. J. Nanomed.* **2020**, *15*, 4717–4737. [CrossRef] [PubMed]
63. Fabiano, A.; Piras, A.M.; Guazzelli, L.; Storti, B.; Bizzarri, R.; Zambito, Y. Impact of Different Mucoadhesive Polymeric Nanoparticles Loaded in Thermosensitive Hydrogels on Transcorneal Administration of 5-Fluorouracil. *Pharmaceutics* **2019**, *11*, 623. [CrossRef] [PubMed]
64. Karpecki, P.; Paterno, M.R.; Comstock, T.L. Limitations of Current Antibiotics for the Treatment of Bacterial Conjunctivitis. *Optom. Vis. Sci.* **2010**, *87*, 908–919. [CrossRef] [PubMed]
65. Khan, D.; Qindeel, M.; Ahmed, N.; Asad, M.I.; Shah, K.U.; Rehman, A.U. Development of an intelligent, stimuli-responsive transdermal system for efficient delivery of Ibuprofen against rheumatoid arthritis. *Int. J. Pharm.* **2021**, *610*, 121242. [CrossRef]
66. Shafique, M.; Sohail, M.; Minhas, M.U.; Khaliq, T.; Kousar, M.; Khan, S.; Hussain, Z.; Mahmood, A.; Abbasi, M.; Aziz, H.C.; et al. Bio-functional hydrogel membranes loaded with chitosan nanoparticles for accelerated wound healing. *Int. J. Biol. Macromol.* **2020**, *170*, 207–221. [CrossRef]
67. Wang, Z.; Liu, X.; Duan, Y.; Huang, Y. Nanoparticle-Hydrogel Systems Containing Platensimycin for Local Treatment of Methicillin-Resistant Staphylococcus aureus Infection. *Mol. Pharm.* **2021**, *18*, 4099–4110. [CrossRef]
68. Saleh, B.; Dhaliwal, H.K.; Portillo-Lara, R.; Shirzaei Sani, E.; Abdi, R.; Amiji, M.M.; Annabi, N. Local Immunomodulation Using an Adhesive Hydrogel Loaded with miRNA-Laden Nanoparticles Promotes Wound Healing. *Small* **2019**, *15*, 1902232. [CrossRef] [PubMed]
69. Negut, I.; Grumezescu, V.; Grumezescu, A.M. Treatment Strategies for Infected Wounds. *Molecules* **2018**, *23*, 2392. [CrossRef]
70. Onaciu, A.; Munteanu, R.A.; Moldovan, C.S.; Berindan-Neagoe, I. Hydrogels Based Drug Delivery Synthesis, Characterization and Administration. *Pharmaceutics* **2019**, *11*, 432. [CrossRef]
71. Jacob, S.; Nair, A.; Shah, J.; Sreeharsha, N.; Gupta, S.; Shinu, P. Emerging Role of Hydrogels in Drug Delivery Systems, Tissue Engineering and Wound Management. *Pharmaceutics* **2021**, *13*, 357. [CrossRef]
72. Aly, U.F.; Aboutaleb, H.A.; Abdellatif, A.A.; Tolba, N.S. Formulation and evaluation of simvastatin polymeric nanoparticles loaded in hydrogel for optimum wound healing purpose. *Drug Des. Dev. Ther.* **2019**, *13*, 1567–1580. [CrossRef]
73. Zeng, Q.; Qian, Y.; Huang, Y.; Ding, F.; Qi, X.; Shen, J. Polydopamine nanoparticle-dotted food gum hydrogel with excellent antibacterial activity and rapid shape adaptability for accelerated bacteria-infected wound healing. *Bioact. Mater.* **2021**, *6*, 2647–2657. [CrossRef] [PubMed]
74. Zhang, Y.; Zhang, J.; Chen, M.; Gong, H.; Thamphiwatana, S.; Eckmann, L.; Gao, W.; Zhang, L. A Bioadhesive Nanoparticle–Hydrogel Hybrid System for Localized Antimicrobial Drug Delivery. *ACS Appl. Mater. Interfaces* **2016**, *8*, 18367–18374. [CrossRef] [PubMed]
75. Mao, K.-L.; Fan, Z.-L.; Yuan, J.-D.; Chen, P.-P.; Yang, J.-J.; Xu, J.; ZhuGe, D.-L.; Jin, B.-H.; Zhu, Q.-Y.; Shen, B.-X.; et al. Skin-penetrating polymeric nanoparticles incorporated in silk fibroin hydrogel for topical delivery of curcumin to improve its therapeutic effect on psoriasis mouse model. *Colloids Surf. B Biointerfaces* **2017**, *160*, 704–714. [CrossRef] [PubMed]
76. Thomas, W.; Trintchina, E.; Forero-Shelton, M.; Vogel, V.; Sokurenko, E.V. Bacterial Adhesion to Target Cells Enhanced by Shear Force. *Cell* **2002**, *109*, 913–923. [CrossRef]
77. dos Santos, A.M.; Carvalho, S.G.; Araujo, V.H.S.; Carvalho, G.C.; Gremião, M.P.D.; Chorilli, M. Recent advances in hydrogels as strategy for drug delivery intended to vaginal infections. *Int. J. Pharm.* **2020**, *590*, 119867. [CrossRef] [PubMed]
78. Zimmermann, E.S.; Ferreira, L.M.; Denardi, L.B.; Sari, M.H.M.; Cervi, V.F.; Nogueira, C.W.; Alves, S.H.; Cruz, L. Mucoadhesive gellan gum hydrogel containing diphenyl diselenide-loaded nanocapsules presents improved anti-candida action in a mouse model of vulvovaginal candidiasis. *Eur. J. Pharm. Sci.* **2021**, *167*, 106011. [CrossRef]
79. Islam, S.; Bhuiyan, M.A.R.; Islam, M.N. Chitin and Chitosan: Structure, Properties and Applications in Biomedical Engineering. *J. Polym. Environ.* **2016**, *25*, 854–866. [CrossRef]
80. Nascimento, M.H.M.; Franco, M.K.K.D.; Yokaichyia, F.; de Paula, E.; Lombello, C.B.; de Araujo, D.R. Hyaluronic acid in Pluronic F-127/F-108 hydrogels for postoperative pain in arthroplasties: Influence on physico-chemical properties and structural requirements for sustained drug-release. *Int. J. Biol. Macromol.* **2018**, *111*, 1245–1254. [CrossRef]
81. Miele, E.; Spinelli, G.P.; Miele, E.; Tomao, F.; Tomao, S. Albumin-bound formulation of paclitaxel (Abraxane ABI-007) in the treatment of breast cancer. *Int. J. Nanomed.* **2009**, *4*, 99–105.
82. Cascone, S.; Lamberti, G. Hydrogel-based commercial products for biomedical applications: A review. *Int. J. Pharm.* **2019**, *573*, 118803. [CrossRef] [PubMed]

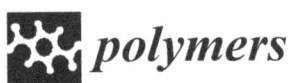

polymers

MDPI

Review

Biofunctionalization and Applications of Polymeric Nanofibers in Tissue Engineering and Regenerative Medicine

Prasanna Phutane [1,*], Darshan Telange [1], Surendra Agrawal [2], Mahendra Gunde [3], Kunal Kotkar [4] and Anil Pethe [1]

1 Department of Pharmaceutics, Datta Meghe Institute of Higher Education and Research, Datta Meghe College of Pharmacy, Wardha 442004, MH, India
2 Department of Pharmaceutical Chemistry, Datta Meghe Institute of Higher Education and Research, Datta Meghe College of Pharmacy, Wardha 442004, MH, India
3 Department of Pharmacognosy, Datta Meghe Institute of Higher Education and Research, Datta Meghe College of Pharmacy, Wardha 442004, MH, India
4 Department of Pharmaceutical Quality Assurance, R.C. Patel Institute of Pharmaceutical Education and Research, Shirpur 425405, MH, India
* Correspondence: prasannaphutane@gmail.com

Abstract: The limited ability of most human tissues to regenerate has necessitated the interventions namely autograft and allograft, both of which carry the limitations of its own. An alternative to such interventions could be the capability to regenerate the tissue in vivo.Regeneration of tissue using the innate capacity of the cells to regenerate is studied under the discipline of tissue engineering and regenerative medicine (TERM). Besides the cells and growth-controlling bioactives, scaffolds play the central role in TERM which is analogous to the role performed by extracellular matrix (ECM) in the vivo. Mimicking the structure of ECM at the nanoscale is one of the critical attributes demonstrated by nanofibers. This unique feature and its customizable structure to befit different types of tissues make nanofibers a competent candidate for tissue engineering. This review discusses broad range of natural and synthetic biodegradable polymers employed to construct nanofibers as well as biofunctionalization of polymers to improve cellular interaction and tissue integration. Amongst the diverse ways to fabricate nanofibers, electrospinning has been discussed in detail along with advances in this technique. Review also presents a discourse on application of nanofibers for a range of tissues, namely neural, vascular, cartilage, bone, dermal and cardiac.

Keywords: nanofiber; scaffold; electrospinning; tissue engineering; regenerative medicine

Citation: Phutane, P.; Telange, D.; Agrawal, S.; Gunde, M.; Kotkar, K.; Pethe, A. Biofunctionalization and Applications of Polymeric Nanofibers in Tissue Engineering and Regenerative Medicine. *Polymers* **2023**, *15*, 1202. https://doi.org/10.3390/polym15051202

Academic Editors: Stephanie Andrade, Joana A. Loureiro and Maria João Ramalho

Received: 5 December 2022
Revised: 7 February 2023
Accepted: 9 February 2023
Published: 27 February 2023

1. Introduction

Humans possess the finite ability to re-grow or regenerate tissues, organs or any part of the body after its resection, except some organs such as liver and lungs possess the good capability to regenerate. Bones and smooth muscles have limited ability to regenerate, while others that scarcely regenerate include the cardiac muscle, lens of the eye, skeletal muscle and nerves. Injury to or resection of many such tissues creates the problem of loss of functionality and unpleasant appearance. Autograft and allograft are the currently available treatments for injury or trauma caused to the tissue. But they carry multiple limitations with them, proposing the researchers to look for better alternatives. The tissues having the capability to regenerate themselves will be the best possible answer, which is evidently not possible in humans rightnow.However, human tissues can be assisted for such regeneration. Such regeneration of tissues applying the principles of life sciences and engineering and using the innate capacity of the cells, is studied under the discipline of tissue engineering and regenerative medicine (TERM).

The key components required for engineering a tissue are regenerative cells, scaffolds and growth-controlling bioactive molecules. These are commonly called as the tissue engineering (TE) triad [1] as shown in Figure 1. Scaffolds have the central role to perform in TERM which is analogous to the role performed by Extracellular Matrix (ECM) in the biological tissues. Scaffold, alike ECM, render structural reinforcement and physical milieu for cells to adhere, multiply, differentiate and migrate. But the scientists are having as enormous variety of choices in scaffolds for TE application. Moreover, mimicking the structure of the ECM at the nanoscale in fabricated scaffold was one of the great limitations in the research area of TE. This limitation was concluded to a great extent by the development of nanofibers. Architecture of ECM consisting of an interwoven fibrous structure in nanoscale range made of an array of multidomain macromolecules inspires the fabrication of scaffolds. Emulation of the structure of ECM to invoke the biological function of the ECM has been one of the central area of research in TE [2].

Nanofibers based systems have been explored for a wide variety of biological [3] as well as non-biological applications [4,5] because of its highly controllable properties. Its biological applications include burn and wound dressing [6,7], facemasks [8], tissue regeneration [9], osteoporosis treatment [10] and drug delivery [11].

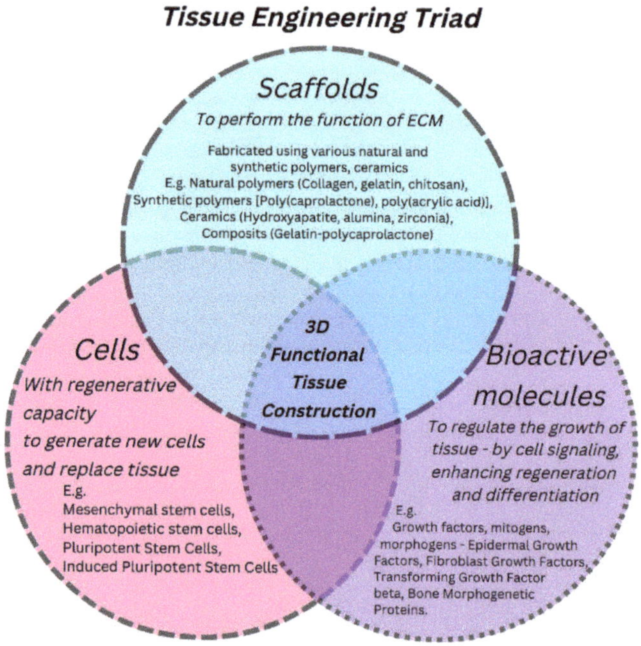

Figure 1. Tissue Engineering Triad.

2. Nanofibers Based Scaffolds in Tissue Engineering

Nanofibers possess some unique features which make it competent candidate for TE application. Some of them are discussed in following paragraphs.

High surface area to volume quotient and consequent high surface energy of nanofibers in comparison to bulk materials results in better attachment of cells, proteins and drug molecules [10]. In comparison to other special kinds of tissue scaffolds such as foam and gel films, fabrication of nanofibers furnishesan opportunity of achieving higher surface area for an equal volume.

High porosity of the scaffolds is preferred to allow for migration, attachment and proliferation of cells, for circulation of oxygen, nutrients and disposal of metabolic byproducts. But an inverse relation has been observed between porosity and tensile strength [12]. Thus, it became difficult for the researchers to achieve mechanical strength, while aiming at high porosity. But nanofibers provide the sufficient mechanical strength, while attaining high porosity. It also showed to possess organized porous architectures and porosity achieved in nanofiber based scaffolds was reported to be more than 90 percent [13]. Nanofibers resemble the porous arrangement of ECM, hence they are favorable for tissue regeneration [10].

Nanofiber commonly demonstrates high aspect ratio, which is a ratio of the length to width of the fiber [9]. High aspect ratio is associated with good tensile strength of the fibrous matrix, due to lengthier nanofibers which impart overall strength to the fibrous matrix [14].

Nanofibers are a peculiar class of material providing biomimetic environment at nanometric level, suitable surface properties and three-dimensional framework on the micrometric level and mechanical performance and physiological acceptability on the macrometric scale [15] Furthermore, many ex-vivo studies on scaffolds, facemasks and wound dressings of nanofibrous origin have established their superiority over their counterparts composed of the same material at micro or macrometric scale [16].

As the attributes of the nanofibers are very sensitive to the properties of the polymer and parameters used in manufacturing techniques, these attributes of the nanofibers can be regulated according to the required application. Thus, flexibility of nanofiber assemblies can be tailored to great magnitude. Some other manageable attributes are diameter of the fiber, flexibility, directional properties, etc. Morphology of nanofibers can be customized to befit various types of tissues or to encapsulate biologically active molecules.In addition, a great range of polymers can be electrospun to serve different applications.

These noteworthy properties make nanofibers idyllic candidate for a broad variety of biomedical and healthcare applications, including TERM.

3. Bio-Degradable Polymers

The selection of material is a central consideration in regulating the utility of the nanofibers for TERM. Biocompatibility is the foremost feature of the polymer to consider for its use in the biomedical applications, followed by biodegradability. There are several polymers which are biocompatible but non-biodegradable. Polymers can be generally categorized based on the susceptibility of their chemical backbone to degradation on exposure to water by the process of hydrolysis, as non-biodegradable and biodegradable. For tissues namely bone, tendons, cartilages, ligaments, and blood vessels, mechanical character is a prime factor. For tissue regeneration of such tissues, non-degradable polymers find a use. Non-biodegradable polymers have also been tried in guided TE application such as directing re-growth of tissues. They have been used for orbital reconstruction, facial reconstruction and rhinoplasty [17]. They also find use in ex-vivo guidance of tissue growth. Some examples of non-degradable polymers used are poly(tetrafluroethylene) (PTFE), extended-PTFE, polyurethane, poly(ethylene terephthalate), poly(ethersulfones), etc. Similarly, gels made from non-degradable polymers such as poly(ethyleneglycol), poly(ethylene oxide) (PEO), poly(vinyl alcohol) and Pluronic (block co-polymers of PEO and poly(propyleneoxide)) have been explored in the domain of engineering of scaffolds.

Another category of biomaterials called tissue adhesives possess the adhesive properties and help to stick the non-adhesive scaffold devices in vivo. Examples of such tissue adhesives are fibrin, albumin and cyanoacrylates. But such tissue adhesives can not be utilized as the proper scaffold to regenerate tissue owing to many of its limitations. To conquer these limitations, adhesive tissue engineering scaffolds (ATESs) have been developed. These ATESs can be secured at the site in vivo without the need of gluing or suturing [18].

With the evolution of TE discipline, the focus has been shifted more on 'functional TE'. There has been a convincing assertion for the use of degradable polymers to fabricate hybrid tissue equivalents. With the overarching need to dissolve the synthetically produced tissue equivalent in situ, in progressive way and in tandem with the process of tissue regeneration, biodegradable polymers came in the spotlight. This review will emphasize primarily on the biodegradable polymers, as it forms the large chunk of the research on TE.

Numerous biodegradable polymers, both natural as well as synthetic, have been utilized in the production of scaffolds based on nanofiber with diverse morphological features. Each class of polymer has its unique set of attributes some of which are expedient for fabrication of scaffolds whereas others are detrimental to overall performance of the scaffold. Table 1 enumerates some of the strengths as well as weaknesses of polymer of natural and synthetic origin for its application in TE. To utilize the strengths of polymers of both natural and synthetic origin, scientists have manufactured composite scaffolds having mechanical properties and bioactivity suitable for regeneration of tissues. Table 2 enlists polymers and polymer composites used to electrospun nanofibers for different TE applications with the results of in vitro/in vivo evaluationtests.

Table 1. Advantages and disadvantages of natural and synthetic polymers.

	Advantages	Disadvantages
Natural polymers	• Inherently bioactive • Possess cell-interactive groups on their backbones • Offer better cell attachment, growth, multiplication and differentiation • Chemically benign degradation products • Elicit low immune response	• Difficult processing • Low cost effectiveness • Poor mechanical properties • Precarious outcome due to batch-to-batch variations • Insufficient mechanical strength • Hydrophilicity • Need of crosslinking to improve strength
Synthetic polymers	• High flexibility in the processing • More economical • Tunable mechanical properties • Higher mechanical strength • Better structural stability	• Lacking bioactivity • May produce intense immune response • Necessitate more modifications compared to natural polymers to impart bioactivity

Table 2. Natural and synthetic polymers used for tissue engineering applications.

Polymer/Polymer Composite	Novel Step	Electrospinning Technique	Application	Result
Collagen nanofibers [19]	Electrospun nanofibers were treated with catecholamines and calcium choride followed by exposure to ammonium carbonate to enable the formation of in situ crosslinked collagen-$CaCO_3$ composite scaffolds.	Electrospinning	Bone tissue engineering	Inclusion of Ca^{2+} into catecholamines containing collagen and ensuing mineralization improved the elastic features, mechanical strength and stiffness. Human Fetal Osteoblasts demonstrated enhanced cell proliferation and osteogenic differentiation in the mineralized composite mats compared to pristine collagen mats.
Gelatin nanofibers [20]	Mild solvents have been utilised to preserve gelatin in a sol state at ambient temperature, for the electrospinning of nanofibers. A model protein reagent, (ALP) was embedded in the gelatin nanofibers to evaluate protein stability	Single nozzle electrospinning	Tissue engineering scaffolds	Mild neutral dipolar aprotic solvents, N,N-dimethylacetamide (DMA), N,N-dimethylformamide (DMF) and N-methyl-2-pyrrolidone (NMP), allowed gelatin to remain in sol state at room temperature. DMA, DMF and NMP conserved the alkaline phosphatase activity substantially, indicating their effectiveness for encapsulating protein reagents while preserving their activities. Swiss 3T3 fibroblasts grew well on the manufactured gelatin nanofibers.

Table 2. *Cont.*

Polymer/Polymer Composite	Novel Step	Electrospinning Technique	Application	Result
Sodium alginate/ polycaprolactone core-shell nanofibers [21]	Emulsion electrospinning of sodium alginate has been tried to fabricate nanofibers with core-shell morphology	Water-in-oil emulsion electrospinning	Tissue engineering scaffolds and controlled drug delivery	Increase in PCL concentration improved the loss and storage moduli and also increases the diameter of the manufactured fibers. Cytotoxicity assay using human dermalfibroblasts indicated no cytotoxicity of the manufactured core-shell nanofibers.
Chitosan/ hydroxyapatite (HA) nanofibers [22]	HA nanopowder was dispersed in chitosan solution to be electrospun to replicate the structure and composition of natural bone tissue. Cross-linking was carried out with exposure to the vapors of a glutaraldehyde	Blend electrospinning	Bone tissue engineering	Addition of HA caused statistically significant reduction in the average fiber diameter and an enhancement in Young's modulus and Ultimate Tensile Strength compared to chitosan nanofiber samples. High cell viability was observed for HA incorporated chitosan nanofibers.
PVA/Hyaluronic acid nanofibers [23]	Incorporated cellulose nanocrystals (CNCs) as nanofiller to improve mechanical properties of the nanofibers. L-arginine was loaded as wound healing accelerator.	Blend electrospinning	Dermal tissue engineering	Inclusion of CNCs into PVA/HA blend substantially augmented mechanical and swelling properties of nanofibers. PVA/HA/CNC/L -arginine nanofibers displayed excellent hemocompatibility, enhanced protein adsorption, remarkable proliferative and adhesive capability.
SF/kappa-carrageenan nanofibers [24]	kappa-carrageenan was blended with SF for electrospinning nanofibers to improve biological properties of SF based nanofibers and to mimic bone ECM structure, while genipin was used for crosslinking agent.	Blend electrospinning	Bone tissue engineering	Blending of kappa-carrageenan in nanofibrous matrix effectively moderated the hydrophobic nature of SF nanofibers, thus enhancing cell survival and proliferation. The scaffold was able to guide the osteogenic differentiation, stimulate mineralization and developement of bone tissue in vitro. Ultimate tensile strength and Young's modulus of the SF mats improved post-crosslinking with genipin.
Poly caprolactone (PCL) electrospun nanofiber [25]	PCL electrospun nanofibrous matrix was combined with hydrogels of polyethylene glycol diacrylate (PEGDA), sodium alginate (SA) and type I collagen (CG1) to fabricate three kinds of scaffolds. Composite scaffold were created using the layers of hydrogel and PCL nanofibers.	Electrospinning	Dermal tissue engineering	Cells were more capable of proliferating and differentiating in the CG1-PCL scaffold compared to PEGDA-PCL and SA-PCL. The mean number of cells proliferated was greater for the CG1-PCL scaffold in comparison to other scaffolds. CG1-PCL also has lower hydrophilicity and degradability compared to PEGDA-PCL and SA-PCL which makes it appropriate as a dermal equivalent.
Polyaniline-co-(polydopamine grafted-poly(D,L-lactide) [PANI-co-(PDA-g-PLA)] electrospun nanofibers [26]	PANI-co-PDA was manufactured using a single -step chemical oxidization approach. Later, D,L-lactide monomer was inserted onto PDA segment using a ring opening polymerization to create PANI-co-(PDA-g-PLA) terpolymer. PANI and PDA were incorporated to improve hydrophobicity and biological activity of PLA.Fabricated terpolymer was electrospun into nanofibers and a conductive nanofibrous matrix was fabricated.	Electrospinning	Bone tissue engineering	The surface wettability of the scaffold was found acceptable for a successful TE application. Manufactured scaffold demonstrated exceptional performance in terms of adhesion, migration and proliferation of the mouse osteoblast MC3T3-E1 cells, primarily because of excellent and accessible binding cites in the scaffold owing to presence of PDA and PLA chains, biocompatible nature of PANI-co-(PDA-g-PLA) nanofibers and communication between the cells via electrical conductive matrix.

Table 2. *Cont.*

Polymer/Polymer Composite	Novel Step	Electrospinning Technique	Application	Result
Polyglycolic acid/gelatin nanofibers [27]	Blend of Gelatin with PGA was electrospun into nanofibers. The polymer blend was utilised to enhance cell attachment, improve survival of the cells of the vasculature, namely endothelial and smooth muscle cells, and to impart appropriate biomechanical properties to the scaffold. Variable weight proportions of gelatin was tried to fabricate electrospun fibrous scaffolds.	Blend electrospinning	Vascular tissue engineering	Incorporation of gelatin substantially improved tensile strength and the Young's modulus of the fiber sheets. Electrospun fibers with PGA and 10 wt% and 30 wt% gelatin had tensile strength values approximating that of natural vein values.Fibers with PGA and 10 wt% gelatin showed enhanced endothelial cells density whilst PGA with 30 wt% gelatin increased smooth muscle cell density with enhanced adhesion and survival compared to other scaffold blends.
Polyvinyl alcohol (PVA) electrospun nanofibers [28]	Epidermal growth factor (EGF) and fibroblast growth factor (FGF) were included into PVA to be co-electrospun into nanofibers for the fabrication of wound dressing. Single, mix, multilayer electrospun nanofibers were fabricated.	Electrospinning	Dermal tissue engineering	Fiber diameter decreased, surface roughness decreased, wettability increased after incorporation of growth factors within the PVA Nanofibers. The GFs incorporation in PVA nanofibers induced cell proliferation and better cell attachemnt compared to PVA control sample. PVA-growth factors nanofibrous matrix demonstrated to be a better scaffold to heal burn-wounds in comparison to PVA only nanofiber.
Dipeptide polyphosphazene-polyester blend nanofibers [29]	Polymeric blend composed of poly[(glycine ethyl glycinato)$_1$ (phenylphenoxy)$_1$ phosphazene] (PPHOS) and poly(lactide-co-glycolide) (PLAGA) in a 25:75 weight ratio was chosen to fabricate the BLEND nanofi bers via electrospinning. Biomimetic scaffolds were fabricated with concentric orientation of fibers with an open central lumen to mimic bone marrow cavity, as well as the lamellar structure of bone.	Electrospinning	Bone tissue engineering	The tensile strength value for BLEND nanofi bers was 25% higher than the tensile strength of trabecular bone. BLEND nanofiber matrices assisted osteoblasts attachement and proliferation and demonstrated an enhanced phenotype expression compared to polyester nanofibers. Additionally, the 3D structure supported osteoblast infiltration and ECM secretion, bridging the spaces in concentric walls in scaffold during in vitro culture. Scaffolds showed similar lamellar ECM organization to that of native bone

In the following paragraphs, the natural and synthetic polymers employed in fabrication of scaffolds are discussed with their strengths and weaknesses.

3.1. Natural Polymers

Polymers of natural origin are biocompatible, biodegradable and show low immunogenicity. They offer the advantage of being similar and many a times identical with the ECM, thus elicits the favorable interaction with the cells. In addition, some possess antimicrobial and anti-inflammatory properties, thus buttressing the course of tissue repair and regeneration. Some of the limitations in fabrication of nanofibers using natural polymers are difficult processing, lowcost effectiveness, poor mechanical properties and precarious outcome. Natural polymers also show variations in its characteristics between different batches and different sources. Insufficient mechanical strength and greater water solubility are the limitations of most of the naturally derived biopolymers used to construct the scaffolds. These limitations are overcome by the way of crosslinking to preserve their architectural cohesion in aqueous medium and to increase mechanical toughness. The issue associated with the crosslinking is the cytotoxicity of the chemicals utilized to crosslink. Enzymatic as well as physical methods have been explored for the crosslinking of the biopolymers along with crosslinking using non-toxic and low-toxic chemicals [30]. But crosslinking using non-toxic chemicals demonstrate the low degree of crosslinking compared to crosslinking with glutaraldehyde and other common toxic chemicals. Some examples of extensively studied natural polymers to manufacture nanofibers for application in TE are collagen, gelatin, alginate, chitosan, hyaluronic acid and silk fibroin.

3.1.1. Collagen

Collagen is the most plentiful protein in humans and animals and is the prime ECM protein which imparts it structural integrity [31]. Thus, it can be inferred that nanofibers made out of collagen will most closely mimic the histological structure of native tissues. Other suitable attributes of the collagen are the induction of very low immunogenic response and its suitability for the regeneration of most body tissues [32]. One of the main deficiencies of the collagen is poor mechanical strength. Scaffolds made out of pure collagen displays inadequate resistance to water and collagenase which results in reduced rigidity to withstand handling while implanting the scaffold [33]. Mechanical strength can be improved via cross-linking, which imparts degradation resistance and increased strength. D-banding observed with the quaternary structure of native collagen type I is pivotal to the mechanical stability of native collagen. Such D-banding is lost in solubilization process during processing to construct scaffolds [34]. Thus, methods which could preserve or recreate D-banding needs to be explored, to overcome the limitation associated with constructs designed out of collagen. Mechanical features can also be improved by deciphering the origins of the unique mechanical attributes of the native collagen fibrils. The mechanical strength of collagen microfibrils originate from the hierarchical structure at the nanometer scale, which, upon application of stress, results in straightening of twisted molecules, followed by stretching at the axis and further molecular uncoiling. Such sequence of deformation mechanisms impart collagen fibrils its' strength, specifically its great extensibility, strain hardening, and toughness. Usage of pure collagen molecules in its primary structure can not provide the broad range of mechanical functionality necessary for physiological functioning of collagenous tissues [35]. The hierarchical structure of the biological collagen fibrils inspires the fabrication of scaffolds which reproduce dimensional aspects and functionality of the native ECM.

A study developed mineralized nanofibrous composite structure similar to bone with electrospun collagen containing catecholamines and Ca^{2+}. Divalent cation induced crosslinking of collagen nanofibers, thus providing constructs with mechanical strength. Further mineralization of construct ammonium carbonate resulted into scaffold with exceptional mechanical strength with Young's modulus nearing the thresholds of cancellous bone. The scaffolds showed excellent biocompatibility with human fetal osteoblast cells and osteogenic efficiency [19].

3.1.2. Gelatin

Gelatin is partially hydrolyzed form of collagen having shorter chains of amino acids. Even though gelatin is technically a form of collagen, gelatin is less expensive, more readily available, presents a reduced immunological risk, provides improved hydrophilicity and cell adhesion [36]. Dissimilarity in the chemical composition of these two proteins makes them each act very differently. Gelatin nanofibers have demonstrated to be efficacious scaffolds for TE application with good cell adhesion activity. Thermo-responsive property of gelatin aqueous solution is a crucial feature which causes its' reversible transformation from sol to gel when temperature lowered below its critical solution temperature [37]. But the same becomes the limitation due its gelation at ambient temperature range. In a study to manufacture electrospun gelatin nanofibers, numerous organic solvents have been screened for their potential to preserve gelatin in a sol state at ambient temperature. Fluorinated alcohols as well as acidic organic solvents are observed to impede gelatinizing at room temperature [20]. The thermo-responsive behavior of gelatin can also be used to achieve desirable viscosity of the gelatin solution for nanofiber making.

3.1.3. Alginate

Alginate, also known as algin or alganic acid, has been one of the materials of choice for TE. But the electrospinning technique has affixed a new dimension to this polymeric material. Alginate is a natural polysaccharides, extracted from the cell walls of brown algae [38]. Alginate possesses some exceptional properties, namely high biocompatibility, fairly low immune response, and unique gel-forming capacity. It also shows structural similarity to proteoglycans, which is a crucial element of the ECM [39]. One of the limitations of alginate is its inability to precisely interact with mammalian cells. Thus, the material must be adapted to support cell adhesion. One of the means is attaching of cell adhesive peptides with covalent bonding to the polysaccharide backbone of alginate [40]. Despite the potential of alginate nanofibers, electrospinning trials of pure alginate nanofibers have not been successful. Dense intra- and intermolecular hydrogen bonding in the alginate was reported to pose a challenge in its electrospinning. To produce continual running and uniform nanofibers from the electrospinning of pure alginate solutions whether aqueous or non-aqueous, is an arduous task [41]. But its' electrospinnability can be improved by mixing it with polymers like polyvinyl alcohol [42] and polyethylene oxide [43]. However, existence of impurities in manufactured scaffold and challenges in bulk production of alginate-based nanofibers are still unresolved challenges for alginate polymer.

In a study, sodium alginate/polycaprolactone core-shell nanofibers were prepared using emulsion electrospinning. Water in oil emulsion was prepared where sodium alginate aqueous solution formed the dispersed phase whereas polycaprolactone in chloroform formed the continuous phase. This core-shell nanofiber was developed to act as promising candidate for incorporating both hydrophilic and hydrophobic bioactive molecules for biomedical application [21].

3.1.4. Chitosan

Chitosan is another polysaccharide that has been extensively examined as a biomaterial for scaffold fabrication for tissue regeneration purpose. Chitosan nanofibers are quite commonly used in the area of TE on account of its' morphological and chemical analogy with natural ECM, thus being biocompatible and biodegradable. In addition, its antimicrobial [44], antiulcer [45] and antitumoral [46] properties has been reported in the literature. It is obtained from a deacetylation reaction of chitin. Some of the strengths of the chitosan are that, it can assume numerous conformations and it can be attached with a broad range of functional groups to meet specific applications [47]. Its cationic nature, becomes the reason of its significance from the biomedical application perspective [48]. Although, obtaining defect-free chitosan nanofibers still represents a major hurdle, this issue has been tackled by the use of several cosolvents and copolymers [49,50]. The presence of amine group in molecule makes it a weak base which adds another limitation of its insolubility at higher pH. Native chitosan also has relatively poor transfection efficiency and lack of some functionalities which are highly desirable for few TE applications. Therefore several chemical alteration techniques have been tried in order to subdue these weaknesses of chitosan. Graft copolymerization is most frequently used technique among others for chitosan [48].

Wang et al. fabricated composite nanofibrous membrane of chitosan and polyvinyl alcohol using electrospinning for application in wound healing. Antibiotic was loaded in nanofibers at different concentrations. These nanofibers were found to have more and larger nanobeads with increasing concentration of chitosan. The nanofibrous composite was observed to be promising candidate for skin tissue regeneration [50].

3.1.5. Hyaluronic Acid

Hyaluronic acid (HA) is a kind of non-protein glycosaminoglycan which is a large water loving, biodegradable and biocompatible molecule. Characters which make it peculiar biopolymer for application in TERM are its unique viscoelastic properties [51]. HA is another main component of ECM besides collagen [52]. Nevertheless, high viscosity and interfacial tension of HA aqueous solutions, even at low concentrations, make electrospinning a challenging task [53]. Furthermore, insufficient drying of nanofibers during electrospinning due to the robust water holding capability of HA may cause troublesome fusing of electrospun nanofibers on the collector [54]. Therefore, the exploration of a solvent system which will facilitate the electrospinning of HA nanofibers is essential.

Hussein et al. fabricated L-arginine loaded polyvinyl alcohol—HAelectrospun nanofibers for wound healing purpose. Polyvinyl alcohol was blended with HA to promote its electrospunability and citric acid was used as cross-linking agent to improve nanofibers' resistance against degradation in aqueous environment. Poor mechanical properties of the nanofibers were found to be significantly improved by incorporating cellulose nanocrystals as nanofiller. Developed nanofibers exhibited excellent heamocompatibilityand prominent wound healing effect [23].

3.1.6. Silk Fibroin

Silk fibroin (SF) is a unique natural protein obtained from silkworm silk. Considering many desirable physiochemical characteristics of SF i.e., excellent biocompatibility, biodegradability, resorbability, low immunogenicity, and tunable mechanical characteristics, it has been explored as a potential biopolymer for TE [55]. Silk fibroin carries the property of tailorable degradation rates providing the functional life to scaffold from hours to years. It also exhibits noteworthy mechanical properties when fabricated into different forms [56]. Manufactured scaffolds possess resistance against tensile and compressive forces and have mechanical performance analogues to biological tissues. Their outstanding mechanical properties includes high elongation at break, great strength and toughness [57]. SF has been suggested as one of the best biomaterials for skeletal tissue regeneration [58]. It also shows desirable permeation ability for the exchange of nutrients and wastes [59].

Electrospun SF/kappa-carrageenan nanofibrous membranes were developed by-Roshanfaret al. for bone regeneration purpose.Genipin was used as crosslinker which facilitated more crystalline and stable structure of SF. Blending of kappa-carrageenan in nanofibersefficiently moderated the hydrophobic nature of SF-based nanofibers, thus enhancing cell survival and proliferation. The scaffold was able to guide the differentiation towards osteogenic lineage, stimulate the mineralization and development of bone tissue in vitro [24].

3.2. *Synthetic Polymers*

Numerous polymers of synthetic origin have also been tested for the fabrication of nanofibers. The excellence of synthetic polymers which explains its use alone or in combination with natural polymer is due to their features such as its fitness to spinning, excellent mechanical strength and cost-efficiency [60]. Synthetic polymers that are broadly investigated in the fabrication of nanofibers for application in TERM are polycaprolactone (PCL), polyvinyl alcohol (PVA), polyethylene oxide (PEO), polylactic acid (PLA), polyglycolic acid (PGA), polyglycerol sebacate and polyurethanes. The characteristics of the individual polymers are decided by their respective composition and molecular architecture including arrangement of side chains. Biodegradability of the polymers is directed by the characteristics such as chain length, degree of branching and crystallinity [61].

3.2.1. Polycaprolactone

PCL is a highly endorsed synthetic biopolymer owing to its FDA approval. It is frequently studied biodegradable polymer which possesses properties such as adequate mechanical strength and tailorable hydrophobicity. Blends, copolymers and composites of PCL with other polymers can be manufactured to achieve desirable physiochemical and mechanical properties [61]. Hydrophobic nature of PCL reduces cell affinity towards PCL surface. Thus, lack of cell-scaffold interactions causes inadequate cells attachment, migration, growth and differentiation, and conclusively results into very slow tissue regeneration [62]. But interfacial characteristics of PCL nanofibers can be altered for TERM usability by making desirable surface alterations in addition to mixing with other polymers [63].

A study aimed at developing dermal equivalent scaffold, fabricated PCL electrospun nanofiber and assembled it with polyethylene glycol diacrylate, sodium alginate and type I collagen (CG1) to fabricate three kinds of dermal equivalent scaffolds. These three group of nanofiber matrices were analyzed for cell viability, adhesion and differentiation and rheological properties, which revealed that the combination of CG1 and PCL is the best suited as dermal equivalent and has potential to be used as graft for dermal regeneration [25].

3.2.2. Polylactic Acid

Wide application of PLA in TE is not only due to its peculiar cytocompatibility and biodegradability [64], but also by the virtue of its chirality. Enantiomers of lactic acid i.e., L- and D-lactic acid can be synthesized having different stereoregularities, which in turn governs the physical and chemical attributes of the polymers, like thermal and mechanical features as well as degradation aspects [65]. Biologically inert and hydrophobic nature of PLA leads to low cell adhesion and lower rate of degradation. Another drawback linked with the usage of PLA is acidic degradation products that causes inflammation at the site of implant [66] These shortcomings hinder PLA's application in tissue-regenerative treatments. Further research is needed to overcome mentioned drawbacks.

A terpolymer having aniline, dopamine and lactide was used to create conductive nanofibrous scaffold for bone tissue engineering. Adequate physicochemical characteristics such as mechanical, conductivity, electroactivity, wettability, and morphology, along withgood biological properties, made the nanofibers made from this terpolymer a budding candidate to manufacture scaffolds for TE applications [26].

3.2.3. Polyglycolic Acid

Apart from biocompatibility and biodegradability, PGA posses features such as predictable bioabsorption and hydrophilic nature [67]. For the electrospun PGA nanofibers, it has been observed that large surface area of nanofibers brings about speedy degradation and faster loss of strength [68]. Thus PGA happens to be wise choice when a scaffold is expected to be tough initially possessing high strength and elasticity but degrades at a faster rate for quick resorption. However, accompanying sharp increase in localized pH caused by high rate of degradation may induce unwanted tissue responses. Such undesirable tissue responses may precipitate if the region is lacking sufficient buffering capacity or enough means for the rapid elimination of metabolites [69]. Absence of a methyl group in molecular structure of PGA compared with the molecular structure of PLA, makes it more hydrophilic and demonstrate lower solubility in organic solvent. In case of PLA, presence of methyl group creates steric hindrance making it less labile to hydrolysis. Absence of such steric hindrance for PGA leads to faster degradation rate [70]. PGA and PLA are stiff materials which render them unsuitable polymers to fabricate matrices for engineering of soft tissues [68]. PGA is commonly copolymerized with PLA to form poly(lactic-co-glycolic acid) (PLGA.) PLGA is one of the extensively utilized biodegradable polymer on account of its adjustable mechanical characteristics and rate of degradation by varying the lactic acid to glycolic acid copolymer ratio [71]. A terpolymer of lactide, glycolide and caprolactone has been utilized to manufacture porous scaffold for TE purpose. This terpolymer has shown to maintain their dimensions, porous microstructure and mechanical strength for 6 weeks in phosphate buffered saline, even after topographical changes at the surface [72], but further exploration of this terpolymer is needed for application TE.

3.2.4. Polyvinyl Alcohol

PVA has been utilized for TE scaffold fabrication owing to its chemo-thermal stability, mechanical efficiency and its aqueous solubility along with excellent biocompatibility and biodegradability [73]. PVA based scaffolds are noted for maintaining mechanical integrity with ability to withstand high tensile stress, exhibiting good percent elongation as well as high flexibility [74]. PVA is obtained from the hydrolysis or alcoholysis of polyvinyl acetate, thus different grades with varied degrees of hydrolysis are available. PVA grades obtained from high degrees of hydrolysis demonstrate low solubility in water, thus offers high water resistance. One of the limitation of the PVA is its' hydrophilicity, and thus its' immediate dissolution on contact with water. This limitation necessitates modification of PVA fibers by chemical or physical crosslinking to enhance its mechanical performance and resistance to water [75]. Another limitation of PVA is poor cell adhesion owing to its low affinity to protein [76], which can be improved using techniques such as blending with macromolecules like chitosan, fibronectin, etc and surface chemical modification such as amination [77].

Asiriet al. fabricated multilayered PVA electrospun nanofibers with epidermal growth factor and fibroblast growth factor to act as biological wound dressing scaffolds. Incorporation of growth factors improved the wettabilty of the PVA nanofibers and stimulated cell adherence and proliferation. This multilayered scaffold showed wound reduction in one week and wound repair in 2–3 weeks, thus exhibiting the potential to be used as biological dressing scaffold [28].

3.2.5. Polyphosphazene

Polyphosphazene symbolizes next generation of biocompatible and biodegradable biomaterials as the excellent design pliability of polyphosphazenes enables the designing of tunable polymers. In addition, such polymers allows to be employed solely or as composites with other polymers to accommodate needs of the application [78]. Chemical groups added to the polyphosphazene backbone chiefly controls the physico-chemical attributes of the polymer [79]. Thus the degradation rate and mechanical stability are controllable with alterations in side groups attached to core molecule. In addition to being biodegradable, the polyphosphazene polymer degrades into products which are non-toxic. Moreover, degradation products do not alter the pH of surrounding tissue because of the buffering capacity of phosphates and ammonia produced during polyphosphazene degradation [80]. Its buffering ability have also been used for neutralization of the acidic byproducts originated from degradation of polymers such as PLGA [81]. Polyphosphazene-polyester blends are drawing attention for TE applications due to non-toxic and neutral pH degradation products along with their controllable degradation pattern [82]. In polyphosphazene, the main chains are flexible due to alternating nitrogen and phosphorus atoms, but it also causes the fiber to shrivel during electrospinning. Surmounting this drawback to get a mechanically sound fiber is a challenge. Many studies which tried to solve this problem aimed at altering molecular structure by addition of large side-groups [83], while others experimented with blend of polyphosphazene with more rigid polymers [82].

Deng et al. fabricated electrospun fibers from dipeptide polyphosphazene-polyetser blend to mimic collagen fibrils. 3D scaffold was designed with concentric alignment of nanofibers with an empty central lumen. These blend nanofibrous scaffolds were shown to support osteoblast adhesion and proliferation and demonstrated an enhanced phenotype expression compared to nanofibers fabricated out of polyester alone. The 3D structure also encourages ECM secretion, indicating its potential for bone regeneration [29].

4. Biofunctionalization of Polymers

As discussed in previous paragraphs, most of the natural polymers used to construct scaffolds retain some form of similarity with the ECM found in tissues, but these polymers lack the required attributes such as mechanical strength, adequate stability in vivo and elasticity for its application in TE. Thus, investigators have incorporated synthetic polymers for their favorable mechanical qualities such as strength and elasticity, along with other desirable features of hydrophobicity and slow degradation rate. But synthetic polymers are also ridden with many drawbacks such as inadequate cellular interaction and nonresponse toward tissue integration. These challenges linked with synthetic polymers are due to the structural differences at molecular level which leads to lack of cell surface recognition sites.

One of the way to get around this barrier is surface alteration with biomolecules, where the bulk properties of the polymer especially elasticity and its ability to withstand stress remain unaffected, although alterations in the surface confer necessary characteristics. Such superficial modifications favor an enhanced cellular adherence, causing a drastic improvement in cellular proliferation and supports faster integration of the implant in vivo [84].

Surface modification using biomolecules has remained one of the preferred methods for the advantages it provides in tissue regeneration. Such biofunctionalization involves immobilization of biomolecules on the polymer matrix surfaces to promote cell adhesion and proliferation. Preferred biosignal molecules used for immobilization are cell-growth-factor proteins, therapeutic proteins and cell-adhesion-factor protein [85,86]. Such biomolecules for immobilization includes growth factors, peptide sequences (RGD), natural ECM proteins (fibronectin, laminin, collagen), heparin, heparin sulfate binding peptides among others [87]. Besides providing structural backbone, the scaffolds modified with ECM components initiate cellular interactions which are decisive for cell attachment, growth and differentiation [88].

Numerous techniques have been worked out for physical or chemical immobilization of such protein molecules. These are grafting, polymer blending and chemically modifying the polymers. To comprehend about the biofunctionalization of polymers, is it necessary to be aware about the composition of the ECM.

ECM is a complex network comprised of a cluster of macromolecules organized according to tissue type. It is composed of two prime families of macromolecules: fibrous proteins and proteoglycans (PGs) [89]. Collagens, elastins, fibronectins and laminins are the fundamental fibrous ECM proteins [90]. Collagen is the principal structural element of the ECM and the most extensive fibrous protein forming the ECM. It makes up about 30% of the total protein weight in animals and perform an array of functions such as providing resistance to breaking under tension, controlling cell adhesion, assisting chemotaxis and directing development of tissues [91]. Collagen is accompanied by elastin, which is another essential ECM fibrous protein. Elastin confers recoiling property to those tissues which undergoes frequent stretching. Fibronectins are engaged in guiding the arrangement of ECM with an essential role in facilitating cell attachment. These proteins are associated together by proteoglycans and makes up the thin fibers of the ECM [92]. Proteoglycans (PGs) are constituted of glycosaminoglycan (GAG) chains linked to a core protein with covalent bonding. Proteoglycans perform an important function of signal transduction by binding various signal molecules and regulate many cellular processes, in addition to being a structural protein [93]. GAGs are highly water loving and adopt immensely extended conformations that lead to development of hydrogels. The matrices formed by GAGs are capable to withstand high compressive forces [90].

Modification of the polymers with ECM proteins and growth factors is a commonly followed strategy. Table 3 summarizes biofunctionalization with a range of bioactive molecules, methods used for biofunctionalization and outcome of biofunctionalization. Forthcoming paragraphs will review biofunctionalization with various molecules and the improvements achieved using such biofunctionalization.

Table 3. Functionalization with bioactive molecules and their applications in tissue engineering.

Bioactive Molecule	Method of Functionalization	Research/Study	Outcome of Biofunctionalization	Cells Used/Tissue to Regenerate
Collagen [94]	Remote plasma treatment followed by immobilization of collagen on the nanofibersurface	PCL nanofibers were electrospun and layered with collagen	Collagen coating improved hydrophilicity and increased cell proliferation compared to non-coated PCL nanofibers	Primary human dermal fibroblasts (HDFs)/ Dermal tissue
Collagen [95]	Coaxial electrospinning technique and by soaking the PCL matrix in collagen solution	PCL nanofibers were electrospun and coated with collagen using two techniques	Density of human dermal fibroblasts on collagen layered PCL nanofibers prepared using coaxial electrospinning increased linearly compared to roughly collagen coated and uncoated PCL nanofibers	Human dermal fibroblasts/Dermal tissue
Gelatin [96]	Air plasma treatment followed by covalent grafting of gelatin molecules	PCL nanofibers were electrospun and grafted with gelatin molecules	Viability and proliferation rate of fibroblast cells increased in biofunctionalized nanofibers compared to tissue culture polystyrene (TCPS)	Fibroblast cells/Tissue engineering
Fibronectin [97]	Three different approaches were used -protein surface entrapment, chemical functionalization and coaxial electrospinning	PCL nanofibers were electrospun and functionalized with fibronectin using three approaches	Improved cell adhesion and proliferation of bone murine stromal cells was observed for scaffolds functionalized using all the three approaches, but sample with the surface entrapment of fibronectin demonstrated better performance.	Bone murine stromal cells/ bone tissue
Fibronectin [98]	Immersing in fibronectin solution overnight.	PCL nanofibers were electrospun with radial alignment and coated with fibronectin	Improved cell adhesion, cell migration and helped in more uniform distribution of cells. Boosted the effect of topographic cues offered by the fiber alignment.	Dural fibroblast cells/dural tissue
RGD [99]	RGD peptide was conjugated on nanofibers using Polyethylene glycol as a spacer.	Polyurethane electrospun matrix was immobilized with RGD peptide.	Improved viability, promoted proliferation of cells in comparison with an unaltered surface.	Human umbilical vein endothelial cells/vascular tissue
RGD [100]	RGD functionalization via strain-promoted azide–alkyne cycloaddition.	PCL aligned nanofibers were electrospun and functionalized with RGD peptide.	RGD functionalization decreased muscular atrophy and hastened sensory recovery. Facilitated regeneration of sciatic nerve in animal model compared to non-functionalized nanofibers.	Rat sciatic nerve repair
RGD [101]	Chemical conjugation of RGD on nanofibers was carried out, after activation of carboxyl groups of polymer	Polybutylene adipate-*co*-terephthalate (PBAT)/gelatin elctrospun nanofibers were loaded with Doxycycline and modified using RGD	RGD functionalized PBAT/gelatin nanofibers showed notably improved wound closure and histopathological results with re-epithelialization and angiogenesis in animal model compared to the control groups.	Dermal wounds
Aspartic acid (ASP) and Glutamic acid (GLU) Templated Peptides [102]	Cold atmospheric plasma (CAP) was used to modify the nanofiber surface and to mediate the conjugation with peptides	PLGA nanofiberswere electrospun and conjugated with peptides	Peptide conjugation improved the osteoinductive capacity of nanofibers. ASP templated peptide conjugation to nanofibers increased the expression of key osteogenic markers and induced cell proliferation more than GLU templated peptide conjugated nanofibers.	Human bone marrow derived mesenchymal stem cells/bone tissue

Table 3. *Cont.*

Bioactive Molecule	Method of Functionalization	Research/Study	Outcome of Biofunctionalization	Cells Used/Tissue to Regenerate
Laminin [103]	Physical coating method and the chemical bonding method used for functionalization of the surfaceof the nanofiber	Slow-degrading silica nanofibers were electrospun and attached with Laminin on the surface	Nanofibers with covalently attached laminin showed significantly longer neurite extensions than those observed on unmodified nanofibers and nanofibers subjected to physical adsorption of laminin.	Rat pheochromocytoma cell line/neuron
Laminin [104]	covalent binding, physical adsorption or blended electrospinning procedures.	PLLA nanofibers were electrospun and modified with laminin.	Functionalized nanofibers were capable of enhancing axonal extensions. In comparison to covalent immobilized and physical adsorbed, blending for electrospinning of laminin and synthetic polymer is a simple and effective method to functionalize nanofibers	Rat pheochromocytoma cell-line PC12 cells/neurons
Laminin [105]	Functionalization with laminin usingcarbodiimide based crosslinking and physical adsorption method	Nanofibers were electrospun from the blends of poly(caprolactone) (PCL) and chitosan and modified with laminin	Number of cells attached and the rate of proliferation on the lamininc oated scaffolds were higher than those of pure PCL and PCL-chitosan scaffolds. Schwann Cell attachment and proliferation were significantly higher on PCL-chitosan scaffolds with crosslinked laminin than the PCL-chitosan nanofibrous matrices with adsorbed laminin.	Schwann Cell/nerve tissue
Avidin-biotin system [106]	Avidin immobilization on nanofibers	Poly(caprolactone-co-lactide)/Pluronic (PLCL/Pluronic) nanofibers were electrospun and modified with avidin. Adipose-derived stem cells (ADSCs) were modified with biotin.	Biotinylated ADSCs showed more rapid attachment onto avidin-treated nanofibrous matrices compared to normal ADSCs adherence on untreated matrices, and the difference of attached cell number between the two groups was notable. It also promoted cell spreading on nanofibrous matrices.	Adipose-derived stem cells (ADSCs)
Fibroblast Growth Factor-2 (FGF-2) [107]	FGF-2 was immobilized on the surface of the nanofibers through avidin-biotin covalent binding.	Gelatin nanofibers were electrospun, crosslinked using glutaraldehyde, and modified with FGF-2	FGF-2 immobilization led to proportionate increase in cell proliferation and adhesion.	Adipose derived stem cells
Insulin [108]	Insulin was bound to carboxylic moieties of the polymer backbone through a standard carbodiimide chemistry	PCL and cellulose acetate micro-nanofibers were electrospun and functionalized with insulin.	Enhanced expression of tendon phenotypic markers by Mesenchymal stem cells (MSCs) akin to observations from insulin supplemented media, indicatedconservation of insulin bioactivity upon functionalization.	MSCs/tendon
Insulin-like Growth Factor-1 (IGF-1) [109]	Physical adsorption of IGF-1 due to soaking into suspension of IGF-1 in PBS and shaking for 4 h	Graphene oxide (GO)-incorporated PLGAnanofibres were electrospun and functionalized with IGF-1	Survival, proliferation, and differentiation of neural stem cells (NSCs) was significantly increased. Higher survival rate of NSCs in the IGF-1 modifed nanofibers compared to unmodifed nanofibers was observed.	NSCs/nerve cells

Table 3. *Cont.*

Bioactive Molecule	Method of Functionalization	Research/Study	Outcome of Biofunctionalization	Cells Used/Tissue to Regenerate
Polydopamine assisted bromelain [110]	Soaking in solution of dopamine and bromelain, with continuoue stirring for 8 h. Dopamine-assisted co-deposition strategy was used.	PCL nanofibers were electrospun and immobilized with bromelain using polydopamine (PDA) to create bromelain-polydopamine-PCL (BrPDA-PCL) nanofibers	BrPDA-PCL fibers exhibited superior biocompatibility compared to PCL fibers PDA coating made scaffold hydrophilic, allowing for better cell attachment and spreading PDA and bromelain both showed anti-bacterial activity.	L929 fibroblast cells/wound healing
Poly norepinephrine (pNE) [111]	Soaking in norepinephrine solution for 15 h	PCLfibers were electrospun andcoated using mussel-inspired pNE.	pNE coating improved the ECM proteins accumulation on the fibers, which supported cell adhesion and proliferation of cells on PCL fibrous membranes.	Skeletal muscle cell line L6/skeletal muscles
pNE mediated collagen [112]	Soaking in norepinephrine solution 16 h, followed by soaking in collagen solution overnight.	Poly(lactic acid-co-caprolactone) (PLCL) nanofibers were electrospun and coated with poly norepinephrine, followed by collagen.	pNE coating assisted in collagen anchoring to improve cell adhesion and to immobilize nerve growth factor to advance differentiation to neurons. pNE–collagen coating was observed to be the better substrate for PC12 cells differentiation.	PC12 cells/neurons
Polyphenol [113]	Blend electrospinning	Polylactic acid/date palm polyphenol nanofibers were electrospun using blend electrospinning.	Both cell proliferation and cell viability were enhanced with increased polyphenol concentration within the scaffolds. Higher polyphenol content resulted into better cell migration	NIH/3T3 fibroblast cell/wound healing
Vascular endothelial growth factor (VEGF) [114]	Blend and co-axial electrospinning	PCL-gelatin nanofibers were electrospun and modified with VEGF.	Functionalization improved proliferation of mesenchymal stem cells, but no significant difference in proliferartion between nanofibers manufactured with both techniques was observed. Expression of cardiac specific proteins enhanced.	Human mesenchymal stem cells/myocardium
VEGF [115]	Covalent coupling to VEGF by forming stable amide bond	PCL nanofibers were electrospun and modified with VEGF.	Biological activity of immobilised VEGF was maintained and functionalised substrates demonstrated to induce a higher cell number compared to non-functionalised scaffolds.	Human umbilical vein endothelial cells
Epidermal growth factor (EGF) and fibroblast growth factor (FGF) [28]	Blend electrospinning	PVAnanofibers were electrospun and modified with EGF and FGF.	GFs incorporated PVA nanofibers induced cell proliferation andenhanced cell survival compared to PVA control sample In in-vivo study, PVA/EGF/FGF nanofibers demonstratedquick recovery of the wounds in contrast to that of only EGF or FGF nanofibers.	Human dermal fibroblasts/wound healing.

Collagen (type I) is the most copious extracellular protein and it exists in a nanorange fibrillar structure. Such fibrillar morphology has been demonstrated to be crucial for attachment of cells, their growth and differentiation [116]. Collagen is one of the most favored bioactive molecules used for coating, as it provides the biomimetic environment for cell life cycle. Duan et al. constructed PCL nanofibers using electrospinning and layered it with collagen to merge the desirable attributes of collagen and PCL. PCL possess superior mechanical characteristics, yet its hydrophobicity and poor cell affinity results into poor cell attachment and proliferation. Collagen was immobilized on PCL nanofibers with the aim to improve the cell affinity of nanofibers after surface modification using remote plasma treatment. This study indicated that collagen immobilization along with plasma treatment offered an significant enhancement in surface hydrophilicity and greatly improved the primary human dermal fibroblasts (HDF) attachment and growth compared with pristine material [94]. In another study, collagen coated PCL nanofibers were prepared using two different methods, using coaxial electrospinning technique to give core-shell structure and by soaking the PCL matrix in collagen solution to form a rough collagen coating over PCL nanofibers. Although both kind of collagen immobilization over PCL nanofibers favored cell proliferation, HDF density found more over the nanofibers with core-shell structure compared to simple collagen coating over nanofibers [95].

The inclusion of Gelatin in scaffolds enhances the characteristics such as cell attachment, cell growth and biomineralization. Coating of the polymer matrices using gelatin resulted into enhanced biocompatibility and mechanical performance [117]. Such coating with gelatin also suppresses the activation of the complement system and opsonization, thus reduces immunogenicity of other polymers in matrix [118]. The presence of gelatin improved cellular proliferation of mouse embryonic fibroblasts (MEF) in electrospun PCL nanofibers blended with gelatin and those coated with gelatin, but the highest improvement was observed for nanofibrous scaffolds prepared using blend of PCL and gelatin [119]. Safaeijavan et al. altered the surface of PCL nanofibers by gelatin grafting to enhance their compatibility with living medium. For grafting, PCL scaffolds were initially given air plasma treatment which adds carboxyl groups on polymer surface. Gelatin molecules were then covalently grafted on nanofiber, which inserted amine functional groups on the surface. Such grafting not only increased the hydrophilicity of the scaffold but also enabled the scaffold to hold fibroblast cells and support their survival and functioning [96].

Fibronectin is large adhesive glycoprotein of the ECM essential for cell functions such as adhesion, spreading and motility. In a study, the functionalization of PCL electrospun fibers with fibronectin was achieved using three different approaches—protein surface entrapment, chemical functionalization and coaxial electrospinning. Improved cell adhesion and proliferation of bone murine stromal cells was obtained for scaffolds functionalized using all the three approaches. But sample with the surface entrapment of fibronectin demonstrated better performance in terms of cell response, which indicated that surface entrapment was the best approach to attain efficient functionalization for electrospun fibers [97]. Xie et al. fabricated scaffolds of PCL nanofibers with radial alignment. Influence of fiber alignment and fibronectin surface layering on cell motility of fibroblasts was studied. It indicated that fibronectin coating was able to boost the effect of topographic cues offered by the fiber alignment on cell morphology. Even in the case of randomly aligned nanofibers coated withfibronectin, cell adherence and distribution were enhancedcompared to the unlayered sample [98].

One of the most frequently employed peptides is RGD (arginine-glycine-aspartic acid) which originates from fibronectin. RGD is the leading integrin-binding domain situated inside various ECM proteins such as fibrinogen, fibronectin, vitronectin, osteopontin, bone sialoprotein as well as in some laminins and collagens [120]. It not only regulates the endothelial cells adhesion, migration and proliferation but also can be utilized to preferentially focus on certain cell lines and bring out specific cell responses. The grafting of short peptide sequences like RGD has some benefits when compared to entire protein molecules, such as greater stability under sterilization processes, storage, heat application, pH alterations and against enzymatic degradation. Short peptides also has lower space requirement, which leads to a higher density packaging of the peptides [121]. But RGD is recognized by numerous integrins, thus acts as a nonspecific peptide [122]. Choi et al. developed electrospun nanofibrous matrix of polyurethane over which RGD peptides were immobilized to enhance affinity of endothelial cells. RGD-immobilized matrix exhibited improved viability of human umbilical vein endothelial cells in comparison with an unaltered surface, proving that immobilization of RGD peptide has benefitted cell proliferation [99]. Besides RGD, several other cell adhesion motifs have been recognized namely DGEA peptide from collagen, GREDVY, KQAGDV peptide from fibronectin, PHSRN, etc. [121]. Thus, the RGD sequence can not be considered as the "universal cell recognition motif", nevertheless it is one-of-a-kind given its broad distribution and usage.

Laminin (LM) is heterotrimeric glycoprotein having high molecular weight. It is an essential constituent of basement membrane lining many tissues. This glycoprotein is necessary for activities like cell attachment, survival, growth, mobility and specialization [123]. Junka et al. developed electrospun nanofibers for tissue regeneration in large-gap peripheral nerve injury. Nanofibrous scaffolds employed blends of PCL and chitosan. Functionalization of the scaffold surface with laminin was done by crosslinking and by using conventional adsorption method. Schwann cell attachment and proliferation rates were found to be significantly greater on laminin crosslinked to PCL-chitosan scaffolds in comparison to scaffolds adsorbed with laminin or scaffolds without laminin [105]. Incorporation of Laminin in scaffolds has been tried for the regeneration of many diverse tissues including intervertebral fibrocartilage, muscles, neurons and blood vessels [123].

The natural adhesion between the ECM and cells generally depends on the creation of integrin-interceded bonds between integrins in the cell membrane and adhesion proteins or motifs in ECM. Here, the presence of cell membrane integrin controls the efficiency of cell adhesion. However, avidin-biotin linkage is an extrinsic, integrin-independent, high affinity receptor-ligand complex. This avidin-biotin system can be utilized for improved seeding of the cells into scaffolds. Given approach is founded on the existence of multiple binding sites on avidin for biotin and the strong non-covalent interaction between them. In TERM applications, biopolymer matrices are conjugated with avidin and cell membranes are attached with biotin to enhance cell interaction with matrices. Pan et.al evaluated avidin-biotin technology with poly(caprolactone-co-lactide)/Pluronic (PLCL/Pluronic) nanofiber based scaffolds for improving cell adhesion. Nanofiber surface is coated with avidin, whereas cellular membrane is attached with biotin. This research showed the improved adhesion and proliferation of stem cells on nanofiber matrix with the aid of technique based on avidin-biotin complex [106].

After in vivo exposure of scaffold, fibronectin and vitronectin gets adsorbed on the surface of scaffold non-specifically. By virtue of such adhered ECM proteins, the cell-scaffold interaction improves. Such interactions are controlled by integrins, which are cell surface receptors principally involved in attachment of cells to ECM [86].

Another commonly exercised approach in TE to bring out cellular differentiation is the utilization of growth factors. Yet, the constraints linked with the use of growth factors, such as rapid blood clearance, large dose requirement and heavy price, have aroused the exploration of growth factor substitutes, including mimicking molecules. Insulin has been examined as a biochemical signal due to its structural alikeness with Insulin Growth Factor-1 (IGF-1) and similarity between their receptors [124]. Ramos et al. developed insulin functionalized scaffolds where insulin was immobilized on polycaprolactone—cellulose acetate electrospun fiber matrices. The cells incubated on insulin conjugated scaffolds presented a rise in tendon markers, indicating potential of its use for tendon repair and regeneration [108]. In another study, a significantly increase in collagen I and III was observed postsurgery where bioactive insulin-immobilized electrospun nanofiber matrices cultured with mesenchymal stem cells were sutured to transected Achilles tendons in animal model. Furthermore, these matrices promoted alignment of collagen fibrils in regenerated tendons [125].

Mussle inspired peptides have attracted significant attention to functionalize material surfaces because it caters a simple and flexible approach and eliminate the requirement of expensive or complex instruments and procedures. Mussle inspired chemistry is founded on catechol-effectuated molecular adhesion [126]. Polydopamine (PDA) is one of the mussle inspired molecule. Chen et al. successfully used PDA to mediate bromelain immobilization on electrospun PCL fibers. Purpose of such immobilization was to apply antibacterial, anti-inflammatory, anti-edematous activities of bromelein and its capability to hydrolyze necrotic tissues to augment rates of wound healing. Bromelain–polydopamine–polycaprolactone (BrPDA-PCL) fibers exhibited superior biocompatibility given the hydrophilicity of the PDA coating which provides a suitable surface for cell adhesion. BrPDA-PCL fibrous membrane was observed to be highly effective wound dressing. It exhibited antibacterial activity, in addition to assist both cellular adhesion and proliferation [110]. In another study, mussel-inspired polynorepinephrine (pNE) was used to coat PCL fibers to improve hydrophilic nature and cellular interaction of hydrophobic surfaces. pNE functionalization created suitable environment both in vitro and in vivo for skeletal muscle cell adhesion and proliferation [111]. pNE coating has been also been utilized to create bio-interface by applying smooth coating of pNE on electrospun Poly(lactic acid-co-caprolactone) fibers. Here, the catechol groups from pNE assisted in collagen anchoring to improve cell adhesion and to immobilize nerve growth factor to advance differentiation to neurons [112].Polyphenol is another biomolecule whose addition in nanofibrous scaffolds increases cell adhesion, proliferation and differentiation, along with exhibiting their antioxidant and antimicrobial activity. Many polyphenols such as curcumin, naringin, apigenin, icarrin have been studied for bone tissue regeneration, which indicates their prospective for use in TE [127].

Along with the improvement in cell adhesion and proliferation with adoption of biofunctionalization using different approaches as seen in earlier paragraphs, further improvement in tissue regeneration can be achieved with the use of various growth factors. The simultaneous deliverance of angiogenesis-related factors and other biomolecules by nanofibrous matrices has demonstrated to boost tissue repair and regeneration [128]. Angiogenesis is of a pivotal occurrence in tissue regeneration which is essential to carry out the functions such as delivery of oxygen, nutrients, growth factors, ligands and disposal of metabolic byproducts. Therefore, numerous bioactive molecules have been incorporated in biomaterials to impart angiogenic activity. Scaffold-based transfer of vascular endothelial growth factor (VEGF) and bone morphogenetic protein 2 (BMP2) is the commonly investigated combination to promote angiogenesis and osteogenesis owing to their respective pro-angiogenic and osteoinductive activities [129,130]. Kai et al. fabricated PCL-gelatin (PG) nanofibers in which VEGF was incorporated using two individual methods namely blending and co-axial electrospinning to induce the cardiac differentiation of cells. The VEGF incorporated nanofibers improved the cell growth and division of mesenchymal stem cells (MSCs), promoted cardiac differentiation of MSCs and helped in enhancing the translation of cardiac-specific proteins [114].

VEGF has reported to be angiogenic and promoted formation of natural bypasses in cases of myocardium infarction by promoting generation of neovasculature and dissolution of existing vasculature [131]. Many recent findings in TERM offer proof that surface immobilization of growth factors helps in induction of activity for prolonged duration. Guex et al. used electrospinning for fabricating PCL nanofibrous constructs and VEGF was covalently bound to it. On evaluation of its effect on cell division of endothelial cells in vitro, it was observed that number of endothelial cells were noticeably increased on VEGF-immobilized scaffolds in comparison to non-functionalized PCL scaffolds, suggesting that biological activity of immobilized VEGF was maintained [115].

Epidermal growth factor (EGF) induces growth, proliferation, differentiation as well as cell survival by binding with its membrane receptor [132] and is considered the frontrunner in advancement of wound healing [133]. EGF facilitate wound healing by improving epidermal and mesenchymal restoration, cell migration, proliferation and ECM regeneration [133]. PVA electrospun nanofibers were fabricated to act as biological wound dressing scaffolds by Asiri et al. EGF and fibroblast growth factor (FGF) were incorporated in the

PVA nanofibers which resulted in the improvement in wettability and surface roughness. Growth factor release from the PVA nanofibers resulted in stimulation of cell adhesion, proliferation and improvement in cell viability. In vivo evaluation showed that GFs added PVA nanofibers expedited the healing process in burn wound by boosting epithelialization and proliferation of dermal fibroblasts [28].

5. Fabrication Techniques of Nanofibers

Diverse ways has been explored to fabricate nanofibers, some of which are template synthesis, phase separation, self assembly, interfacial polymerization and electrospinning [134]. Apart from the selection of the material from the broad range of polymers for fabricating nanofibers, the management of nanofiber diameter is extremely decisive in biomedical applications, as it decides the surface area for cellular interactions. Alongside fiber diameter, other attributes namely fiber morphology, architecture and alignment are also the significant variables instrumental in deciding the cell-fiber interactions for biomedical applications [16,135]. Above mentioned are some of the parameters used in selection of nanofiber fabrication technique. Scalability to the commercial scale is another crucial factor to consider while selecting fabrication technique. Among the mentioned techniques, electrospinning is the extensively experimented nanofiber fabrication technique and it has offered the most encouraging outcomes for TERM applications. Nanofiber synthesis using other techniques for TE application has been studies on relatively limited basis.

In Phase separation technique, nano-fibrous matrices are prepared following a process that involves polymer dissolution, thermally induced gelation, exchange of solvent, freezing and freeze-drying [136]. Gelation is the decisive stage in this technique for the creation of fibrillar matrix. Gelation of polymer solution depends upon solvents used, polymer concentration and gelation temperature. Gelling temperature is another critical element influencing the porous structure of the fibrous matrices. Porosity up to 98.5% had been achieved using this technique [137]. Some of the strengths of phase separation technique are minimum requirement of sophisticated equipments, simplified procedure, ability of the method to produce fibers in nanorange and capability to construct fibrous scaffold matrix in the anatomical shape of the body part using mold.

Biomolecular self-assembly presents an easy way to manufacture functional nanomaterials. The self-assembly mechanisms of biomolecules are based on varied internal interactions, such as hydrogen bonding, electrostatic interactions, hydrophobic interactions, π–π stacking, ligand–receptor binding and DNA base pairing. In addition, self-assembly can be induced using external stimulations like making alterations in solution attributes such as pH, ion concentration and temperature, by addition of organic solvents or enzymes, and with the help of light [138]. Self assembly is a bottom up technique to manufacture nanofibers in which molecules tend to align themselves in specific patterns to generate nanofibers. The structure of the individual molecules taking part in self-assembly and the intermolecular forces involved in molecular interactions decides the morphology of the nanofibers. This technique is capable of producing fibres in nano range. Yet the drawbacks such as low productivity, arduous handling of the fiber dimensions, constricted choices of materials which can self-assemble and being a cumbersome process, this method is of least preference [134].

Template synthesis of nanofibers involves extrusion of polymer precursor solution from the nanoporous membrane into the solidifying solution under pressure. As soon as polymer solution touches solidifying solution, nanofibers are created. Nanofibrous membrane containing cylindrical pores is used as as a template/mold. Aligned nanofibers with different diameters can be fabricated using templates with different pore diameters [139]. Limitation of this technique is the formation of discontinuous fibers having variable diameters.

Interfacial polymerization is another method of generating nanofibers, which is mainly a polycondensation reaction happening at the interface between two kinds of monomers solubilised in two non-miscible solvents. On mixing two distinct phases containing monomers, polymerization happens at the interface of the dispersed phase and dispersion medium of emulsion. Homogeneous nucleated growth is the key determinant in this technique.

5.1. Electrospinning Method

Electrospinning has come out as one of the most rewarding techniques, considering its capability to manufacture fibers in the nanometer range, having morphology which is comparable to the ECM fibrous structure. It also provides control over the fiber diameter, its composition and the porosity of nanofiber meshes [136]. Furthermore, this technique is remarkably simple, robust, and versatile, thus making it the preferred choice for preparing nanofibers. Polymer can be selected from a wide range of materials amenable to electrospinning and this technique has been employed by now to manufacture nanofibers of more than 100 different kinds of polymers [140].This method is adept for scaling up to make production at commercial scale [141]. Since the attributes of electrospun nanofibers are highly manageable and can be customized to befit various tissues or to encapsulate different drugs, such fabricated nanofibers are highly resilient for use in different biomedical applications.

The electrostatic repulsion between polymer molecules help in overcoming surface tension of the polymer solution, drawing a jet from polymer solution drop and further stretching of the jet is the principle on which electrospinning is based. Such electrostatic repulsion between polymer molecules develops with the help of electrical potential difference applied between the electrodes. A symbolic electrospinning setup as shown in Figure 2 comprised of following essential components: a higher voltage potential current, a syringe with needle tip, a pump and a metal collector [142]. On the induction of voltage between the needle and the collector, charges begin to develop on the polymer molecules in the needle. The magnitude of charge in the polymer molecule determines the extent of electrostatic repulsion experienced by the molecules of polymer in polymer solution. Surface area of the polymer solution increases as the electrostatic repulsion between polymer molecules increase [16]. But to produce fibers in nanometer range, charges on the polymer molecules should be dense enough and at the same time it should not that high to cause the solution jet to split into droplets [143]. The electrostatic repulsion developed between polymer molecules opposes the cohesive forces between polymer molecules and led to the formation of Taylor cone. This Taylor cone then turns into charged jet, which stretches, thins out and finally collects on the metal surface. All along the travel of polymer solution from Taylor cone to the collector, solvent evaporation provides rigidity to the fiber. The solvent evaporation mechanism also influences the porosity of the fibers [16].

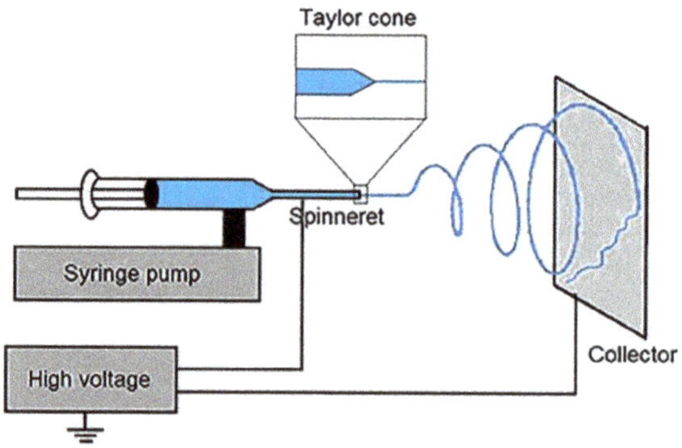

Figure 2. Schematic representation of typical electrospinning system [144] ©2016 Yawen Li and Therese Bou-Akl. Originally published in "Chapter 6 Electrospinning in Tissue Engineering"from Book "Electrospinning—Material, Techniques, and Biomedical Applications" under Creative Commons Attribution License (http://creativecommons.org/licenses/by/3.0, accessed on 4 December 2022). Available from: https://doi.org/10.5772/65836, accessed on 4 December 2022.

The parameters which influences the characteristics of the nanofibers obtained can be broadly divided into parameters related to electrospinning solution, parameters related to processand environmental parameters. Polymer solution concentration controls viscosity, surface tension as well as charge density and is thus the prime parameter deciding fiber diameter [145]. Other solution related parameters includes polymer molecular weight and distribution and solvent or mixture of solvent used. Process parameters include parameters related to equipment set up namely orifice diameter, voltage, solution feed rate, spinning distance, design of collector and motion of collectorthat affects the nanofiber attributes. Apart from these, the ambient factors like temperature and humidity, which are covered under environmental parameters affect the quality of nanofibers obtained. High humidity lengthens the solidification time required by the fibers after their ejection from the needle orifice, whereas low humidity assists in efficient removal of the solvents from the nanofibers. Humidity also influences the surface morphology of the fibers, with increase in humidity has been reported to cause an increase in number and size of the pores in the nanofibers [146,147]. Increase in ambient temperature has been reported to cause reduction in diameter of electrospun fibers [148]. Yet, the influence of the environmental factors on the properties of fibers should be studied on case by case basis.

Fridrikh et al. presented a model that predicts terminal jet diameter based on the availability of information on flow rate, applied voltage, and interfacial tension of the liquid. Fiber formation in electrospinning is a result of counterbalance between polymer adhesive forces due to surface tension and repulsive forces due similar charges on polymer molecules. On this relation, the prediction of fiber diameter has been based [149]. However, electrospinning is also linked with certain shortcomings such as wide range of electrospun fiber diameter, irregular alignment of fibers, and poor mechanical performance of the fiber matrices [135].

Electrospinning method can be classified according to the kind of nozzle used into three classes-single nozzle, coaxial and multi-nozzle electrospinning, whereas according to the kind and number of solutions or melts used, it can be categorized into blend electrospinning, co-axial and emulsion electrospinning as shown in Figure 3.

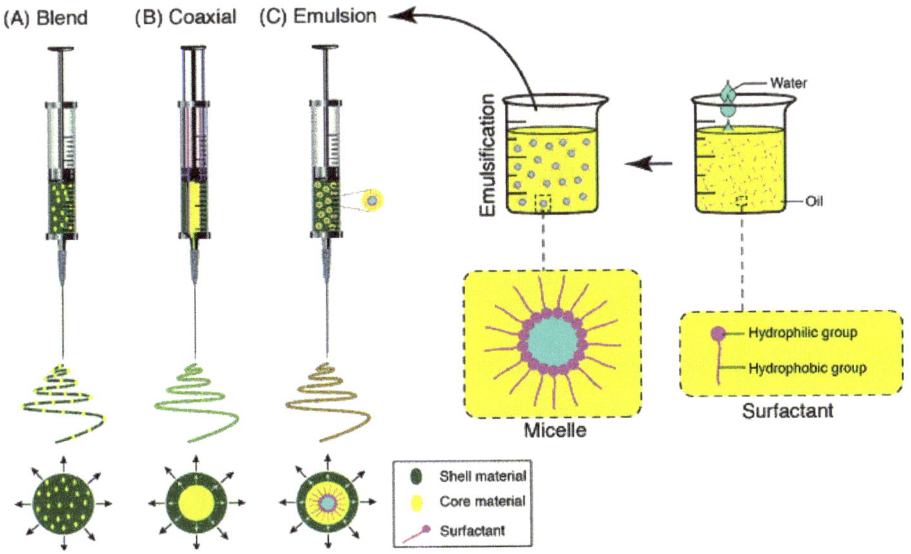

Figure 3. Electrospinning techniques—Blend, Co-axial and emulsion electrospinning. Reproduced from reference [150] with permission from Royal Society of Chemistry.

5.1.1. Single Nozzle Electrospinning

Single nozzle electrospinning uses a nozzle with single aperture through which polymer melt or solution outflows. Composite nanofibers can be manufactured using this mode of electrospinning. Compatible polymers can be employed for electrospinning of polymer blends. Moreover, solid nanoparticles can be embedded within electrospun fibers and liquid phase particles can be electrospun using emulsion electrospinning method. Single nozzle electrospinning method uses either blend of bioactive molecules in polymer dissolved in solvent, melt of polymers or emulsions for electrospinning.

5.1.2. Co-Axial Electrospinning

Co-axial electrospinning produces core-sheath fibers by physically separating them using two co-axial electrospinning needles and two solutions. Co-axial electrospinning uses the concurrent flow of different solutions through two co-axial capillaries to physically separate core and shell fibers. Specific processing parameters such as solution flow rates and solution properties like viscosities and electrical conductivities are typically taken into account while attempting to apply the co-axial approach. For example, the compositions of the fiber's core and sheath may be chosen depending on their ability to provide strength and their ability to support cells respectively. Selection of polymeric material and solvent is of importance for consistent generation of coaxial fibers. Viscoelasticity of the polymer solution forming shell should be sufficient to stabilize the fluid jet to create core-shell morphology of the fibers. A study shows that the morphology of the nanofibers depends on the interaction between the core and shell solutions during co-axial electrospinning, rather than their individual effects. If two highly miscible solutions are used, then partial mixing of those solutions occurs during co-electrospinning, which significantly influences the morphology of resulting nanofibers [151].

This technique can be applied to even non-electrospinnable materials, which can form the core of the core-shell nanofibers, whereas solutions forming the shells should possess spinnability. Moreover, active compounds devoid of fibrous characteristics can also be enclosed in the fiber core. The technique also offers the benefit of building a single drug delivery system from two or more bioactive compounds with varied biological activity and solubilities. Coaxial electrospun fibers with topographical and biochemical features are utilized for TE applications. Drugs are often incorporated in the core and released by shell polymer degradation or sheath pores. A longer-lasting drug release is possible using coaxial electrospinning of drugs and polymers. The sheath barrier effect can stop an initial burst discharge of the drugs [152]. Core–shell nanofibers have also been explored for the dual discharge of growth factors, wherein growth factor incorporated in core followed a time-controlled release compared to the growth factor attached on the shell surface [153]. Dual drug loading has also been achieved using core-shell nanofibers by loading drugs each in the core and the shell. Core exhibited the long-term release, whereas shell showed short-term release of the drugs to improve the tissue regeneration efficiency of the scaffolds fabricated from such fibers [154].

Triaxial electrospun fibers have also been developed with dual drug delivery capability using modified electrospinning technique. In triaxial fibers, interaction between core-intermediate layers and sheath-intermediate layers contributes to the mechanical strength of the fibers [155].Living cells can also be electrospun into the core of the fibers, encapsulated within polymeric shell. But this kind of electrospinning has been discussed in separate section in this review.

Although it requires a complicated setup, coaxial electrospinning offers numerous benefits like one-step technique for encapsulating, ability to make composite nanofibers, and its applicability for a variety of materials. With all of its benefits, coaxial electrospinning has been extensively utilised in the creation of nanofibers for varied applications [156].

5.1.3. Multiple-Jet Electrospinning

The preliminary form of electrospinning uses a single-needle to efflux the polymer solution and to create fibers. Notwithstanding the range of benefits offered by this simple form of electrospinning, the major drawback of the lower production rate with conventional electrospinning restricts the utilization of the process at a commercial scale. Multiple-jet electrospinning technique has been considered to surmount this deficiency of lower rate of production, but creation of multiple jets carries with it the issues including repulsion between jets, non-uniformity of electrical fields, poor control over the process and decline in fiber quality [157]. This necessitates even further development and optimization of the process. The operating principle of multi-nozzle electrospinning is same as that of conventional single-needle electrospinning technique, with the major difference lies in use of multiple nozzles. In a single setup, multiple nozzles are arranged in various configurations to generate multiple jets [158]

5.1.4. Blend Electrospinning

In Blend electrospinning technique, bioactive materials are solubilised or suspended within polymer solution. Physicochemical characteristics of the solution and its interaction with the bioactive materials decides the disposition of bioactive molecules within fibers [150]. Blend electrospinning method is uncomplicated compared to coaxial and emulsion electrospinning, but it also has some drawbacks such as sensitive bioactive agents may get denatured due to presence of the solvents and thus suffer from loss of their bioactivity [159]. Polymers such as poly(ethylene oxide) (PEO) and poly(vinyl alcohol) (PVA) having good water solubility, have been utilized to encapsulate bioactive proteins [160,161]. Surface accumulation of bioactive molecules is commonly observed in nanofibers, because such molecules are charged and they migrate towards the surface of the jet due to repulsion between them, during jet ejection and elongation.

5.1.5. Emulsion Electrospinning

The emulsion electrospinning involves basic set up similar to that of blend electrospinning but comprises spinning of emulsion. This is another unique and simple approach to electrospin core-shell nanofibers using a single nozzle spinneret. In comparison with coaxial electrospinning which employs coaxial needles to manufacture nanofibers with core-shell morphology, emulsion electrospinning uses single nozzle to electrospun nanofibers, therefore making it simple and more conducive for scaling up [21]. Core-shell structure is obtainable in electrospun nanofibers using either water-in-oil (W/O) [21] or oil-in-water (O/W) emulsions [162] to load respectively hydrophilic or hydrophobic compounds into the core of nanofibers [163]. In this method, polymer is solubilised in organic or aqueous solvent to form the dispersion medium whereas bioactive substances are solubilised in organic or aqueous solvent forming dispersed phase. Formulation of emulsion eliminates the requirement for common solvent for polymer and bioactive molecules. Availability of common solvent is considered as a primary necessity of the blend electrospinning technique, which is omitted in emulsion electrospinning. After ejection of jet in emulsion electrospinning, evaporation of the solvent of dispersion medium from ejected jet increases the viscosity of that phase. As a consequence, droplets of the dispersed phase travel to the core of the jet due to viscosity gradient [164]. Mutual dielectrophoresis caused by electric field led to coalescence of the droplets at the centre of the fiber, thus giving fiber a core–shell morphology. Stability of the emulsion is a decisive consideration for emulsion electrospinning, which necessitates addition of emulsifier to prevent emulsion from breaking down. This technique has been developed to incorporate functional elements such as enzymes, bioactive proteins and drugs. This technique is a potential alternative to conventional electrospinning methods because it enables loading of lipophilic drugs using affordable hydrophilic polymers and bypass the requirement of restricted, less safe solvents [150]. Some other crucial determinants of fiber characteristics include the nature of emulsion, strength of applied electric potential difference, conductivity of dispersed phase, interfacial tension exhibited by emulsion, and cooling time among others [165].

5.1.6. Cell-Electrospinning

On account of aforementioned benefits, electrospinning has earned noteworthy attention for applications in biomedical field. Nevertheless, it carries some constraints with it, such as utilization of toxic solvents, low and uncertain cell penetration and non-uniform cell distribution [166]. In addition, to seed cells in scaffold, it needs to be kept in bioreactor for long durations. Even then, there remains the uncertainty about the distribution of the cells within scaffold. To subdue such limitations, a unique approach, termed cell-electrospinning (C-ES), was invented. Cell electrospinning was discovered in around 2005–2006 [167,168]. C-ES originated from the typical electrospinning technique but is capable of embedding living cells inside the fibers. Use of viable cell in the electrospinning process differentiates C-ES it from conventional electrospinning technique. C-ES allows us to construct fully cellularized three-dimensional tissue construct by directly handling the cells. Cell-electrospun fibers demonstrate primacy by directing the cells along the fibers, enabling an effective and quick exchange of nutrients and gases and providing better cell-to-cell interaction compared to cell-embedded bulk 3D construct. But some of its limitations include inadequate mechanical strength, restrictions in development of 3D structures and less control over cell density [166].

In this technique, one of the approach is to encapsulate the biosuspension containing living cells in the core of a fiber, using a coaxial needle, within a shell fabricated out of a biocompatible polymer [168]. Electrical conductivity of biosuspension flowing through the inner needle and polymer solution flowing through the outer needle make an important consideration for the electrospinning. Viscosity, flow rate of both the liquids and the strength of the applied electric field are critical variables to analyze in this technique. Needle with different configuration such as single as well as tri-needle can also be used. The type of needle used decides the core arrangement, which can vary from single to tri-core morphology. Another consideration is that the ground electrode in cell electrospinning is significantly different compared to those used in conventional electrospinning technique. This modification is to avoid dehydration of the encapsulated cells to avert cellular damage or death [167]. Due to the negative effect of the electric field on the viability of the cells in biosuspension, magnitude of the electric field can not be raised above threshold. Dehydration and shear developed during stretching of the fibers are the probable reasons besides the applied electric field for the low viability of the cells during cell electrospinning process.

A study developed active biological microthreads using coaxial electrospinning method. A concentrated living biosuspension was used to form the core and a medical grade poly(dimethylsiloxane) was used to form the shell of the microthreads. Along the length of the microthreads, cell aggregates generated the capsules. Cell viability assay showed that the viability of the cells passed through the electric field to be around 67%, which was not statistically much different from the viability of the control cells. No indications of any harm to encapsulated cells were observed while generating microthreads through cell electrospinning using co-axial needles [168].

Guo et al. developed an electrospinning approach to enclose cellular aggregates into fibrin/polyethylene oxide microfibers. Encapsulated cellular aggregates within fibrinogen microfibers were suspended into a rotating bath containing thrombin to produce fibrin fibrils by thrombin induced polymerization of fibrin. Researcher established that loading cellular aggregates less than 100 μm in size and adjusting process parameters in electrospinning led to improved cell survival [169]. Considering the great interest developed in the area of cell electrospinning owing to the benefits provided, more of the studies are expected in future.

Table 4 discusses the advantages and limitations of different electrospinning methods reviewed in above paragraphs.

Table 4. Different electrospinning techniques to develop Nanofibers—Their Advantages and Limitations.

Electrospinning Technique	Advantages	Limitations
Single nozzle electrospinning	• Simple process with least number of controllable parameters • Most experimented due to simple gear and process used	• Compatibility between polymer/polymer solutions/bioactive molecules is essential to eject mixture through single nozzle. • Involves the use of solvents • For electrospinning of polymer blend, availability of common solvent is essential • Lower rate of production • Denaturation of sensitive bioactive agents or loss of their bioactivity due to presence of the solvents, in case of blend of bioactive with polymer(s)
Co-axial electrospinning	• Actives which are lacking fibrous characteristics can still be enclosed in the nanofiber core. • Two or more bioactive compounds with varied biological activity and solubilities can be encapsulated within single multilayered fiber. • Controlled drug release is possible due to core and shell morphology	• Sophisticated gear required • Use of co-axial needle make electrospinning process complex, involving numerous controllable parameters
Multi-nozzle electrospinning	• Increased production rate • Multiple nozzles can be arranged in various configurations to alter alignment of nanofibers	• Increases complications in process with repulsion between jets, non-uniformity of electrical fields, poor control over the process and decline in fiber quality
Emulsion electrospinning	• Able to fabricate core-shell Nanofibers without use of co-axial nozzle • Hydrophilic or hydrophobic actives can be loaded into core of nanofibers by electrospinning W/O or O/W emulsions respectively	• Formulation of emulsion eliminates the requirement of common solvent for polymer and bioactive molecules
Cell electrospinning	• It includes all the advantages of electrospinning • In addition, it is capable of embedding living cells inside the fibers • encapsulation of cells within fibers enables effective and rapid exchange of nutrients and oxygen • enables excellent interaction between cells and help to achieve uniform cell distribution	• More critical variables to consider to keep cells viable in electrospun fibers • Low mechanical properties • Low control over cell density

6. Applications of Polymeric Nanofibers

Use of nanofibers has been evaluated for range of tissues from a cornea [170], myocardial tissue [171] to skeletal muscles [135]. This review summarizes the applications of nanofibers in regeneration of neural, vascular, cartilage, bone and dermal tissues.

6.1. Neural Tissue Regeneration

The peripheral nerve injury creates a major problem in their repair and restoration. Autografting, allografting and xenografting offers recourse to overcome this difficulty. However, donor site morbidity, the lack of donors and low proficiency in grafting techniques turn up to be limitations of these alternatives [172]. In contrast, the electrospun nanofibers offer multiple benefits, including controlled alignment which provides spatial assistance for neurite outgrowth, axon lengthening [173] and mechanical cues for differentiation of stem cells [174]. Apart from this, aligned nanofibers were noticed for supporting Schwann cells migration and thus assist in reestablishing a growth cone at the tip [175].

Afrash et al. developed a nerve growth factor (NGF) functionalized aligned nanofibrous scaffold based on polycaprolactone/chitosan (PCL/CS) polymers for tissue regeneration of neural cells. NGF was used as a neurotrophin and it was attached to PCL/CS nanofibers with the use of dopamine coating. Polydopamine coating reduced the hydrophilicity of the nanofibers, whereas immobilization of NGF on the nanofibers improved the hydrophilic nature. It was observed that, aligned fibers were more hydrophilic compared to randomly aligned fibers. It established that topography and morphology can control interfacial tension. It also demonstrated that regular alignment of PCL/CS nanofibers could provide desirable conditions for neural cell growth [9]. In another study by Xieet al., the characterization of embryonic stem (ES) cell culture on electrospun PCL nanofibers with regular and irregular alignment, manifested the significance of material topography in cell differentiation. PCL nanofibers seeded with ES cells showed that stem cells specialized to oligodendrocytes and astrocytes along with many other neural lineage cells. In addition, this study demonstrated that regular alignment of nanofibers could retard the specialization and maturation of ES cells into astrocytes, which play a critical role inthe repair of spinal cord traumas [176].

6.2. *Vascular Tissue Regeneration*

An impediment of in vitro fabrication of vascular tissues to fulfill the clinical necessity of tissue grafts is lingering since the dawn of TERM [177].The 1950s saw the development of synthetic tissue-engineered vascular grafts (TEVGs) to restore blocked arteries after surgical complications. TEVGs were used as a remedy to the regular scarcity of allogenic tissue grafting and to mitigate the problem of immunological rejections after transplantation. But these synthetic TEVGs were found to be unable to noticeably reduce overall mortality and morbidity [178]. To solve this issue, the researchers have employed various in vitro strategies to prepare vascular tissue having ability to interact with cells to develop new blood vessels [60].

Shin et al. fabricated PLGA nanofibers with co-functionalization of RGD peptide and graphene oxide (GO) for vascular TE using the electrospinning technique. Surface functionalization with RGD and GO on PLGA nanofiber improved hydrophilicity and facilitated interaction between nanofiber and cells. RGD peptide functionalization greatly increased initial attachment and growth of vascular smooth muscle cells (VSMCs). In addition, GO also supported enhanced proliferation of VSMCs. The study shows the promising potential of RGD-GO-PLGA nanofiber matrices for vascular tissue regeneration [179]. Marelli et al.electrospun SF into fibers with tubular morphology for small diameter vessel grafting. These electrospun tubes were able to resist pressure up to 575 ± 17 mm Hg, which is more than fourfold of normal systolic pressure i.e., 120 mm Hg and more than twice that of pathological upper pressure of 220 mm Hg. SF tubes displayed good cytological compatibility in in vitro analysis. Thus, electrospun tubes designed in this study show bright prospects for small diameter blood vessel grafting [180].

6.3. *Cartilage Tissue Regeneration*

Articular cartilage is a functional connective tissue which covers the ends of bones at the site of junction of two or more moving bones. The ECM in cartilage tissue is dense, while chondrocytes are thinly distributed within matrix. Such entrapment of chondrocytes within dense microenvironment prevents its mobility to adjoining regions within cartilage. Though chondrocytes responds to variety of stimuli, it rarely form cell –to-cell contacts for direct cell transduction. Moreover, limited ability of chondrocytes to replicate is responsible for poor regenerative capability of cartilage in case of an injury [181]. In addition, articular cartilage lacks blood vessels, lymphatics and nerves, limiting its ability to regenerate tissue after injury. Damage in the cartilages necessitates replacement most of the times. To settle this problem, researchers have experimented with many TE strategies, including sponges, hydrogel scaffolds, gelatin microsphere, and collagen sponges. These approaches showed limited improvement in the cartilage healing process. Conversely, the nanofibers synthesized from synthetic, natural, and composite polymers provide good results due to its resemblance with the ECM. Such Nanofibers promote the cell-ECM interaction and chondrogenic differentiation. Very high surface area compared to total volume of the aligned nanofibers manifests the potential of engineering articular cartilage using approaches of TE [60,135].

Semitela et al. synthesized the bioactive polycaprolactone-gelatin nanofibers scaffolds (PCL + GEL) with enhanced pore size and interconnectivity for cartilage tissue repair. Polyethylene glycol (PEG) was incorporated during the electrospinning process and subsequently eliminated to get enlarged pore size. This innovative method was used to subdue two weaknesses of PCL electrospun fibers which are small pore size and lack of bio-inductive property. The scaffolds with improved pore diameter and interconnectivity enabled enhanced cell infiltration and homogeneous cell distribution, thus creating the potential to generate functional tissue [182].

An electrospun composite containing uniformly distributed but distinct fibers of PCL and PEO was developed. In this composite scaffold, fibers of polyethylene oxide formed the removable sacrificial fiber fraction. Both polymers were chosen based on their stability in hydrated environment, such as PCL is slowly degrading, whereas PEO dissolves immediately upon hydration. Although, removal of sacrificial fiber content resulted in reduced mechanical properties, it increased the size of the pores within the scaffold. Increase in sacrificial fiber fraction in the construct augmented cellular infiltration within construct. Construct with 60% PEO fraction, was observed to be fully colonized with seeded cells and was able to direct cell morphology and consequent matrix formation [183].

6.4. Bone Tissue Regeneration

Bone is one of the highly vascularized tissues in the human body. It is categorized into cortical bone and trabecular bone. Cortical bone is a dense, solid bone, extends mechanical support to human body and protection to bone marrow whereas trabecular bone is biologically active, enables joint and limb movement. The bone structure is made up of 69% of inorganic component containing hydroxyapatite and calcium phosphate complex contributing to bone its compactness and stiffness, while organic component composed of collagen and other structural proteins accounts for about 22% [184].

The regeneration of bones is a complex process involving a series of osteoinductive processes. Therefore bone TE demands the utilization of scaffolding, cells, chemical signaling and mechanical forces to create customized tissues. Biomimetic scaffolding for bone repair can include features such as high porosity to aid cell attachment, migration, proliferation and differentiation and biomechanical stress tolerance ability to endure stress generated within body during tissue regeneration [60]. A growing number of bone illnesses including infections, cancer and bone loss, necessitate bone regeneration. The vigorous course of bone TE begins with movement and recruitment of osteoprogenitor cells, and continues with cellular growth, differentiation, matrix production, and bone remodelling. Mechanical characteristics of bones are due to unique structural design of bone that extends from nano range to macro range dimensions, along with specific interconnections. Bone TE focuses on developing three-dimensional scaffolds that can replicate the ECM, offer structural support as well as aid in regeneration of bone. To increase the attachment, viability and mobility of osteogenic cells, scaffolds should have osteo-conductive, osteo-inductive, and osteogenic characteristics. To impart these characteristics, scaffolds are manufactured to provide mechanical and chemical cues that induce osteoblastic lineage formation [185].

Several scientists have attempted to alter the mechanical properties of scaffolds namely stiffness, strength and toughness using various methods and to create nanostructures to imitate bone's natural architecture [186]. Despite many studies focusing on bone TE, much-needed advancements in scaffolds with conceivably superior clinical outcomes are still required. Electrospinning has long been considered an appropriate manufacturing technique for scaffolds by virtue of its multidimensional capacity of making nano- and micro-range fibrous frameworks with configurable fiber features [187].

Using an electrospinning process, PLA fibers encapsulating Fe_3O_4 nanoparticles at concentration of 2 and 5 percent were formed. Bone deformities transplanted with Fe_3O_4/PLA nanofibers displayed a significantly greater rate of bone healing compared to deformities transplanted with plain PLA nanofibers. Furthermore, CT scan demonstrated that the bone defects grafted with Fe_3O_4/PLA nanofibers encapsulating 2 and 5 percent Fe_3O_4 nanoparticles presented 1.9- and 2.3-fold enhancement, respectively, in volume of bone in comparison to the control sample [188]. Miszuket al. fabricated a composite nanofiber based scaffold using polycaprolactone/hydroxyapatite for regeneration of bone using an new thermally induced self-agglomeration (TISA) technique based on electrospinning. High elasticity, porosity even after coating with minerals and easy alteration with the applied pressure to fit to different defect shapes are the reported features making it desirable for application in bone TE. In addition, biomimetic mineral coating on fabricated scaffolds allows simultaneously encapsulation of different types of proteins, small molecules and drugs, under physiologically mild conditions. This study suggested that the innovative nanofiber based composite scaffold, that are press-fit, can be a sound means to deliver multiple drugs along with bone TE [189].

6.5. *Dermis Tissue Regeneration*

Skin lesions usually heal by forming epithelialized scar tissue rather than full skin regeneration. The epidermis has a poor ability to heal compared to the dermis; thus in case of substantial damage to the epidermis, biological regeneration process is insufficient. On the other hand, the dermis has a tremendous ability to rejuvenate. After a skin injury, the scar tissue develops with deficiency of dermis, thus loses the flexibility, elasticity, and toughness of natural dermis [190]. The fibrous structure in native ECM always shows a more intricate design than just straightforward unidirectional alignment. Skin tissue contains collagen fibrils that have a pattern like a basketweave or mesh. As a result, scaffolds with crossed nanofibers performed better than those with random or unidirectionally aligned nanofibers in terms of keratinocyte and fibroblast migration rates.

Collagen in its indigenous form acts as a natural foundation for cell adhesion, division, growth and specialization. Collagen exhibits significant strength in its biological form [191]. In addition, its' biological origin make it the most biomimetic skin substitute created and thus the most preferred material to fabricate nanofibrous scaffold.

Powell and Boyce prepared electrospun submicron fibers using PCL and collagen to design a scaffold. Mechanical performance of nanofibers improved noticeably with mixing of little amount of PCL to collagen without negotiating on the biocompatibility of nanofibers. Keratinocytes and dermal fibroblasts cultured on collagen/PCL nanofiber scaffolds promoted the regeneration of the layered epidermis, dermis and uninterrupted basal layers [192]. Another study intended to evaluate in vivo performance of the PGA/collagen nanofiber on granulation histology and its capability of stimulating new vasculature was conducted out by Sekiya et al. This group of researchers developed PGA/collagen nanofibers using electrospinning technique. When compared to commercially available collagen matrix, histology revealed that fabricated nanofibers demonstrated considerably higher cell density with greater number of migrating cells. These observations indicated the superior ability of the developed nanofibers in relation to cell migration and neovascularization compared to collagen matrix product. This desirable outcome was attributed to the nano-range diameter of fibers and inclusion of PGA [193]. In another study, a highly porous scaffold created out of PCL/chitosan fibers with core–shell morphology were developed using emulsion electrospinning. Presence of high porosity and interconnectivity assisted penetration and proliferation of cells. The scaffold also supported ECM protein translation and in vitro layered epithelialization. Successful incorporation of the scaffold with margins of wound in animal model and rapid healing in around 20 days established the effectiveness of the scaffold as skin graft [194].

6.6. Cardiac Tissue Regeneration

Cardiac tissue has very restricted regenerative capacity, thus cardiac tissue regeneration using the principles of TE is a necessary and appropriate alternative. Some of the challenges associated with cardiac tissue engineering are the choice of polymers for fabricating scaffolds and achieving the required alignment of the microfibrils for guiding the growth of cells and contraction of cardiac muscle cells. Another clinical challenge is the regeneration of heart valves (HVs) because of their complex anatomical structure with leaflets and numerous supporting structures along with having complex, striated ECM [195]. Earlier many attempts to engineer the valves met the failure with disordered ECM and inability to function due to use of isotropic and homogeneous scaffolds [196]. Biomimetic scaffold with heterogeneous and anisotropic characteristics which approach that of inherent heart valve tissue are applicable for regenerating HV tissue. Tissue engineered HVs are contemplated to be capable of adapting to such complexities, indicating its potential as a alternative to existing treatments.

Ahmadi et al. manufactured polyurethane/chitosan/carbon nanotubes (PU/Cs/CNT) composite nanofibrous scaffolds using two techniques. PU/Cs/CNT electrospun scaffolds were manufactured by blending CNT and electrospinning this blend of polyurathane, chitosan and CNT. In other technique, polyurethane/chitosan solution electrospun into nanofibers and CNT were electrosprayed onto nanofibers from the opposite side. The nanofibers were also collected with random and aligned orientation. Addition of CNT substantiallyameliorated the mechanical characteristicsand hydrophilicity of the nanofibers. Improvement of surface properties by hydrophilic chitosan and carboxylated CNTs led to proliferation enhancement of Human umbilical vein endothelial cells in PU/Cs/CNT scaffold compared to PU scaffold. Cardiac rat myoblast cells (H9C2 cells) proliferation on fibrous matrix with electrosprayed CNT was more notable than cell proliferation on PU scaffold. Alamar blue assays demonstrated that number of H9C2 cells on scaffold with electrosprayed CNT, in both aligned and random scaffolds, enhanced notably higher than other scaffolds and control group [197].

To fabricate fibrous scaffold for replicating the anisotropic nature of native heart valves, Xue et al. utilised ring-shaped copper collector for collecting electrospun fibers. This group of researchers fabricated anisotropic fibrous matrices manufactured with (poly(1,3-diamino-2-hydroxypropane-co-glycerol sebacate)-co-poly (ethylene glycol) (APS-co-PEG) and PCL polymer blends that hadadjustable and controllable fiber morphologies and mechanical features. The polymer formulationswere electrospun onto flat aluminum foil and ring-shaped copper wire, producing isotropic and anisotropic fibers, respectively. The scaffolds gathered on flat aluminum foil demonstratedalike mechanical properties in the two perpendicular directions, revealing an isotropic behavior, whereas the scaffolds collected on the ring-shaped collector acted differently in their fiber and cross-fiber directions, indicating mechanical anisotropy. The anisotropic scaffold also showed to possess a Degree of anisotropy (DA) close to that of a porcine aortic valve, indicating its prospectives to be used to regenerate the heart valves [198].

7. Conclusions

We can witness the tremendous work done in the discipline of TERM where nanofibrous scaffolds have been employed as reinforcement to allow regeneration of tissue. Promising outcomes of the TERM research conducted for the variety of tissues and several disorders is increasing hopes for a therapy that will be a better alternative to existing therapies. Among the available therapies, graft surgery is indispensible for numerous health issues. Extensive research in TERM is taking the discipline forward by small but consistent leaps. Kind of physical and chemical cues, their amount and timing for the regulation of cell activities has still puzzled the researchers. Increasing comprehension about the cellular behavior to an array of physical and chemical cues will help the researchers to come up with the right approach for tissue regeneration.

Natural and synthetic polymers have their own set of advantages which it offers to the fabricated scaffolds for the tissue regeneration purpose. Natural polymers offers features such as biocompatibility, biodegradability, low immunogenicity and its ability to elicit the favorable interaction with the cells, whereas synthetic polymers offers characteristics of their fitness to spinning, excellent mechanical strength and cost-efficiency. Features of both natural and synthetic polymers are complementary too each other explaining their often use in combination to bring desirable attributes in the scaffold. Nonetheless, further research in the field will need to focus on innovative polymers with characteristic features for specific tissue regeneration applications.

Electrospun nanofibers have been integrated with 3D printed tissue constructs fabricated using additive manufacturing or 3D printing technique, which is another promising technology, to design composite scaffolds. The merit provided by the nanofibers was the enhancement in infiltration of the cells in the scaffold, whereas 3D printed component of the composite extends its mechanical strength to the scaffold and also enables preparation of complex 3D forms at the macro level. Such alliances between the matrices are significant for regeneration of load bearing tissues requiring mechanical strength of the scaffold such as bone and cartilages. Similarly electrospun nanofibers composites with hydrogels have been experimented to overcome limitation of both of them and to reap the merits provided by both components for the TE purpose.Such combination can be studied more for the purpose of soft tissue regeneration. More of such combinations involving more than one matrix fabricated using different techniques needs to be studied to mimic the biological ECM in various tissues to a maximum extent.

To achieve high scale production of nanofibers is still challenging due to the low yields of existing processes. Centrifugal jet spinning has shown to be capable of producing high quantities of nanofibers in short duration and consuming less power. This technique also has the potential for scaling up to commercial production levels [199,200]. Other unattended challenges associated with existing nanofiber fabrication methodsinclude utilization of toxic solvents and low and uncertain cell penetration into fiber matrices. High cost of the biomedical research also adds to existing obstacles in the path of researchwhich needs to be addressed to reach the feasible as well as affordable solution.Given the potential demonstrated by Nanofibers in TERM, new pursuits in the application of nanofibers in TERM are anticipated to deepen understanding of tissue regeneration, to bring answers to unsolved queries and to offer intervention to apply on wide population to regenerate tissues.

Author Contributions: Conceptualization, P.P.; methodology, A.P.; writing—original draft preparation, P.P., D.T., S.A., M.G. and K.K.; writing—review and editing, P.P. and A.P.; visualization, A.P.; supervision, A.P. and S.A.; project administration, A.P.; funding acquisition, A.P. All authors have read and agreed to the published version of the manuscript.

Funding: This research received no external funding.

Institutional Review Board Statement: Not applicable.

Data Availability Statement: Data sharing is not applicable to this article.

Conflicts of Interest: The authors declare no conflict of interest.

References

1. Chan, B.P.; Leong, K.W. Scaffolding in Tissue Engineering: General Approaches and Tissue-Specific Considerations. *Eur. Spine J.* **2008**, *17*, 467–479. [CrossRef] [PubMed]
2. Vogt, L.; Liverani, L.; Roether, J.A.; Boccaccini, A.R. Electrospun Zein Fibers Incorporating Poly(Glycerol Sebacate) for Soft Tissue Engineering. *Nanomaterials* **2018**, *8*, 150. [CrossRef] [PubMed]
3. Mousa, H.M.; Hussein, K.H.; Sayed, M.M.; Abd El-Rahman, M.K.; Woo, H.-M. Development and Characterization of Cellulose/Iron Acetate Nanofibers for Bone Tissue Engineering Applications. *Polymers* **2021**, *13*, 1339. [CrossRef] [PubMed]

4. Liang, W.; Xu, Y.; Li, X.; Wang, X.-X.; Zhang, H.-D.; Yu, M.; Ramakrishna, S.; Long, Y.-Z. Transparent Polyurethane Nanofiber Air Filter for High-Efficiency PM2.5 Capture. *Nanoscale Res. Lett.* **2019**, *14*, 361. [CrossRef] [PubMed]
5. Ebrahimi Vafaye, S.; Rahman, A.; Safaeian, S.; Adabi, M. An Electrochemical Aptasensor Based on Electrospun Carbon Nanofiber Mat and Gold Nanoparticles for the Sensitive Detection of Penicillin in Milk. *Food Meas.* **2021**, *15*, 876–882. [CrossRef]
6. Alotaibi, B.S.; Shoukat, M.; Buabeid, M.; Khan, A.K.; Murtaza, G. Healing Potential of Neomycin-Loaded Electrospun Nanofibers against Burn Wounds. *J. Drug Deliv. Sci. Technol.* **2022**, *74*, 103502. [CrossRef]
7. Liu, X.; Xu, H.; Zhang, M.; Yu, D.-G. Electrospun Medicated Nanofibers for Wound Healing: Review. *Membranes* **2021**, *11*, 770. [CrossRef]
8. Yu, B.; Chen, J.; Chen, D.; Chen, R.; Wang, Y.; Tang, X.; Wang, H.-L.; Wang, L.-P.; Deng, W. Visualization of the Interaction of Water Aerosol and Nanofiber Mesh. *Phys. Fluids* **2021**, *33*, 092106. [CrossRef]
9. Afrash, H.; Nazeri, N.; Davoudi, P.; Majidi, R.; Ghanbari, H. Development of a Bioactive Scaffold Based on NGF Containing PCL/Chitosan Nanofibers for Nerve Regeneration. *Biointerface Res. Appl. Chem.* **2021**, *11*, 12606–12617. [CrossRef]
10. Ghajarieh, A.; Habibi, S.; Talebian, A. Biomedical Applications of Nanofibers. *Russ. J. Appl. Chem.* **2021**, *94*, 847–872. [CrossRef]
11. Aggarwal, U.; Goyal, A.K.; Rath, G. Development and Characterization of the Cisplatin Loaded Nanofibers for the Treatment of Cervical Cancer. *Mater. Sci. Eng. C* **2017**, *75*, 125–132. [CrossRef]
12. Venugopal, J.; Vadgama, P.; Kumar, T.S.S.; Ramakrishna, S. Biocomposite Nanofibres and Osteoblasts for Bone Tissue Engineering. *Nanotechnology* **2007**, *18*, 055101. [CrossRef]
13. Samadian, H.; Farzamfar, S.; Vaez, A.; Ehterami, A.; Bit, A.; Alam, M.; Goodarzi, A.; Darya, G.; Salehi, M. A Tailored Polylactic Acid/Polycaprolactone Biodegradable and Bioactive 3D Porous Scaffold Containing Gelatin Nanofibers and Taurine for Bone Regeneration. *Sci. Rep.* **2020**, *10*, 13366. [CrossRef]
14. Sharma, P.R.; Zheng, B.; Sharma, S.K.; Zhan, C.; Wang, R.; Bhatia, S.R.; Hsiao, B.S. High Aspect Ratio Carboxycellulose Nanofibers Prepared by Nitro-Oxidation Method and Their Nanopaper Properties. *ACS Appl. Nano Mater.* **2018**, *1*, 3969–3980. [CrossRef]
15. Rošic, R.; Kocbek, P.; Pelipenko, J.; Kristl, J.; Baumgartner, S. Nanofibers and Their Biomedical Use. *Acta Pharm.* **2013**, *63*, 295–304. [CrossRef]
16. Leung, V.; Ko, F. Biomedical Applications of Nanofibers. *Polym. Adv. Technol.* **2011**, *22*, 350–365. [CrossRef]
17. Shastri, V. Non-Degradable Biocompatible Polymers in Medicine: Past, Present and Future. *Curr. Pharm. Biotechnol.* **2003**, *4*, 331–337. [CrossRef]
18. Chen, S.; Gil, C.J.; Ning, L.; Jin, L.; Perez, L.; Kabboul, G.; Tomov, M.L.; Serpooshan, V. Adhesive Tissue Engineered Scaffolds: Mechanisms and Applications. *Front. Bioeng. Biotechnol.* **2021**, *9*, 683079. [CrossRef]
19. Dhand, C.; Ong, S.T.; Dwivedi, N.; Diaz, S.M.; Venugopal, J.R.; Navaneethan, B.; Fazil, M.H.U.T.; Liu, S.; Seitz, V.; Wintermantel, E.; et al. Bio-Inspired in Situ Crosslinking and Mineralization of Electrospun Collagen Scaffolds for Bone Tissue Engineering. *Biomaterials* **2016**, *104*, 323–338. [CrossRef]
20. Aoki, H.; Miyoshi, H.; Yamagata, Y. Electrospinning of Gelatin Nanofiber Scaffolds with Mild Neutral Cosolvents for Use in Tissue Engineering. *Polym. J* **2015**, *47*, 267–277. [CrossRef]
21. Norouzi, M.-R.; Ghasemi-Mobarakeh, L.; Itel, F.; Schoeller, J.; Fashandi, H.; Borzi, A.; Neels, A.; Fortunato, G.; Rossi, R.M. Emulsion Electrospinning of Sodium Alginate/Poly(ε-Caprolactone) Core/Shell Nanofibers for Biomedical Applications. *Nanoscale Adv.* **2022**, *4*, 2929–2941. [CrossRef] [PubMed]
22. Liverani, L.; Abbruzzese, F.; Mozetic, P.; Basoli, F.; Rainer, A.; Trombetta, M. Electrospinning of Hydroxyapatite-Chitosan Nanofibers for Tissue Engineering Applications: Electrospinning of hydroxyapatite-chitosan nanofibers. *Asia-Pac. J. Chem. Eng.* **2014**, *9*, 407–414. [CrossRef]
23. Hussein, Y.; El-Fakharany, E.M.; Kamoun, E.A.; Loutfy, S.A.; Amin, R.; Taha, T.H.; Salim, S.A.; Amer, M. Electrospun PVA/Hyaluronic Acid/L-Arginine Nanofibers for Wound Healing Applications: Nanofibers Optimization and in Vitro Bioevaluation. *Int. J. Biol. Macromol.* **2020**, *164*, 667–676. [CrossRef] [PubMed]
24. Roshanfar, F.; Hesaraki, S.; Dolatshahi-Pirouz, A. Electrospun Silk Fibroin/Kappa-Carrageenan Hybrid Nanofibers with Enhanced Osteogenic Properties for Bone Regeneration Applications. *Biology* **2022**, *11*, 751. [CrossRef] [PubMed]
25. Khandaker, M.; Nomhwange, H.; Progri, H.; Nikfarjam, S.; Vaughan, M.B. Evaluation of Polycaprolactone Electrospun Nanofiber-Composites for Artificial Skin Based on Dermal Fibroblast Culture. *Bioengineering* **2022**, *9*, 19. [CrossRef]
26. Massoumi, B.; Abbasian, M.; Jahanban-Esfahlan, R.; Mohammad-Rezaei, R.; Khalilzadeh, B.; Samadian, H.; Rezaei, A.; Derakhshankhah, H.; Jaymand, M. A Novel Bio-Inspired Conductive, Biocompatible, and Adhesive Terpolymer Based on Polyaniline, Polydopamine, and Polylactide as Scaffolding Biomaterial for Tissue Engineering Application. *Int. J. Biol. Macromol.* **2020**, *147*, 1174–1184. [CrossRef]
27. Hajiali, H.; Shahgasempour; Naimi-Jamal, M.R. Peirovi Electrospun PGA/Gelatin Nanofibrous Scaffolds and Their Potential Application in Vascular Tissue Engineering. *Int. J. Nanomed.* **2011**, 2133–2141. [CrossRef]
28. Asiri, A.; Saidin, S.; Sani, M.H.; Al-Ashwal, R.H. Epidermal and Fibroblast Growth Factors Incorporated Polyvinyl Alcohol Electrospun Nanofibers as Biological Dressing Scaffold. *Sci. Rep.* **2021**, *11*, 5634. [CrossRef]
29. Deng, M.; Kumbar, S.G.; Nair, L.S.; Weikel, A.L.; Allcock, H.R.; Laurencin, C.T. Biomimetic Structures: Biological Implications of Dipeptide-Substituted Polyphosphazene–Polyester Blend Nanofiber Matrices for Load-Bearing Bone Regeneration. *Adv. Funct. Mater.* **2011**, *21*, 2641–2651. [CrossRef]

30. Ehrmann, A. Non-Toxic Crosslinking of Electrospun Gelatin Nanofibers for Tissue Engineering and Biomedicine—A Review. *Polymers* **2021**, *13*, 1973. [CrossRef]
31. Cen, L.; Liu, W.; Cui, L.; Zhang, W.; Cao, Y. Collagen Tissue Engineering: Development of Novel Biomaterials and Applications. *Pediatr. Res.* **2008**, *63*, 492–496. [CrossRef]
32. Zhang, J.; Elango, J.; Wang, S.; Hou, C.; Miao, M.; Li, J.; Na, L.; Wu, W. Characterization of Immunogenicity Associated with the Biocompatibility of Type I Collagen from Tilapia Fish Skin. *Polymers* **2022**, *14*, 2300. [CrossRef]
33. Torres-Giner, S.; Gimeno-Alcañiz, J.V.; Ocio, M.J.; Lagaron, J.M. Comparative Performance of Electrospun Collagen Nanofibers Cross-Linked by Means of Different Methods. *ACS Appl. Mater. Interfaces* **2009**, *1*, 218–223. [CrossRef]
34. Blackstone, B.N.; Gallentine, S.C.; Powell, H.M. Collagen-Based Electrospun Materials for Tissue Engineering: A Systematic Review. *Bioengineering* **2021**, *8*, 39. [CrossRef]
35. Gautieri, A.; Vesentini, S.; Redaelli, A.; Buehler, M.J. Hierarchical Structure and Nanomechanics of Collagen Microfibrils from the Atomistic Scale Up. *Nano Lett.* **2011**, *11*, 757–766. [CrossRef]
36. Mousavi, S.; Khoshfetrat, A.B.; Khatami, N.; Ahmadian, M.; Rahbarghazi, R. Comparative Study of Collagen and Gelatin in Chitosan-Based Hydrogels for Effective Wound Dressing: Physical Properties and Fibroblastic Cell Behavior. *Biochem. Biophys. Res. Commun.* **2019**, *518*, 625–631. [CrossRef]
37. Arun, A.; Malrautu, P.; Laha, A.; Ramakrishna, S. Gelatin Nanofibers in Drug Delivery Systems and Tissue Engineering. *Eng. Sci.* **2021**, *16*, 71–81.
38. Fertah, M.; Belfkira, A.; Dahmane, E.M.; Taourirte, M.; Brouillette, F. Extraction and Characterization of Sodium Alginate from Moroccan *Laminaria digitata* Brown Seaweed. *Arab. J. Chem.* **2017**, *10*, S3707–S3714. [CrossRef]
39. Mokhena, T.C.; Mochane, M.J.; Mtibe, A.; John, M.J.; Sadiku, E.R.; Sefadi, J.S. Electrospun Alginate Nanofibers toward Various Applications: A Review. *Materials* **2020**, *13*, 934. [CrossRef]
40. Rowley, J.A.; Madlambayan, G.; Mooney, D.J. Alginate Hydrogels as Synthetic Extracellular Matrix Materials. *Biomaterials* **1999**, *20*, 45–53. [CrossRef]
41. Taemeh, M.A.; Shiravandi, A.; Korayem, M.A.; Daemi, H. Fabrication Challenges and Trends in Biomedical Applications of Alginate Electrospun Nanofibers. *Carbohydr. Polym.* **2020**, *228*, 115419. [CrossRef] [PubMed]
42. Aadil, K.R.; Nathani, A.; Sharma, C.S.; Lenka, N.; Gupta, P. Fabrication of Biocompatible Alginate-Poly(Vinyl Alcohol) Nanofibers Scaffolds for Tissue Engineering Applications. *Mater. Technol.* **2018**, *33*, 507–512. [CrossRef]
43. Jeong, S.; Krebs, M.; Bonino, C.; Khan, S.; Alsberg, E. Electrospun Alginate Nanofibers with Controlled Cell Adhesion for Tissue Engineering. *Macromol. Biosci.* **2010**, *10*, 934–943. [CrossRef] [PubMed]
44. Goy, R.C.; Morais, S.T.B.; Assis, O.B.G. Evaluation of the Antimicrobial Activity of Chitosan and Its Quaternized Derivative on E. Coli and S. Aureus Growth. *Rev. Bras. Farmacogn.* **2016**, *26*, 122–127. [CrossRef]
45. Ito, M.; Ban, A.; Ishihara, M. Anti-Ulcer Effects of Chitin and Chitosan, Healthy Foods, in Rats. *Jpn. J. Pharmacol.* **2000**, *82*, 218–225. [CrossRef]
46. Abedian, Z.; Moghadamnia, A.A.; Zabihi, E.; Pourbagher, R.; Ghasemi, M.; Nouri, H.R.; Tashakorian, H.; Jenabian, N. Anticancer Properties of Chitosan against Osteosarcoma, Breast Cancer and Cervical Cancer Cell Lines. *Casp. J. Intern. Med.* **2019**, *10*, 439–446. [CrossRef]
47. Brun, P.; Zamuner, A.; Battocchio, C.; Cassari, L.; Todesco, M.; Graziani, V.; Iucci, G.; Marsotto, M.; Tortora, L.; Secchi, V.; et al. Bio-Functionalized Chitosan for Bone Tissue Engineering. *Int. J. Mol. Sci.* **2021**, *22*, 5916. [CrossRef]
48. Thakur, V.K.; Thakur, M.K. Recent Advances in Graft Copolymerization and Applications of Chitosan: A Review. *ACS Sustain. Chem. Eng.* **2014**, *2*, 2637–2652. [CrossRef]
49. Mahoney, C.; Conklin, D.; Waterman, J.; Sankar, J.; Bhattarai, N. Electrospun Nanofibers of Poly(ε-Caprolactone)/Depolymerized Chitosan for Respiratory Tissue Engineering Applications. *J. Biomater. Sci. Polym. Ed.* **2016**, *27*, 611–625. [CrossRef]
50. Wang, M.; Roy, A.K.; Webster, T.J. Development of Chitosan/Poly(Vinyl Alcohol) Electrospun Nanofibers for Infection Related Wound Healing. *Front. Physiol.* **2017**, *7*, 683. [CrossRef]
51. Movahedi, M.; Asefnejad, A.; Rafienia, M.; Khorasani, M.T. Potential of Novel Electrospun Core-Shell Structured Polyurethane/Starch (Hyaluronic Acid) Nanofibers for Skin Tissue Engineering: In Vitro and In Vivo Evaluation. *Int. J. Biol. Macromol.* **2020**, *146*, 627–637. [CrossRef]
52. Fischer, R.L.; McCoy, M.G.; Grant, S.A. Electrospinning Collagen and Hyaluronic Acid Nanofiber Meshes. *J. Mater. Sci. Mater. Med.* **2012**, *23*, 1645–1654. [CrossRef]
53. Li, J.; He, A.; Han, C.C.; Fang, D.; Hsiao, B.S.; Chu, B. Electrospinning of Hyaluronic Acid (HA) and HA/Gelatin Blends. *Macromol. Rapid Commun.* **2006**, *27*, 114–120. [CrossRef]
54. Hsu, F.-Y.; Hung, Y.-S.; Liou, H.-M.; Shen, C.-H. Electrospun Hyaluronate-Collagen Nanofibrous Matrix and the Effects of Varying the Concentration of Hyaluronate on the Characteristics of Foreskin Fibroblast Cells. *Acta Biomater.* **2010**, *6*, 2140–2147. [CrossRef]
55. Sun, W.; Gregory, D.A.; Tomeh, M.A.; Zhao, X. Silk Fibroin as a Functional Biomaterial for Tissue Engineering. *Int. J. Mol. Sci.* **2021**, *22*, 1499. [CrossRef]
56. Kasoju, N.; Bora, U. Silk Fibroin in Tissue Engineering. *Adv. Healthc. Mater.* **2012**, *1*, 393–412. [CrossRef]
57. Koh, L.-D.; Cheng, Y.; Teng, C.-P.; Khin, Y.-W.; Loh, X.-J.; Tee, S.-Y.; Low, M.; Ye, E.; Yu, H.-D.; Zhang, Y.-W.; et al. Structures, Mechanical Properties and Applications of Silk Fibroin Materials. *Prog. Polym. Sci.* **2015**, *46*, 86–110. [CrossRef]

58. MacIntosh, A.C.; Kearns, V.R.; Crawford, A.; Hatton, P.V. Skeletal Tissue Engineering Using Silk Biomaterials. *J. Tissue Eng. Regen. Med.* **2008**, *2*, 71–80. [CrossRef]
59. Chirila, T.V. Oxygen Permeability of Silk Fibroin Membranes: A Critical Review and Personal Perspective. *Biomater. Tissue Technol.* **2017**, *1*, 1–5. [CrossRef]
60. Nemati, S.; Kim, S.; Shin, Y.M.; Shin, H. Current Progress in Application of Polymeric Nanofibers to Tissue Engineering. *Nano Converg.* **2019**, *6*, 36. [CrossRef]
61. Sowmya, B.; Hemavathi, A.B.; Panda, P.K. Poly (ε-Caprolactone)-Based Electrospun Nano-Featured Substrate for Tissue Engineering Applications: A Review. *Prog. Biomater.* **2021**, *10*, 91–117. [CrossRef] [PubMed]
62. Gautam, S.; Dinda, A.K.; Mishra, N.C. Fabrication and Characterization of PCL/Gelatin Composite Nanofibrous Scaffold for Tissue Engineering Applications by Electrospinning Method. *Mater. Sci. Eng. C* **2013**, *33*, 1228–1235. [CrossRef] [PubMed]
63. Nazeer, M.A.; Yilgor, E.; Yilgor, I. Electrospun Polycaprolactone/Silk Fibroin Nanofibrous Bioactive Scaffolds for Tissue Engineering Applications. *Polymer* **2019**, *168*, 86–94. [CrossRef]
64. Santoro, M.; Shah, S.R.; Walker, J.L.; Mikos, A.G. Poly(lactic acid) nanofibrous scaffolds for tissue engineering. *Adv. Drug Deliv. Rev.* **2016**, *107*, 206–212. [CrossRef]
65. Bigg, D.M. Polylactide Copolymers: Effect of Copolymer Ratio and End Capping on Their Properties. *Adv. Polym. Technol.* **2005**, *24*, 69–82. [CrossRef]
66. Maduka, C.V.; Alhaj, M.; Ural, E.; Habeeb, M.O.; Kuhnert, M.M.; Smith, K.; Makela, A.V.; Pope, H.; Chen, S.; Hix, J.M.; et al. Polylactide Degradation Activates Immune Cells by Metabolic Reprogramming. *bioRxiv* **2022**. [CrossRef]
67. Gorth, D.; J Webster, T. 10—Matrices for Tissue Engineering and Regenerative Medicine. In *Biomaterials for Artificial Organs*; Lysaght, M., Webster, T.J., Eds.; Woodhead Publishing Series in Biomaterials; Woodhead Publishing: Sawston, UK, 2011; pp. 270–286. ISBN 978-1-84569-653-5.
68. Wong, W.H.; Mooney, D.J. Synthesis and Properties of Biodegradable Polymers Used as Synthetic Matrices for Tissue Engineering. In *Synthetic Biodegradable Polymer Scaffolds*; Atala, A., Mooney, D.J., Eds.; Birkhäuser: Boston, MA, USA, 1997; pp. 51–82. ISBN 978-1-4612-4154-6.
69. Barnes, C.P.; Sell, S.A.; Boland, E.D.; Simpson, D.G.; Bowlin, G.L. Nanofiber Technology: Designing the next Generation of Tissue Engineering Scaffolds. *Adv. Drug Deliv. Rev.* **2007**, *59*, 1413–1433. [CrossRef]
70. Gilding, D.K.; Reed, A.M. Biodegradable Polymers for Use in Surgery—Polyglycolic/Poly(Actic Acid) Homo- and Copolymers: 1. *Polymer* **1979**, *20*, 1459–1464. [CrossRef]
71. Oh, S.H.; Kang, S.G.; Kim, E.S.; Cho, S.H.; Lee, J.H. Fabrication and Characterization of Hydrophilic Poly(Lactic-Co-Glycolic Acid)/Poly(Vinyl Alcohol) Blend Cell Scaffolds by Melt-Molding Particulate-Leaching Method. *Biomaterials* **2003**, *24*, 4011–4021. [CrossRef]
72. Pamuła, E.; Dobrzyński, P.; Bero, M.; Paluszkiewicz, C. Hydrolytic Degradation of Porous Scaffolds for Tissue Engineering from Terpolymer of L-Lactide, ε-Caprolactone and Glycolide. *J. Mol. Struct.* **2005**, *744–747*, 557–562. [CrossRef]
73. Teixeira, M.A.; Amorim, M.T.P.; Felgueiras, H.P. Poly(Vinyl Alcohol)-Based Nanofibrous Electrospun Scaffolds for Tissue Engineering Applications. *Polymers* **2019**, *12*, 7. [CrossRef]
74. Coelho, D.; Sampaio, A.; Silva, C.J.S.M.; Felgueiras, H.P.; Amorim, M.T.P.; Zille, A. Antibacterial Electrospun Poly(Vinyl Alcohol)/Enzymatic Synthesized Poly(Catechol) Nanofibrous Midlayer Membrane for Ultrafiltration. *ACS Appl. Mater. Interfaces* **2017**, *9*, 33107–33118. [CrossRef]
75. Park, J.-C.; Ito, T.; Kim, K.-O.; Kim, K.-W.; Kim, B.-S.; Khil, M.-S.; Kim, H.-Y.; Kim, I.-S. Electrospun Poly(Vinyl Alcohol) Nanofibers: Effects of Degree of Hydrolysis and Enhanced Water Stability. *Polym. J.* **2010**, *42*, 273–276. [CrossRef]
76. Huang, C.-Y.; Hu, K.-H.; Wei, Z.-H. Comparison of Cell Behavior on Pva/Pva-Gelatin Electrospun Nanofibers with Random and Aligned Configuration. *Sci. Rep.* **2016**, *6*, 37960. [CrossRef]
77. Ino, J.M.; Chevallier, P.; Letourneur, D.; Mantovani, D.; Le Visage, C. Plasma Functionalization of Poly(Vinyl Alcohol) Hydrogel for Cell Adhesion Enhancement. *Biomatter* **2013**, *3*, e25414. [CrossRef]
78. Ogueri, K.S.; Ogueri, K.S.; Allcock, H.R.; Laurencin, C.T. Polyphosphazene Polymers: The next Generation of Biomaterials for Regenerative Engineering and Therapeutic Drug Delivery. *J. Vac. Sci. Technol. B Nanotechnol. Microelectron.* **2020**, *38*, 030801. [CrossRef]
79. Allcock, H. Chemistry and Applications of Polyphosphazenes | Wiley. Available online: https://www.wiley.com/en-in/Chemistry+and+Applications+of+Polyphosphazenes-p-9780471443711 (accessed on 12 August 2022).
80. Deng, M.; Kumbar, S.G.; Wan, Y.; Toti, U.S.; Allcock, H.R.; Laurencin, C.T. Polyphosphazene Polymers for Tissue Engineering: An Analysis of Material Synthesis, Characterization and Applications. *Soft Matter* **2010**, *6*, 3119–3132. [CrossRef]
81. Deng, M.; Nair, L.S.; Nukavarapu, S.P.; Kumbar, S.G.; Brown, J.L.; Krogman, N.R.; Weikel, A.L.; Allcock, H.R.; Laurencin, C.T. Biomimetic, Bioactive Etheric Polyphosphazene-Poly(Lactide-Co-Glycolide) Blends for Bone Tissue Engineering. *J. Biomed. Mater. Res. A* **2010**, *92A*, 114–125. [CrossRef]
82. Deng, M.; Nair, L.S.; Nukavarapu, S.P.; Jiang, T.; Kanner, W.A.; Li, X.; Kumbar, S.G.; Weikel, A.L.; Krogman, N.R.; Allcock, H.R.; et al. Dipeptide-Based Polyphosphazene and Polyester Blends for Bone Tissue Engineering. *Biomaterials* **2010**, *31*, 4898–4908. [CrossRef]

83. Singh, A.; Krogman, N.R.; Sethuraman, S.; Nair, L.S.; Sturgeon, J.L.; Brown, P.W.; Laurencin, C.T.; Allcock, H.R. Effect of Side Group Chemistry on the Properties of Biodegradable L-Alanine Cosubstituted Polyphosphazenes. *Biomacromolecules* **2006**, *7*, 914–918. [CrossRef]
84. Sengupta, P.; Prasad, B.L.V. Surface Modification of Polymers for Tissue Engineering Applications: Arginine Acts as a Sticky Protein Equivalent for Viable Cell Accommodation. *ACS Omega* **2018**, *3*, 4242–4251. [CrossRef] [PubMed]
85. Ito, Y.; Zheng, J.; Qin Liu, S.; Imanishi, Y. Novel Biomaterials Immobilized with Biosignal Molecules. *Mater. Sci. Eng. C* **1994**, *2*, 67–72. [CrossRef]
86. Vasita, R.; Shanmugam, I.K.; Katt, D.S. Improved Biomaterials for Tissue Engineering Applications: Surface Modification of Polymers. *Curr. Top. Med. Chem.* **2008**, *8*, 341–353. [CrossRef] [PubMed]
87. Tallawi, M.; Rosellini, E.; Barbani, N.; Cascone, M.G.; Rai, R.; Saint-Pierre, G.; Boccaccini, A.R. Strategies for the Chemical and Biological Functionalization of Scaffolds for Cardiac Tissue Engineering: A Review. *J. R. Soc. Interface* **2015**, *12*, 20150254. [CrossRef]
88. Lutolf, M.P.; Hubbell, J.A. Synthetic Biomaterials as Instructive Extracellular Microenvironments for Morphogenesis in Tissue Engineering. *Nat. Biotechnol.* **2005**, *23*, 47–55. [CrossRef]
89. Järveläinen, H.; Sainio, A.; Koulu, M.; Wight, T.N.; Penttinen, R. Extracellular Matrix Molecules: Potential Targets in Pharmacotherapy. *Pharm. Rev.* **2009**, *61*, 198–223. [CrossRef]
90. Frantz, C.; Stewart, K.M.; Weaver, V.M. The Extracellular Matrix at a Glance. *J. Cell Sci.* **2010**, *123*, 4195–4200. [CrossRef]
91. Rozario, T.; DeSimone, D.W. The Extracellular Matrix in Development and Morphogenesis: A Dynamic View. *Dev. Biol.* **2010**, *341*, 126–140. [CrossRef]
92. Lotfi, M.; Nejib, M.; Naceur, M. *Cell Adhesion to Biomaterials: Concept of Biocompatibility*; IntechOpen: London, UK, 2013; ISBN 978-953-51-1051-4.
93. Schaefer, L.; Schaefer, R.M. Proteoglycans: From Structural Compounds to Signaling Molecules. *Cell Tissue Res.* **2010**, *339*, 237–246. [CrossRef]
94. Duan, Y.; Wang, Z.; Yan, W.; Wang, S.; Zhang, S.; Jia, J. Preparation of Collagen-Coated Electrospun Nanofibers by Remote Plasma Treatment and Their Biological Properties. *J. Biomater. Sci. Polym. Ed.* **2007**, *18*, 1153–1164. [CrossRef]
95. Zhang, Y.Z.; Venugopal, J.; Huang, Z.-M.; Lim, C.T.; Ramakrishna, S. Characterization of the Surface Biocompatibility of the Electrospun PCL-Collagen Nanofibers Using Fibroblasts. *Biomacromolecules* **2005**, *6*, 2583–2589. [CrossRef]
96. Safaeijavan, R.; Soleimani, M.; Divsalar, A.; Eidi, A.; Ardeshirylajimi, A. Biological Behavior Study of Gelatin Coated PCL Nanofiberous Electrospun Scaffolds Using Fibroblasts. *Arch. Adv. Biosci.* **2014**, *5*, 67–73. [CrossRef]
97. Liverani, L.; Killian, M.S.; Boccaccini, A.R. Fibronectin Functionalized Electrospun Fibers by Using Benign Solvents: Best Way to Achieve Effective Functionalization. *Front. Bioeng. Biotechnol.* **2019**, *7*, 68. [CrossRef]
98. Xie, J.; MacEwan, M.R.; Ray, W.Z.; Liu, W.; Siewe, D.Y.; Xia, Y. Radially Aligned, Electrospun Nanofibers as Dural Substitutes for Wound Closure and Tissue Regeneration Applications. *ACS Nano* **2010**, *4*, 5027–5036. [CrossRef]
99. Choi, W.S.; Bae, J.W.; Lim, H.R.; Joung, Y.K.; Park, J.-C.; Kwon, I.K.; Park, K.D. RGD Peptide-Immobilized Electrospun Matrix of Polyurethane for Enhanced Endothelial Cell Affinity. *Biomed. Mater.* **2008**, *3*, 044104. [CrossRef]
100. Cavanaugh, M.; Silantyeva, E.; Pylypiv Koh, G.; Malekzadeh, E.; Lanzinger, W.D.; Willits, R.K.; Becker, M.L. RGD-Modified Nanofibers Enhance Outcomes in Rats after Sciatic Nerve Injury. *J. Funct. Biomater.* **2019**, *10*, 24. [CrossRef]
101. Varshosaz, J.; Arabloo, K.; Sarrami, N.; Ghassami, E.; Yazdani Kachouei, E.; Kouhi, M.; Jahanian-Najafabadi, A. RGD Peptide Grafted Polybutylene Adipate-*Co*-Terephthalate/Gelatin Electrospun Nanofibers Loaded with a Matrix Metalloproteinase Inhibitor Drug for Alleviating of Wounds: An in Vitro/in Vivo Study. *Drug Dev. Ind. Pharm.* **2020**, *46*, 484–497. [CrossRef]
102. Onak, G.; Şen, M.; Horzum, N.; Ercan, U.K.; Yaralı, Z.B.; Garipcan, B.; Karaman, O. Ozan Karaman Aspartic and Glutamic Acid Templated Peptides Conjugation on Plasma Modified Nanofibers for Osteogenic Differentiation of Human Mesenchymal Stem Cells: A Comparative Study. *Sci. Rep.* **2018**, *8*, 17620. [CrossRef]
103. Chen, W.; Guo, L.; Tang, C.; Tsai, C.; Huang, H.; Chin, T.; Yang, M.-L.; Chen-Yang, Y. The Effect of Laminin Surface Modification of Electrospun Silica Nanofiber Substrate on Neuronal Tissue Engineering. *Nanomaterials* **2018**, *8*, 165. [CrossRef]
104. Koh, H.S.; Yong, T.; Chan, C.K.; Ramakrishna, S. Enhancement of Neurite Outgrowth Using Nano-Structured Scaffolds Coupled with Laminin. *Biomaterials* **2008**, *29*, 3574–3582. [CrossRef]
105. Junka, R.; Valmikinathan, C.M.; Kalyon, D.M.; Yu, X. Laminin Functionalized Biomimetic Nanofibers for Nerve Tissue Engineering. *J. Biomater. Tissue Eng.* **2013**, *3*, 494–502. [CrossRef] [PubMed]
106. Pan, J.; Liu, N.; Shu, L.; Sun, H. Application of Avidin-Biotin Technology to Improve Cell Adhesion on Nanofibrous Matrices. *J. Nanobiotechnol.* **2015**, *13*, 37. [CrossRef]
107. Lee, H.; Lim, S.; Birajdar, M.S.; Lee, S.-H.; Park, H. Fabrication of FGF-2 Immobilized Electrospun Gelatin Nanofibers for Tissue Engineering. *Int. J. Biol. Macromol.* **2016**, *93*, 1559–1566. [CrossRef] [PubMed]
108. Ramos, D.M.; Abdulmalik, S.; Arul, M.R.; Rudraiah, S.; Laurencin, C.T.; Mazzocca, A.D.; Kumbar, S.G. Insulin Immobilized PCL-Cellulose Acetate Micro-Nanostructured Fibrous Scaffolds for Tendon Tissue Engineering. *Polym. Adv. Technol.* **2019**, *30*, 1205–1215. [CrossRef] [PubMed]
109. Qi, Z.; Guo, W.; Zheng, S.; Fu, C.; Ma, Y.; Pan, S.; Liu, Q.; Yang, X. Enhancement of Neural Stem Cell Survival, Proliferation and Differentiation by IGF-1 Delivery in Graphene Oxide-Incorporated PLGA Electrospun Nanofibrous Mats. *RSC Adv.* **2019**, *9*, 8315–8325. [CrossRef]

110. Chen, X.; Wang, X.; Wang, S.; Zhang, X.; Yu, J.; Wang, C. Mussel-Inspired Polydopamine-Assisted Bromelain Immobilization onto Electrospun Fibrous Membrane for Potential Application as Wound Dressing. *Mater. Sci. Eng. C* **2020**, *110*, 110624. [CrossRef]
111. Liu, Y.; Zhou, G.; Liu, Z.; Guo, M.; Jiang, X.; Taskin, M.B.; Zhang, Z.; Liu, J.; Tang, J.; Bai, R.; et al. Mussel Inspired Poly-norepinephrine Functionalized Electrospun Polycaprolactone Microfibers for Muscle Regeneration. *Sci. Rep.* **2017**, *7*, 8197. [CrossRef]
112. Taskin, M.B.; Xu, R.; Zhao, H.; Wang, X.; Dong, M.; Besenbacher, F.; Chen, M. Poly(Norepinephrine) as a Functional Bio-Interface for Neuronal Differentiation on Electrospun Fibers. *Phys. Chem. Chem. Phys.* **2015**, *17*, 9446–9453. [CrossRef]
113. Zadeh, K.M.; Luyt, A.S.; Zarif, L.; Augustine, R.; Hasan, A.; Messori, M.; Hassan, M.K.; Yalcin, H.C. Electrospun Polylactic Acid/Date Palm Polyphenol Extract Nanofibres for Tissue Engineering Applications. *Emerg. Mater.* **2019**, *2*, 141–151. [CrossRef]
114. Kai, D.; Prabhakaran, M.P.; Jin, G.; Tian, L.; Ramakrishna, S. Potential of VEGF-Encapsulated Electrospun Nanofibers for in Vitro Cardiomyogenic Differentiation of Human Mesenchymal Stem Cells. *J. Tissue Eng. Regen. Med.* **2017**, *11*, 1002–1010. [CrossRef]
115. Guex, A.G.; Hegemann, D.; Giraud, M.N.; Tevaearai, H.T.; Popa, A.M.; Rossi, R.M.; Fortunato, G. Covalent Immobilisation of VEGF on Plasma-Coated Electrospun Scaffolds for Tissue Engineering Applications. *Colloids Surf. B Biointerfaces* **2014**, *123*, 724–733. [CrossRef]
116. Strom, S.C.; Michalopoulos, G. Collagen as a Substrate for Cell Growth and Differentiation. *Methods Enzym.* **1982**, *82 Pt A*, 544–555. [CrossRef]
117. Ashwin, B.; Abinaya, B.; Prasith, T.P.; Chandran, S.V.; Yadav, L.R.; Vairamani, M.; Patil, S.; Selvamurugan, N. 3D-Poly (Lactic Acid) Scaffolds Coated with Gelatin and Mucic Acid for Bone Tissue Engineering. *Int. J. Biol. Macromol.* **2020**, *162*, 523–532. [CrossRef]
118. Foox, M.; Zilberman, M. Drug Delivery from Gelatin-Based Systems. *Expert Opin. Drug Deliv.* **2015**, *12*, 1547–1563. [CrossRef]
119. Bikuna-Izagirre, M.; Aldazabal, J.; Paredes, J. Gelatin Blends Enhance Performance of Electrospun Polymeric Scaffolds in Comparison to Coating Protocols. *Polymers* **2022**, *14*, 1311. [CrossRef]
120. Bellis, S.L. Advantages of RGD Peptides for Directing Cell Association with Biomaterials. *Biomaterials* **2011**, *32*, 4205–4210. [CrossRef]
121. Hersel, U.; Dahmen, C.; Kessler, H. RGD Modified Polymers: Biomaterials for Stimulated Cell Adhesion and Beyond. *Biomaterials* **2003**, *24*, 4385–4415. [CrossRef]
122. Kurpanik, R.; Stodolak-Zych, E. Chemical and Physical Modifications of Electrospun Fibers as a Method to Stimulate Tissue Regeneration—Minireview. *Eng. Biomater.* **2021**, 31–41. [CrossRef]
123. Talovic, M.; Marcinczyk, M.; Ziemkiewicz, N.; Garg, K. Laminin Enriched Scaffolds for Tissue Engineering Applications. *Adv. Tissue Eng. Regen. Med. Open Access* **2017**, 2, 194–200.
124. Rechler, M.M.; Nissley, S.P. The Nature and Regulation of the Receptors for Insulin-like Growth Factors. *Annu. Rev. Physiol.* **1985**, *47*, 425–442. [CrossRef]
125. Ramos, D.M.; Abdulmalik, S.; Arul, M.R.; Sardashti, N.; Banasavadi-Siddegowda, Y.K.; Nukavarapu, S.P.; Drissi, H.; Kumbar, S.G. Insulin-Functionalized Bioactive Fiber Matrices with Bone Marrow-Derived Stem Cells in Rat Achilles Tendon Regeneration. *ACS Appl. Bio Mater.* **2022**, *5*, 2851–2861. [CrossRef] [PubMed]
126. Chen, X.; Gao, Y.; Wang, Y.; Pan, G. Mussel-Inspired Peptide Mimicking: An Emerging Strategy for Surface Bioengineering of Medical Implants. *Smart Mater. Med.* **2021**, *2*, 26–37. [CrossRef]
127. Raja, I.S.; Preeth, D.R.; Vedhanayagam, M.; Hyon, S.-H.; Lim, D.; Kim, B.; Rajalakshmi, S.; Han, D.-W. Polyphenols-Loaded Electrospun Nanofibers in Bone Tissue Engineering and Regeneration. *Biomater. Res.* **2021**, *25*, 29. [CrossRef] [PubMed]
128. Rather, H.A.; Patel, R.; Yadav, U.C.S.; Vasita, R. Dual Drug-Delivering Polycaprolactone-Collagen Scaffold to Induce Early Osteogenic Differentiation and Coupled Angiogenesis. *Biomed. Mater.* **2020**, *15*, 045008. [CrossRef] [PubMed]
129. Li, B.; Wang, H.; Qiu, G.; Su, X.; Wu, Z. Synergistic Effects of Vascular Endothelial Growth Factor on Bone Morphogenetic Proteins Induced Bone Formation In Vivo: Influencing Factors and Future Research Directions. *BioMed Res. Int.* **2016**, *2016*, 2869572. [CrossRef]
130. Barati, D.; Shariati, S.R.P.; Moeinzadeh, S.; Melero-Martin, J.M.; Khademhosseini, A.; Jabbari, E. Spatiotemporal Release of BMP-2 and VEGF Enhances Osteogenic and Vasculogenic Differentiation of Human Mesenchymal Stem Cells and Endothelial Colony-Forming Cells Co-Encapsulated in a Patterned Hydrogel. *J. Control. Release* **2016**, *223*, 126–136. [CrossRef]
131. Kutryk, M.J.B.; Stewart, D.J. Angiogenesis of the Heart. *Microsc. Res. Tech.* **2003**, *60*, 138–158. [CrossRef]
132. Esquirol Caussa, J.; Herrero Vila, E. Epidermal Growth Factor, Innovation and Safety. *Med. Clín. (Engl. Ed.)* **2015**, *145*, 305–312. [CrossRef]
133. Liao, J.-L.; Zhong, S.; Wang, S.-H.; Liu, J.-Y.; Chen, J.; He, G.; He, B.; Xu, J.-Q.; Liang, Z.-H.; Mei, T.; et al. Preparation and Properties of a Novel Carbon Nanotubes/Poly(Vinyl Alcohol)/Epidermal Growth Factor Composite Biological Dressing. *Exp. Med.* **2017**, *14*, 2341–2348. [CrossRef]
134. Bayrak, E. *Nanofibers: Production, Characterization, and Tissue Engineering Applications*; IntechOpen: London, UK, 2022; ISBN 978-1-80355-085-5.
135. Vasita, R.; Katti, D.S. Nanofibers and Their Applications in Tissue Engineering. *Int. J. Nanomed.* **2006**, *1*, 15–30. [CrossRef]
136. Sharma, J.; Lizu, M.; Stewart, M.; Zygula, K.; Lu, Y.; Chauhan, R.; Yan, X.; Guo, Z.; Wujcik, E.K.; Wei, S. Multifunctional Nanofibers towards Active Biomedical Therapeutics. *Polymers* **2015**, *7*, 186–219. [CrossRef]
137. Ma, P.X.; Zhang, R. Synthetic Nano-Scale Fibrous Extracellular Matrix. *J. Biomed. Mater. Res.* **1999**, *46*, 60–72. [CrossRef]

138. Wang, L.; Gong, C.; Yuan, X.; Wei, G. Controlling the Self-Assembly of Biomolecules into Functional Nanomaterials through Internal Interactions and External Stimulations: A Review. *Nanomaterials* **2019**, *9*, 285. [CrossRef]
139. Feng, L.; Li, S.H.; Zhai, J.; Song, Y.L.; Jiang, L.; Zhu, D.B. Template Based Synthesis of Aligned Polyacrylonitrile Nanofibers Using a Novel Extrusion Method. *Synth. Met.* **2003**, *135–136*, 817–818. [CrossRef]
140. Agarwal, S.; Wendorff, J.H.; Greiner, A. Progress in the Field of Electrospinning for Tissue Engineering Applications. *Adv. Mater.* **2009**, *21*, 3343–3351. [CrossRef]
141. Ramakrishna, S.; Fujihara, K.; Teo, W.-E.; Lim, T.-C.; Ma, Z. *An Introduction to Electrospinning and Nanofibers*; World Scientific: Singapore, 2005; ISBN 978-981-256-415-3.
142. Alghoraibi, I. Different Methods for Nanofibers Design and Fabrication. In *Handbook of Nanofibers*; Springer: Berlin/Heidelberg, Germany, 2018; ISBN 978-3-319-42789-8.
143. Baumgarten, P.K. Electrostatic Spinning of Acrylic Microfibers. *J. Colloid Interface Sci.* **1971**, *36*, 71–79. [CrossRef]
144. Li, Y.; Bou-Akl, T. Electrospinning in Tissue Engineering. In *Electrospinning—Material, Techniques, and Biomedical Applications*; Haider, S., Haider, A., Eds.; InTech Open: London, UK, 2016; ISBN 978-953-51-2821-2.
145. Li, D.; Wang, Y.; Xia, Y. Electrospinning Nanofibers as Uniaxially Aligned Arrays and Layer-by-Layer Stacked Films. *Adv. Mater.* **2004**, *16*, 361–366. [CrossRef]
146. Rahmati, M.; Mills, D.K.; Urbanska, A.M.; Saeb, M.R.; Venugopal, J.R.; Ramakrishna, S.; Mozafari, M. Electrospinning for Tissue Engineering Applications. *Prog. Mater. Sci.* **2021**, *117*, 100721. [CrossRef]
147. Casper, C.L.; Stephens, J.S.; Tassi, N.G.; Chase, D.B.; Rabolt, J.F. Controlling Surface Morphology of Electrospun Polystyrene Fibers: Effect of Humidity and Molecular Weight in the Electrospinning Process. *Macromolecules* **2004**, *37*, 573–578. [CrossRef]
148. Jabur, A.R.; Abbas, L.K.; Muhi Aldain, S.M. Effects of Ambient Temperature and Needle to Collector Distance on PVA Nanofibers Diameter Obtained From Electrospinning Technique. *Eng. Technol. J.* **2017**, *35*, 340–347. [CrossRef]
149. Fridrikh, S.V.; Yu, J.H.; Brenner, M.P.; Rutledge, G.C. Controlling the Fiber Diameter during Electrospinning. *Phys. Rev. Lett.* **2003**, *90*, 144502. [CrossRef] [PubMed]
150. Nikmaram, N.; Roohinejad, S.; Hashemi, S.; Koubaa, M.; Barba, F.J.; Abbaspourrad, A.; Greiner, R. Emulsion-Based Systems for Fabrication of Electrospun Nanofibers: Food, Pharmaceutical and Biomedical Applications. *RSC Adv.* **2017**, *7*, 28951–28964. [CrossRef]
151. Yan, K.; Le, Y.; Mengen, H.; Zhongbo, L.; Zhulin, H. Effect of Solution Miscibility on the Morphology of Coaxial Electrospun Cellulose Acetate Nanofibers. *Polymers* **2021**, *13*, 4419. [CrossRef] [PubMed]
152. Liu, Z.; Ramakrishna, S.; Liu, X. Electrospinning and Emerging Healthcare and Medicine Possibilities. *APL Bioeng.* **2020**, *4*, 030901. [CrossRef] [PubMed]
153. Cheng, G.; Yin, C.; Tu, H.; Jiang, S.; Wang, Q.; Zhou, X.; Xing, X.; Xie, C.; Shi, X.; Du, Y.; et al. Controlled Co-Delivery of Growth Factors through Layer-by-Layer Assembly of Core–Shell Nanofibers for Improving Bone Regeneration. *ACS Nano* **2019**, *13*, 6372–6382. [CrossRef]
154. He, P.; Zhong, Q.; Ge, Y.; Guo, Z.; Tian, J.; Zhou, Y.; Ding, S.; Li, H.; Zhou, C. Dual Drug Loaded Coaxial Electrospun PLGA/PVP Fiber for Guided Tissue Regeneration under Control of Infection. *Mater. Sci. Eng. C* **2018**, *90*, 549–556. [CrossRef]
155. Nagiah, N.; Murdock, C.J.; Bhattacharjee, M.; Nair, L.; Laurencin, C.T. Development of Tripolymeric Triaxial Electrospun Fibrous Matrices for Dual Drug Delivery Applications. *Sci. Rep.* **2020**, *10*, 609. [CrossRef]
156. Balusamy, B.; Celebioglu, A.; Senthamizhan, A.; Uyar, T. Progress in the Design and Development of "Fast-Dissolving" Electrospun Nanofibers Based Drug Delivery Systems—A Systematic Review. *J. Control. Release* **2020**, *326*, 482–509. [CrossRef]
157. Salehhudin, H.; Mohamad, E.; Mahadi, W.; Afifi, A. Multiple-Jet Electrospinning Methods for Nanofiber Processing: A Review. *Mater. Manuf. Process.* **2017**, *33*, 479–498. [CrossRef]
158. Salehhudin, H.; Mohamad, E.; Mahadi, W.; Afifi, A. Simulation and Experimental Study of Parameters in Multiple-Nozzle Electrospinning: Effects of Nozzle Arrangement on Jet Paths and Fiber Formation. *J. Manuf. Process.* **2021**, *62*, 440–449. [CrossRef]
159. Szentivanyi, A.; Chakradeo, T.; Zernetsch, H.; Glasmacher, B. Electrospun Cellular Microenvironments: Understanding Controlled Release and Scaffold Structure. *Adv. Drug Deliv. Rev.* **2011**, *63*, 209–220. [CrossRef]
160. Zeng, J.; Aigner, A.; Czubayko, F.; Kissel, T.; Wendorff, J.H.; Greiner, A. Poly(Vinyl Alcohol) Nanofibers by Electrospinning as a Protein Delivery System and the Retardation of Enzyme Release by Additional Polymer Coatings. *Biomacromolecules* **2005**, *6*, 1484–1488. [CrossRef]
161. Szentivanyi, A.; Assmann, U.; Schuster, R.; Glasmacher, B. Production of Biohybrid Protein/PEO Scaffolds by Electrospinning. *Mater. Werkst.* **2009**, *40*, 65–72. [CrossRef]
162. Quan, Z.; Xu, Y.; Rong, H.; Yang, W.; Yang, Y.; Wei, G.; Ji, D.; Qin, X. Preparation of Oil-in-Water Core-Sheath Nanofibers through Emulsion Electrospinning for Phase Change Temperature Regulation. *Polymer* **2022**, *256*, 125252. [CrossRef]
163. Zhang, C.; Feng, F.; Zhang, H. Emulsion Electrospinning: Fundamentals, Food Applications and Prospects. *Trends Food Sci. Technol.* **2018**, *80*, 175–186. [CrossRef]
164. Xu, X.; Yang, L.; Xu, X.; Wang, X.; Chen, X.; Liang, Q.; Zeng, J.; Jing, X. Ultrafine Medicated Fibers Electrospun from W/O Emulsions. *J. Control. Release* **2005**, *108*, 33–42. [CrossRef]
165. Sy, J.C.; Klemm, A.S.; Shastri, V.P. Emulsion as a Means of Controlling Electrospinning of Polymers. *Adv. Mater.* **2009**, *21*, 1814–1819. [CrossRef]

166. Hong, J.; Yeo, M.; Yang, G.H.; Kim, G. Cell-Electrospinning and Its Application for Tissue Engineering. *Int. J. Mol. Sci.* **2019**, *20*, 6208. [CrossRef]
167. Jayasinghe, S.N. Cell Electrospinning: A Novel Tool for Functionalising Fibres, Scaffolds and Membranes with Living Cells and Other Advanced Materials for Regenerative Biology and Medicine. *Analyst* **2013**, *138*, 2215. [CrossRef]
168. Townsend-Nicholson, A.; Jayasinghe, S.N. Cell Electrospinning: A Unique Biotechnique for Encapsulating Living Organisms for Generating Active Biological Microthreads/Scaffolds. *Biomacromolecules* **2006**, *7*, 3364–3369. [CrossRef]
169. Guo, Y.; Gilbert-Honick, J.; Somers, S.M.; Mao, H.-Q.; Grayson, W.L. Modified Cell-Electrospinning for 3D Myogenesis of C2C12s in Aligned Fibrin Microfiber Bundles. *Biochem. Biophys. Res. Commun.* **2019**, *516*, 558–564. [CrossRef] [PubMed]
170. Kong, B.; Mi, S. Electrospun Scaffolds for Corneal Tissue Engineering: A Review. *Materials* **2016**, *9*, 614. [CrossRef] [PubMed]
171. Kim, P.-H.; Cho, J.-Y. Myocardial Tissue Engineering Using Electrospun Nanofiber Composites. *BMB Rep.* **2016**, *49*, 26–36. [CrossRef] [PubMed]
172. Malkoc, V.; Chang, L. Applications of Electrospun Nanofibers in Neural Tissue Engineering. *Eur. J. BioMed. Res.* **2015**, *1*, 25. [CrossRef]
173. Liu, W.; Thomopoulos, S.; Xia, Y. Electrospun Nanofibers for Regenerative Medicine. *Adv. Healthc. Mater.* **2012**, *1*, 10–25. [CrossRef]
174. Qian, J.; Lin, Z.; Liu, Y.; Wang, Z.; Lin, Y.; Gong, C.; Ruan, R.; Zhang, J.; Yang, H. Functionalization Strategies of Electrospun Nanofibrous Scaffolds for Nerve Tissue Engineering. *Smart Mater. Med.* **2021**, *2*, 260–279. [CrossRef]
175. Xie, J.; Liu, W.; MacEwan, M.R.; Bridgman, P.C.; Xia, Y. Neurite Outgrowth on Electrospun Nanofibers with Uniaxial Alignment: The Effects of Fiber Density, Surface Coating, and Supporting Substrate. *ACS Nano* **2014**, *8*, 1878–1885. [CrossRef]
176. Xie, J.; Willerth, S.M.; Li, X.; Macewan, M.R.; Rader, A.; Sakiyama-Elbert, S.E.; Xia, Y. The Differentiation of Embryonic Stem Cells Seeded on Electrospun Nanofibers into Neural Lineages. *Biomaterials* **2009**, *30*, 354–362. [CrossRef]
177. Devillard, C.D.; Marquette, C.A. Vascular Tissue Engineering: Challenges and Requirements for an Ideal Large Scale Blood Vessel. *Front. Bioeng. Biotechnol.* **2021**, *9*, 721843. [CrossRef]
178. Nugent, H.M.; Edelman, E.R. Tissue Engineering Therapy for Cardiovascular Disease. *Circ. Res.* **2003**, *92*, 1068–1078. [CrossRef]
179. Shin, Y.C.; Kim, J.; Kim, S.E.; Song, S.-J.; Hong, S.W.; Oh, J.-W.; Lee, J.; Park, J.-C.; Hyon, S.-H.; Han, D.-W. RGD Peptide and Graphene Oxide Co-Functionalized PLGA Nanofiber Scaffolds for Vascular Tissue Engineering. *Regen. Biomater.* **2017**, *4*, 159–166. [CrossRef]
180. Marelli, B.; Alessandrino, A.; Farè, S.; Freddi, G.; Mantovani, D.; Tanzi, M.C. Compliant Electrospun Silk Fibroin Tubes for Small Vessel Bypass Grafting. *Acta Biomater.* **2010**, *6*, 4019–4026. [CrossRef]
181. Sophia Fox, A.J.; Bedi, A.; Rodeo, S.A. The Basic Science of Articular Cartilage. *Sports Health* **2009**, *1*, 461–468. [CrossRef]
182. Semitela, Â.; Girão, A.F.; Fernandes, C.; Ramalho, G.; Bdikin, I.; Completo, A.; Marques, P.A. Electrospinning of Bioactive Polycaprolactone-Gelatin Nanofibres with Increased Pore Size for Cartilage Tissue Engineering Applications. *J. Biomater. Appl.* **2020**, *35*, 471–484. [CrossRef]
183. Baker, B.M.; Shah, R.P.; Silverstein, A.M.; Esterhai, J.L.; Burdick, J.A.; Mauck, R.L. Sacrificial Nanofibrous Composites Provide Instruction without Impediment and Enable Functional Tissue Formation. *Proc. Natl. Acad. Sci. USA* **2012**, *109*, 14176–14181. [CrossRef]
184. Donnaloja, F.; Jacchetti, E.; Soncini, M.; Raimondi, M.T. Natural and Synthetic Polymers for Bone Scaffolds Optimization. *Polymers* **2020**, *12*, 905. [CrossRef]
185. Turnbull, G.; Clarke, J.; Picard, F.; Riches, P.; Jia, L.; Han, F.; Li, B.; Shu, W. 3D Bioactive Composite Scaffolds for Bone Tissue Engineering. *Bioact. Mater.* **2018**, *3*, 278–314. [CrossRef]
186. Gong, T.; Xie, J.; Liao, J.; Zhang, T.; Lin, S.; Lin, Y. Nanomaterials and Bone Regeneration. *Bone Res.* **2015**, *3*, 15029. [CrossRef]
187. Nekounam, H.; Allahyari, Z.; Gholizadeh, S.; Mirzaei, E.; Shokrgozar, M.A.; Faridi-Majidi, R. Simple and Robust Fabrication and Characterization of Conductive Carbonized Nanofibers Loaded with Gold Nanoparticles for Bone Tissue Engineering Applications. *Mater. Sci. Eng. C* **2020**, *117*, 111226. [CrossRef]
188. Lai, W.-Y.; Feng, S.-W.; Chan, Y.-H.; Chang, W.-J.; Wang, H.-T.; Huang, H.-M. In Vivo Investigation into Effectiveness of Fe3O4/PLLA Nanofibers for Bone Tissue Engineering Applications. *Polymers* **2018**, *10*, 804. [CrossRef]
189. Miszuk, J.; Liang, Z.; Hu, J.; Sanyour, H.; Hong, Z.; Fong, H.; Sun, H. Elastic Mineralized 3D Electrospun PCL Nanofibrous Scaffold for Drug Release and Bone Tissue Engineering. *ACS Appl. Bio Mater.* **2021**, *4*, 3639–3648. [CrossRef] [PubMed]
190. Biazar, E. Application of Polymeric Nanofibers in Soft Tissues Regeneration: Nanofibers in Medical Sciences. *Polym. Adv. Technol.* **2016**, *27*, 1404–1412. [CrossRef]
191. Venugopal, J.; Ramakrishna, S. Biocompatible Nanofiber Matrices for the Engineering of a Dermal Substitute for Skin Regeneration. *Tissue Eng.* **2005**, *11*, 847–854. [CrossRef] [PubMed]
192. Powell, H.M.; Boyce, S.T. Engineered Human Skin Fabricated Using Electrospun Collagen–PCL Blends: Morphogenesis and Mechanical Properties. *Tissue Eng. A* **2009**, *15*, 2177–2187. [CrossRef] [PubMed]
193. Sekiya, N.; Ichioka, S.; Terada, D.; Tsuchiya, S.; Kobayashi, H. Efficacy of a Poly Glycolic Acid (PGA)/Collagen Composite Nanofibre Scaffold on Cell Migration and Neovascularisation in Vivo Skin Defect Model. *J. Plast. Surg. Hand Surg.* **2013**, 498–502. [CrossRef]
194. Pal, P.; Srivas, P.K.; Dadhich, P.; Das, B.; Maulik, D.; Dhara, S. Nano-/Microfibrous Cotton-Wool-Like 3D Scaffold with Core–Shell Architecture by Emulsion Electrospinning for Skin Tissue Regeneration. *ACS Biomater. Sci. Eng.* **2017**, *3*, 3563–3575. [CrossRef]

195. Gumpangseth, T.; Lekawanvijit, S.; Mahakkanukrauh, P. Histological Assessment of the Human Heart Valves and Its Relationship with Age. *Anat. Cell Biol.* **2020**, *53*, 261–271. [CrossRef]
196. Tseng, H.; Puperi, D.S.; Kim, E.J.; Ayoub, S.; Shah, J.V.; Cuchiara, M.L.; West, J.L.; Grande-Allen, K.J. Anisotropic Poly(Ethylene Glycol)/Polycaprolactone Hydrogel–Fiber Composites for Heart Valve Tissue Engineering. *Tissue Eng. A* **2014**, *20*, 2634–2645. [CrossRef]
197. Ahmadi, P.; Nazeri, N.; Derakhshan, M.A.; Ghanbari, H. Preparation and Characterization of Polyurethane/Chitosan/CNT Nanofibrous Scaffold for Cardiac Tissue Engineering. *Int. J. Biol. Macromol.* **2021**, *180*, 590–598. [CrossRef]
198. Xue, Y.; Ravishankar, P.; Zeballos, M.A.; Sant, V.; Balachandran, K.; Sant, S. Valve Leaflet-inspired Elastomeric Scaffolds with Tunable and Anisotropic Mechanical Properties. *Polym. Adv. Technol.* **2020**, *31*, 94–106. [CrossRef]
199. Ravishankar, P.; Khang, A.; Laredo, M.; Balachandran, K. Using Dimensionless Numbers to Predict Centrifugal Jet-Spun Nanofiber Morphology. *J. Nanomater.* **2019**, *2019*, 4639658. [CrossRef]
200. Rogalski, J.J.; Bastiaansen, C.W.M.; Peijs, T. Rotary Jet Spinning Review—A Potential High Yield Future for Polymer Nanofibers. *Nanocomposites* **2017**, *3*, 97–121. [CrossRef]

polymers

MDPI

Review

Angiopep-2-Modified Nanoparticles for Brain-Directed Delivery of Therapeutics: A Review

Saffiya Habib and Moganavelli Singh *

Nano-Gene and Drug Delivery Group, Discipline of Biochemistry, School of Life Sciences, University of KwaZulu-Natal, Private Bag X54001, Durban 4000, South Africa; saffiya.habib@gmail.com
* Correspondence: singhm1@ukzn.ac.za; Tel.: +27-31-2607170

Abstract: Nanotechnology has opened up a world of possibilities for the treatment of brain disorders. Nanosystems can be designed to encapsulate, carry, and deliver a variety of therapeutic agents, including drugs and nucleic acids. Nanoparticles may also be formulated to contain photosensitizers or, on their own, serve as photothermal conversion agents for phototherapy. Furthermore, nano-delivery agents can enhance the efficacy of contrast agents for improved brain imaging and diagnostics. However, effective nano-delivery to the brain is seriously hampered by the formidable blood–brain barrier (BBB). Advances in understanding natural transport routes across the BBB have led to receptor-mediated transcytosis being exploited as a possible means of nanoparticle uptake. In this regard, the oligopeptide Angiopep-2, which has high BBB transcytosis capacity, has been utilized as a targeting ligand. Various organic and inorganic nanostructures have been functionalized with Angiopep-2 to direct therapeutic and diagnostic agents to the brain. Not only have these shown great promise in the treatment and diagnosis of brain cancer but they have also been investigated for the treatment of brain injury, stroke, epilepsy, Parkinson's disease, and Alzheimer's disease. This review focuses on studies conducted from 2010 to 2021 with Angiopep-2-modified nanoparticles aimed at the treatment and diagnosis of brain disorders.

Keywords: Angiopep-2; nanoparticles; transcytosis; drug delivery; brain; targeting

Citation: Habib, S.; Singh, M. Angiopep-2-Modified Nanoparticles for Brain-Directed Delivery of Therapeutics: A Review. *Polymers* **2022**, *14*, 712. https://doi.org/10.3390/polym14040712

Academic Editors: Stephanie Andrade, Joana A. Loureiro and Maria João Ramalho

Received: 19 January 2022
Accepted: 8 February 2022
Published: 12 February 2022

1. Introduction

The blood–brain barrier (BBB) is a selectively permeable network of capillary endothelial cells, astroglia, pericytes, and perivascular mast cells, which stringently regulates the exchange of molecules between the blood and the cerebral tissue. The system functions in protecting the central nervous system (CNS), providing nutrients to the brain, maintaining homeostasis, and regulating communication to and from the CNS [1]. The former protective capability is conferred by the presence of tight intercellular junctions that prevent the entry of pathogens and toxins [2].

In line with the exclusion of foreign substances, the BBB is also a significant impediment to the delivery of therapeutic and diagnostic agents to the brain. Moreover, ATP-binding cassette (ABC) transporters of BBB endothelial cells can expel compounds that may traverse the barrier back into the bloodstream [3]. Consequently, brain disorders are notoriously difficult to diagnose and treat both by conventional methods and nanotechnology.

Understanding the natural routes of transport, such as receptor-mediated transcytosis (RMT), across the BBB has led to the 'trojan horse' concept being widely investigated. This strategy involves modifying nanoparticles (NPs) with ligands that can bind specific receptors at the apical membrane of brain endothelial cells and promote endocytosis. In this way, the entry of the NP is masked through recognition of the ligand. Angiopep-2 is one such ligand [1,4].

Angiopep-2 (TFFYGGSRGKRNNFKTEEY, molecular weight 2.4 kDa) is a 19-amino-acid-long oligopeptide that binds to the low-density lipoprotein receptor-related protein-1

(LRP1) [5]. Identified by Demeule and coworkers [6] as part of a family of Kunitz-domain-derived peptides, Angiopep-2 showed greater transcytosis ability and parenchyma accumulation than the protease inhibitor, aprotinin. Aprotinin was used as it possesses a Kunitz protease inhibitor (KPI) domain, which renders it a good substrate for the low-density lipoprotein receptor-related protein (LRP), which facilitates transport across the BBB.

This study established the framework for the application of Angiopep-2 in brain-directed therapeutics. A representation of RMT of Angiopep-2 is provided in Figure 1. Angiopep-2 has since been appended to anticancer drugs [7,8], a variety of NPs [9–13] and has even been investigated in clinical trials. As a recent example, ANG1005, which consists of three paclitaxel residues linked to Angiopep-2, showed patient benefits in a phase II study of adults with recurrent brain metastases arising from breast cancer [14].

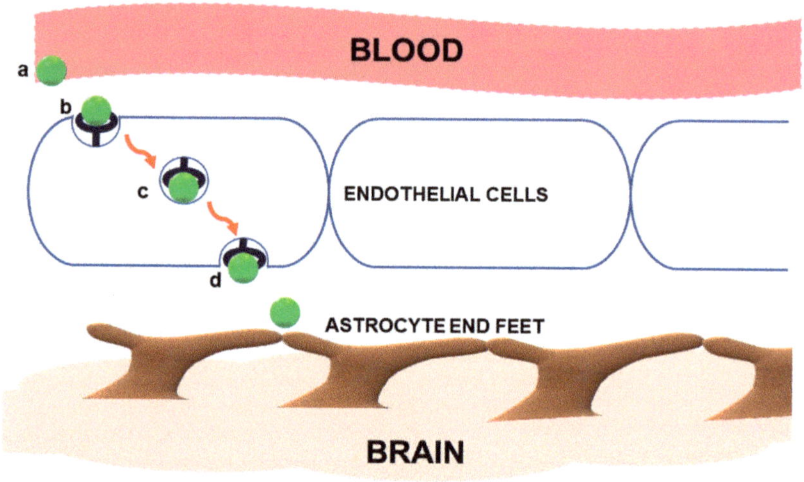

Figure 1. Schematic representation of receptor-mediated transcytosis (RMT) of Angiopep-2. When introduced into the bloodstream, the peptide (a) binds to the low-density lipoprotein receptor-related protein-1 (LRP1) (b) on the apical membrane of brain endothelial cells and initiates invagination of the plasma membrane. The receptor–ligand complex is endocytosed via the intracellular vesicular network (c) and routed to the basolateral membrane, where membrane fusion permits the release of the vesicle contents (d). Angiopep-2 detaches from the receptor and reaches brain cells. Adapted from [15,16].

Although drug–ligand conjugates have demonstrated efficacy [17], the association of drugs with BBB ligand-decorated NPs has, in theory, greater benefits. Not only can the drug-loaded nanostructure traverse the BBB but also it has the potential to improve circulation time, encourage cellular uptake, lower the effective dose required, and reduce drug-induced side effects [18]. Interestingly, Angiopep-2-modified nanoparticles were recently shown to enhance transcytosis across the intestinal epithelium with potential for the design of oral delivery systems [19].

In addition to drugs [20], nanotechnology has opened up the possibility of directing other therapeutic agents to the brain. Functional therapeutic gene segments can be introduced via appropriately designed nanoparticles [21,22]. Similarly, small interfering RNA (siRNA), which function in regulating gene expression via RNA-inference (RNAi) can be introduced to inhibit the expression of disease-causing genes in the brain [23,24].

In other instances, micro RNA (miRNA) technology in the form of miRNA mimics and anti-miRNA oligonucleotides (anti-miRs) can be applied to either restore the function of beneficial miRNA or attenuate that of disease-causing miRNA, such as onco-miRs [25]. In the medical field, nanodevices to transport and deliver contrast agents are important diagnostic tools, improving the efficacy of current imaging systems [26]. This review explores the prospects for nanotherapeutics directed towards the brain, which involve Angiopep-2 as a homing device. The major potential applications of Angiopep-2 decorated NPs is broadly outlined in Figure 2.

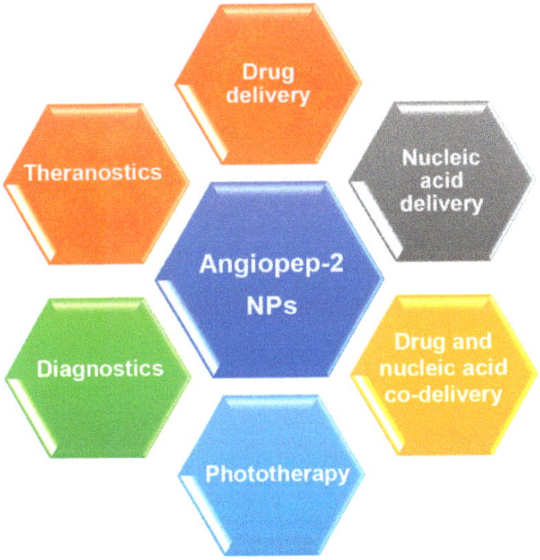

Figure 2. Potential medical usages of Angiopep-2 NPs.

2. Angiopep-2-Decorated Nanoparticles

Angiopep-2 has been appended to a wide variety of nanostructures for the delivery of therapeutic agents to treat brain disorders, which include cancer, brain injury, stroke, epilepsy, fungal infections, Alzheimer's disease (AD), and Parkinson's disease (PD). These encompass both organic and inorganic nanoparticles (Figure 3).

Liposomes are arguably the most common nano-delivery agents. First described in the 1960s by Bangham and coworkers [27], liposomes are spherical lipid vesicles composed of a phospholipid bilayer that can encapsulate therapeutic agents within the aqueous core. Liposomes are versatile in that they are amenable to several useful modifications, including the appendage of ligands, such as Angiopep-2, on the surface. Danyu and coworkers [28] showed that Angiopep-2 liposomes loaded with the anticancer drug, doxorubicin, had glioma targeting therapeutic effects with reduced toxicity.

Cationic lipids can be incorporated into liposome formulations to confer a net positive charge that permits convenient electrostatic binding of nucleic acids [29]. Conveniently, cationic liposomes are amenable to carrying both drugs and genes [30]. Angiopep-2-functionalized cationic liposomes were shown to effectively deliver siRNA against Golgi phosphoprotein 3 (GOLPH3) specifically to glioma and inhibit its growth in U87-GFP-Luci-bearing BALB/c mouse models [31].

Stealth properties can be conferred using polymer shrouds, such as polyethylene glycol (PEG). Xuan and colleagues [32] demonstrated that the encapsulation of dibenzazepine in PEGylated Angiopep-2-modified liposomes enhanced its cytotoxicity against glioblastoma stem cells. More recently, PEGylated Angiopep-2-modified liposomes were shown to promote the anti-glioma effect of arsenic trioxide [33].

An alternative to liposomes is solid lipid NPs (SLNPs). SLNPs are prepared from emulsifier-stabilized lipids that are solid at room temperature [34]. Angiopep-2-grafted SLNPs encapsulating the chemotherapeutic drug, docetaxel showed selective targeting and higher accumulation in the brain than the marketed drug formulation [35].

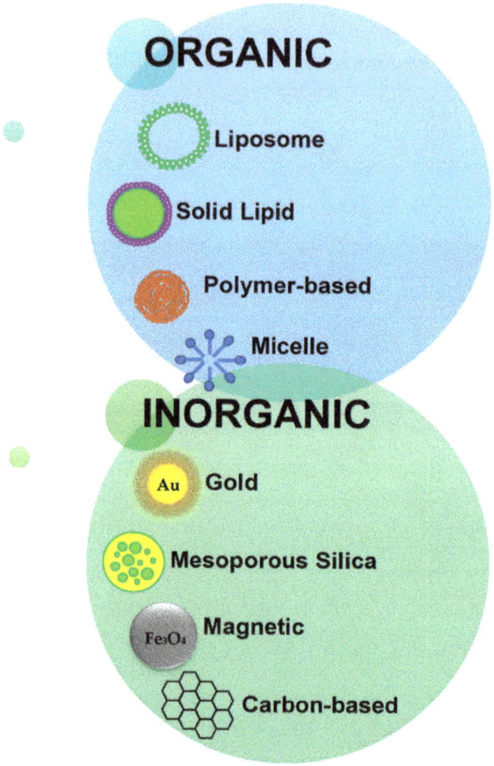

Figure 3. Outline of major nanoparticle classes to which Angiopep-2 has been appended.

Angiopep-2 has also been appended to polymer-based NPs. Polymer-based NPs are colloidal systems formulated from natural, synthetic, or semi-synthetic polymers. Polymeric NPs differ in characteristics based on the type of polymer employed. However, they are generally stable in biological fluids and are versatile enough to regulate the stimuli-induced controlled release of their therapeutic payloads [36].

Parashar and coworkers [37] recently reported on an Angiopep-2-anchored lipoprotein-coated e-caprolactone nanoparticle to deliver the anti-epilepsy drug, carbamazepine, as an alternative to the standard oral and intravenous routes. Functionalization of a trileucine-stabilized β-poly(l-malic acid) nanoplatform with Angiopep-2 resulted in more effective infiltration of the brain parenchyma than those modified with the brain-shuttle peptide, MiniAp-4, and the transferrin receptor ligands, cTfRL and B6 [38].

Micelles are made up of amphiphilic macromolecules, notably polymers, which form through self-assembly in solution. The hydrophobic segments converge to form a core, while the hydrophilic components form an outer shell. Such micelles are intrinsically stealth NPs capable of evading the reticuloendothelial system [39]. Radiolabeled Angiopep-2-anchored poly (ethylene glycol)-block-poly (d,l-lactide acid) (PEG-PLA) micelles demonstrated high brain accumulation for up to 24 h after intravenous administration in mice [40].

Similarly, Angiopep-2-modified PEG-co-poly(ϵ-caprolactone) (PEG-PCL) NPs accumulated at higher levels in the brain cortical layer, lateral ventricle, third ventricles and hippocampus than did unmodified nanoparticles [41]. In the category of polymer NPs, are the dendrimers. Dendrimers are radially symmetrical hyperbranched artificial macromolecules typified by a combination of many functional groups and a compact structure. Surface modification, such as the introduction of the Angiopep-2 moiety, is relatively simple. Moreover, a high level of control can be exerted over dendritic architectures, making them suitable carriers in biomedical applications.

Polyamidoamine (PAMAM) dendrimers are the most common class of dendrimers utilized to date to deliver nucleic acids [42–45]. Dendrimers possess an inner alkyl-diamine core and a peripheral shell of tertiary amine branches [46,47]. Angiopep-2-modified PAMAM dendrimers have demonstrated efficacy in delivering doxorubicin to glioma cells [48,49]. Due to the cationic centers of PAMAM at physiological pH, it has also been used for the binding and transport of DNA to the brain in mouse models when associated with Angiopep-2 [50,51].

In recent years, there has been a surge in the investigation of inorganic nanomaterials as carriers of therapeutic agents. These include gold, mesoporous silica, magnetic and carbon-based nanomaterials, and organic/inorganic hybrids.

Turkevich first reported that the reduction of gold salts in the presence of a reducing agent initiates the nucleation of gold ions [52]. Gold NPs exist in a broad size range of 1 nm to 8 μm and exhibit varying morphology, including nanospheres, nanoclusters, nanorods, nanoshells, nanostars, and nanoprisms [53]. For biomedical applications, the gold core nanostructure is typically modified with an organic monolayer to permit solubility in aqueous environments and control intermolecular interactions of the nanoparticle [54].

Key to the tunable characteristics of the monolayer is the appendage of targeting ligands, such as Angiopep-2. In this way, gold nanospheres [55], nanorods [56], and nanoprisms [57] have been directed to the brain in animal models. For example, Angiopep-2-modified hypoxic lipid radiosensitizer-coated gold NPs were shown to enhance the effects of radiation therapy on brain tumor growth in vivo [58].

Another well-studied class of inorganic carriers is the mesoporous silica NP (MSN). MSNs are a specialized form of silica NPs with well-defined porosity and morphology. The porous honeycomb-like structure accounts for a high drug loading capacity and aids in controlled release [59,60]. They are reportedly non-toxic, do not affect healthy tissues, can be imbued with stimuli-responsive features [61], and can be modified to mediate chemo-photodynamic therapy [62].

Their behavior in biological systems can be attenuated by controlling the surface chemistry and size [63]. Consequently, in addition to appending Angiopep-2, MSNs directed to the brain have been modified with lipids [64,65] and polymers [66]. In a recent study, Angiopep-2-modified lipid-coated MSNs efficiently loaded paclitaxel, increased glioma cell apoptosis, and prolonged the survival of C6 glioma bearing rats [64].

Carbon-based nanomaterials (CBNs) have received extensive attention in biotechnology owing to their tunable surface characteristics and mechanical, electrical, optical, and chemical properties. CBNs, which include graphene oxide, carbon nanotubes, and carbon nanodots, have been functionalized with Angiopep-2. Graphene oxide, as the name suggests, is the oxidized form of graphene, a flat monolayer composed of sp^2 hybridized carbon in two-dimensional sheets of a hexagonally arranged honeycomb lattice [67].

Graphene is considered superior to other CBNs because it has lower levels of metallic impurities and requires purification processes that are less time-consuming [68]. The modification of graphene oxide with Angiopep-2 increased doxorubicin uptake in U87 MG cells over that of unmodified graphene oxide-doxorubicin and free doxorubicin [69]. Carbon nanotubes (CNTs) are formed by rolling the graphene sheet in a cylindrical structure within a specified nano-diameter [70].

PEGylated oxidized multi-walled carbon nanotubes modified with Angiopep-2 demonstrated a combined dual targeting effect in the delivery of doxorubicin to glioma [71]. Carbon nanodots are zero-dimensional spherical allotropes of carbon and are below 10 nm in size. They have great potential for biomedical application due to their biocompatibility, low toxicity, water-solubility, eco-friendly synthesis, conductivity, and desirable optical properties [72]. Angiopep-2 anchored PEGylated carbon nanodots was shown to target C6 glioma cells more effectively than PEGylated carbon nanodots [73].

A relatively recent addition to the growing plethora of inorganic NPs is the superparamagnetic iron oxide NPs (SPIONs). This has been described as one of the most promising tools in theranostic applications. Such NPs typically consist of single or multiple iron oxide cores and are surface modified to promote biocompatibility and stability in biological systems [74]. Hence, brain-targeted magnetic NPs are comprised of hybrid materials. These include Angiopep-2 decorated iron gold alloy NPs [75] and magnetic lipid-polymer hybrid NPs [76]. Of great interest is the possibility of utilizing an external magnetic field to promote deposition of the NP at the desired locality and, in this way, modulating the release of the therapeutics [74]. The advantages and limitations of the major classes of Angiopep-2-modified NPs are summarized in Table 1.

Table 1. Advantages and disadvantages of some Angiopep-2-modified nanoparticles.

Nanoparticle	Advantages	Disadvantages	References
Liposomes 	Biocompatible and easy to prepare. Can deliver anionic and cationic material. Tunable composition. Can be modified for cell-specific targeting.	High production cost Short shelf-life Possibility of drug leakage	[77]
Polymeric Nanoparticles 	Biocompatible and biodegradable. Long half-life. High drug stability.	Complicated synthesis methods. Purification is difficult. Difficult to produce on a large-scale.	[78]
Micelles 	Biocompatible. Can potentiate controlled drug release. Tunable chemical and physical properties.	Low drug loading capacity, which is dependent on micelle concentration	[79]
Dendrimers 	High drug loading capacity. Chemistry can be easily modified. Ability to penetrate biological barriers. Can be modified for cell-specific targeting.	Some dendrimers can be cytotoxic.	[80]
Gold 	Biocompatible. Small size with high surface area to volume ratio. Surface can be easily modified. Potential for green synthesis.	Potential toxicity if retained in the body over a long period.	[53,81]
Mesoporous silica 	Biocompatible and biodegradable. High surface area with large pore sizes. Has well-defined surface properties.	Preparation can be complex. Varied size distributions can occur.	[59]
Carbon-based Nanoparticles 	Can be functionalized. Potential for photothermal therapy (graphene oxide and carbon nanotubes). Carbon nanodots have potential for imaging.	Toxic if not functionalized.	[82–85]
Magnetic Nanoparticles 	Biocompatible and easy to synthesize. Surface can be modified. Magnetic properties can be exploited for targeting and controlled release of therapeutic.	Has no internal loading capacity.	[86]

3. Drug Delivery

Most Angiopep-2-modified drug delivery agents have been designed with a view to treat cancers of the central nervous system. Glioblastoma or glioblastoma multiforme (GBM) is the most common and aggressive malignant brain tumor [87]. The current treatment involves a combination of surgery, radiation therapy, and chemotherapy. However, the disease remains highly resistant to treatment [88].

The conjugation of Angiopep-2 to NPs is reported to have a dual-targeting effect. Not only does the peptide act as a shuttle to promote transport across the BBB but it is also selective for glioma cells due to the overexpression of LRP1 on their surfaces [55]. Using this concept, several nanosystems have been designed to improve the efficacy of existing chemotherapeutic agents, including doxorubicin, paclitaxel, and docetaxel. Encouragingly, many have demonstrated efficacy in animal models (Table 2).

For example, treatment of glioma-bearing rats with Angiopep-2 decorated polymersomes loaded with doxorubicin prolonged the survival time compared with unmodified polymersomes and the free drug [89]. In addition, the incorporation of drugs in Angiopep-2 NPs has been linked with the attenuation of side effects. For example, the histopathological analysis of Angiopep-2 decorated nanocarbon tubes carrying doxorubicin suggested lower cardiac toxicity than the free drug.

On the other hand, using a second ligand to bypass the blood-tumor barrier and encourage NP uptake in glioma cells has also been reported. NPs modified with both Angiopep-2 and an activatable cell-penetrating peptide were shown to localize in gliomas with greater efficiency than NPs with a single ligand [90]. Furthermore, docetaxel-loaded Angiopep-2 and TAT functionalized tandem nanomicelles were shown to have a prolonged blood circulation time in mice and inhibited orthotopic U87MG human glioma better than the Angiopep-2 single peptide-functionalized counterpart [91].

In the same year, Kim and colleagues [92] demonstrated that the conjugation of both Angiopep-2 and anti-CD133 monoclonal antibody to a liposome was effective in delivering temozolomide to glioblastoma stem cells through the BBB. Dual targeting efficiency was also demonstrated with Angiopep-2 and an AS1411 aptamer covalently linked to a doxorubicin-loaded lipid-capped PLGA NP [93].

The incorporation of statins within Angiopep-2-decorated NPs increased LRP-1 expression in brain microvascular endothelial cells and brain metastatic tumor cells. The systemic administration of Angiopep-2-functionalized PEGylated PLGA-PLL NPs co-encapsulating simvastatin and doxorubicin displayed an extended median survival of mice bearing brain metastases due to enhanced BBB transcytosis and the effective targeting of brain metastases [94].

Angiopep-2 has also been incorporated into the design of "smart" nanodrugs that are stimuli-responsive to overcome problems that include incomplete drug release or non-site-specific drug deposition. A common strategy involves exploiting unique features of the tumor microenvironment. Ruan and coworkers [55] tethered doxorubicin to an Angiopep-2-modified PEGylated gold NP via a hydrazone bond to permit drug release upon exposure to the acidic tumor locality. More recently, polyacrylic acid was incorporated as part of liposome-silica hybrid nanovesicles to allow the acid-triggered release of arsenic trioxide [95].

Recently, the matrix metalloproteinase-1 (MMP1)-rich niche of breast cancer brain metastases (BCBMs) was exploited in the design of NPs that can escape abluminal LRP-1-mediated clearance. PLGA-PLL NPs co-carrying doxorubicin and lapatinib were modified with a MMP-1 sensitive fusion peptide containing HER2-targeting KAAYSL and LRP-1-targeting Angiopep-2. MMP1-triggered cleavage removed Angiopep-2 for augmented accumulation in BCBMs-bearing brains [96]. NPs may also be engineered such that drug release is induced via an externally applied stimulus. Luo and colleagues [97] reported on Angiopep-2-decorated PLGA hybrid NPs that encapsulated an ultrasound contrast agent and doxorubicin.

High-intensity focused ultrasound (HIFU) was applied to trigger on-demand doxorubicin release at glioblastoma sites resulting in a mean survival time of 56 days for glioblastoma-bearing mice and minimal traces of tumor cells evident in pathological slices. Table 2 summarizes Angiopep-2-decorated nanodrug delivery systems applicable to the treatment of brain disorders.

Table 2. Angiopep-2-modified nanoparticles that have been used for drug delivery.

Nanoparticle	Drug/s	Disorder Treated	Test System	Reference
Liposome-silica hybrid	Arsenic trioxide	Glioma	C6 glioma-bearing rats	[95]
PAMAM dendrimer	Doxorubicin	Glioma	C6 glioma cells	[48]
poly(dimethylsiloxane)-poly(2-methyloxazoline) (PDMS-PMOXA) diblock copolymer	Doxorubicin	Glioblastoma	U87MG glioblastoma cells	[98]
Carboxymethyl chitosan nanogel	Doxorubicin	Glioblastoma	-	[99]
lipid-poly-(metronidazoles) hypoxic radiosensitized-polyprodrug	Doxorubicin	Glioma	C6 glioma cells Glioma-bearing ICR mice	[100]
lipid-poly (hypoxic radiosensitized polyprodrug)	temozolomide	Glioblastoma	C6 glioma cells Glioma-bearing ICR mice	[101]
PEG-*b*-poly(ε-caprolactone) (PEG-*b*-PCL)	Doxorubicin	Primary CNS lymphoma	SU-DHL-2-LUC lymphoma xenograft mice model	[20]
PEG-*co*- poly(ε-caprolactone) polymersome	Doxorubicin	Glioma	C6 glioma cells C6 glioma-bearing rats	[89]
PCL-PEG	Ginsenoside-Rg3	Glioma	C6 glioma cells	[102]
PEGylated gold	Doxorubicin	Glioma	C6 glioma cells C6 glioma-bearing mice	[55]
Poly (lactic-co-glycolic acid) (PLGA)-based mesoporous silica	Doxorubicin Paclitaxel	Glioma	Human brain micro-vascular endothelial cells BBB model	[66]
PEGylated PLGA-PLL	Doxorubicin Simvastatin	Brain metastases	-	[94]
Biomimetic nanoparticles	Doxorubicin Lexiscan	Glioblastoma	U87MG human glioblastoma tumor-bearing nude mice	[103]
Graphene oxide	Doxorubicin	Glioma	U87 MG cells/ mouse xenograft	[69]
PEGylated oxidized multi-walled carbon nanotubes	Doxorubicin	Glioma	C6 glioma cells C6 glioma bearing mice	[71]
HIFU-responsive PLGA hybrid	Doxorubicin	Glioblastoma	Glioblastoma-bearing mice	
PLGA Gold	Docetaxel	Glioma	-	[104]
Solid lipid nanoparticles	Docetaxel	Glioblastoma	U87MG glioblastoma cells GL261 mouse glioma Glioblastoma- induced C57BL/6 mouse model	[35]
PEG-PCL	Paclitaxel	Glioma	3D glioma tumor spheroids Intracranial glioma mouse model	[105]
Lipid-coated mesoporous silica nanoparticles	Paclitaxel	Glioma	C6 glioma cells C6 glioma-bearing rats	[64]
Phospholipid-functionalized mesoporous silica	Paclitaxel	Glioma	HBMEC cells C6 glioma cells	[65]

Table 2. *Cont.*

Nanoparticle	Drug/s	Disorder Treated	Test System	Reference
PEGylated poly propyleneimine (PPI) dendrimers	Paclitaxel	Glioblastoma	C6 glioma cells Co-culture BCECs model	[106]
redox-responsive virus-mimicking polymersome	Saporin	Glioblastoma	U-87 MG glioblastoma cells U-87 MG human-glioblastoma mouse model	[107]
PEG-PE polymeric micelles	Amphotericin B	Meningo-encephalitis	Immunosuppressive murine *Cryptococcus neooformans* meningo-encephalitis model	[108]
PE-PEG polymeric micelle	Amphotericin B	CNS fungal infections	-	[109]
Ceria	Edaravone	Ischemic stroke	-	[110]
PEG-PLGA	Tanshinone IIA	Ischemic stroke	-	[111]
PEG-PAMAM nanoparticle	Scutellarin	Ischemic stroke	-	[112]
Electro-responsive hydrogel	Phenytoin sodium	Epilepsy	Amygdala kindling seizure model	[113]
Lipoprotein-coated e-caprolactone	Carbamazepine	Epilepsy	Adult male albino rats	[37]
PEGylated 2-methoxy estradiol micelle	2-Methoxy estradiol	Cerebral ischemia-reperfusion injury	PC12 cells	[114]

4. Nucleic Acid Delivery

Gene therapy, the application of nucleic acids to treat disease, promises to revolutionize how brain disorders are addressed. It is theoretically capable of curing the disease rather than merely treating symptoms. Initially envisaged as the introduction of functional gene segments to replace defective genes, gene therapy encompasses more than just the use of therapeutic DNA. Other types of therapeutic nucleic acids applicable to diseases of the brain include small RNA molecules, such as small interfering RNA (siRNA) and micro RNA (miRNA).

However, using these nucleic acids as medicine necessitates their association with biocompatible carriers to encapsulate, protect, and facilitate cellular entry at the correct site. This synergy created between gene therapy and nanomedicine may be a significant association that can benefit the treatment of various disorders [115]. Angiopep-2 has been appended to various nanostructures for the reliable transport of therapeutic nucleic acids.

DNA-based Angiopep-2-modified NPs have been reported for the treatment of brain cancer and Parkinson's disease. As an example, Gao and coworkers reported on the delivery of a suicide gene via Angiopep-2 conjugated cationic PEI-PLL-PEG NPs, which penetrated the BBB and accumulated in the striatum and cortex via systemic administration. The system achieved a remarkable anti-tumor effect and survival benefit in an invasive orthotopic human glioblastoma mouse model by inhibiting proliferation and inducing apoptosis [116].

Angiopep-2-conjugated dendrigraft poly-L-lysine delivered a therapeutic gene encoding human glial cell line-derived neurotrophic factor in a chronic Parkinsonian model. Pharmacodynamic data revealed that rats in the group with five injections of targeted DNA-bound NPs improved in locomotor activity and apparent recovery of dopaminergic neurons compared to those in other groups [117]. In a proof of principle study, Angiopep-2 and TAT dual modified magnetic lipid-polymer hybrid NPs delivered a reporter gene effectively in C6 cells in a magnetic field [76]. Angiopep-2 NPs have been applied to the delivery of siRNA against genes involved in brain cancer progression and survival. These include GOLPH3, Polo-like kinase 1 (PLK1), vascular endothelial growth factor (VEGF), and vascular endothelial growth factor receptor (VEGFR) genes.

Zheng and coworkers [118] introduced a polymer capable of stabilizing siRNA by electrostatic hydrogen bonds and hydrophobic interactions. Given that ROS is enriched in cancer cells, the polymer was designed with a ROS-responsive feature to trigger on-site siRNA release.

With Angiopep-2 functionalization, the polymer successfully delivered siRNA against PLK1 and vascular endothelial growth factor receptor-2 (VEGFR2), leading to effective suppression of tumor growth and significantly improved survival time in mice bearing orthotopic GBM brain tumors. Another Angiopep-2-modified ROS-responsive nanosystem successfully delivered VEGF siRNA into glioma cells. VEGF silencing was accompanied by angiogenesis inhibition and suppressed expression of caveolin-1, which is involved in BBB functional regulation in the occurrence and treatment of glioblastoma [119].

Angiopep-2 was also used to functionalize a biomimetic three-layer core-shell nanostructure to deliver siRNA to glioma cells. The nanostructure was designed to release siRNA in the endo/lysosome by charge conversion from negative to positive. This led to highly potent target-gene silencing with a strong anti-GBM effect and minimal side effects [120].

Angiopep-2 has also been involved in miRNA-directed nanotherapy. Liu and coworkers [121] used polymeric NPs to simultaneously supplement the function of miR-124 and inhibit the function of miR-21 to treat glioblastoma. Co-delivery of a miR-124 mimic and anti-miR-21 regulated the mutant RAS/PI3K/PTEN/AKT signaling pathway in tumor cells. This was accompanied by anti-tumor effects, which included reduction of tumor cell proliferation, migration, invasion and angiogenesis, tumor growth suppression, and improved survival time. Table 3 provides an overview of Angiopep-2-functionalised NPs investigated for the delivery of nucleic acids.

Table 3. Angiopep-2-modified nanoparticles for nucleic acid delivery.

Nanoparticle	Nucleic Acid	Nucleic Acid Details	Disease	Test System	Reference
PAMAM-PEG	DNA	pORF-TRAIL	Glioma	C6 glioma cells	[51]
PAMAM-PEG	DNA	pEGFP-N2	-	BCEC Balb/c mice	[50]
PEI-PLL-PEG	DNA	Herpes simplex virus type I TK gene	Glioblastoma multiforme	Human GBM mouse model	[116]
dendrigraft PLL	DNA	Gene encoding human glial cell line-derived neurotrophic factor	Parkinson's	Rotenone-induced chronic model of Parkinson's disease	[117]
Cationic liposome	siRNA	GOLPH3 siRNA	Glioma	U87-GFP-Luc-bearing BALB/c mouse models	[31]
Polymeric	siRNA	siPLK1 siVEGFR2	Glioblastoma	GBM brain tumor mouse model	[118]
Polyplex	siRNA	-	Glioma	Glioma mouse model	[122]
Chimeric polymersomes	siRNA	siPLK1	Glioblastoma	U-87 MG cells Glioblastoma mouse model	[123]
Biomimetic nanoparticles	siRNA	-	Glioblastoma	U87MG- Luc human glioblastoma mouse model	[120]
ROS cleavable thioketal-linked glycolipid-like nanocarriers	siRNA	siVEGF	Glioblastoma	U87 MG cells	[119]
Polymeric	miRNA	miR-124 anti-miR-21	Glioblastoma	U87MG-Luc human glioblastoma tumor mouse model	[121]

5. Drug and Nucleic Acid Co-Delivery

The idea of treating cancer through drug and nucleic acid combination therapy is receiving significant attention. This strategy affords the ability to target more than one mechanism governing the growth and survival of tumors, giving rise to synergistic anti-cancer effects. Appropriately designed NPs can overcome the challenges associated with the delivery of two therapeutic agents with markedly different physiological properties.

Nucleic acids are hydrophilic, anionic, high-molecular-weight entities, while the most commonly used chemotherapy drugs are small hydrophobic molecules, thus necessitating different mechanisms for encapsulation [124]. In general, nucleic acids are electrostatically associated with cationic components of the NP, while small molecule drugs are enclosed within them by hydrophobic force, electrostatic interactions, or chemical conjugation [125].

Angiopep-2 has served as an essential component of several nanoplatforms for dual agent delivery to the brain. Liposomes, being archetypal delivery systems, have been utilized for multimodal intervention. For example, an Angiopep-2-modified cationic liposome co-carrying the therapeutic gene encoding the human tumor necrosis factor-related apoptosis-inducing ligand (pEGFP-hTRAIL) and paclitaxel was reported to achieve greater apoptosis of glioma cells than single medication systems and the unmodified co-delivery system [126].

Liposomes were also dual-functionalized with Angiopep-2 and the tLyP-1 peptide, which targets the neuropilin-1 receptor on glioma cells, for simultaneous anti-angiogenic and apoptotic effects through the delivery of vascular endothelial growth factor (VEGF) siRNA and docetaxel. This system demonstrated superior anti-tumor effects after both intracranial and systemic administration in mice with U87 MG tumors without activating system-associated toxicity or the innate immune response [127]. Sun and colleagues [128] reported on cationic liposomes modified with Angiopep-2 and an aptamer that binds to CD133.

The co-delivery of survivin siRNA and paclitaxel using this nanocarrier was minimally toxic to brain capillary endothelial cells but selectively caused apoptosis of CD133+ glioma stem cells and improved the differentiation of CD133+ glioma stem cells' into non-stem-cell lineages. In addition, the system inhibited tumorigenesis, induced CD133+ glioma cell apoptosis, and prolonged survival in intracranial glioma tumor-bearing nude mice. Recently, synergistic tumor-inhibitory effects were also noted with an Angiopep-2 decorated cationic liposome that simultaneously delivered doxorubicin, yes-associated protein (YAP) siRNA, and gold nanorods [129].

Wang and colleagues [130] designed Angiopep-2-modified PLGA NPs to encapsulate doxorubicin and siRNA against the EGFR. This co-delivery nanosystem was shown to cause apoptosis of the glioma tissue and prolong lifespan in glioma-bearing mice. Another combinatory anti-glioma system involved the dual release of Gefitinib and GOLPH3 siRNA from an Angiopep-2-modified cationic lipid-PLGA NP.

This system achieved synergistic anti-EGFR activity in that Gefitinib markedly inhibited EGFR signaling, while GOLPH3 silencing promoted EGFR and p-EGFR degradation [131]. In line with the use of polymers for the design of dual agent nanostructures, Wen and coworkers [132] reported on an Angiopep-2 decorated glycolipid-like co-polymeric micelle for the simultaneous delivery of VEGF siRNA and paclitaxel in vivo.

Moreover, the nanovector was designed with a redox-responsive feature to trigger the intracellular release of its payload. In the same year, a combination of Temozolomide and PLK1 siRNA by Angiopep-2-modified PEG-PEI-PCL micelles produced enhanced drug efficacy in glioma [133].

6. Phototherapy

In addition to drug and nucleic acid-mediated therapy, brain cancer can be treated via phototherapy. Phototherapy can be subdivided into two main branches, namely, photodynamic and photothermal therapies. Photodynamic therapy (PDT) uses light-sensitive molecules known as photosensitizers, which produce cytotoxic ROS once exposed to a specific wavelength [134]. PDT is minimally invasive because photosensitizers are only cytotoxic when activated in tumor regions. However, PDT cannot treat advanced cancers due to the difficulty of light delivery and the limited penetration depth.

Photothermal therapy (PTT) is an alternative for advanced tumors, in which photosensitizers absorb near-infrared (NIR) light and release vibrational energy in the form of heat to destroy cancer cells, independent of oxygen [135]. The amalgamation of phototherapy and nanotherapy has led to the construction of NPs that can direct photosensitizers to the tumor site. In other instances, they possess inherent photothermal conversion capability. The latter is true of NPs composed of magnetic and carbon-based materials [136].

Oleic acid-coated upconversion NPs (UCNPs) were conjugated with PEG/Angiopep-2 for the co-delivery of the photothermal agent, IR-780, and photodynamic sensitizer, 5,10,15,20-tetrakis(3-hydroxyphenyl) chlorin (mTHPC) in brain astrocytoma tumors. The photoactivated dual therapies resulted in extensive apoptosis and necrosis of brain tumors, translating into an extended median survival of tumor-bearing mice compared to non-targeted NPs [137]. Recently, Angiopep-2 decorated nanostructured lipid carriers of the photosensitizer, chlorin e6, were evaluated for PDT efficacy in vitro against a glioblastoma model [138].

Phototherapy may also be combined with chemotherapy for synergistic anti-tumor effects. Lu and colleagues [139] developed a multicomponent nanoplatform made up of self-assembled pH-responsive nanodrugs derived from amino acid-conjugated camptothecin and canine dyes coated with an Angiopep-2-conjugated copolymer. The combination of chemotherapy and PTT improved the therapeutic effect with a longer survival time and reduced toxic side effects in orthotopic glioblastoma tumor-bearing nude mice.

7. Diagnostic and Theranostic Applications

Brain cancers are often difficult to detect due to tumors being located deep in brain tissue. More often than not, diagnosis is delayed, which further impedes the success of the treatment administered [140]. Furthermore, accurate tumor imaging is of immense importance in the pre-operative stage and the location of tumor margins [141]. In this regard, Angiopep-2-modified nanosystems have demonstrated great potential.

Magnetic resonance imaging (MRI) is an imaging technique that uses magnetic fields to assess the morphological structure of organs in the body. It has emerged as a dominant imaging modality in brain cancer diagnosis and clinical staging [142]. However, one of the drawbacks is low sensitivity, which reduces its potential in molecular-level detection. Hence, increasing the contrast between healthy and diseased tissues is of the utmost importance [143]. In this regard, the inherent magnetic properties of iron oxide NPs render them suitable alternatives to conventional contrast agents for MRI [144,145].

Chen and colleagues [146] reported on Pluronic®F127-modified water-dispersible poly (acrylic acid)-bound iron oxide NPs modified with Angiopep-2 as brain-directed diagnostic agents. The system demonstrated negligible cytotoxicity, better cellular uptake, and higher $T2$-weighted image enhancement than non-targeted NPs in U87 cells.

Du and colleagues [147] presented the first report on Angiopep-2 conjugated ultra-small superparamagnetic iron oxide NPs (USPIONs) as $T1$-weighted positive MR contrast agents for intracranial targeted glioblastoma imaging. The nanoprobe showed promise for efficient pre-operative tumor diagnosis and the targeted surgical resection of intracranial glioblastomas.

Optical imaging using fluorescent NPs is another alternative. Features that include strong signal strength, resistance to photobleaching, tunable fluorescence emissions, and high sensitivity are the impetus for the application of fluorescent NPs in cancer diagnosis [148]. Additionally, fluorescent NPs display stronger fluorescent brightness, better photostability, water dispersibility, and biocompatibility compared with conventional fluorescent dyes. Fluorescent carbonaceous nanodots were prepared from glucose and glutamic acid with long excitation/emission wavelengths to overcome the limitations associated with shorter wavelengths in imaging diseased tissue.

Decoration with Angiopep-2 resulted in a glioma/normal brain (G/N) ratio of 1.76 [73]. Moreover, the developed system showed good serum stability, hemocompatibility, and low cytotoxicity. Recently, Ren and colleagues [149] designed Angiopep-2-modified Er-based lanthanide NPs with strong NIR IIb fluorescence for imaging-guided surgery of orthotopic glioma. NPs were delivered to gliomas in mice via focused ultrasound sonication to temporarily open the BBB. The highest tumor-to-background ratio (TBR = 12.5) was reported in the targeted NIR IIb fluorescence imaging of small orthotopic glioma through intact skull and scalp was obtained.

Xie and coworkers [141] constructed a MRI/NIR fluorescence dual-modal imaging nanoprobe by combining superparamagnetic iron oxide NPs (SPIONs) with the fluorescent dye indocyanine. This was further modified with the retro-enantiomer of Angiopep-2 to prevent its degradation by enzymes of the blood and cells. In keeping with the idea of dual-modal imaging, Wei and colleagues [150] introduced small-sized iron oxide NPs (SIONs), which were surface modified with Angiopep-2 and the photosensitizer, chlorin e6, to boost fluorescence imaging to support MRI results.

Angiopep-2 has also served as an essential component of nanosystems that seek to integrate active agents for therapy and diagnosis. Such nanoplatforms, categorized under the broad category of theranostics, promise to significantly benefit the diagnosis, treatment and management of brain cancer. Angiopep-2 was appended to pegylated bubble liposomes at the distal ends of PEG chains. The nanosystem was shown to be capable of encapsulating ultrasound contrast gas and nucleic acids. Systemic administration could serve as a useful device for brain-targeted delivery and ultrasound imaging [151].

Crosslinked hyaluronic acid NPs were decorated with Angiopep-2 and formulated to encapsulate gadolinium-diethylenetriamine penta-acetic acid (Gd-DTPA) and the chemotherapeutic agent, irinotecan. The nanosystem showed improved MRI capability, improved uptake in U87 and GS-102 cells, and reduced the irinotecan time response [152]. Lin and colleagues [153] constructed Angiopep-2 coupled bovine serum albumin NPs containing superparamagnetic iron oxide (SPIO), indocyanine green, and the drug, Carmustine.

The nanoprobes were capable of dual MRI and fluorescence imaging and effective drug delivery. In a deviation from the non-viral NPs discussed thus far, a theranostic NP based on the MS bacteriophage capsid was reported. Angiopep-2 was appended to the external surface, while the interior space was loaded with Mn^{2+} via a porphyrin ring to enable detection via MRI. The inner space can further encapsulate therapeutic agents. Systemic introduction of NPs resulted in dose-dependent, non-toxic accumulation in the midbrain [154].

Iron–gold alloy NPs were also conjugated with Angiopep-2 as a minimally invasive theranostic system. These superparamagnetic NPs enhanced negative Glioma image contrast and exhibited a 12 °C temperature elevation when magnetically stimulated. Angiopep-2 modification resulted in a 1.5-fold higher uptake by glioma cells than fibroblasts, and magnetic field induced hyperthermia decreased cell viability by 90%. Furthermore, treatment resulted in a five-fold decrease in tumor volume and extended survival time in vivo [70].

A system integrating targeted brain imaging and chemo- and phototherapy was put forward by Hao and colleagues. PLGA NPs were loaded with indocyanine green as a NIR imaging and phototherapy agent and the anticancer drug, docetaxel. Once modified with Angiopep-2, NIR image-guided chemo-phototherapy resulted in glioma cell death and prolonged survival of glioma xenograft-bearing mice [155].

Lipid NPs containing Angiopep-2, a hypoxia-responsive poly(nitroimidazole) 25, indocyanine green, and doxorubicin were proposed for fluorescence-guided surgery chemotherapy, PDT, and PTT combination multitherapy strategies targeting glioma. The study suggested that this nanoplatform may be useful in preventing the post-surgical recurrence of glioma [156].

8. Discussion and Conclusions

Tang and colleagues [140] commented that *"BBB-crossing nanotechnology is expected to make a revolutionary impact on conventional brain cancer management"*. In this regard, strategies that exploit Angiopep-2-mediated transport are increasingly important. At present, most Angiopep-2-functionalized nanomedicines have been directed towards the treatment of brain cancers, in particular glioblastoma, which is highly aggressive and responds poorly to the current therapy.

Advantageously, Angiopep-2 modification has been reported to have BBB and blood-tumor barrier (BTB) dual-penetrating ability. To our knowledge, little comparative data is available with respect to the performance of Angiopep-2 versus other cell-penetrating peptides. Interestingly, enhanced NP functioning has been reported through co-modification with Angiopep-2 and other cell-penetrating peptides. It is worth noting, however, that functionalization of the chemo-therapeutic agent, PAPTP, with either Angiopep-2 or the $TAT_{48\text{-}61}$ peptide, permitted similar delivery to the brain in mice [157].

In the past five years, the as-functionalized nanosystems have also demonstrated potential for the delivery of agents to treat other brain disorders, including fungal infections, epilepsy, stroke, brain injury, Parkinson's disease, and Alzheimer's disease. For example, gold nanorods functionalized with Angiopep-2 and the D1 peptide that recognizes toxic aggregates of β-amyloid showed efficacy in a *Caenorhabditis elegans* model of Alzheimer's disease [158].

Over the time period considered for this review (2010–2021), Angiopep-2-decorated nanostructures have been employed to deliver a vast array of therapeutic and diagnostic agents. These include chemical compounds; the nucleic acids DNA, siRNA, and miRNA; photosensitizers; and contrast agents. In addition, Angiopep-2-modified nanoparticles are applicable to immunotherapy. Wang and colleagues [159] reported that Angiopep-2 and IP10-EGFRvIIIscFv fusion protein-modified NPs can recruit activated CD8+ T lymphocytes to glioblastoma cells.

While the application of Angiopep-2-modified NPs in phototherapy is well documented, another novel physical method of destroying brain cancer cells, sonodynamic therapy (SDT), based on ultrasound stimulation, has been reported. Qu and colleagues [160] designed an innovative "all-in-one" nanosensitizer platform by combining the sonoactive chlorin e6 and an autophagy inhibitor, hydroxychloroquine, in Angiopep-2-modified liposomes to simultaneously induce apoptosis and inhibit mitophagy in glioma cells.

As in the aforementioned study, Angiopep-2 NPs, are amenable to the integration of dual- and multimodal therapy. Angiopep-2 nanosystems can also be engineered to behave in a stimuli-responsive fashion to permit a controlled and sustained release of their therapeutic cargo. Moreover, their potential in theranostics has been highlighted in recent years. Encouragingly, there is a growing body of in vivo data to support the further design of multifunctional Angiopep-2-modified nanomedicines. Overall, there is a great need to translate the in vitro and in vivo achievements of BBB-crossing nanotherapeutics to the clinic.

In addition to the impact of NP shape, size, and charge on BBB-transcytosis, the number of Angiopep-2 residues displayed on the surface may have a significant influence. The multimeric association between Angiopep-2 peptides and the LRP1 was shown to increase the intracerebral uptake of NPs significantly [161]. The local flow environment is also a necessary consideration for in vitro modelling of the performance of NPs functionalized with Angiopep-2. Studies with Angiopep-2-labelled liposomes suggested that blood flow can influence the binding and BBB penetration of NPs [162].

In summary, this review highlighted the role of Angiopep-2-modified NPs in the diagnosis and treatment of brain disorders. With greater streamlining of NP design, advances in BBB modelling and further in vivo testing, it is envisaged that Angiopep-2-based nanosystems may make their way into the clinic for the routine assessment and treatment of brain cancer and other disorders of the brain in the years to come.

Author Contributions: Conceptualization, S.H. and M.S.; software, S.H.; resources, M.S.; writing—original draft preparation, S.H.; writing—review and editing, M.S.; supervision, M.S.; project administration, M.S.; funding acquisition, M.S. All authors have read and agreed to the published version of the manuscript.

Funding: This review received no external funding, but research in this area was funded by the National Research Foundation of South Africa, grant numbers 120455, 129263 and 138470.

Institutional Review Board Statement: Not applicable.

Informed Consent Statement: Not applicable.

Data Availability Statement: Not applicable.

Conflicts of Interest: The authors declare no conflict of interest.

References

1. Bellettato, C.M.; Scarpa, M. Possible strategies to cross the blood–brain barrier. *Ital. J. Pediatr.* **2018**, *44*, 127–133. [CrossRef] [PubMed]
2. O'Keeffe, E.; Campbell, M. Modulating the paracellular pathway at the blood–brain barrier: Current and future approaches for drug delivery to the CNS. *Drug Discov. Today Technol.* **2016**, *20*, 35–39. [CrossRef] [PubMed]
3. Brzica, H.; Abdullahi, W.; Ibbotson, K.; Ronaldson, P.T. Role of transporters in central nervous system drug delivery and blood-brain barrier protection: Relevance to treatment of stroke. *J. Cent. Nerv. Syst. Dis.* **2017**, *9*, 1179573517693802. [CrossRef]
4. Ceña, V.; Játiva, P. Nanoparticle crossing of blood–brain barrier: A road to new therapeutic approaches to central nervous system diseases. *Future Med.* **2018**, *13*, 1513–1516. [CrossRef]
5. Demeule, M.; Currie, J.C.; Bertrand, Y.; Ché, C.; Nguyen, T.; Régina, A.; Gabathuler, R.; Castaigne, J.-P.; Béliveau, R. Involvement of the low-density lipoprotein receptor-related protein in the transcytosis of the brain delivery vector Angiopep-2. *J. Neurochem.* **2008**, *106*, 1534–1544. [CrossRef] [PubMed]
6. Demeule, M.; Regina, A.; Che, C.; Poirier, J.; Nguyen, T.; Gabathuler, R.; Castaigne, J.-P.; Béliveau, R. Identification and design of peptides as a new drug delivery system for the brain. *J. Pharmacol. Exp. Ther.* **2008**, *324*, 1064–1072. [CrossRef]
7. Ché, C.; Yang, G.; Thiot, C.; Lacoste, M.-C.; Currie, J.-C.; Demeule, M.; Régina, A.; Béliveau, R.; Castaigne, J.-P. New Angiopep-modified doxorubicin (ANG1007) and etoposide (ANG1009) chemotherapeutics with increased brain penetration. *J. Med. Chem.* **2010**, *53*, 2814–2824. [CrossRef]
8. Regina, A.; Demeule, M.; Che, C.; Lavallee, I.; Poirier, J.; Gabathuler, R.; Béliveau, R.; Castaigne, J.-P. Antitumour activity of ANG1005, a conjugate between paclitaxel and the new brain delivery vector Angiopep-2. *Br. J. Pharmacol.* **2008**, *155*, 185–197. [CrossRef]
9. Hoyos-Ceballos, G.P.; Ruozi, B.; Ottonelli, I.; Da Ros, F.; Vandelli, M.A.; Forni, F.; Daini, E.; Vilella, A.; Zoli, M.; Tosi, G.; et al. PLGA-PEG-ANG-2 Nanoparticles for Blood–Brain Barrier Crossing: Proof-of-Concept Study. *Pharmaceutics* **2020**, *12*, 72. [CrossRef]
10. Chen, W.; Zuo, H.; Zhang, E.; Li, L.; Henrich-Noack, P.; Cooper, H.; Qian, Y.; Xu, Z.P. Brain targeting delivery facilitated by ligand-functionalized layered double hydroxide nanoparticles. *ACS Appl. Mater. Interfaces* **2018**, *10*, 20326–20333. [CrossRef]
11. Mei, L.; Zhang, Q.; Yang, Y.; He, Q.; Gao, H. Angiopep-2 and activatable cell penetrating peptide dual modified nanoparticles for enhanced tumor targeting and penetrating. *Int. J. Pharm.* **2014**, *474*, 95–102. [CrossRef] [PubMed]
12. Wei, H.; Liu, T.; Jiang, N.; Zhou, K.; Yang, K.; Ning, W.; Yu, Y. A novel delivery system of cyclovirobuxine D for brain targeting: Angiopep-conjugated polysorbate 80-coated liposomes via intranasal administration. *J. Biomed. Nanotech.* **2018**, *14*, 1252–1262. [CrossRef]
13. Huile, G.; Shuaiqi, P.; Zhi, Y.; Shijie, C.; Chen, C.; Xinguo, J.; Shun, S.; Zhiqing, P.; Yu, H. A cascade targeting strategy for brain neuroglial cells employing nanoparticles modified with angiopep-2 peptide and EGFP-EGF1 protein. *Biomaterials* **2011**, *32*, 8669–8675. [CrossRef] [PubMed]
14. Kumthekar, P.; Tang, S.-C.; Brenner, A.J.; Kesari, S.; Piccioni, D.E.; Anders, C.K.; Carrillo, J.; Chalasani, P.; Kabos, P.; Puhalla, S.L.; et al. ANG1005, a brain-penetrating peptide–drug conjugate, shows activity in patients with breast cancer with leptomeningeal carcinomatosis and recurrent brain metastases. *Clin. Cancer Res.* **2020**, *26*, 2789–2799. [CrossRef] [PubMed]
15. Georgieva, J.V.; Hoekstra, D.; Zuhorn, I.S. Smuggling drugs into the brain: An overview of ligands targeting transcytosis for drug delivery across the blood–brain barrier. *Pharmaceutics* **2014**, *6*, 557–583. [CrossRef]
16. Pulgar, V.M. Transcytosis to cross the blood brain barrier, new advancements and challenges. *Front. Neurosci.* **2019**, *12*, 1019. [CrossRef] [PubMed]
17. Nabi, B.; Rehman, S.; Khan, S.; Baboota, S.; Ali, J. Ligand conjugation: An emerging platform for enhanced brain drug delivery. *Brain Res. Bull.* **2018**, *142*, 384–393. [CrossRef]
18. Patra, J.K.; Das, G.; Fraceto, L.F.; Campos, E.V.R.; del Pilar Rodriguez-Torres, M.; Acosta-Torres, L.S.; Diaz-Torres, L.A.; Grillo, R.; Swamy, M.K.; Sharma, S.; et al. Nano based drug delivery systems: Recent developments and future prospects. *J. Nanobiotechnol.* **2018**, *16*, 71. [CrossRef]

19. Liu, X.; Wu, R.; Li, Y.; Wang, L.; Zhou, R.; Li, L.; Xiang, Y.; Wu, J.; Xing, L.; Huang, Y. Angiopep-2-functionalized nanoparticles enhance transport of protein drugs across intestinal epi-thelia by self-regulation of targeted receptors. *Biomater. Sci.* **2021**, *9*, 2903–2916. [CrossRef]
20. Shi, X.-X.; Miao, W.-M.; Pang, D.-W.; Wu, J.-S.; Tong, Q.-S.; Li, J.-X.; Luo, J.-Q.; Li, W.-Y.; Du, J.-Z.; Wang, J. Angiopep-2 conjugated nanoparticles loaded with doxorubicin for the treatment of primary central nervous system lymphoma. *Biomater. Sci.* **2020**, *8*, 1290–1297. [CrossRef]
21. Gong, C.; Li, X.; Xu, L.; Zhang, Y.-H. Target delivery of a gene into the brain using the RVG29-oligoarginine peptide. *Biomaterials* **2012**, *33*, 3456–3463. [CrossRef]
22. Das, M.; Wang, C.; Bedi, R.; Mohapatra, S.S.; Mohapatra, S. Magnetic micelles for DNA delivery to rat brains after mild traumatic brain injury. *Nanomed. Nanotechnol. Biol. Med.* **2014**, *10*, 1539–1548. [CrossRef] [PubMed]
23. Zhou, Y.; Zhu, F.; Liu, Y.; Zheng, M.; Wang, Y.; Zhang, D.; Anraku, Y.; Zou, Y.; Li, J.; Wu, H.; et al. Blood-brain barrier–penetrating siRNA nanomedicine for Alzheimer's disease therapy. *Sci. Adv.* **2020**, *6*, eabc7031. [CrossRef] [PubMed]
24. Zheng, M.; Tao, W.; Zou, Y.; Farokhzad, O.C.; Shi, B. Nanotechnology-based strategies for siRNA brain delivery for disease therapy. *Trends Biotechnol.* **2018**, *36*, 562–575. [CrossRef]
25. Wen, M.M. Getting miRNA therapeutics into the target cells for neurodegenerative diseases: A mini-review. *Front. Mol. Neurosci.* **2016**, *9*, 129. [CrossRef] [PubMed]
26. Chithrakumar, T.; Thangamani, M. Exploring nanotechnology for diagnostic, therapy and medicine. *IOP Conf. Ser. Mater. Sci. Eng.* **2021**, *1091*, 012062. [CrossRef]
27. Bangham, A.; Standish, M.; Watkens, J. Liposomes by film hydration technique. *JC J. Mol. Biol.* **1965**, *13*, 238. [CrossRef]
28. Danyu, M.; Huile, G.; Wei, G.; Zhiqing, P.; Xinguo, J.; Jun, C. Anti glioma effect of doxorubicin loaded liposomes modified with angiopep-2. *Afr. J. Pharmacy Pharmacol.* **2011**, *5*, 409–414. [CrossRef]
29. Felgner, P.L.; Gadek, T.R.; Holm, M.; Roman, R.; Chan, H.W.; Wenz, M.; Northrop, J.P.; Ringold, G.M.; Danielsen, M. Lipofection: A highly efficient, lipid-mediated DNA-transfection procedure. *Proc. Natl. Acad. Sci. USA* **1987**, *84*, 7413–7417. [CrossRef]
30. Jinka, S.; Rachamalla, H.K.; Bhattacharyya, T.; Sridharan, K.; Jaggarapu, M.M.C.S.; Yakati, V.; Banerjee, R. Glucocorticoid receptor-targeted liposomal delivery system for delivering small molecule ESC8 and anti-miR-Hsp90 gene construct to combat colon cancer. *Biomed. Mater.* **2021**, *16*, 024105. [CrossRef]
31. Yuan, Z.; Zhao, L.; Zhang, Y.; Li, S.; Pan, B.; Hua, L.; Wang, Z.; Ye, C.; Lu, J.; Yu, R.; et al. Inhibition of glioma growth by a GOLPH3 siRNA-loaded cationic liposomes. *J. Neurooncol.* **2018**, *140*, 249–260. [CrossRef] [PubMed]
32. Xuan, S.; Shin, D.H.; Kim, J.-S. Angiopep-2-conjugated liposomes encapsulating c-secretase inhibitor for targeting glioblastoma stem cells. *J. Pharm. Investig.* **2014**, *44*, 473–483. [CrossRef]
33. Xu, H.; Li, C.; Wei, Y.; Zheng, H.; Zheng, H.; Wang, B.; Piao, J.-G.; Li, F. Angiopep-2-modified calcium arsenite-loaded liposomes for targeted and pH-responsive delivery for anti-glioma therapy. *Biochem. Biophys. Res. Commun.* **2021**, *551*, 14–20. [CrossRef]
34. Ghasemiyeh, P.; Mohammadi-Samani, S. Solid lipid nanoparticles and nanostructured lipid carriers as novel drug delivery systems: Applications, advantages and disadvantages. *Res. Pharm Sci.* **2018**, *13*, 288–303. [CrossRef] [PubMed]
35. Kadari, A.; Pooja, D.; Gora, R.H.; Gudem, S.; Kolapalli, V.R.M.; Kulhari, H.; Sistla, R. Design of multifunctional peptide collaborated and docetaxel loaded lipid nanoparticles for antiglioma therapy. *Eur. J. Pharm. Biopharm.* **2018**, *132*, 168–179. [CrossRef]
36. Gagliardi, A.; Giuliano, E.; Eeda, V.; Fresta, M.; Bulotta, S.; Awasthi, V.; Cosco, D. Biodegradable polymeric nanoparticles for drug delivery to solid tumors. *Front. Pharmacol.* **2021**, *12*, 17. [CrossRef]
37. Parashar, A.K.; Kurmi, B.; Patel, P. Preparation and characterization of ligand anchored polymeric nanoparticles for the treatment of epilepsy. *PharmAspire* **2021**, *13*, 1–5.
38. Israel, L.L.; Braubach, O.; Galstyan, A.; Chiechi, A.; Shatalova, E.S.; Grodzinski, Z.; Ding, H.; Black, K.L.; Ljubimova, J.Y.; Holler, E. A combination of tri-leucine and angiopep-2 drives a polyanionic polymalic acid nanodrug platform across the blood–brain barrier. *ACS Nano* **2019**, *13*, 1253–1271. [CrossRef]
39. Alexander-Bryant, A.A.; Vanden Berg-Foels, W.S.; Wen, X. Bioengineering Strategies for Designing Targeted Cancer Therapies. *Adv. Cancer Res.* **2013**, *118*, 1–59. [CrossRef]
40. Shen, J.; Zhan, C.; Xie, C.; Meng, Q.; Gu, B.; Li, C.; Zhang, Y.; Lu, W. Poly (ethylene glycol)-block-poly (D, L-lactide acid) micelles anchored with angiopep-2 for brain-targeting delivery. *J. Drug Target.* **2011**, *19*, 197–203. [CrossRef]
41. Xin, H.; Sha, X.; Jiang, X.; Chen, L.; Law, K.; Gu, J.; Chen, Y.; Wang, X.; Fang, X. The brain targeting mechanism of Angiopep-conjugated poly (ethylene glycol)-co-poly (ϵ-caprolactone) nanoparticles. *Biomaterials* **2012**, *33*, 1673–1681. [CrossRef]
42. Mbatha, L.S.; Maiyo, F.; Daniels, A.; Singh, M. Dendrimer-coated Gold Nanoparticles for Efficient Folate-Targeted mRNA Delivery in vitro. *Pharmaceutics* **2021**, *13*, 900. [CrossRef] [PubMed]
43. Pillay, N.S.; Daniels, A.; Singh, M. Folate-targeted transgenic activity of Dendrimer functionalized Selenium Nanoparticles in vitro. *Int. J. Mol. Sci.* **2020**, *21*, 7177. [CrossRef]
44. Mbatha, L.S.; Maiyo, F.C.; Singh, M. Dendrimer Functionalized Folate-Targeted Gold Nanoparticles for Luciferase Gene Silencing in vitro: A Proof of Principle Study. *Acta Pharm.* **2019**, *69*, 49–61. [CrossRef] [PubMed]
45. Mbatha, L.S.; Singh, M. Starburst Poly(amidoamine) Dendrimer Grafted Gold Nanoparticles as a Scaffold for Folic Acid-Targeted Plasmid DNA Delivery in vitro. *J. Nanosci. Nanotechnol.* **2019**, *19*, 1959–1970. [CrossRef] [PubMed]

46. Abbasi, E.; Aval, S.F.; Akbarzadeh, A.; Milani, M.; Nasrabadi, H.T.; Joo, S.W.; Hanifehpour, Y.; Nejati-Koshki, K.; Pashaei-Asl, R. Dendrimers: Synthesis, applications, and properties. *Nanoscale Res. Lett.* **2014**, *9*, 247. [CrossRef] [PubMed]
47. Carvalho, M.; Reis, R.; Oliveira, J.M. Dendrimer nanoparticles for colorectal cancer applications. *J. Mater. Chem. B* **2020**, *8*, 1128–1138. [CrossRef]
48. Xu, Z.; Wang, Y.; Ma, Z.; Wang, Z.; Wei, Y.; Jia, X. A poly (amidoamine) dendrimer-based nanocarrier conjugated with Angiopep-2 for dual-targeting function in treating glioma cells. *Polym. Chem.* **2016**, *7*, 715–721. [CrossRef]
49. Han, S.; Zheng, H.; Lu, Y.; Sun, Y.; Huang, A.; Fei, W.; Shi, X.; Xu, X.; Li, J.; Li, F. A novel synergetic targeting strategy for glioma therapy employing borneol. combination with angiopep-2-modified, DOX-loaded PAMAM dendrimer. *J. Drug Target.* **2018**, *26*, 86–94. [CrossRef] [PubMed]
50. Ke, W.; Shao, K.; Huang, R.; Han, L.; Liu, Y.; Li, J.; Kuang, Y.; Ye, L.; Lou, J.; Jiang, C. Gene delivery targeted to the brain using an Angiopep-conjugated polyethyleneglycol-modified polyamidoamine dendrimer. *Biomaterials* **2009**, *30*, 6976–6985. [CrossRef]
51. Huang, S.; Li, J.; Han, L.; Liu, S.; Ma, H.; Huang, R.; Jiang, C. Dual targeting effect of Angiopep-2-modified, DNA-loaded nanoparticles for glioma. *Biomaterials* **2011**, *32*, 6832–6838. [CrossRef] [PubMed]
52. Stevenson, P.C.; Turkevich, J.; Hillier, J. A study of the nucleation and growth processes in the synthesis of. *Discuss. Faraday Soc.* **1951**, *11*, 55–75. [CrossRef]
53. Khan, A.; Rashid, R.; Murtaza, G.; Zahra, A. Gold nanoparticles: Synthesis and applications in drug delivery. *Trop. J. Pharm. Res.* **2014**, *13*, 1169–1177. [CrossRef]
54. Arvizo, R.; Bhattacharya, R.; Mukherjee, P. Gold nanoparticles: Opportunities and challenges in nanomedicine. *Expert Opin. Drug Deliv.* **2010**, *7*, 753–763. [CrossRef] [PubMed]
55. Ruan, S.; Yuan, M.; Zhang, L.; Hu, G.; Chen, J.; Cun, X.; Zhang, Q.; Yang, Y.; He, Q.; Gao, H. Tumor microenvironment sensitive doxorubicin delivery and release to glioma using angiopep-2 decorated gold nanoparticles. *Biomaterials* **2015**, *37*, 425–435. [CrossRef]
56. Velasco-Aguirre, C.; Morales-Zavala, F.; Salas-Huenuleo, E.; Gallardo-Toledo, E.; Andonie, O.; Muñoz, L.; Rojas, X.; Acosta, G.; Sánchez-Navarro, M.; Giralt, E.; et al. Improving gold nanorod delivery to the central nervous system by conjugation to the shuttle Angiopep-2. *Nanomedicine* **2017**, *12*, 2503–2517. [CrossRef]
57. Tapia-Arellano, A.; Gallardo-Toledo, E.; Ortiz, C.; Henríquez, J.; Feijóo, C.G.; Araya, E.; Sierpe, R.; Kogan, M.J. Functionalization with PEG/Angiopep-2 peptide to improve the delivery of gold nanoprisms to central nervous system: In vitro and in vivo studies. *Mater. Sci. Eng. C* **2021**, *121*, 111785. [CrossRef]
58. Zhao, Z.; Xu, H.; Li, S.; Han, Y.; Jia, J.; Han, Z.; Zhang, D.; Zhang, L.; Yu, R.; Liu, H. Hypoxic radiosensitizer-lipid coated gold nanoparticles enhance the effects of radiation therapy on tumor growth. *J. Biomed. Nanotech.* **2019**, *15*, 1982–1993. [CrossRef]
59. Tang, F.; Li, L.; Chen, D. Mesoporous silica nanoparticles: Synthesis, biocompatibility and drug delivery. *Adv. Mater.* **2012**, *24*, 1504–1534. [CrossRef]
60. Moodley, T.; Singh, M. Current Stimuli-responsive Mesoporous Silica Nanoparticles for Cancer Therapy. *Pharmaceutics* **2021**, *13*, 71. [CrossRef]
61. Vallet-Regí, M. Our contributions to applications of mesoporous silica nanoparticles. *Acta Biomater.* **2022**, *137*, 44–52. [CrossRef]
62. Yang, Y.; Chen, F.; Xu, N.; Yao, Q.; Wang, R.; Xie, X.; Zhang, F.; He, Y.; Shao, D.; Dong, W.-F.; et al. Red-light-triggered self-destructive mesoporous silica nanoparticles for cascade-amplifying chemo-photodynamic therapy favoring antitumor immune responses. *Biomaterials* **2022**, *281*, 121368. [CrossRef] [PubMed]
63. Jafari, S.; Derakhshankhah, H.; Alaei, L.; Fattahi, A.; Varnamkhasti, B.S.; Saboury, A.A. Mesoporous silica nanoparticles for therapeutic/diagnostic applications. *Biomed. Pharmacother.* **2019**, *109*, 1100–1111. [CrossRef] [PubMed]
64. Zhu, J.; Zhang, Y.; Chen, X.; Zhang, Y.; Zhang, K.; Zheng, H.; Wei, Y.; Zheng, H.; Zhu, J.; Wu, F.; et al. Angiopep-2 modified lipid-coated mesoporous silica nanoparticles for glioma targeting therapy overcoming BBB. *Biochem. Biophys. Res. Commun.* **2021**, *534*, 902–907. [CrossRef] [PubMed]
65. Wang, G.-W.; Fei, W.-D.; Zhang, R.-R.; Guo, M.-M.; Xu, J.-J.; Li, F.-Z. Preparation and in Vitro Evaluation of Paclitaxel-loaded Core-Shell Structural Phospholipid-Functionalized Mesoporous Silica Nanoparticles Modified with Angiopep-2. *Chin. Pharm. J.* **2015**, *24*, 775–783.
66. Heggannavar, G.B.; Vijeth, S.; Kariduraganavar, M.Y. Development of dual drug loaded PLGA based mesoporous silica nanoparticles and their conjugation with Angiopep-2 to treat glioma. *J. Drug Deliv. Sci. Technol.* **2019**, *53*, 101157. [CrossRef]
67. Novoselov, K.S.; Geim, A.K.; Morozov, S.V.; Jiang, D.; Zhang, Y.; Dubonos, S.V.; Grigorieva, I.V.; Firsov, A.A. Electric field effect in atomically thin carbon films. *Science* **2004**, *306*, 666–669. [CrossRef]
68. Shadjou, N.; Hasanzadeh, M. Graphene and its nanostructure derivatives for use in bone tissue engineering: Recent advances. *J. Biomed. Mater. Res. A* **2016**, *104*, 1250–1275. [CrossRef]
69. Zhao, Y.; Yin, H.; Zhang, X. Modification of graphene oxide by angiopep-2 enhances anti-glioma efficiency of the nanoscaled delivery system for doxorubicin. *Aging* **2020**, *12*, 10506. [CrossRef]
70. Heersche, H.B.; Jarillo-Herrero, P.; Oostinga, J.B.; Vandersypen, L.M.; Morpurgo, A.F. Bipolar supercurrent in graphene. *Nature* **2007**, *446*, 56–59. [CrossRef]
71. Ren, J.; Shen, S.; Wang, D.; Xi, Z.; Guo, L.; Pang, Z.; Qian, Y.; Sun, X.; Jiang, X. The targeted delivery of anticancer drugs to brain glioma by PEGylated oxidized multi-walled carbon nanotubes modified with angiopep-2. *Biomaterials* **2012**, *33*, 3324–3333. [CrossRef]

72. Cohen, E.N.; Kondiah, P.P.; Choonara, Y.E.; du Toit, L.C.; Pillay, V. Carbon dots as nanotherapeutics for biomedical application. *Curr. Pharm. Des.* **2020**, *26*, 2207–2221. [CrossRef]
73. Ruan, S.; Qian, J.; Shen, S.; Chen, J.; Zhu, J.; Jiang, X.; He, Q.; Yang, W.; Gao, H. Fluorescent carbonaceous nanodots for noninvasive glioma imaging after angiopep-2 decoration. *Bioconj. Chem.* **2014**, *25*, 2252–2259. [CrossRef] [PubMed]
74. Gul, S.; Khan, S.B.; Rehman, I.U.; Khan, M.A.; Khan, M. A comprehensive review of magnetic nanomaterials modern day theranostics. *Front. Mater.* **2019**, *6*, 179. [CrossRef]
75. Hsu, S.P.; Dhawan, U.; Tseng, Y.-Y.; Lin, C.-P.; Kuo, C.-Y.; Wang, L.-F.; Chung, R.-J. Glioma-sensitive delivery of Angiopep-2 conjugated iron gold alloy nanoparticles ensuring simultaneous tumor imaging and hyperthermia mediated cancer theranostics. *Appl. Mater. Today* **2020**, *18*, 100510. [CrossRef]
76. Qiao, L.; Qin, Y.; Wang, Y.; Liang, Y.; Zhu, D.; Xiong, W.; Li, L.; Bao, D.; Zhang, L.; Jin, X. A brain glioma gene delivery strategy by angiopep-2 and TAT-modified magnetic lipid-polymer hybrid nanoparticles. *RSC Adv.* **2020**, *10*, 41471–41481. [CrossRef]
77. Yadav, D.; Sandeep, K.; Pandey, D.; Dutta, R.K. Liposomes for drug delivery. *J. Biotechnol. Biomater.* **2017**, *7*, 276. [CrossRef]
78. Parveen, S.; Sahoo, S.K. Polymeric nanoparticles for cancer therapy. *J. Drug Target.* **2008**, *16*, 108–123. [CrossRef]
79. Kahraman, E.; Güngör, S.; Özsoy, Y. Potential enhancement and targeting strategies of polymeric and lipid-based nanocarriers in dermal drug delivery. *Ther. Deliv.* **2017**, *8*, 967–985. [CrossRef]
80. Chis, A.A.; Dobrea, C.; Morgovan, C.; Arseniu, A.M.; Rus, L.L.; Butuca, A.; Juncan, A.M.; Totan, M.; Vonica-Tincu, A.L.; Cormos, G.; et al. Applications and limitations of dendrimers in biomedicine. *Molecules* **2020**, *25*, 3982. [CrossRef]
81. Sztandera, K.; Gorzkiewicz, M.; Klajnert-Maculewicz, B. Gold nanoparticles in cancer treatment. *Mol. Pharm* **2018**, *16*, 1–23. [CrossRef]
82. Sajjadi, M.; Nasrollahzadeh, M.; Jaleh, B.; Soufi, G.J.; Iravani, S. Carbon-based nanomaterials for targeted cancer nanotherapy: Recent trends and future prospects. *J. Drug Target.* **2021**, *7*, 716–741. [CrossRef]
83. Simon, J.; Flahaut, E.; Golzio, M. Overview of carbon nanotubes for biomedical applications. *Materials* **2019**, *12*, 624. [CrossRef] [PubMed]
84. Sivasankarapillai, V.S.; Kirthi, A.V.; Akksadha, M.; Indu, S.; Dharshini, U.D.; Pushpamalar, J.; Karthik, L.; Arivarasan, V.K.; Janarthanan, P. Recent advancements in the applications of carbon nanodots: Exploring the rising star of nanotechnology. *Nanoscale Adv.* **2020**, *2*, 1760–1773. [CrossRef]
85. Ostadhossein, F.; Pan, D. Functional carbon nanodots for multiscale imaging and therapy. *Wiley Interdiscip. Rev. Nanomed. Nanobiotechnol.* **2017**, *9*, e1436. [CrossRef] [PubMed]
86. Revia, R.A.; Zhang, M. Magnetite nanoparticles for cancer diagnosis, treatment, and treatment monitoring: Recent advances. *Mater. Today* **2016**, *19*, 157–168. [CrossRef]
87. Skalli, O.; Wilhelmsson, U.; Örndahl, C.; Fekete, B.; Malmgren, K.; Rydenhag, B.; Pekny, M. Astrocytoma grade IV (glioblastoma multiforme) displays 3 subtypes with unique expression profiles of intermediate filament proteins. *Hum. Pathol.* **2013**, *44*, 2081–2088. [CrossRef]
88. Henson, J.W. Treatment of glioblastoma multiforme: A new standard. *Arch. Neurol.* **2006**, *63*, 337–341. [CrossRef]
89. Lu, F.; Pang, Z.; Zhao, J.; Jin, K.; Li, H.; Pang, Q.; Zhang, L.; Pang, Z. Angiopep-2-conjugated poly (ethylene glycol)-co-poly (ε-caprolactone) polymersomes for dual-targeting drug delivery to glioma in rats. *Int. J. Nanomed.* **2017**, *12*, 2117. [CrossRef]
90. Gao, H.; Zhang, S.; Cao, S.; Yang, Z.; Pang, Z.; Jiang, X. Angiopep-2 and activatable cell-penetrating peptide dual-functionalized nanoparticles for systemic glioma-targeting delivery. *Mol. Pharm.* **2014**, *11*, 2755–2763. [CrossRef]
91. Zhu, Y.; Jiang, Y.; Meng, F.; Deng, C.; Cheng, R.; Zhang, J.; Feijen, J.; Zhong, Z. Highly efficacious and specific anti-glioma chemotherapy by tandem nanomicelles co-functionalized with brain tumor-targeting and cell-penetrating peptides. *J. Control. Release* **2018**, *278*, 1–8. [CrossRef] [PubMed]
92. Kim, J.S.; Shin, D.H.; Kim, J.-S. Dual-targeting immunoliposomes using angiopep-2 and CD133 antibody for glioblastoma stem cells. *J. Control. Release* **2018**, *269*, 245–257. [CrossRef] [PubMed]
93. Wang, S.; Zhao, C.; Liu, P.; Wang, Z.; Ding, J.; Zhou, W. Facile construction of dual-targeting delivery system by using lipid capped polymer nanoparticles for anti-glioma therapy. *RSC Adv.* **2018**, *8*, 444–453. [CrossRef]
94. Guo, Q.; Zhu, Q.; Miao, T.; Tao, J.; Ju, X.; Sun, Z.; Li, H.; Xu, G.; Chen, H.; Han, L. LRP1-upregulated nanoparticles for efficiently conquering the blood-brain barrier and targetedly suppressing multifocal and infiltrative brain metastases. *J. Control. Release* **2019**, *303*, 117–129. [CrossRef]
95. Tao, J.; Fei, W.; Tang, H.; Li, C.; Mu, C.; Zheng, H.; Li, F.; Zhu, Z. Angiopep-2-conjugated "core–shell" hybrid nanovehicles for targeted and pH-triggered delivery of arsenic trioxide into glioma. *Mol. Pharm.* **2019**, *16*, 786–797. [CrossRef]
96. Khan, N.U.; Ni, J.; Ju, X.; Miao, T.; Chen, H.; Han, L. Escape from abluminal LRP1-mediated clearance for boosted nanoparticle brain delivery and brain metastasis treatment. *Acta Pharm. Sin. B* **2021**, *11*, 1341–1354. [CrossRef] [PubMed]
97. Luo, Z.; Jin, K.; Pang, Q.; Shen, S.; Yan, Z.; Jiang, T.; Zhu, X.; Yu, L.; Pang, Z.; Jiang, X. On-demand drug release from dual-targeting small nanoparticles triggered by high-intensity focused ultrasound enhanced glioblastoma-targeting therapy. *ACS Appl. Mater. Interfaces* **2017**, *9*, 31612–31625. [CrossRef]
98. Figueiredo, P.; Balasubramanian, V.; Shahbazi, M.-A.; Correia, A.; Wu, D.; Palivan, C.G.; Hirvonen, J.T.; Santos, H.A. Angiopep2-functionalized polymersomes for targeted doxorubicin delivery to glioblastoma cells. *Int. J. Pharm.* **2016**, *511*, 794–803. [CrossRef]
99. Song, P.; Song, N.; Li, L.; Wu, M.; Lu, Z.; Zhao, X. Angiopep-2-modified carboxymethyl chitosan-based pH/reduction dual-stimuli-responsive nanogels for enhanced targeting glioblastoma. *Biomacromolecules* **2021**, *22*, 2921–2934. [CrossRef]

100. Hua, L.; Wang, Z.; Zhao, L.; Mao, H.; Wang, G.; Zhang, K.; Liu, X.; Wu, D.; Zheng, Y.; Lu, J.; et al. Hypoxia-responsive lipid-poly-(hypoxic radiosensitized polyprodrug) nanoparticles for glioma chemo-and radiotherapy. *Theranostics* **2018**, *8*, 5088. [CrossRef]
101. Zong, Z.; Hua, L.; Wang, Z.; Xu, H.; Ye, C.; Pan, B.; Zhao, Z.; Zhang, L.; Lu, J.; Liu, H.; et al. Self-assembled angiopep-2 modified lipid-poly (hypoxic radiosensitized polyprodrug) nanoparticles delivery TMZ for glioma synergistic TMZ and RT therapy. *Drug Deliv.* **2019**, *26*, 34–44. [CrossRef]
102. Su, X.; Zhang, D.; Zhang, H.; Zhao, K.; Hou, W. Preparation and characterization of angiopep-2 functionalized ginsenoside-Rg3 loaded nanoparticles and the effect on C6 glioma cells. *Pharm. Dev. Technol.* **2020**, *25*, 385–395. [CrossRef] [PubMed]
103. Zou, Y.; Liu, Y.; Yang, Z.; Zhang, D.; Lu, Y.; Zheng, M.; Xue, X.; Geng, J.; Chung, R.; Shi, B. Effective and targeted human orthotopic glioblastoma xenograft therapy via a multifunctional biomimetic nanomedicine. *Adv. Mater.* **2018**, *30*, 1803717. [CrossRef]
104. Hao, Y.; Zhang, B.; Zheng, C.; Ji, R.; Ren, X.; Guo, F.; Sun, S.; Shi, J.; Zhang, H.; Zhang, Z.; et al. The tumor-targeting core–shell structured DTX-loaded PLGA@ Au nanoparticles for chemo-photothermal therapy and X-ray imaging. *J. Control. Release* **2015**, *220*, 545–555. [CrossRef]
105. Xin, H.; Sha, X.; Jiang, X.; Zhang, W.; Chen, L.; Fang, X. Anti-glioblastoma efficacy and safety of paclitaxel-loading Angiopep-conjugated dual targeting PEG-PCL nanoparticles. *Biomaterials* **2012**, *33*, 8167–8176. [CrossRef]
106. Parashar, A.K.; Gupta, A.K.; Jain, N.K. Synthesis and characterization of Agiopep-2 anchored PEGylated poly propyleneimine dendrimers for targeted drug delivery to glioblastoma multiforme. *J. Drug Deliv. Ther.* **2018**, *8*, 74–79.
107. Jiang, Y.; Yang, W.; Zhang, J.; Meng, F.; Zhong, Z. Protein toxin chaperoned by LRP-1-targeted virus-mimicking vesicles induces high-efficiency glioblastoma therapy in vivo. *Adv. Mater.* **2018**, *30*, 1800316. [CrossRef] [PubMed]
108. Shao, K.; Wu, J.; Chen, Z.; Huang, S.; Li, J.; Ye, L.; Lou, J.; Zhu, L.; Jiang, C. A brain-vectored angiopep-2 based polymeric micelles for the treatment of intracranial fungal infection. *Biomaterials* **2012**, *33*, 6898–6907. [CrossRef] [PubMed]
109. Shao, K.; Huang, R.; Li, J.; Han, L.; Ye, L.; Lou, J.; Jiang, C. Angiopep-2 modified PE-PEG based polymeric micelles for amphotericin B delivery targeted to the brain. *J. Control. Release* **2010**, *147*, 118–126. [CrossRef] [PubMed]
110. Bao, Q.; Hu, P.; Xu, Y.; Cheng, T.; Wei, C.; Pan, L.; Shi, J. Simultaneous blood–brain barrier crossing and protection for stroke treatment based on edaravone-loaded ceria nanoparticles. *ACS Nano* **2018**, *12*, 6794–6805. [CrossRef]
111. Li, Y.; Dang, Y.; Han, D.; Tan, Y.; Liu, X.; Zhang, F.; Xu, Y.; Zhang, H.; Yan, X.; Zhang, X.; et al. An Angiopep-2 functionalized nanoformulation enhances brain accumulation of tanshinone IIA and exerts neuroprotective effects against ischemic stroke. *N. J. Chem.* **2018**, *42*, 17359–17370. [CrossRef]
112. Liu, X.; Li, Y.-T.; Liu, W.; Zhang, F.-M.; Chen, Z.-Z.; Zeng, Z.-Y.; Xu, M.S.; Sun, X.J. Neuroprotective effect of Angiopep-2 peptide modified scutellarin-loaded PEGylated PAMAM dendrimer nanoparticles on ischemic stroke by modulating the Toll-like receptors-dependent MyD88/IKK/NF-κB signaling pathway. *Chin. J. Pharmacol. Toxicol.* **2016**, *10*, 1019–1020.
113. Ying, X.; Wang, Y.; Liang, J.; Yue, J.; Xu, C.; Lu, L.; Xu, Z.; Gao, J.; Du, Y.; Chen, Z. Angiopep-conjugated electro-responsive hydrogel nanoparticles: Therapeutic potential for epilepsy. *Angew. Chem.* **2014**, *126*, 12644–12648. [CrossRef]
114. Hu, L.; Wang, Y.; Zhang, Y.; Yang, N.; Han, H.; Shen, Y.; Cui, D.; Guo, S. Angiopep-2 modified PEGylated 2-methoxyestradiol. micelles to treat the PC12 cells with oxygen-glucose deprivation/reoxygenation. *Colloids Surf. B Biointerfaces* **2018**, *171*, 638–646. [CrossRef] [PubMed]
115. Jagaran, K.; Singh, M. Nanomedicine for Neurodegenerative Disorders: Focus on Alzheimer's and Parkinson's Diseases. *Int. J. Mol. Sci.* **2021**, *22*, 9082. [CrossRef] [PubMed]
116. Gao, S.; Tian, H.; Xing, Z.; Zhang, D.; Guo, Y.; Guo, Z.; Zhu, X.; Chen, X. A non-viral suicide gene delivery system traversing the blood brain barrier for non-invasive glioma targeting treatment. *J. Control. Release* **2016**, *243*, 357–369. [CrossRef]
117. Huang, R.; Ma, H.; Guo, Y.; Liu, S.; Kuang, Y.; Shao, K.; Li, J.; Liu, Y.; Han, L.; Huang, S.; et al. Angiopep-conjugated nanoparticles for targeted long-term gene therapy of Parkinson's disease. *Pharm. Res.* **2013**, *30*, 2549–2559. [CrossRef]
118. Zheng, M.; Liu, Y.; Wang, Y.; Zhang, D.; Zou, Y.; Ruan, W.; Yin, J.; Tao, W.; Park, J.B.; Shi, B. ROS-responsive polymeric siRNA nanomedicine stabilized by triple interactions for the robust glioblastoma combinational RNAi therapy. *Adv. Mater.* **2019**, *31*, 1903277. [CrossRef]
119. Wen, L.; Peng, Y.; Wang, K.; Huang, Z.; He, S.; Xiong, R.; Wu, L.; Zhang, F.; Hu, F. Regulation of pathological BBB restoration via nanostructured ROS-responsive glycolipid-like copolymer entrapping siVEGF for glioblastoma targeted therapeutics. *Nano Res.* **2021**, *15*, 1455–1465. [CrossRef]
120. Liu, Y.; Zou, Y.; Feng, C.; Lee, A.; Yin, J.; Chung, R.; Park, J.B.; Rizos, H.; Tao, W.; Zheng, M.; et al. Charge conversional biomimetic nanocomplexes as a multifunctional platform for boosting orthotopic glioblastoma RNAi therapy. *Nano Lett.* **2020**, *20*, 1637–1646. [CrossRef]
121. Liu, Y.; Zheng, M.; Jiao, M.; Yan, C.; Xu, S.; Du, Q.; Morsch, M.; Yin, J.; Shi, B. Polymeric nanoparticle mediated inhibition of miR-21 with enhanced miR-124 expression for combinatorial glioblastoma therapy. *Biomaterials* **2021**, *276*, 121036. [CrossRef]
122. An, S.; He, D.; Wagner, E.; Jiang, C. Peptide-like polymers exerting effective glioma-targeted siRNA delivery and release for therapeutic application. *Small* **2015**, *11*, 5142–5150. [CrossRef]
123. Shi, Y.; Jiang, Y.; Cao, J.; Yang, W.; Zhang, J.; Meng, F.; Zhong, Z. Boosting RNAi therapy for orthotopic glioblastoma with non-toxic brain-targeting chimaeric polymersomes. *J. Control. Release* **2018**, *292*, 163–171. [CrossRef] [PubMed]
124. Zhao, Z.; Li, Y.; Liu, H.; Jain, A.; Patel, P.V.; Cheng, K. Co-delivery of IKBKE siRNA and cabazitaxel by hybrid nanocomplex inhibits invasiveness and growth of triple-negative breast cancer. *Sci. Adv.* **2020**, *6*, eabb0616. [CrossRef] [PubMed]

125. Dai, X.; Tan, C. Combination of microRNA therapeutics with small-molecule anticancer drugs: Mechanism of action and co-delivery nanocarriers. *Adv. Drug Deliv. Rev.* **2015**, *81*, 184–197. [CrossRef] [PubMed]
126. Sun, X.; Pang, Z.; Ye, H.; Qiu, B.; Guo, L.; Li, J.; Ren, J.; Qian, Y.; Zhang, Q.; Chen, J.; et al. Co-delivery of pEGFP-hTRAIL and paclitaxel to brain glioma mediated by an angiopep-conjugated liposome. *Biomaterials* **2012**, *33*, 916–924. [CrossRef]
127. Yang, Z.-Z.; Li, J.-Q.; Wang, Z.-Z.; Dong, D.-W.; Qi, X.-R. Tumor-targeting dual peptides-modified cationic liposomes for delivery of siRNA and docetaxel to gliomas. *Biomaterials* **2014**, *35*, 5226–5239. [CrossRef]
128. Sun, X.; Chen, Y.; Zhao, H.; Qiao, G.; Liu, M.; Zhang, C.; Cui, D.; Ma, L. Dual-modified cationic liposomes loaded with paclitaxel and survivin siRNA for targeted imaging and therapy of cancer stem cells in brain glioma. *Drug Deliv.* **2018**, *25*, 1718–1727. [CrossRef] [PubMed]
129. Lihuang, L.; Qiuyan, G.; Yanxiu, L.; Mindan, L.; Jun, Y.; Yunlong, G.; Qiang, Z.; Benqiang, S.; Xiumin, W.; Liang-Cheng, L.; et al. Targeted combination therapy for glioblastoma by co-delivery of doxorubicin, YAP-siRNA and gold nanorods. *J. Mater. Sci. Technol.* **2021**, *63*, 81–90. [CrossRef]
130. Wang, L.; Hao, Y.; Li, H.; Zhao, Y.; Meng, D.; Li, D.; Shi, J.; Zhang, H.; Zhang, Z.; Zhang, Y. Co-delivery of doxorubicin and siRNA for glioma therapy by a brain targeting system: Angiopep-2-modified poly (lactic-co-glycolic acid) nanoparticles. *J. Drug Target.* **2015**, *23*, 832–846. [CrossRef]
131. Ye, C.; Pan, B.; Xu, H.; Zhao, Z.; Shen, J.; Lu, J.; Yu, R.; Liu, H. Co-delivery of GOLPH3 siRNA and Gefitinib by cationic lipid-PLGA nanoparticles improves EGFR-targeted therapy for glioma. *J. Mol. Med.* **2019**, *97*, 1575–1588. [CrossRef]
132. Wen, L.; Wen, C.; Zhang, F.; Wang, K.; Yuan, H.; Hu, F. siRNA and chemotherapeutic molecules entrapped into a redox-responsive platform for targeted synergistic combination therapy of glioma. *Nanomed. Nanotechnol. Biol. Med.* **2020**, *28*, 102218. [CrossRef]
133. Shi, H.; Sun, S.; Xu, H.; Zhao, Z.; Han, Z.; Jia, J.; Wu, D.; Lu, J.; Liu, H.; Yu, R. Combined Delivery of Temozolomide and siPLK1 Using Targeted Nanoparticles to Enhance Temozolomide Sensitivity in Glioma. *Int. J. Nanomed.* **2020**, *15*, 3347. [CrossRef] [PubMed]
134. Shi, H.; Sadler, P.J. How promising is phototherapy for cancer? *Br. J. Cancer* **2020**, *123*, 871–873. [CrossRef] [PubMed]
135. Zou, L.; Wang, H.; He, B.; Zeng, L.; Tan, T.; Cao, H.; He, X.; Zhang, Z.; Guo, S.; Li, Y. Current approaches of photothermal therapy in treating cancer metastasis with nanotherapeutics. *Theranostics* **2016**, *6*, 762. [CrossRef]
136. Chitgupi, U.; Qin, Y.; Lovell, J.F. Targeted Nanomaterials for Phototherapy. *Nanotheranostics* **2017**, *1*, 38–58. [CrossRef]
137. Tsai, Y.-C.; Vijayaraghavan, P.; Chiang, W.-H.; Chen, H.-H.; Liu, T.-I.; Shen, M.-Y.; Omoto, A.; Kamimura, M.; Soga, K.; Chiu, H.-C. Targeted delivery of functionalized upconversion nanoparticles for externally triggered photothermal/photodynamic therapies of brain glioblastoma. *Theranostics* **2018**, *8*, 1435–1448. [CrossRef]
138. Pero, M. Nanostructured Lipid Carriers as Effective Delivery Systems of Photosensitizers for Photodynamic Therapy against Gliobastoma Multiforme: Politecnico di Torino 2021. Available online: https://webthesis.biblio.polito.it/19658/ (accessed on 23 October 2021).
139. Lu, L.; Wang, K.; Lin, C.; Yang, W.; Duan, Q.; Li, K.; Cai, K. Constructing nanocomplexes by multicomponent self-assembly for curing orthotopic glioblastoma with synergistic chemo-photothermal therapy. *Biomaterials* **2021**, *279*, 121193. [CrossRef]
140. Tang, W.; Fan, W.; Lau, J.; Deng, L.; Shen, Z.; Chen, X. Emerging blood–brain-barrier-crossing nanotechnology for brain cancer theranostics. *Chem. Soc. Rev.* **2019**, *48*, 2967–3014. [CrossRef] [PubMed]
141. Xie, R.; Wu, Z.; Zeng, F.; Cai, H.; Wang, D.; Gu, L.; Zhu, H.; Lui, S.; Guo, G.; Song, B.; et al. Retro-enantio isomer of angiopep-2 assists nanoprobes across the blood-brain barrier for targeted magnetic resonance/fluorescence imaging of glioblastoma. *Signal. Transduct. Target. Ther.* **2021**, *6*, 1–13. [CrossRef]
142. Leung, D.; Han, X.; Mikkelsen, T.; Nabors, L.B. Role of MRI in primary brain tumor evaluation. *J. Natl. Compr. Cancer Netw.* **2014**, *12*, 1561–1568. [CrossRef] [PubMed]
143. Tocchio, S.; Kline-Fath, B.; Kanal, E.; Schmithorst, V.J.; Panigrahy, A. MRI evaluation and safety in the developing brain. *Semin. Perinatol.* **2015**, *39*, 73–104. [CrossRef]
144. Avasthi, A.; Caro, C.; Pozo-Torres, E.; Leal, M.P.; García-Martín, M.L. Magnetic nanoparticles as MRI contrast agents. *Top. Curr. Chem.* **2020**, *378*, 49–91. [CrossRef]
145. Ramnandan, D.; Mokhosi, S.; Daniels, A.; Singh, M. Chitosan, Polyethylene glycol. and Polyvinyl alcohol. modified MgFe2O4 ferrite magnetic nanoparticles in Doxorubicin delivery: A comparative study in vitro. *Molecules* **2021**, *26*, 3893. [CrossRef] [PubMed]
146. Chen, G.-J.; Su, Y.-Z.; Hsu, C.; Lo, Y.-L.; Huang, S.-J.; Ke, J.-H.; Kuo, Y.-C.; Wang, L.-F. Angiopep-pluronic F127-conjugated superparamagnetic iron oxide nanoparticles as nanotheranostic agents for BBB targeting. *J. Mater. Chem. B* **2014**, *2*, 5666–5675. [CrossRef]
147. Du, C.; Liu, X.; Hu, H.; Li, H.; Yu, L.; Geng, D.; Chen, Y.; Zhang, J. Dual-targeting and excretable ultrasmall SPIONs for T 1-weighted positive MR imaging of intracranial glioblastoma cells by targeting the lipoprotein receptor-related protein. *J. Mater. Chem. B* **2020**, *8*, 2296–2306. [CrossRef]
148. He, J.; Li, C.; Ding, L.; Huang, Y.; Yin, X.; Zhang, J.; Zhang, J.; Yao, C.; Liang, M.; Pirraco, R.P.; et al. Tumor targeting strategies of smart fluorescent nanoparticles and their applications in cancer diagnosis and treatment. *Adv. Mater.* **2019**, *31*, 1902409. [CrossRef]
149. Ren, F.; Liu, H.; Zhang, H.; Jiang, Z.; Xia, B.; Genevois, C.; He, T.; Allix, M.; Sun, Q.; Li, Z.; et al. Engineering NIR-IIb fluorescence of Er-based lanthanide nanoparticles for through-skull targeted imaging and imaging-guided surgery of orthotopic glioma. *Nano Today* **2020**, *34*, 100905. [CrossRef]

150. Wei, R.; Liu, Y.; Gao, J.; Yong, V.W.; Xue, M. Small functionalized iron oxide nanoparticles for dual brain magnetic resonance imaging and fluorescence imaging. *RSC Adv.* **2021**, *11*, 12867–12875. [CrossRef]
151. Endo-Takahashi, Y.; Ooaku, K.; Ishida, K.; Suzuki, R.; Maruyama, K.; Negishi, Y. Preparation of Angiopep-2 peptide-modified bubble liposomes for delivery to the brain. *Biol. Pharm. Bull.* **2016**, *39*, 977–983. [CrossRef]
152. Costagliola di Polidoro, A.; Zambito, G.; Haeck, J.; Mezzanotte, L.; Lamfers, M.; Netti, P.A.; Torino, E. Theranostic design of angiopep-2 conjugated hyaluronic acid nanoparticles (Thera-ANG-cHANPs) for dual targeting and boosted imaging of glioma cells. *Cancers* **2021**, *13*, 503. [CrossRef]
153. Lin, F.; Xiong, X.-L.; Cui, E.-M.; Lei, Y. A novel ANG-BSA/BCNU/ICG MNPs integrated for targeting therapy of glioblastoma. *Res. Sq.* **2021**. [CrossRef]
154. Apawu, A.K.; Curley, S.M.; Dixon, A.R.; Hali, M.; Sinan, M.; Braun, R.D.; Castracane, J.; Cacace, A.T.; Bergkvist, M.; Holt, A.G. MRI compatible MS2 nanoparticles designed to cross the blood–brain-barrier: Providing a path towards tinnitus treatment. *Nanomed. Nanotechnol. Biol. Med.* **2018**, *14*, 1999–2008. [CrossRef]
155. Hao, Y.; Wang, L.; Zhao, Y.; Meng, D.; Li, D.; Li, H.; Zhang, B.; Shi, J.; Zhang, H.; Zhang, Z.; et al. Targeted imaging and chemo-phototherapy of brain cancer by a multifunctional drug delivery system. *Macromol. Biosci.* **2015**, *15*, 1571–1585. [CrossRef]
156. Xu, H.; Han, Y.; Zhao, G.; Zhang, L.; Zhao, Z.; Wang, Z.; Zhao, L.; Hua, L.; Naveena, K.; Lu, J.; et al. Hypoxia-Responsive Lipid–Polymer Nanoparticle-Combined Imaging-Guided Surgery and Multitherapy Strategies for Glioma. *ACS Appl. Mater. Interfaces* **2020**, *12*, 52319–52328. [CrossRef] [PubMed]
157. Parrasia, S.; Rossa, A.; Varanita, T.; Checchetto, V.; De Lorenzi, R.; Zoratti, M.; Paradisi, C.; Ruzza, P.; Mattarei, A.; Szabò, I.; et al. An Angiopep2-PAPTP Construct Overcomes the Blood-Brain Barrier. New Perspec-tives against Brain Tumors. *Pharmaceuticals* **2021**, *14*, 129. [CrossRef]
158. Morales-Zavala, F.; Arriagada, H.; Hassan, N.; Velasco, C.; Riveros, A.; Álvarez, A.R.; Minniti, A.N.; Rojas-Silva, X.; Muñoz, L.L.; Vasquez, R.; et al. Peptide multifunctionalized gold nanorods decrease toxicity of β-amyloid peptide in a Caenorhabditis elegans model of Alzheimer's disease. *Nanomed Nanotechnol. Biol. Med.* **2017**, *13*, 2341–2350. [CrossRef] [PubMed]
159. Wang, X.; Xiong, Z.; Liu, Z.; Huang, X.; Jiang, X. Angiopep-2/IP10-EGFRvIIIscFv modified nanoparticles and CTL synergistical-ly inhibit malignant glioblastoma. *Sci. Rep.* **2018**, *8*, 1–11. [CrossRef]
160. Qu, F.; Wang, P.; Zhang, K.; Shi, Y.; Li, Y.; Li, C.; Lu, J.; Liu, Q.; Wang, X. Manipulation of Mitophagy by "All-in-One" nanosensitizer augments sonodynamic glioma therapy. *Autophagy* **2020**, *16*, 1413–1435. [CrossRef]
161. Gao, X.; Qian, J.; Zheng, S.; Xiong, Y.; Man, J.; Cao, B.; Wang, L.; Ju, S.; Li, C. Up-regulating blood brain barrier permeability of nanoparticles via multivalent effect. *Pharm Res.* **2013**, *30*, 2538–2548. [CrossRef]
162. Papademetriou, I.; Vedula, E.; Charest, J.; Porter, T. Effect of flow on targeting and penetration of angiopep-decorated nanoparticles in a microfluidic model blood-brain barrier. *PLoS ONE* **2018**, *13*, e0205158. [CrossRef] [PubMed]

polymers MDPI

Review

Polysaccharide-Drug Conjugates: A Tool for Enhanced Cancer Therapy

Neena Yadav [1,†], Arul Prakash Francis [1,2,†], Veeraraghavan Vishnu Priya [2], Shankargouda Patil [3], Shazia Mustaq [4], Sameer Saeed Khan [3], Khalid J. Alzahrani [5], Hamsa Jameel Banjer [5], Surapaneni Krishna Mohan [6], Ullas Mony [2] and Rukkumani Rajagopalan [1,*]

1 Department of Biochemistry and Molecular Biology, School of Life Sciences, Pondicherry University, Puducherry 605014, India; neenayadav100@gmail.com (N.Y.); fdapharma@gmail.com (A.P.F.)
2 Centre of Molecular Medicine and Diagnostics (COMManD), Saveetha Institute of Medical & Technical Sciences, Saveetha Dental College and Hospitals, Saveetha University, Chennai 600077, India; vishnupriya@saveetha.com (V.V.P.); ullasmony@gmail.com (U.M.)
3 Department of Maxillofacial Surgery and Diagnostic Sciences, Division of Oral Pathology, College of Dentistry, Jazan University, Jazan 45142, Saudi Arabia; dr.ravipatil@gmail.com (S.P.); samarkhan8@gmail.com (S.S.K.)
4 Dental Health Department, College of Applied Medical Sciences, King Saud University, Riyadh 11451, Saudi Arabia; smushtaqdr@gmail.com
5 Department of Clinical Laboratories Sciences, College of Applied Medical Sciences, Taif University, Taif 21974, Saudi Arabia; ak.jamaan@tu.edu.sa (K.J.A.); h.banjer@tu.edu.sa (H.J.B.)
6 Departments of Biochemistry, Molecular Virology, Research, Clinical Skills & Research Institute & Simulation, Panimalar Medical College Hospital, Varadharajapuram, Poonamallee, Chennai 600123, India; krishnamohan.surapaneni@gmail.com
* Correspondence: ruks2k2@gmail.com; Tel.: +91-(96)-7784-7337
† These authors contributed equally to this work.

Citation: Yadav, N.; Francis, A.P.; Priya, V.V.; Patil, S.; Mustaq, S.; Khan, S.S.; Alzahrani, K.J.; Banjer, H.J.; Mohan, S.K.; Mony, U.; et al. Polysaccharide-Drug Conjugates: A Tool for Enhanced Cancer Therapy. *Polymers* **2022**, *14*, 950. https://doi.org/10.3390/polym14050950

Academic Editors: Stephanie Andrade, Joana A. Loureiro and Maria João Ramalho

Received: 19 January 2022
Accepted: 23 February 2022
Published: 27 February 2022

Publisher's Note: MDPI stays neutral with regard to jurisdictional claims in published maps and institutional affiliations.

Abstract: Cancer is one of the most widespread deadly diseases, following cardiovascular disease, worldwide. Chemotherapy is widely used in combination with surgery, hormone and radiation therapy to treat various cancers. However, chemotherapeutic drugs can cause severe side effects due to non-specific targeting, poor bioavailability, low therapeutic indices, and high dose requirements. Several drug carriers successfully overcome these issues and deliver drugs to the desired sites, reducing the side effects. Among various drug delivery systems, polysaccharide-based carriers that target only the cancer cells have been developed to overcome the toxicity of chemotherapeutics. Polysaccharides are non-toxic, biodegradable, hydrophilic biopolymers that can be easily modified chemically to improve the bioavailability and stability for delivering therapeutics into cancer tissues. Different polysaccharides, such as chitosan, alginates, cyclodextrin, pullulan, hyaluronic acid, dextran, guar gum, pectin, and cellulose, have been used in anti-cancer drug delivery systems. This review highlights the recent progress made in polysaccharides-based drug carriers in anti-cancer therapy.

Keywords: chemotherapy; cancer; polysaccharides; toxicity; drug delivery

1. Introduction

Cancer is the most prevalent deadly disease, next to cardiovascular disease, throughout the world [1]. About 19.3 million new cancer cases and nearly 10.0 million cancer deaths were reported globally in 2020. The global survey on cancer in 2020 reported a higher incidence rate (19%) in men than women. Breast and lung cancer, the most common cancers reported, contributing to 23.1% of the new cases diagnosed in 2020. Colorectal cancer was the third most common cancer, with 1.93 million new cases in 2020 [2]. Approximately 70% of cancer deaths were reported from low- and middle-income countries. The significant causes for cancer include environmental factors, viral or genetic constituents, excessive alcohol intake, tobacco use, excessive body weight, lack of physical exercise, and low intake of vegetables and fruits. In low-income nations, nearly 25% of cancer cases are caused by hepatitis and the human papillomavirus. Genetic lesions in the genes encoding cell cycle proteins and somatic mutations in upstream cell signaling pathways may lead to cancer. Moreover, cancer cell metastasis, heterogeneity, recurrence, and their resistance to chemotherapy and radiotherapy reduce the efficacy of traditional treatment against many malignant tumors [3]. Extensive research has been carried out to develop more effective anti-cancer drugs [4]. Since the existing medications are not effective in treating cancer, the cure from cancer is considered the holy grail. Various treatments are available to treat cancer, such as surgery, radiotherapy, chemotherapy, immunotherapy and hormone therapy [5,6].

Chemotherapy is widely used, but is associated with severe side effects, including nerve problems, hair loss, weight changes, sexual dysfunction, and anemia. Moreover, chemotherapy is required in high doses for the treatment, as it is not specific for the target site. Chemotherapy mainly targets the DNA synthesis and mitosis process to kill cancerous cells [7]. Hence, conventional chemotherapy kills normal healthy cells along with cancer cells, causing severe unwanted side effects. Therefore, it is necessary to develop new chemotherapeutics that target cancerous cells without affecting normal cells. Researchers have focused on polymeric drug delivery carriers since the 1980s to improve the therapeutic index and bioavailability of conventional drugs [8]. In the last two decades, the polymeric drug delivery system has emerged as a key player in cancer treatment [9]. Polymeric carriers can be tuned through chemical modification to target the diseased site and deliver the drug in a controlled manner. Biodegradable and bioabsorbable polymers provide a safe platform for drug delivery that reduces the side effects [10]. Several biocompatible polymers, including carbohydrate-based polymers, and polysaccharides are used to improve the therapeutic efficacy of the cancer drugs and reduce the side effects. Nowadays, various immunological therapies and targeted drug delivery were included in cancer treatment based on carbohydrate polymers. Drug delivery systems based on specific targeting mechanisms will increase the efficacy of therapeutics [11]. Polysaccharide polymers are commonly used as drug carriers because of their acceptance in the body due to their biochemical and structural similarity with human extracellular matrix components. Herein, we review the current research on the polysaccharide-based carriers used in drug delivery systems [12].

2. Properties of Polysaccharides Carriers

Polysaccharides can be defined based on their chemical structure containing monosaccharides units linked together by glycosidic bonds [13]. They can be classified based on the monosaccharide components, chain length, and branching to the chains. The glycosidic linkage between the anomeric carbon atom of both donor and acceptor monosaccharide units distinguish them from proteins and peptides. Polysaccharides naturally possess storage properties that provide physical structure and stability, e.g., cellulose. Further, polysaccharides are differentiated into positively charged (chitosan) and negatively charged (alginate, heparin, hyaluronic acid, and pectin) based on the presence of the functional groups [14]. Chemical modifications including sulfation, phosphorylation, and carboxymethylation alter the biological properties of polysaccharides [15], making them suitable for drug carriers. They have also preferred drug carriers because of their stability, non-toxicity, and biodegradable properties. Moreover, chemical modifications subsequently improve the therapeutic efficacy of the drug through conjugation [16]. Drugs poorly soluble in aqueous solutions result in poor absorption, undergo interaction with food and enzymatic degradation, and may end up with low bioavailability [17]. Polysaccharide drug carriers improve the bioavailability of such small molecules, proteins, and peptides by enhancing their ability to permeate into tissues owing to their subcellular and submicron size. The improved cellular uptake in tissues limits the first-pass metabolism, P-glycoprotein mediated efflux and facilitates intestinal lymphatic transport [18]. The properties of polysaccharides, including their mucoadhesive properties, enhancement in absorption, flexibility to chemical modification, biocompatibility, and reduced toxicity, boost them as effective drug delivery systems. Target-specific delivery of the drug to the diseased site reduces the drug concentration required for treatment and hence enhances the drug efficacy; this, in turn, reduces adverse drug reactions by limiting the drug distribution and thereby bypassing the organs not involved in the diseased state [19].

3. Biopolymer Based Materials

Biopolymers are polymeric biomolecules synthesized by living organisms. Biopolymers take part in many vital functions, such as transmittance of genetic information, cellular construction, cell signalling, and drug delivery. Biopolymers are fabricated into hydrogel, aerogel and composites, which can be used for various applications, including controlled drug release, tissue engineering and wound dressing [20]. Hydrogels are 3D networks fabricated through the cross-linking of natural or synthetic polymeric materials and capable of absorbing a substantial volume of water [21]. Hydrogels are used in drug delivery systems because of their admirable properties, such as biocompatibility, biodegradability, and non-toxicity [20,22]. Nanoparticle addition into the polymeric hydrogels provides unique properties such as enhanced mechanical, thermal and magnetic properties, selectivity, and high swelling rate. Polymer-based hydrogels are widely used in biomedical applications such as tissue engineering, cell bioreactors, and drug delivery [23]. Aerogels, consisting of an ultra-lightweight structure with a porous network, can be obtained from hydrogels by removing the pore liquid using appropriate dehydration methods without any significant modification in the network structure. Biopolymers overcome the drawbacks of traditional materials used for aerogel preparation that limit their use in biomedical applications [24,25]. A study revealed the gastro-retentive properties of bendamustine hydrochloride loaded cellulose nanofiber aerogels prepared using freeze-drying. The nanofiber aerogels showed a higher area-under-the-curve (AUC) in the pharmacokinetics study in comparison with the free drug [26]. The large surface area of aerogels with an interconnected pore network facilitates the contact with the aqueous milieu and hence enhances the drug solubility [27]. Polysaccharide composites made of chitosan, alginate, cellulose, and starch are commonly used in the drug delivery system. A chitosan–polyethylene glycol (PEG) hydrogel fabricated using a silane crosslinker was pH-responsive and helpful in controlled drug delivery [28]. A pectin-based composite prepared using calcium, ethylcellulose and hydroxy propyl methyl cellulose revealed an efficient colon-specific drug delivery [29]. Carbon nanocomposites prepared using carbon nanotubes with biopolymers such as starch and chitosan have been used for drug delivery. A study reported on hollow starch nanoparticles showed an efficient delivery of DOX into hepatocellular liver cells [30]. A redox-sensitive hydroxyethyl starch doxorubicin (HES-SS-DOX) conjugate developed for anti-cancer drug delivery showed a prolonged plasma half-life time and hence exhibited better anti-tumor efficacy and reduced toxicity in comparison to free DOX [31]. Biopolymer conjugates provide a sustained drug release and help to develop time-increased release systems [32].

4. Polysaccharides Based Drug Carriers

Polysaccharides are carbohydrates that comprise more than ten monosaccharide units which are interlinked through a glycosidic bond. Among the natural polymers, polysaccharides have unique properties such as biocompatibility, stability, safety, adhesive properties, affinity towards the specific receptors, and non-toxicity, which make them a vital candidate in the drug delivery system [9]. Polysaccharides show a substantial structural and chemical diversity and possess a variety of functional groups that support various chemical modifications, enhancing their stability, solubility, encapsulation, and target specificity. On the other hand, the presence of a high number of the hydroxyl group in the backbone of polysaccharides encourages the inclusion of specific ligands resulting in functionalized colloidal systems. The anti-cancer potential of polysaccharides from the toxins of *B. prodigiosus* (*Serratia marcescens*) was first reviewed by Nauts et al. in 1946, and they suggested that the bacterial toxins could induce remission in cancer patients [33,34]. Polysaccharide K (PSK) and polysaccharide peptide (PSP) are protein-bound polysaccharides isolated from *Coriolus versicolor* that have been used in adjunctive immunotherapy for various types of cancer, including lung, breast, colorectal and gastrointestinal cancers [35]. Various polysaccharides used in drug delivery systems are listed in Table 1, including chitosan, alginates, cyclodextrin, pullulan, hyaluronic acid, dextran, guar gum, pectin, and cellulose.

Table 1. Polysaccharides used for drug delivery [36,37].

Polysaccharides	Sources	Physicochemical Properties	Applications and Benefits
Chitosan	Shells of crab, shrimp, and krill	Soluble in weak acids, mucoadhesive, reacts with negatively charged surfaces	Tissue regenerative medicine, pulmonary delivery, ionotropic gelation
Alginates	Marine brown algae	Water soluble, anionic coacervation with ions and polycations	pH-dependent swelling nontoxic, diffusion, erosion, in situ forming hydrogels
Cyclodextrin	Degraded starch derived from potato, corn, rice., etc.	Water soluble, nontoxic	Nanocarrier for controlled drug release, gene and drug delivery
Pullulan	Bacterial Homopolysaccharides produced from starch by *Aureobasidium pullalans*	Neutral polymer underivatized pullulan has high water solubility	Emulsifier sustained-release preparations
Hyaluronic acid	Vertebrate organisms	Biodegradable, bioactive, nonimmunogenic	Anti-cancer drug delivery, wound healing and skin regeneration
Dextran	Bacterial strains, cell-free supernatant	Neutral polymer, solubility depends on degree of polymerization	Colon-targeted delivery
Guar gum	Seeds of *Cyamopsis tetragonoloba*	Water soluble, non-ionic, galactomannan forms a thixotropic solution, stable at pH 4–10.5	Controlled release, colon-targeted release, thermoreversible
Pectin	Plant cell wall	Negatively charged molecule	In Situ gelling, sustained delivery, drug delivery in colorectal carcinoma

4.1. Chitosan

Chitosan is a naturally occurring positively charged polysaccharide intensively used in biomedical research. Chitosan is the principal derivative of chitin, found in the cell walls of fungi, mollusk shells, and crustaceans' exoskeletons. Chitosan is derived by deacetylation of chitin under certain conditions, and the degree of deacetylation varies from 60 to 100%. The molecular weight of commercially derived chitosan is between 3800 and 20,000 Daltons [38,39]. Chitosan contains (1 $\rightarrow$ 4)-2-acetamido-2-deoxy-β-D-glucan (*N*-acetyl D-glucosamine) and (1 $\rightarrow$ 4)-2-amino-2-deoxy-β-D-glucan (D-glucosamine) units [40]. Chitosan is a poorly water-soluble natural polysaccharide, and is soluble in a low pH solution. Modified forms of chitin, such as carboxymethyl chitin, fluorinated chitin, and sulfated glycol chitin, have been developed to improve its water solubility [40,41]. Several chemical modifications have been made to obtain many derivatives of chitosan for the controlled drug delivery system [42,43]. Chitosan shows antibacterial, antimicrobial, and anti-coagulation properties, and it also speeds up wound healing [44]. Chitosan with low molecular weight suppresses the growth of tumors and displays anti-tumor activity with a lesser toxic effect on normal growing cells [45,46]. Hence, low-molecular weight chitosan (LMWC), when used as a drug carrier, can induce synergistic effects. The cytotoxic activities of chitosan derivatives were reported in various cancer cell lines including MCF-7, HeLa and HEK293 tumor cell lines [47]. Chitosan has several features, such as cell permeability and mucoadhesive properties, which improve ocular, transdermal, and nasal drug delivery efficiency. Chitosan with a positive charge was co-assembled with negatively charged compounds to provide various types of drug carrier systems [48]. Hence, chitosan plays a major role as a drug delivery agent for low-molecular weight drugs and biomacromolecules.

Chitin and chitosan derivatives are promising candidates as polymeric carriers of anti-cancer agents. The solubility and bioavailability of chitosan are improved by derivatizing chitosan through chemical modification [43]. Previous studies have reported that the derivatization of chitosan with an acetamido residue and amino group increases the solubility of encapsulated drug molecules [12]. Some cancer cells are resistant to some anti-cancer drugs such as docetaxel (DTX), methotrexate (MTX), cisplatin, and 5-fluorouracil [49]. Currently, conventional medicines cause toxic effects on different body parts, such as the gonads, bone marrow, and gastrointestinal lining [50]. LMWC, due to its higher positively charged amino group, is highly attracted towards the cancer cell membrane with a greater negative charge than in normal cells. In addition, chitosan was reported to attack cancer cells through electrostatic interaction with the tumor cell membrane. Moreover, chitosan-drug nanoparticles can be used as alternatives to conventional drugs because of their selectivity towards the cancer cells and their biocompatibility [45,47].

4.1.1. Chitosan with Doxorubicin (DOX)

DOX is an anti-cancer drug used to treat various types of cancer such as bladder cancer, breast cancer, and lymphoma. Still, DOX causes cardiotoxicity and some side effects in the human body [51]. To reduce the toxicity, DOX has been linked with chitosan by a cross-linking method that improves the anti-cancer efficacy of DOX at a lower dose [52]. DOX has been linked with chitosan by the cross-linking technique to enhance anti-cancer efficacy. The drug loading efficiency was increased by adding tripolyphosphate with a chitosan–DOX complex. The diameter of the chitosan–DOX complex was in the range of 130–160 μm. The chitosan–DOX complex showed improved anti-cancer activity in VX2 cells compared to free DOX [53]. DOX-loaded magnetic nanoparticles coated with chitosan readily entered the MCF-7 cells, accumulated around the nucleus, and delivered doxorubicin successfully. DOX-loaded nanoparticles were used as a pH-dependent drug delivery system. As they release the doxorubicin at pH 4.2, the drug can be released into the tumor environment [54]. DOX-loaded cholesterol-modified glycol chitosan micelles conjugated with folic acid significantly induced cytotoxicity against FR-positive HeLa cells [55].

4.1.2. Chitosan with Paclitaxel (PTX)

PTX, obtained from the Taxus brevifolia, is an effective anti-cancer drug for breast, lung, ovarian, and stomach cancer. However, the hydrophobic nature and high hemolytic toxicity of PTX are significant issues during drug delivery. Conjugation or loading of PTX with the biocompatible and biodegradable chitosan can increase the aqueous solubility of PTX and reduce the problems associated with its hydrophobic nature [56]. The PTX–chitosan complex is more efficient in cancer treatment than free PTX. The anti-cancer effect of the PTX–chitosan complex has been tested in triple-negative breast cancer cell lines (MDA-MB-231) in vitro [57]. PTX conjugated to LMWC through cleavable succinic anhydride linker enhanced its water solubility and exhibited equipotent cytotoxicity in cancer cell lines (NCIH358, SK-OV-3, MDA MB231) in comparison with free PTX. In addition, an orally administered aqueous solution of the PTX–LMWC conjugates (Figure 1) inhibited tumor growth significantly in mice bearing xenograft or allograft tumors [58]. Trimethyl chitosan PTX conjugates enhanced the mucoadhesion and intestinal transport of PTX. Additionally, the folic acid functionalization boosted the anti-cancer efficacy of the conjugate, resulting from elevated cellular uptake and intratumor accumulation [59].

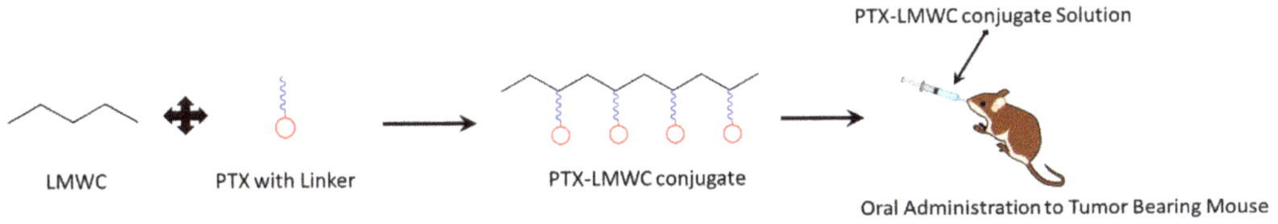

Figure 1. Schematic representation of the PTX–LMWC conjugate and its oral administration to tumor bearing mouse.

4.1.3. Chitosan with DTX

DTX is a chemotherapeutic drug used to treat different types of cancer, such as stomach, breast, non-small cell lung, and prostate cancer. Chitosan was modified into glycol chitosan, and DTX was loaded on the modified chitosan by means of the dialysis method. The anti-cancer action of this conjugate was tested in A549 lung cancer cell-bearing mice. The DTX-loaded glycol chitosan nanoparticles showed lesser toxicity than free DTX. The tumor-suppressing ability was also high for DTX-loaded nanoparticles than the free DTX in the case of lung cancer-bearing mice [60]. Various studies have shown that the drug-loaded glycol chitosan nano-conjugate is a promising nanoformulation in cancer treatment. DTX-loaded chitosan nanoparticles in a water-in-oil nanoemulsion system showed an enhanced cytotoxic potential in a human breast cancer cell line [61]. DTX encapsulated chitosan nanoparticles revealed higher Bax (a pro-apoptotic factor) expression over the BCL-2 (an anti-apoptotic factor) compared to the cells treated with free DTX [62].

4.1.4. Chitosan with MTX

MTX was used for the first time in 1947 to treat cancer. It is used to treat breast cancer, leukemia, and lymphomas, but it causes non-specific side effects. MTX has been conjugated with natural polymers such as chitosan to reduce toxicity. Glutaraldehyde acts as a cross-linking agent to link MTX with chitosan. The ionic gelation process between the MTX, conjugated chitosan, and sodium triphosphate forms the chitosan–MTX–TPP complex. The anti-cancer efficiency of the chitosan–MTX conjugate, investigated in MCF-7, revealed that the conjugated form of MTX was more effective and less toxic than free MTX [63]. Erlotinib-loaded MTX–chitosan magnetic nanoparticles showed a thermo- and pH-dependent drug release. Selective uptake of the MTX–chitosan magnetic nanoparticles via folate receptors promoted them as a smart carrier for targeted treatment in FR-positive solid tumors [64].

4.1.5. Chitosan with Curcumin

Curcumin is an effective anti-cancer agent, but curcumin's major problem is its poor pharmacokinetics profile. Chitosan-curcumin conjugate (Figure 2) obtained through the chemical conjugation of curcumin to chitosan significantly through imine formation improves the solubility and stability of curcumin. The conjugation takes places through imine bonding between the amino group of chitosan and carbonyl group present in chitosan under microwave irradiation [65]. However, a recent study showed that the introduction of 1-ethyl-3 (3-dimethylaminopropyl) carbodiimide spacer reduced the steric hindrance in curcumin conjugation, and the degree of substitution is increased by using acetate as catalyst [66]. The anti-cancer efficacy of curcumin could be increased by encapsulating the curcumin in the chitosan nanoparticles. Encapsulation of curcumin in chitosan nanoparticles increases the bioavailability of curcumin, and it works better than free curcumin. The anti-cancer efficacy of the curcumin-loaded chitosan complex has been tested in lung cancer in vitro [67]. Some in vitro studies have shown that curcumin-loaded chitosan nanoparticles were more efficient in suppressing tumor growth than free curcumin in breast cancer, hepatocellular carcinoma and colorectal cancer [68–70].

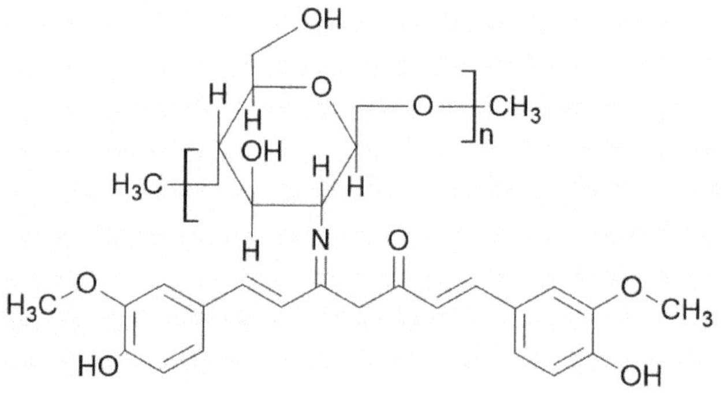

Figure 2. Structure of chitosan-curcumin conjugate.

4.1.6. Chitosan with Oxaliplatin

Oxaliplatin is a platinum-containing chemotherapeutic drug used to treat rectal and colon cancer. It suppresses or slows down the growth of the tumor. Oxaliplatin was loaded on chitosan to make a pH-sensitive nanocarrier for target specific delivery in tumor cells. Oxaliplatin-loaded chitosan nanoparticles increase the therapeutic efficiency of oxaliplatin in cancer treatment. The pH sensitive chitosan-oxaliplatin releases oxaliplatin much more rapidly at pH 4.5 than pH 7.4, which is essential for tumor-targeted drug delivery [71]. The anti-cancer activity of the complex was tested in MCF-7 cell lines. The chitosan–oxaliplatin complex affects the apoptotic pathway, and it increases the expression of cytochrome C, Bax, Bik, and caspases 3 and 9 in breast cancer cells. So, the oxaliplatin–chitosan complex could be a clever approach in cancer treatment. Overall, chitosan-based carriers could be a promising strategy for delivery of anti-tumor drugs.

4.2. Alginate

Alginates are linear unbranched anionic polysaccharides found in the cell walls of brown algae, including *Laminaria* and *Ascophyllum*, consisting of (1 $\rightarrow$ 4′)-linked β-d-mannuronic acid and α-l-guluronic acid residues. Alginate is mainly used in the food industry, pharmaceutical, and industrial applications. Due to its properties including bioavailability, biocompatibility, low toxicity, relatively lesser cost, and moderate gelation by addition of divalent cations, alginate could be a good drug delivery agent [72]. Alginate has broad applications in cell transplantation, wound healing, and acts as a carrier for delivering proteins and small chemical drugs.

The synthesis of alginate hydrogels is straightforward and is performed by the cross-linking method. Alginate hydrogels are easily attached to mucosal surfaces due to their bio-adhesive nature and can be orally given or injected into the body directly. Such properties allow the extensive use of alginate in the pharmaceutical industry [72]. Nanoparticles of alginate can be stabilized by adding cationic polyelectrolytes [73]. The properties of alginate, including as a thickening, stabilizing, and gel-forming agent, promote it as a useful drug carrier.

Controlled release drug delivery systems are considered to give a steady and kinetically expectable drug release. Alginate, as with hydrocolloids, has played a significant role in the design of controlled-release products. Many drugs have been loaded into alginate matrices in the form of microsphere, tablets, beads, and tablets for control release therapies. Alginate nanoparticles have been used to deliver various drugs such as metformin [74], DOX [75], ethionamide [76] and several other drugs. Various administration routes of alginate nanoparticles are pulmonary [77], oral [75], nasal [78,79], intravenous [80], vaginal [81] and ocular [82].

4.2.1. Alginate with DOX

Alginate is used in anti-cancer drug delivery systems in several forms, such as nanogels, hydrogels, and nanoparticles, due to its biocompatible and biodegradable nature [83]. Alginate nanoparticles were synthesized by the gelation of alginate with divalent cations such as calcium and loaded with DOX [84]. DOX shows a greater affinity towards alginate polymers. Alginate nanoparticles are used as a drug carrier due to their high drug loading capacity for DOX. This could be more than 50 mg per 100 mg of alginate. Experimental studies have reported that DOX-loaded nanoparticles are used to treat liver metastasis in mice. DOX-loaded alginic acid/poly(2-(diethylamino)-ethyl methacrylate) nanoparticles were tested in H22 tumor-bearing mice. Alginate nanoparticles showed more permeability and retention effect, targeting the tumor cells passively, which was confirmed by the near-infrared (NIR) fluorescence imaging technique. An intravenously administered DOX-loaded alginic acid/poly(2-(diethylamino) ethyl methacrylate) (ALG-PDEA) nanoparticle solution was more effective in H22 tumor-bearing mice than free DOX (Figure 3) [85].

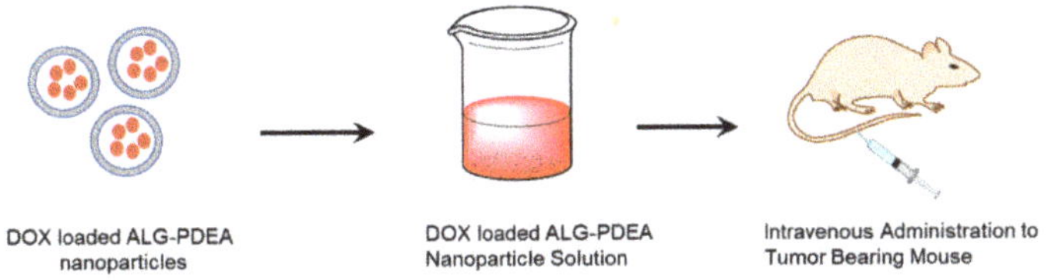

Figure 3. Schematic representation of DOX-loaded nanoparticles and their administration to tumor-bearing mouse.

4.2.2. Alginate with Curcumin

Curcumin is a naturally occurring polyphenolic compound used in the treatment of different types of cancers, such as cervical, bladder, prostate and breast cancer [86]. However, the major problem is its low water solubility and poor bioavailability. This can be overcome by the entrapment of curcumin in hydrophilic calcium alginate nanoparticles and delivering it to the target site. The percentage entrapment efficiency of alginate for curcumin was reported as 49.3 ± 4.3. The curcumin-loaded calcium alginate nanoparticles were synthesized by emulsification and cross-linking method. The particle size was found to be 12.53 ± 1.06 nm. An alginate nanoformulation was checked in prostate cancer, and it showed cytotoxic effects on DU145 prostate cancer cells in vitro [87]. Moreover, the aqueous solubility of curcumin was significantly enhanced through conjugation with alginate (Figure 4). The conjugates were developed via the esterification reaction between the hydroxyl group of curcumin and C-6 carboxylate group of sodium alginate. The cytotoxic studies using L-929 mouse fibroblast cells revealed that the cytotoxic potential of curcumin was retained even after conjugation [88].

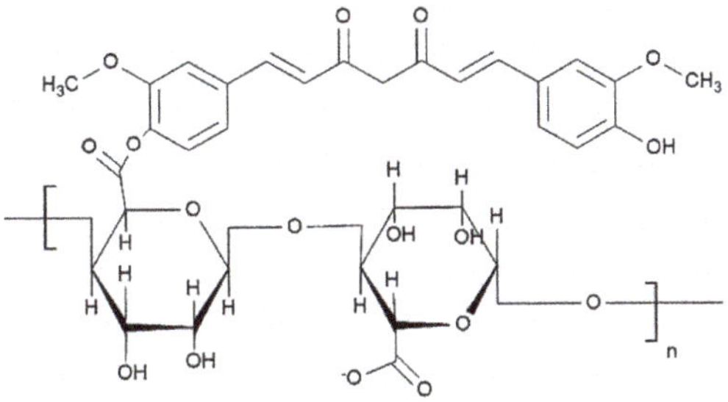

Figure 4. Structure of the alginate–curcumin conjugate.

4.2.3. Alginate with Exemestane (EXE)

EXE is a hydrophobic steroid aromatase inhibitor with high lipophilicity. It has shown excellent solubility in organic solvents. EXE is an oral chemotherapeutic drug mainly used to treat breast cancer by inhibiting the synthesis of estrogen [89]. It is loaded in alginate nanoparticles by simple controlled gelation methods. The loading and unloading were confirmed by XRD, SEM, and FTIR techniques. The XRD studies were performed to check the EXE encapsulation [90].

The anti-cancer activity of the formulation has been tested in vitro using DLA cells (Dalton's lymphoma ascites). EXE has been released from the nanoformulation at pH 7.4 in vitro, and the side effects of chemotherapy are reduced by the controlled release of EXE loaded in alginate nanoparticles. Thus, in vitro studies have shown that the EXE–alginate nanoformulation could be an excellent anti-cancer agent.

4.2.4. Alginate with Tamoxifen (TMX)

TMX has been widely used to treat breast cancer [91]. It shows poor solubility in water, which restricts the oral administration of the drug. This problem can be overcome by means of the use of a nanoparticle carrier system for drug delivery. Nanoparticles with bovine serum albumin and thiolate alginate have been prepared by the coacervation method and loaded with TMX. In vitro release studies of this formulation showed 45–52% TMX release after up to 25 h. Cellular uptake of TMX-loaded nanoparticles was examined in monoculture of MCF7 cells and HeLa cells [92]. The TMX-loaded alginate nanoparticles were found to be effective in both cell lines. Folate targeted alginate–silver nanoparticles loaded with TMX provides better cytotoxic effects in breast cancer cell lines. The cytotoxicity was achieved by inducting ROS, down regulating the survival oncogenic genes (BCL-2 and survivin) and arresting G2/M phase [93].

4.3. *Pectin*

Pectin is a linear polysaccharide found in the middle lamella of plants in higher concentrations [94]. Pectin comprises D-galacturonic acid (GalA) units, which are mainly joined by an α $(1 \rightarrow 4)$ glycosidic bond (homopolymer of $(1 \rightarrow 4)$ a-D-galactopyranosyluronic acid units with varying degrees of carboxyl groups methyl esterified). Rhamnose is also present in the pectin backbone, and galactose, xylose, and arabinose are present in the side chain [94,95]. The most crucial property of pectin is its gel-forming ability. Due to this property, it has been used in the food and pharmaceutical industries. The primary source of pectin is plant residue after juice extraction—the peel of citrus fruits mainly contains around 20–30%, and 10–15% is present in apples. Pectin is also obtained from other sources such as mango waste, legumes, cabbage, sunflower heads, and sugar manufacturing waste. Pectin has been widely used as a drug carrier due to its gel-forming property in acidic media. The gel-forming ability varies according to the molecular composition, molecular weight, and source from where it obtained. Pectin is generally used as a food additive, and it is entirely safe as a food additive.

Nowadays, pectin is used as a drug carrier owing to its advantageous properties such as non-toxicity, bioavailability, and low manufacturing cost. There are different modes of delivery, including administration through the nasal and oral cavity. Because of its carboxylic acid functional group, pectin is easily conjugated to the amino group of the anti-cancer drugs [96]. The anti-cancer drug from the pectin conjugate is quickly released in the tumor cell because the lysosomal enzymes easily hydrolyze the amide bond.

4.3.1. Pectin with Curcumin

Curcumin is obtained from the turmeric rhizomes and used as an anti-inflammatory, antimicrobial, antiviral, and anti-cancer agent [97]. Curcumin shows excellent hepatoprotective effects [98]. Curcumin is an ideal anti-cancer drug candidate due to its antioxidant and anti-inflammatory properties, but its poor bioavailability is the major problem during oral administration. This problem can be overcome by forming the appropriate curcumin nanoformulation with pectin. Curcumin works against human colon cancer, inhibiting cancer cell growth by modulating the NF-κB signaling pathway [99]. Curcumin-loaded pectin showed a more cytotoxic effect in colon cancer (HCT116) than free curcumin. Pectin maintains curcumin's integrity and bioactivity and delivers it to the target site in the colon cancer cell. Pectin-type B gelatin curcumin conjugates have been more effective for curcumin delivery through oral administration in anti-cancer agents [100]. Pectin and curcumin were conjugated via the esterification reaction between carboxylic groups of pectin and phenolic –OH group of curcumin (Figure 5). Cytotoxicity studies of the conjugates showed significant inhibition of KYSE-30 cell lines compared to free curcumin [101].

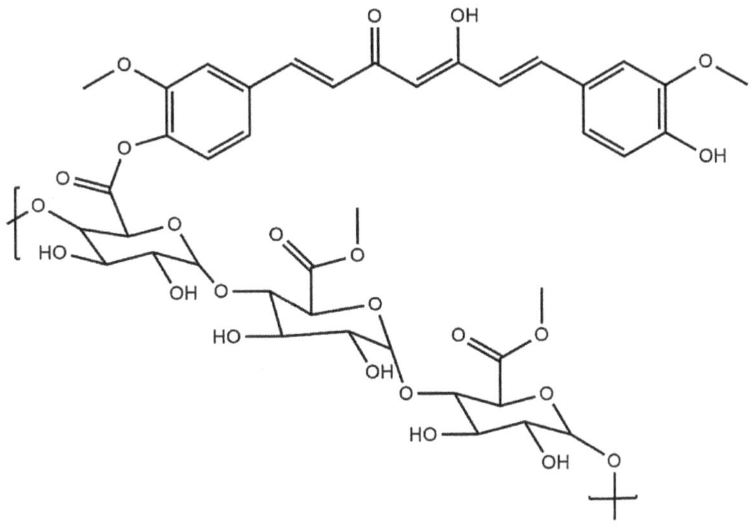

Figure 5. Structure of the pectin–curcumin conjugate.

4.3.2. Pectin with DOX

Pectin has been used as a carrier for anti-cancer drugs such as DOX. Pectin has been linked with DOX in thiolate form to obtain thiolate pectin–DOX conjugate and tested for its anti-cancer activity in human prostate cancer, human bone osteosarcoma cells and colon cancer in vitro. The thiolate pectin–DOX conjugate showed more anti-cancer activity than free DOX in the case of 143B and CT26 cells, but no significant difference was seen between the free DOX and thiolate pectin-DOX in the case of the prostate cancer cells. Thiolated pectin–DOX conjugates have been used in targeted drug delivery in CT26 cells [102]. Self-assembled DOX-conjugated hydrophilic pectin nanoparticles loaded with hydrophobic dihydroartemisinin revealed a quick release of DOX and DHA in a weakly acidic environment. The formulation significantly reduced the tumor growth in a female C57BL/6 mouse model [103].

4.3.3. Pectin with Cisplatin

Cisplatin is used as an anti-cancer drug against various cancers, including lung cancer, head and neck cancer, ovarian cancer, and bladder cancer [104]. Cisplatin can be given alone or along with other cancer therapy such as radiotherapy. Pectin conjugates with cisplatin were tested in B16 cells (murine melanoma) in vitro. These in vivo studies reported the antitumor efficacy in mice. The anti-tumor efficacy was checked by measuring tumor growth. Cisplatin reduced the tumor size effectively, as studied by tumor regression studies [105]. Nano-conjugates of pectin and cisplatin were more effective when given along with radiotherapy than cisplatin. Pectin–cisplatin nano-conjugates revealed a prolonged blood retention profile in mice, which was confirmed by the presence of cisplatin in the circulation after 24 h. The J-774 cells incubated with the nano-conjugates tagged with FITC showed an uptake in 40% cells followed by 30 min of incubation [106].

4.4. Guar Gum

Guar gum is a naturally occurring high-molecular weight and uncharged carbohydrate mainly available in *Cyamopsis tetragonolobus*. Guar gum is made up of galactomannan, which consists of galactose and mannose. The presence of multiple hydroxyl groups makes it an excellent candidate for derivatization. Guar gum is mainly used as a stabilizing and emulsifying agent. The vital characteristic of guar gum is its swelling property, used to control drug release. Guar gum is soluble in water but insoluble in hydrocarbons, fats, alcohol, esters, and ketones [107]. It generally forms a viscous colloidal solution in both hot and cold water. Guar gum has many pharmaceutical applications, mainly in drug delivery, due to its stability, non-toxicity, and biodegradability. Guar gum and its derivative have been used as drug carriers in targeted drug delivery systems. Guar gum has been used to control the release profile of the drugs that are highly water-soluble and have difficulty in delivering at the targeted site [108].

4.4.1. Guar Gum with 5-Fluorouracil (5FU)

5FU is the drug used to treat colon cancer [109]. 5FU is poorly absorbed in oral administration and hence shows low bioavailability. To improve the bioavailability, a microsphere that consists of guar gum and sodium borate was prepared by the emulsification cross-linking method. This helped to control the release of the drug for 24 h, specifically in the colon site, and it worked better against colorectal cancer. The microspheres loaded with 5FU showed more stability, and the release of 5FU followed zero-order kinetics with both degradation and erosion mechanisms. The microspheres could release the maximum amount of drug in colon cancer in a controlled manner. The significant advantages of targeting the colon for 5FU drug delivery are the pH range, which is around 5.5–7 in the colon, and the low digestive enzymatic activity, which supports more drug absorption. The effectiveness of formulation in colon-targeted oral drug delivery systems has been studied in vivo [110].

4.4.2. Guar Gum with Tamoxifen Citrate (TMX)

TMX shows an antagonistic effect in breast cancer cells and is used for the treatment of ER (+) tumors [111]. Guar gum containing TMX nanoparticles was formed by the emulsion process. Guar gum has been used as a drug carrier for different anti-cancer drugs, but very little information is reported about this nanoformulation. Guar gum nanoparticles loaded with TMX was analyzed in albino mice for two days. After two days, nanoparticles were found in both mammary and ovary tissue, but the uptake and retention of nanoparticles were found to be more in mammary glands [112,113].

4.4.3. Guar Gum with MTX

Guar gum microspheres were prepared and used to treat colorectal cancer [107]. The entrapment efficiency of MTX-loaded microspheres was found to be 75.7%. Guar gum microspheres delivered the maximum amount of drug at the targeted site in colon cancer. The formulation was given orally to albino rats and checked for drug release in different parts, such as the stomach, colon, and small intestine, at different time intervals. The study revealed that guar gum microspheres released MTX at the target site in the colon. Thus, guar gum microspheres have been used as an effective system for MTX delivery in the case of colorectal cancer [114].

4.4.4. Guar Gum with Curcumin

Curcumin has been used as an anti-cancer drug due to its antioxidant properties, but its poor absorption in GIT is the major problem [115]. Curcumin has been formulated with guar gum of different concentrations to overcome this problem, and the formulations were investigated for their effect on colon cancer. The drug release of the formulation with 40% guar gum containing preparation was 91.1%, while the 50% guar gum containing formulation showed a drug release of 82.1%, which is lesser in comparison to the preparation containing 40% guar gum [115]. Guar gum's effectiveness and drug release assays against colonic bacteria have been studied with rat cecal contents. Guar gum could be a promising drug carrier for curcumin delivery in colon cancer [108].

4.5. Dextran

Natural polymers have been used as carriers in drug delivery systems due to their non-toxicity and low production cost. Dextran is one of the natural polymers in targeted drug delivery systems [116], which contains a monomeric α-D-glucose unit and has α-(1 → 6) glycosidic linkage in the backbone. Dextran is widely used in pharmaceutical, food, and chemical industries as a carrier, stabilizer, and emulsifying agent [117,118]. Due to its properties such as biocompatibility, non-immunogenicity, non-toxicity, it has been used in drug delivery systems. It can also be easily modified and hence widely used in drug delivery systems. It increases the stability of the drug and prevents the drug from accumulating in the blood [119]. Dextran inhibits the growth of the tumor, and it reduces the toxic effects induced by the drug in the body. Dextran can be a promising drug delivery agent for anti-cancer drugs.

4.5.1. Dextran with DOX

Dextran is a superior carrier in drug delivery systems, increasing the stability of DOX [120]. Mixtures of dextran–DOX and dextran–CPT showed remarkable anti-cancer activity against 4T1 cell line and 4T1 tumor-bearing mice compared to the free drugs [121]. A biocompatible poly pro-drug based on dextran–DOX prodrug (DOXDT) was formed by one-step atom transfer radical polymerization (ATRP). The drug loading capacity of DOXDT prodrug is higher than other lipid-based drug delivery systems, at up to 23.6%. The DOXDT shows improved cytotoxicity against 4T1 and HeLa cells. DOXDT was also studied for tumor-suppressive effects. The DOXDT-based delivery system suppressed the growth of tumor cells [122].

4.5.2. Dextran with PTX

PTX, or Taxol (TXL), is used as an anti-cancer drug in different cancers such as breast, ovarian, and lung cancer. TXL and its derivatives were linked with aminated dextran to form dextran–TXL conjugates and were analyzed for the anti-cancer activity in HeLa-KB cells in vitro. Dextran–TXL conjugates showed two to three times greater anti-cancer effects when linked with folic acid. So, conjugation of TXL with dextran and folic acid may improve the anti-cancer efficacy of Taxol [123]. A Dex–SS–PTX conjugate showed significant cytotoxicity in BT-549 and MCF-7 cells [124].

4.5.3. Dextran with Phenoxodiol (PXD)

PXD is a synthetic analog of the plant isoflavone genistein with improved anti-cancer efficacy [125]. PXD, an anti-tumor agent, was conjugated with dextran to improve its effectiveness [126]. The anti-proliferative activity of this conjugate was tested in glioblastoma, breast cancer MDA-MB-231, and neuroblastoma SKN-BE (2)C cells. The anti-cancer activity was also tested in HMEC-1 (human microvascular endothelial cells) and non-malignant human lung fibroblast MRC-5 cells. This conjugate was more stable, effective and less toxic than the free drug [127].

4.5.4. Dextran with MTX

MTX is used to treat cancer and other hematological diseases. MTX is conjugated with dextran by covalent linkage to obtain a better formulation of the drug [128]. The anti-cancer activity of the conjugate was evaluated in human brain tumors (H80) and 9L gliosarcoma in rat brains. The conjugate killed the tumor cells more effectively than free MTX [129].

4.5.5. Dextran with Curcumin

Curcumin is actively used in cancer treatment due to its well-known biological properties. A study on carboxymethyl dextran-coated liposomal curcumin revealed improved cytotoxicity and enhanced cellular uptake in HeLa cells [130]. Dextran-curcumin micelles fabricated (Figure 6) using self-assembly of dextran curcumin conjugate displayed significant cytotoxicity in cancer cells owing to the enhanced solubility and efficient cellular internalization compared to free curcumin [131]. Curcumin has been conjugated with dextran successfully by a free radical reaction. The anti-cancer activity of curcumin–dextran conjugate has been tested in MCF-7 and adenocarcinomas gastric cell line by MTT assay. The results show that curcumin–dextran conjugates have anti-proliferative effects in MCF-7 cell lines and human gastric adenocarcinoma cells [132]. Curcumin–dextran conjugates act as a carrier for MTX and act synergistically to enhance the anti-cancer efficacy in MCF-7 cell line [133].

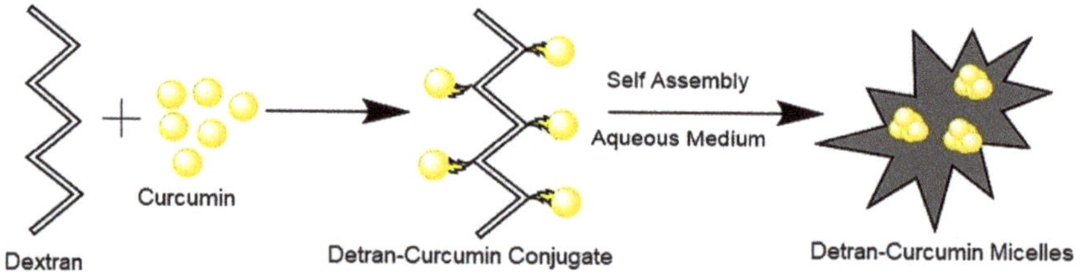

Figure 6. Schematic diagram depicting self-assembly of curcumin–dextran micelles.

4.6. *Hyaluronic Acid (HA)*

HA is a mucopolysaccharide that consists of two saccharide units, glucuronic acid and *N*-acetylglucosamine [134]. HA can be easily modified due to its hydroxyl and carboxylic groups, as well as an *N*-acetyl group. Different drugs can be directly linked with HA to generate a new conjugate with more anti-cancer activity. HA is more specific to several tumor cells, particularly tumor-initiating cells. Tumor cells overexpress the CD44, LYVE-1 receptors, which are HA-binding receptors, and low expression was also seen in the surface of epithelial hematopoietic and neural cells [135]. Due to its properties such as biocompatibility, high viscoelasticity, and biodegradability, HA is used as a drug carrier in drug delivery systems in the form of hydrogel and micelles. The intracellular uptake of HA-drug conjugates was facilitated via CD44 caveolae-mediated endocytosis on tumor cells, thus enhancing the targeted drug delivery efficiency [136,137].

4.6.1. HA with PTX

PTX is obtained from the bark of Pacific yew, and it shows antimitotic activity. PTX is used as an anti-cancer agent, and it promotes tubulin assembly. PTX has been linked with hyaluronic acid through esterification. It inhibits the replication of cells in the late G2/M phase [138]. Hyaluronic acid was first dissolved in DMSO with polyethylene glycol (PEG) then conjugated with PTX by ester linkage without any modification [139]. The conjugate of HA-PTX forms micelles in an aqueous solution. Tumor cells over-express the hyaluronic acid receptor, and hence the HA–PTX conjugate shows more binding and cytotoxic effects in tumor cells than normal cells. The anti-tumor activity of this conjugate has been proved in MCF-7 and HCT-116 cancer cells in vitro [140]. Chitosan-coated HA–PTX nanoparticles were prepared by coating chitosan onto the surface of self-assembled HA–PTX conjugates (Figure 7). HA–PTX NPs displayed higher cellular uptake than free PTX in HepG2 cell lines [141].

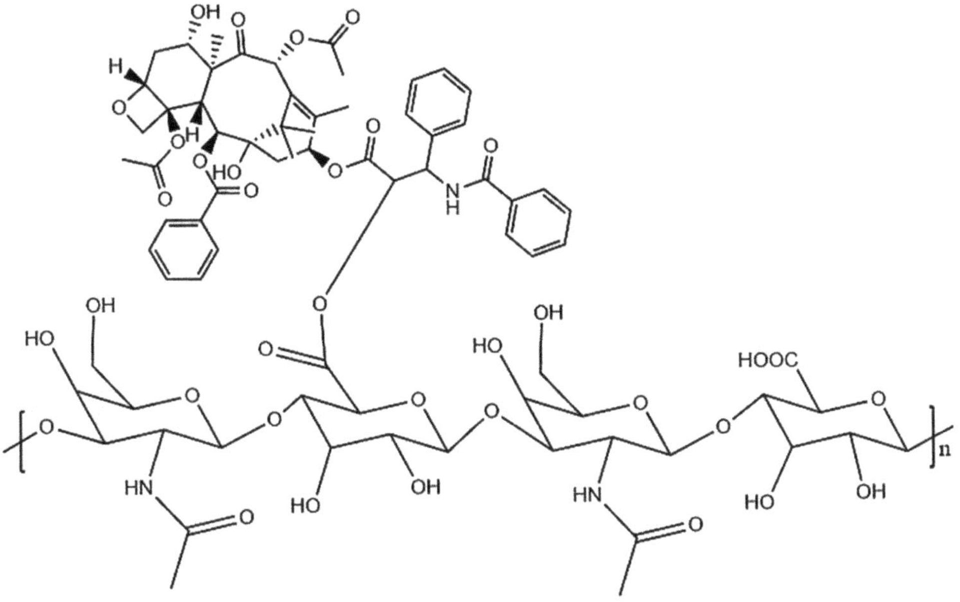

Figure 7. Structure of the HA–PTX conjugate.

4.6.2. HA with DOX

DOX is used effectively to treat ovarian, breast, multiple myeloma, and pediatric solid tumors [142]. DOX has been conjugated with hyaluronic acid through amide linkage between the carboxylic acid group of HA and the amine of Dox to form an amide bond to form a nano-conjugate with improved anti-cancer activity. The anti-cancer activity of the hyaluronan–DOX nano-conjugate has been tested in MDA-MB-231 breast cancer cell lines. The nano-conjugate exhibited improved cytotoxicity in breast cancer cell lines. The conjugate showed minimum toxicity to the normal cells. The nano-conjugate also inhibited breast cancer in vivo and improved the survival rate. The nano-conjugate inhibited tumor growth at early stages in breast cancer [143].

4.6.3. HA with Cisplatin

Cisplatin is generally used to treat bone and blood vessel cancer [68]. Cisplatin is an effective anti-cancer agent, but its use is limited due to its toxic effect on the nervous system. In order to reduce the harmful impact of cisplatin, it has been linked with hyaluronic acid to form a less toxic conjugate with improved tumor-targeted delivery. The conjugate has been delivered in squamous cell carcinoma of the head and neck in vivo to evaluate the anti-tumor activity. The results show the improved anti-cancer efficacy of the conjugate with lesser toxicity. The anti-cancer activity of the conjugate was also shown in dogs with soft tissue carcinoma [144].

4.6.4. HA with Camptothecin (CPT)

CPT is obtained from *Camptotheca acuminate*, a Chinese tree, and it shows anti-tumor activity by inhibiting the nuclear enzymes. CPT is an alkaloid that shows poor solubility in water, so the cellular uptake is less. Several analogs of CPT have been developed to overcome the problem and improve water solubility [145,146]. 7-ethyl-10-hydroxy camptothecin is a CPT analog that has been linked with hyaluronic acid, and the resulting compound is ONCOFID-S. Its anti-cancer activity has been tested in breast cancer, esophageal, ovarian, gastric, and lung cancer [147]. Recently the anti-cancer activity of ONCOFID-S has been proved in mice with peritoneal carcinomatosis from esophageal, colorectal, and gastric adenocarcinomas [148].

4.7. Cyclodextrin

Cyclodextrin (CD), made of cyclic oligosaccharides, contains a hydrophilic outer layer with (α-1,4)-linked α-d-glucopyranose units and a lipophilic central cavity [149]. Cyclodextrins are biocompatible have low toxicity and low immunogenicity. The molecular weight of cyclodextrins is in the range of 1000 to 2000 Da. The example of some natural cyclodextrins is α-cyclodextrin (αCD) with six glucopyranose units, β-cyclodextrin (βCD) with seven glucopyranose units, and γ-cyclodextrin (γCD) with eight glucopyranose units. The most abundant cyclodextrin is β-cyclodextrin, which is used in the pharmaceutical industry due to its low production cost, bio-availability and perfect size cavity. Cyclodextrins form water-soluble complexes with many poorly soluble compounds. Cyclodextrins are hollow truncated, with a cavity that is slightly hydrophobic inside and hydrophilic outside. The drugs with hydrophobic nature can be easily encapsulated into the cavity of cyclodextrins to form an inclusion complex without any chemical reaction. As a result, encapsulating the drug in the cyclodextrin improves the hydrophobic drug's stability and aqueous solubility. Cyclodextrin provides the shielding effect and reduces the side effects of the drug on the human body [150].

4.7.1. Cyclodextrin with DOX

DOX, an anti-cancer drug, was entrapped in pegylated liposomes and conjugated with γ-cyclodextrin and was tested for the anti-cancer effect in BALB/c mice bearing colon-26 tumor cells. The anti-tumor activity was compared in different conjugates such as pegylated liposomes entrapping DOX, γ-CD, and the binary system of liposomes with both DOX and CD and free DOX. Various preparations have been tested in mice, and the results show that the complex-in-liposome with both DOX and CD provided high DOX levels in plasma and solid tumors compared with the other formulations. Furthermore, the results obtained from the above studies show that pegylated liposomes entrapped with DOX and γ-CD retarded tumor growth and improved the survival rate of mice, hence increasing the anti-cancer efficacy of DOX [52].

4.7.2. Cyclodextrin with CPT

CPT is a hydrophobic anti-cancer drug, but its clinical application is less widespread due to some properties such as instability in physiological conditions and poor solubility in aqueous solutions. The solubility and stability have been increased by nanoparticulate systems of amphiphilic cyclodextrins, poly-epsilon-caprolactone, or poly(lactide-co-glycolide) (PLGA) [151]. The drug-loading capacity and anti-cancer efficacy were found to be greater for amphiphilic cyclodextrin nanoparticles compared to others. The anti-tumor activity of nanoparticles was tested in the breast cancer cell line MCF-7. CPT-loaded cyclodextrin nanoparticles acted as an excellent carrier system for the effective delivery of CPT [152].

4.7.3. Cyclodextrin with Curcumin

Curcumin is used as an anti-cancer agent, but its poor oral bioavailability is the major problem that can be overcome by conjugating with cyclodextrin [153]. The curcumin–cyclodextrin complex has been investigated in different anti-cancer studies. In one study, curcumin was encapsulated in the β-cyclodextrin cavity using the saturated aqueous solution method. Cyclodextrin increased the delivery of curcumin, and its therapeutic value increased in vitro. It regulated various pathways such as up-regulated p53/p21 pathway, down-regulated CyclinE-CDK2 combination, increased Bax/caspase 3 expressions MAPK/NF-κB pathway and CD15. The cyclodextrin–curcumin complex has been used to improve the curcumin anti-cancer efficacy and delivery in lung cancer [154].

4.7.4. Cyclodextrin with PTX

PTX is used to treat lung, breast, esophageal, bladder, and ovarian cancer. Its solubility in an aqueous solution is very low, which is a major problem in treatment. The problem can be overcome by conjugation of PTX with cyclodextrin. The anti-cancer activity of the PTX–cyclodextrin conjugate has been proved in MDA-MB-231 breast cancer cells [155]. PTX-loaded cyclodextrin peptide (R8-CMβCD) enhanced the cellular uptake of PTX by reducing the efflux of PTX in tumor cells by inhibiting P-gp efflux pump. PTX-conjugated β-cyclodextrin polyrotaxane exhibits significantly superior potency in reducing tumor growth, and enhances the lifetime of tumor-bearing mice [156,157].

4.8. Pullulan

Pullulan is a biopolymer made of maltotriose units joined via α $(1 \rightarrow 4)$ glycosidic bond, and successive maltotriose units were linked by α $(1 \rightarrow 6)$ glycosidic bond. Pullulan was obtained from the fermentation medium of the *Aureobasidium pullulans* [158] Pullulan is used as a drug carrier broadly due to its high aqueous solubility. A number of derivatives of pullulan were produced by some chemical modifications with different solubilities. Pullulan has not shown any toxicity to cells, and it is a non-immunogenic polymer that is useful in biomedical applications. Pullulan showed higher degradation than dextran in serum. Pullulan has been used in different ways due to its elasticity and thermal stability. The derivatization of pullulan has been performed either with DOX or DOX and folic acid. The activation of pullulan was achieved by means of periodate oxidation and was functionalized by reductive conjugation with cysteamine and PEG $(NH_2)_2$. The DOX–pullulan bioconjugates have been used in passive tumor targeting. Pullulan has been conjugated with different anti-cancer drugs for improving the anti-cancer efficiency of the drugs [159].

4.8.1. Pullulan with DOX

DOX is used as an anti-cancer drug that inhibits the synthesis of DNA in cancer cells. Pullulan can be used as a drug carrier due to its properties such as biocompatibility and non-immunogenicity. Pullulan has a number of functional groups, which is a useful feature in drug delivery [160,161]. DOX shows toxic effects on normal cells that can be reduced by the encapsulation of DOX. Some studies reported that encapsulated DOX nanoparticles holding folic acid were more efficient in suppressing the tumor than free DOX. Folic-acid-conjugated pullulan-g-poly (Lactide-co-glycolide) (PLGA) copolymer has been synthesized to deliver DOX at the target site in tumor cells. Therefore, pullulan and the different pullulan derivatives such as pullulan/PLGA graft copolymer would be the perfect candidates for the fabrication of drug targeting carriers. The anti-cancer activity of the conjugate was shown in KB tumor cells in vitro [162].

4.8.2. Pullulan with Mitoxantrone

Mitoxantrone is an anti-cancer drug that inhibits topoisomerase and can intercalate DNA. Due to its toxicity, the uses of mitoxantrone are limited [163]. A modified pullulan was used for loading mitoxantrone. Pullulan nanoparticles hydrophobically modified with cholesterol were used to load mitoxantrone, and improved mitoxantrone delivery was observed using cholesterol substituted pullulan polymers (CHPs) [164]. The different kind of CHPs were synthesized on the basis of the degree of cholesterol substitution and diameter. The larger the size of the nanoparticles, the greater the drug release capacity will be. Drug-loaded CHP nanoparticles with the largest size showed more anti-cancer activity in bladder cancer cells, which was confirmed by flow cytometry. The drug-loaded nanoparticles could inhibit the migration of MB49 cells [165].

4.8.3. Pullulan with Curcumin

Pullulan is hydrophilic in nature, so it could be used to deliver the hydrophilic substances, but not hydrophobic substances [166]. This problem could be resolved by forming pullulan acetate particles with amphiphilic nature by the process of acetylation. Pullulan acetate could be used as a carrier for delivering the drugs. The nanoparticle synthesis of curcumin-loaded pullulan acetate was performed to improve curcumin's stability and physiochemical properties. It displayed improved biocompatibility and hemocompatibility in the embryo of zebrafish in vitro. The curcumin–pullulan acetate nanoparticle complex was used as a hepatoprotective agent. The conjugate enhances the solubility, pH stability, and photostability of curcumin [167]. A Galactosylated pullulan–curcumin conjugate (Figure 8) was synthesized for targeted delivery of curcumin to hepatocarcinoma. Galactosylated pullulan was conjugated with curcumin using succinic anhydride introducing acid functionalities. Galactosylated pullulan–curcumin conjugate shows higher toxicity and internalization towards HepG2 cells via asialoglycoprotein-mediated endocytosis compared to its non-galactosylated counterpart. The results suggest that the higher uptake of galactosylated pullulan–curcumin conjugate may take place through asialoglycoprotein receptor (ASGPR)-mediated endocytosis [168].

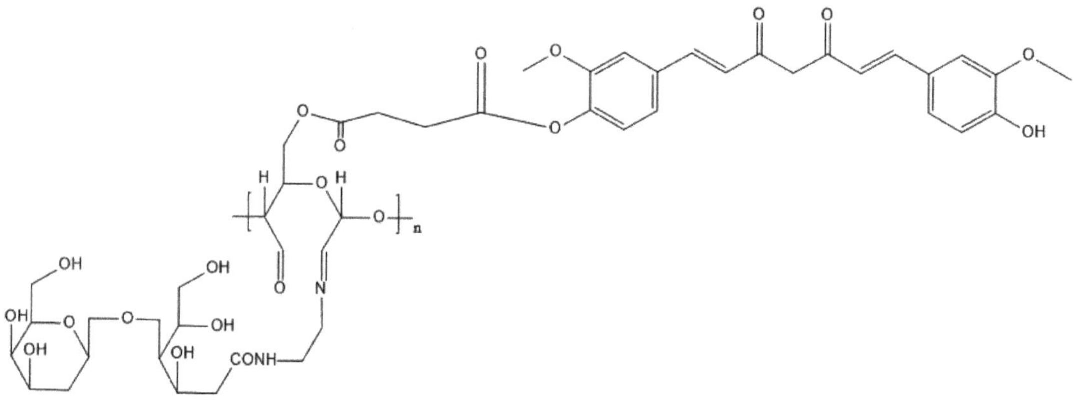

Figure 8. Structure of galactosylated pullan–curcumin conjugate.

5. Potentials and Prospective of Polysaccharides in Cancer Drug Delivery

The utilization of traditional chemotherapy drugs was limited by the side effects and physiological barriers of drug delivery (blood circulation, tumor accumulation, tumor tissue penetration, endocytosis, and drug release) in the human body. The development of new materials and the modification of existing materials are the critical factors in constructing efficient carriers. Natural polysaccharides with good biocompatibility and unique physicochemical properties are considered ideal for drug delivery applications (Table 2). Meanwhile, choosing suitable polysaccharides for drug delivery poses several challenges. Polysaccharides with diverse functional groups enable chemical modification and help to conjugate or load desired drug compounds. The drug molecule can be released to the diseased site through simple diffusion or cleavable cross-linking. Polysaccharides possess biocompatibility and biodegradable properties, and they are often used as a suitable polymer matrix for designing new carriers for drug delivery. Polysaccharides can be quickly degraded under biological conditions and removed by renal clearance compared to other materials. In addition, polysaccharide-based carriers can easily be linked with targeting ligands aiming to deliver the drug to the diseased site without affecting the healthy cells. Drug leakage in the blood, resistance and prolonged blood circulation of the drug are the crucial factors to be considered while designing the carrier. Furthermore, the carrier should provide high encapsulation efficiency, and sustained drug release to achieve the synergistic effect.

Table 2. Anti-cancer studies using polysaccharide–drug formulation.

Formulation	Model Used	Biological Changes	Reference
PTX-trimethyl chitosan conjugates	H22 tumor-bearing mice	Enhanced mucoadhesion and intestinal transport of PTX, increased tumor retardation and survival rate	[59]
Erlotinib-loaded MTX-chitosan magnetic nanoparticles	OVCAR-3 cell lines,	Improved cellular uptake, greater cytotoxicity and target specific delivery in FR-positive cancer cell lines	[64]
Curcumin-loaded chitosan nanoparticles	Swiss albino mice	Inhibiting the B[a]P-induced lung carcinogenesis, overexpression of p65 in the nuclei, reduced the overexpression of proliferating cell nuclear antigen	[67]
Curcumin-loaded folate-modified-chitosan-nanoparticles	MCF7 cell lines, L929 cell lines	Target specific uptake of curcumin into cancerous cells	[68]
Alginate nanoparticles with curcumin and resveratrol	DU145 prostate cancer cells	Increased cell uptake and enhanced cytotoxicity in cancer cells	[87]
EXE-loaded alginate nanoparticles	Dalton's lymphoma ascites cells	Improved cytotoxicity	[90]
Ag/Alg-TMX-PEG/FA core shell nanocomposite	MCF7 cell lines	Inducting reactive oxygen species (ROS), downregulation of survival oncogenic genes, G2/M phase arrest	[93]
Pectin–curcumin composite	KYSE-30 cell lines	Release of curcumin from the composite at acidic pH, enhanced cytotoxicity	[101]
Dihydroartemisinin-loaded DOX–pectin conjugate	MCF-7 cell lines, C57BL/6 mouse	Intranuclear uptake in MCF-7 cell lines, significant reduction in tumor growth	[103]
MTX-loaded guar gum microspheres	Albino rats	Target specific delivery to the colon	[114]
Dextran–DOX micelles	Balb/C mice bearing 4T1 tumors	Acid-sensitive drug release minimize systemic toxicity in normal tissues Selective accumulation in tumor and enhanced tumor-suppressive efficiency	[122]
MTX-loaded dextran–CUR nanoparticles	MCF-7 cell lines	Rapid internalization and enhanced cytotoxicity	[133]
PTX–HA conjugate	MCF-7 cell lines BALB/c nude mice	Cellular internalization, and tumor targeting via CD44 caveolae-mediated endocytosis, target specific drug release in the presence of GSH	[137]
Paclitaxel-loaded cyclodextrin–polypeptide conjugates	MCF-7 and 4T1 cell lines	Enhanced cellular uptake, inhibit P-gp efflux pumps	[157]
Mitoxantrone loaded modified pullulan nanoparticles	MB49 cells	Inhibit the growth migration of MB49 cells	[164]

6. Conclusions

Natural and synthetic polymer-based drug carriers take advantage of the unique delivery mechanism, such as target-specific delivery and prolonged circulation, by circumventing the immune systems. Application of biocompatible polymers, including their use as delivery systems for several potent anti-cancer drugs, are reported. This review highlighted the drug delivery applications of various polysaccharides, including alginate, dextran, chitosan, and hyaluronic acid. The polysaccharides with diverse functional groups can be modified with other chemical groups to make the conjugation of preferred therapeutic molecules easier. The drug moiety can be released through diffusion or by degradable cross-linking. Biocompatibility, biodegradability, and well-defined molecular weight are the crucial factors for a material to be used for therapeutic applications. Polysaccharides naturally possessing these properties can act as an appropriate polymer matrix for designing new materials. Polysaccharide-based nanomaterials have degraded into harmless derivatives under biological conditions and are easily removed from the body. Moreover, polysaccharides can be linked to targeting ligands such as folic acid to achieve effective drug delivery to the diseased site, provide cytotoxic specificity, and prevent toxic effects on healthy cells. Polysaccharide-based drug delivery systems can carry multiple drugs and deliver the drug by sequential or simultaneous release to achieve the synergistic effect. Polysaccharides loaded with proteins, peptides, drugs, and growth factors open up a new pathway for enhanced cancer treatment.

Author Contributions: Conceptualization, A.P.F. and R.R.; writing—original draft preparation, N.Y. and A.P.F.; writing—review and editing, A.P.F., V.V.P., U.M. and R.R.; resources, S.P., S.M., S.S.K., K.J.A., H.J.B. and S.K.M. All authors have read and agreed to the published version of the manuscript.

Funding: This study was funded by University Grants Commission, New Delhi, India, [F.No.16-9 (June 2018)/2019(NET/CSIR) and No.F.4-2/2006 (BSR)/BL/18-19/0428].

Institutional Review Board Statement: Not applicable.

Informed Consent Statement: Not applicable.

Data Availability Statement: Not applicable.

Acknowledgments: The authors N.Y., A.P.F. and R.R. acknowledge the financial support by the University Grants Commission, New Delhi, India.

Conflicts of Interest: The author(s) declare no potential conflicts of interest with respect to the research, authorship, and/or publication of this article.

References

1. Dagenais, G.R.; Leong, D.P.; Rangarajan, S.; Lanas, F.; Lopez-Jaramillo, P.; Gupta, R.; Diaz, R.; Avezum, A.; Oliveira, G.B.F.; Wielgosz, A.; et al. Variations in common diseases, hospital admissions, and deaths in middle-aged adults in 21 countries from five continents (PURE): A prospective cohort study. *Lancet* **2020**, *395*, 785–794. [CrossRef]
2. Sung, H.; Ferlay, J.; Siegel, R.L.; Laversanne, M.; Soerjomataram, I.; Jemal, A.; Bray, F. Global Cancer Statistics 2020: GLOBOCAN Estimates of Incidence and Mortality Worldwide for 36 Cancers in 185 Countries. *CA Cancer J. Clin.* **2021**, *71*, 209–249. [CrossRef] [PubMed]
3. Yahya, E.B.; Alqadhi, A.M. Recent trends in cancer therapy: A review on the current state of gene delivery. *Life Sci.* **2021**, *269*, 119087. [CrossRef] [PubMed]
4. Plummer, M.; de Martel, C.; Vignat, J.; Ferlay, J.; Bray, F.; Franceschi, S. Global burden of cancers attributable to infections in 2012: A synthetic analysis. *Lancet Glob. Health* **2016**, *4*, e609–e616. [CrossRef]
5. Chakraborty, S.; Rahman, T. The difficulties in cancer treatment. *ECancerMedicalScience* **2012**, *6*, ed16. [CrossRef]
6. Liang, C.; Zhang, X.; Yang, M.; Dong, X. Recent Progress in Ferroptosis Inducers for Cancer Therapy. *Adv. Mater.* **2019**, *31*, e1904197. [CrossRef]
7. Senapati, S.; Mahanta, A.K.; Kumar, S.; Maiti, P. Controlled drug delivery vehicles for cancer treatment and their performance. *Signal Transduct. Target. Ther.* **2018**, *3*, 7. [CrossRef]
8. Liechty, W.B.; Kryscio, D.R.; Slaughter, B.V.; Peppas, N.A. Polymers for Drug Delivery Systems. *Annu. Rev. Chem. Biomol. Eng.* **2010**, *1*, 149–173. [CrossRef]
9. Sood, A.; Gupta, A.; Agrawal, G. Recent advances in polysaccharides-based biomaterials for drug delivery and tissue engineering applications. *Carbohydr. Polym. Technol. Appl.* **2021**, *2*, 100067. [CrossRef]
10. Sung, Y.K.; Kim, S.W. Recent advances in polymeric drug delivery systems. *Biomater. Res.* **2020**, *24*, 12. [CrossRef]
11. Ai, J.-W.; Liao, W.; Ren, Z.-L. Enhanced anticancer effect of copper-loaded chitosan nanoparticles against osteosarcoma. *RSC Adv.* **2017**, *7*, 15971–15977. [CrossRef]

12. Miao, T.; Wang, J.; Zeng, Y.; Liu, G.; Chen, X. Polysaccharide-Based Controlled Release Systems for Therapeutics Delivery and Tissue Engineering: From Bench to Bedside. *Adv. Sci.* **2018**, *5*, 1700513. [CrossRef] [PubMed]
13. Muhamad, I.I.; Lazim, N.A.M.; Selvakumaran, S. Natural polysaccharide-based composites for drug delivery and biomedical applications. In *Natural Polysaccharides in Drug Delivery and Biomedical Applications*; Hasnain, M.S., Nayak, A.K., Eds.; Academic Press: Cambridge, MA, USA, 2019; pp. 419–440; ISBN 9780128170557.
14. Liu, Z.; Jiao, Y.; Wang, Y.; Zhou, C.; Zhang, Z. Polysaccharides-based nanoparticles as drug delivery systems. *Adv. Drug Deliv. Rev.* **2008**, *60*, 1650–1662. [CrossRef] [PubMed]
15. Chen, Y.; Yao, F.; Ming, K.; Wang, D.; Hu, Y.; Liu, J. Polysaccharides from Traditional Chinese Medicines: Extraction, Purification, Modification, and Biological Activity. *Molecules* **2016**, *21*, 1705. [CrossRef]
16. Ngwuluka, N.C. Responsive polysaccharides and polysaccharides-based nanoparticles for drug delivery. In *Stimuli Responsive Polymeric Nanocarriers for Drug Delivery Applications*; Salam, A., Makhlouf, H., Abu-Thabit, N.Y., Eds.; Woodhead Publishing: Cambridge, MA, USA, 2018; Volume 1, pp. 531–554; ISBN 9780081019979.
17. Ghadi, R.; Dand, N. BCS class IV drugs: Highly notorious candidates for formulation development. *J. Control. Release* **2017**, *248*, 71–95. [CrossRef]
18. Ravi, P.R.; Vats, R.; Balija, J.; Adapa, S.P.N.; Aditya, N. Modified pullulan nanoparticles for oral delivery of lopinavir: Formulation and pharmacokinetic evaluation. *Carbohydr. Polym.* **2014**, *110*, 320–328. [CrossRef]
19. Park, J.H.; Saravanakumar, G.; Kim, K.; Kwon, I.C. Targeted delivery of low molecular drugs using chitosan and its derivatives. *Adv. Drug Deliv. Rev.* **2010**, *62*, 28–41. [CrossRef]
20. Shin, Y.; Kim, D.; Hu, Y.; Kim, Y.; Hong, I.K.; Kim, M.S.; Jung, S. pH-Responsive Succinoglycan-Carboxymethyl Cellulose Hydrogels with Highly Improved Mechanical Strength for Controlled Drug Delivery Systems. *Polymers* **2021**, *13*, 3197. [CrossRef]
21. Ahmed, E.M. Hydrogel: Preparation, characterization, and applications: A review. *J. Adv. Res.* **2015**, *6*, 105–121. [CrossRef]
22. Gholamali, I.; Hosseini, S.N.; Alipour, E. Doxorubicin-loaded oxidized starch/poly (vinyl alcohol)/CuO bio-nanocomposite hydrogels as an anticancer drug carrier agent. *Int. J. Polym. Mater. Polym. Biomater.* **2020**, *70*, 967–980. [CrossRef]
23. Gholamali, I.; Yadollahi, M. Bio-nanocomposite Polymer Hydrogels Containing Nanoparticles for Drug Delivery: A Review. *Regen. Eng. Transl. Med.* **2021**, *7*, 129–146. [CrossRef]
24. Yahya, E.B.; Jummaat, F.; Amirul, A.A.; Adnan, A.S.; Olaiya, N.G.; Abdullah, C.K.; Rizal, S.; Mohamad Haafiz, M.K.; Abdul Khalil, H.P.S. A Review on Revolutionary Natural Biopolymer-Based Aerogels for Antibacterial Delivery. *Antibiotics* **2020**, *9*, 648. [CrossRef]
25. García-González, C.A.; Budtova, T.; Durães, L.; Erkey, C.; Del Gaudio, P.; Gurikov, P.; Koebel, M.; Liebner, F.; Neagu, M.; Smirnova, I. An Opinion Paper on Aerogels for Biomedical and Environmental Applications. *Molecules* **2019**, *24*, 1815. [CrossRef]
26. Bhandari, J.; Mishra, H.; Mishra, P.K.; Wimmer, R.; Ahmad, F.; Talegaonkar, S. Cellulose nanofiber aerogel as a promising biomaterial for customized oral drug delivery. *Int. J. Nanomed.* **2017**, *12*, 2021–2031. [CrossRef] [PubMed]
27. García-González, C.A.; Sosnik, A.; Kalmár, J.; De Marco, I.; Erkey, C.; Concheiro, A.; Alvarez-Lorenzo, C. Aerogels in drug delivery: From design to application. *J. Control. Release* **2021**, *332*, 40–63. [CrossRef]
28. Atta, S.; Khaliq, S.; Islam, A.; Javeria, I.; Jamil, T.; Athar, M.M.; Shafiq, M.I.; Ghaffar, A. Injectable biopolymer based hydrogels for drug delivery applications. *Int. J. Biol. Macromol.* **2015**, *80*, 240–245. [CrossRef]
29. Liu, L.; Fishman, M.L.; Kost, J.; Hicks, K.B. Pectin-based systems for colon-specific drug delivery via oral route. *Biomaterials* **2003**, *24*, 3333–3343. [CrossRef]
30. Yang, J.; Li, F.; Li, M.; Zhang, S.; Liu, J.; Liang, C.; Sun, Q.; Xiong, L. Fabrication and characterization of hollow starch nanoparticles by gelation process for drug delivery application. *Carbohydr. Polym.* **2017**, *173*, 223–232. [CrossRef] [PubMed]
31. Hu, H.; Li, Y.; Zhou, Q.; Ao, Y.; Yu, C.; Wan, Y.; Xu, H.; Li, Z.; Yang, X. Redox-Sensitive Hydroxyethyl Starch–Doxorubicin Conjugate for Tumor Targeted Drug Delivery. *ACS Appl. Mater. Interfaces* **2016**, *8*, 30833–30844. [CrossRef] [PubMed]
32. Jacob, J.; Haponiuk, J.T.; Thomas, S.; Gopi, S. Biopolymer based nanomaterials in drug delivery systems: A review. *Mater. Today Chem.* **2018**, *9*, 43–55. [CrossRef]
33. Zhang, M.; Cui, S.; Cheung, P.; Wang, Q. Antitumor polysaccharides from mushrooms: A review on their isolation process, structural characteristics and antitumor activity. *Trends Food Sci. Technol.* **2007**, *18*, 4–19. [CrossRef]
34. Nauts, H.C.; Swift, W.E.; Coley, B.L. The treatment of malignant tumors by bacterial toxins as developed by the late William B. Coley, M.D., reviewed in the light of modern research. *Cancer Res.* **1946**, *6*, 205–216. [PubMed]
35. Fritz, H.; Kennedy, D.A.; Ishii, M.; Fergusson, D.; Fernandes, R.; Cooley, K.; Seely, D. Polysaccharide K and Coriolus versicolor Extracts for Lung Cancer: A Systematic Review. *Integr. Cancer Ther.* **2015**, *14*, 201–211. [CrossRef] [PubMed]
36. Chen, L.; Liu, X.; Wong, K.-H. Novel nanoparticle materials for drug/food delivery-polysaccharides. *Phys. Sci. Rev.* **2016**, *1*, 20160053. [CrossRef]
37. Lengyel, M.; Kállai-Szabó, N.; Antal, V.; Laki, A.J.; Antal, I. Microparticles, Microspheres, and Microcapsules for Advanced Drug Delivery. *Sci. Pharm.* **2019**, *87*, 20. [CrossRef]
38. Babu, A.; Ramesh, R. Multifaceted Applications of Chitosan in Cancer Drug Delivery and Therapy. *Mar. Drugs* **2017**, *15*, 96. [CrossRef]
39. Yogeshkumar, N.G.; GuravAtul, S.; Adhikrao, V.Y. Chitosan and Its Applications: A Review of Literature. *Int. J. Res. Pharm. Biomed. Sci.* **2013**, *4*, 312–331.
40. Rinaudo, M. Chitin and Chitosan: Properties and Applications. *Prog. Polym. Sci.* **2006**, *31*, 603–632. [CrossRef]

41. Nahar, K.; Hossain, K.; Khan, T.A. Alginate and Its Versatile Application in Drug Delivery. *J. Pharm. Sci. Res.* **2017**, *9*, 606–617.
42. Lee, M.; Nah, J.; Kwon, Y.; Koh, J.J.; Ko, K.S.; Kim, S.W. Water-Soluble and Low Molecular Weight Chitosan-Based Plasmid DNA Delivery. *Pharm. Res.* **2001**, *18*, 427–431. [CrossRef]
43. Kumar, S.; Dutta, J.; Dutta, P. Preparation and characterization of N-heterocyclic chitosan derivative based gels for biomedical applications. *Int. J. Biol. Macromol.* **2009**, *45*, 330–337. [CrossRef] [PubMed]
44. Gim, S.; Zhu, Y.; Seeberger, P.H.; Delbianco, M. Carbohydrate-based nanomaterials for biomedical applications. *Wiley Interdiscip. Rev. Nanomed. Nanobiotechnol.* **2019**, *11*, e1558. [CrossRef] [PubMed]
45. Wimardhani, Y.S.; Suniarti, D.F.; Freisleben, H.J.; Wanandi, S.I.; Siregar, N.C.; Ikeda, M.-A. Chitosan exerts anticancer activity through induction of apoptosis and cell cycle arrest in oral cancer cells. *J. Oral Sci.* **2014**, *56*, 119–126. [CrossRef] [PubMed]
46. Park, J.K.; Chung, M.J.; Na Choi, H.; Park, Y.I. Effects of the Molecular Weight and the Degree of Deacetylation of Chitosan Oligosaccharides on Antitumor Activity. *Int. J. Mol. Sci.* **2011**, *12*, 266–277. [CrossRef] [PubMed]
47. Adhikari, H.S.; Yadav, P.N. Anticancer Activity of Chitosan, Chitosan Derivatives, and Their Mechanism of Action. *Int. J. Biomater.* **2018**, *2018*, 2952085. [CrossRef] [PubMed]
48. Quiñones, J.P.; Peniche, H.; Peniche, C. Chitosan Based Self-Assembled Nanoparticles in Drug Delivery. *Polymers* **2018**, *10*, 235. [CrossRef]
49. Andreadis, C.; Vahtsevanos, K.; Sidiras, T.; Thomaidis, I.; Antoniadis, K.; Mouratidou, D. 5-Fluorouracil and cisplatin in the treatment of advanced oral cancer. *Oral Oncol.* **2003**, *39*, 380–385. [CrossRef]
50. Remesh, A. Toxicities of anticancer drugs and its management. *Int. J. Basic Clin. Pharmacol.* **2012**, *1*, 2–12. [CrossRef]
51. Zhao, L.; Zhang, B. Doxorubicin induces cardiotoxicity through upregulation of death receptors mediated apoptosis in cardiomyocytes. *Sci. Rep.* **2017**, *7*, 44735. [CrossRef]
52. Arima, H.; Hagiwara, Y.; Hirayama, F.; Uekama, K. Enhancement of antitumor effect of doxorubicin by its complexation with γ-cyclodextrin in pegylated liposomes. *J. Drug Target.* **2006**, *14*, 225–232. [CrossRef]
53. Park, J.M.; Lee, S.Y.; Lee, G.H.; Chung, E.Y.; Chang, K.M.; Kwak, B.K.; Kuh, H.-J.; Lee, J. Design and characterisation of doxorubicin-releasing chitosan microspheres for anti-cancer chemoembolisation. *J. Microencapsul.* **2012**, *29*, 695–705. [CrossRef] [PubMed]
54. Unsoy, G.; Khodadust, R.; Yalcin, S.; Mutlu, P.; Gunduz, U. Synthesis of Doxorubicin loaded magnetic chitosan nanoparticles for pH responsive targeted drug delivery. *Eur. J. Pharm. Sci.* **2014**, *62*, 243–250. [CrossRef] [PubMed]
55. Yu, J.; Xie, X.; Wu, J.; Liu, Y.; Liu, P.; Xu, X.; Yu, H.; Lu, L.; Che, X. Folic acid conjugated glycol chitosan micelles for targeted delivery of doxorubicin: Preparation and preliminary evaluation in vitro. *J. Biomater. Sci. Polym. Ed.* **2013**, *24*, 606–620. [CrossRef] [PubMed]
56. Jordan, M.A.; Wilson, L. Microtubules as a target for anticancer drugs. *Nat. Rev. Cancer* **2004**, *4*, 253–265. [CrossRef]
57. Trickler, W.J.; Nagvekar, A.A.; Dash, A.K. A Novel Nanoparticle Formulation for Sustained Paclitaxel Delivery. *AAPS PharmSciTech* **2008**, *9*, 486–493. [CrossRef]
58. Lee, E.; Lee, J.; Lee, I.-H.; Yu, M.; Kim, H.; Chae, S.Y.; Jon, S. Conjugated Chitosan as a Novel Platform for Oral Delivery of Paclitaxel. *J. Med. Chem.* **2008**, *51*, 6442–6449. [CrossRef]
59. He, R.; Yin, C. Trimethyl chitosan based conjugates for oral and intravenous delivery of paclitaxel. *Acta Biomater.* **2017**, *53*, 355–366. [CrossRef]
60. Hwang, H.-Y.; Kim, I.-S.; Kwon, I.C.; Kim, Y.-H. Tumor targetability and antitumor effect of docetaxel-loaded hydrophobically modified glycol chitosan nanoparticles. *J. Control. Release* **2008**, *128*, 23–31. [CrossRef]
61. Jain, A.; Thakur, K.; Kush, P.; Jain, U.K. Docetaxel loaded chitosan nanoparticles: Formulation, characterization and cytotoxicity studies. *Int. J. Biol. Macromol.* **2014**, *69*, 546–553. [CrossRef]
62. Mirzaie, Z.H.; Irani, S.; Mirfakhraie, R.; Atyabi, S.M.; Dinarvand, M.; Dinarvand, R.; Varshochian, R.; Atyabi, F. Docetaxel-Chitosan nanoparticles for breast cancer treatment: Cell viability and gene expression study. *Chem. Biol. Drug Des.* **2016**, *88*, 850–858. [CrossRef]
63. Wu, P.; He, X.; Wang, K.; Tan, W.; He, C.; Zheng, M. A Novel Methotrexate Delivery System Based on Chitosan-Methotrexate Covalently Conjugated Nanoparticles. *J. Biomed. Nanotechnol.* **2009**, *5*, 557–564. [CrossRef] [PubMed]
64. Fathi, M.; Barar, J.; Erfan-Niya, H.; Omidi, Y. Methotrexate-conjugated chitosan-grafted pH- and thermo-responsive magnetic nanoparticles for targeted therapy of ovarian cancer. *Int. J. Biol. Macromol.* **2020**, *154*, 1175–1184. [CrossRef] [PubMed]
65. Saranya, T.; Rajan, V.; Biswas, R.; Jayakumar, R.; Sathianarayanan, S. Synthesis, characterisation and biomedical applications of curcumin conjugated chitosan microspheres. *Int. J. Biol. Macromol.* **2018**, *110*, 227–233. [CrossRef] [PubMed]
66. Ni, J.; Tian, F.; Dahmani, F.Z.; Yang, H.; Yue, D.; He, S.; Zhou, J.; Yao, J. Curcumin-carboxymethyl chitosan (CNC) conjugate and CNC/LHR mixed polymeric micelles as new approaches to improve the oral absorption of P-gp substrate drugs. *Drug Deliv.* **2016**, *23*, 3424–3435. [CrossRef] [PubMed]
67. Vijayakurup, V.; Thulasidasan, A.T.; Shankar, G.M.; Retnakumari, A.P.; Nandan, C.D.; Somaraj, J.; Antony, J.; Alex, V.V.; Vinod, B.S.; Liju, V.B.; et al. Chitosan Encapsulation Enhances the Bioavailability and Tissue Retention of Curcumin and Improves its Efficacy in Preventing B[a]P-induced Lung Carcinogenesis. *Cancer Prev. Res.* **2019**, *12*, 225–236. [CrossRef]
68. Esfandiarpour-Boroujeni, S.; Bagheri-Khoulenjani, S.; Mirzadeh, H.; Amanpour, S. Fabrication and study of curcumin loaded nanoparticles based on folate-chitosan for breast cancer therapy application. *Carbohydr. Polym.* **2017**, *168*, 14–21. [CrossRef]

69. Chuah, L.H.; Roberts, C.; Billa, N.; Abdullah, S.; Rosli, R. Cellular uptake and anticancer effects of mucoadhesive curcumin-containing chitosan nanoparticles. *Colloids Surf. B Biointerfaces* **2014**, *116*, 228–236. [CrossRef]
70. Duan, J.; Zhang, Y.; Han, S.; Chen, Y.; Li, B.; Liao, M.; Chen, W.; Deng, X.; Zhao, J.; Huang, B. Synthesis and in vitro/in vivo anti-cancer evaluation of curcumin-loaded chitosan/poly(butyl cyanoacrylate) nanoparticles. *Int. J. Pharm.* **2010**, *400*, 211–220. [CrossRef]
71. Vivek, R.; Thangam, R.; Nipunbabu, V.; Ponraj, T.; Kannan, S. Oxaliplatin-chitosan nanoparticles induced intrinsic apoptotic signaling pathway: A "smart" drug delivery system to breast cancer cell therapy. *Int. J. Biol. Macromol.* **2014**, *65*, 289–297. [CrossRef]
72. Lee, K.Y.; Mooney, D.J. Alginate: Properties and biomedical applications. *Prog. Polym. Sci.* **2012**, *37*, 106–126. [CrossRef]
73. Rajaonarivony, M.; Vauthier, C.; Couarraze, G.; Puisieux, F.; Couvreur, P. Development of a New Drug Carrier Made from Alginate. *J. Pharm. Sci.* **1993**, *82*, 912–917. [CrossRef] [PubMed]
74. Kumar, S.; Bhanjana, G.; Verma, R.K.; Dhingra, D.; Dilbaghi, N.; Kim, K.-H. Metformin-loaded alginate nanoparticles as an effective antidiabetic agent for controlled drug release. *J. Pharm. Pharmacol.* **2017**, *69*, 143–150. [CrossRef] [PubMed]
75. Kirtane, A.R.; Narayan, P.; Liu, G.; Panyam, J. Polymer-surfactant nanoparticles for improving oral bioavailability of doxorubicin. *J. Pharm. Investig.* **2017**, *47*, 65–73. [CrossRef]
76. Abdelghany, S.; Alkhawaldeh, M.; AlKhatib, H.S. Carrageenan-stabilized chitosan alginate nanoparticles loaded with ethionamide for the treatment of tuberculosis. *J. Drug Deliv. Sci. Technol.* **2017**, *39*, 442–449. [CrossRef]
77. Zahoor, A.; Sharma, S.; Khuller, G. Inhalable alginate nanoparticles as antitubercular drug carriers against experimental tuberculosis. *Int. J. Antimicrob. Agents* **2005**, *26*, 298–303. [CrossRef] [PubMed]
78. Dehghan, S.; Kheiri, M.T.; Abnous, K.; Eskandari, M.; Tafaghodi, M. Preparation, characterization and immunological evaluation of alginate nanoparticles loaded with whole inactivated influenza virus: Dry powder formulation for nasal immunization in rabbits. *Microb. Pathog.* **2018**, *115*, 74–85. [CrossRef]
79. Hefnawy, A.; Khalil, I.A.; El-Sherbiny, I.M. Facile development of nanocomplex-in-nanoparticles for enhanced loading and selective delivery of doxorubicin to brain. *Nanomedicine* **2017**, *12*, 2737–2761. [CrossRef]
80. Najafabadi, A.H.; Azodi-Deilami, S.; Abdouss, M.; Payravand, H.; Farzaneh, S. Synthesis and evaluation of hydroponically alginate nanoparticles as novel carrier for intravenous delivery of propofol. *J. Mater. Sci. Mater. Electron.* **2015**, *26*, 145. [CrossRef]
81. Wong, T.W.; Dhanawat, M.; Rathbone, M.J. Vaginal drug delivery: Strategies and concerns in polymeric nanoparticle development. *Expert Opin. Drug Deliv.* **2014**, *11*, 1419–1434. [CrossRef]
82. Motwani, S.K.; Chopra, S.; Talegaonkar, S.; Kohli, K.; Ahmad, F.; Khar, R.K. Chitosan–sodium alginate nanoparticles as submicroscopic reservoirs for ocular delivery: Formulation, optimisation and in vitro characterisation. *Eur. J. Pharm. Biopharm.* **2008**, *68*, 513–525. [CrossRef]
83. Lakkakula, J.R.; Gujarathi, P.; Pansare, P.; Tripathi, S. A comprehensive review on alginate-based delivery systems for the delivery of chemotherapeutic agent: Doxorubicin. *Carbohydr. Polym.* **2021**, *259*, 117696. [CrossRef] [PubMed]
84. Goh, C.H.; Heng, P.W.S.; Chan, L.W. Alginates as a useful natural polymer for microencapsulation and therapeutic applications. *Carbohydr. Polym.* **2012**, *88*, 1–12. [CrossRef]
85. Cheng, Y.; Yu, S.; Wang, J.; Qian, H.; Wu, W.; Jiang, X. In vitro and in vivo Antitumor Activity of Doxorubicin-Loaded Alginic-Acid-Based Nanoparticles. *Macromol. Biosci.* **2012**, *12*, 1326–1335. [CrossRef]
86. Tomeh, M.A.; Hadianamrei, R.; Zhao, X. A Review of Curcumin and Its Derivatives as Anticancer Agents. *Int. J. Mol. Sci.* **2019**, *20*, 1033. [CrossRef] [PubMed]
87. Saralkar, P.; Dash, A.K. Alginate Nanoparticles Containing Curcumin and Resveratrol: Preparation, Characterization, and In Vitro Evaluation Against DU145 Prostate Cancer Cell Line. *AAPS PharmSciTech* **2017**, *18*, 2814–2823. [CrossRef] [PubMed]
88. Dey, S.; Sreenivasan, K. Conjugation of curcumin onto alginate enhances aqueous solubility and stability of curcumin. *Carbohydr. Polym.* **2014**, *99*, 499–507. [CrossRef] [PubMed]
89. Dixon, J.M. Exemestane: A potent irreversible aromatase inactivator and a promising advance in breast cancer treatment. *Expert Rev. Anticancer Ther.* **2002**, *2*, 267–275. [CrossRef]
90. Jayapal, J.J.; Dhanaraj, S. Exemestane loaded alginate nanoparticles for cancer treatment: Formulation and in vitro evaluation. *Int. J. Biol. Macromol.* **2017**, *105*, 416–421. [CrossRef]
91. Legha, S.S. Tamoxifen. Use in treatment of metastatic breast cancer refractory to combination chemotherapy. *JAMA* **1979**, *242*, 49–52. [CrossRef]
92. Martínez, A.; Benito-Miguel, M.; Iglesias, I.; Teijón, J.M.; Blanco, M.D. Tamoxifen-loaded thiolated alginate-albumin nanoparticles as antitumoral drug delivery systems. *J. Biomed. Mater. Res. Part A* **2012**, *100A*, 1467–1476. [CrossRef]
93. Ibrahim, O.M.; El-Deeb, N.M.; Abbas, H.; Elmasry, S.M.; El-Aassar, M. Alginate based tamoxifen/metal dual core-folate decorated shell: Nanocomposite targeted therapy for breast cancer via ROS-driven NF-κB pathway modulation. *Int. J. Biol. Macromol.* **2020**, *146*, 119–131. [CrossRef] [PubMed]
94. Sriamornsak, P. Application of pectin in oral drug delivery. *Expert Opin. Drug Deliv.* **2011**, *8*, 1009–1023. [CrossRef] [PubMed]
95. Features, K. The chemistry and technology of pectin. In *Food Science and Technology*; Academic Press: Cambridge, MA, USA, 1991; ISBN 9780080926445.
96. Morris, G.A.; Samil Kök, M.; Harding, S.E.; Adams, G.G. Polysaccharide Drug Delivery Systems Based on Pectin and Chitosan. Biotechnol. *Genet. Eng. Rev.* **2010**, *27*, 257–284.

97. Ghosh, S.; Banerjee, S.; Sil, P.C. The beneficial role of curcumin on inflammation, diabetes and neurodegenerative disease: A recent update. *Food Chem. Toxicol.* **2015**, *83*, 111–124. [CrossRef]
98. Huang, S.; Beevers, C.S. Pharmacological and clinical properties of curcumin. *Bot. Targets Ther.* **2011**, *1*, 5–18. [CrossRef]
99. Giordano, A.; Tommonaro, G. Curcumin and Cancer. *Nutrients* **2019**, *11*, 2376. [CrossRef]
100. Shih, F.-Y.; Su, I.-J.; Chu, L.-L.; Lin, X.; Kuo, S.-C.; Hou, Y.-C.; Chiang, Y.-T. Development of Pectin-Type B Gelatin Polyelectrolyte Complex for Curcumin Delivery in Anticancer Therapy. *Int. J. Mol. Sci.* **2018**, *19*, 3625. [CrossRef]
101. Mundlia, J.; Ahuja, M.; Kumar, P.; Pillay, V. Pectin–curcumin composite: Synthesis, molecular modeling and cytotoxicity. *Polym. Bull.* **2019**, *76*, 3153–3173. [CrossRef]
102. Cheewatanakornkool, K.; Niratisai, S.; Manchun, S.; Dass, C.R.; Sriamornsak, P. Thiolated pectin–doxorubicin conjugates: Synthesis, characterization and anticancer activity studies. *Carbohydr. Polym.* **2017**, *174*, 493–506. [CrossRef]
103. Tao, Y.; Zheng, D.; Zhao, J.; Liu, K.; Liu, J.; Lei, J.; Wang, L. Self-Assembling PH-Responsive Nanoparticle Platform Based on Pectin-Doxorubicin Conjugates for Codelivery of Anticancer Drugs. *ACS Omega* **2021**, *6*, 9998–10004. [CrossRef]
104. Dasari, S.; Tchounwou, P.B. Cisplatin in cancer therapy: Molecular mechanisms of action. *Eur. J. Pharmacol.* **2014**, *740*, 364–378. [CrossRef] [PubMed]
105. Verma, A.K.; Chanchal, A.; Chutani, K. Augmentation of anti-tumour activity of cisplatin by pectin nano-conjugates in B-16 mouse model: Pharmacokinetics and in-vivo biodistribution of radio-labelled, hydrophilic nano-conjugates. *Int. J. Nanotechnol.* **2012**, *9*, 872–886. [CrossRef]
106. Verma, A.; Sachin, A. Novel Hydrophilic Drug Polymer Nano-Conjugates of Cisplatin Showing Long Blood Retention Profile: Its Release Kinetics, Cellular Uptake and Bio-Distribution. *Curr. Drug Deliv.* **2008**, *5*, 120–126. [CrossRef] [PubMed]
107. Patel, J.J.; Karve, M.; Patel, N.K. View of Guar Gum: A Versatile Material for Pharmaceutical Industries. *Int. J. Pharm. Pharm. Sci.* **2014**, *6*, 13–19.
108. Prabaharan, M. Prospective of guar gum and its derivatives as controlled drug delivery systems. *Int. J. Biol. Macromol.* **2011**, *49*, 117–124. [CrossRef]
109. Pardini, B.; Kumar, R.; Naccarati, A.; Novotny, J.; Prasad, R.B.; Forsti, A.; Hemminki, K.; Vodicka, P.; Bermejo, J.L. 5-Fluorouracil-based chemotherapy for colorectal cancer and MTHFR/MTRR genotypes. *Br. J. Clin. Pharmacol.* **2011**, *72*, 162–163. [CrossRef]
110. Kamal, T.; Sarfraz, M.; Arafat, M.; Mikov, M.; Rahman, N. Cross-linked guar gum and sodium borate based microspheres as colon-targeted anticancer drug delivery systems for 5-fluorouracil. *Pak. J. Pharm. Sci.* **2017**, *30*, 2329–2336.
111. Lashley, M.R.; Niedzinski, E.J.; Rogers, J.M.; Denison, M.S.; Nantz, M.H. Synthesis and estrogen receptor affinity of a 4-hydroxytamoxifen-Labeled ligand for diagnostic imaging. *Bioorgan. Med. Chem.* **2002**, *10*, 4075–4082. [CrossRef]
112. Krishnaiah, Y.; Karthikeyan, R.; Satyanarayana, V. A three-layer guar gum matrix tablet for oral controlled delivery of highly soluble metoprolol tartrate. *Int. J. Pharm.* **2002**, *241*, 353–366. [CrossRef]
113. Soppirnath, K.S.; Aminabhavi, T.M. Water transport and drug release study from cross-linked polyacrylamide grafted guar gum hydrogel microspheres for the controlled release application. *Eur. J. Pharm. Biopharm.* **2002**, *53*, 87–98. [CrossRef]
114. Chaurasia, M.; Chourasia, M.K.; Jain, N.K.; Jain, A.; Soni, V.; Gupta, Y.; Jain, S.K. Cross-linked guar gum microspheres: A viable approach for improved delivery of anticancer drugs for the treatment of colorectal cancer. *AAPS PharmSciTech* **2006**, *7*, E143–E151. [CrossRef] [PubMed]
115. Elias, E.J.; Anil, S.; Ahmad, S.; Daud, A. Colon Targeted Curcumin Delivery Using Guar Gum. *Nat. Prod. Commun.* **2010**, *5*, 915–918. [CrossRef] [PubMed]
116. Huang, S.; Huang, G. Preparation and drug delivery of dextran-drug complex. *Drug Deliv.* **2019**, *26*, 252–261. [CrossRef] [PubMed]
117. Díaz-Montes, E. Dextran: Sources, Structures, and Properties. *Polysaccharides* **2021**, *2*, 554–565. [CrossRef]
118. Heinze, T.; Liebert, T.; Heublein, B.; Hornig, S. Functional Polymers Based on Dextran. *Adv. Polym. Sci.* **2006**, *205*, 199–291. [CrossRef]
119. Huang, G.; Huang, H. Application of dextran as nanoscale drug carriers. *Nanomedicine* **2018**, *13*, 3149–3158. [CrossRef]
120. Zhang, Y.; Wang, H.; Mukerabigwi, J.F.; Liu, M.; Luo, S.; Lei, S.; Cao, Y.; Huang, X.; He, H. Self-organized nanoparticle drug delivery systems from a folate-targeted dextran–doxorubicin conjugate loaded with doxorubicin against multidrug resistance. *RSC Adv.* **2015**, *5*, 71164–71173. [CrossRef]
121. Cao, D.; He, J.; Xu, J.; Zhang, M.; Zhao, L.; Duan, G.; Cao, Y.; Zhou, R.; Ni, P. Polymeric prodrugs conjugated with reduction-sensitive dextran–camptothecin and pH-responsive dextran–doxorubicin: An effective combinatorial drug delivery platform for cancer therapy. *Polym. Chem.* **2016**, *7*, 4198–4212. [CrossRef]
122. Zhang, X.; Zhang, T.; Ma, X.; Wang, Y.; Lu, Y.; Jia, D.; Huang, X.; Chen, J.; Xu, Z.; Wen, F. The design and synthesis of dextran-doxorubicin prodrug-based pH-sensitive drug delivery system for improving chemotherapy efficacy. *Asian J. Pharm. Sci.* **2019**, *15*, 605–616. [CrossRef]
123. Nakamura, J.; Nakajima, N.; Matsumura, K.; Hyon, S.-H. Water-soluble taxol conjugates with dextran and targets tumor cells by folic acid immobilization. *Anticancer Res.* **2010**, *30*, 903–909.
124. Kanwal, S.; Naveed, M.; Arshad, A.; Arshad, A.; Firdous, F.; Faisal, A.; Yameen, B. Reduction-Sensitive Dextran-Paclitaxel Polymer-Drug Conjugate: Synthesis, Self-Assembly into Nanoparticles, and in Vitro Anticancer Efficacy. *Bioconjugate Chem.* **2021**, *32*, 2516–2529. [CrossRef] [PubMed]

125. Georgaki, S.; Skopeliti, M.; Tsiatas, M.; Nicolaou, K.A.; Ioannou, K.; Husband, A.; Bamias, A.; Dimopoulos, M.A.; Constantinou, A.I.; Tsitsilonis, O.E. Phenoxodiol, an anticancer isoflavene, induces immunomodulatory effects in vitro and in vivo. *J. Cell. Mol. Med.* **2009**, *13*, 3929–3938. [CrossRef]
126. Gamble, J.R.; Xia, P.; Hahn, C.; Drew, J.J.; Drogemuller, C.J.; Brown, D.; Vadas, M.A. Phenoxodiol, an experimental anticancer drug, shows potent antiangiogenic properties in addition to its antitumour effects. *Int. J. Cancer* **2006**, *118*, 2412–2420. [CrossRef] [PubMed]
127. Yee, E.M.H.; Cirillo, G.; Brandl, M.B.; Black, D.S.; Vittorio, O.; Kumar, N. Synthesis of Dextran–Phenoxodiol and Evaluation of Its Physical Stability and Biological Activity. *Front. Bioeng. Biotechnol.* **2019**, *7*, 183. [CrossRef] [PubMed]
128. Nevozhay, D.; Budzynska, R.; Jagiello, M.; Kanska, U.; Omar, M.S.; Opolski, A.; Wietrzyk, J.; Boratynski, J. The effect of the substitution level of some dextran-methotrexate conjugates on their antitumor activity in experimental cancer models. *Anticancer Res.* **2006**, *26*, 2179–2186.
129. Dang, W.; Colvin, O.M.; Brem, H.; Saltzman, W.M. Covalent coupling of methotrexate to dextran enhances the penetration of cytotoxicity into a tissue-like matrix. *Cancer Res.* **1994**, *54*, 1729–1735. [PubMed]
130. Huang, Q.; Zhang, L.; Sun, X.; Zeng, K.; Li, J.; Liu, Y.-N. Coating of carboxymethyl dextran on liposomal curcumin to improve the anticancer activity. *RSC Adv.* **2014**, *4*, 59211–59217. [CrossRef]
131. Raveendran, R.; Bhuvaneshwar, G.; Sharma, C.P. Hemocompatible curcumin–dextran micelles as pH sensitive pro-drugs for enhanced therapeutic efficacy in cancer cells. *Carbohydr. Polym.* **2016**, *137*, 497–507. [CrossRef] [PubMed]
132. Zare, M.; Sarkati, M.N.; Tashakkorian, H.; Partovi, R.; Rahaiee, S. Dextran-immobilized curcumin: An efficient agent against food pathogens and cancer cells. *J. Bioact. Compat. Polym.* **2019**, *34*, 309–320. [CrossRef]
133. Curcio, M.; Cirillo, G.; Tucci, P.; Farfalla, A.; Bevacqua, E.; Vittorio, O.; Iemma, F.; Nicoletta, F.P. Dextran-Curcumin Nanoparticles as a Methotrexate Delivery Vehicle: A Step Forward in Breast Cancer Combination Therapy. *Pharmaceuticals* **2020**, *13*, 2. [CrossRef]
134. Alaniz, L.; Cabrera, P.V.; Blanco, G.G.; Ernst, G.; Rimoldi, G.; Alvarez, E.; Hajos, S.E. Interaction of CD44 with Different Forms of Hyaluronic Acid. Its Role in Adhesion and Migration of Tumor Cells. *Cell Commun. Adhes.* **2002**, *9*, 117–130. [CrossRef]
135. Jong, A.; Wu, C.-H.; Gonzales-Gomez, I.; Kwon-Chung, K.J.; Chang, Y.C.; Tseng, H.-K.; Cho, W.-L.; Huang, S.-H. Hyaluronic Acid Receptor CD44 Deficiency Is Associated with Decreased Cryptococcus neoformans Brain Infection. *J. Biol. Chem.* **2012**, *287*, 15298–15306. [CrossRef] [PubMed]
136. Widjaja, L.K.; Bora, M.; Chan, P.N.P.H.; Lipik, V.; Wong, T.T.L.; Venkatraman, S.S. Hyaluronic acid-based nanocomposite hydrogels for ocular drug delivery applications. *J. Biomed. Mater. Res. Part A* **2014**, *102*, 3056–3065. [CrossRef] [PubMed]
137. Yin, S.; Huai, J.; Chen, X.; Yang, Y.; Zhang, X.; Gan, Y.; Wang, G.; Gu, X.; Li, J. Intracellular delivery and antitumor effects of a redox-responsive polymeric paclitaxel conjugate based on hyaluronic acid. *Acta Biomater.* **2015**, *26*, 274–285. [CrossRef] [PubMed]
138. Weaver, B.A. How Taxol/paclitaxel kills cancer cells. *Mol. Biol. Cell* **2014**, *25*, 2677–2681. [CrossRef] [PubMed]
139. Lee, H.; Lee, K.; Park, T.G. Hyaluronic Acid−Paclitaxel Conjugate Micelles: Synthesis, Characterization, and Antitumor Activity. *Bioconj. Chem.* **2008**, *19*, 1319–1325. [CrossRef]
140. Wickens, J.M.; Alsaab, H.O.; Kesharwani, P.; Bhise, K.; Amin, M.C.I.M.; Tekade, R.K.; Gupta, U.; Iyer, A.K. Recent advances in hyaluronic acid-decorated nanocarriers for targeted cancer therapy. *Drug Discov. Today* **2017**, *22*, 665–680. [CrossRef] [PubMed]
141. Li, J.; Huang, P.; Chang, L.; Long, X.; Dong, A.; Liu, J.; Chu, L.; Hu, F.; Liu, J.; Deng, L. Tumor targeting and pH-responsive polyelectrolyte complex nanoparticles based on hyaluronic acid-paclitaxel conjugates and Chitosan for oral delivery of paclitaxel. *Macromol. Res.* **2013**, *21*, 1331–1337. [CrossRef]
142. Thorn, C.F.; Oshiro, C.; Marsh, S.; Hernandez-Boussard, T.; McLeod, H.; Klein, T.E.; Altman, R.B. Doxorubicin pathways: Pharmacodynamics and adverse effects. *Pharm. Genom.* **2011**, *21*, 440–446. [CrossRef]
143. Cai, S.; Thati, S.; Bagby, T.R.; Diab, H.-M.; Davies, N.M.; Cohen, M.S.; Forrest, M.L. Localized doxorubicin chemotherapy with a biopolymeric nanocarrier improves survival and reduces toxicity in xenografts of human breast cancer. *J. Control. Release* **2010**, *146*, 212–218. [CrossRef]
144. Florea, A.-M.; Büsselberg, D. Cisplatin as an Anti-Tumor Drug: Cellular Mechanisms of Activity, Drug Resistance and Induced Side Effects. *Cancers* **2011**, *3*, 1351–1371. [CrossRef] [PubMed]
145. Venditto, V.J.; Simanek, E.E. Cancer Therapies Utilizing the Camptothecins: A Review of the in Vivo Literature. *Mol. Pharm.* **2010**, *7*, 307–349. [CrossRef] [PubMed]
146. Pizzolato, J.F.; Saltz, L.B. The camptothecins. *Lancet* **2003**, *361*, 2235–2242. [CrossRef]
147. Serafino, A.; Zonfrillo, M.; Andreola, F.; Psaila, R.; Mercuri, L.; Moroni, N.; Renier, D.; Campisi, M.; Secchieri, C.; Pierimarchi, P. CD44-targeting for antitumor drug delivery: A new SN-38-hyaluronan bioconjugate for locoregional treatment of peritoneal carcinomatosis. *Curr. Cancer Drug Targets* **2011**, *11*, 572–585. [CrossRef] [PubMed]
148. Montagner, I.M.; Merlo, A.; Zuccolotto, G.; Renier, D.; Campisi, M.; Pasut, G.; Zanovello, P.; Rosato, A. Peritoneal Tumor Carcinomatosis: Pharmacological Targeting with Hyaluronan-Based Bioconjugates Overcomes Therapeutic Indications of Current Drugs. *PLoS ONE* **2014**, *9*, e112240. [CrossRef] [PubMed]
149. Di, L.; Kerns, E.H. Formulation. In *Drug-Like Properties Concepts, Structure Design and Methods from ADME to Toxicity Optimization*; Di, L., Kerns, E.H., Eds.; Academic Press: Cambridge, MA, USA, 2016; pp. 497–510.
150. Sharma, N.; Baldi, A. Exploring versatile applications of cyclodextrins: An overview. *Drug Deliv.* **2016**, *23*, 739–757. [CrossRef]
151. Çırpanlı, Y.; Allard, E.; Passirani, C.; Bilensoy, E.; Lemaire, L.; Çalış, S.; Benoit, J.-P. Antitumoral activity of camptothecin-loaded nanoparticles in 9L rat glioma model. *Int. J. Pharm.* **2011**, *403*, 201–206. [CrossRef] [PubMed]

152. Çirpanli, Y.; Bilensoy, E.; Doğan, A.L.; Çaliş, S. Comparative evaluation of polymeric and amphiphilic cyclodextrin nanoparticles for effective camptothecin delivery. *Eur. J. Pharm. Biopharm.* **2009**, *73*, 82–89. [CrossRef] [PubMed]
153. Ghalandarlaki, N.; Alizadeh, A.M.; Ashkani-Esfahani, S. Nanotechnology-Applied Curcumin for Different Diseases Therapy. *BioMed Res. Int.* **2014**, *2014*, 394264. [CrossRef]
154. Zhang, L.; Man, S.; Qiu, H.; Liu, Z.; Zhang, M.; Ma, L.; Gao, W. Curcumin-cyclodextrin complexes enhanced the anti-cancer effects of curcumin. *Environ. Toxicol. Pharmacol.* **2016**, *48*, 31–38. [CrossRef]
155. Jing, J.; Szarpak-Jankowska, A.; Guillot, R.; Pignot-Paintrand, I.; Picart, C.; Auzély-Velty, R. Cyclodextrin/Paclitaxel Complex in Biodegradable Capsules for Breast Cancer Treatment. *Chem. Mater.* **2013**, *25*, 3867–3873. [CrossRef]
156. Yu, S.; Zhang, Y.; Wang, X.; Zhen, X.; Zhang, Z.; Wu, W.; Jiang, X. Synthesis of Paclitaxel-Conjugated β-Cyclodextrin Polyrotaxane and Its Antitumor Activity. *Angew. Chem. Int. Ed.* **2013**, *52*, 7272–7277. [CrossRef] [PubMed]
157. Yusheng, S.; Chenjun, M.; Yingying, H.; Tiantian, W.; Liefeng, Z. Multifunctional nanoparticles of paclitaxel and cyclodextrin–polypeptide conjugates with in vitro anticancer activity. *Pharm. Dev. Technol.* **2020**, *25*, 1071–1080. [CrossRef] [PubMed]
158. Mishra, B.; Suneetha, V.; Ramalingam, C. An overview of Mechanistic Characterization and optimization of Pullulan producing microorganism. *South Asian J. Exp. Biol.* **2011**, *1*, 147–151. [CrossRef]
159. Mishra, B.; Vuppu, S.; Rath, K. The Role of Microbial Pullulan, a Biopolymer in Pharmaceutical Approaches: A Review. *J. Appl. Pharm. Sci.* **2011**, *1*, 45–50.
160. Jeans, A. Dextrans and pullulans: Industrially significant α-d-glucans. In *Encyclopedia of Polymer Science and Technology*; Jeans, A., Ed.; John Wiley and Sons Inc: New York, NY, USA, 1966; Volume 4, pp. 819–821.
161. Yuen, S. Pullulan and Its Applications. *Process Biochem.* **1974**, *22*, 7–9.
162. Gao, F.; Li, L.; Liu, T.; Hao, N.; Liu, H.; Tan, L.; Li, H.; Huang, X.; Peng, B.; Yan, C.; et al. Doxorubicin loaded silica nanorattles actively seek tumors with improved anti-tumor effects. *Nanoscale* **2012**, *4*, 3365–3372. [CrossRef]
163. Posner, L.E.; Dukart, G.; Goldberg, J.D.; Bernstein, T.; Cartwright, K. Mitoxantrone: An overview of safety and toxicity. *Investig. New Drugs* **1985**, *3*, 123–132. [CrossRef]
164. Tao, X.; Tao, T.; Wen, Y.; Yi, J.; He, L.; Huang, Z.; Nie, Y.; Yao, X.; Wang, Y.; He, C.; et al. Novel Delivery of Mitoxantrone with Hydrophobically Modified Pullulan Nanoparticles to Inhibit Bladder Cancer Cell and the Effect of Nano-drug Size on Inhibition Efficiency. *Nanoscale Res. Lett.* **2018**, *13*, 345. [CrossRef]
165. Falvo, E.; Malagrinò, F.; Arcovito, A.; Fazi, F.; Colotti, G.; Tremante, E.; Di Micco, P.; Braca, A.; Opri, R.; Giuffrè, A.; et al. The presence of glutamate residues on the PAS sequence of the stimuli-sensitive nano-ferritin improves in vivo biodistribution and mitoxantrone encapsulation homogeneity. *J. Control. Release* **2018**, *275*, 177–185. [CrossRef]
166. Singh, R.S.; Kaur, N.; Kennedy, J.F. Pullulan and pullulan derivatives as promising biomolecules for drug and gene targeting. *Carbohydr. Polym.* **2015**, *123*, 190–207. [CrossRef] [PubMed]
167. Ganeshkumar, M.; Ponrasu, T.; Subamekala, M.K.; Janani, M.; Suguna, L. Curcumin loaded on pullulan acetate nanoparticles protects the liver from damage induced by DEN. *RSC Adv.* **2016**, *6*, 5599–5610. [CrossRef]
168. Sarika, P.; James, N.R.; Nishna, N.; Kumar, P.A.; Raj, D.K. Galactosylated pullulan–curcumin conjugate micelles for site specific anticancer activity to hepatocarcinoma cells. *Colloids Surf. B Biointerfaces* **2015**, *133*, 347–355. [CrossRef] [PubMed]

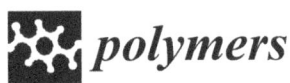

polymers

MDPI

Review

Polymeric Nanoparticles in Brain Cancer Therapy: A Review of Current Approaches

Chad A. Caraway [1], Hallie Gaitsch [1,2], Elizabeth E. Wicks [1,3], Anita Kalluri [1], Navya Kunadi [1] and Betty M. Tyler [1,*]

[1] Hunterian Neurosurgical Research Laboratory, Department of Neurosurgery, Johns Hopkins University School of Medicine, Baltimore, MD 21205, USA; ccarawa2@jhmi.edu (C.A.C.); hgaitsc1@jhmi.edu (H.G.); ewicks2@jhmi.edu (E.E.W.); akallur1@jhmi.edu (A.K.); navyakunadi@gmail.com (N.K.)

[2] NIH-Oxford-Cambridge Scholars Program, Wellcome—MRC Cambridge Stem Cell Institute and Department of Clinical Neurosciences, University of Cambridge, Cambridge CB2 1TN, UK

[3] University of Mississippi School of Medicine, University of Mississippi Medical Center, Jackson, MS 39216, USA

* Correspondence: btyler@jhmi.edu; Tel.: +1-410-502-8197

Abstract: Translation of novel therapies for brain cancer into clinical practice is of the utmost importance as primary brain tumors are responsible for more than 200,000 deaths worldwide each year. While many research efforts have been aimed at improving survival rates over the years, prognosis for patients with glioblastoma and other primary brain tumors remains poor. Safely delivering chemotherapeutic drugs and other anti-cancer compounds across the blood–brain barrier and directly to tumor cells is perhaps the greatest challenge in treating brain cancer. Polymeric nanoparticles (NPs) are powerful, highly tunable carrier systems that may be able to overcome those obstacles. Several studies have shown appropriately-constructed polymeric NPs cross the blood–brain barrier, increase drug bioavailability, reduce systemic toxicity, and selectively target central nervous system cancer cells. While no studies relating to their use in treating brain cancer are in clinical trials, there is mounting preclinical evidence that polymeric NPs could be beneficial for brain tumor therapy. This review includes a variety of polymeric NPs and how their associated composition, surface modifications, and method of delivery impact their capacity to improve brain tumor therapy.

Keywords: brain; cancer therapy; drug delivery; glioblastoma multiforme; nanoparticles; polymers

Citation: Caraway, C.A.; Gaitsch, H.; Wicks, E.E.; Kalluri, A.; Kunadi, N.; Tyler, B.M. Polymeric Nanoparticles in Brain Cancer Therapy: A Review of Current Approaches. *Polymers* **2022**, *14*, 2963. https://doi.org/10.3390/polym14142963

Academic Editors: Stephanie Andrade, Joana A. Loureiro and Maria João Ramalho

Received: 2 June 2022
Accepted: 18 July 2022
Published: 21 July 2022

1. Introduction

Malignant tumors of the central nervous system (CNS), the vast majority of which originate in the brain [1], are the 13th leading cause of cancer mortality worldwide according to GLOBOCAN 2020 [2]. Glioblastoma (GBM) is derived from astrocytes and accounts for 49% of all malignant CNS tumors, making it the most common form of CNS cancer [3]. Despite decades of work aimed at developing new therapies to target GBM, its prognosis remains poor. Worldwide, the median length of survival for GBM patients is approximately 8 months, and even surgical intervention followed by rapid initiation of radiation and chemotherapy standard-of-care treatment only increases the median survival length to 14 months [1,4] with a five-year survival rate of 5–10% [3]. Similarly, the five-year survival rate for individuals with any form of primary malignant brain tumor is just 20% [3]. The challenges associated with successfully treating brain cancers are numerous. Brain tumors often recur following surgical resection. They also frequently occur in areas that are too difficult or dangerous for gross total resection, necessitating the use of alternative or combination treatment strategies. Additionally, most drugs are incapable of crossing the blood–brain barrier (BBB) and blood–brain tumor barrier (BBTB) in sufficient quantities to halt tumor growth. Furthermore, while stereotactic radiosurgery-based approaches are effective in ablating a variety of brain tumors that are visible on MRI and other neuroimaging modalities, these methods are not as effective in treating tumors with high recurrence rates, including metastatic brain cancers and GBM [5].

Numerous approaches aimed at overcoming these limitations have been developed over the years, with variable success [6]. Some of the most promising techniques in development involve the use of nanocarrier systems to bypass the BBB and BBTB, selectively target brain cancer cells, and release anti-cancer compounds into diseased tissue while limiting toxicity to systemic and healthy brain tissue. Nanoparticles (NPs)—carriers ranging from 10–1000 nm in diameter—can be loaded with chemotherapeutic drugs, nucleic acids, antibodies, and other proteins and peptides. These carriers can be engineered from a variety of materials, with metal, lipid, and polymer-based NPs being the most tested in neurological disease research [7]. The general structure of polymeric NPs consists of a core polymer with therapeutic agents either surface-bound or encapsulated and coated with targeting and/or hydrophilic molecules to increase circulation half-life and specific delivery [8].

In recent years, polymeric NPs developed for CNS tumor treatment have been modified with various moieties capable of interacting with the BBB and tumor cells. Appropriate subtypes can be selected by assessing the polymer pharmacokinetics, modifiability, and payload delivery best suited for any given study. Additionally, polymers can be engineered to form other nanomaterials that may be useful for therapeutic development. For example, a recent review detailed the ability to produce optical nanofibers from polymers that can aid in phototherapy, drug delivery, sensing, and more [9]. Others have explored the potential of novel noninvasive delivery methods, like loading polymeric NPs into neutrophils or monocytes, to enhance their transport to brain tumors [10]. Though many challenges remain, mounting evidence from promising preclinical studies utilizing a variety of approaches suggests polymeric NPs may prove effective in treatment of CNS malignancies. This review provides a summary of the major types of polymeric NPs used in brain cancer studies, the ways in which these NPs can be modified and delivered in order to bypass the BBB, and strategies for specific targeting of NPs to cancer cells.

2. Major Polymers in Nanoparticle-Based Brain Cancer Research

2.1. Polyanhydride

Polyanhydride is formed by carboxylic acid polymerization with anhydride linkages [11] and is an exceptionally well-characterized and proven biocompatible and biodegradable polymer for cancer therapy. Before discussing the current state of polyanhydride in nanomedicine, it is imperative to address its role in current therapeutic treatment for GBM and other primary brain malignancies. The 1,3-bis(2-chloroethyl)-1-nitrosourea (BCNU) wafer, more commonly known as the Gliadel® wafer, is a biodegradable carmustine-loaded polyanhydride-based implant that was first approved by the US Federal Drug Administration (FDA) in 1996 for patients with recurrent GBM as an adjunct to surgery [12]. In 2003, that approval was expanded to treatment for patients with newly diagnosed high-grade malignant gliomas as an adjunct to surgery and radiation therapy [12].

Gliadel® wafers were the first FDA-approved method of delivering local, sustained-release chemotherapy to brain tumors [13]. In combination with temozolomide and radiotherapy, Gliadel® wafers have become part of the gold standard for treatment of GBM and have significantly improved median patient survival times [14]. A 2022 observational study of 506 malignant glioma patients receiving adjuvant treatment with Gliadel® wafers reported a median overall survival of 18.0 months, with 39.8% and 31.5% of patients surviving two and three years, respectively [15]. Building on this progress, additional approaches utilizing NP-based formulations encapsulating chemotherapeutic drugs and other compounds have been increasingly studied in an attempt to further improve brain tumor treatments and outcomes.

With the success of the Gliadel® wafer, polyanhydride NPs have been investigated as a potential platform for novel brain cancer therapeutics. Polyanhydride NPs are highly tunable and their degradation profile can be modified from days to months, depending on their copolymer composition [16]. Like other polymeric NPs, surface ligands can also be added to polyanhydrides to enhance targeted delivery [17]. Polyanhydrides are highly hydrophobic, which allows their rate of release and erosion to be largely constant and predictable [18]. However, their erosion characteristics may limit their targeting potential due to inadequate ligand retention times [19]. Additionally, polyanhydrides are relatively difficult to synthesize compared to other polymers utilized in nanomedicine and their potential to acylate nucleophiles can result in limited stability of loaded peptides and proteins [20]. Nevertheless, Brenza et al. reported that polyanhydride NPs could successfully cross the BBB via cell-based delivery and transcytosis [21]. While further studies examining polyanhydride NPs as effective carriers for brain tumor therapy have been rather limited, these findings combined with the proven history of Gliadel® wafers suggest polyanhydrides could be an effective platform for nanomedicine.

2.2. Poly (lactic-co-glycolic acid)

Poly (lactic-co-glycolic acid) (PLGA) is an FDA and European Medicine Agency (EMA)-approved biodegradable anionic polymer that is widely used to encapsulate chemotherapy drugs, anti-inflammatory drugs, antibiotics, and proteins for the treatment of a variety of conditions [7,22]. In addition to its frequent use in microsphere and microparticle drug delivery systems [23], PLGA is a common platform for NP-based therapies due to the unique properties of PLGA NPs, including their simple biodegradability, relative ease of synthesis, tunability, commercial availability, sustained drug-release properties, and biocompatibility [24]. A recent study by Maksimenko et al. utilizing doxorubicin-loaded PLGA NPs coated with poloxamer 188, which is a copolymer surfactant capable of repairing function in damaged cells [25], reported the carriers could penetrate both the BBB and intracranial tumors to result in significant anti-tumor efficacy in vivo [25]. Another group reported that PLGA NPs loaded with the chemotherapeutic drug morusin and conjugated to chlorotoxin, which is a peptide specific for particular chloride channels expressed in glioma cells, resulted in significant anti-tumor effects against two human glioblastoma cell lines in vitro [26]. Similar in vitro findings using various modified forms of PLGA NPs have been published in recent years [27–29], and studies investigating their use against brain tumor-bearing rodent models have also been largely encouraging [30,31]. There is considerable literature documenting the potential of PLGA-based nanocarrier systems to treat CNS tumors via specific surface modifications and loading contents. A general overview of these tunable features for PLGA and other polymeric NPs is shown in Figure 1.

PLGA NPs are degraded into lactate and glycolate, compounds that can be further metabolized through the Krebs cycle [32]. The overall hydrophobicity and degradation rate of PLGA copolymers depends on their ratio of poly (glycolic acid) (PGA), which is hydrophilic, to poly (lactic acid) (PLA), a hydrophobic polymer [32]. Increased hydrophobicity results in a slower rate of degradation and, as a result, a slower rate of drug release [32,33]. PLGA polymers are negatively charged; therefore, cellular uptake through negatively charged cell membranes is limited. However, surface modifications can result in neutral or positively charged PLGA NPs that may penetrate the blood–brain barrier more effectively [7,34]. One challenge in the use of PLGA NPs in therapeutic contexts is their poor drug loading efficiency [34]. Another challenge to the use of PLGA NPs is that high burst release from PLGA NPs, which may be due to drug adsorption to NP surfaces [22], results in low levels of drugs reaching target cells or tissues [34]. Additionally, the production of acids following degradation—a common drawback of biodegradable polymers—can destabilize acid-sensitive drugs and peptides carried in PLGA NPs [35], though there have been many efforts to limit this issue [24,35].

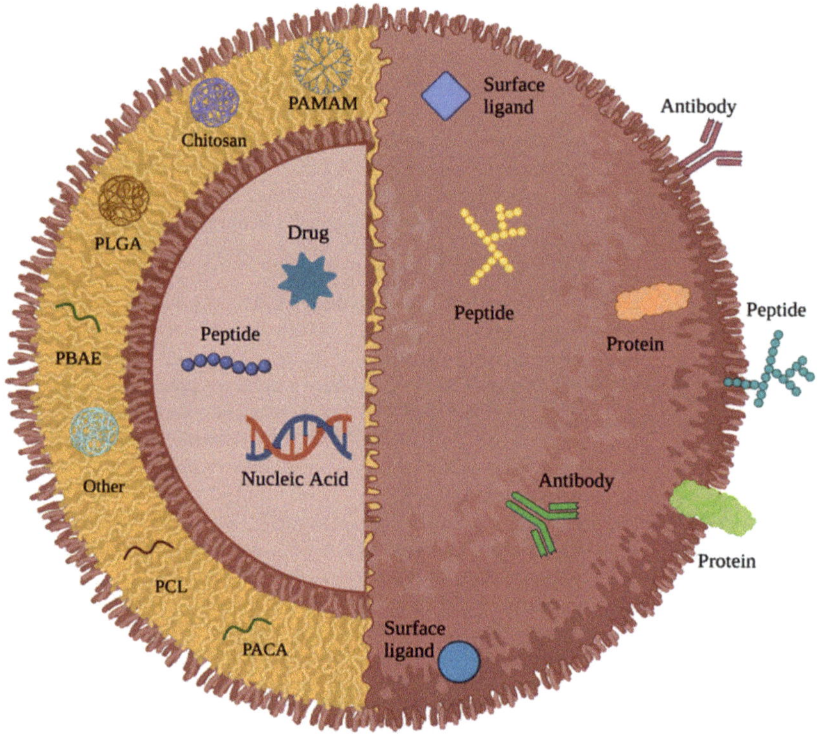

Figure 1. General structure of a polymeric nanoparticle.

2.3. Poly (β-amino ester)

Poly (β-amino ester) (PBAE) is an easily synthesized, biodegradable, and biocompatible cationic polymer commonly used to construct NPs that can deliver polynucleotides and other acid-labile compounds [36]. PBAE NPs are uniquely suited to carrying these types of cargo due to their high efficacy [37] and polyamine nature, which allows these polymers to act as pH buffers [38]. This buffering capacity has been found to enhance the ability of PBAE NPs to escape from endosomes following endocytosis, allowing expression of nucleic acids within target cells [39]. PBAE polymers have an established safety profile [36] and also maintain low cytotoxicity compared to other cationic polymers such as polyethylenimine (PEI) [40] due to their rapid degradation under physiological conditions, which further enhances nucleic acid delivery from NPs to cells [36,37]. Although clinical usage of PBAEs has historically been limited due to their rapid hydrolysis and substantial cationic properties—both of which contribute to PBAE polymer instability in the blood [41]—surface modification studies have attempted to mitigate this disadvantage [42]. Additionally, an extensive library of more than 2300 PBAE polymers was introduced by Anderson et al. in 2003 to determine ideal polymers for targeting particular cell types and tissues [43] and newer libraries have since been developed to expand upon these findings [40,44]. However, while some studies have shown that PBAE NPs can preferentially transfect GBM cells over normal brain tissue in vitro and in vivo [44], others have reported that PBAE and other cationic NPs encounter adhesive interactions with the extracellular matrix (ECM) that limit their ability to achieve widespread gene transfer in brain tumor tissue [45].

2.4. Chitosan

Chitosan is a biodegradable polymer created by deacetylation of the widely abundant, naturally occurring polymer chitin [46]. Given its primary and secondary hydroxyl groups, and its amino group, a range of structures can be derived from chitosan by N-linked and O-linked modifications [47]. NPs constructed using chitosan and its derivatives often possess mucoid and cationic properties that facilitate adherence to mucous membranes and sustained drug release [46]. Furthermore, the adherent properties of chitosan allow it to pass transcellularly through the endothelium and epithelium, making it a promising candidate for crossing the BBB [48] and for nose-to-brain delivery [46] via chitosan-based polymeric NPs. While one potential drawback of chitosan-based drug delivery is its low solubility at physiological pH, this may allow for preferential release in a tumor acidic environment or in intracellular endosomes [47].

Chitosan can also be used as a polymer coating and in hybrid NP carriers. In vitro and in vivo studies have demonstrated the potential for these NP drug delivery systems for GBM. Shevtsov et al. reported that hybrid chitosan-dextran superparamagnetic NPs demonstrated enhanced internalization in U87 and C6 glioma cells compared to those coated with dextran alone, and further demonstrated accumulation of these particles in orthotopic C6 gliomas in rats [49]. Successful tumor growth reduction via magnetically guided delivery of folate-grafted, chitosan-coated magnetic NPs containing doxorubicin to human U87 GBM cells in a subcutaneous tumor model in mice has also been demonstrated [50].

2.5. Poly(amidoamine) Dendrimers

Poly(amidoamine) (PAMAM) dendrimers are flexible, non-toxic, and biocompatible branched NPs that were first synthesized in 1985 [51]. The structure of PAMAMs consists of an initiator core that anchors dendrimer growth, interior dendrimer layers and branches, and terminal functionalized branches in an outer layer [52] that gives PAMAMs an extremely high surface-area-to-volume ratio [51]. Advantages of PAMAMs in drug delivery applications include their high solubility, stability, small size, and presence of readily modifiable surfaces [53]. While PAMAMs have been associated with cytotoxicity, their cytotoxic effects can be modulated by selective modifications to the terminal functional branches in the outer layer [51].

The cationic nature of PAMAMs, coupled with their large hydrophilic surface area, has made them of particular interest in drug delivery [52]. Sarin et al. demonstrated successful crossing of the BBTB by functionalized dendrimers with diameters of less than 11.7 to 11.9 nm in rodents with orthotopic RG-2 malignant glioma, and further noted the accumulation of dendrimers with long half-lives within glioma cells [54]. Additionally, Moscariello et al. demonstrated that a PAMAM dendrimer bioconjugate with streptavidin adapter was capable of transcytosis across the BBB both in vitro and in vivo [55].

2.6. Poly(caprolactone)

Poly(caprolactone) (PCL) is a biodegradable and non-toxic polymer characterized as a semi-crystalline aliphatic polyester that is obtained from a monomer ε-caprolactone ring opening [56]. PCL is a promising, FDA-approved polymer in the development of NP therapies. This is primarily due to its versatility as the combination of PCL with other polymers directly influences its crystallinity, solubility, and rate of degradation, allowing PCL-based drug delivery systems to be utilized for a variety of different approaches [56]. One study highlighting the potential of PCL utilized paclitaxel-loaded PCL NPs conjugated to Angiopep-2, which is a ligand that binds to the low-density lipoprotein receptor related protein (LRP) [57]. LRP is overexpressed on BBB and glioma cells [58], and Xin et al. reported significantly higher penetration, distribution, accumulation, and anti-glioblastoma efficacy in tumor-bearing mice compared to NPs without Angiopep-2 [57].

PCL is a highly stable polymer, due in part to its strong hydrophobicity and crystallinity [59], that requires two to four years for complete degradation [60]. The two stages of PCL degradation include non-enzymatic breakage of ester linkages, followed by enzymatic fragmentation [59]. While this process is slow, hydrophilic polymers can be added to shorten the degradation time [61]. PCL permits modification of its physical, chemical, and ionic properties; therefore, it can be designed to fit the intended properties for specific drug deliveries, ultimately improving therapeutic efficacy. Although most of the current formulations used in drug delivery are satisfied by these PCL modifications, the hydrophobicity of PCL is a drawback that limits its use [59].

2.7. Poly(alkyl cyanoacrylate)

Poly(alkyl cyanoacrylate) (PACA) NPs are typically made from alkyl cyanoacrylate monomers and their mixtures through a one-step mini-emulsion process that results in varying levels of PACA particle degradability [62]. Particles with longer alkyl chains typically degrade at a slower rate [63]. PACA NPs have shown significant promise in delivering drugs across the blood–brain barrier, as well as in infiltrating solid tumor structures [64]. The intracellular drug availability provided by PACA delivery is primarily affected by the degradation of its NPs, which occurs through surface erosion and hydrolysis of esters (with or without esterases), allowing for the release of hydrophobic drugs [65]. PACA has been studied in a variety of physicochemical environments and several studies have utilized PACA-based NPs for hydrogels and delivering nucleic acids and peptides in vivo [66]. One such study by Baghirov et al. reported their novel PACA NPs could be transported across the BBB and into brain tissue with ultrasound-mediated delivery [67]. Similarly, Andrieux et al. summarized several different PACA NP-based approaches and concluded they could readily cross the BBB in animal and cell models with appropriate surface modifications [68]. One such modification is the addition of polysorbate 80, a surfactant that may enhance NP delivery via transcytosis across the BBB [69]. Additionally, the biodegradability of PACA allows for continuous drug delivery rather than in bursts, which are present in traditional cancer treatment methods such as chemotherapy [66]. Furthermore, PACA NPs are reportedly capable of overcoming multidrug resistance, which allows tumors to resist chemotherapeutic drugs such as doxorubicin due to P-glycoprotein overexpression [70]. The mechanistic explanation for this involves the formation of an ion pair between PACA degradation products and doxorubicin [68]. This suggests the PACA polymer base could be a suitable choice for chemotherapeutic approaches in the field of neurosurgery.

3. General Modifications

3.1. Polyethylene Glycol

Polyethylene glycol (PEG) is a hydrophilic polymer that can be covalently attached to NPs and other therapeutics to increase their systemic circulation time [71]. PEG is classified as Generally Regarded as Safe (GRAS) by the FDA and several protein therapeutics coated with PEG, or "PEGylated", have been FDA approved since 1990 [72]. The conjugation of PEG to NP surfaces reduces their recognition by immune cells through minimizing protein adsorption via steric hindrance, thereby increasing bioavailability [8]. More specifically, PEGylated NPs have been shown to exhibit fewer surface interactions with plasma proteins and cell membranes than non-PEGylated controls, meaning they are more resistant to aggregation, opsonization, and phagocytosis [71,72]. Consistent with its FDA categorization, PEGylation of NPs has not been shown to increase toxicity [7,8] and is one of the most popular modifications used to enhance the effects of nanotherapeutics. A table summarizing the utilization of PEG, other common modifications, and polymer subtypes above can be seen in Table 1.

Table 1. Comparison of polymer subtypes and modifications.

Polymer Type	Common Synthesis Techniques	Advantages	Disadvantages	Specific Uses Cited
Polyanhydride	Most often via polycondensations from diacids or diacyl anhydrides; can also be prepared via solvent evaporation from emulsion, or thiol-ene 'click' polymerization, or melt condensation; NP synthesis via nanoprecipitation	Well-characterized; biocompatible; biodegradable; modifiable (depending on copolymer and surface ligand composition); hydrophobic; predictable rate of release/erosion	Rapid erosion can lead to inadequate ligand retention times; difficult to synthesize; limited stability of loaded peptides and proteins due to nucleophile acylation	Gliadel® (BCNU) wafer for local, sustained-release chemotherapy [14]; drug delivery across the BBB [21]; delivery of non-proteinaceous cargo [16]
Poly (lactic-co-glycolic acid)	Co-polymerization of cyclic dimers of glycolic acid and lactic acid; NP synthesis via emulsification-evaporation, nanoprecipitation, phase-inversion, and solvent diffusion; emulsification-evaporation and nanoprecipitation are most commonly used when loading hydrophobic moieties	Widely used; biocompatible; simple biodegradability; easily synthesized; modifiable charge, hydrophobicity, and degradation rate; sustained drug-release; good BBB/tumor penetration	Poor drug loading efficiency; poor drug target delivery efficiency due to high burst release; destabilization of acid-sensitive drugs/peptides	Encapsulation of chemotherapeutics with toxicity profiles indicating sustained, low dosing [7]; microsphere and microparticle drug delivery systems [23]
Poly (β-amino ester)	Conjugate addition of amines to bis(acrylamides) and copolymerization; NP synthesis via solvent/anti-solvent formulation	Established safety profile; biocompatible; biodegradable; easily synthesized; high efficacy; pH buffering capacity; able to escape endosomes and allow intracellular expression of nucleic acids	Instability in blood (rapid hydrolysis) without surface modifications; limited ability to achieve widespread gene transfer due to adhesive interactions with ECM	Delivery of polynucleotides and other acid-labile compounds [36]; delivery of nucleic acids to cells [44]
Chitosan	Enzymatic or chemical deacetylation of chitin, usually through hydrolysis, produces chitosan; NP synthesis via emulsification and crosslinking, microemulsion, precipitation, or ionic gelation	Biodegradable; capable of mucous membrane adherence and transcytosis; sustained drug release; putative preferential release in tumor acidic environment	Low solubility at physiological pH; tendency to aggregate	Nose-to-brain delivery (via mucous membrane adherence) [46]; in situ gelation [73]; tumor targeting via differential pH [47]
Poly(amidoamine) dendrimers	Convergent (beginning with exterior and adding end groups while working towards the core) or divergent synthesis (beginning with core and adding end groups towards the exterior); end group additions via conjugate addition	Biocompatible; flexible, non-toxic; stable; highly soluble; small; modifiable; large hydrophilic surface area; presence of cavities; resistance to denaturation after freezing/thawing	Associated with (modifiable) cytotoxicity; synthesis can lead to heterogeneous mixture of dendrimers unless additional purification steps are completed	Precision-targeting [52]; delivery across the BBB [54]; encapsulating particularly insoluble contents [53]

Table 1. *Cont.*

Polymer Type	Common Synthesis Techniques	Advantages	Disadvantages	Specific Uses Cited
Poly(caprolactone)	Polycondensation of 6-hydroxyhexanoic acid, or ring-opening polymerization of ε-caprolactone; NP synthesis via nanoemulsification, supercritical fluid extraction of emulsion, or solvent evaporation	Biodegradable; non-toxic; modifiable; stable	High hydrophobicity (slow degradation rate of months/years)	Combination with other copolymers to tailor NP suitability to cargo [56]
Poly(alkyl cyanoacrylate)	Free radical, anionic, and zwitterionic polymerization; NP synthesis via polymerization in aqueous acidic phase or through interfacial emulsion polymerization	Biodegradable; modifiable; enhanced intracellular penetration; capable of overcoming multidrug resistance	BBB translocation ability remains controversial	Hydrogel-incorporated drug delivery [66]; delivery of nucleic acids and peptides [66]; continuous drug delivery (vs. bursts) [66]; instances of multidrug resistance [70]
Polymer Modification		**Advantages**	**Disadvantages**	**General Uses**
Polyethylene glycol		Widely used; classified as GRAS; increases systemic circulation time of NPs; reduces recognition of NPs by immune cells; decreases NP aggregation, opsonization, and phagocytosis	Reduced cellular uptake of PEGylated NPs	Modify NP to reduce immunogenicity
pH		Can improve selective tumor targeting via triggered drug release	Limits the types of cargo able to be carried within the NP	Modify NP to selectively target tumor tissue and spare surrounding parenchyma
Size		Can increase NP stability; can potentially increase BBB/BBTB penetration and brain parenchymal spread	Conflicting in vitro/in vivo results on ideal size of NPs for BBB/BBTB penetrance, brain tissue spread, and cellular uptake	Modify NP to increase intra-tumoral spread
Shape		Can modulate NP circulation time, cellular uptake, and BBB penetration	Certain shapes promote accumulation in non-target organs; ideal shape, depending on delivery mechanism, requires further investigation	Modify NP to maximize efficacy based on delivery mechanism (e.g., nose-to-brain vs. across BBB)

3.2. *pH*

The extracellular pH in solid tumors is more acidic compared to normal tissue [74]. Normal tissues maintain a pH of ~7.4, whereas tumor microenvironments exhibit a pH of ~6.5 and can thereby be targeted by pH-responsive nanocarrier modifications and triggered drug release [75]. Development of tumor pH-sensitive drug release systems has been studied extensively. Research has shown incorporation of weak acids and other pH-sensitive compounds into polymeric NPs can result in drug carriers that are stable at physiological pH while destabilized and precipitated in the tumor microenvironment, resulting in drug delivery [76]. One group investigating NPs for brain tumor therapy attached $H_7K(R_2)_2$, a pH responsive peptide, to the surface of PLGA-based, PEGylated NPs in order to enhance targeting of malignant glioma cells in vivo [77]. While $H_7K(R_2)_2$, remained unexposed under physiological conditions due to hydrophobic interactions between PLGA and the H_7 residues, the acidic tumor environment protonated the imidazole ring of H_7, thus making the $H_7K(R_2)_2$ more hydrophilic and selectively exposing the cell-penetrating ligand to glioma cells [77]. Similar approaches have been used to modify various polymer cores to selectively "activate" NPs in acidic environments [76,78]. Additionally, multiple studies have explored how utilizing pH-responsive nanomaterials can enhance selective targeting of brain tumor cells by polymeric NPs [77]. Laboratories have investigated how other stimuli, such as temperature, redox gradients, enzyme concentration, and magnetic field, can be exploited alongside pH to enhance nanocarrier delivery and release in tumor environments [78]. Taken together, designing polymeric NPs that take advantage of tumor-specific stimuli to enhance targeted delivery and release may prove to be a viable strategy for CNS tumor therapy.

3.3. Size

While NPs are 10–1000 nm in diameter by definition, the optimal NP size for clinical brain tumor therapy remains unclear for a variety of reasons. Size plays a key role in NP stability, ability to pass through the BBB/BBTB and spread throughout the brain parenchyma, and likelihood of being endocytosed for cell-mediated delivery [7,71]. Systemically administered NPs with diameters <5 nm are cleared via renal filtration, while larger NPs (>200 nm in diameter) cannot effectively reach the BBB due to splenic sequestration [7]. Fortunately, polymeric NPs can be engineered to precise size specifications [79]. One study of PEG-coated PLGA NPs found that 100 nm particles had longer circulation time and enhanced penetration of the brain parenchyma compared to 200 nm and 800 nm particles in a traumatic brain injury mouse model [80]. Another study utilizing PEG and vitamin E-coated polystyrene NPs of various sizes reported the smallest diameters had the highest brain uptake levels (25 > 50 > 100 > 500 nm) in a rat model [81]. However, conversely, Nowak et al. found spherical polystyrene NPs with diameters of 200 nm crossed the BBB more effectively than 100 and 500 nm spheres in a microfluidic model [82]. Furthermore, other studies have found NP size to have limited or no effect on their ability to penetrate the BBB [83]. These contrasting findings highlight the most common challenges in developing brain tumor therapeutics, as the BBB, BBTB, extracellular space, pore size, and overall physiology varies substantially across in vivo and in vitro models, especially compared to humans [71,82].

NP size also impacts their ability to spread throughout the brain parenchyma, thus affecting therapeutic delivery to tumor regions after crossing the BBB [71]. Similar to BBB studies, accurately replicating the human brain parenchyma and extracellular space remains a significant challenge in bringing NP therapeutics to the clinical trials [71]. Thorne et al. published findings in 2006 suggesting the extracellular space pores in rat brains is up to 64 nm, meaning larger NPs would be unable to further penetrate brain tissue after crossing the BBB [84]. However, Nance et al. later reported in 2012 that larger particles (e.g., up to 114 nm) could still diffuse through the brain extracellular space in both rats and humans with PEG or carboxyl moiety (COOH) coatings [85]. Furthermore, Thorne et al. also suggests that there may be around 25% or more pores in the human brain extracellular space with diameters equal to or larger than 100 nm, with some even exceeding 200 nm [84]. These findings have made clear that more research is required to elucidate the ideal NP size for delivery throughout brain tissue.

3.4. Shape

Polymeric NPs can be engineered in a variety of shapes, though spherical particles are the most studied form. Shape has been shown to influence NP pharmacokinetics, cellular uptake, and BBB penetration [86,87]. Spherical NPs may more readily interact with cellular surfaces and thereby have enhanced uptake and clearance by the spleen before reaching tumor regions compared to long, cylindrical NPs, as summarized by Truong et al. [86]. Similarly, Christian et al. reported enhanced circulation time of flexible filamentous micelles ("filomicelles") compared to spherical micelles in mice [87]. Furthermore, both micelle forms were loaded with paclitaxel and mice treated with filomicelles had enhanced tumor shrinkage and tumor cell apoptosis. Other studies have also reported findings of increased circulation time for various rod-shaped NPs, suggesting that shape is an important characteristic in the development of polymeric NPs for effective brain tumor therapy [88,89]. One particular study investigated endothelial uptake of modified polystyrene nanospheres and nanorods in the brain [90]. Kolhar et al. found the nanorods to have a sevenfold higher accumulation in mouse brains, though they also reported increased accumulation of these particles in the lungs, kidneys, heart, and spleen. Similarly, Nowak et al. reported significantly enhanced transport across a BBB microfluidic model of polystyrene rod-shaped NPs compared to spheres [82]. Indeed, while spherical NPs are most popular, considerable debate regarding the ideal shape for CNS treatment remains. Furthermore, it likely also depends on the mechanism of delivery, as approaches not dependent on crossing the BBB may prefer different conformations than those that must travel through the systemic circulation.

4. Receptor Targeting for Blood–Brain Barrier Penetration

Delivering NPs to the brain for tumor therapy remains a significant challenge due to the BBB. In order to cross the BBB, NPs can be engineered to take advantage of transport processes such as adsorptive-mediated transcytosis (AMT), receptor-mediated transcytosis (RMT), and cell-based delivery (described in the Mechanisms of Delivery section below) [91]. AMT involves electrostatic interactions between positively charged ligands and negatively charged brain capillary endothelial cell membranes [92]. While NPs may be able to take advantage of this process, it is unclear if AMT factors significantly into the delivery of endogenous compounds through the BBB [91]. Conversely, targeting endothelial cells present on the BBB for RMT is among the most common approaches to enhance drug delivery to the brain parenchyma [93]. RMT involves receptor-mediated endocytosis on the luminal side of the BBB, followed by trafficking and sorting through endothelial cells, and concluding with the release of contents to the brain parenchyma [94]. Specific targets for RMT include the transferrin receptor (TfR) [95], insulin receptor [96], low density lipoprotein (LDL) receptor (LDLR) [94], melanotransferrin [94], CD98 [97], and various others [7,94,98]. An overview of potential transport pathways for NPs to penetrate the BBB is shown in Figure 2.

Some receptors, like TfR, are also overexpressed on GBM cells and may be suitable targets for enhancing delivery of NPs through both the BBB and ultimately to brain tumor tissue [99]. TfR is the most commonly targeted protein for enhancing delivery of therapeutics through the BBB via RMT [94,95,97]. One reason for this is that TfR is expressed on brain capillary endothelial cells, but not on endothelial cells in other parts of the body [100]. Additionally, there are multiple potential ligands that can be conjugated to NPs for targeting the Tfr, including transferrin (Tf), antibodies, and targeting peptides [101]. Ramalho et al. reported enhanced internalization of TMZ-loaded PLGA NPs coated with monoclonal antibodies for the TfR (OX26 type) in GBM cells through receptor-mediated endocytosis [99]. Another study, conducted by Kuang et al. [102], investigated the use of dendrigraft poly-L-lysine-based NPs conjugated to a peptide capable of targeting TfR on the BBB and glioma cells to deliver RNA and doxorubicin. They reported high tumor targeting efficiency in vitro and increased cellular uptake, slower tumor growth, improved median survival time, and additional anti-tumor effects in vivo.

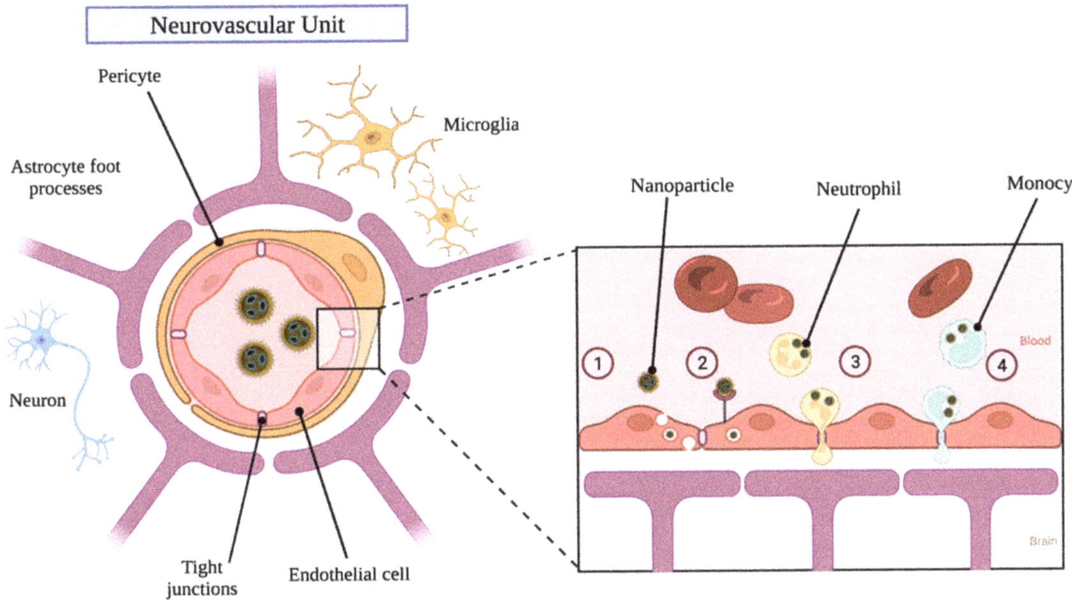

Figure 2. Nanoparticle transport across the blood–brain barrier. (1) Adsorptive-Mediated Transport. (2) Receptor-Mediated Transport. (3) Neutrophil-Mediated Delivery. (4) Monocyte-Mediated Delivery.

Various groups have investigated targeting the insulin receptor, which is expressed on BBB endothelial cells, as well [94]. A study led by Shilo et al. demonstrated five times greater brain localization of insulin-targeted gold NPs compared to controls in a mouse model two hours after intravenous injection [103]. Betzer et al. expanded on these findings to show that similar gold NPs coated with insulin could promote NP localization to specific brain regions in mice [104]. Another group found that anti-insulin receptor antibodies covalently attached to human serum albumin NPs resulted in increased delivery across the BBB compared to controls [105], again suggesting the insulin receptor may be a viable target for BBB penetration.

Conjugating ligands capable of binding LDLR to NPs is another potential strategy for enhancing brain delivery [94] as this molecule is expressed on BBB endothelial cells [98]. Apolipoprotein E and B (ApoE and ApoB) can bind these receptors, and various in vivo and in vitro studies have reported increased diffusion of ApoE-coated NPs across the BBB and BBB models, respectively [106,107]. Other studies have reported successful transcytosis through the BBB of therapeutics containing ApoB fragments as well [108]; however, there is limited data associated with the use of these fragments in NP delivery. Nevertheless, others have demonstrated enhanced delivery of NPs bound to peptides with high affinity for LDLR across the BBB and subsequent glioma localization in vivo [109].

Melanotransferrin, or melanoma tumor antigen p97, binds to LDL receptor related protein 1 (LRP1) and is a type of iron-binding transferrin protein [94]. Karkan et al. [110] demonstrated that covalently linking melanotransferrin to chemotherapy drugs, such as paclitaxel and doxorubicin, could dramatically increase drug delivery to gliomas in vivo compared to controls. Another group had similar results with melanotransferrin conjugated to trastuzumab to treat breast cancer brain metastases compared to controls in mice [111]. These findings suggest that melanotransferrin could be yet another protein that may improve NP RMT through the BBB.

CD98 is a transmembrane, glycoprotein heterodimer composed of CD98 heavy chain (CD98hc) and various CD98 light chains [112] and is expressed in various tissues, including endothelial cells on the BBB, where it functions as an amino acid transporter [113]. CD98 is also overexpressed in tumor and inflammatory cells [114] and has been targeted by various NP formulations to amplify their internalization in diseased tissues [112], though studies targeting CD98 for neurodegenerative disease and brain cancer therapy has been limited. One study led by Zuchero et al. investigated the use of bispecific antibodies targeting CD98hc and β-secretase 1 (BACE1) to reduce amyloid beta production in mice and found that levels of anti-CD98hc/BACE1 in the brain were significantly higher than anti-TfR/BACE1 antibodies following intravenous injection [97]. Others have reported on the enhanced expression of CD98 in astrocytic neoplasms [115], suggesting this complex could be a viable target for NP-based brain tumor therapy. However, while targeting CD98 may be a viable way to enhance RMT of NPs and targeting of cancer cells, the potential associated disruption of amino acid transport across the BBB is a concern that must be studied further [94,113].

5. Receptor Targeting for Delivery to Brain Cancer Cells

5.1. Vascular Endothelial Growth Factor

Vascular endothelial growth factor (VEGF) is a mitogen for endothelial cells and promotes angiogenesis. It is expressed by various cell types, including endothelial cells and several types of tumor cells [116]. While VEGF is a secreted protein, it commonly localizes to cellular membranes and intracellular matrices [117]. Overexpression of VEGF on tumor cells is critical for tumor growth and metastasis through enhanced angiogenesis, and antibodies against VEGF have long been a target for cancer therapies, including many forms of brain cancer [116].

Bevacizumab is a monoclonal antibody against VEGF that has been approved by the FDA for GBM patients whose condition did not improve following treatment with TMZ, a staple chemotherapy drug for treating GBM [31]. Bevacizumab has since been incorporated into several NP-based brain tumor therapy studies, with most focusing on loading NPs with bevacizumab for more effective delivery to tumor cells [31,118]. Some of these studies are promising, including one that reported reduced tumor growth and higher anti-angiogenic effect from bevacizumab-loaded PLGA NPs compared to free drug in mice after 14 days [31]. Others have investigated the potential for conjugating bevacizumab or other VEGF antibodies to NP surfaces for more effective cancer cell targeting [117]. For example, Abakumov et al. used a PEG linker to conjugate VEGF monoclonal antibodies to magnetic NPs for intracranial visualization of glioma cells with MRI in vitro [117]. Similarly, liposomal NPs were conjugated to a novel VEGF monoclonal antibody through a PEG linker and intravenously injected into rats with C6 gliomas by Shein et al. [119]. They reported both specific accumulation of these NPs in tumor tissues and engulfment by glioma cells. Studies utilizing similar NPs loaded with cisplatin and conjugated to antibodies against VEGF and its receptor type II (VEGFR2) have since demonstrated enhanced drug delivery and uptake by glioma cells in vivo [120].

5.2. Epidermal Growth Factor Receptor

Epidermal growth factor receptor (EGFR) is one of four types of receptor tyrosine kinases within the ErbB protein family and is overexpressed in many cancers, including in 40–50% of patients with GBM [71,121]. Several therapeutics for cancers throughout the body targeting EGFR and its variants have been developed over the years, including the FDA-approved EGFR tyrosine kinase inhibitors afatinib and dacomitinib, and monoclonal antibodies against EGFR such as cetuximab, panitumumab, and nimotuzumab [122]. However, many of these therapeutics have not yet been shown to be clinically effective against GBM and other brain cancers, due, in part, to challenges in delivery across the BBB [123]. Westphal et al. published a large summary of various EGFR-targeted therapeutics for GBM in 2017, again citing brain delivery as the most common pitfall [124].

NPs conjugated to EGFR-targeting antibodies or loaded with EGFR inhibitors have been increasingly studied in recent years [71] with the aim of enhancing tumor-specific delivery and inhibition. For example, Mortensen et al. conjugated cetuximab to immuno-liposomal NPs via PEG linkage and reported enhanced uptake and accumulation in an intracranial U87 MG xenograft mouse model [125]. In another study conducted by Erel-Akbaba et al., solid lipid NPs loaded with siRNA against human EGFR were found to exhibit dose-dependent EGFR knockdown when tested in human U87 glioma cells [126]. Furthermore, they found significantly decreased EGFR mRNA in tumor tissue after retro-orbital injection of solid lipid NPs loaded with EGFR siRNA at 8, 9, and 11 days post-GL261 tumor implantation in mice. Other studies have investigated nanosyringes conjugated to EGFR binding peptides for localized delivery specifically to tumor regions in intracranial xenograft mouse models [71]. Additionally, a Phase I clinical trial using weekly administration of novel nanocellular compounds loaded with doxorubicin and conjugated to panitumumab over 8 weeks following standard therapy (radiation and TMZ) demonstrated no dose limiting toxicity in 14 patients with recurrent GBM [121]. Other NP studies have reported similar findings in various rodent and canine models [71], leading to additional clinical trials investigating EGFR-based NPs. These include a recently-completed Phase I study led by groups from EnGeneIC Limited and Johns Hopkins University to investigate EGFR-targeting 400 nm minicells loaded with doxorubicin in the context of GBM; results currently pending [127].

6. Mechanisms of Delivery

6.1. Focused Ultrasound

Focused ultrasound (FUS), is a promising technique that uses a minimally invasive approach to treat a variety of diseases, including those related to the brain [128]. FUS has the unique ability to safely and reversibly open the BBB, thus increasing the ability of drugs to reach normally inaccessible brain sites [129]. This finding has been demonstrated in multiple recent clinical trials [8] with no adverse radiological findings at three month follow-up.

The BBB, which is primarily formed by capillary endothelial cells, astrocyte end-feet, and pericytes, serves as a biological barrier that only allows channel-mediated and small molecules to pass through, under normal conditions. While the BBB primarily functions to protect the brain, it also restricts the types of drugs that can reach the brain. In combination with the use of microbubbles, which are similar to red blood cells in size at ~10 μm in diameter [130], FUS improves drug delivery, allowing for more effective passage of drugs across the BBB [131]. In this way, the use of FUS prior to NP dosing results in enhanced delivery of NPs to cancerous cells [71].

Several studies in recent years have investigated the use of FUS for polymeric NP delivery across the BBB. Yang et al. recently reported successful delivery of lipid-polymer hybrid NPs loaded with clustered regularly interspaced short palindromic repeats (CRISPR)-associated protein 9 (CRISPR/Cas9) plasmids targeting a temozolomide drug-resistance gene to glioblastoma cells in vivo [132]. They subsequently found significantly inhibited tumor growth and prolonged survival times in tumor-bearing mice. Separately, Nance et al. reported pressure-dependent delivery of PLGA-based NPs into regions of the brain parenchyma of a rat model where FUS was used to temporarily disrupt the BBB [133]. Other studies have published similar findings of enhanced NP transport across BBB and localization to target regions via FUS [134,135]. An illustration of FUS and other approaches that can be used to deliver NPs to brain tumor tissue is shown in Figure 3.

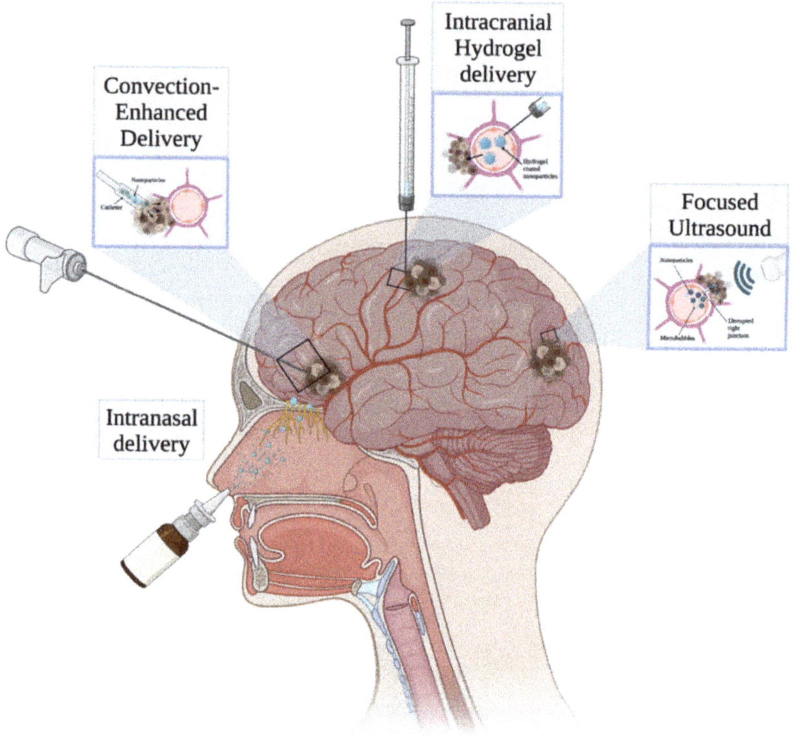

Figure 3. Methods of nanoparticle administration.

Through the combined use of FUS to generate oscillation openings and microbubbles—most commonly Optison, Definity, and Sonovue—to transport drugs, medications are more likely to reach their target sites in the brain and have increased efficacy [136]. Despite these potential benefits, however, it is important to note that FUS-mediated drug delivery presents certain risks that must continue to be investigated, including acute complications such as microhemorrhages and vacuolation of pericytes and other cells at the BBB following sonication, likely due to the temporary BBB disruption [129,137].

6.2. *Convection-Enhanced Delivery*

Convection-enhanced delivery (CED) was developed at the National Institutes of Health in the early 1990s as a way to bypass the BBB for localized drug delivery [138]. CED involves stereotactic placement of one or more catheters and an external pressure gradient to pump drugs via fluid convection directly into regions of interest within the brain [138,139]. This method also results in a greater volume of distribution compared to diffusion, thereby allowing drugs to spread further throughout the brain parenchyma [140]. CED may be particularly beneficial for delivering drug-loaded NPs with tumor cell-targeting properties to remaining cancer cells after surgical resection in clinical settings [139]. Indeed, there are multiple ongoing in vivo studies and clinical trials aimed at using CED to deliver various chemotherapeutic drugs to glioma cells [71,141]. However, no trials involving CED have established a definitive increase in glioma survival time compared to the current standard of care [141,142]. Additionally, with CED comes potential downsides, such as risk of infection, catheter obstruction, and limited therapeutic administration windows compared to systemic treatments [71].

One example illustrating the potential of CED in brain tumor nanomedicine investigated the delivery of PBAE NPs complexed with DNA plasmids coding for the luciferase protein (for fluorescent tracking) and p53 tumor suppressor protein to the brain parenchyma in tumor-bearing rats [143]. Mastorakos et al. reported that the PBAE NPs, which were also conjugated to polyethylene glycol to avoid adhesive trapping in the brain, rapidly traversed the brain parenchyma and orthotopic brain tumors. Furthermore, rats treated with these same DNA-loaded NPs displayed effective gene transfection of tumor cells and significantly enhanced survival compared to similar NPs without the polyethylene glycol coating. While these findings are indeed promising, the study did not investigate any potential complications resulting from CED. Another study utilizing similar DNA-loaded NPs administered via CED resulted in transgene expression in both healthy striatal tissue and malignant glioma cells in a rat model [144]. Nevertheless, significantly more replicated studies and research into potential CED delivery of NPs are required before potentially reaching the clinical trial phase.

6.3. Nose-to-Brain Delivery

The nose-to-brain technique offers a minimally invasive and relatively convenient option for NP delivery [145]. Drugs delivered nasally to the CNS are directly absorbed via the olfactory and trigeminal pathways, circumventing the blood brain barrier [146], and can exhibit favorable pharmacokinetics and pharmacodynamics [147]. One challenge in nasal delivery of drugs includes the presence of various degrading enzymes in these pathways, including cytochrome P450, as well as mucociliary clearance that reduces drug retention and absorption [148]. While it is known that various therapeutic compounds delivered via nose-to-brain are able to spread from the olfactory and trigeminal pathways to brain parenchyma and CSF [149], the exact mechanism allowing for this is unclear and resulting bioavailability is often low [150].

Studies have shown modified NPs are capable of protecting their contents from enzymatic degradation and clearance via nasal cilia [148]. While most molecules are not likely to reach therapeutic doses in the CNS via intranasal delivery, polymeric NPs including chitosan, PLGA, and PLA have been studied for potential delivery via this method [151]. Craparo et al. demonstrated that mPEG-PLGA labeled with Rhodamine-B were suitable for drug delivery via the nose-to-brain route [151]. Separately, Chu et al. successfully targeted glioblastoma in a rodent model using the nose-to-brain technique to deliver modified TMZ-loaded PLGA particles [152]. In addition to the significant glioma cell cytotoxicity found in rats treated with nasal delivery of these modified NPs, fluorescence imaging showed lower buildup of NPs in other organs and higher brain distribution compared to the same carriers delivered intravenously after four hours [152]. Another study reported prolonged life and decreased tumor growth in tumor-bearing rats after nose-to-brain delivery of modified PCL nano-micelles containing therapeutic compounds [153]. Kanazawa et al. followed up on these findings by attaching a peptide that binds specifically to gastrin-releasing peptide receptor (GRPR) to similar PCL nano-micelles in a subsequent study [149]. GRPR is upregulated in various tumor cells, including glioblastoma [154], and the researchers found selective cellular uptake of and cytotoxicity in C6 glioma cells treated with these PCL nano-micelles. It remains to be seen if these promising preliminary rodent findings and others will be translated to successful human clinical trials, but nose-to-brain delivery may indeed serve as a powerful technique for minimally invasive delivery of brain tumor therapeutics.

6.4. Intracranial Hydrogel Delivery

Hydrogels are three-dimensional, hydrophilic gel polymer networks that provide the benefits of minimal toxicity, localized and stimuli-responsive drug delivery, and passively controlled drug release [155]. These same potential advantages can be found in NP-based therapeutic systems, and, indeed, many studies have investigated the use of NP-loaded hydrogel hybrids as a means for managing brain tumors [156]. However, one common obstacle in the use of hydrogels for this purpose is that most chemotherapeutic drugs are hydrophobic, making them largely incompatible with the hydrophilic nature of the gels [157]. Polymeric NPs are highly tunable and can be constructed with hydrophilic shells loaded with hydrophobic drugs, potentially circumventing this issue [7]. NPs can be incorporated into hydrogels in a variety of ways before and after the gelation process, or even by being incorporated into the gel matrix itself as many gel matrices have been shown to be stable environments for NPs [155].

The combined use of NPs and hydrogels has emerged as a potentially powerful tool for improving brain tumor therapy through safe and localized delivery of hydrophobic drugs. Using this combination strategy, drugs can be injected intratumorally or implanted following surgical resection [158]. Still, both mechanisms of delivery are invasive, prompting additional studies aimed at developing nanoscale hydrogels—so-called "nanogels"— for intravenous approaches [159]. These nanogels have the potential to add NP-based advantages, such as the ability to easily cross the BBB, be internalized by cells, and efficiently encapsulate drugs, to hydrogel systems [160]. Combining these with hydrogel swelling capabilities, or ability to expand in aqueous solution [161], and hydrophilicity makes nanogels a powerful tool for drug delivery [158]. Many recent in vivo studies have confirmed the ability of various nanogels to cross the BBB and selectively target tumor cells [162–164]; however, further investigation is required to determine whether these systems can be safely and effectively incorporated into clinical practice.

6.5. Cell-Based Delivery

NPs can be internalized by monocytes, neutrophils, and stem cells for delivery to brain tumor tissue. The utilization of immune cells such as monocytes and neutrophils as vectors for NP delivery is especially appealing as these cells readily cross the BBB to sites of injury, inflammation, and tumor growth [10]. This cell-based approach potentially circumvents the need for NPs to traverse the BBB and brain parenchyma themselves. Additionally, among the greatest challenges of delivering systemically-injected NPs to brain tumor sites remains their rapid uptake and removal from circulation by the reticuloendothelial system, consisting of monocytes/macrophages in the liver, spleen, and other fixed tissues [7,71]. In theory, monocytes/macrophages and neutrophils could be obtained from patients and loaded with drug-containing NPs in vitro before reintroducing them to the bloodstream via intravenous injection, allowing them to eventually migrate into and around tumors [165].

The potential for cell-based delivery of NPs in brain tumor therapy has been increasingly studied in recent years. In 2020, Ibarra et al. reported the efficient delivery of monocytes loaded with conjugated polymer NPs to GBM spheroids and an orthotopic mouse model [166]. In another study investigating neutrophil-based delivery, Wu et al. reported that certain neutrophils internalized with magnetic mesoporous silica NPs containing doxorubicin could actively target inflamed brain tumor tissue following intravenous injection after surgical resection in a mouse model [75]. They noted no effect on neutrophil viability after NP loading and tumor-specific cytotoxicity resulting from NPs accumulating in extracellular traps from neutrophils (NETs) that were then internalized by glioma cells. Another study by Xue et al. had similar findings with liposomes, noting that intravenously-injected neutrophils containing paclitaxel-loaded liposomes could accumulate in the brain and suppress glioma cell growth in mice following surgical resection [167].

Mesenchymal stem cells (MSCs) have also been increasingly studied as potential vectors for glioma treatment, as they are hypoimmunogenic and known to migrate toward tumor cells [168]. In a proof-of-concept study, Roger et al. demonstrated that PLA NPs and lipid nanocapsules could be internalized into MSCs without affecting cell viability and differentiation [169]. In the same study, they also showed that those MSCs could migrate towards human glioma cells in a mouse model following intra-arterial and intracranial injection. Similarly, Clavreul et al. reported that a particular subpopulation of MSCs containing lipid nanocapsules with an organometallic complex injected into the striatum resulted in increased survival time in U87MG-bearing mice compared to mice receiving striatum injections of the organometallic complex-containing lipid nanocapsules alone [170]. Building upon these findings, a 2018 study found injection of MSCs loaded with paclitaxel-containing PLGA NPs into the brains of an orthotopic glioma rat model resulted in paclitaxel resulted in significantly longer survival than rats injected with paclitaxel-primed MSCs or paclitaxel-containing PLGA NPs alone [168]. Furthermore, the same group demonstrated that the same MSCs could transfer paclitaxel to glioma cells and induce tumor cell death in vitro. While there are a limited number of studies successfully showing cell-based delivery of NPs for brain tumor treatment, this approach may be able to alleviate many of the complications in delivering NPs to brain tumor tissue in clinical practice and requires further investigation. A comparison of the various mechanisms of delivery discussed here can be found in Table 2.

Table 2. Comparison of polymeric NP mechanisms of delivery.

Mechanism of Delivery	Type(s) of Polymeric NPs Used	Advantages	Limitations
Focused Ultrasound	PLGA [133]	Can reversibly open BBB; targeted delivery; safety supported via clinical trials; minimal systemic effects	Acute complications such as microhemorrhages reported; invasive
Convection-Enhanced Delivery	PLGA [171], PBAE [143], Chitosan [172], PAMAM [173], PCL [174]	High volume of distribution reported; targeted delivery; multiple ongoing clinical trials; potential for use post-resection; minimal systemic effects	No definitive increase in glioma patient survival time reported; infection; limited therapeutic administration windows; invasive
Nose-to-Brain Delivery	PLGA [151], Chitosan [46], PCL [153]	Minimally invasive; easier to study in vivo; bypasses BBB; minimal systemic effects	Exact delivery mechanism and clearance pathways unclear; non-targeted delivery; bioavailability can be low compared to other delivery mechanisms; limited NP clinical studies
Intracranial Hydrogel Delivery	PLGA [155], Chitosan [73], PCL [175]	Potential for use post-resection; targeted delivery; passively controlled drug release; variety of potential approaches; minimal systemic effects	Difficult to use with hydrophobic NPs; invasive; non-targeted delivery
Cell-Based Delivery	PLGA [168]	Minimally invasive; limited clearance via reticuloendothelial system compared to other systemic delivery approaches	Limited NP clinical studies; non-targeted delivery

7. Discussion and Future Directions

The increasing number of encouraging findings from preclinical studies suggests polymeric NPs may ultimately serve as one of the next great innovations in clinical CNS cancer therapy. Nevertheless, questions regarding their ideal size, surface characteristics, loaded contents, and mechanism of delivery remain, which continues to limit their translation to clinical practice. NP-based systems continue to encounter roadblocks such as low delivery to brain tissue due to mononuclear phagocyte-mediated clearance, despite the widespread adoption of modifications like PEGylation to reduce this barrier [94]. Finding an ideal balance between NP stability and degradation while ensuring the contents are released specifically to target regions also remains elusive. Standardization of NP synthesis and development of more representative preclinical models is essential in progressing towards clinical trials and more comprehensive strategies to mitigate this have been explored [176].

The most effective NP formulations for treating CNS malignancies may include hybrid systems, such as those combining gold NPs coated with PEG for improved magnetic resonance imaging of brain tumor periphery [7]. Similarly, NP formulations containing multiple therapeutic components may be beneficial in overcoming intra-tumor heterogeneity in metastases such as GBM [71]. An example of this was shown by Xu et al., who reported that paclitaxel and temozolomide co-loaded PLGA-based NPs resulted in greater inhibition against two glioma cell lines in vivo compared to either single drug [177]. Furthermore, polymeric NPs designed with dual targeting of the BBB and tumor cells may result in greater and more specific delivery to target sites throughout the brain parenchyma. Many moieties overexpressed on both of these were previously described above. This may, thereby, allow for polymeric NPs that could be further modified to target tumor microenvironmental stimuli, such as pH, while maintaining a simpler composition.

Intravenous injection is the least invasive method for NP-based therapies and minimizes risk of complications such as infection and compromised BBB integrity. However, concerns of limited delivery across the BBB and BBTB, as well as systemic toxicity, remain. Much of the recent CNS-targeted NP literature investigates alternative delivery mechanisms like nose-to-brain, FUS, CED, hydrogels or nanogels, and cell-based delivery to overcome these challenges. Of course, these approaches must continue to be studied to ensure their safe use and effectiveness in clinical practice. Nevertheless, these delivery techniques are proving to be a powerful tool for nanotherapeutics. Bringing an effective polymeric NP-based therapy to CNS malignancies will require carefully constructed polymers loaded with biotherapeutics that can selectively target tumor cells while minimizing toxicity to healthy brain tissue and other organs.

8. Conclusions

While numerous studies utilizing polymer-based NPs for brain tumor therapy have reported promising findings, few approaches have made it to clinical trial, and none have proven to be effective replacements or additions to the current therapeutic regimens. Pitfalls, including low therapeutic delivery and inconsistent study results, must be overcome if polymeric NP-based therapeutics are to be leveraged in the treatment of brain cancer. While PLGA is the most widely-used polymeric platform in studies of nanocarrier application to the treatment of neurological disease, there is no consensus on the most effective polymer for use in brain cancer treatment, as their variable composition influences the specific targeting ligands, contents, mechanisms of delivery, tumor characteristics, and other factors selected for each study. Furthermore, the ideal NP size and shape is dependent on those same factors and similar studies have reported variable findings in BBB and tumor localization. The differences in physiology and permeability in various in vivo and in vitro models compared to humans has been documented and likely also contributes to the challenges in standardization of effective polymeric NP characteristics. While further investigation is warranted, polymeric NPs remain a promising therapeutic method. Ongoing clinical trials and future studies will hopefully shed more light on the optimal polymeric NP composition for managing brain cancer.

Author Contributions: Conceptualization and methodology, C.A.C. and B.M.T.; writing—original draft preparation, C.A.C., A.K. and N.K.; writing—review and editing, C.A.C., H.G. and E.E.W.; visualization, E.E.W.; supervision and project administration, B.M.T. All authors have read and agreed to the published version of the manuscript.

Funding: This research received no external funding.

Data Availability Statement: Not applicable.

Acknowledgments: Figures were created using BioRender.com accessed on 30 May 2022.

Conflicts of Interest: Betty Tyler has research funding from NIH and is a co-owner for Accelerating Combination Therapies*. Ashvattha Therapeutics Inc. has also licensed one of her patents (*includes equity or options).

References

1. Ostrom, Q.T.; Ostrom, Q.T.; Patil, N.; Cioffi, G.; Waite, K.; Kruchko, C.; Barnholtz-Sloan, J.S. CBTRUS Statistical Report: Primary Brain and Other Central Nervous System Tumors Diagnosed in the United States in 2013–2017. *Neuro-Oncology* **2020**, *22* (Suppl. S2), iv1–iv96. [CrossRef] [PubMed]
2. Sung, H.; Ferlay, J.; Siegel, R.L.; Laversanne, M.; Soerjomataram, I.; Jemal, A.; Bray, F. Global Cancer Statistics 2020: GLOBOCAN Estimates of Incidence and Mortality Worldwide for 36 Cancers in 185 Countries. *CA Cancer J. Clin.* **2021**, *71*, 209–249. [CrossRef] [PubMed]
3. Visser, O.; Ardanaz, E.; Botta, L.; Sant, M.; Tavilla, A.; Minicozzi, P.; Hackl, M.; Zielonke, N.; Oberaigner, W.; Van Eycken, E.; et al. Survival of adults with primary malignant brain tumours in Europe; Results of the EUROCARE-5 study. *Eur. J. Cancer* **2015**, *51*, 2231–2241. [CrossRef]
4. Tang, L.; Feng, Y.; Gao, S.; Mu, Q.; Liu, C. Nanotherapeutics Overcoming the Blood-Brain Barrier for Glioblastoma Treatment. *Front. Pharmacol.* **2021**, *12*, 786700. [CrossRef] [PubMed]
5. Roy, S.; Lahiri, D.; Maji, T.; Biswas, J. Recurrent Glioblastoma: Where we stand. *S. Asian J. Cancer* **2015**, *4*, 163. [CrossRef] [PubMed]
6. Aldape, K.; Brindle, K.M.; Chesler, L.; Chopra, R.; Gajjar, A.; Gilbert, M.R.; Gottardo, N.; Gutmann, D.H.; Hargrave, D.; Holland, E.C.; et al. Challenges to curing primary brain tumours. *Nat. Rev. Clin. Oncol.* **2019**, *16*, 509–520. [CrossRef] [PubMed]
7. Zhang, W.; Mehta, A.; Tong, Z.; Esser, L.; Voelcker, N.H. Development of Polymeric Nanoparticles for Blood–Brain Barrier Transfer—Strategies and Challenges. *Adv. Sci.* **2021**, *8*, 2003937. [CrossRef]
8. Fisher, D.; Price, R.J. Recent Advances in the Use of Focused Ultrasound for Magnetic Resonance Image-Guided Therapeutic Nanoparticle Delivery to the Central Nervous System. *Front. Pharmacol.* **2019**, *10*, 1348. [CrossRef]
9. Wang, Y.; Huang, Y.; Bai, H.; Wang, G.; Hu, X.; Kumar, S.; Min, R. Biocompatible and Biodegradable Polymer Optical Fiber for Biomedical Application: A Review. *Biosensors* **2021**, *11*, 472. [CrossRef]
10. Yu, H.; Yang, Z.; Li, F.; Xu, L.; Sun, Y. Cell-mediated targeting drugs delivery systems. *Drug Deliv.* **2020**, *27*, 1425–1437. [CrossRef]
11. Lluch, C.; Lligadas, G.; Ronda, J.C.; Galià, M.; Cadiz, V. "Click" Synthesis of Fatty Acid Derivatives as Fast-Degrading Polyanhydride Precursors. *Macromol. Rapid Commun.* **2011**, *32*, 1343–1351. [CrossRef] [PubMed]
12. Parikh, P.M.; Panigrahi, M.; Das, P.K. Brain tumor and Gliadel wafer treatment. *Indian J. Cancer* **2011**, *48*, 11–17. [CrossRef] [PubMed]
13. Gabikian, P.; Abel, T.; Ryken, T.; Lesniak, M. Gliadel for brain metastasis. *Surg. Neurol. Int.* **2013**, *4* (Suppl. S4), S289–S293. [CrossRef]
14. Qi, Z.-Y.; Xing, W.-K.; Shao, C.; Yang, C.; Wang, Z. The role of Gliadel wafers in the treatment of newly diagnosed GBM: A meta-analysis. *Drug Des. Dev. Ther.* **2015**, *9*, 3341–3348. [CrossRef]
15. Iuchi, T.; Inoue, A.; Hirose, Y.; Morioka, M.; Horiguchi, K.; Natsume, A.; Arakawa, Y.; Iwasaki, K.; Fujiki, M.; Kumabe, T.; et al. Long-term effectiveness of Gliadel implant for malignant glioma and prognostic factors for survival: 3-year results of a postmarketing surveillance in Japan. *Neuro-Oncol. Adv.* **2022**, *4*, vdab189. [CrossRef]
16. Brenza, T.M.; Ghaisas, S.; Ramirez, J.E.V.; Harischandra, D.; Anantharam, V.; Kalyanaraman, B.; Kanthasamy, A.G.; Narasimhan, B. Neuronal protection against oxidative insult by polyanhydride nanoparticle-based mitochondria-targeted antioxidant therapy. *Nanomed. Nanotechnol. Biol. Med.* **2017**, *13*, 809–820. [CrossRef] [PubMed]
17. Chavez-Santoscoy, A.V.; Roychoudhury, R.; Pohl, N.L.; Wannemuehler, M.J.; Narasimhan, B.; Ramer-Tait, A. Tailoring the immune response by targeting C-type lectin receptors on alveolar macrophages using "pathogen-like" amphiphilic polyanhydride nanoparticles. *Biomaterials* **2012**, *33*, 4762–4772. [CrossRef] [PubMed]
18. Binnebose, A.M.; Haughney, S.L.; Martin, R.J.; Imerman, P.M.; Narasimhan, B.; Bellaire, B.H. Polyanhydride Nanoparticle Delivery Platform Dramatically Enhances Killing of Filarial Worms. *PLOS Negl. Trop. Dis.* **2015**, *9*, e0004173. [CrossRef]
19. Schlichtmann, B.W.; Kalyanaraman, B.; Schlichtmann, R.L.; Panthani, M.G.; Anantharam, V.; Kanthasamy, A.G.; Mallapragada, S.K.; Narasimhan, B. Functionalized polyanhydride nanoparticles for improved treatment of mitochondrial dysfunction. *J. Biomed. Mater. Res. Part B Appl. Biomater.* **2021**, *110*, 450–459. [CrossRef]

20. Geraili, A.; Mequanint, K. Systematic Studies on Surface Erosion of Photocrosslinked Polyanhydride Tablets and Data Correlation with Release Kinetic Models. *Polymers* **2020**, *12*, 1105. [CrossRef]
21. Brenza, T.M.; Schlichtmann, B.W.; Bhargavan, B.; Ramirez, J.E.V.; Nelson, R.D.; Panthani, M.G.; McMillan, J.M.; Kalyanaraman, B.; Gendelman, H.E.; Anantharam, V.; et al. Biodegradable polyanhydride-based nanomedicines for blood to brain drug delivery. *J. Biomed. Mater. Res. Part A* **2018**, *106*, 2881–2890. [CrossRef]
22. Kumari, A.; Yadav, S.K.; Yadav, S.C. Biodegradable polymeric nanoparticles based drug delivery systems. *Colloids Surf. B Biointerfaces* **2010**, *75*, 1–18. [PubMed]
23. Mirakabad, F.S.T.; Nejati-Koshki, K.; Akbarzadeh, A.; Yamchi, M.R.; Milani, M.; Zarghami, N.; Zeighamian, V.; Rahimzadeh, A.; Alimohammadi, S.; Hanifehpour, Y.; et al. PLGA-Based Nanoparticles as Cancer Drug Delivery Systems. *Asian Pac. J. Cancer Prev.* **2014**, *15*, 517–535. [CrossRef]
24. Makadia, H.K.; Siegel, S.J. Poly lactic-co-glycolic acid (PLGA) As biodegradable controlled drug delivery carrier. *Polymers* **2011**, *3*, 1377–1397. [CrossRef] [PubMed]
25. Maksimenko, O.; Malinovskaya, J.; Shipulo, E.; Osipova, N.; Razzhivina, V.; Arantseva, D.; Yarovaya, O.; Mostovaya, U.; Khalansky, A.; Fedoseeva, V.; et al. Doxorubicin-loaded PLGA nanoparticles for the chemotherapy of glioblastoma: Towards the pharmaceutical development. *Int. J. Pharm.* **2019**, *572*, 118733. [CrossRef]
26. Agarwal, S.; Mohamed, M.S.; Mizuki, T.; Maekawa, T.; Kumar, D.S. Chlorotoxin modified morusin–PLGA nanoparticles for targeted glioblastoma therapy. *J. Mater. Chem. B* **2019**, *7*, 5896–5919. [CrossRef]
27. Banstola, A.; Duwa, R.; Emami, F.; Jeong, J.-H.; Yook, S. Enhanced Caspase-Mediated Abrogation of Autophagy by Temozolomide-Loaded and Panitumumab-Conjugated Poly(lactic-*co*-glycolic acid) Nanoparticles in Epidermal Growth Factor Receptor Overexpressing Glioblastoma Cells. *Mol. Pharm.* **2020**, *17*, 4386–4400. [CrossRef]
28. Eivazi, N.; Rahmani, R.; Paknejad, M. Specific cellular internalization and pH-responsive behavior of doxorubicin loaded PLGA-PEG nanoparticles targeted with anti EGFRvIII antibody. *Life Sci.* **2020**, *261*, 118361.
29. Younis, M.; Faming, W.; Hongyan, Z.; Mengmeng, T.; Hang, S.; Liudi, Y. Iguratimod encapsulated PLGA-NPs improves therapeutic outcome in glioma, glioma stem-like cells and temozolomide resistant glioma cells. *Nanomed. Nanotechnol. Biol. Med.* **2019**, *22*, 102101. [CrossRef]
30. Mao, J.; Meng, X.; Zhao, C.; Yang, Y.; Liu, G. Development of transferrin-modified poly(lactic-co-glycolic acid) nanoparticles for glioma therapy. *Anti-Cancer Drugs* **2019**, *30*, 604–610. [CrossRef]
31. Sousa, F.; Dhaliwal, H.K.; Gattacceca, F.; Sarmento, B.; Amiji, M.M. Enhanced anti-angiogenic effects of bevacizumab in glioblastoma treatment upon intranasal administration in polymeric nanoparticles. *J. Control Release* **2019**, *309*, 37–47. [CrossRef]
32. Rezvantalab, S.; Drude, N.; Moraveji, M.K.; Güvener, N.; Koons, E.K.; Shi, Y.; Lammers, T.; Kiessling, F. PLGA-Based Nanoparticles in Cancer Treatment. *Front. Pharmacol.* **2018**, *9*, 1260. [CrossRef] [PubMed]
33. Visan, A.; Popescu-Pelin, G.; Socol, G. Degradation Behavior of Polymers Used as Coating Materials for Drug Delivery—A Basic Review. *Polymers* **2021**, *13*, 1272. [CrossRef] [PubMed]
34. Danhier, F.; Ansorena, E.; Silva, J.M.; Coco, R.; Le Breton, A.; Préat, V. PLGA-based nanoparticles: An overview of biomedical applications. *J. Control Release* **2012**, *161*, 505–522. [CrossRef] [PubMed]
35. Houchin, M.; Neuenswander, S.; Topp, E. Effect of excipients on PLGA film degradation and the stability of an incorporated peptide. *J. Control Release* **2007**, *117*, 413–420. [CrossRef]
36. Choi, J.; Rui, Y.; Kim, J.; Gorelick, N.; Wilson, D.R.; Kozielski, K.; Mangraviti, A.; Sankey, E.; Brem, H.; Tyler, B.; et al. Nonviral polymeric nanoparticles for gene therapy in pediatric CNS malignancies. *Nanomed. Nanotechnol. Biol. Med.* **2020**, *23*, 102115. [CrossRef]
37. Sunshine, J.; Peng, D.Y.; Green, J.J. Uptake and Transfection with Polymeric Nanoparticles Are Dependent on Polymer End-Group Structure, but Largely Independent of Nanoparticle Physical and Chemical Properties. *Mol. Pharm.* **2012**, *9*, 3375–3383. [CrossRef]
38. Liu, J.; Huang, Y.; Kumar, A.; Tan, A.; Jin, S.; Mozhi, A.; Liang, X.J. pH-sensitive nano-systems for drug delivery in cancer therapy. *Biotechnol. Adv.* **2014**, *32*, 693–710. [CrossRef]
39. Fornaguera, C.; Guerra-Rebollo, M.; Lazaro, M.A.; Cascante, A.; Rubio, N.; Blanco, J.; Borrós, S. In Vivo Retargeting of Poly(beta aminoester) (OM-PBAE) Nanoparticles is Influenced by Protein Corona. *Adv. Healthc. Mater.* **2019**, *8*, e1900849. [CrossRef]
40. Park, H.J.; Lee, J.; Kim, M.J.; Kang, T.J.; Jeong, Y.; Um, S.H.; Cho, S.W. Sonic hedgehog intradermal gene therapy using a biodegradable poly(beta-amino esters) nanoparticle to enhance wound healing. *Biomaterials* **2012**, *33*, 9148–9156. [CrossRef]
41. Feng, R.; Chen, Q.; Zhou, P.; Wang, Y.; Yan, H. Nanoparticles based on disulfide-containing poly(beta-amino ester) and zwitterionic fluorocarbon surfactant as a redox-responsive drug carrier for brain tumor treatment. *Nanotechnology* **2018**, *29*, 495101. [CrossRef] [PubMed]
42. Xu, Y.; Liu, D.; Hu, J.; Ding, P.; Chen, M. Hyaluronic acid-coated pH sensitive poly (beta-amino ester) nanoparticles for co-delivery of embelin and TRAIL plasmid for triple negative breast cancer treatment. *Int. J. Pharm.* **2020**, *573*, 118637. [CrossRef] [PubMed]
43. Anderson, D.G.; Lynn, D.M.; Langer, R. Semi-Automated Synthesis and Screening of a Large Library of Degradable Cationic Polymers for Gene Delivery. *Angew. Chem. Int. Ed. Engl.* **2003**, *42*, 3153–3158. [CrossRef]
44. Guerrero-Cázares, H.; Tzeng, S.Y.; Young, N.P.; Abutaleb, A.O.; Quiñones-Hinojosa, A.; Green, J.J. Biodegradable Polymeric Nanoparticles Show High Efficacy and Specificity at DNA Delivery to Human Glioblastoma In Vitro and In Vivo. *ACS Nano* **2014**, *8*, 5141–5153. [CrossRef] [PubMed]

45. Negron, K.; Zhu, C.; Chen, S.-W.; Shahab, S.; Rao, D.; Raabe, E.H.; Eberhart, C.G.; Hanes, J.; Suk, J.S. Non-adhesive and highly stable biodegradable nanoparticles that provide widespread and safe transgene expression in orthotopic brain tumors. *Drug Deliv. Transl. Res.* **2020**, *10*, 572–581. [CrossRef] [PubMed]
46. Mohammed, M.A.; Syeda, J.T.M.; Wasan, K.M.; Wasan, E.K. An Overview of Chitosan Nanoparticles and Its Application in Non-Parenteral Drug Delivery. *Pharmaceutics* **2017**, *9*, 53. [CrossRef] [PubMed]
47. Rizeq, B.R.; Younes, N.N.; Rasool, K.; Nasrallah, G.K. Synthesis, Bioapplications, and Toxicity Evaluation of Chitosan-Based Nanoparticles. *Int. J. Mol. Sci.* **2019**, *20*, 5776. [CrossRef]
48. Caprifico, A.E.; Foot, P.J.S.; Polycarpou, E.; Calabrese, G. Overcoming the Blood-Brain Barrier: Functionalised Chitosan Nanocarriers. *Pharmaceutics* **2020**, *12*, 1013. [CrossRef]
49. Shevtsov, M.; Nikolaev, B.; Marchenko, Y.; Yakovleva, L.; Skvortsov, N.; Mazur, A.; Tolstoy, P.; Ryzhov, V.; Multhoff, G. Targeting experimental orthotopic glioblastoma with chitosan-based superparamagnetic iron oxide nanoparticles (CS-DX-SPIONs). *Int. J. Nanomed.* **2018**, *13*, 1471–1482. [CrossRef]
50. Yang, C.-L.; Chen, J.-P.; Wei, K.-C.; Chen, J.-Y.; Huang, C.-W.; Liao, Z.-X. Release of Doxorubicin by a Folate-Grafted, Chitosan-Coated Magnetic Nanoparticle. *Nanomaterials* **2017**, *7*, 85. [CrossRef]
51. Fana, M.; Gallien, J.; Srinageshwar, B.; Dunbar, G.L.; Rossignol, J. PAMAM Dendrimer Nanomolecules Utilized as Drug Delivery Systems for Potential Treatment of Glioblastoma: A Systematic Review. *Int. J. Nanomed.* **2020**, *15*, 2789–2808. [CrossRef] [PubMed]
52. Araújo, R.V.D.; Santos, S.D.S.; Igne Ferreira, E.; Giarolla, J. New Advances in General Biomedical Applications of PAMAM Dendrimers. *Molecules* **2018**, *23*, 2849. [CrossRef]
53. Florendo, M.; Figacz, A.; Srinageshwar, B.; Sharma, A.; Swanson, D.; Dunbar, G.L.; Rossignol, J. Use of Polyamidoamine Dendrimers in Brain Diseases. *Molecules* **2018**, *23*, 2238. [CrossRef]
54. Sarin, H.; Kanevsky, A.S.; Wu, H.; Brimacombe, K.R.; Fung, S.H.; Sousa, A.A.; Auh, S.; Wilson, C.M.; Sharma, K.; Aronova, A.M.; et al. Effective transvascular delivery of nanoparticles across the blood-brain tumor barrier into malignant glioma cells. *J. Transl. Med.* **2008**, *6*, 80. [CrossRef]
55. Moscariello, P.; Ng, D.; Jansen, M.; Weil, T.; Luhmann, H.J.; Hedrich, J. Brain Delivery of Multifunctional Dendrimer Protein Bioconjugates. *Adv. Sci.* **2018**, *5*, 1700897. [CrossRef] [PubMed]
56. Thomas, V.; Jagani, S.; Johnson, K.; Jose, M.V.; Dean, D.R.; Vohra, Y.K.; Nyairo, E. Electrospun Bioactive Nanocomposite Scaffolds of Polycaprolactone and Nanohydroxyapatite for Bone Tissue Engineering. *J. Nanosci. Nanotechnol.* **2006**, *6*, 487–493. [CrossRef] [PubMed]
57. Xin, H.; Sha, X.; Jiang, X.; Zhang, W.; Chen, L.; Fang, X. Anti-glioblastoma efficacy and safety of paclitaxel-loading Angiopep-conjugated dual targeting PEG-PCL nanoparticles. *Biomaterials* **2012**, *33*, 8167–8176. [CrossRef]
58. Belykh, E.; Shaffer, K.V.; Lin, C.; Byvaltsev, V.A.; Preul, M.C.; Chen, L. Blood-Brain Barrier, Blood-Brain Tumor Barrier, and Fluorescence-Guided Neurosurgical Oncology: Delivering Optical Labels to Brain Tumors. *Front. Oncol.* **2020**, *10*, 739. [CrossRef]
59. Gou, M.; Wei, X.; Men, K.; Wang, B.; Luo, F.; Zhao, X.; Wei, Y.; Qian, Z. PCL/PEG Copolymeric Nanoparticles: Potential Nanoplatforms for Anticancer Agent Delivery. *Curr. Drug Targets* **2011**, *12*, 1131–1150. [CrossRef]
60. Sanchez-Gonzalez, S.; Diban, N.; Urtiaga, A. Hydrolytic Degradation and Mechanical Stability of Poly(epsilon-Caprolactone)/Reduced Graphene Oxide Membranes as Scaffolds for In Vitro Neural Tissue Regeneration. *Membranes* **2018**, *8*, 12. [CrossRef]
61. Liu, J.; Zeng, Y.; Shi, S.; Xu, L.; Zhang, H.; Pathak, J.L.; Pan, Y. Design of polyaspartic acid peptide-poly (ethylene glycol)-poly (epsilon-caprolactone) nanoparticles as a carrier of hydrophobic drugs targeting cancer metastasized to bone. *Int. J. Nanomed.* **2017**, *12*, 3561–3575. [CrossRef] [PubMed]
62. Sulheim, E.; Baghirov, H.; Von Haartman, E.; Bøe, A.; Åslund, A.K.O.; Mørch, Y.; Davies, C.D.L. Cellular uptake and intracellular degradation of poly(alkyl cyanoacrylate) nanoparticles. *J. Nanobiotechnol.* **2016**, *14*, 1. [CrossRef] [PubMed]
63. Müller, R.H.; Lherm, C.; Herbert, J.; Couvreur, P. In vitro model for the degradation of alkylcyanoacrylate nanoparticles. *Biomaterials* **1990**, *11*, 590–595. [CrossRef]
64. Kante, B.; Couvreur, P.; Dubois-Krack, G.; De Meester, C.; Guiot, P.; Roland, M.; Mercier, M.; Speiseru, P. Toxicity of Polyalkylcyanoacrylate Nanoparticles I: Free Nanoparticles. *J. Pharm. Sci.* **1982**, *71*, 786–790. [CrossRef]
65. Vauthier, C.; Dubernet, C.; Fattal, E.; Pinto-Alphandary, H.; Couvreur, P. Poly(alkylcyanoacrylates) as biodegradable materials for biomedical applications. *Adv. Drug Deliv. Rev.* **2003**, *55*, 519–548. [CrossRef]
66. Vauthier, C. A journey through the emergence of nanomedicines with poly(alkylcyanoacrylate) based nanoparticles. *J. Drug Target.* **2019**, *27*, 502–524. [CrossRef] [PubMed]
67. Baghirov, H.; Snipstad, S.; Sulheim, E.; Berg, S.; Hansen, R.; Thorsen, F.; Mørch, Y.; Davies, C.D.L.; Åslund, A.K.O. Ultrasound-mediated delivery and distribution of polymeric nanoparticles in the normal brain parenchyma of a metastatic brain tumour model. *PLoS ONE* **2018**, *13*, e0191102. [CrossRef] [PubMed]
68. Andrieux, K.; Couvreur, P. Polyalkylcyanoacrylate nanoparticles for delivery of drugs across the blood-brain barrier. *Wiley Interdiscip. Rev. Nanomed. Nanobiotechnol.* **2009**, *1*, 463–474. [CrossRef]
69. Li, Y.; Wu, M.; Zhang, N.; Tang, C.; Jiang, P.; Liu, X.; Yan, F.; Zheng, H. Mechanisms of enhanced antiglioma efficacy of polysorbate 80-modified paclitaxel-loaded PLGA nanoparticles by focused ultrasound. *J. Cell. Mol. Med.* **2018**, *22*, 4171–4182. [CrossRef]
70. Vauthier, C.; Dubernet, C.; Chauvierre, C.; Brigger, I.; Couvreur, P. Drug delivery to resistant tumors: The potential of poly(alkyl cyanoacrylate) nanoparticles. *J. Control Release* **2003**, *93*, 151–160. [CrossRef]

71. McCrorie, P.; Vasey, C.E.; Smith, S.J.; Marlow, M.; Alexander, C.; Rahman, R. Biomedical engineering approaches to enhance therapeutic delivery for malignant glioma. *J. Control Release* **2020**, *328*, 917–931. [CrossRef] [PubMed]
72. Suk, J.S.; Xu, Q.; Kim, N.; Hanes, J.; Ensign, L.M. PEGylation as a strategy for improving nanoparticle-based drug and gene delivery. *Adv. Drug Deliv. Rev.* **2016**, *99 Pt A*, 28–51. [CrossRef]
73. Wang, W.; Zhang, Q.; Zhang, M.; Lv, X.; Li, Z.; Mohammadniaei, M.; Zhou, N.; Sun, Y. A novel biodegradable injectable chitosan hydrogel for overcoming postoperative trauma and combating multiple tumors. *Carbohydr. Polym.* **2021**, *265*, 118065. [CrossRef] [PubMed]
74. Helmlinger, G.; Yuan, F.; Dellian, M.; Jain, R.K. Interstitial pH and pO2 gradients in solid tumors in vivo: High-resolution measurements reveal a lack of correlation. *Nat. Med.* **1997**, *3*, 177–182. [CrossRef] [PubMed]
75. Wu, M.; Zhang, H.; Tie, C.; Yan, C.; Deng, Z.; Wan, Q.; Liu, X.; Yan, F.; Zheng, H. MR imaging tracking of inflammation-activatable engineered neutrophils for targeted therapy of surgically treated glioma. *Nat. Commun.* **2018**, *9*, 1–13. [CrossRef] [PubMed]
76. Cheng, R.; Meng, F.; Deng, C.; Klok, H.-A.; Zhong, Z. Dual and multi-stimuli responsive polymeric nanoparticles for programmed site-specific drug delivery. *Biomaterials* **2013**, *34*, 3647–3657. [CrossRef]
77. Zhou, M.; Jiang, N.; Fan, J.; Fu, S.; Luo, H.; Su, P.; Zhang, M.; Shi, H.; Huang, Y.; Li, Y.; et al. H7K(R2)2-modified pH-sensitive self-assembled nanoparticles delivering small interfering RNA targeting hepatoma-derived growth factor for malignant glioma treatment. *J. Control Release* **2019**, *310*, 24–35. [CrossRef]
78. Mura, S.; Nicolas, J.; Couvreur, P. Stimuli-responsive nanocarriers for drug delivery. *Nat. Mater.* **2013**, *12*, 991–1003. [CrossRef]
79. Hickey, J.W.; Santos, J.L.; Williford, J.-M.; Mao, H.-Q. Control of polymeric nanoparticle size to improve therapeutic delivery. *J. Control Release* **2015**, *219*, 536–547. [CrossRef]
80. Cruz, L.J.; Stammes, M.A.; Que, I.; van Beek, E.R.; Knol-Blankevoort, V.T.; Snoeks, T.J.; Chan, A.; Kaijzel, E.L.; Löwik, C.W. Effect of PLGA NP size on efficiency to target traumatic brain injury. *J. Control Release* **2016**, *223*, 31–41. [CrossRef]
81. Kulkarni, S.A.; Feng, S.-S. Effects of Particle Size and Surface Modification on Cellular Uptake and Biodistribution of Polymeric Nanoparticles for Drug Delivery. *Pharm. Res.* **2013**, *30*, 2512–2522. [CrossRef] [PubMed]
82. Nowak, M.; Brown, T.D.; Graham, A.; Helgeson, M.E.; Mitragotri, S. Size, shape, and flexibility influence nanoparticle transport across brain endothelium under flow. *Bioeng. Transl. Med.* **2020**, *5*, e10153. [CrossRef] [PubMed]
83. Voigt, N.; Henrich-Noack, P.; Kockentiedt, S.; Hintz, W.; Tomas, J.; Sabel, B.A. Surfactants, not size or zeta-potential influence blood-brain barrier passage of polymeric nanoparticles. *Eur. J. Pharm. Biopharm.* **2014**, *87*, 19–29. [CrossRef] [PubMed]
84. Thorne, R.G.; Nicholson, C. In vivo diffusion analysis with quantum dots and dextrans predicts the width of brain extracellular space. *Proc. Natl. Acad. Sci. USA* **2006**, *103*, 5567–5572. [CrossRef]
85. Nance, E.A.; Woodworth, G.F.; Sailor, K.A.; Shih, T.Y.; Xu, Q.; Swaminathan, G.; Xiang, D.; Eberhart, C.; Hanes, J. A dense poly(ethylene glycol) coating improves penetration of large polymeric nanoparticles within brain tissue. *Sci. Transl. Med.* **2012**, *4*, 149ra119. [CrossRef]
86. Truong, N.; Whittaker, M.; Mak, C.W.; Davis, T.P. The importance of nanoparticle shape in cancer drug delivery. *Expert Opin. Drug Deliv.* **2015**, *12*, 129–142. [CrossRef]
87. Christian, D.A.; Cai, S.; Garbuzenko, O.B.; Harada, T.; Zajac, A.L.; Minko, T.; Discher, D.E. Flexible Filaments for in Vivo Imaging and Delivery: Persistent Circulation of Filomicelles Opens the Dosage Window for Sustained Tumor Shrinkage. *Mol. Pharm.* **2009**, *6*, 1343–1352. [CrossRef]
88. Huang, X.; Li, L.; Liu, T.; Hao, N.; Liu, H.; Chen, D.; Tang, F. The Shape Effect of Mesoporous Silica Nanoparticles on Biodistribution, Clearance, and Biocompatibility In Vivo. *ACS Nano* **2011**, *5*, 5390–5399. [CrossRef]
89. Doshi, N.; Prabhakarpandian, B.; Rea-Ramsey, A.; Pant, K.; Sundaram, S.; Mitragotri, S. Flow and adhesion of drug carriers in blood vessels depend on their shape: A study using model synthetic microvascular networks. *J. Control Release* **2010**, *146*, 196–200. [CrossRef]
90. Kolhar, P.; Anselmo, A.C.; Gupta, V.; Pant, K.; Prabhakarpandian, B.; Ruoslahti, E.; Mitragotri, S. Using shape effects to target antibody-coated nanoparticles to lung and brain endothelium. *Proc. Natl. Acad. Sci. USA* **2013**, *110*, 10753–10758. [CrossRef]
91. Grabrucker, A.M.; Ruozi, B.; Belletti, D.; Pederzoli, F.; Forni, F.; Vandelli, M.A.; Tosi, G. Nanoparticle transport across the blood brain barrier. *Tissue Barriers* **2016**, *4*, e1153568. [CrossRef]
92. Lu, W. Adsorptive-Mediated Brain Delivery Systems. *Curr. Pharm. Biotechnol.* **2012**, *13*, 2340–2348. [CrossRef] [PubMed]
93. Simonneau, C.; Duschmalé, M.; Gavrilov, A.; Brandenberg, N.; Hoehnel, S.; Ceroni, C.; Lassalle, E.; Kassianidou, E.; Knoetgen, H.; Niewoehner, J.; et al. Investigating receptor-mediated antibody transcytosis using blood–brain barrier organoid arrays. *Fluids Barriers CNS* **2021**, *18*, 43. [CrossRef] [PubMed]
94. Terstappen, G.C.; Meyer, A.H.; Bell, R.D.; Zhang, W. Strategies for delivering therapeutics across the blood–brain barrier. *Nat. Rev. Drug Discov.* **2021**, *20*, 362–383. [CrossRef] [PubMed]
95. Niewoehner, J.; Bohrmann, B.; Collin, L.; Urich, E.; Sade, H.; Maier, P.; Rueger, P.; Stracke, J.O.; Lau, W.; Tissot, A.C.; et al. Increased Brain Penetration and Potency of a Therapeutic Antibody Using a Monovalent Molecular Shuttle. *Neuron* **2014**, *81*, 49–60. [CrossRef]
96. Boado, R.J.; Zhang, Y.; Zhang, Y.; Pardridge, W.M. Humanization of anti-human insulin receptor antibody for drug targeting across the human blood–brain barrier. *Biotechnol. Bioeng.* **2007**, *96*, 381–391. [CrossRef]

97. Zuchero, Y.J.Y.; Chen, X.; Bien-Ly, N.; Bumbaca, D.; Tong, R.K.; Gao, X.; Zhang, S.; Hoyte, K.; Luk, W.; Huntley, M.A.; et al. Discovery of Novel Blood-Brain Barrier Targets to Enhance Brain Uptake of Therapeutic Antibodies. *Neuron* **2016**, *89*, 70–82. [CrossRef]
98. Dehouck, B.; Fenart, L.; Dehouck, M.-P.; Pierce, A.; Torpier, G.; Cecchelli, R. A New Function for the LDL Receptor: Transcytosis of LDL across the Blood–Brain Barrier. *J. Cell Biol.* **1997**, *138*, 877–889. [CrossRef]
99. Ramalho, M.; Sevin, E.; Gosselet, F.; Lima, J.; Coelho, M.; Loureiro, J.; Pereira, M. Receptor-mediated PLGA nanoparticles for glioblastoma multiforme treatment. *Int. J. Pharm.* **2018**, *545*, 84–92. [CrossRef]
100. Johnsen, K.B.; Moos, T. Revisiting nanoparticle technology for blood–brain barrier transport: Unfolding at the endothelial gate improves the fate of transferrin receptor-targeted liposomes. *J. Control Release* **2016**, *222*, 32–46. [CrossRef]
101. Ramalho, M.J.; Loureiro, J.A.; Coelho, M.A.N.; Pereira, M.C. Transferrin Receptor-Targeted Nanocarriers: Overcoming Barriers to Treat Glioblastoma. *Pharmaceutics* **2022**, *14*, 279. [CrossRef] [PubMed]
102. Kuang, Y.; Jiang, X.; Zhang, Y.; Lu, Y.; Ma, H.; Guo, Y.; Zhang, Y.; An, S.; Li, J.; Liu, L.; et al. Dual Functional Peptide-Driven Nanoparticles for Highly Efficient Glioma-Targeting and Drug Codelivery. *Mol. Pharm.* **2016**, *13*, 1599–1607. [CrossRef] [PubMed]
103. Shilo, M.; Motiei, M.; Hana, P.; Popovtzer, R. Transport of nanoparticles through the blood–brain barrier for imaging and therapeutic applications. *Nanoscale* **2014**, *6*, 2146–2152. [CrossRef] [PubMed]
104. Betzer, O.; Shilo, M.; Opochinsky, R.; Barnoy, E.; Motiei, M.; Okun, E.; Yadid, G.; Popovtzer, R. The effect of nanoparticle size on the ability to cross the blood–brain barrier: An in vivo study. *Nanomedicine* **2017**, *12*, 1533–1546. [CrossRef]
105. Ulbrich, K.; Knobloch, T.; Kreuter, J. Targeting the insulin receptor: Nanoparticles for drug delivery across the blood–brain barrier (BBB). *J. Drug Target.* **2011**, *19*, 125–132. [CrossRef]
106. Tamaru, M.; Akita, H.; Kajimoto, K.; Sato, Y.; Hatakeyama, H.; Harashima, H. An apolipoprotein E modified liposomal nanoparticle: Ligand dependent efficiency as a siRNA delivery carrier for mouse-derived brain endothelial cells. *Int. J. Pharm.* **2014**, *465*, 77–82. [CrossRef]
107. Neves, A.R.; Queiroz, J.F.; Lima, S.A.C.; Reis, S. Apo E-Functionalization of Solid Lipid Nanoparticles Enhances Brain Drug Delivery: Uptake Mechanism and Transport Pathways. *Bioconj. Chem.* **2017**, *28*, 995–1004. [CrossRef]
108. Spencer, B.; Verma, I.; Desplats, P.; Morvinski, D.; Rockenstein, E.; Adame, A.; Masliah, E. A Neuroprotective Brain-penetrating Endopeptidase Fusion Protein Ameliorates Alzheimer Disease Pathology and Restores Neurogenesis. *J. Biol. Chem.* **2014**, *289*, 17917–17931. [CrossRef]
109. Zhang, B.; Sun, X.; Mei, H.; Wang, Y.; Liao, Z.; Chen, J.; Zhang, Q.; Hu, Y.; Pang, Z.; Jiang, X. LDLR-mediated peptide-22-conjugated nanoparticles for dual-targeting therapy of brain glioma. *Biomaterials* **2013**, *34*, 9171–9182. [CrossRef]
110. Karkan, D.; Pfeifer, C.; Vitalis, T.Z.; Arthur, G.; Ujiie, M.; Chen, Q.; Tsai, S.; Koliatis, G.; Gabathuler, R.; Jefferies, W.A. A unique carrier for delivery of therapeutic compounds beyond the blood-brain barrier. *PLoS ONE* **2008**, *3*, e2469. [CrossRef]
111. Nounou, M.I.; Adkins, C.E.; Rubinchik, E.; Terrell-Hall, T.B.; Afroz, M.; Vitalis, T.; Gabathuler, R.; Tian, M.M.; Lockman, P.R. Anti-cancer Antibody Trastuzumab-Melanotransferrin Conjugate (BT2111) for the Treatment of Metastatic HER2+ Breast Cancer Tumors in the Brain: An In-Vivo Study. *Pharm. Res.* **2016**, *33*, 2930–2942. [CrossRef] [PubMed]
112. Song, H.; Canup, B.S.B.; Ngo, V.L.; Denning, T.L.; Garg, P.; Laroui, H. Internalization of Garlic-Derived Nanovesicles on Liver Cells is Triggered by Interaction with CD98. *ACS Omega* **2020**, *5*, 23118–23128. [CrossRef] [PubMed]
113. Feral, C.C.; Nishiya, N.; Fenczik, C.A.; Stuhlmann, H.; Slepak, M.; Ginsberg, M.H. CD98hc (SLC3A2) mediates integrin signaling. *Proc. Natl. Acad. Sci. USA* **2005**, *102*, 355–360. [CrossRef]
114. Xiao, B.; Laroui, H.; Viennois, E.; Ayyadurai, S.; Charania, M.A.; Zhang, Y.; Zhang, Z.; Baker, M.T.; Zhang, B.; Gewirtz, A.T.; et al. Nanoparticles with Surface Antibody Against CD98 and Carrying CD98 Small Interfering RNA Reduce Colitis in Mice. *Gastroenterology* **2014**, *146*, 1289–1300.e19. [CrossRef]
115. Nawashiro, H.; Otani, N.; Shinomiya, N.; Fukui, S.; Nomura, N.; Yano, A.; Shima, K.; Matsuo, H.; Kanai, Y. The Role of CD98 in Astrocytic Neoplasms. *Hum. Cell* **2002**, *15*, 25–31. [CrossRef]
116. Kim, K.J.; Li, B.; Winer, J.; Armanini, M.; Gillett, N.; Phillips, H.S.; Ferrara, N. Inhibition of vascular endothelial growth factor-induced angiogenesis suppresses tumour growth in vivo. *Nature* **1993**, *362*, 841–844. [CrossRef] [PubMed]
117. Abakumov, M.A.; Nukolova, N.V.; Sokolsky-Papkov, M.; Shein, S.A.; Sandalova, T.O.; Vishwasrao, H.M.; Grinenko, N.F.; Gubsky, I.L.; Abakumov, A.M.; Kabanov, A.V.; et al. VEGF-targeted magnetic nanoparticles for MRI visualization of brain tumor. *Nanomed. Nanotechnol. Biol. Med.* **2015**, *11*, 825–833. [CrossRef]
118. Alves, A.D.C.S.; Lavayen, V.; Dias, A.D.F.; Bruinsmann, F.A.; Scholl, J.N.; Cé, R.; Visioli, F.; Battastini, A.M.O.; Guterres, S.S.; Figueiró, F.; et al. EGFRvIII peptide nanocapsules and bevacizumab nanocapsules: A nose-to-brain multitarget approach against glioblastoma. *Nanomedicine* **2021**, *16*, 1775–1790. [CrossRef]
119. Shein, S.A.; Nukolova, N.V.; Korchagina, A.A.; Abakumova, T.; Kiuznetsov, I.I.; Abakumov, M.; Baklaushev, V.P.; Gurina, O.I.; Chekhonin, V.P. Site-Directed Delivery of VEGF-Targeted Liposomes into Intracranial C6 Glioma. *Bull. Exp. Biol. Med.* **2015**, *158*, 371–376. [CrossRef]
120. Shein, S.A.; Kuznetsov, I.I.; Abakumova, T.; Chelushkin, P.S.; Melnikov, P.A.; Korchagina, A.A.; Bychkov, D.A.; Seregina, I.F.; Bolshov, M.A.; Kabanov, A.V.; et al. VEGF- and VEGFR2-Targeted Liposomes for Cisplatin Delivery to Glioma Cells. *Mol. Pharm.* **2016**, *13*, 3712–3723. [CrossRef]

121. Whittle, J.R.; Lickliter, J.D.; Gan, H.K.; Scott, A.M.; Simes, J.; Solomon, B.J.; MacDiarmid, J.A.; Brahmbhatt, H.; Rosenthal, M.A. First in human nanotechnology doxorubicin delivery system to target epidermal growth factor receptors in recurrent glioblastoma. *J. Clin. Neurosci.* **2015**, *22*, 1889–1894. [CrossRef] [PubMed]
122. An, Z.; Aksoy, O.; Zheng, T.; Fan, Q.-W.; Weiss, W.A. Epidermal growth factor receptor and EGFRvIII in glioblastoma: Signaling pathways and targeted therapies. *Oncogene* **2018**, *37*, 1561–1575. [CrossRef] [PubMed]
123. Reardon, D.A.; Nabors, L.; Mason, W.P.; Perry, J.R.; Shapiro, W.; Kavan, P.; Mathieu, D.; Phuphanich, S.; Cseh, A.; Fu, Y.; et al. Phase I/randomized phase II study of afatinib, an irreversible ErbB family blocker, with or without protracted temozolomide in adults with recurrent glioblastoma. *Neuro-Oncology* **2015**, *17*, 430–439. [CrossRef] [PubMed]
124. Westphal, M.; Maire, C.L.; Lamszus, K. EGFR as a Target for Glioblastoma Treatment: An Unfulfilled Promise. *CNS Drugs* **2017**, *31*, 723–735. [CrossRef]
125. Mortensen, J.H.; Jeppesen, M.; Pilgaard, L.; Agger, R.; Duroux, M.; Zachar, V.; Moos, T. Targeted Antiepidermal Growth Factor Receptor (Cetuximab) Immunoliposomes Enhance Cellular Uptake In Vitro and Exhibit Increased Accumulation in an Intracranial Model of Glioblastoma Multiforme. *J. Drug Deliv.* **2013**, *2013*, 209205. [CrossRef] [PubMed]
126. Erel-Akbaba, G.; Carvalho, L.A.; Tian, T.; Zinter, M.; Akbaba, H.; Obeid, P.J.; Chiocca, E.A.; Weissleder, R.; Kantarci, A.G.; Tannous, B.A. Radiation-Induced Targeted Nanoparticle-Based Gene Delivery for Brain Tumor Therapy. *ACS Nano* **2019**, *13*, 4028–4040. [CrossRef]
127. A Study to Evaluate the Safety, Tolerability and Immunogenicity of EGFR(V)-EDV-Dox in Subjects with Recurrent Glioblastoma Multiforme (GBM). Available online: https://www.clinicaltrials.gov/ct2/show/NCT02766699 (accessed on 23 May 2022).
128. Arif, W.M.; Elsinga, P.H.; Gasca-Salas, C.; Versluis, M.; Martínez-Fernández, R.; Dierckx, R.A.; Borra, R.J.; Luurtsema, G. Focused ultrasound for opening blood-brain barrier and drug delivery monitored with positron emission tomography. *J. Control Release* **2020**, *324*, 303–316. [CrossRef]
129. McDannold, N.; Vykhodtseva, N.; Hynynen, K. Use of Ultrasound Pulses Combined with Definity for Targeted Blood-Brain Barrier Disruption: A Feasibility Study. *Ultrasound Med. Biol.* **2007**, *33*, 584–590. [CrossRef]
130. Sirsi, S.; Borden, M. Microbubble Compositions, Properties and Biomedical Applications. *Bubble Sci. Eng. Technol.* **2009**, *1*, 3–17. [CrossRef]
131. Burgess, A.; Dubey, S.; Yeung, S.; Hough, O.; Eterman, N.; Aubert, I.; Hynynen, K. Alzheimer Disease in a Mouse Model: MR Imaging–guided Focused Ultrasound Targeted to the Hippocampus Opens the Blood-Brain Barrier and Improves Pathologic Abnormalities and Behavior. *Radiology* **2014**, *273*, 736–745. [CrossRef]
132. Yang, Q.; Zhou, Y.; Chen, J.; Huang, N.; Wang, Z.; Cheng, Y. Gene Therapy for Drug-Resistant Glioblastoma via Lipid-Polymer Hybrid Nanoparticles Combined with Focused Ultrasound. *Int. J. Nanomed.* **2021**, *16*, 185–199. [CrossRef] [PubMed]
133. Nance, E.; Timbie, K.; Miller, G.W.; Song, J.; Louttit, C.; Klibanov, A.L.; Shih, T.-Y.; Swaminathan, G.; Tamargo, R.J.; Woodworth, G.F.; et al. Non-invasive delivery of stealth, brain-penetrating nanoparticles across the blood-brain barrier using MRI-guided focused ultrasound. *J. Control Release* **2014**, *189*, 123–132. [CrossRef] [PubMed]
134. Timbie, K.F.; Afzal, U.; Date, A.; Zhang, C.; Song, J.; Miller, G.W.; Suk, J.S.; Hanes, J.; Price, R.J. MR image-guided delivery of cisplatin-loaded brain-penetrating nanoparticles to invasive glioma with focused ultrasound. *J. Control Release* **2017**, *263*, 120–131. [CrossRef] [PubMed]
135. Mead, B.P.; Mastorakos, P.; Suk, J.S.; Klibanov, A.L.; Hanes, J.; Price, R.J. Targeted gene transfer to the brain via the delivery of brain-penetrating DNA nanoparticles with focused ultrasound. *J. Control Release* **2016**, *223*, 109–117. [CrossRef] [PubMed]
136. Wang, J.B.; Di Ianni, T.; Vyas, D.B.; Huang, Z.; Park, S.; Hosseini-Nassab, N.; Aryal, M.; Airan, R.D. Focused Ultrasound for Noninvasive, Focal Pharmacologic Neurointervention. *Front. Neurosci.* **2020**, *14*, 675. [CrossRef]
137. Hynynen, K.; McDannold, N.; Sheikov, N.A.; Jolesz, F.A.; Vykhodtseva, N. Local and reversible blood–brain barrier disruption by noninvasive focused ultrasound at frequencies suitable for trans-skull sonications. *NeuroImage* **2005**, *24*, 12–20. [CrossRef]
138. Mehta, A.M.; Sonabend, A.M.; Bruce, J.N. Convection-Enhanced Delivery. *Neurotherapeutics* **2017**, *14*, 358–371. [CrossRef]
139. Saucier-Sawyer, J.K.; Seo, Y.-E.; Gaudin, A.; Quijano, E.; Song, E.; Sawyer, A.J.; Deng, Y.; Huttner, A.; Saltzman, W.M. Distribution of polymer nanoparticles by convection-enhanced delivery to brain tumors. *J. Control Release* **2016**, *232*, 103–112. [CrossRef]
140. Bobo, R.H.; Laske, D.W.; Akbasak, A.; Morrison, P.F.; Dedrick, R.L.; Oldfield, E.H. Convection-enhanced delivery of macromolecules in the brain. *Proc. Natl. Acad. Sci. USA* **1994**, *91*, 2076–2080. [CrossRef]
141. Kunwar, S.; Chang, S.; Westphal, M.; Vogelbaum, M.; Sampson, J.; Barnett, G.; Shaffrey, M.; Ram, Z.; Piepmeier, J.; Prados, M.; et al. Phase III randomized trial of CED of IL13-PE38QQR vs. Gliadel wafers for recurrent glioblastoma. *Neuro-Oncology* **2010**, *12*, 871–881. [CrossRef]
142. Vogelbaum, M.; Healy, A. Convection-enhanced drug delivery for gliomas. *Surg. Neurol. Int.* **2015**, *6* (Suppl. S1), S59–S67. [CrossRef] [PubMed]
143. Mastorakos, P.; Zhang, C.; Song, E.; Kim, Y.E.; Park, H.W.; Berry, S.; Choi, W.K.; Hanes, J.; Suk, J.S. Biodegradable brain-penetrating DNA nanocomplexes and their use to treat malignant brain tumors. *J. Control Release* **2017**, *262*, 37–46. [CrossRef] [PubMed]
144. Negron, K.; Khalasawi, N.; Lu, B.; Ho, C.-Y.; Lee, J.; Shenoy, S.; Mao, H.-Q.; Wang, T.-H.; Hanes, J.; Suk, J.S. Widespread gene transfer to malignant gliomas with In vitro-to-In vivo correlation. *J. Control Release* **2019**, *303*, 1–11. [CrossRef]
145. Ansari, M.A.; Chung, I.-M.; Rajakumar, G.; Alzohairy, M.A.; Alomary, M.; Thiruvengadam, M.; Pottoo, F.H.; Ahmad, N. Current Nanoparticle Approaches in Nose to Brain Drug Delivery and Anticancer Therapy—A Review. *Curr. Pharm. Des.* **2020**, *26*, 1128–1137. [CrossRef]

146. Dufes, C.; Olivier, J.-C.; Gaillard, F.; Gaillard, A.; Couet, W.; Muller, J.-M. Brain delivery of vasoactive intestinal peptide (VIP) following nasal administration to rats. *Int. J. Pharm.* **2003**, *255*, 87–97. [CrossRef]
147. Erdő, F.; Bors, L.A.; Farkas, D.; Bajza, Á.; Gizurarson, S. Evaluation of intranasal delivery route of drug administration for brain targeting. *Brain Res. Bull.* **2018**, *143*, 155–170. [CrossRef]
148. Su, Y.; Sun, B.; Gao, X.; Dong, X.; Fu, L.; Zhang, Y.; Li, Z.; Wang, Y.; Jiang, H.; Han, B. Intranasal Delivery of Targeted Nanoparticles Loaded With miR-132 to Brain for the Treatment of Neurodegenerative Diseases. *Front. Pharmacol.* **2020**, *11*, 1165. [CrossRef] [PubMed]
149. Kanazawa, T.; Taki, H.; Okada, H. Nose-to-brain drug delivery system with ligand/cell-penetrating peptide-modified polymeric nano-micelles for intracerebral gliomas. *Eur. J. Pharm. Biopharm.* **2020**, *152*, 85–94. [CrossRef] [PubMed]
150. Upadhaya, P.G.; Pulakkat, S.; Patravale, V.B. Nose-to-brain delivery: Exploring newer domains for glioblastoma multiforme management. *Drug Deliv. Transl. Res.* **2020**, *10*, 1044–1056. [CrossRef]
151. Craparo, E.F.; Musumeci, T.; Bonaccorso, A.; Pellitteri, R.; Romeo, A.; Naletova, I.; Cucci, L.M.; Cavallaro, G.; Satriano, C. mPEG-PLGA Nanoparticles Labelled with Loaded or Conjugated Rhodamine-B for Potential Nose-to-Brain Delivery. *Pharmaceutics* **2021**, *13*, 1508. [CrossRef]
152. Chu, L.; Wang, A.; Ni, L.; Yan, X.; Song, Y.; Zhao, M.; Sun, K.; Mu, H.; Liu, S.; Wu, Z.; et al. Nose-to-brain delivery of temozolomide-loaded PLGA nanoparticles functionalized with anti-EPHA3 for glioblastoma targeting. *Drug Deliv.* **2018**, *25*, 1634–1641. [CrossRef] [PubMed]
153. Kanazawa, T.; Morisaki, K.; Suzuki, S.; Takashima, Y. Prolongation of Life in Rats with Malignant Glioma by Intranasal siRNA/Drug Codelivery to the Brain with Cell-Penetrating Peptide-Modified Micelles. *Mol. Pharm.* **2014**, *11*, 1471–1478. [CrossRef] [PubMed]
154. Cornelio, D.; Roesler, R.; Schwartsmann, G. Gastrin-releasing peptide receptor as a molecular target in experimental anticancer therapy. *Ann. Oncol.* **2007**, *18*, 1457–1466. [CrossRef] [PubMed]
155. Gao, W.; Zhang, Y.; Zhang, Q.; Zhang, L. Nanoparticle-Hydrogel: A Hybrid Biomaterial System for Localized Drug Delivery. *Ann. Biomed. Eng.* **2016**, *44*, 2049–2061. [CrossRef]
156. Bastiancich, C.; Vanvarenberg, K.; Ucakar, B.; Pitorre, M.; Bastiat, G.; Lagarce, F.; Préat, V.; Danhier, F. Lauroyl-gemcitabine-loaded lipid nanocapsule hydrogel for the treatment of glioblastoma. *J. Control Release* **2016**, *225*, 283–293. [CrossRef]
157. Pillai, J.J.; Thulasidasan, A.K.T.; Anto, R.J.; Chithralekha, D.N.; Narayanan, A.; Kumar, G.S.V. Folic acid conjugated cross-linked acrylic polymer (FA-CLAP) hydrogel for site specific delivery of hydrophobic drugs to cancer cells. *J. Nanobiotechnol.* **2014**, *12*, 25. [CrossRef]
158. Basso, J.; Miranda, A.; Nunes, S.; Cova, T.; Sousa, J.; Vitorino, C.; Pais, A. Hydrogel-Based Drug Delivery Nanosystems for the Treatment of Brain Tumors. *Gels* **2018**, *4*, 62. [CrossRef]
159. Eckmann, D.M.; Composto, R.J.; Tsourkas, A.; Muzykantov, V.R. Nanogel carrier design for targeted drug delivery. *J. Mater. Chem. B* **2014**, *2*, 8085–8097. [CrossRef]
160. Park, H.; Guo, X.; Temenoff, J.S.; Tabata, Y.; Caplan, A.I.; Kasper, F.K.; Mikos, A.G. Effect of Swelling Ratio of Injectable Hydrogel Composites on Chondrogenic Differentiation of Encapsulated Rabbit Marrow Mesenchymal Stem Cells In Vitro. *Biomacromolecules* **2009**, *10*, 541–546. [CrossRef]
161. Žuržul, N.; Ilseng, A.; Prot, V.E.; Sveinsson, H.M.; Skallerud, B.H.; Stokke, B.T. Donnan Contribution and Specific Ion Effects in Swelling of Cationic Hydrogels are Additive: Combined High-Resolution Experiments and Finite Element Modeling. *Gels* **2020**, *6*, 31. [CrossRef]
162. Shatsberg, Z.; Zhang, X.; Ofek, P.; Malhotra, S.; Krivitsky, A.; Scomparin, A.; Tiram, G.; Calderon, M.; Haag, R.; Satchi-Fainaro, R. Functionalized nanogels carrying an anticancer microRNA for glioblastoma therapy. *J. Control Release* **2016**, *239*, 159–168. [CrossRef] [PubMed]
163. Chen, Z.; Liu, F.; Chen, Y.; Liu, J.; Wang, X.; Chen, A.T.; Deng, G.; Zhang, H.; Liu, J.; Hong, Z.; et al. Targeted Delivery of CRISPR/Cas9-Mediated Cancer Gene Therapy via Liposome-Templated Hydrogel Nanoparticles. *Adv. Funct. Mater.* **2017**, *27*, 1703036. [CrossRef] [PubMed]
164. Qin, M.; Zong, H.; Kopelman, R. Click Conjugation of Peptide to Hydrogel Nanoparticles for Tumor-Targeted Drug Delivery. *Biomacromolecules* **2014**, *15*, 3728–3734. [CrossRef] [PubMed]
165. Madsen, S.J.; Baek, S.-K.; Makkouk, A.R.; Krasieva, T.; Hirschberg, H. Macrophages as Cell-Based Delivery Systems for Nanoshells in Photothermal Therapy. *Ann. Biomed. Eng.* **2012**, *40*, 507–515. [CrossRef] [PubMed]
166. Ibarra, L.E.; Beaugé, L.; Arias-Ramos, N.; Rivarola, V.A.; Chesta, C.A.; López-Larrubia, P.; Palacios, R.E. Trojan horse monocyte-mediated delivery of conjugated polymer nanoparticles for improved photodynamic therapy of glioblastoma. *Nanomedicine* **2020**, *15*, 1687–1707. [CrossRef]
167. Xue, J.; Zhao, Z.; Zhang, L.; Xue, L.; Shen, S.; Wen, Y.; Wei, Z.; Wang, L.; Kong, L.; Sun, H.; et al. Neutrophil-mediated anticancer drug delivery for suppression of postoperative malignant glioma recurrence. *Nat. Nanotechnol.* **2017**, *12*, 692–700. [CrossRef]
168. Wang, X.; Gao, J.-Q.; Ouyang, X.; Wang, J.; Sun, X.; Lv, Y. Mesenchymal stem cells loaded with paclitaxel–poly(lactic-co-glycolic acid) nanoparticles for glioma-targeting therapy. *Int. J. Nanomed.* **2018**, *13*, 5231–5248. [CrossRef]
169. Roger, M.; Clavreul, A.; Venier-Julienne, M.-C.; Passirani, C.; Sindji, L.; Schiller, P.; Montero-Menei, C.; Menei, P. Mesenchymal stem cells as cellular vehicles for delivery of nanoparticles to brain tumors. *Biomaterials* **2010**, *31*, 8393–8401. [CrossRef]

170. Clavreul, A.; Lautram, N.; Franconi, F.; Passirani, C.; Montero-Menei, C.; Menei, P.; Tetaud, C.; Montagu, A.; Laine, A.-L.; Vessieres, A. Targeting and treatment of glioblastomas with human mesenchymal stem cells carrying ferrociphenol lipid nanocapsules. *Int. J. Nanomed.* **2015**, *10*, 1259–1271. [CrossRef]
171. Arshad, A.; Yang, B.; Bienemann, A.S.; Barua, N.U.; Wyatt, M.J.; Woolley, M.; Johnson, D.E.; Edler, K.; Gill, S.S. Convection-Enhanced Delivery of Carboplatin PLGA Nanoparticles for the Treatment of Glioblastoma. *PLoS ONE* **2015**, *10*, e0132266. [CrossRef]
172. Danhier, F.; Messaoudi, K.; Lemaire, L.; Benoit, J.P.; Lagarce, F. Combined anti-Galectin-1 and anti-EGFR siRNA-loaded chitosan-lipid nanocapsules decrease temozolomide resistance in glioblastoma: In vivo evaluation. *Int. J. Pharm.* **2015**, *481*, 154–161. [CrossRef] [PubMed]
173. Lesniak, W.G.; Oskolkov, N.; Song, X.; Lal, B.; Yang, X.; Pomper, M.; Laterra, J.; Nimmagadda, S.; McMahon, M.T. Salicylic Acid Conjugated Dendrimers Are a Tunable, High Performance CEST MRI NanoPlatform. *Nano Lett.* **2016**, *16*, 2248–2253. [CrossRef] [PubMed]
174. Young, J.S.; Bernal, G.; Polster, S.; Nunez, L.; Larsen, G.F.; Mansour, N.; Podell, M.; Yamini, B. Convection-Enhanced Delivery of Polymeric Nanoparticles Encapsulating Chemotherapy in Canines with Spontaneous Supratentorial Tumors. *World Neurosurg.* **2018**, *117*, e698–e704. [CrossRef]
175. Panja, S.; Dey, G.; Bharti, R.; Mandal, P.; Mandal, M.; Chattopadhyay, S. Metal Ion Ornamented Ultrafast Light-Sensitive Nanogel for Potential in Vivo Cancer Therapy. *Chem. Mater.* **2016**, *28*, 8598–8610. [CrossRef]
176. Hare, J.I.; Lammers, T.; Ashford, M.B.; Puri, S.; Storm, G.; Barry, S.T. Challenges and strategies in anti-cancer nanomedicine development: An industry perspective. *Adv. Drug Deliv. Rev.* **2017**, *108*, 25–38. [CrossRef] [PubMed]
177. Xu, Y.; Shen, M.; Li, Y.; Sun, Y.; Teng, Y.; Wang, Y.; Duan, Y. The synergic antitumor effects of paclitaxel and temozolomide co-loaded in mPEG-PLGA nanoparticles on glioblastoma cells. *Oncotarget* **2016**, *7*, 20890–20901. [CrossRef] [PubMed]

polymers

MDPI

Review

Approaches to Improve Macromolecule and Nanoparticle Accumulation in the Tumor Microenvironment by the Enhanced Permeability and Retention Effect

Victor Ejigah [1], Oluwanifemi Owoseni [1], Perpetue Bataille-Backer [1], Omotola D. Ogundipe [1], Funmilola A. Fisusi [1,2] and Simeon K. Adesina [1,*]

1 Department of Pharmaceutical Sciences, College of Pharmacy, Howard University, Washington, DC 20059, USA; victor.eligah@bison.howard.edu (V.E.); oluwanifemi.owoseni@bison.howard.edu (O.O.); perpetue.batailleba@bison.howard.edu (P.B.-B.); omotola.ogundipe@bison.howard.edu (O.D.O.); funmilolaadesodun.fi@howard.edu (F.A.F.)
2 Faculty of Pharmacy, Obafemi Awolowo University, Ile-Ife 220005, Nigeria
* Correspondence: simeon.adesina@howard.edu; Tel.: +1-202-250-5304; Fax: +1-202-806-7805

Abstract: Passive targeting is the foremost mechanism by which nanocarriers and drug-bearing macromolecules deliver their payload selectively to solid tumors. An important driver of passive targeting is the enhanced permeability and retention (EPR) effect, which is the cornerstone of most carrier-based tumor-targeted drug delivery efforts. Despite the huge number of publications showcasing successes in preclinical animal models, translation to the clinic has been poor, with only a few nano-based drugs currently being used for the treatment of cancers. Several barriers and factors have been adduced for the low delivery efficiency to solid tumors and poor clinical translation, including the characteristics of the nanocarriers and macromolecules, vascular and physiological barriers, the heterogeneity of tumor blood supply which affects the homogenous distribution of nanocarriers within tumors, and the transport and penetration depth of macromolecules and nanoparticles in the tumor matrix. To address the challenges associated with poor tumor targeting and therapeutic efficacy in humans, the identified barriers that affect the efficiency of the enhanced permeability and retention (EPR) effect for macromolecular therapeutics and nanoparticle delivery systems need to be overcome. In this review, approaches to facilitate improved EPR delivery outcomes and the clinical translation of novel macromolecular therapeutics and nanoparticle drug delivery systems are discussed.

Keywords: enhanced permeability and retention effect; nanotechnology; tumor microenvironment; Zwitterionic polymers; synthetic microbe; macrophages; liposomes

Citation: Ejigah, V.; Owoseni, O.; Bataille-Backer, P.; Ogundipe, O.D.; Fisusi, F.A.; Adesina, S.K. Approaches to Improve Macromolecule and Nanoparticle Accumulation in the Tumor Microenvironment by the Enhanced Permeability and Retention Effect. *Polymers* **2022**, *14*, 2601. https://doi.org/10.3390/polym14132601

Academic Editor: Stephanie Andrade

Received: 26 May 2022
Accepted: 18 June 2022
Published: 27 June 2022

1. Introduction

Improved clinical outcomes in the treatment of some cancers have not been observed despite tremendous efforts in the development of new chemotherapeutic agents. This is largely due to the failure to deliver these agents selectively to tumors and the lack of precision in targeting cancer cells [1]. This poor targeting to tumors greatly increases the propensity for debilitating off-target consequences and minimizes the magnitude of therapeutic efficacy. Innovations in the field of nanotechnology, immunology, chemistry, and pharmaceutical technology have led to the development of a variety of drug delivery systems aimed at improving the plasma half-life, biodistribution, and the target-site accumulation of chemotherapeutic drugs [2,3].

Nanoparticles for drug delivery are carrier systems in the nanometer size range, composed of biocompatible and biodegradable natural and/or synthetic polymers, self-assembled lipids, or inorganic materials, capable of carrying payloads such as small molecule therapeutics, peptides, proteins, or nucleic acids and delivering their cargoes in a controlled manner at the target site [4]. These delivery systems have shown promise in preclinical trials, and some nanoparticles, such as Doxil^{TM} (doxorubicin/liposome) and Abraxane^{TM} (paclitaxel/albumin), designed to modify the pharmacokinetics of existing chemotherapeutic agents, are already routinely used in clinical practice [2,5].

The enthusiasm around nanocarrier systems has led to a plethora of publications demonstrating the efficacy of different constructs in a variety of animal cancer models; however, a meta-analysis of published studies [6] revealed that in the past decade, only about 0.7% tumor site accumulation was achieved for particles with sizes below 100 nm in solid tumors. While this may appear to be low, nanomedicines have demonstrated substantially higher delivery efficiencies than most conventional chemotherapeutic formulations in relative terms [7,8]. It has been reported that the dense extracellular matrix, high interstitial fluid pressure, and non-uniform blood perfusion limit nanoparticle accumulation in solid tumors [7,9]. The preferential accumulation of drug-loaded nanoparticles in neoplastic tissues is referred to as passive targeting [9,10]. Passive targeting is facilitated by the enhanced permeability and retention (EPR) effect [11], i.e., the mechanism by which high molecular weight drug carriers accumulate in the tumor microenvironment (TME) due to increased vascular permeability and the nanometer size of nanoparticles [12]. The TME also possesses impaired lymphatic drainage that prevents the efficient removal of these macromolecules or nanoparticles, thus enhancing their retention within neoplastic tissues [13–15].

The concept of EPR originated from a landmark study in 1986 by Matsumura and Maeda on the mechanisms of the tumoritropic accumulation of proteins and chemotherapeutic agents [16]. In this study, an increase in the accumulation and uptake of a derivatized styrene-maleic acid polymer loaded with neocarzinostatin (SMANCS) was observed in tumor cells relative to the native neocarzinostatin. SMANCS has a molecular size of 16 kDa and can bind serum albumin (67 kDa) to become a larger molecule [17]. The superior accumulation of SMANCS in tumor tissues offered prolonged duration of action and increased therapeutic efficacy. The authors opined that EPR is made possible by an increased vascular permeability, a dysfunctional lymphatic drainage system, and the relative size of nanoparticles. They showed that very small molecules will traverse biological barriers easily, while larger molecules like nanoparticles may be filtered through highly vascularized and permeable barriers. The dependence of EPR on molecular size was further demonstrated via the encapsulation of SMANCS in liposomes (ethiodized poppyseed oil); an outcome that stimulated research on the use of liposomes as chemotherapeutic drug delivery systems [12].

In contrast to tumor blood supply, blood supply to healthy tissues is well organized with tight endothelial junctions. Healthy tissues also possess a functional lymphatic drainage system that rids the extracellular matrix of nonresident molecules. The difference between healthy and neoplastic blood supply forms the basis of the selectivity inherent in the EPR effect. The effectiveness of EPR has been validated [18,19] and has led to the discovery and development of carrier-based therapeutic products [20] that are currently in clinical use. With advances in the field of nanoparticle drug delivery, it has become apparent that the therapeutic efficacy of passively targeted nanomedicines is vastly influenced by the heterogeneity of the intensity of the EPR effect within a tumor, at different stages of a tumor, and among individual tumors [21], as well as other physiological barriers including the reticuloendothelial system.

In this review, we will discuss the principle of EPR as a function of the features of the tumor microenvironment. We will also summarize the challenges of EPR-based passive tumor targeting. Our goal is to explore strategies and approaches for enhancing the accumulation of macromolecules and nanoparticles within the tumor microenvironment for improved therapeutic outcomes.

2. Principle of EPR

The unique vascular physiology of solid tumors is the basis for the EPR effect. First, the microvasculature in solid tumor (Figure 1) tissues lacks basement membrane support, making it unresponsive to physiological cues and stimuli that regulate blood flow [10]. Second, rapid metabolism in tumor cells leads to a high demand for nutrients and oxygen, which drives angiogenesis at high rates and prevents adequate vessel maturation [22]. Taken together, these factors create a hyperpermeable state with fenestrations in the endothelial cell lining of newly formed tumor blood vessels [23]. The leaky, heterogenous, and disorganized nature of neoplastic blood vessels allows the escape of macromolecules and nanoparticles into tumor tissues. In addition, nanoparticle entrapment and retention in the tumor interstitium occur because of the tumor's dysfunctional lymphatic drainage system, preventing proper drainage [3].

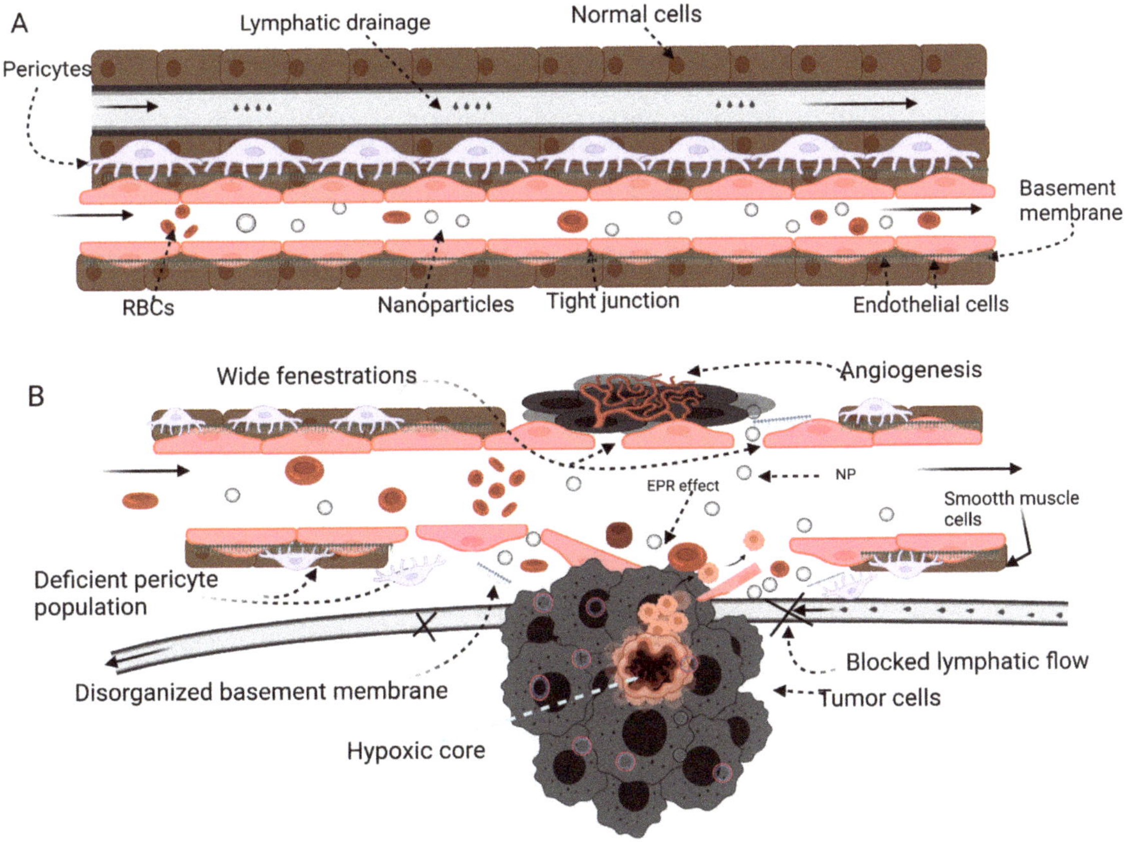

Figure 1. Comparison of the microenvironment of (**A**) healthy and (**B**) tumor tissues. The tumor microenvironment shows the disorganized components (hypoxic core, blocked lymphatic drainage, deficient pericyte population, disorganized basement membrane and wide fenestration) that are exploited for enhancing EPR effect. Created with BioRender.com Accessed on 25 July 2021.

Thus, tumor-selective anticancer drug delivery can be achieved by attaching, conjugating, or encapsulating low-MW antineoplastic drugs to macromolecular carriers (dendrimers, liposomes, polymers, and micelles) in single or multiple-drug formulations [24].

The EPR effect-mediated tumor accumulation of macromolecules and nanoparticles has been demonstrated in several studies [25] with carrier molecular sizes of more than

40 kDa and particle sizes of 6–8 nm or larger [26]. For passive targeting via the EPR effect, several criteria must be met. First, the macromolecular carrier must remain in systemic circulation for a considerable length of time because the accumulation at the tumor site via the EPR effect is a time-dependent occurrence [18]. Second, macromolecular carriers must possess critical quality attributes such as: (i) appropriate size to allow extravasation and accumulation in tumors via the wide fenestrations found only in tumor blood supply; (ii) an ideal surface charge to avoid opsonin aggregation; (iii) an optimum surface chemistry that allows tissue penetration; (iv) stability of macromolecule integrity to avoid nonspecific release; and (v) biocompatibility [24]. Finally, the nanoparticles must have a uniform size distribution to avoid aggregation due to a high surface to volume ratio which predisposes to recognition by the reticuloendothelial system, especially the Kupffer cells, liver sinusoidal endothelial cells, and liver stellate cells that are responsible for the elimination of 30–99% of injected nanoparticles and macromolecules [27].

3. Essential Considerations for EPR Effect

The translation of various macromolecular and nanocarrier constructs from successful preclinical experiments to clinical practice has been largely unsuccessful, despite all the efforts made to optimize nanoparticle design to promote active or passive targeting [28]. A meta-analysis by Wilhelm [6] reported that for all the published articles evaluated, only 0.7% of the injectable dose (ID) on average was found in the tumor. Of all the parameters analyzed, only hydrodynamic size < 10 nm (0.7%; p = 0.0001), neutral zeta potential (0.7%; p = 0.0068), spherical shape (0.8%; p = 0.00479), and orthotopic tumor models (1.1%; p = 0.001) correlated significantly to the highest percentage delivered. Upon stratification to determine delivery efficiency based on type of nanomaterials, organic nanomaterials showed significantly superior efficiency compared to inorganic materials for the parameters mentioned above. Therefore, it is logical to optimize these attributes in organic nanocarriers to engender greater delivery efficiency and increase accumulation of macromolecules and nanoparticles within the TME.

The differences that exist between tumor xenografts in mice models and tumors in man must be considered in translating preclinical successes to clinical settings [2,29–31]. Most cell line grafts in animal models are much less heterogeneous than human tumors because the experimental models are usually standardized—same genetic background, origin, and age—to allow valid statistical analyses. Human tumors, on the other hand, can be highly heterogeneous, i.e., ranging from 1 mm to as large as 100 mm or more, and are observed in individuals of different ages, lifestyles, and genetic backgrounds. In addition, the possibility that currently used xenograft models are not suitable for direct comparison and extrapolation to human tumors must also be considered [2,20,32,33]. For instance, the rate of development of tumors in animal models is faster than in humans; this rapid rate of tumor growth results in accelerated angiogenesis, leading to unusually disorganized vascular walls which are amenable to EPR [3]. Similarly, the rate of metabolism in mice is faster than that in humans; this allows a more aggressive dosing of macromolecules [34].

The total amount of nanoparticles required to deliver the desired payload is another factor to be taken into consideration. Data from the study by Wilhelm et al. shows that to achieve IC_{50} in a tumor volume of 0.5 cm^3 for a mouse of about 20 g body weight, a total of 1.2×10^{12} nanoparticles, or a dose of 6.5 mg kg^{-1}, must be injected, provided the nanoparticles encapsulate 20 wt% of the drug. They also posited that an increase in this dose to a total of 2.8×10^{12} nanoparticles or 15.7 mg kg^{-1} will be necessary if drugs are loaded on the particle surface at a surface density of one drug molecule per nm^2. This dose is feasible for preclinical administration in mice. Translating this to man on similar metrics would require a dose of 2.7×10^{14} drug-encapsulated nanoparticles, or 6.4×10^{14} surface-loaded nanoparticles, based on the surface-area dosing strategy [35]. This is a challenging proposition, as it would require scaling up production of nanoparticles, which may lead to issues of colloidal instability, aggregation, short shelf life, systemic toxicity, and poor bioavailability due to elimination by the reticuloendothelial system [6]. Other limitations in the translation of experimental results of EPR in animal models to the clinic include differences in genetics, immunology, syngeneic attributes, and non-orthotopic tumor grafting [36].

Additional differences between tumor xenografts in mice models and tumors in man are seen in the relative size of tumors to the host body weight. Mice tumors are usually grown to more than 10% of the animal's total body weight before treatment is administered. The high tumor volume relative to the total body weight in mice allows for significant contact with circulating drug loaded nanoparticles, leading to better efficacy outcomes [3]. In contrast, some human tumors constitute just about 0.005% of the total body weight of a 70 kg man. Therefore, for a chance to significantly encounter circulating drug loaded nanoparticles, a tumor requires considerable exposure, i.e., of an average of ten days or more [3]. Hence, a loaded nanoparticle has a higher propensity to There appear to be a plethora of challenges in achieving adequate accumulation of nanoparticles in the TME based on the essential factors discussed above. Therefore, to successfully translate preclinical efforts to human subjects, the fundamental strategy will be to improve the delivery efficiency of nanoparticles to tumors [6]. Part of this strategy would involve creating tumor models that would properly replicate the human TME to produce outcomes that are reproducible post-translation. To achieve this, some studies have attempted to slow down the rate of angiogenesis in mice by using anti-angiogenic factors to encounter and extravasate into a mice xenograft tumor by EPR than it does a human tumor produce vessels that are less permeable to large nanoparticles [22]. Other novel experimental tumor models, including the use of laboratory methods, organ-on-a-chip methods, ex vivo systems, and dynamic organoids are becoming increasingly popular [37].

The meta-analysis by Wilhelm et al. has not gone unchallenged. A recent publication [23] asserted that the analysis did not take into consideration important factors such as tumor size, effectiveness of drug delivery at target site, and tumor heterogeneity. They observed that the analysis focused on percentage injected dose while ignoring the ratio of drug concentration in the tumor to that in blood. Similarly, another study [38] performed a reanalysis on the same data set used by Wilhelm et al. and concluded that based on traditional pharmacokinetic (PK) evaluation, the %ID in tumor was poorly correlated with standard PK metrics that describe nanoparticle tumor delivery (AUC_{tumor}/AUC_{blood} ratio) and is only moderately associated with maximal tumor concentration. The author proposed that a better interpretation of the finding of Wilhelm et al. should be that an average of 0.67% of ID was found in the tumor per hour interval throughout the entire PK evaluation period. Using the same dataset and based on the more appropriate AUC_{tumor}/AUC_{blood} ratio metric, Price et al. showed that the exposure of tumors to overall plasma nanoparticles (AUC_{blood}) was 76.12%, i.e., a 100-fold greater value than that based on %ID. If this were the case, then it is expected that preclinical trials should translate successfully to clinical settings, but that is not the experience so far. Therefore, the imperative to explore further and improve EPR cannot be overemphasized.

4. Approaches and Techniques to Improve EPR Effect

4.1. Modification of Physicochemical Properties of Nanoparticles and Macromolecules

4.1.1. Particle Size

Particle size is a fundamental characteristic of nanoparticles that significantly influences the efficiency of tumor targeted delivery systems with respect to circulation, biodistribution, tumor penetration/accumulation, and cellular uptake [39]. The EPR effect is substantially dependent on the particle size of nanoparticles. In general, small molecules or particles with sizes less than the renal glomerular filtering threshold of 40 kDa or 6–8 nm are largely removed via renal excretion or via the liver by the stellate cells, and are eliminated in urine or feces [40]. The optimum nanoparticle size is largely dependent on the type of tumor; a small nanoparticle size does not necessarily translate to improved tumor delivery via the EPR effect [39]. It has been demonstrated that small size is critical to tumor tissue penetration [20]; however, small-sized macromolecules and nanoparticles are easily extruded by the interstitial pressure within the tumor environment [41]. On the flip side, large molecules are poor at penetrating tumor cells, but when effectively delivered, are more likely to be retained [42].

The large size of nano-based drugs, therefore, plays a critical role in EPR-dependent drug accumulation in tumors, because they are less likely to be eliminated [24]. Most nanomedicines are designed in the range of 10–200 nm in diameter, but even particle sizes up to 1–2 μm (bacteria cells) could also accumulate in tumor cells via the EPR while sparing healthy tissues [43]. Such macromolecules show prolonged circulation time (increased $t_{1/2}$) and high area under the concentration-time curve (AUC) in plasma, thus enabling gradual permeation and accumulation within tumor tissues [19].

Striking the right balance between tissue permeation and retention of nanoparticles has been a subject of research in the recent past. In one study, an evaluation of the permeation and retention ability of nanoparticles of different sizes showed that 60% of nanoparticles in the range of 100–200 nm circulated longer in plasma compared to 20% of nanoparticles with sizes less than 50 nm and 20% of nanoparticles greater than 250 nm. Another study reported a four-fold uptake of nanoparticles between 100–200 nm compared to nanoparticles with sizes less than 50 nm and greater than 300 nm [44]. These results clearly show that nanoparticles with small particle sizes may not be ideal in providing adequate tumor accumulation in some circumstance, considering how quickly they could be extruded from the tumor microenvironment.

A strategy designed to strike the balance between tissue penetration and retention of nanoparticles has been demonstrated in the formulation of large gelatin-based nanoparticles of 100 nm that can be degraded by tumor-associated matrix metalloproteinases (MMPs) to 10 nm upon extravasation to promote tissue penetration [45].

4.1.2. Surface Charge

The surface charge of nanoparticles and macromolecules is a critical physicochemical property that should be tailored to prolong circulation half-life and enhance accumulation in tumors via the EPR effect [24]. This is because surface charge is associated with solubility, aggregation, biocompatibility, and the ability of nanoparticles to move across biological barriers [33]. The vascular endothelial luminal surface and membranes of the cells of the liver and spleen have very high negative charge, so nanoparticles with positive charges or cationic polymers are easily bound by these surfaces, resulting in a rapid decrease in plasma concentration due to removal by the reticuloendothelial system (RES) [46]. Besides, strong electrostatic interaction between cell membranes and positively charged nanoparticles has been shown to cause cytotoxicity [47]. That notwithstanding, some studies have shown that positively charged nanoparticles can still achieve good cell uptake depending on the type of cell. For instance, positively charged nanoparticles were used to induce a significant adaptive immunologic response to a pulmonary vaccine compared to a negatively charged control. Positively charged nanoparticles were more associated to dendritic cells that are required for adaptive immunity, unlike negatively charged nanoparticles, that were more associated with alveolar macrophages [48]. In another instance, cationic nanoparticles have been shown to attract proteins which form complexes that enhance cell uptake. The mechanism is largely due to the nature of the receptors found on specific cells. The proteins that adsorb unto cationic polymers interact with phagocytic cells that have an abundance of receptors on their cell membranes that facilitate significant internalization. When cationic polymers were used in an environment that lacks opsonins, the extent of cell binding and uptake was significantly reduced [49]. Furthermore, uptake of inorganic nanoparticles modified with cationic polymer was considerably increased in SK-BR-3 cell lines compared to nanoparticles with negative charges [50]. It should be noted that positively charged particles may facilitate endosomal escape via the 'proton sponge effect' [51], a strategy that circumvents the degradative effect of the acidic and enzyme-rich endo-lysosomal compartment on drug cargo [19].

Strongly negatively charged particles are also rapidly removed by the RES; hence the importance of considering the effect of surface charge during the development of nanoparticles [24]. Research has shown that nanoparticles with neutral and slightly negatively charged surfaces adsorb less plasma opsonins and demonstrate low non-specific cell uptake [52]. To demonstrate this concept, a study evaluated phagocytic and non-phagocytic cell uptake of positively charged rhodamine B (RhB)-labeled carboxymethyl chitosan conjugated nanoparticles (RhB-CMCNP) and negatively charged chitosan hydrochloride conjugated nanoparticles (RhB-CHNP) in various cell lines (L02, SMMC-7721, HEK 293, 786-O, HFL-I or A549 cells). The result showed a high uptake of positively charged nanoparticles (14.8–34.6 mV) by phagocytic cells compared to negatively charged nanoparticles (−13.2 to −38.4 mV). On the flipside, slightly negatively charged nanoparticles were significantly internalized by non-phagocytic cells compared to either positively charged or highly negatively charged nanoparticles [53]. The superior uptake of nanoparticles with small negative charge was likely due to the reduced repulsive forces between the cell membrane and charged species on nanoparticles, as observed from the progressive decrease in cell uptake with increasing negative zeta potential. This result was validated in a biodistribution study of RhB-CMCNP and RhB-CHNP in H-22 tumor bearing mice.

Thus, for effective nanoparticle delivery to tumors, one would desire a neutral or slightly negative nanoparticle surface charge upon intravenous administration, but a switch to positive charge upon arrival at the tumor site [13,19], as demonstrated by the design of a switchable zwitterionic nanoparticle based on TME cues [54]. In this instance, a docetaxel loaded co-polymer was linked to a negatively charged group, i.e., dimethyl maleic acid (DMA), via a pH sensitive amide linker to avoid adsorption of opsonins in plasma. However, at the low pH within the tumor environment, the amide crosslinker was cleaved to release DMA, thereby exposing the positively charged amine groups to enhance adsorptive interaction with the cell membrane.

4.1.3. Shape

The shape of nanoparticles is one of the most researched physicochemical properties (after size) that can be modulated to improve the delivery efficiency of nanocarriers [55]. The shape of nanoparticles has been shown to have remarkable impact on tissue targeting, internalization, immune cell association, cell adhesion, and uptake [56]. The meta-analyses by Wilhelm et al. showed the importance of shape to the overall delivery efficiency of nanoparticles. Their data showed that rod shaped (0.8%) nanoparticles had the highest delivery efficiency compared to spherical shape (0.7%) and others [6]. However, other researchers have explored discoidal shaped nanoparticles because of their unique tumbling and margination dynamics that favor vessel wall interactions considerably more than spherical particles, with implications for better extravasation into the TME [57].

It has also been reported that the shape of nanoparticles is critical to the extent of clearance by macrophages via phagocytosis. This is because geometric parameters, such as curvature and aspect ratio, affect uptake [13]. The kinetics of phagocytosis reveal that particles possessing a length of normalized curvature (designated as Ω) $\leq 45°$, as observed with spherical particles, undergo faster internalization than particles with $\Omega \geq 45°$. The influence of shape on tumor internalization has led to further evaluations of ellipsoidal, cylindrical, and discoidal shaped nanoparticles [13]. The ability of filamentous polymicelles to align with blood flow has been exploited due to their high aspect ratios (>10) and longitudinal length (10µm) that ensures successful retention in the blood for up to a week [53]. This filamentous property enabled the delivery and accumulation of high levels of paclitaxel in tumors when compared to spherical micelles [58].

4.1.4. Elasticity

Elasticity or deformability and biodegradability should also be considered in a bid to enhance EPR. This is because organs like the liver and spleen have fenestrated endothelia that filter rigid particles with diameters that exceed the cut-off limit of their inter-endothelial fenestrae [13]. It has been reported that by decreasing the nanoparticle elastic modulus by eight-fold, the blood circulation half-life thereof can be increased by a factor of thirty [27], hence extending the residence time needed for efficient EPR. In a study that evaluated the influence of elasticity on cellular and tumor accumulation of a PEG-based hydrogel nanoparticle formulation, softer nanoparticles with elastic modulus of 10 kpa demonstrated longer circulation time and enhanced tumor targeting compared to hard nanoparticles with high elastic modulus (3000 kpa). The authors reported a 3.5-fold uptake of hard nanoparticles by macrophages, which explains their short circulation time [59]. Guo et al. [60] investigated the impact of elasticity on cellular uptake of nanoliposomes (NLP), uncrosslinked nanolipogel, and cross-linked nanolipogel (NLG) with varying Young's elastic moduli (NLP: +45 kpa; and NLG 1.6 ± 0.6 MPa to 19 ± 5 MPa). Their data demonstrate that NLPs and soft NLGs accumulated significantly more in tumors, whereas NLGs with high elastic moduli preferentially accumulated in the liver [60]. Evidence from the work by Hui et al. [61] demonstrated higher uptake of soft silica nanoparticles (560 kPa) compared to hard silica nanoparticles (1.18 GPa) by SKOV3 cell lines, while hard silica nanoparticles were taken up faster by RAW264.7 phagocytic cells [61]. A possible explanation for the increased phagocytosis of hard particles is that very soft particles can potentially undergo deformation in response to the forces of phagocytosis, which can lead to changes in the particle radius of curvature. Deformation that leads to changes such as elongation of nanoparticle shape can reduce susceptibility to phagocytosis. Shape elongation drastically reduces the radius of curvature of nanoparticles, which potentially diminishes the ease of phagocytosis by macrophages [60]. The accumulation of nanoparticles via the EPR effect in these studies seemed to largely depend on the long circulation time of the nanoparticles within the blood.

The impact of deformability on uptake of soft nanoparticles by macrophages may also pose a problem for uptake by non-phagocytic cells. This concern is largely due to evidence from studies that show higher cellular uptake with hard nanoparticles. In one report, hydrogel nanoparticles with intermediate Young's moduli of 35 and 136 kPa were significantly internalized compared to those with elastic moduli of 18 kPa and 211 kPa [62]. This suggests that an optimum elastic modulus may be essential to improve the EPR effect. The advantages of soft nanoparticles have been contested by other reports. In a recent study on the role of elasticity in the uptake of silicon oxide nanocapsules by HeLa cells, the authors reported that increased elasticity resulted in higher cellular uptake of stiff SiO_2 nanocapsules by about nine-fold compared to soft nanocapsules [63]. Mechanistic investigation showed that hard nanocapsules were internalized via clathrin mediated endocytosis, while soft nanocapsules were taken up by either the caveolae dependent pathway or via micropinocytosis, as observed in soft nanoparticles with extremely low elastic modulus [60,64].

4.2. *Enhancing Nanoparticle Navigation in Systemic Circulation*

4.2.1. Circumventing Opsonization

Opsonization is an immune process that involves the adsorption of serum protein fragments (opsonins) to foreign pathogens for recognition and elimination by phagocytes [65]. In the absence of opsonins, the negatively charged cell walls on both pathogens and phagocytes will repel each other, thus enabling the pathogen to replicate uncontrollably while avoiding destruction.

Nanoparticles may attain high concentrations in neoplastic tissues via the EPR effect only if they are able to evade the cells of the reticuloendothelial system [10]. The journey of macromolecules and nanoparticles begins with their injection and continues through different stages of circulation, extravasation, accumulation, endocytosis, endosomal escape, intracellular localization, and pharmacological action [3]. Nanoparticles and macromolecules are particularly prone to opsonization due to their nanosize, that confers a large surface area. The high surface area to volume ratio of nanoparticles generates very high surface energies that engender unusual behavior [48], such as the adsorption of plasma proteins, i.e., serum albumin, apolipoproteins, components of the complement system and immunoglobulins, on the surface of circulating nanoparticles [66]. Opsonization leads to the formation of a protein corona around nanoparticles, a process that is dependent on several factors, such as, nanoparticle size, surface charge, hydrophobicity, and surface chemistry [48]. With the adsorption of proteins, the surface of nanoparticles is primed for attachment to specific receptors on the surface of phagocytes. Phagocytosis then occurs, and the nanoparticles are internalized, transported to phagosomes, and fused with lysosomes [13]. In addition to increasing uptake by the RES, opsonization is detrimental to active-targeting strategies for nanoparticles, since the protein adsorption creates a corona mask that diminishes the binding affinity of targeting ligands, resulting in a marked reduction in specificity [13].

To overcome the challenge of opsonization, hydrophilic polymers that confer stealth properties have been used to drastically reduce the adsorption of opsonins to nanoparticles. The stealth property of nanoparticles implies their ability, via various modification strategies, to navigate biological systems without detection and destruction by the immune system. Modification of the nanoparticle surface to confer stealth property prolongs in vivo circulation time and enhances passive targeting via the EPR effect [67].

The most common strategy to confer stealth property is PEGylation. Using this approach, polyethylene glycol (PEG) is usually grafted, adsorbed, or covalently bonded to the nanoparticle surface [3,13]. Polyethylene glycol is believed to provide a steric barrier, i.e., a hydration zone, around nanocarriers because of its hydrophilicity. This reduces the adsorption of opsonins on the surfaces of nanocarriers, thereby reducing nanoparticle uptake by the cells of the RES in the liver and spleen. This reduced uptake leads to prolonged blood circulation time [68]. The graft density of PEG on the nanoparticle surface is critical to the effectiveness of the PEG coating in resisting protein adsorption. It has been reported that low PEG density leaves regions of the nanoparticle surface exposed to binding by opsonins, while high density may constrain PEG and decimate its ability to push adsorbing proteins away, instead becoming a new surface for protein adhesion [3].

The use of PEG to reduce adsorption of opsonins is not without challenge. Studies [68] have shown that a second dose of PEGylated liposomes in rats or rhesus monkeys was cleared very rapidly from circulation when the interval between the first and second injection was between 5 and 21 days, largely due to enhanced accumulation in the liver. This phenomenon, commonly referred to as 'accelerated blood clearance', is responsible for the sudden decrease in nanocarrier concentration after subsequent injections of PEGylated nanoparticles and macromolecules. This leads to short circulation time and low accumulation of nanoparticles at tumor sites, thus compromising efficacy. The rational design of nanoparticles requires that more attention be paid to the ABC effect and its overall impact on EPR.

To avoid PEG-induced ABC effect, various hydrophilic synthetic coatings (polyvinylpyrrolidone, PVP; polyphosphoesters, PPEs; polyelectrolytes, and zwitterionic polymers) and natural polymeric coatings (polynucleotides, polypeptides, dextran, and chitosan) have been explored with the goal of preserving the physicochemical properties, surface properties, and functional integrity of nanoparticles within biological systems, to prolong blood circulation. These agents mitigate ABC by preventing protein corona formation, warding off RES cells, and averting nanoparticle agglomeration, as well as preventing other bio-nanoparticle interactions that serve as barriers for effective nanoparticle drug delivery [69]. Another strategy involves the use of polyglycerol (PG) lipids for modification of nanocarrier surfaces. Application of PG to a liposome nanocarrier was shown to produce neither an anti-polymer immune response nor the ABC phenomenon upon repeated administration, resulting in enhanced therapeutic efficacy of encapsulated doxorubicin in a tumor-bearing mouse model [68]. Similarly, a novel cleavable PEG lipid derivative (mPEG-Hz-CHEMS) in which the PEG moiety was linked to cholesterol by two ester bonds and one pH-sensitive hydrazone molecule has been reported as a promising PEG alternative [70]. The authors revealed that liposomes functionalized with this novel PEG-lipid derivative demonstrated prolonged blood circulation characteristics, and upon repeated administration, showed no ABC phenomenon compared with liposomes modified with mPEG-CHEMS lipid derivative.

Hydrophilic synthetic materials such as zwitterionic polymers have received a lot of attention in recent years. Zwitterionic polymers are made of moieties with both cationic and anionic groups, characterized by high dipole moments from highly charged species, and yet maintain charge neutrality [71]. Zwitterionic materials are biocompatible, resist nonspecific protein adsorption in the blood, and do not induce immunological response in vivo [72]. The mechanism by which zwitterionic materials prevent nonspecific protein adsorption involves strong electrostatic interactions between highly charged groups that induce hydration around nanoparticles, thus preventing biofouling. Although both PEG and zwitterionic polymers induce hydration, besides the non-induction of immunologic responses, zwitterionic materials differ in their structure and extent of hydration. They have been shown in molecular dynamics simulations to possess lower free energy of hydration, which translates to stronger hydration. Lower free energy of hydration leads to low water mobility and wider dipole moments that together repulse adsorption of proteins and other charged species, respectively [71].

Zwitterionic materials are made from small molecule zwitterions like phospholipids, betaine, amino acids, and their derivatives. Polymeric zwitterions are formed from the surface derivatization of nanoparticles, proteins, and hydrogels, and include polycarboxybetain acrylamide (polyCBAA), polycarboxybetaine methacrylate (polyCBMA), and polysulfobetaine methacrylate (polySBMA) [73]. The efficiency of zwitterion polymers in enhancing drug delivery and EPR has been demonstrated in several studies. In one study, the ability of a uricase-loaded pCB nanocarrier to prevent humoral immune response (from either uricase or polymer) and reduction in efficacy after repeated administration was investigated. The result showed that in a clinical rat model of gout, repeated administration of pCB loaded carrier was superior to pegylated uricase, as indexed from a lack of immune response and sustained efficacy [74]. In a different study, the biocompatibility and circulation time of a pCB-based nanocarrier were comparable to those of a pegylated system [75]. The conjugation of polycarboxybetain polymers to organophosphate hydrolase increased its pulmonary delivery due to a significant increase in bioavailability (5% to 53%) [76]. The co-conjugation of docetaxel and curcumin onto a polycarboxybetaine (pCB) polymer led to a significantly enhanced EPR effect in multidrug resistant MCF-7/Adr cell lines due to the antifouling effect of pCB. The result showed cell cytotoxicity at IC_{50} of 5.87 μg/mL for docetaxel-curcumin-pCB, compared to either docetaxel (437.2 μg/mL) or pCB-Dox (14.1 μg/mL) [77]. Delivery of docetaxel via EPR effect in a zwitterionic shielded, pH-responsive folate conjugated polymer has also been reported [78]. In this construct, a zwitterionic co-polymer was synthesized via reversible fragmentation chain transfer (RAFT) copolymerization from 2-(methacryloyloxy)ethyl phosphorylcholine and polyethylene glycol methacrylate ester benzaldehyde. The authors reported that upon conjugation of the docetaxel-loaded zwitterionic co-polymer to folate and exposure to HeLa cells, there was rapid and efficient internalization due to the presence of folate, and strong interaction between multivalent phosphorylcholine (PC) groups and cell membranes. Such efficient uptake has potential to enhance the EPR effect in vivo.

4.2.2. Overcoming the Impact of the Reticuloendothelial System

The reticuloendothelial system (RES) is a diverse collection of phagocytic cells expressed in systemic tissues as part of the innate immune system that are actively involved in the elimination of particles and soluble substances in both blood circulation and tissues. The RES consists broadly of Kupffer cells of the liver, microglia of the brain, alveolar macrophages, bone marrow, lymph nodes, and macrophages of the intestine and other tissues [79]. Typically, less than 5% of an injected dose of nanoparticles is delivered to the cancer tissue, resulting in an extremely low drug delivery efficiency and, consequently, a poor therapeutic outcome [80]. Elimination of the bulk of nanoparticles occurs via the

RES organs of the liver and spleen [81]. The injection of large amounts of drug-loaded nanoparticles to compensate for this loss raises toxicity concerns with respect to decreased RES function and the risk of nonspecific drug release [80]. Therefore, reducing the uptake of nanoparticles by the RES is a strategic approach to enhance the magnitude of the EPR effect that could lead to high accumulation of macromolecules at the site of action.

Prominent among the RES apparatus are the Kupffer cells positioned in liver sinusoids as part of the body's innate immunity [82]. They are specialized macrophages formed from liver adhering circulating monocytes that polarize into cells with highly differentiated surface receptors which facilitate the binding and/or uptake of foreign materials [27]. The extent of nanoparticle uptake and retention in Kupffer cells is strongly associated with the nanoparticle's surface charge, ligand chemistry, and size, with particles of highly cationic and anionic surface charge being cleared faster than those with neutral charge [83]. Similarly, large nanoparticles are more likely to be sequestered and destroyed by the Kupffer cells [20]. The fate of nanoparticles is further complicated by the fact that smaller monodispersed nanomaterials may be taken up by liver sinusoidal epithelial cells (LSECs) to a higher degree [27]. This suggests that escape from the Kupffer cells does not imply freedom to circulate. Hence, the importance of the rational design and optimization of the physicochemical properties of nanoparticles should be obvious.

Enhancing Nanoparticle Delivery by Silencing or Depleting Kupffer Cells

A strategy to improve the EPR effect is to silence the Kupffer cells, since they are responsible for the bulk of nanoparticle sequestration. This approach is based on the principle that when liposome loaded with clodronate or other agents are taken up by macrophages, the phospholipid bilayers of liposomes are digested by lysosomal phospholipases to release clodronate, which inhibits ADP/ATP translocase in the mitochondria and ultimately triggers the apoptosis of macrophages [23]. The advantage of this strategy is that only phagocytic cells are targeted, and the remaining clodronate are eliminated via the renal system, leading to an extremely short half-life in the bloodstream. Nevertheless, the drawback of this strategy is that at high doses, the depletion of splenic macrophages may result in splenomegaly that may predispose patients to sepsis [84].

Tavares et al. [85] were able to increase nanoparticle delivery to tumor 150-fold by removing all or a portion of Kupffer cells. They reported a series of experiments that involved dose-dependent reduction of macrophages using dichloromethylene diphosphonic acid liposomes (clodronate liposomes), followed by i.v. administration of various nanoparticles (gold-, silver-, silica nanoparticles, and liposomes) in different xenograft models (ovarian, breast, skin, prostate, and lung cancer) in a two-step dosing schedule spread over 48 h. Their data showed that tumor accumulation of nanoparticles increased 150-fold for 50-nm gold nanoparticles, while there was 100-fold increase in tumor accumulation in animals with PC3 orthotopic xenografts for 100-nm gold nanoparticles 48 h after i.v. administration of clodronate liposomes. The data showed that although there were high blood levels of nanoparticles with almost 98% bioavailability, only 2% was delivered to the tumor site.

Minimizing the sequestration of nanoparticles by Kupffer cells is expected to dramatically increase tumor accumulation of nanoparticles. The low tumor accumulation in a situation of diminished Kupffer cell activity begs the question of the contribution of other organs such as the skin, lymph nodes, spleen, and lungs, as well as the pathophysiology of the tumor to nanoparticle sequestration and drug delivery [85]. Tavares et al. also evaluated the possibility of infection during periods of Kupffer cell depletion by using a polymicrobial model of sepsis to determine the outcome of infection in a depleted Kupffer cell situation. The result showed that for acute infection during the period of Kupffer cell depletion therapy, the prognosis will be less favorable. In addition, the study by Tavares et al. revealed that tumor accumulation of nanoparticles, although low, is largely dependent on nanoparticle size, material composition, and tumor type. Thus, optimizing the physicochemical properties of nanoparticles is a rational and strategic initial step towards enhancing the EPR effect.

Other research groups have used gadolinium chloride ($GdCl_3$), a Kupffer cell deactivator, to inhibit the function of Kupffer cells by suppressing phagocytosis via inhibition of calcium transport across the cell membrane [86–89]. In one such example, it was demonstrated that a significant number of liver Kupffer cells were inactivated after pretreatment with systemic administration of $GdCl_3$, leading to an increase in the circulatory half-life of Quantum dots, and consequently, a 50% increase in tumor-specific uptake [87]. Clodronate and gadolinium chloride have no effect on liver sinusoidal epithelial cell uptake and may be more beneficial to inhibiting liver sequestration of larger nanomaterials. While attractive, these transient depletion strategies are not well-characterized for their safety, and studies investigating dose–efficacy relationships and concurrent effect on innate immunity are rare [23].

Another research group proposed the use of intralipid 20%, an FDA-approved fat emulsion used for parenteral nutrition to temporarily blunt the phagocytic capacity of Kupffer cells by decreasing the accumulation of nanoparticles in the liver and spleen, thus increasing the bioavailability of nanodrugs via the EPR effect [90]. This strategy stems from a report that infusion of Intralipid 20% impedes Kupffer cell function by inhibiting peritoneal clearance, hence impairing their phagocytic activity [91]. Their data showed that in rodents, intralipid reduced Kupffer cell uptake by approximately 50%, leading to an increase in the blood half-life of nano- and micron-sized super paramagnetic iron-oxide particles by ~three-fold. They also demonstrated that a single clinical dose (2 g/kg) of intralipid 20% could decrease the accumulation of platinum nanoparticles in the liver by 20.4%, in the spleen by 42.5%, and in the kidney by 39.3% after 24 h post nanodrug administration. Subsequently, the bioavailability of the platinum-nanodrug increased by 18.7% during the first 5 h and by 9.4% after 24 h, respectively.

Most strategies aimed at silencing the impact of the RES have focused solely on Kupffer cells. Earlier, it had been established that small-sized nanoparticles are more likely to be trapped within the liver sinusoidal endothelial cells (LSEC), which could affect the overall bioavailability of these particles. The major receptors involved in the sequestering action of liver sinusoidal epithelial cells include mannose, Fcγ, collagen-alpha receptor, and the hyaluronan scavenger receptors [27]. Kupffer cells equally express mannose and Fcγ receptors. These two receptors could be targeted in both Kupffer cells and LSECs by pretreating with a combination of different sized nanoparticles bearing inhibitors of both cell types, designed to fit the size threshold of Kupffer cells ($\geq$100 nm) and LSECs (<30 nm).

Enhancing Nanoparticle Delivery by Saturating Kupffer Cells

The saturation of the receptors of Kupffer cells with bait and nontoxic unloaded nanoparticles prior to dosing of nanotherapeutics may enhance nanoparticle accumulation in neoplastic tissues [27]. In one study, liposomes made of phosphatidylcholine and cholesterol were used to saturate phagocytosis by macrophages [91]. The subsequent inhibition of phagocytosis occurred within 90 minutes after dosing with liposomes, resulting in an increased intratumoral accumulation that persisted for 48 h. The application of this strategy led to a two-fold increase in the accumulation of PEGylated nanoparticles in a human prostate cancer xenograft model after a single dose, compared to controls [91]. Unlike the Kupffer cell-depleting strategy mentioned in the previous section, the Kupffer cell saturation approach is safe and does not damage the innate immunity, as reflected in the lack of weight loss, non-impairment of liver function, and unchanged host defense in the experimental animals used [91]. It should be noted, however, that this strategy is limited by the fact that the phagocytic function of Kupffer cells was not fully inhibited in this study, judging from the impact of only a two-fold improvement in tumor accumulation. Further development may require the use of nanoparticles made of materials that have slower degradation rates [27], or the titration of the blank decoy nanoparticles to evaluate the kinetics of optimum accumulation.

A technique which overwhelms the uptake rate of nanoparticles by Kupffer cells resulting in decreased hepatic clearance has been used to establish high tumor accumulation of nanoparticles. In a study which focused on evaluating the relationship between nanoparticle dose and liver clearance [92], it was hypothesized that the proportion of nanoparticles taken up by the Kupffer cells in the liver would decrease considerably if the dose were increased beyond the uptake rate of Kupffer cells. The aim of this strategy was to find a threshold nanoparticle dose that would minimize liver clearance without compromising liver function. In line with the desire to translate preclinical findings to the clinic, the authors sought to describe the administered dose in terms of an equivalent number of nanoparticles rather than the dose derived from pharmacological allometry. To determine the relative number of nanoparticles required to saturate Kupffer cells, in vivo experiments [92,93] revealed an estimate of 10 million Kupffer cells in the mouse liver and a projected total clearance rate limit of about 1 trillion nanoparticles per 24 h. Therefore, a gradual increase in the dose of nanoparticles above 1 trillion particles is expected to result in a progressive saturation and decrease in Kupffer cell activity. Liver mass extrapolation estimates 8, 63, and 1.5 quadrillion nanoparticles for rat, rabbit, and a 70-kg man whose livers weigh 8, 63, and 1500 times that of the mouse, respectively. At nanoparticle doses below these thresholds, rapid liver elimination would occur, leading to suboptimal accumulation in the tumor [92].

To prove this hypothesis, Ouyang et al. [92] intravenously injected 4T1 tumor-bearing BALB/c mice (mammary carcinoma cell mouse model) with variable quantities of 50 nm-sized polyethylene glycol-conjugated (PEGylated) gold nanoparticles (ranging from 50 billion to 50 trillion nanoparticles). Their data showed that the liver sequestered less pegylated gold nanoparticles with increasing dose, leading to an increase in the blood half-life from 2 min to 8 h. This strategy was repeated by the same authors in a proof-of-concept study by pretreating xenograft mice models with blank pegylated liposomes (Figure 2) to overwhelm the uptake rate of Kupffer cells before administering Caelyx® (PEGylated doxorubicin loaded liposomes). The result showed a 12% accumulation of injected dose at the tumor site without any record of death, despite the known toxicity of doxorubicin. The reader is referred to the study by Ouyang et al. for a detailed description of their technique.

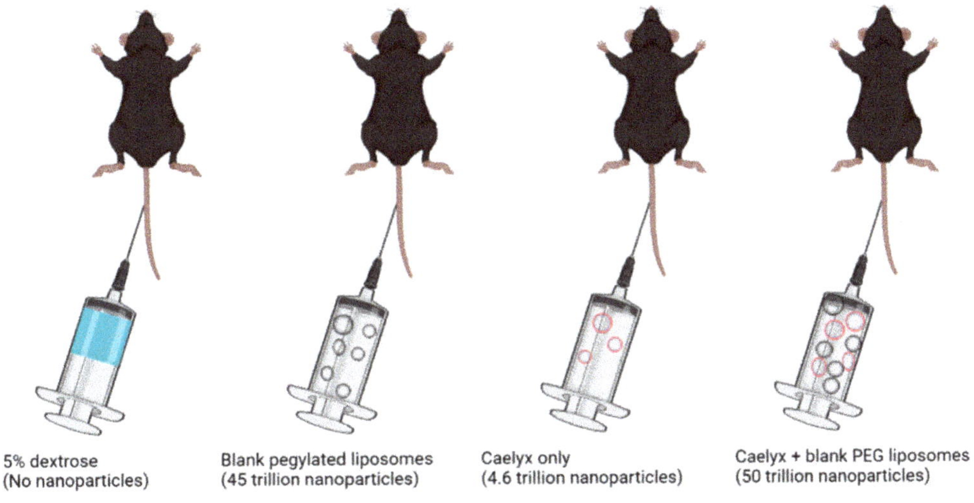

Figure 2. Scaled dosing of nanoparticles according to threshold required to overwhelm Kupffer cells. Adapted from [92].

To reiterate the importance of optimal dosing in overcoming nanoparticle sequestration and improved tumor accumulation, it has been reported that investigational agents, i.e., BIND-014 and NK105, that failed in clinical trials, were dosed below the 1.5 quadrillion threshold, at 1.0 and 0.9 quadrillion nanoparticles per patient, respectively [93], despite being optimized for size, ligand density, drug encapsulation, and release kinetics [94]. Possibly, high sequestration by Kupffer cells and suboptimal delivery to tumors may have been responsible for their therapeutic failure.

4.3. *Promoting Nanoparticle Delivery with Circulating Cells*

Despite improvements seen with the implementation of passive- and active targeting techniques to increase delivery and tumor accumulation of macromolecules, a large percentage of these particles are still cleared by the RES, with a small fraction reaching the tumor microenvironment, thus limiting the clinical translation of nanoformulations. The development of strategies to evade the RES is therefore a key element in fashioning delivery systems that can remain in the blood circulation for long periods to facilitate the efficient accumulation of nanoparticles within solid tumors by the EPR effect [23]. To achieve improved accumulation, several approaches have been proposed, such as 'back packing', 'cellular hitchhiking', and 'Trojan Horse' strategies, that exploit the natural ability of cells within the circulatory system to evade the immune system and be transported via natural tropism to specific, vascular, or systemic locations around the body while crossing biological barriers that are otherwise nearly impermeable [95]. Hence, modeling nanocarrier designs from nature or other bioinspired approaches have been evaluated for reducing nanoparticle uptake by the cells of the RES in the liver [27]. These techniques use, *inter alia*, circulating red blood cells, leukocytes, and monocytes that differentiate into macrophages.

4.3.1. Nanodelivery with Red Blood Cells (RBCs)

Red blood cells (RBCs) are of particular interest due to their safety, abundance, and life span of approximately 120 days [96]. A group of researchers evaluated the use of organic nanoparticles as RBC hitchhikers by developing a smart RBC system containing doxorubicin and bovine serum albumin nanocomplexes for the chemo- and photothermal therapy of glioblastoma cells [95]. These RBCs were further functionalized with RGD peptide (arginine-glycine-aspartic acid) to target the integrins of the endothelium of tumor blood supply. The challenge with using RBCs is that they lack phagocytotic properties (making it difficult to incorporate nanoparticles into them), and they are prone to membrane disturbances especially during processing (adsorption, electrostatic or covalent interactions) with therapeutic agents or carrier molecules, leading to agglutination, stiffness, increased sensitivity to osmosis, and mechanical and oxidative stress that increases the chances of exposure of membrane phosphatidylserine [97]. Research has shown that exposure of phophatidylserine and other 'eat me' signals are responsible for the recognition and phagocytosis of perturbed and dying cells by the RES [97]. By applying the Trojan Horse strategy, a research group successfully loaded nanoparticles into RBC vesicles via a co-extrusion method. This was done by decorating the surface of RBC vesicles with negatively charged sialyl residues via polysaccharide linkers to confer charge asymmetry on the RBC membrane. This charge asymmetry facilitated the fusion of nanoparticles with RBC vesicles when subjected to co-extrusion [98] (Figure 3).

4.3.2. Nanodelivery with Monocytes and Macrophages

Leukocytes (granulocytes, monocytes and, lymphocytes) are found in large quantities in the blood (4−10 billion) and have circulation times of about three weeks [95]. They constitute the body's adaptive and innate immunity, and are widely investigated as nanoparticle vehicles due to their natural tropism that allows movement through endothelial barriers to sites of disease and hypoxia, like the TME [23,99]. Monocytes are produced in the bone marrow and differentiate into macrophages in deep tissues and organs, where they detect and phagocytose necrotic cells, pathogens, and sundry foreign particles through their specialized membrane receptors [23]. Macrophages can also circulate to inflammatory and hypoxic regions, as well as cross the blood−brain barrier via diapedesis and chemotaxis, thus making them a versatile ally in the delivery of nanoparticles [99].

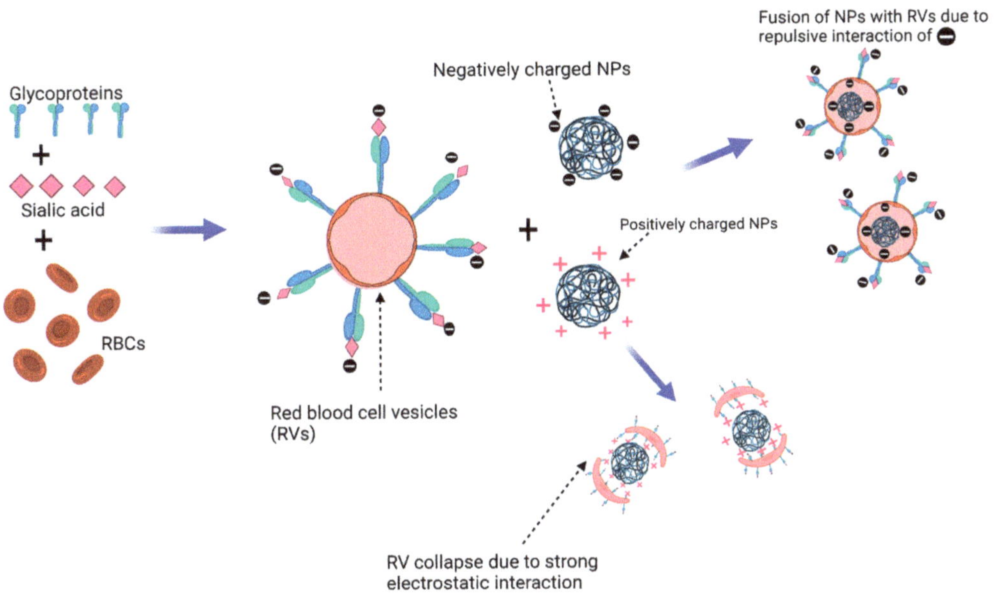

Figure 3. Preparation of RBC vesicle-coated nanoparticles. Schema of electrostatic interactions between negatively and asymmetrically charged RVs (red blood cell vesicles) with negatively charged nanoparticles on one side and positively charged ones on the other. First, the RBCs are designed to form negatively charged vesicles by the addition of glycoproteins and sialic acid. Secondly, the RVs are then mixed with positively and negatively charged nanoparticles, respectively, with their embedded payloads. Naturally, the negatively charged nanoparticles strongly repulsed the negatively charged RVs; however, fusion was achieved by co-extrusion, where mechanical force drives the nanoparticles through the lipid bilayer to fuse with the intracellular membrane side of RBCs. The strong affinity between the positively charged nanoparticles and the negatively charged RVs led to the collapse of the lipid bilayer, which prevented nanoencapsulation. Modified from (Ref. [98]) and Created with BioRender.com Accessed on 24 October 2021.

Notwithstanding the promise of efficient nanoparticle delivery by macrophages, the loading of large quantities of nanoparticles into macrophages remains a challenge. A typical strategy employed in loading nanoparticles into macrophages is the 'backpack' approach, which involves conjugating drug molecules or drug-loaded nanoparticles onto the plasma membrane [100]. This "backpack" approach has been used to load therapeutic nanoparticles onto stem cells, leukocytes, red blood cells, and T cells, with varying degrees of success [96,100]. In this approach, however, since the plasma membrane is essential for cell function and plasticity, conjugated nanoparticles may adversely affect cell signal transduction, adhesion, and migration. In addition, the number of nanoparticles that can be loaded on the membrane at a time is limited. Furthermore, monocytes and macrophages can readily engulf the nanoparticles loaded on the surface of their plasma membrane, thus diminishing the overall efficiency of the strategy [100].

The alternative strategy of loading drugs into the cell cytosol ('Trojan Horse strategy') is also difficult, because most anticancer drugs are highly toxic to macrophages, and hence, the encapsulation of high concentrations of drugs in macrophages could induce immediate cell death, while the encapsulation of low drug concentrations may lead to insufficient drug loading and sub-lethal dosing [100].

The direct conjugation of nano-constructs to the surface of macrophages and the Trojan Horse strategy are usually performed ex vivo, starting with the harvesting of plasma, followed by surface modification or nanoencapsulation, before reintroducing the modified macrophages back to the animal. In a bid to resolve the challenges associated with loading macrophages with nanoparticles ex vivo, a research group developed drug-silica nanocomplexes (DSN-Mf) with negative charges to increase interaction with loaded drugs, reduce the propensity of drug release and thereby decrease the possibility of death of carrier macrophages [100]. The silica nanoparticle was designed in a manner that retarded the release of the drug payload over a period of 48 h. The incubation of macrophages with nanoparticles was limited to 2 h for optimum nanoparticle loading without overwhelming the macrophages. In a U87MG subcutaneous tumor xenograft, the macrophage-loaded silica nanoparticles showed an impressive tumor growth inhibition rate of 62.66% on day 14, with a substantial extension of the animal median survival to 26 days, compared to 14 days in the control group.

In a different study, doxorubicin-containing echogenic liposomes were loaded into polycation (polyallylamine hydrochloride (PAH), polydiallyldimethylammonium chloride (PDAC)) and polyanion (polyacrylic acid (PAA) or polystyrene sulfonate (SPS)) films in a layer- by-layer assembly before 'back packing' to the surface of monocytes [101]. Silicon wafers were used as substrates to create a polyelectrolyte multilayer film by incubating it with one polycation (PAH or PDAC) and one polyanion (PAA or SPS) layer. The polycation was used as the first and last layer to increase the number of positive groups on the film surface to engender firm interaction with liposomes that are naturally electronegative. This multilayered film was then mixed with liposomes, followed by a second layer of multilayered film to create a vesicle of liposome-loaded doxorubicin sandwiched between both layers. The multilayered liposomal construct was then backpacked to the surface of mouse monocytes to evaluate its efficacy. The result showed that use of echogenic liposomes for drug encapsulation into backpacks enabled up to three times DOX loading compared to backpacks without echogenic liposomes. In vitro cytotoxicity evaluation revealed that monocyte backpack conjugates remain viable even after 72 h, demonstrating their promise as a drug delivery vehicle.

4.3.3. Cellular Hitchhiking with Macrophages

Hitchhiking involves the use of live macrophages in vivo without modifying the membrane. The previously discussed methods (Section 4.3.2) of conjugating and encapsulating macrophages are expensive and limited in the amount of drug that can be loaded. However, by targeting circulating monocytes, its natural phagocytic properties can be harnessed for nanoparticle loading, with the idea that upon homing to the tumor site, they can differentiate into macrophages and serve as "Trojan Horses" to release their cargo deep within the hypoxic tumor in a controlled manner [23]. For example, Yang et al. [102] used live circulating monocytes to hitchhike a docetaxel-loaded polymeric-micelle formulation synthesized from chitosan and stearic acid. The authors chose chitosan for its biocompatible and biodegradable properties and stearic acid for its cell membrane compatibility that promotes cell uptake. To enhance uptake by monocytes, the micelle batch with particle size of 86 nm and positive zeta potential (23 mv) was chosen for further investigation. This is because positive zeta potential values promote adsorption of opsonins which, in turn, promotes uptake by circulating monocytes. Tumor delivery was achieved by exocytosis of the drug loaded micelle upon differentiation of monocytes to macrophages at the tumor site [102].

In an innovative strategy, Zhang et al. [103] exploited circulating macrophages to target CD47-rich tumor cells in a 4T1 murine breast cancer model. These authors designed drug-loaded silicon nanoparticles and decorated the surface with calreticulin and anti-CD4 antibody to enhance the EPR effect. The mechanism of this strategy is to diminish the 'don't eat me' signal of CD47-rich tumor cells with anti-CD47 antibodies while promoting 'eat me signal' with calreticulin. The findings show that the simultaneous application of anti-phagocytic and pro-phagocytic signals can significantly enhance macrophage-mediated cancer delivery [103]. The reader is referred to the review by Izci et al. [23] for a detailed description of other hitchhiking strategies with circulating monocytes and macrophages.

4.3.4. Enhancing Drug Delivery with Other Cell-Based Strategies

Platelets are circulating cells that are devoid of a nucleus but are useful as delivery vehicles because of their sensitivity toward inflamed tissues upon activation. Such sensitivity can be harnessed for the delivery of therapeutic payloads to platelet-activating tumors [101,104]. In one report, to overcome cancer recurrence after surgical resection, anti-PDL-1 antibodies were conjugated to the cell surface of mice-derived platelets via a maleimide linker for delivery to the freshly resectioned sites. The administration of platelet bound anti-PDL1 considerably prolonged overall mouse survival after surgery by reducing the risk of cancer regrowth and metastatic spread [105].

Extracellular vesicles (EVs) derived from mesenchymal stem cells (MSC) have been used for nanodelivery to improve the EPR effect [106]. EVs are lipid bilayer membrane vesicles that are produced by eukaryotic cells from several regulatory processes involving endocytosis, fusion, and efflux. They are less immunogenic, relatively non-toxic, and can penetrate tumors and inflamed tissues [107]. The RES clearance threshold for EVs is relatively high, making them useful for stealth delivery. Wei et al. [107] exploited this attribute to deliver microRNA (miR21)-loaded CD47-EVs to inflammatory cardiac cells in a mouse model of acute myocardial infarction and ischemia-reperfusion. The result showed longer circulation time (120 vs. 30 min) and preferential accumulation of CD47-EVs in inflamed cardiac tissues compared to unmodified EVs.

Cancer cell membranes are known to specifically recognize homologous cells, persist for a long time in blood circulation, and possess the capacity to evade phagocytic cells. As such, they are useful in the formulation of nanocarriers for improving the EPR effect [108]. Fang et al. [109] demonstrated the EPR enhancing effect of cancer cell membranes for drug delivery in a construct of PLGA nanoparticles coated with B16−F10 mouse melanoma cells without intracellular content. The data showed that when exposed to MDA-MB-243 cell line, cancer cell membrane-coated nanoparticles (CCNP) were significantly internalized compared to either RBC coated PLGA or plain PLGA. In addition, the CCNP induced a cancer-directed immune response [108]. This construct may be further optimized by encapsulating antineoplastic agents within the PLGA core for dual effect. Chen et al. [110] present evidence of enhanced EPR effect with an indocyanine green-loaded and cancer cell membrane-coated PLGA-PEG nanoparticle. The cancer cell membrane was derived from MCF-7 cancer cells. The study show that biomimetic nanoparticles significantly promote cell endocytosis and homologous-targeting of MCF-7 tumor xenograft, leading to substantial accumulation in vivo when indexed against plain indocyanine green and PLGA. Moreover, the biomimetic nanoparticles persisted in circulation for a long time at 7–14-fold concentration of both plain indocyanine green and PLGA due to reduced sequestration and elimination by the liver and kidney [110].

4.4. Enhancing Nanodelivery via the "Don't Eat Me Strategy"

Another evasive technique to avoid cells of the RES is the 'active stealth' approach explored by Rodriguez and coworkers [111]. In their experiment, CD47 (a putative self-marker) was used to decorate the surface of nanoparticles, making them biomimetic and recognizable by the macrophages of the blood and Kupffer cells as 'self', thus avoiding phagocytic clearance (Figure 4). In this study, CD47 'self' peptides were computationally designed, synthesized, and attached to 160-nm paclitaxel-encapsulated nanobeads, followed by administration to NSG- (NOD) severe combined immunodeficient *IL2rγ null* mice. The self-peptide (CD47) functionalized nanobeads substantially prolonged nanoparticle circulation by impairing phagocytic clearance by the RES. An in vivo evaluation of nanoparticles derivatized with the CD47 'self-peptide' showed superior accumulation in A549 tumors within 10 min of administration, followed by release of the encapsulated paclitaxel, resulting in significant tumor shrinkage compared to the conventional Cremophor EL formulation of the drug [111].

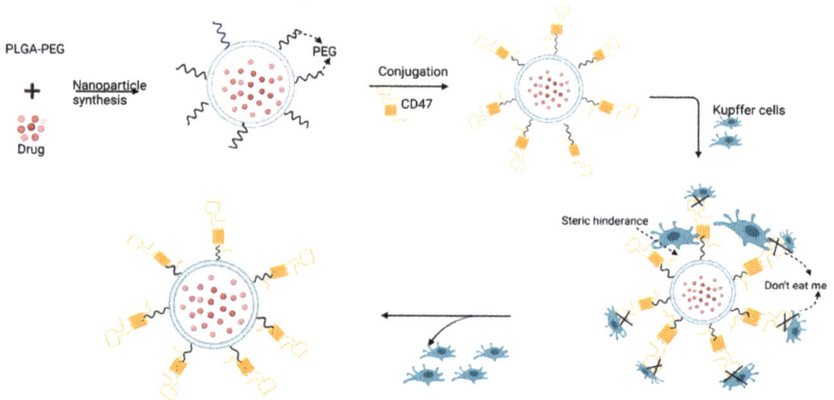

Figure 4. Preparation of a PLGA-PEG nanoparticle loaded with drug molecules and coated with CD47 peptides to prevent phagocytosis by Kupffer cells. Created with BioRender.com Accessed on 30 January 2022.

To increase tumor accumulation, nanoparticles derivatized with CD47 peptides may also be decorated with targeting moieties that e home in specifically on exclusive targets within the TME. Alternatively, CD47 peptides may be conjugated to nanoparticles via a cleavable pH labile hydrazone bond that can only be cleaved in a highly hypoxic environment, thus ensuring exclusive delivery while preventing non-specific distribution [68,112,113].

4.5. The Synthetic Microbe Strategy

Intracellular pathogens have evolved with unique instruments to evade the immune system and thrive within host cells, making them difficult to treat [114]. An understanding of these microbial evasive mechanisms provides an opportunity to apply such survival strategies for drug delivery. The repurposing of bacterial effectors in an appropriate combination for the development of a macrophage-based delivery system for the conveyance and controlled delivery of therapeutic agents packaged in a "synthetic microbe" has been proposed [114].

In this proposed scheme (Figure 5), the drug is first incorporated into two layers of nonbiodegradable but biocompatible nanoparticles with built-in membrane-escaping agents (LLO, pore forming listeriolysin-O, actin inducing protein (ActA), and actin polymerizing complex (ARP2/3) that mimic microbial effectors such as those found in *Listeria monocytogenes*. The microbial effectors in *Listeria monocytogenes* allow the expulsion of mature *Listeria monocytogenes* cells without macrophage destruction. The surface of this drug-loaded construct is decorated with opsonins to promote phagocytosis, and then incubated with macrophages together with wortmannin, chloroquine, and concanamycin to arrest phagosome maturation, thus preventing intracellular degradation. Macrophage encapsulation of the synthetic microbe is meant to proceed for a maximum of 1 h to avoid overloading the macrophages. The loaded macrophages are thereafter injected into experimental models for in vivo evaluation. It is expected that natural tropism and chemotactic mobility will drive the migration of the synthetic microbe to tumor cells. Final drug release will be dependent on the built-in release mechanism. For instance, if temperature sensitive poly-N-isopropylacrylamide nanoparticles are used, then an increase in temperature can be used to trigger release. Likewise, ultrasound may be used to trigger release if microbubbles are used (Figure 6). In terms of safety, drugs may be released prematurely from macrophages but are less likely to be expelled from the nanoparticles, except at the site of external delivery support.

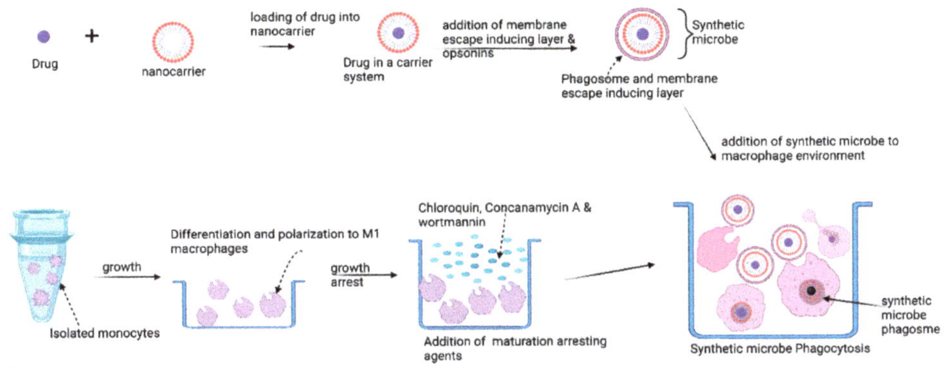

Figure 5. Formulation of the proposed synthetic microbe. Drugs are first loaded into an appropriate nanocarrier and coated with a membrane escape-inducing layer, similar to macrophage escape effector proteins in micro-organisms. The membrane-inducing layer is further coated with opsonins (a phagosome-inducing layer) to facilitate uptake by macrophages. The next step involves monocyte isolation and differentiation into M1 macrophages. They will be left to grow, but maturation will be arrested by the addition of chloroquine, concanamycin, and wortmannin. Maturation arrest is required to prevent the premature release of the synthetic microbes from the macrophages in vivo. The last step is the introduction of the synthetic microbe in the microphage environment to activate phagocytosis. Adapted from (Ref. [114]) and Created with BioRender.com Accessed on 30 March 2022.

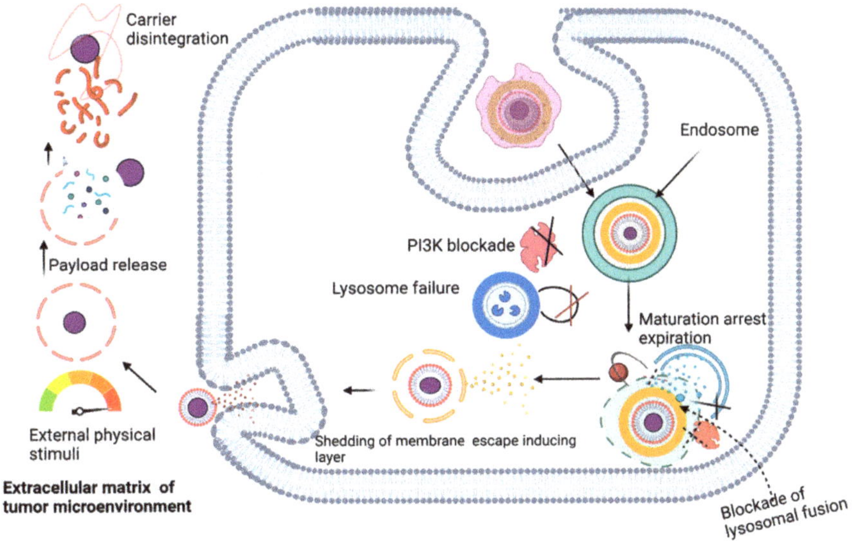

Figure 6. Mechanism of drug release from synthetic microbe. The microbe undergoes endocytosis but is not degraded due to the presence of effectors that block the attachment of PIK3 and lysosomes. Upon expiration of the macrophage maturation arrest, the macrophage degrades the membrane escape-inducing layer (effectors), thereby releasing the drug-containing nanoparticles, which expel the drug upon application of an external physical stimuli at the tumor site. Modified from Ref. [114].

4.6. Exploiting Glutathione-Mediated Biotransformation

Glutathione efflux from hepatocytes into liver sinusoidal endothelial cells mediates the biotransformation of small foreign molecules, leading to their elimination from the body through the renal or hepatobiliary system [115]. Although the sequestration and clearance of nanoparticles from systemic circulation by cells of the RES has been well studied [27], the mechanism of glutathione biotransformation remains unclear. A recent study [115] reported prolonged blood circulation time of gold nanoparticles with a substantial improvement in EPR effect in a mouse model of human breast cancer (MCF-7) mediated by a glutathione biotransformation mechanism. In this study, glutathione-coated gold (Au25) nanoclusters (GS-Au25) were conjugated to indocyanine green (ICG_4) to promote protein adhesion (opsonization), thus ensuring delivery of the nano-construct to the liver for biotransformation. Opsonization also increases the hydrodynamic size of GS-Au25 and prevents renal elimination of the fraction that goes straight to the kidney.

The small size of the nano-construct ensures that despite an increase in the hydrodynamic size due to opsonization, the overall diameter remains below the kupffer cell phagocytic threshold thus escaping sequestration and migrating preferentially to liver sinusoidal epithelial cells where it engages in disulfide exchange with the local high concentration of glutathione and cysteine. Upon biotransformation, some, or all the ICG_4 on the surface of Au25 may be shed to reduce the protein-binding affinity on Au25 (Figure 7). The ICG-GS dislocated as part of the disulfide exchange reaction is taken up by hepatocytes and eliminated through the hepatobiliary pathway while the transformed Au25 nanocarrier goes into circulation for tumor targeting together with the untransformed ICG_4-GS-Au25 via the EPR effect. The outcome of this strategy showed that tumor targeting improved 27-fold after 24 h compared to the ICG control while the EPR effect of the ICG-conjugated construct was 2.3 times higher than the non-conjugated control.

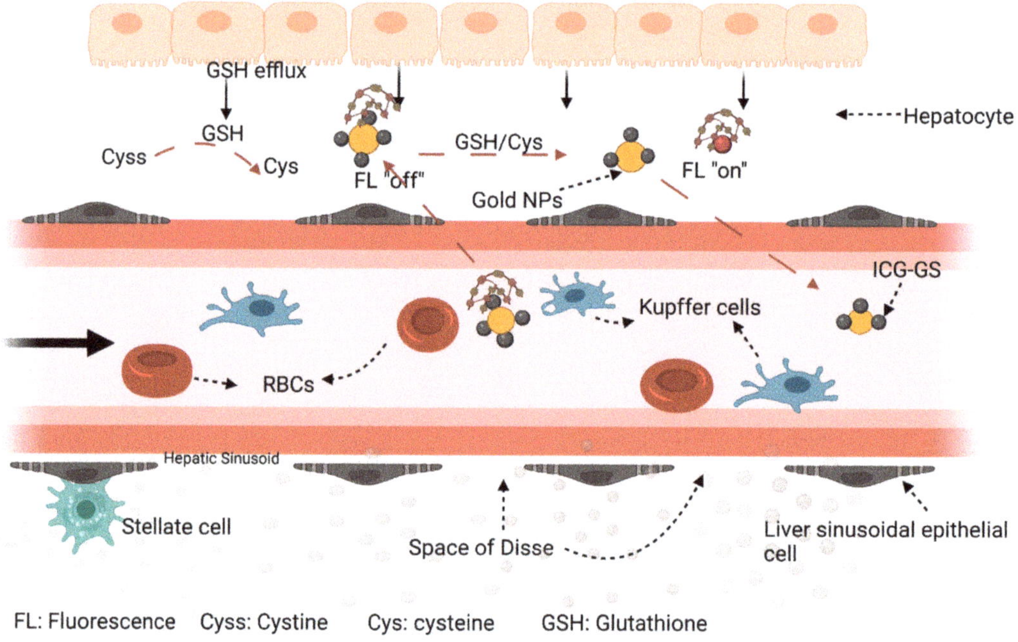

Figure 7. GSH secreted from hepatocytes converts extracellular cystine (Cyss) to cysteine (Cys), which together with GSH, undergo disulfide bond interaction with ICG_4-GS-Au25 conjugates thereby displacing protein bound ICG-GS from the surface of gold nanoparticles (Au25). The displaced ICG-GS (whose fluorescence was suppressed while bound to gold nanoparticles) regains fluorescence and become susceptible to renal clearance. After periods of residence within the body, enough time for a chance to extravasate into the tumor microenvironment, untransformed ICG_4-GS-Au25 is eventually broken down and eliminated. Modified from (Ref. [115]) and created with BioRender.com Accessed on 15 April 2022.

This strategy can be applied to nanoparticles between 5 nm–100 nm if they are designed to interact efficiently with liver sinusoidal GSH. This assumes that biotransformation occurs side by side with the actions of the RES and a design that ensures rapid biotransformation will reduce phagocytic uptake by the RES. Tumor accumulation can be further improved with the addition of ligands that targets specific components of the tumor microenvironment if such addition does not compromise the efficiency of the biotransformation process.

4.7. Exploiting the TME (Tumor Microenvironment)

It has been previously established that the tumor microenvironment consists of fibrous extracellular matrix (ECM) proteins, proteoglycan, growth factor receptors, and transmembrane receptors [5,21]. With the rapid growth associated with tumors, the tumor core is mostly hypoxic due to anaerobic respiration [14]. Hypoxia promotes intratumoral heterogeneity, inhibits innate and adaptive immunity, enhances metastasis, and promotes tumor resistance to ROS-generating cancer therapies (photodynamic therapy, sonodynamic therapy, chemotherapy, and radiation) thus leading to ineffective therapeutic outcomes. Besides, there is a high interstitial fluid pressure [2] within tumors that serves as a barrier to nanoparticle penetration hence drastically reducing vascular transport into the core of the tumor tissue. These obstacles make tissue penetration a daunting task for nanoparticles. To achieve adequate tumor accumulation via the EPR effect, there is a need to exploit the opportunities provided by the TME.

4.7.1. Degradation of ECM to Improve EPR Effect

To improve accumulation and penetration of nanoparticles in neoplastic tissues, degradation of the fibrous ECM with hyaluronidase, an enzyme that breaks down collagen, is a strategy used in combination with other techniques to enhance the EPR effect [5]. For nanoparticles to gain access to the core of tumor tissues, it is imperative that the barriers that prevent accumulation and penetration of nanoparticles in tumors be removed. The challenge of degrading the ECM is the risk of increasing metastasis as the tumor cells become easily mobile when the ECM is compromised [5]. Phesgo®, a combination of two monoclonal antibodies, pertuzumab and transtuzumab for treatment of HER 2+ breast cancer, incorporates hyaluronidase as part of a tripartite strategy to target HER2 receptors in the treatment of HER2+ tumors [116]. Both transtuzumab and pertuzumab are included in this product for their capacity to actively target domains IV and II of the epidermal growth factor receptors (EGFR) respectively, that are overexpressed in several cancers like breast, cervical and intestinal cancers [117,118].

4.7.2. Integrin Receptors as Target for Improving EPR

Integrin $\alpha v\beta 3$ receptors are an established tumor-specific marker of angiogenic activity in the ECM that are overexpressed in rapidly proliferating blood vessels, playing key roles in tumor growth and metastasis [21]. Targeting integrin $\alpha v\beta 3$ leads nanoparticles to the rapidly forming and leaky tumor vessels, thereby increasing extravasation and accumulation of nanoparticles in the TME via the EPR effect [21]. The affinity of the peptide ligand Arg-Gly-Asp (RGD) for integrin $\alpha v\beta 3$ is well established and its use in research is richly documented [15,21]. In a published article, epirubicin loaded near-infrared (NIR) fluorescent nanoparticles assembled from multiple units of cyclic peptides, cyclo [-(D-Ala-L-Glu-D-Ala-L-Trp)2-] were decorated with RGD peptide to evaluate its targeting efficiency to integrin receptors [119]. The authors reported an increased accumulation of the nano construct in tumor tissues compared to healthy tissues. The selective binding of RGD decorated nanoparticles was attributed to the presence of integrin $\alpha_v\beta_3$ receptors on the tumor cells. RGD selective binding in this study led to a substantial improvement in EPR which enhanced therapeutic efficacy.

4.7.3. Exploiting Hypoxia within Tumors

Tumor tissue hypoxia can be harnessed to improve EPR in several ways. One such way is by targeting molecular markers such as phosphatidylserine (overexpressed in hypoxic regions) with Sapocin C, a lysosomal protein that binds to it specifically in hypoxic environment [120]. Selective targeting of tumor phosphatidylserine by Sapocin C spares phosphatidylserine expressed on healthy cells. By designing drug loaded nanoparticles with linkers (azobenzene or 2-nitroimidazole) that can be degraded in a hypoxic environment like tumor tissues, EPR can be greatly improved [121].The reader is referred to reviews by Milane [14,120] for details on other hypoxia based techniques.

4.8. The Use of EPR-Adaptive Delivery Strategies

Apart from modification of the physicochemical properties of nanocarriers, various chemical and physical approaches have also been used to modify the TME for enhanced accumulation of macromolecules via the EPR effect. Modification of the TME can be achieved using external physical or chemical delivery strategies. Chemical EPR-adaptive delivery strategies involve using EPR enhancement factors to adjust the tumor vasculature while external physical inducements typically applied are photodynamic therapy, radiation, sonoporation, and hyperthermia to enhance tumor vascular permeability [122,123]. For in-depth review of these methods, the reader is referred to: Fang [19] and Golombek [2] for PDT; Golombek et al. [2] Park et al. [124] for radiation therapy; Fang et al. [19] and Park et al. [21] for hyperthermia; and Iwanaga et al. [125], Duan et al. [126] and Theek et al. [127] for sonoporation. A recently published review dwelt extensively on physical and pharmacological strategies to improve the EPR effect and the interested reader is referred [128]

4.8.1. Physical Methods

Physical techniques have been developed by various groups to enhance the EPR effect towards improving the therapeutic efficacy of nanomedicines [2,19,21,119,126,127].

Photodynamic Therapy (PDT)

Photodynamic therapy (PDT) as one such example refers to the treatment of tissues, typically tumors, with a photosensitizing agent, followed by activation via locally applied light of specific wavelength [129]. The principle is based on the formation of reactive oxygen species (ROS), such as singlet oxygen species which damages nucleic acids and protein resulting in apoptosis of cancer cells [2]. The decrease in mass of the tumor creates an enabling environment for accumulation of nanoparticles via EPR due to the decrease in interstitial fluid pressure and overall ECM mass. PDT is limited by penetration depth of the applied light (max. 1–2 cm), and the short migration distance of the produced oxygen radicals, and this reduces the impact on the tumor core [2]. One study utilized PDT to target tumor blood vessels by tagging photosensitizers with RGD to enable interaction with endothelial integrin. Application of PDT increased vessel permeability and consequently, enhanced EPR [130]. The authors reported markedly increased tumor accumulation of doxil®. Pretreatment of tumor tissues with photodynamic therapy (PDT) before administering nanomedicine has been demonstrated to improve therapeutic efficacy due to a more efficient accumulation of nanocarriers within the tumor [131]. For example, a 5-fold increase in the tumor accumulation of liposomes loaded with daunorubicin was observed in EGFR-positive A431 epidermoid carcinoma cells coexisting with a small fraction of EGFR-negative Balb-3T3 embryonic fibroblasts after pre-treatment with EGFR-targeted photodynamic therapy [31].

Radiation Therapy (RT)

Radiation therapy (RT) is commonly used alone or with chemotherapeutic agents in cancer treatment. Ionizing irradiation used in radiation therapy has the ability to decrease IFP (interstitial fluid pressure) by generating cytotoxic radicals leading to a decrease in cell density within tumors [132]. Ionizing radiation can also increase tumor vascular leakiness through upregulation of VEGF expression and fibroblast growth factor [124]. Taken together, the various effect of radiotherapy increases the accumulation of low-molecular-weight drugs and nanomedicine formulations in the tumor microenvironment. When radiotherapy is combined with nano-based chemotherapy, there is an increase in the anti-cancer treatment effectiveness. For instance, verteporfin-loaded nanoparticles injected into mice with rhabdomyosarcoma tumors and exposed to laser light [635 nm (0.2 mW cm^{-2}) for 1 min] at 15, 30, and 60 min after injection demonstrated enhanced accumulation of nanoparticles at tumor site. The mice exposed longer to the laser light showed significantly less side effect [133]. In another instance, an analysis of the impact of nanoparticles with RGD moieties in conjunction with radiation therapy showed that treatment with radiation improved the therapeutic response to chemotherapy [119] most likely because of improved EPR.

Hyperthermia

The use of high temperatures (between 39 °C and 42 °C) is another EPR enhancing strategy that induces tissue ablation and promotes perfusion, vasodilation, and vascular permeability and extravasation (i.e., EPR effect) [19,21]. This high temperature strategy is referred to as hyperthermia. Hyperthermia induced by high-intensity focused ultrasound (HIFU) has been used to increase tumor perfusion and vascular permeability within neoplastic tissues [21]. It can be used to increase nanomedicine accumulation, especially in non-leaky tumors which exhibit low EPR effect [2]. Rhodamine-labelled liposomes (100 nm) demonstrated enhanced accumulation in tumor interstitium after intravenous injection into xenograft human ovarian cancer models that had been subjected to gradual temperature increase (between 39 °C and 42 °C) for one hour. The tumors were previously impervious to the rhodamine liposomes at ambient temperature [134]. The ability of HIFU to enhance the EPR effect has been evaluated in various preclinical studies and is close to clinical translation [122]. Temperature-sensitive nanoformulations e.g., ThermoDox®, have become the focus of new research due to the universal relevance of vascular heat shock response after exposure of vessels to increased temperature [21].

Ultrasonication

Ultrasonication using micro/nanobubbles as contrast agents has been utilized in imaging techniques; however, it can also be exploited to increase vessel perfusion and permeability to improve EPR effect [125]. Acoustic transducers are used to generate transverse or longitudinal waves with frequencies greater than 20 kHz, which causes nanobubble oscillation or implosion, leading to vascular cavitation that results in the extravasation of loaded nanobubbles into the tumor microenvironment [126]. Drugs may be encapsulated in the microbubble core or conjugated to the microbubble either directly or indirectly to suitable nanocarriers [135]. Other techniques involve the co-delivery of microbubbles and nanoparticles. This codelivery strategy allows the nanoparticles to take advantage of the cavitation created at the tumor site via an increase in microbubble size in response to transducers [126].

Ultrasound waves have a greater ability to penetrate tissue compared to light waves in PDT, and therefore, are more useful for deep-seated tumors [21]. Ultrasound imaging can produce sonoporation within cell membranes due to the generation of reactive oxygen species by ultrasound waves. When sonoporation is controlled with an ultrasound contrasting agent, temporary pores can be produced within the endothelial layer, causing increased vascular permeability which improves EPR [126]. Micro/nano-bubbles (MNBs), which are used as ultrasound contrast agents, have been employed to control temporary increases in tumor vascular leakiness for improved nanomedicine accumulation within tumor tissues [126,136].

When exposed to ultrasound, micro-nanobubbles undergo oscillation which generates a fluid flow that increases the permeation and accumulation of encapsulated payload in tumor tissues by the EPR effect [126]. In one report, transferrin conjugated nanobubbles were used to traffic drug loaded nanoparticles to the tumor site. Upon application of external ultrasound waves, vascular agitation by the waves led to an increase in the permeability of the nascent blood vessels, resulting in the enhanced delivery of nanoparticles to deep sites within the neoplastic tissue, thus enhancing EPR [137]. The combination of ultrasound, microbubbles, and gemcitabine was evaluated in a phase II clinical trial. The result showed that there was no added toxicity in the combined regimen, and patients tolerated an increased number of gemcitabine cycles compared with historical controls. In addition, the result demonstrated a progressive decrease in tumor size and an increase in the median survival time (17.6 months) in the experimental arm compared to the historical control arm (8.9 months) [138]. Likewise, nanoparticles loaded with doxorubicin were reported to demonstrate enhanced accumulation and distribution within tumors by low-intensity focused ultrasound [139]. Ultrasound waves can be transformed into heat energy due to the friction generated as they propagate through tissue. The heat energy produced may enhance nanocarrier extravasation in tumor tissues by altering tumor hemodynamics. Micronanobubbles have been applied for theranostic purposes by using a phase conversion strategy to deliver ultrasound aided contrast agents via the EPR effect as shown in Figure 8 [126].

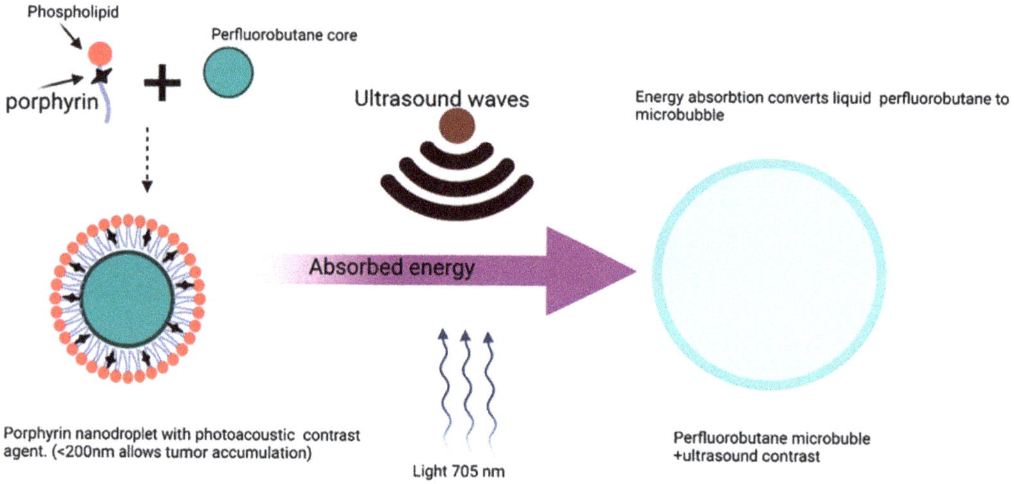

Figure 8. Phase conversion strategy of microbubble ultrasound-based delivery. Porphyrin nanodroplets provides photoacoustic contrast due to the strong optical absorption of porphyrin. The absorption of photons or acoustic energy induces a phase transition of perfluorobutane from liquid to gas microbubble. This system can be loaded with a payload for delivery through cavitations induced by the hyper-echogenic and nonlinear acoustic properties of microbubbles. Modified from Ref. [126] and created with BioRender.com Accessed on 15 March 2022.

4.8.2. Chemical EPR-Adaptive Delivery Strategies

Chemical strategies involve modifying the blood vessels with endogenous cytokines and pharmacological agents to overcome factors associated with the poor accumulation of nanoparticles in tumor cells. These factors include poor tumor perfusion, irregular leakiness, and compressed blood vessels that together hinder the accumulation of adequate concentrations of nanocarriers at the tumor site. Several vasomodulators have been used to enhance tumor blood perfusion, with the expectation that such perfusion will translate to better drug delivery. These agents include bradykinin, tumor necrosis factor-alpha (TNFα), serotonin, and histamine [19,140]. However, toxicity from the systemic administration of cytokines has limited their application. Pharmacological agents that produce a similar vasomodulatory effect to that of cytokines are nitric oxide, angiotensin II blockers, angiotensin converting enzyme inhibitors (ACEI), and nitroglycerin [141–143]. The use of pharmacological agents relies on their capacity to reduce peripheral resistance, which translates to enhanced tissue perfusion. Chemical EPR-adaptive delivery strategies have been discussed in-depth by several authors and are outside the scope of this review. The interested reader is referred to reviews by Ikeda-Imafuku [128] and Ojha et al. [144] for chemical strategies to improve EPR.

5. Development of Rapid Quantitative EPR-Imaging Technologies

Imaging technologies which make it possible to visualize and quantify the extent of EPR-based tumor targeting can provide critical information for future nanocarrier design, predictions of nanomedicine accumulation, and in the pre-selection of patients who would respond to EPR-based therapies [5,126,145,146]. Single photon emission computed tomography (SPECT), positron emission tomography (PET), computed tomography (CT), and MNBs-enhanced (micro nanobubbles) ultrasound imaging have been used to quantify nanomedicines accumulation and distribution in human tumors [20,147]. A radiolabeled analog of a nanomedicine could be given to a cohort of patients individually, and its accumulation quantified in each patient. Patients showing evidence of accumulation (EPR positive group) could then be treated with the nanomedicine, following standard protocols, while the patients with no evidence/or suboptimal level of nanomedicine accumulation could be moved to a different type of treatment [32]. Miller and colleagues showed that ferumoxytol containing super paramagnetic iron oxide nanoparticles could be used as a substitute or companion particle to predict the extravasation, distribution, and accumulation of a PLGA-PEG-based nanomedicine in tumors [148]. More recently, a vascular multiphase tumor growth model that can pre-check the effects of EPR factors such as tumor lymphatic drainage or the size and permeability of vascular endothelial cell pores on the biodistribution of different sizes of nanoparticles in the TME was developed and tested [30].

6. Conclusions

The EPR effect can be considered a hallmark mechanism that differentiates a tumor from healthy tissue and exploits the anatomical and pathophysiological defects in the tumor vasculature to achieve selective anticancer nanomedicine delivery. The desire to improve therapeutic efficacy and safety of clinically relevant and newly developed antineoplastics has led to the advancement of several strategies to precisely deliver these drugs to tumors. Passive targeting strategies based on the EPR effect have shown great therapeutic potential in various preclinical animal models. However, the therapeutic outcome of passively targeted nanomedicines in clinical practice is not encouraging, mainly due to the inherent heterogeneity of the EPR effect. In addition, it has been recognized that for most actively targeted techniques, an effective EPR strategy is still a sine qua non to efficient targeting. Despite the best effort of researchers, the delivery and accumulation of nanoparticles at tumor sites remain very low. Although the notion of the poor delivery of nanoparticles with previous strategies has been disputed, and some published studies have shown tremendous improvement in terms of percent delivery, it remains to be seen if these strategies will translate to clinical success. Thus, understanding and manipulating the factors contributing to the EPR effect can further improve the selective targeting of anticancer nanomedicines to tumors.

Our review differs from a recent review of strategies to improve the EPR effect by Ikeda-Imafuku et al. [128]. In their review, they extensively explored the literature on physical and pharmacological techniques to improve EPR. While there are areas of overlap in general background information, our review focuses on molecular dynamics and challenges in the odyssey of nanomedicines and macromolecular carriers. Our goal was to explore existing and novel nanoformulation strategies that have been demonstrated to improve the EPR effect.

In this review, the critical issues that are essential to improving the EPR effect for increased nanoparticle and macromolecule accumulation at tumor sites are discussed. Since all anticancer nanomedicine delivery systems benefit from the EPR effect, passive and active targeting strategies should be combined in the design and development of nanomedicines to facilitate EPR-based tumor accumulation for better therapeutic efficacy and reduced adverse effects.

Author Contributions: Review: V.E., O.O., P.B.-B., O.D.O., F.A.F., S.K.A.; Editing: V.E., S.K.A.; Visualization, V.E.; Supervision, S.K.A. All authors have read and agreed to the published version of the manuscript.

Funding: This project was supported in part by the Health Resources and Services Administration (HRSA) Center of Excellence Grant under award number D34HP16042.

Institutional Review Board Statement: Not applicable.

Informed Consent Statement: Not applicable.

Data Availability Statement: Not applicable.

Conflicts of Interest: The authors declare no conflict of interest.

References

1. Kreuter, J. Nanoparticles—A historical perspective. *Int. J. Pharm.* **2007**, *331*, 1–10. [CrossRef]
2. Golombek, S.K.; May, J.-N.; Theek, B.; Appold, L.; Drude, N.; Kiessling, F.; Lammers, T. Tumor targeting via EPR: Strategies to enhance patient responses. *Adv. Drug Deliv. Rev.* **2018**, *130*, 17–38. [CrossRef]
3. Nichols, J.W.; Bae, Y.H. Odyssey of a cancer nanoparticle: From injection site to site of action. *Nano Today* **2012**, *7*, 606–618. [CrossRef] [PubMed]
4. Biswas, S.; Torchilin, V.P. Nanopreparations for organelle-specific delivery in cancer. *Adv. Drug Deliv. Rev.* **2014**, *66*, 26–41. [CrossRef] [PubMed]
5. Danhier, F. To exploit the tumor microenvironment: Since the EPR effect fails in the clinic, what is the future of nanomedicine? *J. Control. Release* **2016**, *244*, 108–121. [CrossRef] [PubMed]
6. Wilhelm, S.; Tavares, A.J.; Dai, Q.; Ohta, S.; Audet, J.; Dvorak, H.F.; Chan, W.C.W. Analysis of nanoparticle delivery to tumours. *Nat. Rev. Mater.* **2016**, *1*, 16014. [CrossRef]
7. Greish, K. Enhanced permeability and retention (EPR) effect for anticancer nanomedicine drug targeting. *Methods Mol. Biol.* **2010**, *624*, 25–37.
8. Maeda, H. Nitroglycerin enhances vascular blood flow and drug delivery in hypoxic tumor tissues: Analogy between angina pectoris and solid tumors and enhancement of the EPR effect. *J. Control. Release* **2010**, *142*, 296–298. [CrossRef]
9. Maeda, H.; Matsumura, Y. Tumoritropic and lymphotropic principles of macromolecular drugs. *Crit. Rev. Ther. Drug Carrier. Syst.* **1989**, *6*, 193–210.
10. Bazak, R.; Houri, M.; Achy, S.E.; Hussein, W.; Refaat, T. Passive targeting of nanoparticles to cancer: A comprehensive review of the literature. *Mol. Clin. Oncol.* **2014**, *2*, 904–908. [CrossRef]
11. Shi, Y.; van der Meel, R.; Chen, X.; Lammers, T. The EPR effect and beyond: Strategies to improve tumor targeting and cancer nanomedicine treatment efficacy. *Theranostics* **2020**, *10*, 7921–7924. [CrossRef]
12. Maheshwari, N.; Atneriya, U.K.; Tekade, M.; Sharma, M.C.; Elhissi, A.; Tekade, R.K. Guiding Factors and Surface Modification Strategies for Biomaterials in Pharmaceutical Product Development. In *Biomaterials and Bionanotechnology*; Elsevier: Amsterdam, The Netherlands, 2019; pp. 57–87. [CrossRef]
13. Blanco, E.; Shen, H.; Ferrari, M. Principles of nanoparticle design for overcoming biological barriers to drug delivery. *Nat. Biotechnol.* **2015**, *33*, 941–951. [CrossRef]
14. Milane, L.; Ganesh, S.; Shah, S.; Duan, Z.; Amiji, M. Multi-modal strategies for overcoming tumor drug resistance: Hypoxia, the Warburg effect, stem cells, and multifunctional nanotechnology. *J. Control. Release* **2011**, *155*, 237–247. [CrossRef]
15. Liu, X.; Jiang, J.; Meng, H. Transcytosis—An effective targeting strategy that is complementary to "EPR effect" for pancreatic cancer nano drug delivery. *Theranostics* **2019**, *9*, 8018–8025. [CrossRef] [PubMed]
16. Matsumura, Y.; Maeda, H. A new concept for macromolecular therapeutics in cancer chemotherapy: Mechanism of tumoritropic accumulation of proteins and the antitumor agent smancs. *Cancer Res.* **1986**, *46*, 6387–6392. [PubMed]
17. Maeda, H. Tumor-Selective Delivery of Macromolecular Drugs via the EPR Effect: Background and Future Prospects. *Bioconjug. Chem.* **2010**, *21*, 797–802. [CrossRef] [PubMed]
18. Maeda, H. Vascular permeability in cancer and infection as related to macromolecular drug delivery, with emphasis on the EPR effect for tumor-selective drug targeting. *Proc. Jpn. Acad. Ser. B* **2012**, *88*, 53–71. [CrossRef]
19. Fang, J.; Islam, W.; Maeda, H. Exploiting the dynamics of the EPR effect and strategies to improve the therapeutic effects of nanomedicines by using EPR effect enhancers. *Adv. Drug Deliv. Rev.* **2020**, *157*, 142–160. [CrossRef]
20. Rosenblum, D.; Joshi, N.; Tao, W.; Karp, J.M.; Peer, D. Progress and challenges towards targeted delivery of cancer therapeutics. *Nat. Commun.* **2018**, *9*, 1410. [CrossRef]

21. Park, J.; Choi, Y.; Chang, H.; Um, W.; Ryu, J.H.; Kwon, I.C. Alliance with EPR Effect: Combined Strategies to Improve the EPR Effect in the Tumor Microenvironment. *Theranostics* **2019**, *9*, 8073–8090. [CrossRef]
22. Jain, R.K. Normalization of Tumor Vasculature: An Emerging Concept in Antiangiogenic Therapy. *Science* **2005**, *307*, 6. [CrossRef] [PubMed]
23. Izci, M. The Use of Alternative Strategies for Enhanced Nanoparticle Delivery to Solid Tumors. *Chem. Rev.* **2021**, *121*, 1746–1803. [CrossRef] [PubMed]
24. Nakamura, H.; Jun, F.; Maeda, H. Development of next-generation macromolecular drugs based on the EPR effect: Challenges and pitfalls. *Expert Opin. Drug Deliv.* **2015**, *12*, 53–64. [CrossRef] [PubMed]
25. Kalyane, D.; Raval, N.; Maheshwari, R.; Tambe, V.; Kalia, K.; Tekade, R.K. Employment of enhanced permeability and retention effect (EPR): Nanoparticle-based precision tools for targeting of therapeutic and diagnostic agent in cancer. *Mater. Sci. Eng. C* **2019**, *98*, 1252–1276. [CrossRef]
26. Noguchi, Y.; Wu, J.; Duncan, R.; Strohalm, J.; Ulbrich, K.; Akaike, T.; Maeda, H. Early Phase Tumor Accumulation of Macromolecules: A Great Difference in Clearance Rate between Tumor and Normal Tissues. *Jpn. J. Cancer Res.* **1998**, *89*, 307–314. [CrossRef]
27. Zhang, Y.-N.; Poon, W.; Tavares, A.J.; McGilvray, I.D.; Chan, W.C.W. Nanoparticle–liver interactions: Cellular uptake and hepatobiliary elimination. *J. Control. Release* **2016**, *240*, 332–348. [CrossRef]
28. Batchelor, T.T.; Gerstner, E.R.; Emblem, K.E.; Duda, D.G.; Kalpathy-Cramer, J.; Snuderl, M.; Ancukiewicz, M.; Polaskova, P.; Pinho, M.C.; Jennings, D.; et al. Improved tumor oxygenation and survival in glioblastoma patients who show increased blood perfusion after cediranib and chemoradiation. *Proc. Natl. Acad. Sci. USA* **2013**, *110*, 19059–19064. [CrossRef]
29. Muntimadugu, E.; Kommineni, N.; Khan, W. Exploring the Potential of Nanotherapeutics in Targeting Tumor Microenvironment for Cancer Therapy. *Pharmacol. Res.* **2017**, *126*, 109–122. [CrossRef]
30. Wirthl, B.; Kremheller, J.; Schrefler, B.A.; Wall, W.A. Extension of a multiphase tumour growth model to study nanoparticle delivery to solid tumours. *PLoS ONE* **2020**, *15*, e0228443. [CrossRef]
31. Sano, K.; Nakajima, T.; Choyke, P.L.; Kobayashi, H. The Effect of Photoimmunotherapy Followed by Liposomal Daunorubicin in a Mixed Tumor Model: A Demonstration of the Super-Enhanced Permeability and Retention Effect after Photoimmunotherapy. *Mol. Cancer Ther.* **2014**, *13*, 426–432. [CrossRef]
32. Natfji, A.A.; Ravishankar, D.; Osborn, H.M.; Greco, F. Parameters Affecting the Enhanced Permeability and Retention Effect: The Need for Patient Selection. *J. Pharm. Sci.* **2017**, *106*, 3179–3187. [CrossRef] [PubMed]
33. Navya, P.N.; Kaphle, A.; Srinivas, S.P.; Bhargava, S.K.; Rotello, V.M.; Daima, H.K. Current trends and challenges in cancer management and therapy using designer nanomaterials. *Nano Converg.* **2019**, *6*, 23. [CrossRef] [PubMed]
34. Zhu, J.; Powis de Tenbossche, C.G.; Cané, S.; Colau, D.; van Baren, N.; Lurquin, C.; Schmitt-Verhulst, A.-M.; Liljeström, P.; Uyttenhove, C.; Van den Eynde, B.J. Resistance to cancer immunotherapy mediated by apoptosis of tumor-infiltrating lymphocytes. *Nat. Commun.* **2017**, *8*, 1404. [CrossRef]
35. Reagan-Shaw, S.; Nihal, M.; Ahmad, N. Dose translation from animal to human studies revisited. *FASEB J.* **2008**, *22*, 659–661. [CrossRef]
36. Maeda, H. Toward a full understanding of the EPR effect in primary and metastatic tumors as well as issues related to its heterogeneity. *Adv. Drug Deliv. Rev.* **2015**, *91*, 3–6. [CrossRef] [PubMed]
37. Osman, N.M.; Sexton, D.W.; Saleem, I.Y. Toxicological assessment of nanoparticle interactions with the pulmonary system. *Nanotoxicology* **2020**, *14*, 21–58. [CrossRef] [PubMed]
38. Price, L.S.L.; Stern, S.T.; Deal, A.M.; Kabanov, A.V.; Zamboni, W.C. A reanalysis of nanoparticle tumor delivery using classical pharmacokinetic metrics. *Sci. Adv.* **2020**, *6*, eaay9249. [CrossRef]
39. Yu, W.; Liu, R.; Zhou, Y.; Gao, H. Size-Tunable Strategies for a Tumor Targeted Drug Delivery System. *ACS Cent. Sci.* **2020**, *6*, 100–116. [CrossRef]
40. Soo Choi, H.; Liu, W.; Misra, P.; Tanaka, E.; Zimmer, J.P.; Itty Ipe, B.; Bawendi, M.G.; Frangioni, J.V. Renal clearance of quantum dots. *Nat. Biotechnol.* **2007**, *25*, 1165–1170. [CrossRef]
41. Liu, X.; Chen, Y.; Li, H.; Huang, N.; Jin, Q.; Ren, K.; Ji, J. Enhanced retention and cellular uptake of nanoparticles in tumors by controlling their aggregation behavior. *ACS Nano* **2013**, *7*, 6244–6257. [CrossRef]
42. Perrault, S.D.; Walkey, C.; Jennings, T.; Fischer, H.C.; Chan, W.C.W. Mediating tumor targeting efficiency of nanoparticles through design. *Nano Lett.* **2009**, *9*, 1909–1915. [CrossRef] [PubMed]
43. Zhao, M.; Yang, M.; Li, X.-M.; Jiang, P.; Baranov, E.; Li, S.; Xu, M.; Penman, S.; Hoffman, R.M. Tumor-targeting bacterial therapy with amino acid auxotrophs of GFP-expressing Salmonella typhimurium. *Proc. Natl. Acad. Sci. USA* **2005**, *102*, 755–760. [CrossRef] [PubMed]
44. Raza, K.; Kumar, P.; Kumar, N.; Malik, R. Pharmacokinetics and biodistribution of the nanoparticles. In *Advances in Nanomedicine for the Delivery of Therapeutic Nucleic Acids*; Elsevier: Amsterdam, The Netherlands, 2017; pp. 165–186. [CrossRef]
45. Wong, C.; Stylianopoulos, T.; Cui, J.; Martin, J.; Chauhan, V.P.; Jiang, W.; Popovic, Z.; Jain, R.K.; Bawendi, M.G.; Fukumura, D. Multistage nanoparticle delivery system for deep penetration into tumor tissue. *Proc. Natl. Acad. Sci. USA* **2011**, *108*, 2426–2431. [CrossRef] [PubMed]

46. Campbell, R.B.; Fukumura, D.; Brown, E.B.; Mazzola, L.M.; Izumi, Y.; Jain, R.K.; Torchilin, V.P.; Munn, L.L. Cationic Charge Determines the Distribution of Liposomes between the Vascular and Extravascular Compartments of Tumors. *Cancer Res.* **2002**, *62*, 6831–6836.
47. Nel, A.E.; Mädler, L.; Velegol, D.; Xia, T.; Hoek, E.M.V.; Somasundaran, P.; Klaessig, F.; Castranova, V.; Thompson, M. Understanding biophysicochemical interactions at the nano–bio interface. *Nat. Mater* **2009**, *8*, 543–557. [CrossRef]
48. Fromen, C.A.; Rahhal, T.B.; Robbins, G.R.; Kai, M.P.; Shen, T.W.; Luft, J.C.; DeSimone, J.M. Nanoparticle surface charge impacts distribution, uptake and lymph node trafficking by pulmonary antigen-presenting cells. *Nanomedicine* **2016**, *12*, 677–687. [CrossRef]
49. Fleischer, C.C.; Payne, C.K. Nanoparticle surface charge mediates the cellular receptors used by protein-nanoparticle complexes. *J. Phys. Chem. B* **2012**, *116*, 8901–8907. [CrossRef]
50. Cho, E.C.; Xie, J.; Wurm, P.A.; Xia, Y. Understanding the Role of Surface Charges in Cellular Adsorption versus Internalization by Selectively Removing Gold Nanoparticles on the Cell Surface with a I2/KI Etchant. *Nano Lett.* **2009**, *9*, 1080–1084. [CrossRef]
51. Behra, R.; Sigg, L.; Clift, M.J.D.; Herzog, F.; Minghetti, M.; Johnston, B.; Petri-Fink, A.; Rothen-Rutishauser, B. Bioavailability of silver nanoparticles and ions: From a chemical and biochemical perspective. *J. R. Soc. Interface* **2013**, *10*, 20130396. [CrossRef]
52. Alexis, F.; Pridgen, E.; Molnar, L.K.; Farokhzad, O.C. Factors Affecting the Clearance and Biodistribution of Polymeric Nanoparticles. *Mol. Pharm.* **2008**, *5*, 505–515. [CrossRef]
53. He, C.; Hu, Y.; Yin, L.; Tang, C.; Yin, C. Effects of particle size and surface charge on cellular uptake and biodistribution of polymeric nanoparticles. *Biomaterials* **2010**, *31*, 3657–3666. [CrossRef] [PubMed]
54. Yuan, Y.-Y.; Mao, C.-Q.; Du, X.-J.; Du, J.-Z.; Wang, F.; Wang, J. Surface Charge Switchable Nanoparticles Based on Zwitterionic Polymer for Enhanced Drug Delivery to Tumor. *Adv. Mater.* **2012**, *24*, 5476–5480. [CrossRef] [PubMed]
55. Anselmo, A.C.; Zhang, M.; Kumar, S.; Vogus, D.R.; Menegatti, S.; Helgeson, M.E.; Mitragotri, S. Elasticity of Nanoparticles Influences Their Blood Circulation, Phagocytosis, Endocytosis, and Targeting. *ACS Nano* **2015**, *9*, 3169–3177. [CrossRef]
56. Anselmo, A.C.; Modery-Pawlowski, C.L.; Menegatti, S.; Kumar, S.; Vogus, D.R.; Tian, L.L.; Chen, M.; Squires, T.M.; Sen Gupta, A.; Mitragotri, S. Platelet-like Nanoparticles: Mimicking Shape, Flexibility, and Surface Biology of Platelets to Target Vascular Injuries. *ACS Nano* **2014**, *8*, 11243–11253. [CrossRef]
57. Gentile, F.; Chiappini, C.; Fine, D.; Bhavane, R.C.; Peluccio, M.S.; Cheng, M.M.-C.; Liu, X.; Ferrari, M.; Decuzzi, P. The effect of shape on the margination dynamics of non-neutrally buoyant particles in two-dimensional shear flows. *J. Biomech.* **2008**, *41*, 2312–2318. [CrossRef] [PubMed]
58. Christian, D.A.; Cai, S.; Garbuzenko, O.B.; Harada, T.; Zajac, A.L.; Minko, T.; Discher, D.E. Flexible Filaments for in Vivo Imaging and Delivery: Persistent Circulation of Filomicelles Opens the Dosage Window for Sustained Tumor Shrinkage. *Mol. Pharm.* **2009**, *6*, 1343–1352. [CrossRef] [PubMed]
59. Anselmo, A.C.; Mitragotri, S. Impact of particle elasticity on particle-based drug delivery systems. *Adv. Drug Deliv. Rev.* **2017**, *108*, 51–67. [CrossRef]
60. Guo, P.; Liu, D.; Subramanyam, K.; Wang, B.; Yang, J.; Huang, J.; Auguste, D.T.; Moses, M.A. Nanoparticle elasticity directs tumor uptake. *Nat. Commun.* **2018**, *9*, 130. [CrossRef]
61. Hui, Y.; Yi, X.; Wibowo, D.; Yang, G.; Middelberg, A.P.J.; Gao, H.; Zhao, C.-X. Nanoparticle elasticity regulates phagocytosis and cancer cell uptake. *Sci. Adv.* **2020**, *6*, eaaz4316. [CrossRef]
62. Banquy, X.; Suarez, F.; Argaw, A.; Rabanel, J.-M.; Grutter, P.; Bouchard, J.-F.; Hildgen, P.; Giasson, S. Effect of mechanical properties of hydrogel nanoparticles on macrophage cell uptake. *Soft Matter* **2009**, *5*, 3984–3991. [CrossRef]
63. Ma, X.; Yang, X.; Li, M.; Cui, J.; Zhang, P.; Yu, Q.; Hao, J. Effect of Elasticity of Silica Capsules on Cellular Uptake. *Langmuir* **2021**, *37*, 11688–11694. [CrossRef]
64. Yao, C.; Akakuru, O.U.; Stanciu, S.G.; Hampp, N.; Jin, Y.; Zheng, J.; Chen, G.; Yang, F.; Wu, A. Effect of elasticity on the phagocytosis of micro/nanoparticles. *J. Mater. Chem. B* **2020**, *8*, 2381–2392. [CrossRef] [PubMed]
65. Thau, L.; Asuka, E.; Mahajan, K. Physiology, Opsonization. In *StatPearls*; StatPearls Publishing: Treasure Island, FL, USA, 2022.
66. Tenzer, S.; Docter, D.; Kuharev, J.; Musyanovych, A.; Fetz, V.; Hecht, R.; Schlenk, F.; Fischer, D.; Kiouptsi, K.; Reinhardt, C.; et al. Rapid formation of plasma protein corona critically affects nanoparticle pathophysiology. *Nat. Nanotechnol.* **2013**, *8*, 772–781. [CrossRef]
67. Adeyemi, S.A.; Kumar, P.; Choonara, Y.E.; Pillay, V. Stealth Properties of Nanoparticles Against Cancer: Surface Modification of NPs for Passive Targeting to Human Cancer Tissue in Zebrafish Embryos. In *Surface Modification of Nanoparticles for Targeted Drug Delivery*; Pathak, Y.V., Ed.; Springer International Publishing: New York, NY, USA, 2019; pp. 99–124. [CrossRef]
68. Lila, A.S.A. The accelerated blood clearance (ABC) phenomenon: Clinical challenge and approaches to manage. *J. Control. Release* **2013**, *172*, 38–47. [CrossRef] [PubMed]
69. Schöttler, S.; Becker, G.; Winzen, S.; Steinbach, T.; Mohr, K.; Landfester, K.; Mailänder, V.; Wurm, F.R. Protein adsorption is required for stealth effect of poly(ethylene glycol)- and poly(phosphoester)-coated nanocarriers. *Nat. Nanotechnol.* **2016**, *11*, 372–377. [CrossRef]
70. Chen, D.; Liu, W.; Shen, Y.; Mu, H.; Zhang, Y.; Liang, R.; Wang, A.; Sun, K.; Fu, F. Effects of a novel pH-sensitive liposome with cleavable esterase-catalyzed and pH-responsive double smart mPEG lipid derivative on ABC phenomenon. *Int. J. Nanomed.* **2011**, *6*, 2053. [CrossRef]
71. Shao, Q.; Jiang, S. Molecular understanding and design of zwitterionic materials. *Adv. Mater* **2015**, *27*, 15–26. [CrossRef]

72. Harijan, M.; Singh, M. Zwitterionic polymers in drug delivery: A review. *J. Mol. Recognit.* **2022**, *35*, e2944. [CrossRef] [PubMed]
73. Xiong, Z.; Shen, M.; Shi, X. Zwitterionic Modification of Nanomaterials for Improved Diagnosis of Cancer Cells. *Bioconjug Chem.* **2019**, *30*, 2519–2527. [CrossRef]
74. Li, B.; Yuan, Z.; Zhang, P.; Sinclair, A.; Jain, P.; Wu, K.; Tsao, C.; Xie, J.; Hung, H.-C.; Lin, X.; et al. Zwitterionic Nanocages Overcome the Efficacy Loss of Biologic Drugs. *Adv. Mater.* **2018**, *30*, 1705728. [CrossRef] [PubMed]
75. Lin, W.; Ma, G.; Ji, F.; Zhang, J.; Wang, L.; Sun, H.; Chen, S. Biocompatible long-circulating star carboxybetaine polymers. *J. Mater. Chem. B* **2014**, *3*, 440–448. [CrossRef] [PubMed]
76. Tsao, C.; Yuan, Z.; Zhang, P.; Liu, E.; McMullen, P.; Wu, K.; Hung, H.-C.; Jiang, S. Enhanced pulmonary systemic delivery of protein drugs via zwitterionic polymer conjugation. *J. Control. Release* **2020**, *322*, 170–176. [CrossRef]
77. Zhao, G.; Sun, Y.; Dong, X. Zwitterionic Polymer Micelles with Dual Conjugation of Doxorubicin and Curcumin: Synergistically Enhanced Efficacy against Multidrug-Resistant Tumor Cells. *Langmuir* **2020**, *36*, 2383–2395. [CrossRef]
78. Li, L.; Song, Y.; He, J.; Zhang, M.; Liu, J.; Ni, P. Zwitterionic shielded polymeric prodrug with folate-targeting and pH responsiveness for drug delivery. *J. Mater. Chem. B* **2019**, *7*, 786–795. [CrossRef] [PubMed]
79. Kalkanis, A.; Judson, M.A.; Kalkanis, D.; Vavougios, G.D.; Malamitsi, J.; Georgou, E. Reticuloendothelial system involvement in untreated sarcoidosis patients as assessed by 18F-FDG PET scanning. *Sarcoidosis Vasc. Diffus. Lung Dis.* **2016**, *33*, 423–425.
80. Zhou, Y.; Dai, Z. New Strategies in the Design of Nanomedicines to Oppose Uptake by the Mononuclear Phagocyte System and Enhance Cancer Therapeutic Efficacy. *Chem. Asian J.* **2018**, *13*, 3333–3340. [CrossRef]
81. Poon, W.; Zhang, Y.; Ouyang, B.; Kingston, B.R.; Wu, J.L.Y.; Wilhelm, S.; Chan, W.C.W. Elimination Pathways of Nanoparticles. *ACS Nano* **2019**, *13*, 5785–5798. [CrossRef]
82. Bertrand, N.; Leroux, J.-C. The journey of a drug-carrier in the body: An anatomo-physiological perspective. *J. Control. Release* **2012**, *161*, 152–163. [CrossRef]
83. Di, J.; Gao, X.; Du, Y.; Zhang, H.; Gao, J.; Zheng, A. Size, shape, charge and "stealthy" surface: Carrier properties affect the drug circulation time in vivo. *Asian J. Pharm. Sci.* **2021**, *16*, 444–458. [CrossRef]
84. Ohara, Y.; Oda, T.; Yamada, K.; Hashimoto, S.; Akashi, Y.; Miyamoto, R.; Kobayashi, A.; Fukunaga, K.; Sasaki, R.; Ohkohchi, N. Effective delivery of chemotherapeutic nanoparticles by depleting host Kupffer cells. *Int. J. Cancer* **2012**, *131*, 2402–2410. [CrossRef]
85. Tavares, A.J.; Poon, W.; Zhang, Y.-N.; Dai, Q.; Besla, R.; Ding, D.; Ouyang, B.; Li, A.; Chen, J.; Zheng, G.; et al. Effect of removing Kupffer cells on nanoparticle tumor delivery. *Proc. Natl. Acad. Sci. USA* **2017**, *114*, E10871–E10880. [CrossRef] [PubMed]
86. Andrés, D.; Sánchez-Reus, I.; Bautista, M.; Cascales, M. Depletion of Kupffer cell function by gadolinium chloride attenuates thioacetamide-induced hepatotoxicity. Expression of metallothionein and HSP70. *Biochem. Pharmacol.* **2003**, *66*, 917–926. [CrossRef]
87. Diagaradjane, P.; Deorukhkar, A.; Gelovani, J.G.; Maru, D.M.; Krishnan, S. Gadolinium Chloride Augments Tumor-Specific Imaging of Targeted Quantum Dots in vivo. *ACS Nano* **2010**, *4*, 4131–4141. [CrossRef] [PubMed]
88. Opperman, K.S.; Vandyke, K.; Clark, K.C.; Coulter, E.A.; Hewett, D.R.; Mrozik, K.M.; Schwarz, N.; Evdokiou, A.; Croucher, P.I.; Psaltis, P.J.; et al. Clodronate-Liposome Mediated Macrophage Depletion Abrogates Multiple Myeloma Tumor Establishment In Vivo. *Neoplasia* **2019**, *21*, 777–787. [CrossRef] [PubMed]
89. Usynin, I.; Khar'kovsky, A.; Balitskaya, N.; Panin, L. Gadolinium Chloride-Induced Kupffer Cell Blockade Increases Uptake of Oxidized Low-Density Lipoproteins by Rat Heart and Aorta. *Biochem. Biokhimiia* **1999**, *64*, 620–624.
90. Liu, L.; Ye, Q.; Lu, M.; Lo, Y.-C.; Hsu, Y.-H.; Wei, M.-C.; Chen, Y.-H.; Lo, S.-C.; Wang, S.-J.; Bain, D.J.; et al. A New Approach to Reduce Toxicities and to Improve Bioavailabilities of Platinum-Containing Anti-Cancer Nanodrugs. *Sci. Rep.* **2015**, *5*, 10881. [CrossRef]
91. Liu, T.; Choi, H.; Zhou, R.; Chen, I.-W. RES blockade: A strategy for boosting efficiency of nanoparticle drug. *Nano Today* **2015**, *10*, 11–21. [CrossRef]
92. Ouyang, B.; Poon, W.; Zhang, Y.-N.; Lin, Z.P.; Kingston, B.R.; Tavares, A.J.; Zhang, Y.; Chen, J.; Valic, M.S.; Syed, A.M.; et al. The dose threshold for nanoparticle tumour delivery. *Nat. Mater.* **2020**, *19*, 1362–1371. [CrossRef]
93. Bouwens, L.; Baekeland, M.; De Zanger, R.; Wisse, E. Quantitation, tissue distribution and proliferation kinetics of Kupffer cells in normal rat liver. *Hepatology* **1986**, *6*, 718–722. [CrossRef]
94. Fujiwara, Y.; Mukai, H.; Saeki, T.; Ro, J.; Lin, Y.-C.; Nagai, S.E.; Lee, K.S.; Watanabe, J.; Ohtani, S.; Kim, S.B.; et al. A multi-national, randomised, open-label, parallel, phase III non-inferiority study comparing NK105 and paclitaxel in metastatic or recurrent breast cancer patients. *Br. J. Cancer* **2019**, *120*, 475–480. [CrossRef]
95. Su, Y.; Xie, Z.; Kim, G.B.; Dong, C.; Yang, J. Design Strategies and Applications of Circulating Cell-Mediated Drug Delivery Systems. *ACS Biomater. Sci. Eng.* **2015**, *1*, 201–217. [CrossRef] [PubMed]
96. Anselmo, A.C.; Mitragotri, S. Cell-mediated delivery of nanoparticles: Taking advantage of circulatory cells to target nanoparticles. *J. Control. Release* **2014**, *190*, 531–541. [CrossRef] [PubMed]
97. Wu, Y.; Tibrewal, N.; Birge, R.B. Phosphatidylserine recognition by phagocytes: A view to a kill. *Trends Cell Biol.* **2006**, *16*, 189–197. [CrossRef]
98. Xia, Q. Red blood cell membrane-camouflaged nanoparticles_ a novel drug delivery system for antitumor application. *Acta Pharm. Sin. B* **2019**, *9*, 675–689. [CrossRef]

99. Combes, F.; Meyer, E.; Sanders, N.N. Immune cells as tumor drug delivery vehicles. *J. Control. Release* **2020**, *327*, 70–87. [CrossRef] [PubMed]
100. Zhang, N.; Zhang, J.; Wang, P.; Liu, X.; Huo, P.; Xu, Y.; Chen, W.; Xu, H.; Tian, Q. Investigation of an antitumor drug-delivery system based on anti-HER2 antibody-conjugated BSA nanoparticles. *Anti-Cancer Drugs* **2018**, *29*, 307–322. [CrossRef]
101. Polak, R.; Lim, R.M.; Beppu, M.M.; Pitombo, R.N.M.; Cohen, R.E.; Rubner, M.F. Liposome-Loaded Cell Backpacks. *Adv. Healthc. Mater.* **2015**, *4*, 2832–2841. [CrossRef]
102. Yang, X.; Lian, K.; Tan, Y.; Zhu, Y.; Liu, X.; Zeng, Y.; Yu, T.; Meng, T.; Yuan, H.; Hu, F. Selective uptake of chitosan polymeric micelles by circulating monocytes for enhanced tumor targeting. *Carbohydr. Polym.* **2020**, *229*, 115435. [CrossRef]
103. Zhang, Y.; Luo, J.; Zhang, J.; Miao, W.; Wu, J.; Huang, H.; Tong, Q.; Shen, S.; Leong, K.W.; Du, J.; et al. Nanoparticle-Enabled Dual Modulation of Phagocytic Signals to Improve Macrophage-Mediated Cancer Immunotherapy. *Small* **2020**, *16*, 2004240. [CrossRef]
104. Wang, Y.; Zhang, P.; Wei, Y.; Shen, K.; Xiao, L.; Miron, R.J.; Zhang, Y. Cell-Membrane-Display Nanotechnology. *Adv. Healthc. Mater.* **2021**, *10*, e2001014. [CrossRef]
105. Wang, C.; Sun, W.; Ye, Y.; Hu, Q.; Bomba, H.N.; Gu, Z. In situ activation of platelets with checkpoint inhibitors for post-surgical cancer immunotherapy. *Nat. Biomed. Eng.* **2017**, *1*, 11. [CrossRef]
106. Wei, Z.; Chen, Z.; Zhao, Y.; Fan, F.; Xiong, W.; Song, S.; Yin, Y.; Hu, J.; Yang, K.; Yang, L.; et al. Mononuclear phagocyte system blockade using extracellular vesicles modified with CD47 on membrane surface for myocardial infarction reperfusion injury treatment. *Biomaterials* **2021**, *275*, 121000. [CrossRef] [PubMed]
107. Wei, Z.; Qiao, S.; Zhao, J.; Liu, Y.; Li, Q.; Wei, Z.; Dai, Q.; Kang, L.; Xu, B. miRNA-181a over-expression in mesenchymal stem cell-derived exosomes influenced inflammatory response after myocardial ischemia-reperfusion injury. *Life Sci.* **2019**, *232*, 116632. [CrossRef] [PubMed]
108. Fang, R.H.; Kroll, A.V.; Gao, W.; Zhang, L. Cell Membrane Coating Nanotechnology. *Adv. Mater.* **2018**, *30*, e1706759. [CrossRef] [PubMed]
109. Fang, R.H.; Hu, C.-M.J.; Luk, B.T.; Gao, W.; Copp, J.A.; Tai, Y.; O'Connor, D.E.; Zhang, L. Cancer cell membrane-coated nanoparticles for anticancer vaccination and drug delivery. *Nano Lett.* **2014**, *14*, 2181–2188. [CrossRef]
110. Chen, Z.; Zhao, P.; Luo, Z.; Zheng, M.; Tian, H.; Gong, P.; Gao, G.; Pan, H.; Liu, L.; Ma, A.; et al. Cancer Cell Membrane-Biomimetic Nanoparticles for Homologous-Targeting Dual-Modal Imaging and Photothermal Therapy. *ACS Nano* **2016**, *10*, 10049–10057. [CrossRef]
111. Rodriguez, P.L.; Harada, T.; Christian, D.A.; Pantano, D.A.; Tsai, R.K.; Discher, D.E. Minimal 'Self' Peptides That Inhibit Phagocytic Clearance and Enhance Delivery of Nanoparticles. *Science* **2013**, *339*, 971–975. [CrossRef]
112. Nie, Y.; Günther, M.; Gu, Z.; Wagner, E. Pyridylhydrazone-based PEGylation for pH-reversible lipopolyplex shielding. *Biomaterials* **2011**, *32*, 858–869. [CrossRef]
113. Zhang, M.; Guo, X.; Wang, M.; Liu, K. Tumor microenvironment-induced structure changing drug/gene delivery system forovercoming delivery-associated challenges. *J. Control. Release* **2020**, *323*, 203–224. [CrossRef]
114. Visser, J.G.; Van Staden, A.D.P.; Smith, C. Harnessing Macrophages for Controlled-Release Drug Delivery: Lessons From Microbes. *Front. Pharmacol.* **2019**, *10*, 22. [CrossRef]
115. Jiang, X.; Du, B.; Zheng, J. Glutathione-mediated biotransformation in the liver modulates nanoparticle transport. *Nat. Nanotechnol.* **2019**, *14*, 874–882. [CrossRef] [PubMed]
116. Phesgo FDA 2020. Available online: https://www.fda.gov/news-events/press-announcements/fda-approves-breast-cancer-treatment-can-be-administered-home-health-care-professional (accessed on 18 June 2022).
117. Mondaca, S.; Margolis, M.; Sanchez-Vega, F.; Jonsson, P.; Riches, J.C.; Ku, G.Y.; Hechtman, J.F.; Tuvy, Y.; Ber-ger, M.F.; Shah, M.A.; et al. Phase II study of trastuzumab with modified docetaxel, cisplatin, and 5 fluorouracil in metastatic HER2-positive gastric cancer. *Gastric Cancer* **2019**, *22*, 355–362. [CrossRef]
118. Shu, M.; Yan, H.; Xu, C.; Wu, Y.; Chi, Z.; Nian, W.; He, Z.; Xiao, J.; Wei, H.; Zhou, Q.; et al. A novel anti-HER2 antibody GB235 reverses Trastuzumab resistance in HER2-expressing tumor cells in vitro and in vivo. *Sci. Rep.* **2020**, *10*, 2986. [CrossRef] [PubMed]
119. Fan, Z.; Chang, Y.; Cui, C.; Sun, L.; Wang, D.H.; Pan, Z.; Zhang, M. Near infrared fluorescent peptide nanoparticles for enhancing esophageal cancer therapeutic efficacy. *Nat. Commun.* **2018**, *9*, 2605. [CrossRef] [PubMed]
120. Li, Y.; Jeon, J.; Park, J.H. Hypoxia-responsive nanoparticles for tumor-targeted drug delivery. *Cancer Lett.* **2020**, *490*, 31–43. [CrossRef]
121. Thambi, T.; You, D.G.; Han, H.S.; Deepagan, V.G.; Jeon, S.M.; Suh, Y.D.; Choi, K.Y.; Kim, K.; Kwon, I.C.; Yi, G.R.; et al. Bioreducible Carboxymethyl Dextran Nanoparticles for Tumor-Targeted Drug Delivery. *Adv. Healthc. Mater.* **2014**, *3*, 1829–1838. [CrossRef]
122. Dhaliwal, A.; Zheng, G. Improving accessibility of EPR-insensitive tumor phenotypes using EPR-adaptive strategies: Designing a new perspective in nanomedicine delivery. *Theranostics* **2019**, *9*, 8091–8108. [CrossRef]
123. Attia, M.F.; Anton, N.; Wallyn, J.; Omran, Z.; Vandamme, T.F. An overview of active and passive targeting strategies to improve the nanocarriers efficiency to tumour sites. *J. Pharm. Pharmacol.* **2019**, *71*, 1185–1198. [CrossRef]
124. Park, J.-S.; Qiao, L.; Su, Z.-Z.; Hinman, D.; Willoughby, K.; McKinstry, R.; Yacoub, A.; Duigou, G.J.; Young, C.S.H.; Grant, S.; et al. Ionizing radiation modulates vascular endothelial growth factor (VEGF) expression through multiple mitogen activated protein kinase dependent pathways. *Oncogene* **2001**, *20*, 3266–3280. [CrossRef]

125. Iwanaga, K.; Tominaga, K.; Yamamoto, K.; Habu, M.; Maeda, H.; Akifusa, S.; Tsujisawa, T.; Okinaga, T.; Fukuda, J.; Nishihara, T. Local delivery system of cytotoxic agents to tumors by focused sonoporation. *Cancer Gene Ther.* **2007**, *14*, 354–363. [CrossRef]
126. Duan, L.; Yang, L.; Jin, J.; Yang, F.; Liu, D.; Hu, K.; Wang, Q.; Yue, Y.; Gu, N. Micro/nano-bubble-assisted ultrasound to enhance the EPR effect and potential theranostic applications. *Theranostics* **2020**, *10*, 462–483. [CrossRef] [PubMed]
127. Theek, B.; Gremse, F.; Kunjachan, S.; Fokong, S.; Pola, R.; Pechar, M.; Deckers, R.; Storm, G.; Ehling, J.; Kiessling, F.; et al. Characterizing EPR-mediated passive drug targeting using contrast-enhanced functional ultrasound imaging. *J. Control. Release* **2014**, *182*, 83–89. [CrossRef]
128. Ikeda-Imafuku, M.; Wang, L.L.-W.; Rodrigues, D.; Shaha, S.; Zhao, Z.; Mitragotri, S. Strategies to improve the EPR effect: A mechanistic perspective and clinical translation. *J. Control. Release* **2022**, *345*, 512–536. [CrossRef]
129. Dougherty, T.J.; Gomer, C.J.; Henderson, B.W.; Jori, G.; Kessel, D.; Korbelik, M.; Moan, J.; Peng, Q. Photodynamic Therapy. *J. Natl. Cancer Inst.* **1998**, *90*, 889–905. [CrossRef] [PubMed]
130. Zhen, Z.; Tang, W.; Chuang, Y.-J.; Todd, T.; Zhang, W.; Lin, X.; Niu, G.; Liu, G.; Wang, L.; Pan, Z.; et al. Tumor Vasculature Targeted Photodynamic Therapy for Enhanced Delivery of Nanoparticles. *ACS Nano* **2014**, *8*, 6004–6013. [CrossRef] [PubMed]
131. Mao, C.; Li, F.; Zhao, Y.; Debinski, W.; Ming, X. P-glycoprotein-targeted photodynamic therapy boosts cancer nanomedicine by priming tumor microenvironment. *Theranostics* **2018**, *8*, 6274–6290. [CrossRef]
132. Baronzio, G.; Parmar, G.; Baronzio, M. Overview of Methods for Overcoming Hindrance to Drug Delivery to Tumors, with Special Attention to Tumor Interstitial Fluid. *Front. Oncol.* **2022**, *5*, 165. [CrossRef]
133. Fan, Y.-T.; Zhou, T.-J.; Cui, P.-F.; He, Y.-J.; Chang, X.; Xing, L.; Jiang, H.-L. Modulation of Intracellular Oxygen Pressure by Dual-Drug Nanoparticles to Enhance Photodynamic Therapy. *Adv. Funct. Mater.* **2019**, *29*, 1806708. [CrossRef]
134. Kong, G.; Braun, R.D.; Dewhirst, M.W. Characterization of the Effect of Hyperthermia on Nanoparticle Extravasation from Tumor Vasculature. *Cancer Res.* **2001**, *61*, 3027–3032.
135. Koczera, P.; Appold, L.; Shi, Y.; Liu, M.; Dasgupta, A.; Pathak, V.; Ojha, T.; Fokong, S.; Wu, Z.; van Zandvoort, M.; et al. PBCA-based polymeric microbubbles for molecular imaging and drug delivery. *J. Control. Release* **2017**, *259*, 128–135. [CrossRef]
136. Fu, B.; Shi, L.; Palacio-Mancheno, P.; Badami, J.; Shin, D.W.; Zeng, M.; Cardoso, L.; Tu, R. Quantification of transient increase of the blood–brain barrier permeability to macromolecules by optimized focused ultrasound combined with microbubbles. *Int. J. Nanomed.* **2014**, *9*, 4437. [CrossRef] [PubMed]
137. Zhang, Q.; Wang, N.; Ma, M.; Luo, Y.; Chen, H. Transferrin Receptor-Mediated Sequential Intercellular Nanoparticles Relay for Tumor Deep Penetration and Sonodynamic Therapy. *Adv. Therap.* **2019**, *2*, 1800152. [CrossRef]
138. Dimcevski, G.; Kotopoulis, S.; Bjånes, T.; Hoem, D.; Schjøtt, J.; Gjertsen, B.T.; Biermann, M.; Molven, A.; Sor-bye, H.; McCormack, E.; et al. A human clinical trial using ultrasound and microbubbles to enhance gemcitabine treatment of inoperable pancreatic cancer. *J. Control. Release* **2016**, *243*, 172–181. [CrossRef] [PubMed]
139. Cao, Y.; Chen, Y.; Yu, T.; Guo, Y.; Liu, F.; Yao, Y.; Li, P.; Wang, D.; Wang, Z.; Chen, Y.; et al. Drug Release from Phase-Changeable Nanodroplets Triggered by Low-Intensity Focused Ultrasound. *Theranostics* **2018**, *8*, 1327–1339. [CrossRef] [PubMed]
140. Islam, W.; Fang, J.; Imamura, T.; Etrych, T.; Subr, V.; Ulbrich, K.; Maeda, H. Augmentation of the Enhanced Permeability and Retention Effect with Nitric Oxide–Generating Agents Improves the Therapeutic Effects of Nanomedicines. *Mol. Cancer Ther.* **2018**, *17*, 2643–2653. [CrossRef]
141. Chauhan, V.P.; Martin, J.D.; Liu, H.; Lacorre, D.A.; Jain, S.R.; Kozin, S.V.; Stylianopoulos, T.; Mousa, A.S.; Han, X.; Adstamongkonkul, P.; et al. Angiotensin inhibition enhances drug delivery and potentiates chemotherapy by decompressing tumour blood vessels. *Nat. Commun.* **2013**, *4*, 2516. [CrossRef]
142. Diop-Frimpong, B.; Chauhan, V.P.; Krane, S.; Boucher, Y.; Jain, R.K. Losartan inhibits collagen I synthesis and improves the distribution and efficacy of nanotherapeutics in tumors. *Proc. Natl. Acad. Sci. USA* **2011**, *108*, 2909–2914. [CrossRef]
143. Wu, J. The Enhanced Permeability and Retention (EPR) Effect: The Significance of the Concept and Methods to Enhance Its Application. *J. Pers. Med.* **2021**, *11*, 771. [CrossRef]
144. Ojha, T.; Pathak, V.; Shi, Y.; Hennink, W.E.; Moonen, C.T.; Storm, G.; Kiessling, F.; Lammers, T. Pharmacological and physical vessel modulation strategies to improve EPR-mediated drug targeting to tumors. *Adv. Drug Deliv. Rev.* **2017**, *119*, 44–60. [CrossRef]
145. Lammers, T.; Kiessling, F.; Hennink, W.E.; Storm, G. Drug targeting to tumors: Principles, pitfalls and (pre-) clinical progress. *J. Control. Release* **2012**, *161*, 175–187. [CrossRef] [PubMed]
146. Rezvantalab, S.; Drude, N.I.; Moraveji, M.K.; Güvener, N.; Koons, E.K.; Shi, Y.; Lammers, T.; Kiessling, F. PLGA-Based Nanoparticles in Cancer Treatment. *Front. Pharmacol.* **2018**, *9*, 1260. [CrossRef] [PubMed]
147. Belfiore, L.; Saunders, D.N.; Ranson, M.; Thurecht, K.J.; Storm, G.; Vine, K.L. Towards clinical translation of ligand-functionalized liposomes in targeted cancer therapy: Challenges and opportunities. *J. Control. Release* **2018**, *277*, 1–13. [CrossRef] [PubMed]
148. Miller, M.A.; Gadde, S.; Pfirschke, C.; Engblom, C.; Sprachman, M.M.; Kohler, R.H.; Yang, K.S.; Laughney, A.M.; Wojtkiewicz, G.; Kamaly, N.; et al. Predicting therapeutic nanomedicine efficacy using a companion magnetic resonance imaging nanoparticle. *Sci. Transl. Med.* **2015**, *7*, ra183–ra314. [CrossRef] [PubMed]

MDPI
St. Alban-Anlage 66
4052 Basel
Switzerland
www.mdpi.com

Polymers Editorial Office
E-mail: polymers@mdpi.com
www.mdpi.com/journal/polymers

www.ingramcontent.com/pod-product-compliance
Lightning Source LLC
LaVergne TN
LVHW070228100526
838202LV00015B/2103
* 9 7 8 3 7 2 5 8 0 1 5 2 7 *